A Natural History of Negation

THE DAVID HUME SERIES
PHILOSOPHY AND COGNITIVE SCIENCE REISSUES

The *David Hume Series on Philosophy and Cognitive Science Reissues* consists of previously published works that are important and useful to scholars and students working in the area of cognitive science. The aim of the series is to keep these indispensable works in print.

A NATURAL HISTORY OF NEGATION

LAURENCE R. HORN

THE DAVID HUME SERIES
PHILOSOPHY AND COGNITIVE SCIENCE REISSUES

CSLI PUBLICATIONS

Center for the Study of Language and Information
Leland Stanford Junior University
Printed in the United States
05 04 03 02 01   5 4 3 2 1

Library of Congress Cataloging-in-Publication Data

Horn, Laurence R.
A natural history of negation / Laurence R. Horn.
p. cm.
Originally published: Chicago: University of Chicago Press, c1989.
Includes bibliographical references and indexes.

ISBN 1-57586-336-7 (paper : alk. paper)

1. Grammar, Comparative and general—Negatives. 2. Negation (Logic).
3. Language and languages—Philosophy. I. Title.
P299.N4 H67 2001
415—dc21 2001037822
CIP

∞ The acid-free paper used in this book meets the minimum requirements of the American National Standard for Information Sciences—Permanence of Paper for Printed Library Materials, ANSI Z39.48-1984.

CSLI was founded early in 1983 by researchers from Stanford University, SRI International, and Xerox PARC to further research and development of integrated theories of language, information, and computation. CSLI headquarters and CSLI Publications are located on the campus of Stanford University.

CSLI Publications reports new developments in the study of language, information, and computation. In addition to lecture notes, our publications include monographs, working papers, revised dissertations, and conference proceedings. Our aim is to make new results, ideas, and approaches available as quickly as possible. Please visit our web site at
http://cslipublications.stanford.edu/
for comments on this and other titles, as well as for changes and corrections by the author and publisher.

Contents

A Note on Notation

I adopt the following typographic conventions:

'Single quotes' are used for direct quotes and for glosses.
"Double quotes" are used for quotes within single quotes and as scare quotes (for terms whose utility or applicability I am questioning).
Italics are used for cited expressions.
SMALL CAPS are used for the first relevant mention of a technical term defined explicitly or ostensively in text.
Underlining is used for emphasized or highlighted material.
Boldface is used for several technical abbreviations, with **sans-serif** and **serif** styles distinguished for ease of reference.

I adopt the following standard and nonstandard logical notations:

small boldface letters (**p**, **q**) for proposition variables
capital boldface letters (**P**; $\mathbf{P}_i$, $\mathbf{P}_j$, . . .) for predicate variables
early (α, β) and late (ϕ, ψ) Greek letters also for argument and proposition variables respectively
$\mathbf{p} \wedge \mathbf{q}$ for '**p** and **q**'
$\mathbf{p} \vee \mathbf{q}$ for '**p** or **q**' (inclusive disjunction)
$\mathbf{p}$ w $\mathbf{q}$ for '**p** or **q** but not both' (exclusive disjunction, whose status as a logical operator I reject in chapter 4)
$\mathbf{p} \rightarrow \mathbf{q}$ for 'if **p** then **q**' (material conditional)
$\mathbf{p} \leftrightarrow \mathbf{q}$ for '**p** if and only if **q**' (material equivalence or bi-conditional)
$\forall \mathbf{x} \phi$ for 'for all **x**, ϕ' (universal quantification)
$\exists \mathbf{x} \phi$ for 'for some [at least one] **x**, ϕ' (existential quantification)
$\Box \phi$ for 'necessarily ϕ', 'ϕ is necessary'
$\Diamond \phi$ for 'possibly ϕ', 'ϕ is possible'
$\vdash \mathbf{p}$ (Frege's assertion sign) for '**p** is asserted'
$\mathbf{p} \Vdash \mathbf{q}$ for '**p** (logically or semantically) entails **q**'
$\mathbf{p} \dashrightarrow \mathbf{q}$ for '[the utterance of] **p** conversationally implicates **q**', '**q** follows not logically but pragmatically from **p**'

For negation, $\sim$**p** is used to denote the standard contradictory propositional negation undifferentiated for the internal/external parameter (see chapter 2) and for Aristotle's predicate denial (see chapter 1). $\neg$**p** and $-$**p** are used to denote internal (contrary) and external (contradictory) negation, respectively, when this distinction is relevant (cf. §2.4).

These conventions are overridden when material from Montague's works are cited (in chapter 7), where his conventions are adopted. (Similarly, the conventions employed within Generalized Phrase Structure Grammar are followed when that material is cited.)

As diacritics on sentences, * denotes formal ungrammatically, ? marginal status, and # pragmatic anomaly. Dialectically acceptable examples are annotated with %. **X*(Y)Z** signals that the string **XYZ** is grammatical but the string **XZ** is not, while **X(*Y)Z** signals the opposite. I use {**A**/**B**} as a horizontal variant of $\{\begin{smallmatrix}\mathbf{A}\\ \mathbf{B}\end{smallmatrix}\}$, for '**A** or **B**'. Other nonstandard or semistandard linguistic terminology (e.g., ` and ˇ for falling and fall-rise contours respectively, $\langle \mathbf{P}_i, \mathbf{P}_j, \ldots, \mathbf{P}_n \rangle$ for quantitative scales) are introduced and defined as needed.

Acknowledgments

George Lakoff once (in Davidson and Harman 1972:651) cited a work he dubbed '*Studies in the Semantics of Negation* (L. Horn, doctoral dissertation in preparation).' This reference to the dissertation I didn't write in 1972 is extensionalized in the present study, which has indeed been in preparation ever since. Given both the history and the subject matter of this document, it would be easier, and no doubt fairer, for me to offer negative acknowledgments to all those to whom I am not indebted for help. But I shall not attempt to do so. Rather, I will simply list some whose comments, criticisms, collaboration, or support has been particularly salient—Barbara Abbott, Jay Atlas, Samuel Bayer, Herb Clark, Benoît de Cornulier, Robin Cooper, Warren Cowgill, Alice Davison, David Dowty, Donka Farkas, Georgia Green, Robin Horn, Polly Jacobson, Ruth Kempson, Steve Levinson, Jim McCawley, Jerry Morgan, Barbara Partee, Jerry Sadock—without implicating that the help provided by others was any more dispensable. These colleagues, named and nameless, deserve much of the credit for the virtues of this study and none of the blame for its vices; nor does their perception of which is which necessarily coincide with my own. This book owes its title to Haj Ross and his eponymous (but non-historical) 1972 seminar on negation at MIT.

Support for the research that made this book possible came from the National Endowment for the Humanities (Grant F73-313), the Project on Language Universals, the University of Wisconsin Graduate School, and especially the Morse Fellowship Fund at Yale University.

My wife Judy and my children David and Meryl provided less tangible but no less vital support, vouchsafing me those countless hours to spend with my dusty tomes, Microsoft Word, and basement Mac. This book is dedicated to the memory of my parents, Maxine and Bob Horn, without whom not.

permission of Faber and Faber Ltd. from *The Whitsun Weddings* by Philip Larkin.

Part of chapter 5 appeared as "Short-Circuited Implicature," by L. R. Horn and S. Bayer, in *Linguistics and Philosophy* 7, no. 4 (1984): 397–414. © 1984 by D. Reidel Publishing Co. Reprinted by permission.

Chapter 6 is revised and expanded from "Metalinguistic Negation and Pragmatic Ambiguity," which appeared in *Language* 61 (March 1985): 121–74. Reprinted by permission.

Introduction

> What the world most needs today are negative virtues—not minding people, not being huffy, touchy, irritable or revengeful. Positive ideals are becoming a curse, for they can seldom be achieved without someone being killed, or maimed, or interned. (E. M. Forster, 1939)

All human systems of communication contain a representation of negation. No animal communication system includes negative utterances, and consequently none possesses a means for assigning truth value, for lying, for irony, or for coping with false or contradictory statements (cf. Altmann 1967:353–55). The distinction between the largely digital nature of linguistic representation in human language and the purely analog mechanisms of animal communication (Sebeok 1962) can be argued to result directly from the essential use humans make of negation and opposition. If we are by definition the animals that talk, we are ipso facto the animals that deny, for as Spinoza and Hegel argue, any linguistic determination directly or indirectly involves a negation.

The study of the concept of opposition and of its expression in negative words and statements has engaged the close and often passionate attention of linguists, logicians, metaphysicians, and philosophers of language from Plato and Aristotle to the scholars of today. The explication of negation in natural and formal language has produced some of the most important linguistic discoveries (and arguably some of the most important linguistic errors) of thinkers as diverse as Aristotle, Russell, Frege, Bergson, Jespersen, Wittgenstein, Strawson, and Searle.

One contrast has proved especially provocative: despite the simplicity of the one-place connective of propositional logic ($\sim$**p** is true if and only if **p** is not true) and of the laws of inference in which it participates (e.g., the Law of Double Negation: from $\sim\sim$**p** infer **p**, and vice versa), the form and function of negative statements in ordinary language are far from simple and transparent. In particular, the absolute symmetry definable between affirmative and negative propositions in logic is not reflected by a comparable symmetry in language structure and language use. Much of the speculative, theoretical, and empirical work on negation over the last twenty-three centuries has focused on the relatively marked or complex nature of the negative statement vis-à-vis its affirmative counterpart.

Negation still occupies the crossroads of developments in linguistic theory, psycholinguists, the philosophy of language and of mind, and the history of ideas. For semanticists and pragmaticists, negation must be investigated not only because of its unique position among the constants of classical logic as the one-place truth-functional connective, but also for its complex and systematic interaction with the other logical operators, especially the quantifiers and modals. Its complex behavior in lexical incorporations (*impossible, can't, prevent*) provides vital data on the nature of constraints on word formation. Psychologists and psycholinguists have adduced empirical evidence, based on language acquisition data and on delayed response latencies associated with the processing of overt and inherent negation, for particular conclusions about the mental and linguistic structure of negative lexical items and ultimately about the psychological reality of semantic representation itself. Philosophers and linguists have argued from particular views on the alleged ambiguity of negative sentences (*The number 2 is not red; The king of France is not bald*) to conclusions concerning the existence of logical presuppositions and truth-value gaps and the domains of meaninglessness and ambiguity in semantic theory.

These are among the issues I shall be examining in this study. In addition to providing a historical perspective on the place of negation within classical, traditional, and modern investigations of language and thought, I shall try to show how recent developments in formal theory apply to the analysis of negative statements and how results from work on negation affect (or should affect) current work within semantic and pragmatic theory. We shall see why negation is to the linguist and linguistic philosopher as fruit to Tantalus: waving seductively, alluringly palpable, yet just out of reach, within the grasp only to escape once more.

The history begins in chapter 1 with a sketch of the Aristotelian theory of negation and of the development of that theory and its heirs through 2,300 years of commentary. The major issues introduced in the first chapter will extend through the remainder of the study: the distinction between contradictory and contrary opposition, the purported scopal ambiguity of the negative statement and its relation to existential import (does *Socrates is not well* entail that Socrates exists?), the proper formal treatment of wide-scope negation as a mode of predication (as in Aristotle) or as a one-place external propositional connective (as in the Stoics' model), the definition and scope of the Law of Contradiction and the Law of Excluded Middle, the semantics of general or quantified negative expressions, the nature of subcontrary "opposition" (*Some men are white / Some men are not white, It is possible to go / It is possible not to go*), and the character of the asymmetry between negation and affirmation.

This last issue is important enough in both classical and contemporary negative polemics to earn its own section, §1.2, where I trace a variety of attempts seeking either to eliminate negation entirely from ontology and logic or to reduce contradictory negation to an antecedently understood and purportedly more basic concept (e.g., falsity or contrariety). The historical overview concludes in §1.3 with a survey of the place of negation within the history of ideas in both India and the West.

In chapter 2 I look in greater detail at one of the issues raised in the first chapter: the circumstances under which we can isolate some meaning component of a sentence by its constancy under negation as a SEMANTIC PRESUPPOSITION, such that if that component is not satisfied in the context of utterance the sentence (or the statement it expresses) must be neither true nor false. I shall examine three candidates for the status of presuppositionality: the (positive or negative) statement concerning future contingent events (in §2.1), the (positive or negative) singular statement with a vacuous, nonreferring subject (in §2.2), and the (positive or negative) category mistake or selectional violation (in §2.3).

As we shall see, the claim that presupposition failure results in a loss of bivalence and/or a violation of the law of excluded middle, yielding truth-value gaps or the assignment of a third, nonclassical truth value, appears to be tenable only if a systematic semantic ambiguity is posited for negation. On such an account, a presupposition-preserving INTERNAL or CHOICE negation contrasts with a presupposition-canceling EXTERNAL or EXCLUSION negation. I shall investigate (first in §2.4, and again in chapter 6) a variety of approaches for describing and representing such an ambiguity. Finally, in §2.5, I touch briefly on one contemporary approach to presuppositional phenomena in which no truth-value gaps arise, but negation nevertheless retains its presupposition-theoretic ambiguity. The concerns of chapter 2 return to haunt the last two chapters, when I reconsider the presuppositional evidence and its consequences for the proper treatment of the purported ambiguity of negation.

Chapter 3 picks up where another part of chapter 1 leaves off. In §1.2, I chronicled the campaigns of those seeking to expose the ontological dubiousness, the epistemological worthlessness, or the linguistic inferiority of the negative statement with respect to its positive counterpart. I now scan the empirical record to find similar views being proposed (often in total independence of the rich philosophical tradition) and defended with evidence from the psycholinguistics of language acquisition (summarized in §3.1) and language processing (in §3.2).

The view of the negative statement as the MARKED member of the positive/negative opposition, outlined at the start of chapter 3, is supported by naturalistic observations and experimental data, the latter represented by a series of consistent and interrelated results from studies by H. Clark, P. C.

Wason, and their colleagues. These studies tend to confirm the standard view that negative statements are harder to verify than their affirmative counterparts, that the difficulty posed by negation correlates directly with the implausibility in the context of the corresponding affirmative supposition, and that overt negation presents more problems to the language processor than does inherent or implicit negation.

But just where should this asymmetry between negation and affirmation be situated within a formal theory of language? And how are we to represent the relation between the negative statement and the "affirmative (pre)supposition" taken (by both philosophical and psycholinguistic asymmetricalists) to underlie it? In the last section of chapter 3, the results from these studies are reevaluated and interpreted within the pragmatic theory of nonlogical inference originating in Grice 1975 and reformulated in Horn 1984b. Once the 'presuppositionality' of negatives is properly explained within the pragmatic theory of language use, we can appreciate the divergent goals of the symmetricalists and asymmetricalists and propose at least an armed truce in the polemic warfare between logicians and functionalists which has characterized so much of the history of negation.

The first three chapters are primarily addressed to semantic or psychological questions posed by negative facts, statements, and propositions and their representations within speculative and formal theories of language and thought. In the next three chapters, I turn to questions arising directly from the pragmatics of negation: the function and use of negative (and positive) statements as affected by the context of utterance. The relation between negation and implicature, with which chapter 3 concludes, returns in a variety of guises within the following three chapters, as does the dualistic taxonomy of implicature defined in Horn 1984b and summarized here in §3.3.

Chapter 4 is an update of the core of my 1972 dissertation, *On the Semantics of Logical Operators in English,* and of my 1973 "Greek Grice" paper. The focus is on the relation between the one-sided (less than *p*) and the two-sided (exactly *p*) readings that tend to be available for a scalar operator or predicate *p* in sentences like those in (1), where the scalar value is underlined:

(1)	Chris has three children.	[vs. more than three]
	Some of my friends are linguists.	[vs. most or all]
	It's possible that it will rain today.	[vs. likely or certain]
	Serkin's performance was good.	[vs. excellent]
	Kim is happy.	[vs. ecstatic]

The central questions that arise for the analysis of these sentences and of their negative counterparts are:

(1) What is the relationship between the 'less than *n*' and 'exactly *n*' understandings arising in scalar predications? (Are these two distinct readings of a lexically ambiguous operator, or can one reading be derived from the other by some general, independently motivated pragmatic mechanism?)

(2) Why is negation in the same contexts:

Chris doesn't have three children.	(= 'less than three')
It's not possible that it will rain today.	(= 'less than possible', i.e., impossible)
Kim is not happy.	(= 'less than happy', i.e., unhappy, neutral, or indifferent)

normally interpreted as 'less than *n*', as noted by Jespersen (1917)?

(3) How are we to analyze the marked instances of scalar negation which are not restricted to the less-than interpretation (cf. again Jespersen 1917)? Thus, for example, we can get:

> Chris doesn't have three children, he has four.
> It's not possible that it will rain today, it's downright certain.
> Kim is not happy—she's ecstatic.

It is argued in chapter 4 that the relationship in question in paragraph (1) is that of a generalized conversational implicature (Grice 1975) arising through the exploitation of the Maxim of Quantity. The less-than-*n* readings of scalar negations in (2) thus constitute straightforward contradictory negations. Alternative analyses are considered and argued against, and a new notational convention is introduced—superimposing the quantitative and pragmatic scales of Horn (1972), Ducrot (1973), Fauconnier (1975a, 1975b) and others onto the classical Square of Opposition (cf. chapter 1 of this book)—for depicting the logical and pragmatic relationships obtaining among quantifiers, modals, and related scalar operators. Earlier accounts of scalar terms by pre-twentieth-century logicians and by such linguists as Jespersen and Sapir are presented, evaluated, and compared with the Gricean model advocated here. I also consider some consequences of Barwise and Cooper's (1981) theory of generalized quantifiers for the current treatment. (I defer the issue raised in (3) above until chapter 6.)

Two related topics which receive special treatment in chapter 4 are:

1. The complex scope relations between universal expressions and negation, as evidenced in the ambiguity of *All the boys didn't leave* (= 'all . . . not . . .' vs. 'not all . . .'). This is investigated here in §4.3 and again in the last section of chapter 7.

2. The nature and extent of the conspiracy against the lexicalization—or often even the direct representation—of logical complexes corresponding to the **O** vertex (southeast corner) of the classical Square of Opposition. This constraint, originally proposed in my thesis (Horn 1972: chapter 4), is exemplified by the lack of a simplex realization for *not all, not always, not everybody,* and *not [logically] necessary,* alongside the existence of lexicalizations like *no(ne)* (= 'all . . . not'), *never* (= 'always . . . not'), *nobody* (= 'everybody . . . not'), and *impossible* (= 'not [logically] possible') for complexes corresponding to the **E** vertex (northeast corner) of the Square. This is discussed in §4.5.

In chapter 5, abbreviated versions of which were presented at the December 1985 LSA meeting in Seattle and in talks at Brown and Yale the following spring, I turn to another aspect of the pragmatics of negation and scalar predicates in natural language: the inference to a stronger or more specific interpretation triggered by a certain definable range of scalar negations. A century ago, the Idealist philosopher Bosanquet (1888) observed that 'the essence of formal negation is to invest the contrary with the nature of the contradictory'. While contradictories (*black/not black, odd/even*) exclude any middle term, contraries in principle do not; my shirt may be neither black nor white. But the context may fill in the gap between the contraries, establishing a disjunction of the type normally associated with contradictories. The middle is not so much excluded as pragmatically absorbed, and **p or q** becomes an instance of **p or not-p**. One context triggering this absorption is the black-or-white, centrifugal politics of polarization ('He that is not with me is against me', 'If you're not part of the solution, you're part of the problem'). Applied to scalar predicates, this same polarizing tendency motivates three fundamental processes in the universal morphosyntax of negation, processes which have never been fully explained—in part because they have never been seen as reflecting the same functional dynamic.

(*a*) The simple base **X** of an affixal negation of limited productivity (*un-***X**, *iN-***X**, *dis-***X**) is scalar (gradable) and typically positive, but does not represent an extreme scalar value; the derived form is interpreted as the contrary (antonym) rather than the contradictory of its base:

> *unhappy* vs. **unecstatic*, **unsad*, **unmiserable*
> *dislike* vs. **dislove*, **dishate*

(cf. also the scalar, contrary values of affixal negations like *un-American* [vs. *non-American*], *irrational* [vs. *nonrational*], *uneven* [≠ 'odd'], etc.).

(*b*) The apparently contradictory negation of a second-order predicate takes on a 'neg-raising' (NR) understanding (**not F** = **F** . . . **not** . . .) which can be seen as the assertion of a contrary, but only when that predicate is a weak positive value just above the midpoint of its scale:

*I don't believe that **p*** [= I believe that ~**p**] (no NR with *know, doubt, disbelieve*)
*It's not likely that **p*** [= it's likely that ~**p**] (no NR with *(im)possible, certain*)
*I don't advise you to **VP*** [= I advise you not to **VP**] (no NR with *insist, forbid*)

(*c*) The unincorporated, apparently contradictory negation of a scalar predicate allows a stronger, contrary understanding, but only when the original value is positive and relatively weak:

He's not happy [contrary] vs. *He's not {ecstatic/sad/miserable}* [contradictory]
I don't like you [contrary] vs. *I don't {love/dislike/hate} you* [contradictory]
It's not right [contrary] vs. *It's not {perfect/wrong/sinful}* [contradictory]

The correlation of these processes involves one repeated premise: in a context licensing the pragmatic assumption ***p*** v ***q*** to assert ***p*** is to implicate ***q***. Thus, a formally contradictory negation ***not-P*** tacitly conveys a contrary assertion—but only when ***P***, the focus of negation, is a relatively weak positive scalar predicate, representing the unmarked term in its contrast set. Where the three constructions differ is in the nature and degree of conventionalization governing this pragmatic strengthening process; the inference which is general and exceptionless in the examples of (*c*) is partially fossilized as a SHORT-CIRCUITED IMPLICATURE licensing the NR readings in (*b*) and becomes fully conventionalized in the idiosyncratic lexical examples of (*a*).

The semantic, pragmatic, and morphological properties of affixal negation are discussed in §5.1, with the initial section devoted to the analysts of the semiproductive *un-* prefix and the nonproductive *iN-* prefix in English. Other negative prefixes are considered in §5.1.2, where I undertake a brief cross-linguistic survey of the semantics of negative affixation, following Zimmer 1964. In §5.1.3 I look at so-called logical double negation and in particular the motivation for the use of the *not un-* construction in English. In §5.2, a revised and abbreviated version of Horn 1978b, I reexamine the neg-raising phenomenon (cf. [*b*] above) as an instance of a partially conventionalized pragmatic strengthening process. Finally, in §5.3 a consideration of the (apparently) simple examples of (*c*) above launches my comprehensive account of the phenomenon of the negative strengthening process and the parameters of conventionalization by which it is regulated.

The euphemistic motivation of the strengthening which convert a semantic contradictory to an "acting" contrary in the contexts under consideration here—affixal negation, so-called neg-raising environments, and the litotic understanding of simple negation—is characteristic of the nature of **R**-based implicature. Thus the upper-bounding scalar inference at the heart of chapter 4 and the strengthening inference exemplified in chapter 5 are classic illustrations of the two dialectically opposed patterns of **Q**-based vs. **R**-based inference. Some consequences of this distinction will be brought out in §5.3 and (as they affect the distribution of metalinguistic negation) in chapter 6.

The final panel of the pragmatic triptych, chapter 6, depicts the marked understanding of scalar negation (see (3) above) as one instance of the general phenomenon of the METALINGUISTIC use of negation. As I observe in chapter 2, marked or "external" negation has typically been treated as an additional semantic operator alongside the straightforward truth-functional, presupposition-preserving ordinary ("internal") negation. Alternatively, some have chosen to reject the putative ambiguity of negation, along with the existence of semantic presuppositions, and to collapse the internal and external operators into one unified general truth-conditional operator on propositions.

I argue in chapter 6, a revised and extended version of Horn 1985, that neither of these two approaches (summarized in §6.1) does justice to the differences and kinships between and within these two manifestations of negation in natural language. Marked negation is not reducible to a truth-functional one-place connective with the familiar truth table associated with logical negation, nor is it definable as a distinct logical operator; it represents, rather, a metalinguistic device for registering an objection to the content or form of a previous utterance (not to a proposition) on any grounds whatever, including the implicatures (conventional and conversational), the grammatical and phonetic form, or the register associated with that utterance.

In §6.2 I extend the range of data beyond the presupposition-canceling external negation considered by the multivalued logicians and their monoguist rivals and defend the view that negation is not semantically but PRAGMATICALLY ambiguous, with its metalinguistic use as a signal of objection motivated by the nature of its truth-conditional meaning. Other logical operators are shown to contain their own analogous extended metalinguistic functions. In §6.3 I consider the interaction of metalinguistic negation with both the **Q**-based scalar implicata of chapter 4 and the **R**-based strengthening implicata of chapter 5; an asymmetry in this interaction is described and motivated. The remainder of the chapter is devoted to the investigation of some phonological, morphosyntactic, and distributional correlates of metalinguistic negation, the examination of other recent attempts to explain (or explain away) the phenomena under investigation, and the cross-linguistic realization of the two uses of negation.

In the concluding chapter I attempt to gather together the narrative strands that have been unrolling since my exposition of the Aristotelian theory of predicate denial and term negation in §1.1.1. I investigate a set of interrelated issues concerning the form, function, and meaning of negation in natural language, beginning in §7.1 with a survey of the various parameters affecting the synchronic character of negation in surface sentences and the development of negation over time. I explore and seek to motivate the process known as JESPERSEN'S CYCLE, the tendency for preverbal negation to weaken gradually until it is reinforced by postverbal indefinites or expressions of minimal quantity to which it eventually yields its negative force. The original negation then withers away and disappears, while the new postverbal negation may in its own turn gravitate leftward toward the preverbal position again, either directly or (as in English) indirectly, setting the stage for a repetition of the sequence. This leftward drift of negative morphemes represents another functional principle recognized by Jespersen and investigated here, which I label NEG FIRST: the hypothesis that a negative morpheme tends, for functional reasons, to precede the element on which it focuses. (I invoke this same principle in §5.1 to explain the strong typological preference for prefixal over suffixal negation.)

One striking result of the cross-linguistic investigations into the typology of negation surveyed here (cf. Dahl 1979; Payne 1985) is the extreme rarity of syntactic external negation, that is, the absence of negative morphemes in the position classical propositional logic would lead us to expect to find one-place sentential connectives. I seek to explain this gap in §7.2 by reviving and extending classical term logic, the theory of predication founded by Aristotle and the Peripatetics (§1.1) but repudiated by the Stoics and, since Frege, by practitioners of modern formal logic. In this program, unconjoined sentences retain a canonical subject-predicate (categorical) organization, permitting an accord between logical and grammatical form in the spirit of both Aristotle and, I argue, Montague.

The iterating, external sentential negation of standard propositional logic, absent from both the Organon and Montague English, is systematically excluded from Extended Term Logic; apparent instances of sentential negation involve a metalinguistic use of the negative operator (cf. chapter 6), while ordinary contradictory negation associated with the auxiliary or finite verb is a MODE OF PREDICATION, a syncategorematic operation on subjects and predicates. This analysis of wide-scope descriptive negation is contrasted with the traditional modern view of negation as a one-place "connective" on fully formed propositions, and my version of Extended Term Logic is contrasted with the more radical neo-Aristotelian proposals in the work of Fred Sommers.

In the final section, §7.3, I review some of the problems arising in the interaction of negation with presupposition and the scope of quantified expressions. My approach predicts the full range of scopal interpretations for negation and generalized quantifiers within simple sentences, including a set of apparently impossible scope possibilities. I suggest a functional explanation for the asymmetry between the (virtually) nonexistent wide-scope (Q-NEG) understanding for negation following an existential (*some . . . not*) and the more readily available wide-scope reading for negation following a universal (*all . . . not*). I argue that the apparent scope of an overt negative with respect to quantifiers (as here) and presuppositions (see chapters 1 and 2) is largely determined pragmatically by the context of a given utterance and by the availability of alternative unmarked expressions, rather than being predictable from syntactic and logical form alone.

I seek in this disquisition to show how the study of negation in natural language has been informed by, and how it informs, research into the character of logical form, the nature of implicature and presupposition, and the delineation of the semantics/pragmatics borderline within linguistic theory and the philosophy of language. More broadly, my study is intended both as a synthesis of much of the significant work on negation and related topics from the last 2,500 years and as a current perspective on the roles of negation in natural language. May it also serve as a negative image for the pictures to be developed in the laboratory of tomorrow.

Introduction to the Reissue Edition

Since the initial appearance of *A Natural History of Negation* in 1989, there has been an explosion of interest in the grammar, semantics, pragmatics, and psycholinguistics of natural language negation. Without implicating the R-based principle of *post hoc propter hoc* (cf. Horn 2000b), we can begin by noting that the decade following the publication of *NHN* witnessed the following conferences and workshops devoted to the properties of negation and polarity:

April 1991, Chicago: Chicago Linguistic Society Parasession on Negation (at CLS 27); proceedings in Dobrin et al., eds. (1991)

May 1991, Corsendonk (Belgium): ESPRIT (Dialogue and Discourse) Workshop on Negation in Natural Language (chiefly semantic, pragmatic, and computational aspects)

November 1992, U. of Paris X–Nanterre (France): Colloque sur la Négation (grammar, semantics, rhetoric, psychology); proceedings in Attal, ed. (1994)

April 1993, Boston University: Workshop on Negation in the Kwa Languages of West Africa

June 1994, U. of Groningen (Netherlands): PIONIER Colloquium on Negation and Polarity (mostly formal semantics, with some pragmatics and some syntax)

September 1994, U. of Leipzig (Germany): Conference on Negation (mostly philosophically/logically oriented); proceedings in Wansing, ed. (1996)

May 1995, U. of Ottawa (Canada): Conference on Negation–Syntax and Semantics (mostly Principles & Parameters/Minimalist syntax and formal semantics); proceedings in Forget et al., eds. (1997)

August 1996, U. of Groningen (Netherlands): Conference, Perspectives on Negation; proceedings in Hoeksema, et al., eds. (2001)

I am indebted to Anastasia Giannakidou, Yasuhiko Kato, and Akiko Yoshimura for their comments on an earlier version of these remarks.

August 1997, Krakow (Poland): Conference on Positive-Negative Asymmetry and Reasoning (largely on the social psychology of negation)

November 1997, Tokyo Metropolitan University (Tokyo, Japan), The English Linguistic Society of Japan: Workshop on Negative Polarity in Current Linguistic Theories (Minimalist, HPSG, and logico-semantic approaches)

October 1998, U. of Salford (European Studies Research Centre) Conference on Negation: Syntax, Semantics and Pragmatics; partial proceedings in *Transactions of the Philological Society* 98:1 (2000)

October 1999, Poznan, Poland: Workshop on the Syntax and Semantics of Slavic Negation at the 32nd Poznan Linguistics Meeting

Besides the published proceedings of these conferences and workshops, three other major anthologies, Haegeman, ed. (1993/94), Gabbay & Wansing, eds. (1999) and Horn & Kato, eds. (2000), appeared during this period. In addition, formal or informal sessions on negation and polarity were scheduled on several occasions at the annual meeting of the Linguistic Society of America and the annual SALT (Semantics and Linguistic Theory) conference. Aficionados of negation will also want to take note of the impressive bibliographies compiled by Brütsch et al. (1990) and Seifert & Welte (1987) as well as the valuable descriptive work in Tottie's (1991) empirical study of the manifestations of negation in spoken and written English.

One reason for the surge of interest is the central role of sentential negation in the development of contemporary syntactic theory. In the generative models of negation surveyed in *NHN*, the principal bone of contention was whether the abstract marker of sentential negation originates as a pre-sentential syncategorematic operator (as in Klima), in Comp (as in Lasnik), as a higher predicate (as in generative semantics) or within the auxiliary or predicate (as in GPSG). Within Chomsky's Principles and Parameters model, Pollock (1989) advanced the influential Split or Exploded IP hypothesis, according to which functional elements like negation, agreement and tense are analyzed in terms of a head projecting a full phrasal category (see also Ouhalla 1990, Laka 1990 for early implementations of this idea). Negation is represented in this account as the Neg^0 head of the functional category NegP, and various particulars of negative polarity, negative concord, movement and scope, as well as word order typologies and the diachronic processes related to Jespersen's Cycle of successive movement, weakening, and strengthening of negative particles (see *NHN* §7.1), have been dealt with in these terms.

Landmarks in the study of the syntax of negation over the last decade include the work of Zanuttini (1991, 1997) and Haegeman (1995), who have also collaborated on the development of the Neg Criterion, a well-formedness constraint on the licensing of negation through a Spec-Head relation between a neg operator and a Neg^0 head (Haegeman & Zanuttini 1991). Zanuttini has also stressed the possibility that the "heavier" postverbal negatives in some Romance varieties as well as in English and German may appear as XP adjuncts, accounting for some of the observed cross-linguistic variation in negated structures. These investigations into the expression of sentential negation have helped drive recent developments in syntactic theory and continue to inform our understanding of the nature of the syntax-semantics interface and of its consequences for the nature of parametric variation within Universal Grammar.

Just as Haegeman's and Zanuttini's research was inspired in part by the character of negation in West Flemish and in comparative Romance respectively, many other significant theoretically oriented publications during this period were also informed by the descriptive details of negative structures in particular languages. These include ASL (Copley 1996), Berber (Chaker & Caubet 1996), Chinese (Ernst 1995), English (Baker 1991, Ernst 1992, Potsdam 1997), French (Rowlett 1998, Larrivée 2000), Greek (Giannakidou 1997, 1998), Italian (Acquaviva 1997), Hungarian (Puskás 1998), Japanese (Kato 2000), comparative Romance (Déprez 2000), Russian (Brown 1999), and Serbo-Croatian (Progovac 1994). Ladusaw (1992, 1994, 1996a,b) has also insightfully investigated the interplay between the syntactic analyses of Laka, Zanuttini, et al. and the formal semantic properties of negation, polarity, and concord.

In addition to the P&P/Minimalist approaches outlined above, Kim (2000) treats negation within the competing constraint-based (HPSG) framework, and Drozd (1993) applies the same theory to the grammar of negation in child language, while Payne & Chisarik (2000) advance an Optimality-theoretic account of negation and focus in Hungarian. In fact, the cross-linguistic interaction of focus and the scope of negation has proved to be extremely fruitful area of research within a variety of syntactic and semantic frameworks; in addition to the papers in Hoepelman & Schnitzer, eds. (1991), see inter alia Mufwene (1993), Yeh (1995), Hajičová (1996), Büring (1997), Rohrbaugh (1997), Stroik (1997), and Herbuger (2000).

I should also acknowledge three important comprehensive treatments of the syntax and semantics of negation from the previous decade that appeared in languages other than English and that were inexcusably ignored in the preparation of *NHN*, those of Ota (1980), Bosque (1980) and Jacobs (1982) (cf. also Jacobs 1991).

A related strand of research on negation in the '90s is the pursuit of the grail of negative polarity (see *NHN* §5.3.1, §6.4.2, §6.5.1, and Chapter 7). As with the study of negation proper, the investigation of polarity phenomena in the '90s has been enriched by cross-linguistic results from a wide range of languages, including classical Armenian (Klein 1997), Catalan (Vallduví 1994), Chinese (Lin 1996), Dutch (van der Wouden 1996b), French (Muller 1991), Greek (Giannakidou 1997 et seq.), Hiberno-English (Duffield 1993), Hindi (Lahiri 1998), Hungarian (Tóth 1999), Italian (Tovena 1998), Korean (C. Lee 1996), Japanese (Aoyagi & Ishii 1994, Kawashima & Kitahara 1992, Kato 1994, Kuno 1995), Moroccan Arabic (Benmamoun 1997), Serbo-Croatian (Progovac 1994), and South Asian languages (Bhatia 1995).

As Ladusaw (1996a: 326ff.) has observed (although see von Klopp 1998 for a contrary view), the most fundamental problem in polarity research is to determine the character and membership of the class of negative contexts. To this "licensor question", the answer that Ladusaw and other formal semanticists have developed over the last two decades builds on the insights of earlier scholars into the nature of scalar predication and scale reversal and the property of monotonicity within the formal theory of generalized quantifiers (cf. *NHN* §4.4). Ladusaw identified the set of environments licensing negative polarity items (NPIs) with the semantic notion of downward entailment, the property of licensing inferences from sets to subsets, from the general to the specific, thereby providing content to the [+affective] feature Klima associated arbitrarily with NPI-inducing contexts. Positive polarity items receive far shorter shrift in both syntactic and semantic treatments (but see Progovac 1994, Israel 1996), but are standardly taken to be anti-triggered by the same operators that trigger NPIs.

We have cited work on the formal syntax of negation in the 1990s that offers a configurational account of the distribution of, and restrictions on, NPIs (see also Uribe-Echevarria 1994 for an overview). An earlier syntactic (more properly, syntactico-pragmatic) analysis of polarity was developed by Linebarger, who has continued to refine her theory to address the problems posed by an insufficiently restrictive nature of the implicature component of her approach (Linebarger 1991; see also the critique in Yoshimura 1999). Progovac (1993, 1994) has advanced her own configurational account based on a generalization of binding theory; see Horn & Y. S. Lee (1995) for a critical evaluation. Meanwhile, others have refined and extended the theory of downward-entailing contexts to address the objections and counterexamples to Ladusaw's theory raised by Linebarger and others–cf. Krifka (1991, 1995), Kadmon & Landman (1993), Dowty (1994), Horn (1996, 2001), von Fintel (1999), and the members of what we might call the Groningen school of negative polarity (e.g. Zwarts 1991,

1998, Kas 1993, Sánchez Valencía 1994, Jackson 1994, Hoeksema 1994, Rullmann 1996, van der Wouden 1996b, Giannakidou 1997); see also the papers collected in special issues and anthologies edited by Hoeksema (1995, 2001), Forget et al. (1997), and Horn & Kato (2000).

In the formal semantics of polarity, two complementary developments are (i) the sharpening of the notion of monotonicity into a Boolean algebra of interdefined licensing conditions that correlate with NPI distribution (cf. Kas 1993, van der Wouden 1996b, Atlas 1997, Vasishth 1997) and (ii) the elaboration of the notion of non-veridicality as the core semantic licensing property (cf. Zwarts 1995, Giannakidou 1997, 1998, 1999, Hoeksema 1998, Toth 1999, Pereltsvaig 2000). Within the larger battles on the nature of negative polarity, a lively front has erupted over the proper treatment of "free choice" items and the possibility of developing a unified indefinite account of both NPI and free choice *any*: cf. Kadmon & Landman (1993), Y. S. Lee & Horn (1994), C. Lee (1996), Quer (1998), Giannakidou (1998, to appear), Horn (2000a), and for dissenting universalist views of free choice items, Dayal (1998) and Sæbø (to appear).

The most comprehensive overview of the semantics of polarity and related properties of negation is van der Wouden (1996b), while Haspelmath's (1997) encyclopedic descriptive typology of indefinites overlaps significantly with that of negation and polarity phenomena. (See also Ladusaw 1996a for a superb small-canvas rendering of the state of the art in negation and polarity.) The relationship between negative polarity and negative concord is treated in Ladusaw (1992, 1996b), van der Wouden & Zwarts (1993), Dowty (1994), Przepiórkowski (1999), Przepiórkowski & Kupsc (1999), and Giannakidou (2000). Additional typological and theoretical perspectives on polarity are provided by Yoshimoto (1995), Yoshimura (1992, 1994, 1999), and the papers collected in Forget et al., eds. (1997) and Horn & Kato, eds. (2000). Finally, Israel (1996, to appear) presents a creative foundational study of both negative and positive polarity that combines the downward-entailment insights of Ladusaw and the pragmatic inferences of Linebarger with his own construction grammar-oriented approach to lexical semantics and scalar models. For Israel, as for Tovena, von Klopp, Giannakidou, and Sæbø, the licensee sensitivity question–what are the lexical semantic properties of a given expression that determine its status as a polarity item?–is at least as central as the licensor question, and indeed the two are inextricably linked.

We turn now to developments that more directly relate to the themes explored in *NHN*; I shall proceed chapter by chapter, although in some cases this will prove somewhat arbitrary, since the themes of the chapters are not always discrete.

The logic of negation and its historical development, as explored in §1.1

and revisited in the Extended Term Logic outlined in §7.2, is the focus of the collections in Wansing, ed. (1996) and Gabbay & Wansing, eds. (1999); these, especially the latter, are rigorously formal in character. On the historical side, see now the definitive treatment by Pelletier (1990) of Parmenides and his influence, as well as the comprehensive (if somewhat idiosyncratic) study by Pacitti (1991). Nonexistence and the problems of negative existentials and fictional discourse are the focus of Chakrabarti (1997) and the collection in Everett & Hofweber, eds. (2000). A new edition of Peter of Spain's commentaries on negation (Spruyt, ed. 1989) will also be useful for discerning the medieval picture.

In §1.1.3, I sketched some approaches to the problem of existential import connected with the general statement forms mapped onto the Square of Opposition. Horn (1997) attempts to deal with these issues in a more systematic way, invoking the Brentano-Marty-Kuroda distinction of thetic and categorical judgments (cf. *NHN* §7.3.4) to account for the difference between weak, import-free readings and strong or presuppositional readings of both quantified and indefinite sentences. Ladusaw (1994, 1996b) draws on the same distinction to a different but related end, and much other work in the 1990s addresses this heretofore underappreciated dichotomy in mode of judgment.

Using data from Japanese narrative, Yamada (2000) offers empirical support for the asymmetricalist view of negation described in §1.2. The broader connections between negation and literary and psychoanalytic theory touched on briefly in §1.3 are the focus of some recent publications (Budick & Iser 1989, Fischlin 1994) and of an impressively ambitious study, ranging from Hegel to Freud to syntactic and acquisitional concerns, that was inexplicably overlooked in *NHN* (Ver Eecke 1984). Another synoptic treatment that eluded my web is Lloyd's encyclopedic essay on polarity in ancient philosophy, now reissued as Lloyd (1992). Ruthrof (1997), in his 32-page section on "Negation from Frege to Freud and Beyond", offers a critical rereading of the ground covered in *NHN* §1.2–1.3 (with forays into the later chapters) and advances his own anti-formalist "corporal" perspective on the asymmetry wars, the role of negation in language and consciousness, and the semantics/pragmatics division of labor.

This might be a good point to observe the passing of the highway sign for Exit 6 of the eastbound Cross-Westchester Expressway (I-287) approaching White Plains, N.Y., which stood for years as a bold refutation of the Law of Contradiction:

White Plains No White Plains

–until the powerful neo-Aristotelian lobby evidently pulled some strings with the highway commission and had the sign removed.

On the logic of negation and presupposition (*NHN*, Chapter 2), see now Burton-Roberts (1989), Atlas (1989, Chapter 3), and von Fintel (1999). Conventional implicature, the central topic of §2.5, has not fared well over the intervening years and was finally declared dead in Bach (1999), although Elvisesque sightings are periodically reported.

Among the more important recent treatments of the acquisition of negation (*NHN* §3.1) are Drozd (1993, 1995), Koster & van der Wal (1995), and van der Wal (1995, 1996). Moxey & Sanford (1993) have offered a new approach to the processing of negative (and affirmative) quantified sentences. For some legal applications of the known processing difficulties posed by negation, especially multiple negation, see Horn (1995). The dualistic model of inference (with its distinction of Q-based and R-based implicature and the division of pragmatic labor derived from them) invoked in §3.3.1 and later chapters of *NHN* is further elaborated in Horn (1991, 1993), and a related framework is unveiled in the landmark treatise on generalized conversational implicature by Levinson (2000).

Q-based scalar implicature represents the core of *NHN*'s Chapter 4, which also touches on the Relevance theoretic arguments for the radical underspecification of–and pragmatic intrusion into–propositional content. The 1990s saw a lively give-and-take among neo-Griceans, post-Griceans, and intermediaries in a debate over the proper treatment of scalar predication and other instances of pragmatically determined meaning: see Carston (1988, 1995) and Récanati (1989, 1993) for variants of the "explicature"-based RT approach to what is said, Horn (1992) and Levinson (2000) for modified versions of the classical Gricean line, and Bach (1994, 2001) for a persuasive alternative view on which non-literality is taken as primarily involving implicit components of meaning that are not part of what is said–implic*i*tures as distinct from both the explicatures of RT and the later-computed implic*a*tures of Gricean pragmatics.

The historical roots of scalar implicature, touched on in §4.1 and §4.2, are examined in more detail in Horn (1990a). The natural history of negation's disjunctive cousin is treated impressively in Jennings (1994) and the relationship of modality to negation and scalarity is the focus of van der Auwera (1996) and de Haan (1997). The quantifier-negation scope issues explored in *NHN* §4.3 (and later reprised in §7.3) are insightfully examined in the light of formal semantics, syntax, and intonation in Büring (1997), while the interaction of scalar structures, lexical semantics, and negation, discussed in §4.4, is revisited in Lundquist & Jarvella (1994), Israel (1996), and Schwenter (1999). Herburger (2000) presents a coherent neo-Davidsonian event-structural account of the interaction of scope of negation with focus and quantification. Another topic of discussion in §4.4 is the semantics of *only*, and in particular the question of whether NPs of the form *only a* are or are not monotone decreasing. A flurry of exchanges on this topic blanketed the pages of the *Journal of Semantics* during the 1990's, with Atlas (1991, 1993, 1996) analyzing *only a* as non-monotonic and essentially conjunctive and Horn (1992, 1996) stressing the negative and hence asymmetric properties of *only* phrases (e.g. as licensers of NPIs). It is safe to say that the dispute will grace the 21st century as it did the late 20th (or, indeed, the 13th, as the historical discussion in these papers makes clear).

The functional asymmetry of the Square of Opposition–in particular, the failure or reluctance of O-vertex values (*not all, not necessary*) to lexicalize–is attributed in *NHN* §4.5 to a combination of Gricean pragmatics (in that the two subcontraries Q-implicate each other) and the markedness of negation (whence the preference for I-vertex lexicalizations); see also Horn (1990a). Rival explanations for this asymmetry have since been proposed; cf. Hoeksema (1999) and the discussion and references in van der Wouden (1996b: 100, fn. 47). The reader will not be surprised to learn that I do not find these alternate explanations compelling; they strike me as incapable of generalizing beyond the quantifiers to the full range of lexicalization phenomena addressed in §4.5.

Chapter 5 explored three dimensions of R-based strengthening, the tendency to conceal the meaning of contrary negation within the clothing of contradictory sentential negation. The morphology-pragmatics interface issues explored through the constraints on un-verb formation in §5.1.2 are investigated more carefully in Horn (1988), which also describes the interplay of the Q and R principles in licensing redundant reversative verbs (*unthaw, unloosen, debone, dissever*). The same dualistic model is applied in Horn (1991) to the semantic and rhetorical motivation for "logical double negation" (*not impossible, not unhappy*) as earlier sketched in *NHN* §5.1.3, and both these constructions are reconsidered alongside other instances of the division of pragmatic labor in Horn (1993).

The phenomenon of so-called neg-raising, described under the rubric of short-circuited R-based implicature in §5.2 and §5.3.1, is revisited in Nuyts (1990), Bublitz (1992), Horn (1998a), and Tovena (2001). Litotes, the core theme of §5.3 (and revisited in Horn 1991), is examined insightfully from diachronic, pragmatic, and logical perspectives by Hofmann (1987) and van der Wouden (1996a,b). The various applications of R-based negative strengthening in Chapter 5 are linked in Horn (2000b) to the practice of conditional perfection, the natural (if logically fallacious) tendency to treat an "if p then q" conditional as if it were an "if and only if p, q" biconditional.

The story of metalinguistic negation related in *NHN*, Chapter 6 (itself a revision and expansion of my 1985 *Language* article) has sparked a lively number of exchanges, ranging over the attempt by Burton-Roberts (1989a,b) to enlist MN on behalf of a neo-Strawsonian theory of semantic presupposition (but see the replies by Seuren 1990 and Horn 1990b as well as *NHN* §7.3.2), the critiques by Foolen (1991), van der Sandt (1991), and the especially incisive disentanglement of metalinguistic and contrastive negation in McCawley (1991), the Relevance-theoretic approaches to MN as echoic negation (Carston 1996, Chapman 1996, Yoshimura 1998), a rekindled debate over the presupposition-cancellation cases (Burton-Roberts 1997, 1999 vs. Carston 1998, 1999) and recent accounts of the implications of MN for the logic of negation within dynamic models of semantics (Geurts 1998, Seuren 2000). There has also been a considerable amount of cross-linguistic work on metalinguistic negation; cf. for example Biq (1989), Yeh (1995), and Wible & Chen (2000) on MN in Mandarin or Choi (2000, Chapter 4) on Korean.

The status of metalinguistic negation as one means among several others for induced double processing (or retroactive accommodation) is noted in Horn (1992), which also introduces Saturday Night Live's ironic retro-NOT into the scholarly literature (following the slightly earlier and much more extensive discussion of the phenomenon on Linguist List). As documented in that paper, and more systematically in Sheidlower & Lighter (1993), this supposedly novel device (which strikingly flouts the Neg-First principle defined and exemplified in *NHN*, Chapter 7) actually dates back to the dawn of the 20th century, if not the dusk of the 19th.

The descriptive typology of negation surveyed rather briskly in *NHN* §7.1 has since been supplemented by a dissertation, Honda (1996), two important articles, Dryer (1989) and Croft (1991), and several valuable collections: Kahrel & van den Berg, eds. (1994), Bernini & Ramat (1996) (on which see also Horn 1998b), and Hovdhauven & Mosel, eds. (1999). Another anthology, Tieken-Boom van Ostade et al., eds. (1999), focuses specifically on the diachronic aspects of negative concord and word order in the history of English, with several contributions devoted to the implications of Jespersen's Cycle. (Cf. also Frisch 1997 and Ingham 2000.) A lighter approach to the Cycle is given in Horn (2001), which daringly foresees a future in which the all-purpose English negative particle takes the form of *squat*. Finally, the neo-Aristotelian treatment of negation as a mode of predication supported in §7.2 is extended insightfully by Moser (1992) and Ladusaw (1994, 1996b).

The publication of *A Natural History of Negation* in early 1989 was followed closely by that of its somewhat more accessible companion volume, *Negaholics: How To Overcome Your Negativity and Turn Your Life Around.* In this monograph, Chérie Carter-Scott offered a diagnosis and treatment protocol for those afflicted with negaholism, "the physiological, chemical rush you experience every time you engage in negative thoughts, words, or actions" (Carter-Scott 1989: 8). As it turns out, there were more negaholics at large in the early 1990s than one might have thought, judging from the fact that the first printing of Horn (1989) sold out within six years, despite the reader's burden of negotiating the arduous seven-chapter program prescribed therein. On behalf of my fellow addicts, I am delighted that CSLI has now restored *NHN* to print, making it possible for even the most "confirmed negaholics" among us to "detoxify [ourselves] from the negative demon within" (Carter-Scott 1989: 6).

As we have seen, the fin-de-siècle has been a very good time for natural language negation. (We speak here of *human* natural language, since even the otherwise eloquent bonobo Kanzi–let alone his less gifted common chimp cousins–has proved incapable of dealing with negation, as Savage-Rumbaugh et al. 1998 concede.) On April 23, 1990, just over a year after the first release of *NHN*, Elizabeth Clare Prophet, a.k.a. Guru Ma, speaking on behalf of God, prepared her disciples for "twelve years of intense negative karma" to begin that day, a spell that would culminate with an unwelcome visit by the four horsemen of the apocalypse, probably in the form of a nuclear holocaust (Egan 1990). While the leader of the Church Universal and Triumphant opted to meet this threat by leading her 750 disciples into an immense underground shelter in Paradise Valley, Montana that she described as "Noah's Ark in the earth", my own sense of the upcoming dozen years of intense negative karma was much more sanguine. I am pleased to say in retrospect that we have indeed emerged relatively unscathed from this twelve-year reign, and I trust that the future of negative karma will be equally positive.

No reprint would be complete without an acknowledgment of typos and less excusable errors that managed to creep into the original. An errata sheet–reasonably complete, I hope–appears as Appendix A to this document, followed a list of reviews, review articles, and book notices of *NHN* in Appendix B.

This reissue of *NHN* is dedicated to the memory of James D. McCawley (1938–1999), scholar, epicure, and polymaven nonpareil.

Appendix A

Errata in the original edition of *A Natural History of Negation*

p. xvi, line 12 from bottom: *Semantics* → *Semantic Properties*
p. xix, line 17: to assert ***p*** → to assert ***not-p***
p. xxii, line 5: Q-NEG → NEG-Q
p. 14, line 6: 22b17-18 → 23a17-18
p. 39, (33)(ii)... $(\forall y)(y = \mathbf{RF})$... → ... $(\forall y)(y \in \mathbf{RF})$...
p. 43, line 4 from bottom: given → give
p. 46, line 13: eight → nine
p. 89, line 23: LC → LEM
p. 114, line 1: (28b) → (27b)
p. 120, line 15: predictions → predications
p. 120, line 4 from bottom: reduncant → redundant
p. 124, lines 4–5: concealing → conceding
p. 129, line 2: true → false
p. 133, line 3: ELP → EPL
p. 147, line 26: (77) → (77a)
p. 154, line 7 of first paragraph: of of → of
p. 158, line 12: 5 stands for → |5| stands for
p. 159, line 5 from bottom: *marked* → *unmarked*
p. 165, (9): In Type C of table, under "Assertion", S will do P → S will do A
p. 176, line 6 from bottom: *hot* → *cold*
p. 178, line 13 from bottom: or (in Cornish's terms presupposing) → or (in Cornish's terms) presupposing
p. 186, line 14 from bottom: latter → former
p. 189, last line: catetory → category
p. 208, line 8: (9c) → (9c')
p. 214, line 6 from bottom: (vi) and (vii) → (v) and (vi)
p. 225, lines 12–13: exclusive (***p*** ∨ ***q***) and inclusive (***p*** w ***q***) disjunction → exclusive (***p*** w ***q***) and inclusive (***p*** ∨ ***q***) disjunction
p. 226, (36'): Tout le monde n'est past → ... n'est pas
p. 230, line 9: the pairs in (41) → the pairs in (41)–(44).
p. 238, lines 10–12, replace the sentence:

> While the sum of the values of P and P~ is always zero, the sum of the values of P and ~P (e.g., of 'some' and 'none', of 'all' and 'not all') is always just over |1|, the absolute value of 1.

with these two sentences:

> Note that the sum of the values of P and P~ is always zero. Another intuitively useful observation is that the sum of the absolute values of the values for P and ~P (e.g. of 'some' (=|0.01|) and 'none' (=|-1.00|), or of 'all' (=|1.00|) and 'not all' (=|-0.01|), is always just greater than 1.[1]

p. 238, line 25: the diagnostic tests in (49) → in (51).
p. 241, (64), last parenthesis: Op-Ed pieces → Op-Ed piece
p. 253, line 15 from bottom: the universal negative (A) → (E).
p. 254, (82): ikana katta → ikanakatta.
p. 256, line 15: Old Norse *nē*) → Old Norse) *nē*.
p. 260, line 17: with outer or wide-scope negation over □ → with □ outside the scope of negation
p. 278, line 5: delete 'thus'
p. 278, (12(i)): On its first occurrence, R^{-1} should be explicated: R^{-1} → its converse R^{-1}.
p. 290, line 10: delete 'nonentity,'
p. 291, line 5: classc → classic

---

1. Clarificatory note: the motivating idea here, which doesn't work under the published formulation, was to yield a generalization over the following data:

P		**~P**		**sum of (P + ~P), i.e. diagonal sum**	
all	1.00	not all	-.01	1.00 + (-.01) =	.99
some	.01	none	-1.00	.01 + (-1.00) =	-.99
half	.50	not half	-.51	.50 + (-.51) =	-.01
a majority of	.51	not a majority of	-.50	.51 + (-.50) =	.01

where the value of .01 is used as an arbitrary way of representing ANY value above .00, so e.g. *a majority of* is .51 because it's .01 (an arbitrarily small value) above half, just *as some* is .01 (a value an arbitrarily small amount or number above zero.) The corrected version of the text yields the following appropriate results:

all + not all:	|1.00|	+	|.01|	=	|1.01|
some + none:	|.01|	+	|1.00|	=	|1.01|
half + not half:	|.50|	+	|.51|	=	|1.01|
a majority of + not a majority of:	|.51|	+	|.50|	=	|1.01|

Of course, given the earlier comments, "1.01" really just stands for 'an amount/number which is an arbitrary small amount/number greater than 1'.

p. 307, line 11: that → than
p. 319, (52b), in translation: Hans is coming → Hans is not coming
p. 348, (95b): koro → koto
p. 349, (98): kunu → kuru
p. 361, end of last paragraph, from the fifth line from the bottom, should read as follows:

> ...negation may be strengthened (or 'filled in', à la Bosanquet) to yield a contrary interpretation. Where the three constructions differ is in the nature and degree of conventionalization governing this R-based pragmatic strengthening process: The inference which is general and virtually exceptionless in the (relatively) simple cases of litotes discussed in §5.3 is partially fossilized as a short-circuited implicature or convention of usage in the NR cases of §5.2 (whence the unmediated nature of the inference and the lexical exceptions associated with the NRP), and is partially or fully conventionalized in the lexical affixal negations of §5.1.

p. 363, line 2 from bottom: a subject description → the scope of negation
p. 363, last line: the scope of negation → the scope of the subject description
p. 401, line 4: as → in
p. 437, (140b): insert * in: Alik {*ne byl doma/ne byl doma}, a budet.
p. 442, line 5: former → latter
p. 442, (151a): *Kuruma → Kuruma [the following asterisk is correct]
p. 476, line 17: The form → The forms
p. 481, line 4 from bottom: tenseless and tensed → tensed and tenseless
p. 490, (38'b): were they? → weren't they?
p. 492, (40'a): {*so/neither} → {*neither/so}
p. 527, fn. 16: 1055a10 → 1055b10
p. 534, fn. 67, line 5: *Ergängungsnegation* → *Ergänzungsnegation*
p. 559, line 5: (99'b) → (99'a)
p. 561, fn. 1, fourth line under (iv): classical → classically
p. 574, last line of fn. 17: no is man → no man is
p. 578, fn. 39, (ii): imasu → imasu ga
p. 598, Kempson (1986) reference: Ambiguity, and → Ambiguity, Negation, and

Appendix B

Reviews and book notices of the original edition of *A Natural History of Negation*

Mathematical Reviews, issue 89k, 1989 (book notice by Pierre Kerszberg)
Canadian Philosophical Reviews, 10.5, 1990 (reviewed by Brendan Gillon)
Philosophical Psychology 3: 318–21, 1990 (reviewed by John Snapper)
Lingua 85: 253–61, 1991 (review article by Eva Hajičová)
English Linguistics 8: 190–208, 1991 (Japan, review article by Yasuhiko Kato)
Journal of Symbolic Logic 56: 1104–5, 1991 (reviewed by Jon Barwise)
Philosophy and Rhetoric 24: 164–68, 1991 (reviewed by Glen Helman)
Word 42: 179–82, 1991 (U.S.A., reviewed by Barbara Abbott)
Germanistik, 32: 627–8, 1991 (Germany, book notice by Ewald Lang)
Journal of Pragmatics, 16: 269–87, 1991 (Netherlands, reviewed by Alexis Kalokerinos)
Notes on Linguistics, No. 57: 47–57, May 1992 (reviewed by James K. Watters)
Le français moderne 94: 103–27, May 1992 (France, commentaire critique by Pierre Attal)

Supplemental Bibliography

Additional bibliographic items since the release of the original edition of *A Natural History of Negation*

Acquaviva, Paolo (1997). *The Logical Form of Negation*. New York: Garland.

Aoyagi, Hiroshi and Toru Ishii (1994). On NPI licensing in Japanese. *Japanese/Korean Linguistics* 4: 295–311.

Atlas, Jay David (1989). *Philosophy Without Ambiguity*, Chapter 3. Oxford: Clarendon Press.

——— (1991). Topic/comment, presupposition, logical form and focus stress implicatures: The case of focal particles *only* and *also*. *Journal of Semantics* 8: 127–147.

——— (1993). The importance of being 'only': Testing the neo-Gricean versus neo-entailment paradigms. *Journal of Semantics* 10: 301–18.

——— (1996). 'Only' Noun Phrases, Pseudo-Negative Generalized Quantifiers, Negative Polarity Items, and Monotonicity. *Journal of Semantics* 13: 265–332.

——— (1997). Negative Adverbials, Prototypical Negation, and the De Morgan Taxonomy. *Journal of Semantics* 14: 349–68.

Attal, Pierre, ed. (1994). *La Négation. LINX 1994*, Numéro spécial. Actes de colloque de Paris X- Nanterre, Novembre 1992.

van der Auwera, Johan (1996). Modality: the three-layered scalar square. *Journal of Semantics* 13: 181–95.

Bach, Kent (1994). Conversational impliciture. *Mind & Language* 9: 125–62.

——— (1999). The myth of conventional implicature. *Linguistics and Philosophy* 22: 327–66.

——— (2001). You don't say? *Synthese* 127: 11–31.

Baker, C. L. (1991). The syntax of English *not*: The limits of core grammar. *Linguistic Inquiry* 22: 387–429.

Benmamoun, Elabbas (1997). Licensing of negative polarity items in Moroccan Arabic. *NLLT* 15: 263–87.

von Bergen, Anke and Karl von Bergen (1993). *Negative Polarität im Englischen*. Tübingen: Gunter Naar Verlag.

Bernini, Giuliano and Paolo Ramat (1996). *Negative Sentences in the Languages of Europe*. Berlin: Mouton de Gruyter.

Bhatia, Tej (1995). *Negation in South Asian Languages*. Patiala: Indian Institute of Language Studies.

Biq, Yung-O (1989). Metalinguistic negation in Mandarin. *Journal of Chinese Linguistics* 17: 75–94.

Bosque, Ignacio (1980). *Sobre la negación*. Madrid: Ediciones Cátedra.

Brown, Sue (1999). *The Syntax of Negation in Russian: A Minimalist Approach*. Stanford: CSLI.

Brütsch, Edgar, Markus Nussbaumer and Horst Sitta (1990). *Negation.* (*Studienbibliographien Sprachwissenshaft 1.*). Heidelberg: Julius Groos Verlag.

Bublitz, Wolfram (1992). Transferred negation and modality. *Journal of Pragmatics* 18: 551–577.

Budick, Sanford and Wolfgang Iser (1989). *Languages of the Unsayable: The Play of Negativity in Literature and Literary Theory.* New York: Columbia University Press.

Büring, Daniel (1997). The Great Scope Inversion Conspiracy. *Linguistics and Philosophy* 20: 175–194.

Burton-Roberts, Noel (1989a). *The Limits to Debate: A Revised Theory of Semantic Presupposition.* Cambridge: Cambridge University Press.

——— (1989b). On Horn's dilemma: presupposition and negation. *Journal of Linguistics* 25: 95–125.

——— (1997). On preservation under negation. *Lingua* 101: 65–88.

——— (1999). Presupposition-cancellation and metalinguistic negation: A reply to Carston. *Journal of Linguistics* 35: 347–64.

Callebaut, Bruno, ed. (1992). *Les Négations. Langue française* 94. [Includes a critical commentary by Pierre Attal on Horn 1989.]

Carston, Robyn (1988). Implicature, explicature, and truth-theoretical semantics. In R. Kempson, ed., *Mental Representations*, 155–81. Cambridge: Cambridge University Press.

——— (1995). Quantity maxims and generalized implicature. *Lingua* 96: 213–44.

——— (1996). Metalinguistic negation and echoic use. *Journal of Pragmatics* 25: 309–30.

——— (1998). Negation, 'presupposition' and the semantics/pragmatics distinction. *Journal of Linguistics* 34: 309–50.

——— (1999). Negation, 'presupposition' and metarepresentation: A response to Noel Burton-Roberts. *Journal of Linguistics* 35: 365–89.

Carter-Scott, Chérie (1989). *Negaholics: How to Overcome Negativity and Turn Your Life Around.* New York: Fawcett Crest.

Chaker, Salem and Dominique Caubet (1996). *La négation en Berbère et en Arabe Maghrébin.* Paris: L'Harmattan.

Chakrabarti, Arindam (1997). *Denying Existence: The Logic, Epistemology, and Pragmatics of Negative Existentials and Fictional Discourse.* Dordrecht: Kluwer.

Chapman, Siobhan (1996). Metalinguistic negation: some theoretical implications. *Journal of Linguistics* 32: 387–402.

Choi, Seungja (2000). Topicality, Genericity, and Logophoricity. Yale University dissertation.

Copley, Bridget (1996). The Sign-ificance of Negative Suprasegmentals: Evidence for Term Logic Negation in American Sign Language. Unpublished senior essay, Yale University.

Croft, William (1991). The evolution of negation. *Journal of Linguistics* 27: 1–27.

Dayal, Veneeta (1998). Any as inherently modal. *Linguistics and Philosophy* 21: 433–76.

Déprez, Vivianne (2000). Parallel (a)symmetries and the internal structure of negative expressions. *NLLT* 18: 253–342.

Dobrin, Lise et al., eds. (1991). *The Parasession on Negation. CLS* 27, *Part Two.*

Dowty, David (1994). The role of negative polarity and concord marking in natural language reasoning. *SALT IV*, 114–44.

Drozd, Kenneth (1993). A Unification Categorial Grammar of Child English Negation. University of Arizona dissertation.

——— (1995). Child English pre-sentential negation as metalinguistic exclamatory sentence negation. *Journal of Child Language* 22: 583–610.

Dryer, Matthew (1989). Universals of negative position. In E. Moravcsik, J. Wirth and M. Hammond, eds., *Studies in Syntactic Typology*, 93–124. Amsterdam: John Benjamins.

Duffield, Nigel (1993). On Case-checking and NPI licensing in Hiberno-English. *Rivista di Linguistica* 5: 215–44.

Egan, Timothy (1990). "Guru's Bomb Shelter Hits Legal Snag", *New York Times*, 24 April 1990, A16.

Ernst, Thomas (1992). The phrase structure of English negation. *The Linguistic Review* 9: 109–144.

——— (1995). Negation in Mandarin Chinese. *NLLT* 13: 665–707.

Espinal, M. Teresa (1993). The interpretation of *no-pas* in Catalan. *Journal of Pragmatics* 19: 353–69.

Everett, Anthony and Thomas Hofweber, eds. (2000). *Empty Names, Fiction, and the Puzzles of Non-Existence*. Stanford: CSLI Publications.

von Fintel, Kai (1999). NPI licensing, Strawson-entailment, and context dependency. *Journal of Semantics* 16: 97–148.

Fischlin, Daniel (1994). *Negation, Critical Theory, and Postmodern Textuality*. Dordrecht: Kluwer.

Foolen, Ad (1991). Metalinguistic negation and pragmatic ambiguity: Some comments on a proposal by Laurence Horn. *Pragmatics* 1: 137–57.

Forest, Robert (1993). *Négations: Essai de syntaxe et de typologie linguistique*. Paris: Librarie C. Klincksieck.

Forget, D., P. Hirschbühler et al., eds. (1997). *Negation and Polarity: Syntax and Semantics*. [Proceedings of 1995 Ottawa conference.] Amsterdam: Benjamins.

Frisch, Stefan (1997). The change in negation in Middle English: A NEGP licensing account. *Lingua* 101: 21–64.

Gabbay, Dov and Heinrich Wansing, eds. (1999). *What is Negation?* Dordrecht: Kluwer.

Geurts, Bart (1997). On *no*. *Journal of Semantics* 13: 67–86.

——— (1998). The mechanisms of denial. *Language* 74: 274–307.

Giannakidou, Anastasia (1997). The Landscape of Polarity Items. University of Groningen dissertation.

——— (1998). *Polarity Sensitivity as (Non)veridical Dependency*. Amsterdam: John Benjamins.

——— (1999). Affective dependencies. *Linguistics and Philosophy* 22: 367–421.

——— (2000). Negative...concord? *NLLT* 18: 457–523.

——— (to appear). The meaning of free choice. To appear in *Linguistics and Philosophy*.

de Haan, Ferdinand (1997). *The Interaction of Modality and Negation*. New York: Garland.

Haegeman, Liliane, ed. (1993/1994). *Rivista di Linguistica* 5.2 and 6.1. Special issues on sentential negation.

Haegeman, Liliane (1995). *The Syntax of Negation*. Cambridge: Cambridge University Press.

Haegeman, Liliane and Raffaella Zanuttini (1991). Negative heads and the Neg Criterion. *The Linguistic Review* 8: 233–51.

Hajičová, Eva (1996). Topic-focus articulation—a matter of langue or parole? The case of negation. In R. Sackmann, ed., *Papers in Honour of Hans-Heinrich Lieb on Occasion of his 60th Birthday*, 167–75. Amsterdam: John Benjamins.

Haspelmath, Martin (1997). *Indefinite Pronouns*. New York: Oxford University Press.

Herburger, Elena (2000). *What Counts: Focus and Quantification*. Cambridge: MIT Press.

Hoeksema, Jack (1994). On the grammaticalization of negative polarity items. *BLS* 20. 273–82.

———, ed. (1995). *Linguistic Analysis*, Volume 25, Number 3–4. Special Issue on Negative Polarity.

——— (1998). On the (non)loss of polarity sensitivity: Dutch *ooit*. In R. Hogg and L. van Bergen, eds., *Historical Linguistics* 1995, *Germanic Linguistics*, 101–14. Amsterdam: John Benjamins.

———, et al., eds. (2001). *Perspectives on Negation and Polarity Items*. Amsterdam: John Benjamins.

Hoepelman, Jakob and Rudolf Schnitzer, eds. (1991) Focus in phonetics, syntax, semantics, and pragmatics. Special issue of *Journal of Semantics*, Vol. 8, Nos. 1–2.

Hoffmann, Maria (1987). *Negatio Contrarii: A Study of Latin Litotes*. Assen: van Gorcum.

Honda, Isao (1996). Negation: A Cross-Linguistic Study. SUNY Buffalo dissertation.

Horn, Laurence (1988). Morphology, pragmatics, and the un-verb. *Proceedings of the Fifth Eastern States Conference on Linguistics*, 210–33.

——— (1990a). Hamburgers and truth: Why Gricean inference is Gricean. *BLS* 16, 454–71.

——— (1990b). Showdown at Truth-Value Gap: Burton-Roberts on presupposition. Review article. *Journal of Linguistics*, 26: 483–503.

——— (1991). *Duplex negatio affirmat...*: The economy of double negation. In Dobrin et al., eds., 78–106.

——— (1992). The said and the unsaid. *SALT II: Proceedings of the Second Conference on Semantics and Linguistic Theory*, 163–92. Columbus: Ohio State University Department of Linguistics.

——— (1993). Economy and redundancy in a dualistic model of natural language. In S. Shore and M. Vilkuna, eds., *SKY 1993: 1993 Yearbook of the Linguistic Association of Finland*, 33–72.

——— (1995). Vehicles of meaning: Unconventional semantics and unbearable interpretations. *Washington University Law Quarterly* 73: 1145–52.

——— (1996). Exclusive company: Only and the dynamics of vertical inference. *Journal of Semantics* 13: 1–40.

——— (1997). All John's children are as bald as the King of France: Existential import and the geometry of opposition. *CLS* 33, 155–79.

——— (1998a). Neg-raising revisited: Tinkering with the short circuit. Published as an afterword to the reprint of Horn and Bayer (1984). "Short-circuited implicature: A negative contribution" in A. Kasher, ed., *Pragmatics: Critical Concepts*, Volume IV, 676–82. London: Routledge.

——— (1998b). Review of Bernini and Ramat, Negative Sentences in the Languages of Europe (Berlin: Mouton de Gruyter, 1996). *Journal of Linguistics* 34: 552–54.

——— (2000a). Pick a theory (not just *any* theory): Indiscriminatives and the free-choice indefinite. In Horn and Kato, eds., 147–92.

——— (2000b). From IF to IFF: Conditional perfection as pragmatic strengthening. *Journal of Pragmatics* 32: 289–326.

——— (2001). Flaubert triggers, squatitive negation, and other quirks of grammar. In Hoeksema, et al., eds. (2001), 173–200.

Horn, Laurence and Yasuhiko Kato, eds. (2000). *Negation and Polarity: Syntactic and Semantic Perspectives.* Oxford: Oxford University Press.

Horn, Laurence and Young-Suk Lee (1995). Progovac on polarity. Review article on Progovac (1994). *Journal of Linguistics* 31: 401–24.

Hovdhauven, Even and Ulrike Mosel, eds. (1999). *Negation in Oceanic Languages: Typological Studies.* Muenchen: LINCOM Europa. (LINCOM Studies in Austronesian Linguistics 02.)

Ingham, Richard (2000). Negation and OV order in Late Middle English. *Journal of Linguistics* 36: 13–38.

Israel, Michael (1996). Polarity sensitivity as lexical semantics. *Linguistics and Philosophy* 19: 619–66.

——— (to appear). *Polarity and Scalar Reasoning: The Pragmatic Basis of Grammar.* Cambridge: Cambridge University Press.

Jackson, Eric (1994). Negative polarity and general statements. *Language & Cognition* 4, 97–110. University of Groningen.

Jacobs, Joachim (1982). *Syntax und Semantik der Negation im Deutschen.* München: W. Fink.

——— (1991). Negation. In A. von Stechow and D. Wunderlich, eds., *Semantics: An International Handbook of Contemporary Research,* 560–96. Berlin: Walter de Gruyter.

Jennings, R. E. (1994). *The Genealogy of Disjunction.* New York: Oxford University Press.

Jordan, Michael. (1998). The power of negation in English: Text, context and relevance. *Journal of Pragmatics* 29: 705–752.

Kadmon, Nirit and Fred Landman (1993). Any. *Linguistics and Philosophy* 16: 353–422.

Kahrel, Peter (1996). Aspects of Term Negation. University of Amsterdam dissertation.

Kahrel, Peter and René van den Berg, eds. (1994). *Typological Studies in Negation.* Amsterdam: Benjamins.

Kas, Mark (1993). Essays on Boolean Functions and Negative Polarity. U. of Groningen dissertation.

Kato, Yasuhiko (1994). Negative polarity and movement. *Formal Approaches to Japanese Linguistics* 1 (= *MITWPL 24*), 101–20.

——— (2000). Interpretive asymmetries of negation. In Horn and Kato, eds., 62–87.

Kawashima, Ruriko and Hisatsugu Kitahara (1992). Licensing of negative polarity items and checking theory: A comparative study of English and Japanese. *FLSM* 3, 139–154.

Kim, Jong-Bok (2000). *The Grammar of Negation: A Constraint-Based Approach.* Stanford: CSLI Publications. [Stanford University dissertation]

Klein, Jared (1997). Indefinite pronouns, polarity, and related phenomena in Classical Armenian. *Transactions of the Philological Society* 95: 189–245.

von Klopp, Ana (1998). An alternative view of polarity items. *Linguistics and Philosophy* 21: 393–432.

Koster, Charlotte and Sjoukje van der Wal (1995). Acquiring a negative polarity verb. *Papers from the Dutch-German Colloquium on Language Acquisition*. Amsterdam Series in Child Language Development.

Krahmer, Emiel and Reinhard Muskens (1995). Negation and disjunction in Discourse Representation Theory. *Journal of Semantics* 12: 357–76.

König, Ekkehard (1991). *The Meaning of Focus Particles*. London: Routledge.

Krifka, Manfred (1991). Some remarks on polarity items. *Semantic Universals and Universal Semantics*, D. Zaeferrer, ed., 150–89. Berlin: Foris.

——— (1995). The semantics of weak and strong polarity items in assertion. *SALT* IV, 195–219.

Kuno, Susumu (1995). Negative polarity items in Japanese and English. *Harvard Working Papers in Linguistics* 5, 165–197.

Kuno, Susumu and Ken-ichi Takami (1992). Negation and extraction. *CLS* 28, 297–317.

Ladusaw, William (1992). Expressing negation. *SALT* II, 237–59.

——— (1996a). Negation and polarity items. *Handbook of Contemporary Semantic Theory*, S. Lappin, ed., 321–42. Oxford: Blackwell.

——— (1996b). Negative concord and 'mode of judgment.' In Wansing, ed., 127–44.

Lahiri, Utpal (1998). Focus and negative polarity in Hindi. *Natural Language Semantics* 6: 57–123.

Laka, Itziar (1990). *Negation in Syntax: On the Nature of Functional Categories and Projections*. MIT dissertation.

——— (1993). Negative fronting in Romance: Movement to Σ. *Linguistic Perspectives on the Romance Languages*, W. J. Ashby et al., eds., 315–333. Amsterdam: Benjamins.

Larrivée, Pierre (2000). *L'interprétation des phrases négatives: portée et foyer des négations en français*. Paris: Duculot.

Larrivée, Pierre and Patrick Duffley (1998). *Need, dare*, and negative polarity. *Linguistic Analysis* 28: 89–107.

Lee, Chungmin (1996). Negative polarity items in English and Korean. *Language Sciences* 18: 505–23.

Lee, Young-Suk and Laurence Horn (1994). *Any* as indefinite + *even*. Ms., Yale University.

Levinson, Stephen C. (2000). *Presumptive Meanings: The Theory of Generalized Conversational Implicature*. Cambridge: MIT Press.

Lin, Jo-Wang (1996). Polarity Licensing and Wh-Phrase Quantification in Chinese. University of Massachusetts dissertation.

Linebarger, Marcia (1991). Negative polarity as linguistic evidence. *CLS* 27, Part Two: Papers from the Parasession on Negation, 165–89.

Lloyd, G.E.R. (1992). *Polarity and Analogy: Two Types of Argumentation in Early Greek Thought*. Indianapolis: Hackett Press.

Lundquist, Lita and Robert J. Jarvella (1994). Ups and downs in scalar inferences. *Journal of Semantics* 11: 33–54.

Martin, John (1995). Existence, negation, and abstraction in the neoplatonic hierarchy. *History and Philosophy of Logic* 16: 169–96.

McCawley, James D. (1991). Contrastive negation and metalinguistic negation. In Dobrin et al., eds., 189–206.

Moser, Margaret (1992). The Negation Relation: Semantic and Pragmatic Aspects of a Relational Analysis of Sentential Negation. University of Pennsylvania dissertation.

Moxey, Linda and Anthony Sanford (1993). *Communicating Quantities*. Hove: Lawrence Erlbaum.

Mufwene, Salikoko (1993). Scope of negation and focus in Gullah. In F. Byrne and D. Winford, eds., *Focus and Grammatical Relations in Creole Languages*, 95–116. Amsterdam: John Benjamins.

Muller, Claude (1991). *La négation en français*. Genève: Droz.

Nuyts, Jan (1990). Neg-raising reconsidered: Arguments for a cognitive-pragmatic approach. *Journal of Pragmatics* 14: 559–88.

Ota, Akira (1980) *Hitei no imi: imiron josetsu* [= *The Meaning of Negation: Prolegomena to Semantics*]. Tokyo: Taikushan.

Ouhalla, Jamal (1991). *Functional Categories and Parametric Variation*. London: Routledge.

Pacitti, Domenico (1991). *The Nature of the Negative: Toward an Understanding of Negation and Negativity*. Pisa: Giardini.

Payne, John and Erika Chisarik (2000). Negation and focus in Hungarian: An Optimality Theory account. *Transactions of the Philological Society* 98: 185–230.

Pelletier, F. J. (1990). *Parmenides, Plato, and the Semantics of Not-Being*. Chicago: University of Chicago Press.

Pereltsvaig, Asya (2000). Monotonicity-based vs. veridicality-based approaches to negative polarity: evidence from Russian. In T. H. King and I. A. Sekerina, eds., *Annual Workshop on Formal Approaches to Slavic Linguistics*, 328–346. Ann Arbor: Michigan Slavic Publishers.

Pollock, Jean-Yves (1989). Verb movement, UG, and the structure of IP. *Linguistic Inquiry* 20: 365–424.

Potsdam, Eric (1997). NegP and subjunctive complements in English. *Linguistic Inquiry* 28: 533–41.

Progovac, Ljiljana (1993). Negative polarity: Entailment and binding. *Linguistics and Philosophy* 16: 149–80.

——— (1994). *Negative and Positive Polarity: A Binding Approach*. Cambridge: Cambridge University Press.

Przepiórkowski, Adam (1999). On negative eventualities, negative concord, and negative yes/no questions. *SALT* 9.

Przepiórkowski, Adam and Anna Kupsc (1999). Eventuality negation and negative concord in Polish and Italian. In R. Borsley and A. Przepiórkowski, eds., *Slavic in HPSG*. Stanford: CSLI.

Puskás, Genoveva (1998). On the Neg-criterion in Hungarian. *Acta Linguistica Hungarica* 45: 167–213.

Pustejovsky, James (2000). Events and the semantics of opposition. In C. Tenny and J. Pustejovsky, eds., *Events as Grammatical Objects*, 445–82. Stanford: CSLI.

Quer, Josep (1998). *Mood at the Interface*. The Hague: Holland Academic Graphics.

Récanati, François (1989). The pragmatics of what is said. *Mind & Language* 4: 295–329.

——— (1993). *Direct Reference: From Language to Thought*. Oxford: Blackwell.

Rohrbaugh, Eugene (1997). The role of focus in the licensing and interpretation of negative polarity items. In Forget et al., eds., 311–22.

Rowlett, Paul (1998). *Sentential Negation in French*. Oxford: Oxford University Press.

Rullmann, Hotze (1995). Two types of negative polarity items. *NELS* 26.

Ruthrof, Horst (1997). *Semantics and the Body: Meaning from Frege to the Postmodern*. Toronto: University of Toronto Press.

Sæbø, Kjell Johan (to appear). The semantics of Scandinavian free choice items. To appear in *Linguistics and Philosophy*.

Sánchez Valencia, Victor (1991). *Studies on Natural Logic and Categorial Grammar*. University of Amsterdam dissertation.

van der Sandt, Rob (1991). Denial. In Dobrin et al., eds., 331–44.

Savage-Rumbaugh, Sue et al. (1998). *Apes, Language, and the Human Mind*. New York: Oxford University Press.

Schwenter, Scott (1999). *Pragmatics of Conditional Marking: Implicature, Scalarity, and Exclusivity*. New York: Garland.

Seifert, Stephan and Werner Welte (1987). *A Basic Bibliography on Negation in Natural Language*. Tübingen: G. Narr.

Seuren, Pieter A. M. (1990). Burton-Roberts on presupposition and negation. *Journal of Linguistics* 26: 425–53.

——— (2000). Presupposition, negation and trivalence. *Journal of Linguistics* 36: 261–98.

Sheidlower, Jesse and Jonathan Lighter (1993). A recent coinage (Not!). *American Speech* 68: 213–18.

Spruyt, Joke, ed. (1989). *Peter of Spain on Composition and Negation*, text edited and translated by Spruyt. Nijmegen: Ingenium.

Stroik, Thomas (1997). Negation and predication. *Linguistic Analysis* 27: 245–68.

Tieken-Boom van Ostade, Ingrid (2000). *The Two Versions of Malory's Morte d'Arthur: Multiple Negation and the Editing of the Text*. Cambridge: D.S. Brewer.

Tieken-Boom van Ostade, Ingrid, Gunnel Tottie and Wim van der Wurff, eds. (1999). *Negation in the History of English*. Berlin: Mouton de Gruyter.

Tóth, Ildikó (1999). Negative Polarity Licensing in Hungarian. *Acta Linguistica Hungarica* 46: 119–142.

Tottie, Gunnel (1991). *Negation in English Speech and Writing: A Study in Variation*. New York: Academic Press.

Tovena, Lucia (1998). *The Fine Structure of Polarity Items*. New York: Garland.

——— (2001). Neg-raising: Negation as failure? In Hoeksema, et al., eds. (2001).

Uribe-Etchevarria, Maria (1994). *Interface Licensing Conditions on Negative Polarity Items: A Theory of Polarity and Tense Interactions.* University of Connecticut dissertation.

Vasishth, Shravan (1998). Monotonicity constraints on negative polarity in Hindi. *Ohio State University Working Papers in Linguistics* 51: 147–66.

Vallduví, Enric (1994). Polarity Items, n-words, and minimizers in Catalan and Spanish. *Probus* 6: 263–294.

Ver Eecke, Wilfried (1984). *Saying "No": Its Meaning in Child Development, Psychoanalysis, Linguistics, and Hegel.* Pittsburgh: Duquesne University Press.

van der Wal, Sjoukje (1995). The development of negation and negative polarity in child speech. *CLS* 31.

——— (1996). *Negative Polarity Items & Negation: Tandem Acquisition.* U. of Groningen dissertation.

Wansing, Heinrich, ed. (1996). *Negation: A Notion in Focus.* Berlin: Walter de Gruyter.

Wible, David and Eva Chen (2000). Linguistic limits on metalinguistic negation: Evidence from Mandarin and English. *Language and Linguistics* 1: 233–255..

van der Wouden, Ton (1996a). Litotes and downward monotonicty. In Wansing, ed., 145–68.

——— (1996b). *Negative Contexts: Collocation, Polarity, and Multiple Negation.* London: Routledge.

van der Wouden, Ton and Frans Zwarts (1993). A semantic analysis of negative concord. *SALT* III, 202–19.

Yamada, Masamichi (2000). Negation in Japanese Narratives: A Functional Analysis. Georgetown University dissertation.

Yeh, Ling-Hsia (1995). Focus, metalinguistic negation, and contrastive negation. *Journal of Chinese Linguistics* 23: 42–75.

Yoshimoto, Yasushi (1995). The Terrain and Locality of Negative Polarity Items. University of North Carolina dissertation.

Yoshimura, Akiko (1992). The cognitive structure of negation as an NPI-licensing condition. *English Linguistics* 9: 244–264.

——— (1994). A cognitive constraint on negative polarity phenomena. *BLS* 20, 599–610.

——— (1998). Procedural semantics and metalinguistic negation. In R. Carston and S. Uchida, eds., *Relevance Theory: Applications and Implications*, 105–22. Amsterdam: John Benjamins.

——— (1999). *Hitei Kyokusei Gensho* [*Negative Polarity Phenomena*]. Tokyo: Eihosya.

Zanuttini, Raffaella (1991). Syntactic Properties of Sentential Negation. University of Pennsylvania dissertation.

——— (1997). *Negation and Clausal Structure: A Comparative Study of Romance Languages.* New York: Oxford University Press.

Zwarts, Frans (1991). Negation and generalized quantifiers. In J. van der Does and J. van Eijck, eds., *Generalized Quantifier Theory and Applications.*

——— (1995). Nonveridical contexts. *Linguistic Analysis* 25: 286–312.

——— (1996). A hierarchy of negative expressions. In Wansing, ed., 169–94.

——— (1998). Three types of polarity. In F. Hamm and E. Hinrichs, eds., *Plural Quantification*, 177–238. Dordrecht: Kluwer.

1 Negation and Opposition in Classical Logic

> The *not*-relation is one of the simplest and most fundamental relations known to the human mind. For the study of logic, no more important and fruitful relation is known. (Royce 1917:265)

> No agreement exists as to the possibility of defining negation, as to its logical status, function, and meaning, as to its field of applicability . . . , and as to the interpretation of the negative judgment. (Heinemann 1944:135)

While linguists, philosophers, and psychologists concerned with tracking the pursuit of our quarry may well be interested in the history of that pursuit for its own sake, the record is also worth reviewing as a lesson in the perils of ignorance. Let us begin, then, at the beginning. In both Western and Eastern logical traditions, negation has long occupied a central place at the logical table. Yet in each tradition, negation has been regarded as a suspect guest at that table, if not as a spy from the extralogical domains.

In this first chapter, I set out the parameters of the field of combat, parameters which continue to function today as they did 2,300 years ago when the Aristotelians and Stoics were doing battle. For the Eleatic philosophers in Ancient Greece and the early Buddhists in India, the first explorations of negative concepts were associated with the status of nonbeing in metaphysics and ontology. The study of linguistic negation proper can be said to begin with Plato's *Sophist;* the Stranger in this dialogue seeks to identify negation (the not-***p***) with otherness (that which is distinct from ***p***). Through the Stranger, Plato introduces two of the recurring themes of our history: the view that negation can be eliminated by defining it away in terms of the (putatively) positive concept of otherness or difference, and the observation that negative statements are in some sense less valuable than affirmative ones, in being less specific or less informative.

But it is with Aristotle that the locus of the study of negation leaves the realm of pure ontology and enters the domains of language and logic. Aristotle's theory of negation, presented in discontinuous chunks in the *Categories, De Interpretatione,* the *Prior Analytics,* and the *Metaphysics,* represents the backbone of my entire study. The opposition between contrary and contradictory opposition with which I touch off in §1.1.1 remains a central issue in the psycholinguistics of negation (as we shall see in chapter 3); it returns as the focus of my examination of the pragmatics of contrary and contradictory negation in chapter 5. The psychological, ontological, and/or linguistic asymmetry between negative and affirmative statements, an asymmetry variously stipulated, assumed, and denied at different points within the Organon, is the core problem of §1.2 and of chapter 3. Vacuous singular negative statements and their connection to the putative ambiguity of negation, explored by Aristotle in the *Categories* and *De Interpretatione* (see §1.1.1), represent a major issue in the study of presupposition (see chapter 2); the related contrast in the *Prior Analytics* between wide-scope predicate denial and narrow-scope predicate term negation launches my own attack on the "ambiguity" of negation in chapters 6 and 7. Even when I depart most strongly from the Aristotelian foundations, as in my analysis of weak scalar terms like *possible* and *some* (see chapter 4), it is the Stagirite's lead which I not-follow.

Aristotle, then, is the *primus inter pares;* his (unacknowledged) heirs include Russell, Jespersen, and Montague, as we shall see in this and later chapters. The approach in this most historical chapter of my natural history is partly chronological and partly thematic. I begin by introducing the terms of debate, as Aristotle sets them forth in the Organon: the differentiated types of opposition obtaining among terms and propositions, the identification of negation with contradictory (rather than contrary) opposition, the relation between positive and negative predications, and the analysis of negative quantified and modal expressions mapped onto the classical (but post-Aristotelian) Square of Opposition. I conclude §1.1.1 with an exploration of the formal nature and theoretical status of the two central laws governing opposition, LC (the LAW OF (NON-) CONTRADICTION) and LEM (the LAW OF EXCLUDED MIDDLE).

In §1.1.2, I introduce a rival formal conception of negation. Where the term logic of Aristotle took wide-scope, proposition-level negation to represent a mode of predication, a rule for combining subject and predicate, the Stoics anticipated the modern, Fregean view of negation as a one-place operator on propositions. It is to the Stoics that we owe the first (Occidental) formalization of the law of double negation, as well as the first exposition of a true propositional logic (and as I suggest in chapter 7, the first step in the betrayal of natural language for logical elegance and simplicity).

Two points touched on in the first subsection—the question of which positive and negative statements build in EXISTENTIAL IMPORT (i.e., license an inference of the corresponding existential proposition) and the question of which criteria determine the status of a proposition or predication as positive or negative—are expanded in §1.1.3 and §1.1.4, respectively.

The first section then concludes with a historical overview of the contributions of the Aristotelian legacy to the study of negation; I delineate the different notions of contrariety assumed by Aristotle and his successors and reexamine (not for the last time) the role of Aristotle's two readings of negation.

The central fact about negation for a powerful army of philosophers (Plato, Bacon, Kant, Hegel, Bergson, Strawson), linguists (Apostel, Leech, Ducrot, Givón), and psychologists (Wason, H. Clark) is the asymmetry between positive sentences, statements, and facts and their negative counterparts. Negative statements, for the asymmetricalists, are less primitive, less informative, less objective, less godly, and/or less valuable than their affirmative counterparts. For another, equally impressive array of combatants, including Frege, Geach, and Ayer among their number, this central fact about negation does not exist. Others, including Aristotle and Russell, seem to waver between these two camps. In §1.2 I give an account of this war, although the issues raised here animate much of my later discussion, especially in chapter 3, where I report the news from the psycholinguistic front and essay a pragmatically mediated solution to the conflict.

One traditional conceit of the asymmetricalist forces is the paradox of negative judgment: if a positive statement refers or corresponds to a positive fact, to what does a negative statement refer or correspond? Clearly not to a negative fact, the most notorious mythical beast this side of the unicorn. (Or is it?) The first part of my chronicle, §1.2.1, recounts a variety of attempts to banish negative facts and eliminate negation entirely: the Platonic move to explain it away as positive otherness or difference, the related scheme to absorb apparent sentence-level negation into the predicate, the attempt to reduce negation to falsity. Each of these methods of subjugating negation has its modern champions, yet each proves ultimately circular, incoherent, or otherwise inadequate.

The more measured asymmetricalist positions are examined in more detail in §1.2.2. Here we are variously informed that the negative statement betokens a symptom of fallibility, a necessary evil for a finite mind (cf. Bacon, Kant, and the neo-Hegelians), a device for warding off error (Kant, Givón), a pragmatic faute de mieux (Morris, Russell), a subjective attitude or modality (Bergson, Russell, Apostel), or a speech act of negation (Strawson, Searle, Givón). One common thread in these asymmetricalist manifestos is the view that every negative statement presupposes a corresponding affirmative (although it is not always clear just which affirmative), but not vice versa. Negation is consequently a second-order affirmation: negative statements are about positive statements, while affirmatives are directly about the world. This argument finds proponents in the eighth-century Indian logician Śaṇkara, in Hegel and his followers, in Russell (sometimes, anyway), preeminently in Bergson, and—within our own era—in García and Givón.

We also witness some of the more telling counterthrusts of the symmetricalists: the argument by Frege (later buttressed by Geach and Gale) that there is no distinct negative judgment or speech act, the rejections by Quine and Austin of the view that negation is a second-order concept identifiable with falsity, the observation by Wittgenstein that while negations may presuppose the corresponding affirmation, every affirmative proposition equally presupposes the existence of a corresponding negative, and the complementary critiques by Frege, Ayer, and Geach of the purported epistemological worthlessness (or worth-less-ness), inherent complexity, and/or presuppositionality of the negative statement.

While there is a clear consensus, among the asymmetricalists at least, that negation is less basic than affirmation, it is less clear just how this distinction is manifested. And if negatives do unilaterally "presuppose" positives, under which notion(s) of presupposition? These questions will remain open within this section; I return to their consideration (if not their resolution) in chapter 3.

My natural history enters a broader and somewhat more speculative realm in §1.3, where I consider the place of negation within the history of ideas. My survey begins in the East, where we find that the Buddhist and Nyāya logicians of India, rather than rejecting the Laws of Contradiction, Excluded Middle, and Double Negation out of either sheer perversity or a spirit of illumination (as alleged by adherents and antagonists of these laws, respectively), in fact displayed a full panoply of attitudes to the validity of these laws and to their incorporation within an overall theory of opposition and negation. The development of these approaches, in their multiplicity, their complexity, and their occasional inscrutability, is reminiscent of parallel developments in Western logic. The emergence of an Eastern front in the asymmetricalist wars will similarly come as no surprise to readers of the previous section.

Nor is the Law of Contradiction (LC), that first principle of Aristotelian logic, repudiated by Hegel, as is frequently asserted or assumed. We see in §1.3.2 that the assumption of LC, along with a dynamic conception of Double Negation, is actually a vital step in the motivation of the Hegelian dialectic. In this same subsection, I also touch (with pathetic brevity) on a range of Marxist, Freudian, exegetical, and literary theses on the role of negation in human (and superhuman) consciousness. But here, as in the reductionist battles waged by the asymmetricalists, there is no final solution to the question of the negative statement.

While my account of the analyses of negation may seem discouraging in its revelation of repeated independent rediscoveries of the same observations, the same generalizations, and often the same mistakes, it is also (I hope) instructive. As in other linguistic (and extralinguistic) domains, those who do not learn from the history of ideas are condemned to relive it.

The historical perspective on negation and its roles within the philosophy of language and of mind provided in chapter 1 is designed to introduce the reader to my protean protagonist (along with members of the supporting cast, i.e., the remaining logical operators). This chapter will thus serve as a program for following the rest of my tale, which unfolds to reveal a set of interlocking puzzle boxes concealing the philosophical, psychological, and linguistic clues to the character of negation in the logic of natural language.

1.1 Negation and the Legacy of Aristotle

Never shall this thought prevail, that not-being is:
Nay: keep your mind from this path of investigation.
(Parmenides, cited by Plato [*Sophist:* 258D])

With this Delphic imprecation—more familiar in its short form, *Not-being is not*—Parmenides laid down the gauntlet for 2,500 years of unresolved, and no doubt unresolvable, disputation over the nature of nonexistence in the realm of things and negation in the realm of language.

In his rebuttal (*Sophist* 254C–259B), Plato has his spokesman, the Stranger, respond to Parmenides' dictum by contending that in a real sense, things which <u>are not</u> nevertheless <u>are</u>:

> When we say not-being, we speak, I think, not of something that is the opposite of being, but only of something different.
> (*Sophist* 257B)

For negation cannot in general be read as opposition or contrariety: when we speak of the 'not great' (*mē mega*), we do not pick out 'what is small any more than what is of middle size', rather we refer simply to what is different from the great. Therefore, the Stranger concludes (257B–C),

> When we are told the negative signifies the opposite, we shall not admit it; we shall admit only that the particle "not" [*ou* or *mē*] indicates something different from the words to which it is prefixed, or rather from the things denoted by the words that follow the negative.

The not-beautiful (*mē kalon*) is simply that which is other than beautiful and is just as much a part of being as the beautiful by which it is defined; similarly for the not-great, the not-just, and so on (257D–258B).

If not-X is identified with mere otherness or difference, the fundamental question of whether pure not-being exists can be finessed (258E), and the Stranger can successfully banish any Parmenidean nightmares (or Quinean qualms). But the bogeyman of Not-Being, from which Parmenides averted his eyes and around which Plato's Stranger executed his deft end run, was not so easily dispelled. Plato's argument was finally no more compelling than Parmenides' injunction in settling the matter; indeed, one of the threads I shall pick up below is the treatment of nonbeing and the ontology of negative facts and events from its classical sources through the metaphysical schools of the East (the Indian traditions to be examined in §1.3.1) and the West (the Hegelian tradition to be discussed in §1.3.2). But the focus of my attention rests elsewhere, its gaze concentrating inevitably on the imposing figure of Aristotle.

1.1.1 The Footprints of Aristotle

> At this point, as it often happens in philosophy, we suddenly realize that the path of inquiry we hoped to open is already marked by the footprints of Aristotle. (Vendler 1967: 194)

Aristotle transformed the study of negation by shifting the issue from the domain of ontology to that of logic and language (cf. Wood 1933; Matilal 1968; Englebretsen 1981a, 1981b). The picture of negation and opposition presented in the *Categories*, the *De Interpretatione*, the *Prior Analytics*, and the *Metaphysics* remains as vivid today as it has ever been; analysts from Aristotle's Stoic opponents to his Peripatetic followers, from his Arab, Jewish, and Christian commentators of the scholastic period to his modern interpreters, defenders, and critics—logicians, philosophers, and linguists of every conceivable persuasion owe him an incalculable (if often unacknowledged) debt. The questions and distinctions first raised in the *Organon*—contradictory vs. contrary negation, the effect of negation on quantified and modal expressions, the truth conditions for negative propositions with vacuous subject terms (and those exhibiting category mistakes), the law of excluded middle and its application to future contingent propositions—are as central and as controversial today as when Aristotle broached them. Indeed, the detailed elaboration of these issues has generated much of the light and not a little of the heat emitted during the last twenty-three centuries of philosophic cross fire.

Aristotle's theory of negation has its roots within his system of oppositions between pairs of terms. Four species of opposition are distinguished in the *Categories* (*Cat.* 11b17):

(1) CORRELATION (between two RELATIVES), e.g., *double* vs. *half*
CONTRARIETY (between two CONTRARIES), e.g., *good* vs. *bad*

PRIVATION (PRIVATIVE to POSITIVE), e.g., *blind* vs. *sighted*
CONTRADICTION (AFFIRMATIVE to NEGATIVE), e.g., *He sits* vs. *He does not sit*

Diagnostics are offered for determining which of these 'senses of "opposite"' a given pair falls under (11b23ff.).

Correlation, corresponding to the modern notion of converseness, involves the interdependence of reference: A is the double of B if and only if B is the half of A. Privatives and positives apply to the same subject:

> We say that that which is capable of some particular faculty or possession has suffered privation when the faculty or possession in question is in no way present in that in which, and at the time in which, it should be naturally present. We do not call that toothless which has not teeth, or that blind which has not sight, but rather that which has not teeth or sight at the time when by nature it should. (12a28–33)

On this understanding, a newborn kitten is no more blind than is a chair, and a baby does not count as toothless.[1]

While it is later made clear that two contraries cannot both apply at once—*A* may not be simultaneously good and bad, or black and white—Aristotle seems to take this property as following automatically from the encompassing genus of opposition under which contrariety falls as a species. He thus concentrates in his earlier exposition (12a1–25) on the distinction between what (following Boethius) we can call MEDIATE VS. IMMEDIATE contraries:

> Those contraries which are such that the subjects in which they are naturally present, or of which they are predicated, must necessarily contain either the one or the other of them, have no intermediate. Thus disease and health are naturally present in the body of an animal, and it is necessary that one or the other should be present in the body of an animal.[2] (*Cat.* 12a1–7)

Sickness and health are hence immediate contraries, as are even and odd: every number (i.e., every integer) must be either even or odd, while none can be both. But other contraries do allow an intermediate, including *white* vs. *black* (since bodies to which the terms naturally apply, i.e., colored things, men, etc., may be neither white nor black) and *good* vs. *bad* ('It is not true to say that everything that may be good or bad must be either good or bad'—12a17).[3] Sometimes the nonexcluded middle between mediate contraries has a name (e.g., 'grey and sallow and all the other colours that come between white and black') but sometimes it does not, and 'we must define it as that which is not either extreme, as in the case of that which is neither good nor bad, neither just nor unjust' (12a20–25).[4]

Contradictory opposition

What distinguishes the contradictory relation—'statements opposed to each other as affirmation and negation'—from the other classes of oppositions is a twofold criterion. First, unlike correlation, contrariety, and privation, contradiction is restricted to statements or propositions: terms are never, for Aristotle, related as contradictories. Secondly, 'in this case, and in this case only, it is necessary for the one to be true and the other false' (13b2–3).

It is evident that opposition between terms cannot involve truth or falsity, since only statements (subject–predicate combinations) can be true or false (13b3–12). But two statements may be members of a contrary or privative opposition. Crucially, however, in these cases both members of the opposition may be simultaneously false, although (as with contradictories) they may not be simultaneously true.

The most striking aspect of the exposition for a modern reader lies in Aristotle's selection of illustrative material. Rather than being given an uncontroversial example involving mediate contraries (e.g., *This man is white / This man is black; Socrates is good / Socrates is bad*), we are offered a pair of sentences containing immediate contraries: *Socrates is ill / Socrates is well.* Aristotle insists that it is not the case that one of this pair must be true if the other is false, even though every man is either ill or well:

> For if Socrates exists, one will be true and the other false, but if he does not exist, both will be false; for neither 'Socrates is ill' nor 'Socrates is well' is true, if Socrates does not exist at all.
> (13b17–19)

The argument from contraries is extended to pairs of statements related as privative and positive, for example, *Socrates is blind / Socrates has sight.* Again, both statements can be simultaneously false when Socrates is nonexistent—and they are also both false 'when he is not yet able to acquire the power of vision' (13b20–26), that is, when Socrates is a baby (in accord with Aristotle's empirically inadequate theory of anatomy).[5]

When we move from contrariety and privation to contradiction, the case is altered:

> But in the case of affirmation and negation, whether the subject exists or not, one is false and other true. For manifestly, if Socrates exists, one of the two propositions 'Socrates is ill', 'Socrates is not ill' is true, and the other false. This is likewise the case if he

> does not exist, for if he does not exist, to say that he is ill is false, to say that he is not ill is true. (13b26–32)

> Thus it is that for contradictory opposites ('statements opposed as affirmation and negation'), and for these alone, 'the rule holds good that one of the pair must be true and the other false'. (13b35)

Schematically, the truth conditions assigned by Aristotle work out as follows:

(2)

	if Socrates exists		if Socrates does not exist
a. Socrates is ill.	**T**	**F**	**F**
b. Socrates is well.	**F**	**T**	**F**
c. Socrates is not ill.	**F**	**T**	**T**

This analysis never enjoyed unanimous approval; commentators as early as Boethius (fifth century A.D.) took (2a) and (2b) to be contradictories, and—as we shall see—logicians have long dueled over the assignment of truth conditions to negative propositions like (2c), whose subject term fails to refer (among the more prominent recent warriors can be reckoned Frege, Russell, and Strawson; cf. §2.2). There is some evidence that Aristotle himself was not entirely convinced of its correctness (Ackrill 1963:111). At *De Interpretatione* 21a26–28 it is argued that from *Homer is a poet* it does not follow that *Homer is,* since *is* is used incidentally of Homer in the former example. However, Aristotle might (although he did not) offer the rejoinder that the two cases are distinguishable by the nature of the predication; in English the difference shows up in our willingness to assert *Homer is a poet* in the present tense as against, for example, *Homer is ill* (or *blind*).

In the *Categories,* then, the property that the corresponding affirmative and negative members of a given pair of statements 'divide truth and falsity between them', as the medievals were to put it, represented a distinguishing characteristic of contradictory opposites. The defining criterion of contradiction, however, seems to be syntactic, depending on the fact that members of such typical pairs as (2a)–(2c) or *He sits / He does not sit* are formally identical except for the negation. In Aristotelian terms, a predicate is AFFIRMED of the subject in one case and DENIED of the same subject in the other.

In the *De Interpretatione*, this criterion is first formulated explicitly and later severely constrained.

> An affirmation is a positive assertion of something about something, a denial a negative assertion. . . . It is plain that every affirmation has an opposite denial, and similarly every denial an opposite affirmation. We will call such a pair of propositions a pair of contradictories. Those positive and negative propositions are said to be contradictory which have the same subject and predicate.[6] (*De Int.* 17a25–35)

But this criterion, satisfied simply enough in the case of singular expressions such as those discussed above, loses its utility when we turn to quantified expressions, both those which 'signify universally' (*every man, no man*) and those which do not (*some man, not every man*).[7] The facts here, as Aristotle judges them, are as follows:

> An affirmation is opposed to a denial in the sense which I denote as 'contradictory' when, while the subject remains the same, the affirmation is of universal character and the denial is not. The affirmation 'every man is white' is the CONTRADICTORY of the denial 'not every man is white', or again, the proposition 'no man is white' is the CONTRADICTORY of the proposition 'some man is white'. But propositions are opposed as CONTRARIES when both the affirmation and the denial are universal, as in the sentences 'every man is white', 'no man is white'. (17b16–23)

The result is the logical figure illustrated by the familiar square of opposition (actually first employed eight hundred years later by the commentators Apuleius and Boethius; cf. Kneale and Kneale 1962; Sullivan 1967):

(3)

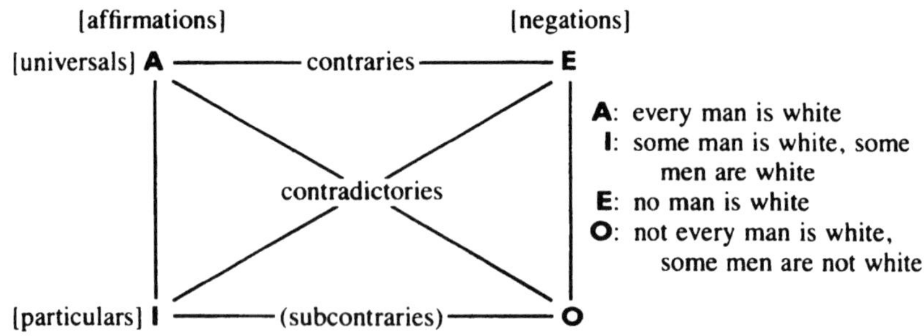

The horizontal axis represents a distinction in QUALITY, the vertical axis a distinction in QUANTITY. The vertex labels derive from the vowels in the Latin verbs ***a**ffirmo* 'I affirm' and *n**e**g**o*** 'I deny'.

Notice that Aristotle has shifted to a semantically based definition of opposition. The **A/O** and **I/E** pairs are contradictories because in any state of affairs one member of each must be true and the other false. Similarly, the **A/E** pair represents contrary opposition because of its truth conditions:

> We see that in a pair of this sort both propositions cannot be true, but the contradictories of a pair of contraries can sometimes be true with reference to the same subject; for instance, 'not every man is white' and 'some men are white' are both true.
>
> (17b23–25)

Note that the last-mentioned opposition, that between the contradictories of contraries (i.e., between **I** and **O**) is a peculiar opposition indeed. This is brought out nicely in a passage from a later text:

> Verbally four kinds of opposition are possible, viz. universal affirmative to universal negative [**A/E**], universal affirmative to particular negative [**A/O**], particular affirmative to universal negative [**I/E**], and particular affirmative to particular negative [**I/O**]: but really there are only three: for the particular affirmative is only verbally opposed to the particular negative. Of the genuine opposites I call those which are universal CONTRARIES, e.g., 'every science is good', 'no science is good'; the others I call CONTRADICTORIES. (*Prior Analytics* 63b21–30)[8]

The same truth-conditional criteria are brought to bear when Aristotle considers the arrangement of contradiction and contrariety among the modal propositions (*De Int.*, chapters 12 and 13; *Pr. An.* 32b4ff.). The question of the logical relations among the modals is treated gingerly ('for the subject is not without difficulty', as Aristotle drily concedes at *De Int.* 21a37), but the original working hypothesis, that the contradictory of a modal proposition is formed by negating the verb rather than the modal operator, is discarded on the grounds that the "contradictories" that would result, for example:

(4) It may be / It may [not be]

(5) It is possible for X to be cut / It is possible for X not to be cut

would be mutually consistent:

> Everything that may be cut or may walk may also escape cutting and refrain from walking. . . . But since it is impossible that contradictory propositions should both be true of the same subject, it follows that 'it may *not* be' is not the contradictory of 'it may be'.
>
> (*De Int.* 21b10–18)

Rather, in the language of the medieval commentators (e.g., Cajetan, book 2, lesson 10, in Oesterle 1962), the contradictory of a modal proposition (e.g., *possible to be*) is formed by adding negation to the MODE (*not possible to be*) and not to the verb (*possible not to be*). The resultant analysis may be represented, as pointed out by Cajetan (Oesterle 1962:207) and other medieval commentators, in the form of a square of modal opposition superimposable onto the standard square in (3) above:

(6)

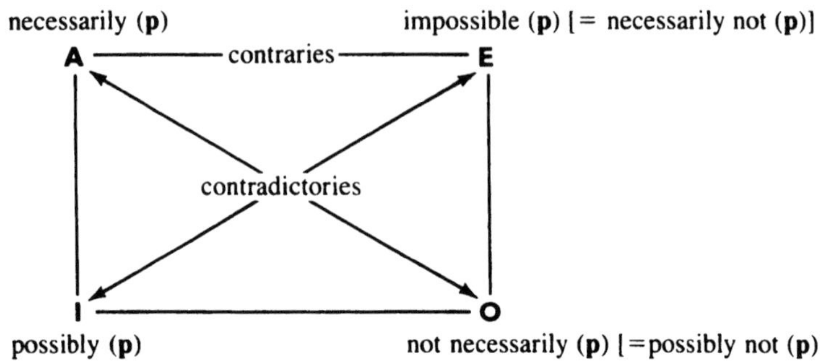

While Aristotle's treatment of contradiction and contrariety for singular, quantified, and modal propositions is relatively clear and consistent, some troublesome questions remain. First, what relation, if any, obtains between a given universal (or necessary) predication and the corresponding particular (or possible) predication? That is, what is the opposition relating **A** and **I** (or **E** and **O**) propositions? Secondly, just what is the relation Aristotle takes to obtain between the contradictories-of-contraries, the **I/O** relation later known—based on the topographic fact that these vertices are literally below the contraries—as SUBCONTRARIETY?

On the first question Aristotle generally maintains a discreet silence (especially for the ASSERTORIC, nonmodal propositions), the possible motivation for which I shall return to in §1.1.3 below. He does, however, note at *De Interpretatione* 22b12 that 'when it is necessary that a thing should be, it is possible that it should be', which might suggest that the relation between the **A** and **I** vertices, at least for the modals, is logical entailment (known as SUBALTERNATION in medieval commentaries on the Square). But this claim runs up against the observation in the earlier passage that whatever is capable of being cut or walking is also capable of not being cut or not walking. Indeed, the latter observation yields Aristotle's law of COMPLEMENTARY CONVERSION (*Pr. An.* I, chap. 32), licensing the inference of an **O**-type modal proposition from the corresponding **I**-type proposition, and vice versa. In the same spirit, we have seen that for Aristotle the corresponding **I**-type and **O**-type particular statements (*Some As are B* / *Some As are not B [= Not all As are B]*) are taken to be merely 'verbally opposed' (*Pr. An.* 63b27–28).

Whence the difficulty from which modal logic did not fully emerge until the modern development of formal pragmatics. Based on the passage at *De Interpretatione* 22b12 (whatever is necessary is possible), Aristotle could identify the possible as the simple contradictory of the impossible; this is what we now know as ONE-SIDED POSSIBILITY. Or, as in the *Prior Analytics*, he could choose the narrower, TWO-SIDED reading:[9]

> I use the terms 'to be possible' and 'the possible' of that which is not necessary but, being assumed, results in nothing impossible. . . . That which is possible then will not be necessary.
> (*Pr. An.* 32a18–28)

Aristotle here leans noticeably toward the two-sided definition (cf. also *De Int.* 22b20: 'If a thing may be, it may also not be'), yet in the same *Prior Analytics* passage (at 32a20) he concedes that 'we say indeed ambiguously of the necessary that it is possible', and while this one-sided reading is rejected here, it figures essentially in the modal syllogisms of *De Interpretatione*, chapters 12 and 13.[10] Thus for Aristotle there are two situations in which something is said to be possible, as the scale in (7) suggests (cf. Hintikka 1960 and Horn 1972, 1973 for elaboration):

(7)

If impossibility and necessity are taken as mediate contraries, as Aristotle assumes, then two-sided possibility is their intermediate; one-sided possibility is the true contradictory of impossibility. But the ambiguity of Aristotle's possibility is more pernicious than it is systematic. In particular, the incompatibility of the *necessary* $\Vdash$ *possible* entailment (*De Int.* 22b11) and the *possible* $\dashrightarrow$ *possible not* (= *not necessary*) conversion is ignored, resulting in the apparent chaos laid out in the argument in (8):[11]

(8) (i) $\Box Fa \rightarrow \Diamond Fa$ [*De Int.* 22b11; *Pr. An.* 32a20]
(ii) $\Diamond Fa \rightarrow \Diamond{\sim}Fa$ [*De Int.* 21b12, 21b35, 22b20; *Pr. An.* 32a29]
(iii) $\therefore \Box Fa \rightarrow \Diamond{\sim}Fa$ (by detachment)
(or equivalently $\Box Fa \rightarrow {\sim}\Box Fa$, given that *not necessary* = *possible not*)

Thus, whatever is necessarily true (*Socrates is Socrates*, 2 + 2 = 4) would be possibly not true, that is, not necessarily true—a modal logician's nightmare. Aristotle himself was not unaware of this result of his reasoning: 'It comes about, therefore, that the thing which must necessarily be need not be, which is absurd' (*De Int.* 22b16); 'And thus it would follow that a thing which must necessarily be may possibly not be; which is false' (22b34). Unfortunately, all he could extricate from the morass was the observation that 'That which is necessary is also possible, though not in every sense in which the word is used' (22b17–18). The problem is dismissed rather than solved.

But concealed within the vagueness of this summation is the germ of a tenable Aristotelian solution to the puzzle (cf. Ackrill 1963: 152): the entailment in (8i) applies to one-sided possibility alone, the bilateral sense being incompatible with complementary conversion (as (7) suggests), while the conversion in (8ii) is restricted to two-sided possibility. From

(9) (i) $\Box \mathbf{Fa} \rightarrow \Diamond_1 \mathbf{Fa}$
(ii) $\Diamond_2 \mathbf{Fa} \rightarrow \Diamond_2 {\sim}\mathbf{Fa}$

no unwelcome inferences emerge.

It will be observed that only one-sided possibility ($\Diamond_1 \mathbf{Fa}$) truly maps onto the **I** (southwest) corner of the Square in (6); the bilateral sense ($\Diamond_2 \mathbf{Fa}$) represents a conjunction of the **I** and **O** vertices. Theophrastus (The Old Peripatetic, second century B.C.), who—in rejecting the conversion principle along with the two-sided possibility modal engaging it—struck the mold for virtually all post-Aristotelian modal logicians, anticipated modern practice by defining a secondary modal notion of CONTINGENCY along just these lines:

(10) contingent (p) $=_{df} \Diamond \mathbf{p} \wedge \Diamond {\sim}\mathbf{p}$ [where $\Diamond$ is equivalent to Aristotle's $\Diamond_1$]

Thus, for Theophrastus and his heirs, the possible has become strictly the not impossible, and the *necessary* ⊩ *possible* entailment (but not complementary conversion) can be retained.

Aristotle's two negations

Throughout his exposition of contradiction, contrariety, and (what we now know as) subcontrariety, Aristotle assumes a framework based on a logic of TERMS, rather than the now far more familiar logic of PROPOSITIONS. In term logic (cf. Sommers 1970, Englebretsen 1981a, 1981b), all statements are categorical, consisting of something (the SUBJECT) about which something (the PREDICATE) is affirmed or denied.

Subject and predicate may be complex, allowing for internal connectives (e.g., conjunction and disjunction, along with negation), but there is no provision for external operators. The negative, hypothetical, and disjunctive judgments at the heart of nineteenth- and early-twentieth-century treatises on logic, along with the unary and binary propositional connectives of modern (post-Fregean) symbolic logic, simply do not exist in the term logic of Aristotle or his Peripatetic followers. Nevertheless, the modern dichotomy of INTERNAL VS. EXTERNAL NEGATION can be traced back directly to the Stagirite.

The solution to this apparent paradox involves the exploitation of one of the oldest weapons in the philosopher's arsenal: scope distinctions. We have seen that predicate denial—*A is not B*—has for Aristotle the appropriate semantics for contradictory negation; it is true if and only if the corresponding affirmation—*A is B*—is false. It is for this reason that denying a predicate of a nonexistent subject results in a true statement (e.g., *Socrates is not ill, The king of France is not bald*). Yet Aristotle was as aware as Russell was, two millennia later, that statements of this type are not always judged true when their subject phrases fail to denote. Under these circumstances, we seem to be ascribing a negative property (e.g., being not-ill) to a subject (Socrates) that does not exist—and negative attributes are no more ascribable to nonexistent subjects than are positive attributes.

I shall discuss Russell's solution in §2.2. Aristotle's analysis begins with the premise that in a copular sentence, the predicate (e.g., *is ill*) consists of a predicate term (*ill*) together with the copula. We have seen that the entire predicate may be affirmed or denied of the subject, resulting in the latter case in contradictory negation (*Socrates is not ill*). But alongside ordinary predicate denial, Aristotle acknowledges the existence of TERM NEGATION, in which a negative predicate term (*not-ill*) is affirmed of a subject.

In addition to ordinary subject and predicate names we must therefore countenance INFINITE OF INDEFINITE names, consisting of a term negation and the term it negates, for example, *not-man, not-ill, not-recovers*. This approach yields a means for distinguishing a false proposition involving term negation (e.g., (11b)) from the closely related true proposition (11a) involving predicate denial:

(11)	a. Socrates is not ill.	(i.e., Socrates [is not] ill.)
	b. Socrates is not-ill.	(i.e., Socrates is [not ill].)

It should be borne in mind that for Aristotle and his Greek- and Latin-writing commentators, the distinction between affirming a negative term and denying a positive term (or predicate) was not signaled by hyphens or brackets but by word order—that is, syntactic scope. Thus the real contrast cited above[12] was, for the ancients, more literally that between (11′a, b):[13]

(11′)	a. Socrates ill not is.	(normal word order, **T** if Socrates does not exist)
	b. Socrates not ill is.	(marked word order, **F** if Socrates does not exist)

Given both predicate denial and term negation, we thus obtain four possible statement types: 'The verb "is" is added either to the term "just" or to the term "not-just", and two negative propositions are formed in the same way' (*De Int.* 19b24–25). Based on the four, Aristotle defines a generalized square of opposition for affirmation and negation, which I can render as follows (*De Int.* 19b18–30; cf. McCall 1967a: 121; Englebretsen 1976:535):

(12)

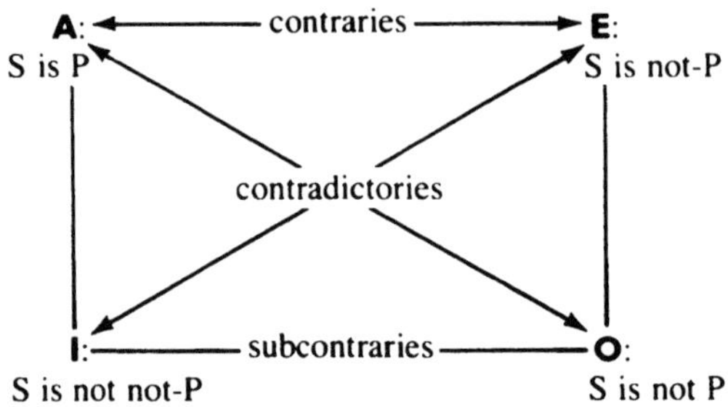

As required, a proposition forms its contradictory opposite by ordinary predicate denial, while a given positive predication and the corresponding infinite predication represent contraries rather than contradictories, since they may be simultaneously false, namely, when the subject term fails to denote or when it is not the sort of thing which can be characterized by the predicate (cf. §2.3 below).[14]

But what sort of predication is *S is not-P,* and what is its relation to the corresponding predicate denial, *S is not P?* The key passage (*Pr. An.* I, 46:51b36–52a17) begins as follows:

> In establishing or refuting, it makes some difference whether we suppose the expressions 'not to be this' and 'to be not-this' are identical or different in meaning, e.g., 'not to be white' and 'to be not-white'. For they do not mean the same thing, nor is 'to be not-white' the negation of 'to be white', but 'not to be white' [is].
>
> (*Pr. An.* 51b5–10)

For Aristotle, (13a) is to (13b) as (13′a) is to (13′b)

(13) a. It is white.
b. It is not-white.

(13′) a. He {can/is able to} walk.
b. He {can/is able to} not-walk.

But (13′b) is clearly not the (contradictory) negation of (13′a), since if it were, 'capacity to walk and incapacity to walk will belong at the same time to the same person', given that 'the same man can both walk and not walk', whereas the definition of contradictory opposition requires that 'an affirmation and a denial which are opposed to one another do not belong at the same time to the same thing'. Given his stipulated correspondence between the two pairs above, Aristotle concludes that (13a, b), like (13′a, b), cannot be contradictory opposites.[15] Furthermore, an infinite negative term like *not-equal* corresponds directly to the privative *unequal,* which clearly does not reduce to the predicate denial *not equal,* for 'not everything is either equal or unequal, but everything is equal or is not equal'.[16]

The next argument is based directly on the scope of negation:

> The expression 'it is a not-white log' and 'it is not a white log' do not imply one another's truth. For if 'it is a not-white log', it must be a log: but that which is not a white log need not be a log at all. (*Pr. An.* 51b28–32)

Socrates is clearly not a white log, but he is not a not-white log either.

In conclusion, then, 'It is clear that "it is not-good" is not the denial of "it is good"' (51b32). But then what is it? Since every (declarative) statement is either an affirmation or a negation, 'if it is not a negation, it must be in some sense an affirmation'. But every affirmation has a corresponding negation. The negation then of 'it is not-good' is 'it is not not-good' (51b33–36), whence the arrangement of the generalized square in (12). Notice also that while predicate denials (*It is not good, Socrates is not ill*) cannot be negated in Aristotle's system, predicates with negative terms (*is not-good, is not-ill*) can themselves be denied as well as affirmed of a subject. But the denial of a negative does not reduce to a simple affirmation.

While *A is not-B* and *A is not B* do not 'imply one another's truth' (51b29), the former does imply the latter:

> If it is true to say 'it is not-white', it is true also to say 'it is not white': for it is impossible that a thing should simultaneously be white and be not-white; if the affirmation does not belong, the denial must belong. (51b42–52a4)

But as we have seen, the converse does not hold: *A* may be neither *B* nor *not-B*. *B* and *not-B* can both fail to apply to a given subject, but cannot both apply at the same time. Thus the unilateral SUBALTERN relation, corresponding to one-way logical entailment, can be established between the **E** and **O** vertices of (12), the generalized square.

The system of opposition described in *Prior Analytics* I, chapter 46 and summarized here is both insightful and internally consistent; its echoes can be heard in Jespersen's distinction between NEXAL negation (*not happy*) and SPECIAL negation (*unhappy*), Von Wright et al.'s distinction of WEAK (contradictory) vs. STRONG (contrary) negation, and Jackendoff's semantic revision of Klima's categories of SENTENTIAL VS. CONSTITUENT negation.

In each case, a negative marker whose scope is narrower than the proposition determines a statement which is, as Aristotle observes, 'in some sense an affirmation', rather than a simple negation or proposition-level denial.

But some inconsistency does arise. Consider the two exchanges below (*De Int.* 20a16–30):

(14)	(i) Is Socrates wise?	(14′)	(i) Is every man wise?
	(ii) No.		(ii) No.
	(iii) Then Socrates is not-wise.		(iii) Then every man is not-wise.

The conclusion in (14′iii) clearly does not follow; instead, all that can properly be inferred is the weaker *Not every man is wise:* 'This last is the contradictory, the former [14′iii] the contrary [of the universal affirmative *Every man is wise*]' (20a30). So far so good; but in the earlier exchange, (14), the conclusion (14iii) is held to follow (20a25–27). How can this be, in the light of the argument in *Pr. An.* I, 46 (or, for that matter, the argument here at *De Int.* 19b18–30) that *A is not-B* constitutes the contrary rather than the contradictory of *A is B?* In particular, how can (14iii) be warranted from (14i, ii) if Socrates does not exist (cf. *Cat.* 13b17–19)? Or, given Aristotle's semantics, if Socrates is a newborn infant, incapable of being either wise or not-wise?

The best guess is probably that of Thompson (1953:256, n. 8): 'We must interpret this remark as assuming that the question (i.e. [14i]) would not be asked if Socrates were nonexistent'. Or perhaps (and this suggestion is not incompatible with Thompson's) Aristotle simply loses sight of the fallacy in (14) in his eagerness to focus on the essential point for the current discussion, the salient contrast between singular expressions (e.g., *Socrates*) and general or quantified expressions (e.g., *every man*).[17]

LC and LEM: Toward a formal definition of contradiction and contrariety

The twin foundations of Aristotle's logic of opposition are the two principles known today as the LAW OF CONTRADICTION (LC) and the LAW OF (THE) EXCLUDED MIDDLE (LEM).[18] These are taken to be basic, undemonstrable principles; as his two instances of 'the common doctrines from which all men prove something' (*Met.* 996b18–30), Aristotle gives LC ('It is impossible to be and not to be at the same time') and LEM ('In every case we must either affirm or deny'). 'The most certain principle of all' is LC, the rule that 'the same thing cannot at the same time both belong and not belong to the same object and in the same respect' (*Met.* 1005b19–23).[19] Crucially, LC applies to both contradictory and contrary oppositions:

> Since it is impossible for contradictories to be truly said of the same object at the same time, it is evident that neither can contraries belong to the same object at the same time.[20]
>
> (*Met.* 1011b17–19)

There are some who would challenge the status of LC as an undemonstrable axiom—or even as a valid law. For these individuals Aristotle reserves his harshest vituperation. Those who stubbornly demand a proof of LC do so simply 'because they lack education': since 'a demonstration of everything is impossible', resulting as it would in infinite regress, at least some principles or axioms (*axiomata*) must be taken as primitive rather than derived from other propositions—and what principle more merits this status than LC? (1006a6–12).[21]

Aristotle acknowledges that some, including Sophists, Pythagoreans, and 'even many physicists' claim that it is possible for the same thing to be and not to be at the same time and in the same respect.[22] But such a position self-destructs 'if only our opponent says something', since as soon as he opens his mouth to make an assertion, any assertion, he must accept LC. But what if he does not open his mouth? Against such an individual 'it is ridiculous to seek an argument', for such a man—insofar as he responds with silence—is no more than a vegetable (1006a1–15).

The same point is made in a later section of Book Γ: one may choose to reject the basic criterion of contradictory opposition, namely, 'that the denial is false whenever the affirmation is true, and the affirmation is false whenever the denial is true', believing instead 'that all speak alike falsely and truly'. (Judge not, lest ye be judged.) But such a man, however noble and generous his spirit, 'can neither speak nor mean anything': 'If he has no belief of anything, how would he differ from a plant?' (1008a35–b12).[23]

Contradictory opposition is governed by LEM as well as LC: the denial is true whenever the affirmation is false, and the affirmation is true when the denial is false. In other words, a corresponding affirmation and denial cannot both be true, by LC, but neither can they both be false, by LEM: 'There cannot be anything between two contradictories, but of any one subject, one thing must either be affirmed or denied' (1011b23–24; cf. 1057a35). Thus, in general, for any two contradictories, 'one of the two must be true and the other false' (*De Int.* 18a31), or at least this is so if we ignore the case of contingent propositions about the future (cf. §2.1). Of course while LC applies to both contradictory and contrary oppositions, LEM holds only for contradictories: 'Nothing can exist between two contradictories, but something may exist between contraries' (*Met.* 1055b2).

LEM is one of Aristotle's first principles, if perhaps not as first a principle as LC. Just as Heraclitus's anti-LC position, 'that everything is and is not, seems to make everything true', so too Anaxagoras's anti-LEM stance, 'that an intermediate exists between two contradictories, makes everything false' (*Met.* 1012a25–29).

Before closing this section, I might try to determine how these two indemonstrable principles governing Aristotle's negation are to be represented in modern notation. The standard versions of LC and LEM (as in the *Principia,* Whitehead and Russell 1910; cf. Russell 1940: 259) are given as in (15a) and (15b) respectively:

(15) a. $\sim(\mathbf{p} \wedge \sim\mathbf{p})$ $[LC_{prop}]$
b. $\mathbf{p} \vee \sim\mathbf{p}$ $[LEM_{prop}]$

A semantic version of these laws is offered by Lukasiewicz (1922):

(15′) a. Two contradictory sentences are not true together. $[LC_{prop/sem}]$
b. Two contradictory sentences are not false together. $[LEM_{prop/sem}]$

But, as Rescher (1969: 149) and Geach ([1972] 1980: 74–75) independently point out, these formulations, employing the 'rather sophisticated' notion of propositional negation (Geach [1972] 1980: 75), are less basic—and less faithful to the traditional conception of the laws—than the quantified (term-based) versions in (16):[24]

(16) a. $\sim\exists \mathbf{x}(\mathbf{Px} \wedge \sim\mathbf{Px})$ $[LC_{term}]$ (or its equivalent, $\forall\mathbf{x}\sim(\mathbf{Px} \wedge \sim\mathbf{Px})$)
b. $\forall\mathbf{x}(\mathbf{Px} \vee \sim\mathbf{Px})$ $[LEM_{term}]$

Thus, LC is read not as the (propositional logic) principle that no statement can be true simultaneously with its negation, but as the (term logic) law that nothing can be both **P** and not **P**. In the same way, LEM is not the principle that every statement is either true or has a true negation, but the law that everything is either **P** or **not P** (≠ **not-P**). For any object **x,** either **x** is red or **x** is not red (but **x** may be neither red nor not-red: if, for instance, **x** is a unicorn or a prime number).

Notice that **P** acts like a predicate variable rather than a predicate constant in the formulas of (16). We can remedy this by taking LC and LEM as statements of second-order predicate logic, following Barnes (1969) and Lear (1980), who offer (16′a) as a representation of the traditional Law of Contradiction; the corresponding version of LEM would appear as in (16′b).

(16′) a. $(\forall\mathbf{P})(\forall\mathbf{x})\sim(\mathbf{Px} \wedge \sim\mathbf{Px})$
b. $(\forall\mathbf{P})(\forall\mathbf{x})(\mathbf{Px} \vee \sim\mathbf{Px})$

The version of LC implied in at least one of Aristotle's instantiations of the law—that at *Metaphysics* 1006b33–34, whose literal equivalent is rendered by Dancy (1975: 162) as 'It is not possible for it to be true to say at the same time of the same thing that it is a man and not a man'—requires a more complex formalization, one involving operators for possibility and truth and allowing quantification over time.[25] But whatever formulation we choose, we should be aware that any translation of the term logic operation of predicate denial into the one-place truth-functional connective of propositional (or sentence) negation cannot faithfully render Aristotle's vision.

1.1.2 The Stoic Opposition

Aristotle's analysis of negation, as we have seen, hinges on the distinction between contradictory opposition, characterized by LC and LEM, and contrary opposition, characterized by LC but not LEM. Predicate denial—in which the entire predicate is negated—results in contradictory negation. Predicate term negation—in which a negative verb is affirmed of the subject—results in a contrary affirmation. Thus, there is strictly speaking no EXTERNAL, propositional negation as such, but two syntactically and semantically distinct types of INTERNAL negation.

Propositional negation originated with the Stoics. While Aristotle and his followers of the Peripatetic school practiced a term logic, using term variables and restricting themselves to categorical (subject/predicate) statements, their Stoic rivals developed the first propositional logic, employing logical constants and propositional variables in the style of modern formal logic and allowing both hypothetical (*if **p** then **q***) and disjunctive (***p** or **q***) propositions. The modern Fregean theory of logical deduction traces back, not to Aristotle's syllogistic, but to the Stoic model of syllogism, formulated as rules of inference, via the medieval theory of consequences derived from the Stoics (cf. Lukasiewicz 1934; Mates 1953).

Alongside the standard binary connectives in the Stoic inventory of truth-functional propositional operators figures the one-place negative connective.[26] In fact, the Stoics—in work transmitted by Diogenes Laertius (Mates 1953: 31ff.; cf. Sullivan 1967: 41–42)—distinguished three varieties of negation, none of them corresponding exactly to Aristotle's predicate denial:

(17) a. DENIAL (*arnētikon*), composed of a 'denying particle' and a predicate: *No one is walking.*
b. PRIVATION (*sterētikon*), formed from an atomic proposition by reversing the predicate: *This man is unkind.* (cf. Aristotle's predicate term negation)
c. NEGATION (*apophatikon*), a negative proposition formed from an atomic or complex proposition by prefixing *oukhi* 'not': *Not: it is day.*

As a contradictory operator, the Stoics' *apophatikon* might be thought to be a notational variant of the predicate denial of term logic. But there are important differences. While Aristotle certainly countenanced multiple negation, generating even such unlikely sequences as *Not man is not not-just* (*De Int.* 19b36), each proposition may contain only one instance of predicate denial (juxtaposed here with both a negated subject term and a negated predicate term). The reason is obvious: each proposition may contain only one predicate. Contradictory negation is essentially introduced syncategorematically by Aristotle, in such a way as to be incapable of applying to its own output.

The Stoics' negation, on the other hand, is an external operator which can be attached iteratively, as its wielders were well aware. Two negations will indeed cancel out, although a negation will not cancel a privation, any more than a predicate denial will cancel a predicate term negation in the Organon (*Socrates is just* entails, but is not entailed by, *Socrates is not not-just*). True double negation is always the negation of a negation, which posits the corresponding simple positive proposition (*Duplex negatio affirmat*). For the Stoic Alexander of Aphrodisias, '*Not: not: it is day* differs from *it is day* only in manner of speech' (Mates 1953: 126). Thus the LAW OF DOUBLE NEGATION (LDN) is born in the propositional logic of the Stoics and not in the term logic of the Aristotelians.[27]

The Stoics can also be credited with the first discussion of the scope of negation (although this notion is implicit in Aristotle, as we have seen), if not of scope simpliciter, in Western logic (cf. Kneale and Kneale 1962: 147). Sextus Empiricus recognizes that for one proposition to constitute the contradictory negation of another it is not sufficient that it merely exceed the other by a negative element. Rather, the two are contradictories only 'if the following condition is satisfied: the negative is prefixed to the proposition in question, for in that case the negative has scope over [or GOVERNS—*kyrieuei*] the whole proposition'. Hence, the negation of *it is day and it is light* must be *not: it is day and it is light* rather than *it is day and it is not light*. In the latter case, 'the negative does not have scope enough to negate the whole proposition, since it is inside the proposition' (*Adv. Math.* 8, 89ff., cited in Mates 1953:95).

The prefixed *oukhi* (or *ouk*) + S formula constituting the Stoics' propositional negation may or may not have been a theoretical fiction. Mates (1953:31) argues that this form is more natural in Greek than is its English equivalent, but Geach ([1972] 1980:75) begs to differ:

> The negation Aristotle was interested in was predicate-negation; propositional negation was as foreign to ordinary Greek as to ordinary English, and he never attained to a distinct conception of it.

> The Stoics did reach such a convention, but in doing so they violated accepted Greek usage; their use of an initial *οὐχί* must have read just as oddly as sentences like 'Not: the sun is shining' do in English.

Unfortunately, no classical Greek informants survive to support either Mates and the Stoics or Geach and the Peripatetics.

From the monasteries of the medieval period to the academic corridors of our own, the debate has raged on over the relative priority of the term logic and categorical syllogisms of Aristotle vs. the propositional logic and hypothetical syllogisms of the Stoics. One contemporary philosopher has no trouble reaching a verdict: 'We know today that propositional logic is logically prior to the logic of terms' (Lukasiewicz 1934:79). The two-valued logic of propositions, 'founded by the Stoics, carried on by the Scholastics, and axiomatized by Frege' (1934:87) has indeed carried the field, as Lukasiewicz points out. But others remain unconvinced: the work of Sommers and Englebretsen offers a spirited defense of term logic, and I shall essay my own in chapter 7 below.

The standard concept of negation in modern symbolic logic is substantially a direct extension of the Stoics' treatment of *apophatikon;* Aristotle's predicate denial and term negation—along with the Stoics' denial and privation operators—have been banished into Parmenides' forbidden realm of nonbeing. Whether external propositional negation is sufficient—or, dare we ask, necessary—for dealing with the varied panoply of negation in natural language is a question to which I must return in this and later chapters.

1.1.3 Existential Import and the Square of Opposition

One thread left hanging in my exploration of the Square in §1.1.1 (represented in (3), repeated here) was the relation between the universal proposition and the corresponding particular (i.e., between **A** and **I**, and between **E** and **O**).

(3)

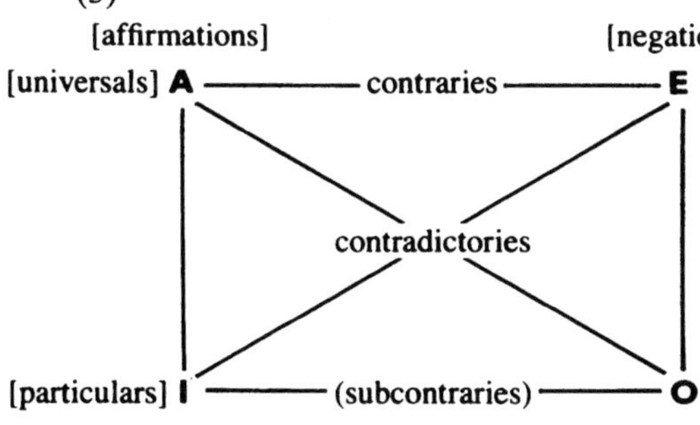

A: every man is white
I: some man is white, some men are white
E: no man is white
O: not every man is white, some men are not white

For the corresponding modal oppositions (cf. (6)), Aristotle did posit entailment between necessity and possibility (at least some of the time). But the parallel entailment from *All S is P* to *Some S is P* is never stated in the *Organon;* the first citation of this SUBALTERN relation is given by Apuleius in his *Peri Hermenias* (2d century A.D.) and is then taken up by Boethius (5th century) and later commentators (cf. Sullivan 1967 for discussion).

One reason for this apparent oversight may have been the issue of EXISTENTIAL IMPORT, as it has come to be known: what are the truth conditions for **A**, **I**, **E**, and **O** propositions when their subject terms fail to denote? A proposition has existential import if and only if it entails the corresponding existential proposition based on its subject term. If there are no unicorns, does *All unicorns are equine* entail *Some unicorns are equine?* If Socrates does not exist, is it really true that he is not bald? There are, as it happens, at least four distinct ways of answering such questions:

(18) (i) Existential import is determined by the quality of the proposition; affirmative (**A** and **I**) propositions entail existence, while negative ones (**E** and **O**) do not.
(ii) Existential import is determined by the quantity of the proposition: universals (**A** and **E**) have no existential import, while particulars (**I** and **O**) do.
(iii) Existential import corresponds to a presupposition associated with **A**, **E**, **I**, and **O** propositions.
(iv) The question of existential import is entirely absent from the Square of Opposition.

I shall sketch each of these positions in turn.

The qualitative approach (18i) has its roots in Aristotle, as Thompson (1953:257) and Sullivan (1967:42) point out. The existence of Socrates, as we have seen, is a necessary condition for the truth of any singular proposition concerning him (e.g., *Socrates is ill*), while his nonexistence is a sufficient condition for the truth of the corresponding (contradictory) negation (*Socrates is not ill*). It is natural to extend this observation to the generalization that the existence of what is denoted by the subject term should be a necessary condition for the truth of any affirmative proposition about that subject (and its nonexistence a sufficient condition for the truth of the corresponding negation). By this reasoning, *All unicorns are equine* and *Some unicorns are equine* come out false, and their respective contradictories, *Not all unicorns are equine* and *No unicorns are equine,* come out true. By adopting this position, Apuleius was thus able to fill out the Square with its now standard subaltern relations (**A** → **I** and **E** → **O**). But the matter was by no means settled; Apuleius (and the Stoics who joined him in this move) had merely begun, or more accurately had resumed from Parmenides and Plato, a lively debate on the status of negative existentials (*X does not exist*) that has continued through the Scholastic era into our own.

The linguistic form adopted for negative (**E** and **O**) propositions is tied to the decision taken on existential import, but writers have not always opted for consistency. Aristotle's (somewhat confusing, if not confused) argument in the last chapter of *De Interpretatione* (23a28ff.) for why the real contrary of *Every man is just* should be *No man is just,* rather than *Every man is unjust,* may have been motivated by considerations based on existential import (Thompson 1953:258). Apuleius, while supporting existential import only for affirmative propositions, nevertheless took negation to be expressed canonically in the predicate (rather than incorporated as *nullus* or *non omnis*). His square thus was spelled out in the manner of (19):[28]

(19)

A:
omnis voluptas
bonum (est)

E:
omnis voluptas
non est bonum

('{all/some} pleasure {is/is not} good')

I:
quaedam voluptas
bonum (est)

O:
quaedam voluptas
non est bonum

Apuleius, however, was well aware of the equivalences we know today as the Laws of Quantifier Negation. Following the Stoics, Apuleius argued that the placement of a negative particle in front of a proposition converts it into its contradictory opposite. The Stoics, curiously, had never generalized their rule for syntactically external negation to quantified formulas; indeed their (extant) texts contain no mention of universal affirmatives (Mates 1952:32). Apuleius remedied this defect, supporting equivalences of the form $\sim$**A** $\leftrightarrow$ **O**, so that *Non omnis voluptas bonum* is taken as logically (although not syntactically) identical to the contradictory of the universal affirmative, namely, *Quaedam voluptas non est bonum* (Sullivan 1967:71, 148–49).[29]

Boethius, whose system of logic in general and of the Square in particular seems to have been either borrowed directly (and without credit) from Apuleius or developed independently (three centuries later) from a (lost) common post-Aristotelian source (see Sullivan for extensive discussion), constructed a Square formally identical to that in (19), except that the **E** proposition is given in the form *Nulla voluptas bonum est,* with incorporated negation.

The first detailed consideration of the interrelationship of negation, scope, and existential import appears in Abelard's *Dialectica* (early twelfth century; Abelard 1956:177–78). Noting that the contradictory of an **I** proposition like *Quidam homo est iustus* 'Some man is just' must be *Non quidam homo est iustus* rather than *Quidam homo non est iustus,* Abelard generalizes the observation into an (apparently) independent rediscovery of the Stoic doctrine: 'The proper negation of *any* proposition results from the placing of the negation in front of the whole' (Kneale and Kneale 1962: 210; emphasis mine). Thus, even a singular proposition like *Socrates est homo* must be negated as *Non Socrates est homo,* rather than the far more natural *Socrates non est homo.*

Abelard's uniform treatment of propositional negation yields the following pairs of contradictories:

(20)	*omnis homo est albus* (every man is white)	:	*non omnis homo est albus*
	quidam homo est albus (some man is white)	:	*non quidam homo est albus*
	Socrates est albus	:	*non Socrates est albus*
	homo est albus ((a) man is white)	:	*non homo est albus*

Consistency has been achieved—at the cost of naturalness. The same strategy is applied to the issue of existential import. Since *omnis* involves existence—that is, an expression of the form *Omnis A est B* entails the existence of at least one A—even *Omnis homo est homo* is false if there are no men. There is thus a crucial truth-conditional distinction for Abelard between a proper **O** expression of the form *Non omnis A est B* and the corresponding particular negative *Quidam A non est B:* the former is true and the latter false if there is nothing satisfying the subject term A. This follows from Abelard's adoption of a neo-Stoic (or proto-Fregean) approach to sentential negation, in which the contradictory of any proposition **p** is by definition ***non-p*** (Abelard 1956:183).

For Aristotle, *Not every man is white* was indeed taken to be the canonical contradictory of *Every man is white* (*De Int.* 24b6), but there is no suggestion that it is not considered to be equivalent to *Some man is not white;* for Apuleius and Boethius, these two forms were explicitly taken to be notational variants. Abelard's results, despite the consistency of his argumentation, were apparently too counterintuitive to be taken seriously; later medieval (and modern) logicians almost without exception rejected this distinction between *non omnis* and *quidam non.*

Among the later Scholastics, Buridan (14th century) most closely echoes Apuleius's qualitative view of existential import. For Buridan, the motivating factor is the truth-conditional relation of contradictories:

> Contradictories are such that one is affirmative and the other negative and such that it is necessary that one be true and the other

> false. . . . Whatever is required for the truth of the affirmatives is required for the falsity of the negative. And likewise, whatever suffices for the falsity of the affirmative, suffices for the truth of the contradictory negative. (Buridan 1966: chap. 2, concl. 11)

Following Moody (1953: 39), we can attribute to Buridan the two equivalences in (21):

(21) $\mathbf{T(p) \leftrightarrow F({\sim}p)}$
$\mathbf{T({\sim}p) \leftrightarrow F(p)}$

Buridan's starting point on existential import is essentially that of Apuleius: 'Every affirmative proposition whose subject or predicate stands for nothing is false'; universal affirmatives are specifically included in this general statement (chapter 1, concl. 15 in Buridan 1966: 72). Indeed, someone who points to a stone and announces 'This man is a substance' speaks falsely, since the subject (*this man*) is empty.

Moody (1953: 51) notes that Buridan and his contemporaries 'made constant use of the principle that universal affirmatives [of the form $\mathbf{\forall x(Fx \rightarrow Gx)}$] have two conditions of falsity', namely (in modern notation), either $\mathbf{{\sim}\exists xFx}$ or $\mathbf{\exists x(Fx \wedge {\sim}Gx)}$. *All unicorns are equine* is equally falsifiable by the nonexistence of unicorns or by the existence of a feline one. These are, by the same token, sufficient conditions for the truth of the particular negative. Since an **O**-type statement is true if and only if either of these conditions holds, such a statement corresponds more exactly to the nonentailing *Not every F is a G* than to the entailing *Some F is not a G*. The former is automatically true but the latter false in a state of *F*-lessness. (For singular expressions, however, a contradictory negation for Buridan and his cohort was formed by adding *not* to the affirmative copula, as in Aristotle, rather than by an external propositional negation in the manner of the Stoics and Abelard.)

Given the principle of existential import for all and only affirmative sentences, the version of the square employed by Buridan and others would translate into modern notation as follows (Moody 1953: 51–52):

(22)

A	**E**
$\mathbf{\exists xFx \wedge \forall x(Fx \rightarrow Gx)}$	$\mathbf{{\sim}(\exists x)(Fx \wedge Gx)}$
I	**O**
$\mathbf{\exists x(Fx \wedge Gx)}$	$\mathbf{{\sim}(\exists x)Fx \vee \exists x(Fx \wedge {\sim}Gx)}$

For Aristotle, as we have seen, a singular proposition has existential import if it is affirmative (*S is (not-)P*) but not if it is negative (*S is not (not-)P*). For Apuleius, Boethius, Abelard, and Buridan, any proposition has existential import if and only if it is affirmative; hence **A** and **I**, but not **E** or **O**, statements can be true only if something exists which satisfies the subject term. The qualitative view of existential import, which was the received position for the ancients and medievals, has more recently been endorsed by Brentano (Vandamme 1972:83), Peirce (1933:440), and Thompson (1953, 1954).

But the qualitative approach rapidly lost ground with the development of modern formal logic. In the predicate calculus, particular statements—*Some F {is / is not} G*—are explicitly treated as existential, while universal quantification does not build in existential import (although the existential proposition could always be added as a separate conjunct, as Moody does in his representation of Buridan's semantics, (22) above). What results is the quantitative approach, (18ii): particulars have existential import, universals do not.

The straightforward translations of the classical propositional forms into standard first-order quantificational logic work out as follows:

(23) **A**: All Fs are G $\forall x(Fx \rightarrow Gx)$
I: Some Fs are G $\exists x(Fx \wedge Gx)$
(= at least one F is G)
E: No Fs are G $\forall x{\sim}(Fx \wedge Gx)$ [or the equivalent ${\sim}\exists x(Fx \wedge Gx)$]
O: Some Fs are not G, Not every F is G $\exists x(Fx \wedge {\sim}Gx)$ [or the equivalent ${\sim}\forall x(Fx \rightarrow Gx)$]

It will be noticed that no distinction is drawn here between the roles of subject and predicate; *F* and *G* are merely two different predicate names. More relevant to our purposes is the fact that the universal affirmative (*All Fs are G*) is vacuously true if there are no Fs, as is the universal negative. The latter position is unarguable, at least when the **E** proposition is given in its incorporated form (*No unicorns are equine*). The former claim, however, has often struck critics as a reductio of the proposal that the predicate calculus can adequately model natural language semantics. Do we really want to count *All unicorns are equine* as true? And as true because of the fact that no unicorns exist? Further, the classical Square of Opposition must be modified or abandoned on the quantitative view, since a proposition with the logical form of (23**A**) does not entail the corresponding proposition with the form of (23**I**): if nothing satisfies *F*, the former is true and the latter false.

The two remaining approaches focus on a perceived inadequacy of accounts on which a sentence like *All ogres are wicked* comes out either false (under (18i)) or vacuously true (under (18ii)) merely because no ogres exist. Strawson (1952: 163–79) and Hart (1951) seek to rescue the traditional interpretation of the Square by taking existential import to be a presupposition rather than an entailment. Strawson and Hart, the prime advocates of (18iii), reject the idea of preserving the Square of Opposition (in particular, the **A** ⊩ **I** subaltern entailment) by building in existential import in the manner of the modern adherents to the qualitative view, who analyze a simple universal like *All men are white* into a conjunction (as in (22)) so that it may continue to entail the corresponding existential.

The villain of the presuppositionalists' piece is the 'logician's prejudice'—whose source, as we shall see, is Aristotle—that 'on every occasion of use a meaningful [declarative] sentence must be true or false' (Hart 1951: 204–5). A sentence is (or is not) meaningful; on a particular occasion of use a meaningful sentence may (or may not) express a statement which is true or false. If its presuppositions are unsatisfied, the question of the statement's truth or falsity simply fails to arise. Such is the case with *Smith has stopped beating his wife,* but also with (the normal use of) *All ogres are wicked:* 'The A form in the absence of a special indication "presupposes" or "strongly suggests" the truth of the existential form' (Hart 1951: 207). 'The existence of members of the subject class is to be presupposed' (Strawson 1952: 176).

For Hart, conjoining an **A**-form proposition to the rejection of its existential import, as in the familiar Oxonian example

(24) All Smith's children are girls—but he has none.

produces neither the logical contradiction that would result on a theory like that of (18i) nor the automatic verification that would result from a minimalist approach like that of (18ii), but rather the continuation 'would cancel the original remark, rendering it pointless', the typical diagnostic of an annulled presupposition.

On this option, we do indeed end up with a modification of the Square, or more exactly a restriction on the domain of its application. We can no longer say, with Aristotle, that of any two contradictories (**A**/**O**, **I**/**E**), one must be true simpliciter, but only that if either is true, the other is false (and vice versa): 'The existence or nonexistence of members of the subject class determines not the truth or falsity of statements of these forms but only the prior question whether the question of their truth or falsity can arise' (Hart 1951: 209).

Like Thompson (1953) on behalf of (18i), Hart assimilates to his position both Aristotle and the vox (tacita) populi: 'Questions of existence lie outside the scope of the square and those who constructed it along those lines accurately interpreted the ordinary usages of these forms' (Hart 1951: 209).

With this last remark, the presuppositionalist account of Strawson and Hart begins to shade perceptibly into the agnostic theory (18iv) advocated by Nelson, the latter's disclaimers to the contrary notwithstanding. For Nelson, the Square of Opposition can—and must—be abstracted away from questions of existential import. The laws of subalternation (the **A** ⊩ **I** and **E** ⊩ **O** entailments), contrariety, and contradiction are semantically unrestricted in their application, but must be evaluated against the 'appropriate realm of discourse'—mythological, religious, mathematical, zoological, or whatever. Universals cannot be claimed to have or to lack existential import; they are simply neutral. Thus any **A**-form proposition—even *All ogres are wicked*—will entail the corresponding **I** proposition (*Some ogres are wicked*), since the question of import (are there any ogres?) need never be broached; the question of existential import is entirely absent from the Square of Opposition. Like Thompson, whom he seeks to rebut, Nelson claims that Aristotle, as well as ordinary language, is really on his side. To the extent that his position requires a shorter leap of faith than that of his rivals, since Aristotle never directly addresses the topic of existential import in general (as opposed to singular) statements, at least Nelson's historical claim may well be correct.[30]

Evidently, which side of the existential import question Professor X comes down on tells us more about X's logical and philosophical Weltanschauung than it does about the proper approach to existence, negation, quantification, and the Square of Opposition. The Square thus continues to serve effectively not only as a device for ordering our thoughts about contradiction, contrariety, subcontrariety, and entailment, but also as a litmus test for linguistic philosophers—albeit one whose results are sometimes hard to read. The second moral that can be drawn concerns the ghost of Aristotle and the spirit of ordinary usage, both of which prove to be rather more elusive than their would-be recruiters seem to imagine.

1.1.4 What Is a Negative Proposition?

Twenty-five centuries of dispute over the nature of negative propositions—what is the relation between negation and affirmation? what is the canonical form of negative propositions, and what existential (and other) inferences can be drawn from them? how many different forms of negation must be countenanced?—have not settled the most basic question of all: just what is a negative proposition, and how can we tell?

For Aristotle, the question as such does not arise. Aristotle's is a term logic; my use of notions like 'negative proposition' in my rendering of the Organon is more a function of my own interests than of his. As we have seen, the closest equivalent of the negative proposition within this system is predicate denial, in which a predicate (which may itself contain a negative term, the 'in(de)finite' ***not-P*** or the 'privative' ***un-P***) is denied of a subject *S*. Speaking somewhat freely, then, (25a) would count as a negative proposition for Aristotle, while (25b) would not.

(25) a. (Not-)**S** is not ({not-/un-})**P**
  b. (Not-)**S** is ({not-/un-})**P**

This distinction, however, does not automatically generalize to quantified expressions, where the relations of contradiction and contrariety must, as we have seen, be treated semantically.[31]

Pinning down the class of negative propositions first emerged as a major goal for the Stoics. As noted in §1.1.2, the Stoics' logic, like Frege's, was propositional in nature, leading to their formulation of the Law of Double Negation. For the Stoics, a denied (or negative) proposition was simply that proposition which begins with a negative particle (*ouk, oukhi*). Hence, contra Aristotle, propositions like those in (26a, b) cannot be negations (*apophatika*), and neither can (26c).

(26) a. Pleasure is not good.
  b. Some pleasure is not good.
  c. It is light and it is not good.

For the Stoics, as for Aristotle and the Peripatetics, every proposition is either negative or affirmative, so the propositions in (26) count as affirmative.

Apuleius, accepting this as his starting point, distinguishes the *ABDICATIVA* (negative) proposition from the *DEDICATIVA* (assertive). He may have been the first to recognize what was to become a recurring motif for the developers of the negative theme: the observation that an affirmative proposition may be logically equivalent to a negative counterpart, as in the pair

(27) a. It is not the case that some pleasure is not good.
  b. Every pleasure is good.

The more constrained approach of the Stoics, Apuleius, and Abelard eventually gave way to a looser tradition within the classical and medieval periods, in which denial or contradictory negation was simply any affirmation containing *not* within the predicate, as for example in (26a). This approach was implicitly or explicitly favored by Porphyry, Boethius, and al-Fārābi, among the early commentators.

For Mill ([1843] 1919:87), following Aristotle, an affirmative proposition is 'one in which the predicate is affirmed of the subject' and a negative proposition 'one in which the predicate is denied of the subject' (cf. *De Int.* 17a25). In a singular expression (*Caesar is / is not dead*), the copula consists of *is* (the 'sign of affirmation') or *is not* (the 'sign of negation'). In this way, Mill seeks to preserve Aristotle's requirement that the two contradictories share the same subject (*Caesar*) and predicate (*dead*), although Mill employs the latter term in a non-Aristotelian way.[32]

Mill's position is set off against that of Hobbes, according to which the copula is simply *is*, and the negative sign—simply *not*—is attached to the predicate. The Hobbesian view essentially collapses the two Aristotelian negations; a negative proposition is one in which the predicate is a negative name. For Mill, this move is a mere fudge, an evasion of reality: 'The fundamental distinction is between a fact and the non-existence of that fact. To put things together and to put or keep them asunder, will remain different operations, whatever tricks we may play with language' (Mill [1843] 1919:87).

Not until our own century did philosophers begin to question the universal assumption behind the debate over the form of the negative proposition or (after Kant) the 'negative judgment'. Is there in fact a coherent class of negative propositions? Is there even a negative judgment per se? Frege (1919:125) entertained these questions, and replied in the negative: 'People speak of affirmative and negative judgments; even Kant does so. Translated into my terminology, this would be a distinction between affirmative and negative thoughts. For logic, at any rate, such a distinction is wholly unnecessary . . . I know of no logical principle whose verbal expression makes it necessary, or even preferable, to use these terms'.

Nor is it easy—or perhaps possible—to determine just which judgments, thoughts, or propositions would count as negative and which as affirmative. For Kant ([1787] 1964:B97/A72), *The soul is not mortal* is unquestionably a negative judgment, while its affixal counterpart, *The soul is nonmortal (nichtsterblich)*, is an affirmative judgment insofar as its logical form is concerned. This is the orthodox Aristotelian line. But, Frege bids us, consider a paradigm like that in (28),

(28) a. Christ is immortal.
b. Christ lives forever.
c. Christ is not immortal.
d. Christ is mortal.
e. Christ does not live forever.

and try to give a coherent, non–ad hoc reply to the obvious question: 'Now which of the thoughts we have here is affirmative, and which is negative'?

According to Peirce (extrapolating from his answer to a similar poser), the judgments corresponding to (28a, b, d) would count as affirmative and those corresponding to (28c, e) as negative, at least within the ancient and traditional model. But Quantity, on this account, 'is an affair of the mode of expression solely' (Peirce 1933:440, §4.552), thereby begging the Fregean question.

Negation within the predicate phrase of a sentence constitutes for Frege neither a necessary nor a sufficient criterion for producing a negative judgment: 'A negation may occur anywhere in a sentence without making the thought indubitably negative' (1919: 125). If the notion 'negative judgment' is not simply a linguistic concept (identifiable with the negative sentence) illegitimately smuggled into philosophical discourse, its deployment is at best premature (in the absence of clear diagnostics for cases like those in (28) and at worst pointless, resulting in 'endless disputes, carried on with the greatest subtlety, and nevertheless essentially sterile' (Frege 1919: 125–26).

Royce (1917) offers a different argument leading to the same conclusion. Citing the symmetry of contradictory negation ('of **p**, **not-p**, exactly one must be true') and the Law of Double Negation ('every proposition is the negation of its own negation'), Royce argues that there cannot be a coherent class of negative (or affirmative) propositions. Every denial is ipso facto an affirmation, and vice versa, since 'to affirm is to deny the contradictory of whatever one affirms' (Royce 1917: 265–66).

Ayer ([1952] 1963) considers this issue in greater depth. He begins by distinguishing the relatively straightforward task of defining the NEGATION-OF relation from the much more difficult (if not impossible) aim of determining what makes a statement negative. In the first case, 'It seems a fair reflection of ordinary usage to identify the negation of S with any statement T which is so related to S that if either is true it follows that the other is false'; Ayer ([1952] 1963: 42) here echoes the philosophers of yesterday and anticipates the linguists (e.g., Jackendoff 1969) of today. By this criterion, affixal negation (*un-*, *iN-*, *-less*) does not yield (true) negation. In reaching this result, Ayer endorses what is for once a unanimous verdict (or near-unanimous: cf. Frege 1919, cited in §1.2.2 below): from the Peripatetics and the Stoics, for whom so-called privative statements—*A is un-B*—were affirmative in nature, to Sigwart (1895: 138), Strawson (1952: 7), Zimmer (1964), and H. Clark (1974), it is universally agreed that affixal negation—well, at least *most* affixal negation (cf. §5.1)—produces a contrary affirmation, a 'polarization' (Vandamme 1972: 69), rather than a true contradictory.

But knowing when a proposition or statement T is the contradictory negation of another proposition or statement S tells us nothing about whether T—or for that matter S—is or is not itself negative, just as knowing when one man B is taller than another man A does not enlighten us as to whether A or B is himself tall. Ayer recognizes this point, which hinges on the difference between relational and absolute knowledge, and he experiments with diverse syntactic criteria for settling the issue, leading him to a rediscovery of Frege's point: different forms can express the same statement. He offers the example of the pair in (29):

(29) a. Everest is the highest mountain in the world.
b. There is no mountain in the world higher than Everest.

Perhaps, then, there are psychological grounds for sequestering negations as a special class of statements used only for rebuttals or denials? But any statement can be so used (Ayer [1952] 1963:38). In the end, the difference between affirmative and negative statements is reduced to one of approach: 'A statement is negative if it states that an object lacks a certain property rather than stating that it possesses the complementary property: a statement is negative if it states that a certain property is not instantiated, rather than stating that the complementary property is universally instantiated (Ayer [1952] 1963:61). But it is by no means clear that this criterion, or its subsequent winnowing via Ayer's notion of specificity—by which the negative statement will always be the less specific member of a pair of contradictories—can be successfully defended.[33]

One moral to be drawn from the discussions in Frege 1919, Royce 1917, and Ayer 1952 is that the literature on negative propositions is racked with the confusion of statements and sentences focused on by Strawson and others. Kissin (1969:5), adding his own pair of semantically equivalent sentences distinguished by the presence vs. absence of negation:

(30) a. He's staying.
b. He's not leaving.

to the inventory established by Apuleius (27a, b), Frege (28c, d, e), and Ayer (29a, b), points out that the traditional criteria for negativity—the presence of a negative particle, its appearance in a specified syntactic location, and so forth—apply only to sentences, not to statements or propositions.[34] In fact, for Kissin, there is no such animal as a negative statement or proposition. AFFIRMATION (AFFIRMATIVE) refers to what people do with sentences; its opposite number is DENIAL. NEGATION (NEGATIVE) is a property of sentences; as its counterpart, Kissin proposes NONNEGATION (NONNEGATIVE). But having drawn the appropriate distinction between these two pairs of terms, Kissin proceeds to restrict his dissertation to the latter set, ignoring the more general and philosophically more potent pair entirely.

One final point on the status of the negative proposition is worth making. Wittgenstein warns us, in his characteristically cryptic Tractarian fashion, not to be misled by our notation into believing that ~ (or *not*) has the power to render a proposition negative:

> In '~p' it is not '~' that negates; it is rather what is common to all the signs of this notation that negate *p*. That is to say the rule that governs the construction of '~p', '~~~p', '~p v ~p',

> '∼p · ∼p', etc. etc. (ad inf.). And this common factor mirrors [*spiegelt*] negation. (1922: §5.512)

As we shall see in §1.2, the fact that no clear criteria have been adduced for defining a class of negative propositions has not deterred centuries of scholars from debating the true nature of the negative proposition. Nor did the one-to-one correspondence between affirmative and negative propositions (or sentences?) stipulated by Aristotle, Royce, and Wittgenstein (and challenged by others, as we shall also see) dissuade their contemporaries (or, for that matter, themselves) from taking negatives to be inherently asymmetrical with, and in some sense inferior to, their affirmative counterparts.

1.1.5 Negation and the Legacy of Aristotle: Retrospect and Prospect

> Now if the oppositions are contradiction, privation, contrariety, and relatives, and since of these contradiction is primary, and nothing can exist between two contradictories but something may exist between two contraries, it is clear that contradictories and contraries are not the same. (Aristotle, *Metaph.* 1055b1–4)

Contradiction is the primary opposition, since two contradictories—in the language of the great tenth-century Arab commentator Avicenna (aka ibn-Sīnā; cf. Madkour 1934: 178–84)—'divide the true and the false between them'. Indeed, in the Arabic tradition developed by Avicenna and others, the same term *tanâqoḍ* translated not only *antiphasis* (Aristotle's 'contradiction') but also *antithesis* ('opposition'), although the contradiction/contrariety distinction itself was preserved intact.

It has often been observed that Aristotle was the first to consistently identify negation (denial) with contradiction, but the supposition (by, inter alia, Kissin 1969) that Plato's negation was basically the affirmation of a contrary cannot be substantiated. As I have noted, the Stranger's proof (*Sophist* 257B) that negation cannot be reduced to pure nonbeing or opposition hinges on the fact that *mē mega,* literally 'not big, not great', cannot be read as a contrary affirmation: that which is not big need not therefore be small. Plato (through the Stranger) argues explicitly that negation is not *enantion* (contrary) but *heteron* (other). The Platonic concept of negation as a mark of DIFFERENCE assimilates neither to contrariety nor to contradiction. This approach—if it is not entirely circular (cf. §1.2)—is unfortunately insufficient for dealing with the range of negative terms, predications, and propositions Aristotle and his successors undertook to describe; cf. Gale (1976) for extended discussion.

Contraries and contraries—Merry black-and-white contrary,
how do your meanings grow?

How many contraries can a given term or proposition have? In a broad sense, every pair of incompatible terms or statements are contraries of each other under the Aristotelian definitions. *Black* and *white* are contrary terms, since they cannot inhere in the same thing at the same time (cf. Cajetan, lesson 13, in Oesterle 1962:234). But then *white* will have as many contraries as there are distinct colors, since nothing can be simultaneously white and red, or white and green, or white and purple. On this understanding, both *excessive* and *insufficient* are contrarily opposed to *moderate*, as well as to each other (*De Int.*, chapters 11 and 14); contrariety is not an exclusive club.

Further: if any two mutually inconsistent statements are ipso facto contraries, any **A**-form statement will have (at least) two contraries, the corresponding **E** statement and the corresponding **O** statement (its contradictory). *All dogs have fleas* is as inconsistent with its contradictory (*Not all dogs have fleas, Some dogs do not have fleas*) as with its (true) contrary (*No dogs have fleas*). Aristotle at times does indeed seem to endorse the view that contradictories are somehow more contrary than (mere) contraries are (cf. *De Int.*, chapter 14 and Cajetan's commentary, lesson 13, Oesterle 1962:254).

On any account in which contradictories must satisfy LC and LEM and contraries must satisfy LC, any two contradictories emerge as ipso facto contraries. Thus, compare Strawson's definitions of these terms based on the relation of INCONSISTENCY:

> To say of two statements that they are contradictories is to say that they are inconsistent with each other [i.e., that they obey LC] and that no statement is inconsistent with both of them [i.e., that they obey LEM]. To say of two statements that they are contraries is to say that they are inconsistent with each other, while leaving open the possibility that there is some statement inconsistent with both of them. (Strawson 1952:16)

A bit further on (p. 25), Strawson notes that 'two statements are contraries when it is logically impossible for them both to be true; subcontraries when it is logically impossible for them both to be false'. But then any two contradictories (e.g., *Every man is a chauvinist / Not every man is a chauvinist; Socrates is ill / Socrates is not ill*), by virtue of 'splitting the true and the false between them', are automatically contraries and subcontraries as well.

If this result seems more troublesome to us than it apparently did to Aristotle, Cajetan, or Strawson (or Englebretsen 1976:536), we could always limit the class of contrary statement pairs by fiat to those statements which are not contradictories. Contradiction, as the primary opposition, should be capable of relegating other oppositions to an 'elsewhere' clause.[35]

While a given term or proposition may have more than one contrary in the weak sense of the term (where any two incompatible terms or propositions count as contraries), Aristotle more often seems to assume a different, stronger sense of contrariety, one which I shall call POLAR contrariety:

> Since things which differ from one another may do so to a greater or a less degree, there exists also a greatest difference, and this I call 'contrariety'. (*Met.* 1055a4–6)

> Contrariety is complete difference. . . . This being so, it is evident that each contrary cannot have more than one contrary; for neither can there be anything more extreme than the extreme, nor can there be more than two extremes for one interval. (*Met.* 1055a17–28)

> We call contraries . . . those which differ most in the same genus. (*De Int.* 23b23)

The image of polar contraries as extremes along an interval is especially important for my purposes (cf. the discussion of scales and antonymy in chapters 4 and 5); Aristotle invokes the same metaphor elsewhere:

> It seems that in defining contraries of every kind men have recourse to a spatial metaphor, for they say that those things are contraries which, within the same class, are separated by the greatest possible distance. (*Categories* 6a15–19)

Aristotle depicts polar contraries as endpoints on a continuum, corresponding to **A** and **C** below:

(31) $\mathbf{A}\quad \mathbf{B}_1\quad \mathbf{B}_2 \ldots \mathbf{B}_n\quad \mathbf{C}$

$\longleftrightarrow$

An intermediate or middle term is any point $\mathbf{B}_1 \ldots \mathbf{B}_n$ through which one passes in traveling from **A** to **C** (or from **C** to **A**); thus in moving from *white* to its polar contrary *black*, one must first pass through the intermediate *gray* (*Met.* 1057a24).

Saint Thomas (lesson 11, in Oesterle 1962:90) points out that for any two polar contraries which by definition cannot simultaneously inhere in the same thing (e.g., *white* and *black*), their contradictories (*not white, not black*) can (i.e., when something is pallid or yellow). Thus the contradictories of contraries define an intermediate term. The same applies on the level of propositions: the particular affirmative (**I**) and particular negative (**O**) each function as a 'mean between contraries' (*medium inter contraria*).

Aristotle may have the same point in mind when he concludes *De Interpretatione* (24b1) by observing that 'contraries are those which enclose their opposites'. So translates Ackrill, who comments (1963:155) that 'it is natural to think of A and E as extremes and of I and O as lying within them'. Other translators and commentators omit this scalar interpretation, but it is one which is very much in the spirit of the 'spatial metaphor' noted above.

The notion of polar contrariety outlined here and its differentiation from the weaker notion of simple incompatibility were well known to Aristotle's medieval and modern commentators, although the relevant terms have gone under a variety of labels. Cajetan (lesson 13, in Oesterle 1962:237) distinguishes ABSOLUTE (= polar) contraries (e.g., *black / white*) from REDUCTIVE (simple) contraries (e.g., *black / red*). Mill (1867:516–21) criticizes Hamilton's overgeneralized application of contrariety, maintaining that each term or statement can have but one contrary, its 'extreme opposite'; otherwise we are dealing not with true CONTRARIA, but with DISPARATA. For Sigwart (1895:137), only 'those terms in a series which are the farthest apart' can properly be contraries (i.e., all contraries are polar contraries); other mutually exclusive terms (e.g., *red / yellow*) are DISJUNCT.

Just as any given term may have at most one immediate or logical contrary (e.g., for *white, not-white*), any given term may also have at most one polar contrary (for *white, black*), and indefinitely many simple contraries (for *white,* any other color: *red, green, blue,* and—depending on the precise definition of simple contrariety—*not-white* and *black* as well).[36] The unique polar contrary and the unique immediate contrary of a given term will not in general coincide. Thus, given one possible scale for *white* and *black,*

(32)

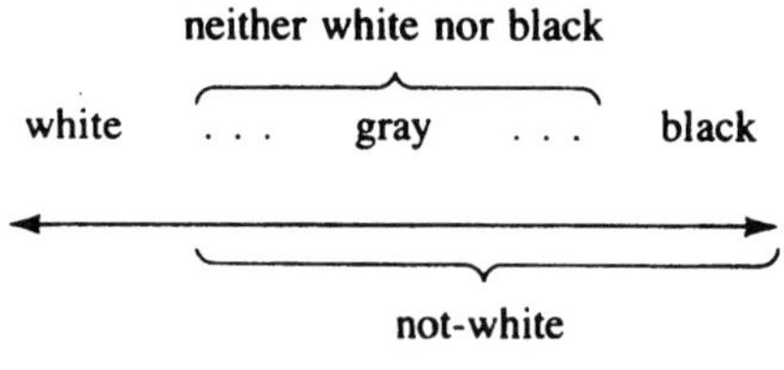

we see by inspection that *white* and *black* are polar contraries, but not immediate contraries, since gray—and indeed, according to Aristotle's somewhat idiosyncratic theory of color, all hues—mediate between them. The immediate contrary of *white* is *not-white* (or *nonwhite*); and in general the immediate contrary of any term will be defined by the operation of predicate term negation (cf. §1.1.1). There may be a simple lexical equivalent of the negative term, however, as in the case of immediate contrary pairs like *odd* (= *not-even*) / *even* (= *not-odd*), or *well* (= *not-ill*) / *ill* (= *not-well*).

Let me define the three notions of contrariety more formally, adapting for my purposes the schema of Barnes (1969). I follow Barnes in writing **R** for the range of a predicate (the RANGE of **P** is the set of things that can be either **P** or **not-P**, equivalent to Sommers's notion (1965, 1970, 1982) of the SPAN of a predicate) and $\mathbf{C_i(F, G)}$ for '**G** is contrary$_i$ to **F**', or '**F** and **G** are predicates in contrary$_i$ opposition'. Then we get these definitions:

(33) (i) $\mathbf{C_1(F, G)} =_{df} [\mathbf{RF} = \mathbf{RG} \wedge (\forall \mathbf{x})\Box \sim (\mathbf{Fx} \wedge \mathbf{Gx})]$

(ii) $\mathbf{C_2(F, G)} =_{df} [\mathbf{RF} = \mathbf{RG} \wedge (\forall \mathbf{x})\Box \sim (\mathbf{Fx} \wedge \mathbf{Gx}) \wedge (\forall \mathbf{y})(\mathbf{y} = \mathbf{RF} \rightarrow \Box(\mathbf{Fy} \vee \mathbf{Gy}))]$

(iii) $\mathbf{C_3(F, G)} =_{df} [\mathbf{RF} = \mathbf{RG} \wedge (\forall \mathbf{x})\Box \sim (\mathbf{Fx} \wedge \mathbf{Gx}) \wedge (\forall \mathbf{H})((\mathbf{RH} = \mathbf{RF} \wedge \mathbf{H} \neq \mathbf{F} \wedge \mathbf{H} \neq \mathbf{G}) \rightarrow (\mathbf{H}\text{ is between } \mathbf{F}\text{ and }\mathbf{G}))]$

$\mathbf{C_1}$ is the relation of contrariety simpliciter (INCOMPATIBILITY for Barnes), $\mathbf{C_2}$ that of immediate contrariety (Barnes's CONTRARY PREDICATES), and $\mathbf{C_3}$ that of polar contrariety. Notice that the scalar relation of 'betweenness' is invoked, but not defined, in (33iii). In addition, we must insure (as Barnes does not) that the variable *y* utilized in (33ii) can range over actual existents only, since we need to allow for both immediate contraries failing to hold when the subjects are within the range of the predicate but not in the class of existent objects (since Socrates is neither ill nor not-ill, the largest prime number is neither odd nor even, etc.).

The system of opposition among terms that emerges from these definitions can now be schematized as follows:[37]

(34)

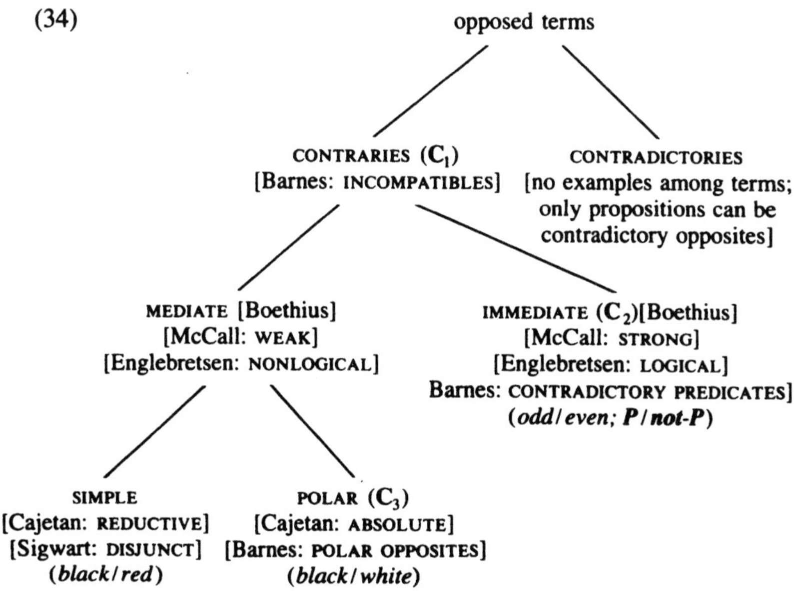

Corresponding notions of propositional contrariety can be defined accordingly. In particular, if two terms ***F*** and ***G*** satisfy $\mathbf{C}_i$, then the two propositions ***a*** *is* ***F*** and ***a*** *is* ***G*** satisfy $\mathbf{C}_i$ as well. In addition, we need to secure Saint Thomas's insight (which may or may not also have been Aristotle's insight) that *All S is P* and *No S is P* represent polar (rather than simple or immediate) contraries. But is it in fact the case that for each proposition there will be exactly one immediately contrary proposition, just as there is exactly one immediately contrary predicate term (***not-P*** or any of its equivalents) for any arbitrary predicate term ***P****?* The answer, especially when I turn to the consideration of general propositions (with quantified subject terms) appears to be no, as we shall see later in this section. Before I return to the question of how to define contrariety over propositions, I need to fill in some background on Aristotle's two negations (predicate denial and term negation), and their fate in the course of later interpretation.

The two negations revisited

A great deal of the post-Aristotelian history of logical negation can be read as an extended commentary on the distinction in the *Prior Analytics* between predicate denial (the operation of contradictory negation taking scope over the entire predication) and predicate term negation (a contrariety-producing operation with scope restricted to the 'infinite' or 'indefinite' term thereby produced).

Avicenna used morphological, syntactic, and semantic criteria to distinguish the NEGATIVE JUDGMENT, *A is not B,* from the INDEFINITE JUDGMENT, *A is not-B* (Madkour 1934: 169). The former is expressed in Arabic by the particle *laysa,* yielding Madkour's gloss 'A n'[est] pas B', the latter by *ghayr* (A [est] non B), often incorporated into the predicate term (e.g., *Zayd ghayr baçir* 'Zayd is nonclairvoyant'). In tripartite (copular) sentences, the place of negation depends on the type of negation, precopular for the negative judgment, between copula and verb for the indefinite (exactly as described by Aristotle). In bipartite (noncopular) sentences, there is no syntactic differentiation, but the semantic distinction remains: for Avicenna, as for Aristotle, an indefinite judgment can be true only if the subject exists, while the negative judgment is automatically true if the subject term fails to denote.

In Latin, as in Greek, there is no morphological differentiation of true negation (affecting the proposition for the Stoics, the subject-predicate connection for the Aristotelians) and infinite (indefinite, infinitive, or term) negation. But the medieval commentators tended to respect the distinction nonetheless. Here, for example, is Burleigh: 'The negation *non* can be taken either merely negatively or infinitively. When it is taken merely negatively, it always negates some propositional complex. But when it is taken infinitively, then it negates some nominal element in the proposition, namely the subject or predicate' (Henry 1972:79–80).[38]

The medieval Jewish logical position on negation, as summarized in Spinoza's *Ethics* (Spinoza 1934:134–35), is somewhat more complex. PARTICULAR negation (Spinoza's PRIVATIO) represents a contingent fact (e.g., *Balaam does not see*), ABSOLUTE negation (Spinoza's NEGATIO) represents a more generally true negative proposition (e.g., *The wall does not see*). 'Privation . . . is denying of a thing something which we think belongs to its nature; negation . . . is denying of a thing something because it does not belong to its nature' (Spinoza 1934:134–35).

When Spinoza, in a letter of 1674, delivered his celebrated edict, *Determinatio est negatio,* 'Determination is negation', he was explicitly referring to the restricted notion of negatio: every description necessarily implies a limitation or exclusion. It will be observed that Spinoza's senses of negation and privation cannot be directly assimilated to those of Aristotle (§1.1.1) or the Stoics (§1.1.2).

Leibniz (1966:18; cf. Castañeda 1976:483–84; Englebretsen 1981a: 13–15) acknowledges both denial negation: *x non est P,* in which the negation sign signifies the mode of predication, and privative or term negation: *x est non P,* in which a negative term is affirmed of the subject. Apparent propositional negation is explained away as a predication of falsehood rather than a negation of truth: 'If B is a proposition, not-B is the same as that B is false' (Leibniz 1966:58). Nor does an initial negation form a constituent with an immediately following quantifier:

> 'Not every' and 'not some' may not properly occur in propositions; for they only negate the proposition affected by the sign 'every' and 'some', and do not make a new sign, 'not-every' or 'not-some'. Thus, if I say 'Not, some man is an animal' [Non, quidam homo est animal], this is the same as that it is false that some man is an animal. (Leibniz 1966:185)

Thus, contrary negation corresponds to a negative predicate, but contradictory negation has two canonical forms: predicate denial and apparent propositional negation (= assertion of falsity). My own neo-Leibnizian position will unfold along similar lines in later chapters.

The fall (and rise?) of contrariety in modern logic

While Leibniz, operating in the extended shadow of Aristotle, maintained the subject/predicate split as a crucial ingredient of syntactic and semantic analysis, the rival Stoic-Abelardian tradition was to gain rapid ascendancy with the birth of modern axiomatic propositional logic. Frege (1919) explicitly disavows the subject-predicate division as an illegitimate importation from the treacherous realms of natural language, to be replaced with function-argument analysis. Predicate denial gives way to the external propositional operator in the representation of contradictory negation, and contrary (term) negation is eliminated. Frege follows Aristotle in observing that 'for every thought there is a contradictory thought'. But, contrary to what the surface syntax suggests, an entire sentence can be (contradictorily) negated by combining a negative element, affix or particle, with a single constituent. Thus, in saying (35b), as in (35a),

(35) a. The man is not celebrated.
b. The man is uncelebrated.

'we indicate the falsity of the thought that he is celebrated' (Frege 1919: 131).

But, pace Frege, it is by no means clear that Aristotle's privative or predicate term negation can in general be assimilated to contradictory propositional negation in this manner. In saying *The man is unhappy,* do we thereby simply indicate the falsity of the thought that the man is happy? Philosophers from Aristotle and Sigwart to Strawson and Drange, and linguists from Sweet and Jespersen to Klima, Zimmer, and Jackendoff, have marshaled syntactic and semantic evidence against this claim; I shall return to this evidence in later chapters.

The same objection can be raised against Madkour (1934: 170), who dismisses Avicenna's distinction between negative and indefinite judgments (later adopted by Kant and others; cf. §1.2 and §2.4) on the grounds that (36a, b) express identical thoughts,

(36) a. The tree is not dry.
b. The tree is not-dry (non-dry).

both representable (in the manner of the Stoics, Abelard, and Frege) by means of an external operator, as *Not: the tree is dry* or *It-is-not-the-case-that the tree is dry.* Madkour argues that the indefinite judgment (*A is not-B*) is a linguistic artifact; while this may be partly true, the same point could be (and has been) made against the external propositional operator itself (see chapters 6 and 7).[39] That Frege's and Madkour's arguments appear as convincing as they do is largely a property of the architects' skillful example construction: the oppositions *not celebrated / uncelebrated, not dry / not-dry,* and (to take another instance from Madkour) *not finite / infinite* all involve privative or term negations which form immediate rather than mediate contraries: there is no unexcluded middle here. Even if this finesse is granted, of course, we may choose to side with Aristotle in distinguishing immediate contraries from true contradictories on the grounds that a man who does not exist is neither celebrated nor uncelebrated,[40] justice is neither dry nor not-dry, and a unicorn is neither odd nor even.[41]

The subject-predicate-based term logic of Aristotle, with its two modes of internal negation, and the function-argument-based propositional logic of Frege, with its all-purpose external negation operator, may themselves represent contrary positions on a continuum rather than mutually exhaustive contradictories. In the late nineteenth century, Brentano and Marty developed a theory of propositions in which the CATEGORICAL judgment, conforming to the subject-predicate paradigm, is distinguished from the THETIC judgment, which is 'simply the recognition or rejection of material of a judgment' (Kuroda 1972). For the categorical judgment, two separate acts are required: a subject is recognized, and a predicate is affirmed or denied of that subject. The thetic judgment is simple, involving just one act, often expressed as an existential (*God exists; There is/are . . .*) or an impersonal (*It's snowing*); they may have subject-predicate surface form but are always essentially subjectless in logical form. Thus, negative categoricals will look Aristotelian and negative thetics Fregean. Kuroda (1972) argues that Japanese typically marks categorical judgments with *-wa* and thetic judgments with *-ga,* and Babby (1980) presents extensive evidence that the categorical-thetic split correlates with distinct patterns of case-marking in negative sentences in Russian. I return to Kuroda's analysis in §7.3.

Deviations aside, it is indisputable that the Fregean model has carried the day. The syntax of negation in the first-order predicate calculus is simply **~p**, where **p** is any proposition. The semantics is equally straightforward, at least if presuppositional phenomena are ignored (cf. chapter 2): **~p** is true if and only if **p** is false.[42] Yet the passionate defense of subject-predicate logic by Sommers and Englebretsen, as well as the evenhanded interpretation of traditional and modern logic offered by Strawson (1952) and Geach (1970, 1972), suggest that Lukasiewicz's requiem for Aristotelian term logic and for the contrary or term negation formulable within it may be somewhat premature. Indeed, within the framework of multivalued logic (in the very footsteps of Lukasiewicz), McCall (1967a) and Rescher (1969) have worked out treatments of contrariety which draw on the joint resources of term-based and proposition-based logic; cf. also the related system of Von Wright (1959), summarized in §2.4 below.

For my purposes, the crucial factor differentiating the Aristotelian logic of terms from the Stoic (and modern) logic of propositions is that 'although both logics include and given formal recognition to the relation of contradiction, only the former, and not the latter, takes account of the relation of contrariety' (McCall 1967a: 121). McCall attempts to fill this void by formalizing a non-truth-functional contrariety operator ***R*** on propositions such that the falsity of ***Rp*** follows from the truth of ***p***, but neither the truth nor falsity of ***Rp*** follows from the falsity of ***p***.[43] The fact that contrary negation is stronger than contradictory negation (in that the former unilaterally entails the latter) can be expressed simply by the axiom ***CRpNp***, where ***N*** is the (Polish-notated) ordinary (contradictory) propositional negation.

But Geach finds such a position untenable: 'Contradictory negation may be thought of as operating upon entire propositions or as operating upon predicables; contrariety can be treated as an operator only upon predicables, not upon entire propositions' (Geach 1972:73). In the case of propositions containing quantified expressions, contrariety cannot be regarded as a function: 'We may speak of the contradictory of a proposition, since no proposition has two (non-equivalent) contradictories, but a proposition may well have more than one contrary in the square-of-opposition sense of the word' (Geach 1972:71–72).

Thus, Geach bids us consider a doubly quantified example like (37) and the two candidates for its contrary opposite, (37′a, b):

(37) Every cat detests every dog.

(37′) a. No cat detests every dog.
b. There is no dog every cat detests.

The two propositions (37′a, b) are each logically inconsistent with (37) and hence represent joint contraries of it, but they are not logically equivalent to each other. In the language I employed earlier in this section, Geach's argument supports the position that some propositions have no immediate contraries, and hence a contrariety operator on propositions would at best yield a rather than the (unique) contrary of the original proposition.[44]

Geach's verdict on propositional contrariety is (independently) upheld by one of today's staunchest advocates of term logic, George Englebretsen, who charges McCall with failure to recognize that the strong contrariety operator *R* is in reality an operator on predicates, not on sentences or propositions:

> The contrariety between two propositions derives from the contrariety (incompatibility) between their predicates. Two propositions are contraries if and only if they are exactly alike except that their predicates are contraries. . . . A logic of contrariety must be a logic of analyzed propositions, i.e. a term logic.
>
> (Englebretsen 1974:614)

But Geach's objection against McCall—that a propositional contrariety operator comes a cropper when it is applied to quantified expressions—can be turned even more tellingly on Englebretsen, whose definition seems to predict incorrectly that *Something is red* and *Something is not-red* are a pair of contrary statements (rather than subcontraries, both possibly true but not possibly false at the same time), and to incorrectly exclude the classic pair *Everything is red* and *Nothing is red.*

In any case, as Geach and Englebretsen argue, contrary negation is certainly more at home within term logic. Indeed, as I shall argue in chapter 7, even contradictory negation—as it surfaces in natural language—may demand a term-logic-based model. While much has undeniably been gained in clarity, explicitness, and logical power in the years since Frege inaugurated twentieth-century mathematical logic with his rejection of the subject-predicate distinction and his homogenization of all negative signs into a single propositional connective, much may also have been lost—or at least buried. This emerges clearly when I examine (as I shall in chapter 2) the price Russell must pay to acknowledge the apparent ambiguity of singular negative statements (*The king of France is not bald*) while attempting to keep the twin Aristotelian ghosts of predicate denial and term negation safely interred (cf. also Henry 1972; Sommers 1970, 1982; Englebretsen 1976, 1981a, 1981b).

Nor have modern philosophers and linguists been notably successful in laying a more ancient ghost, that of the (putative) inferiority of the negative statement vis-à-vis its affirmative counterpart. It is to this dispute that I shall now turn.

1.2 The Paradox of Negative Judgment: Negation and Its Discontents

You've got to
  Accentuate the positive,
  Eliminate the negative
  Latch on to the affirmative,
  Don't mess with Mr. In-Between. (Arlen and Mercer 1944)

Negation may or may not be the most basic, most debated, or most ancient of the logical connectives, but it is without doubt the most maligned. Logicians from Parmenides to Russell have sought to banish logical negation entirely, to explain it away, or to relegate it to a secondary and inferior status. In the battle between the assailers and defenders of negative facts, judgments, and statements, several separate but overlapping fronts can be descried. Consider the following theses on the purported asymmetry between affirmation and negation:

(38) a. Affirmation is logically prior, negation secondary.
  b. Affirmation is ontologically prior, negation secondary.

c. Affirmation is epistemologically prior, negation secondary.
d. Affirmation is psychologically prior, negation secondary
e. Affirmation is basic and simplex, negation complex.
f. Affirmation is essential, negation eliminable.
g. Affirmation is objective, negation subjective.
h. The affirmative sentence describes a fact about the world, the negative sentence a fact about the affirmative.
i. In terms of information, the affirmative sentence is worth more, the negative worth less (if not worthless).

Moderate asymmetricalists seek to support various subsets of these positions, while hard-line asymmetricalists (Parmenides, Bergson, Givón) would argue for all of them.

Before, during, and after evaluating the eight theses in (38), I must address a bevy of related questions:

(39) What, if anything, is a negative judgment?
In what sense, if any, does negation presuppose affirmation?
When, if ever, is negation equivalent to falsity?
Is there a speech act of negation?
Can negation be reduced to speaker denial or rejection?

Again, as with the theses of (38), agreement on the answers to these questions is rare.

Included in the asymmetricalist faction as leaders, proselytizers, acolytes, and camp followers are Parmenides, Plato, Aristotle (sometimes), Saint Thomas Aquinas, Kant, Goethe, Hegel, Bergson, the neo-Hegelian Idealists, Russell (usually), Strawson, Tesnière, Givón, and a variety of psycholinguists; the symmetricalist camp claims Aristotle (sometimes), Frege, Royce, Russell (occasionally), Wittgenstein (perhaps), Ayer, and Geach. It will be observed that the two pivotal figures in the history of negation, Aristotle and Russell, stand astride the field with one foot in each camp (the latter shifting easily during his career from a symmetricalist dove to an asymmetricalist hawk, and perhaps partway back again).

I shall attempt here to report from the various battle fronts, describing the aims and tactics of the warring sides, counting the casualties, and—in later chapters—proposing a strategy for a face-saving, negotiated settlement of the conflict. I shall begin once again with Aristotle.

It is traditional for parties to a logical conflict to cite Aristotelian precedent, and in this case each side may aptly claim justification. The Stagirite is infuriatingly Delphic on the question: 'The first class of propositions is the simple affirmation, the next, the simple denial' (*De Int.* 17a8). But first, in what way? Logically? Epistemologically? The first to be mentioned here?

In his commentary on this passage, Ackrill (1963: 127) suggests that affirmation may be first because negation (i.e., predicate denial) is realized through the addition of a negative marker: 'the negative presupposes, in that it involves adding something to, the affirmation'. But this is just a guess, and there have been others. Aristotle himself implies elsewhere that he is thinking more of epistemological than logical or ontological priority for affirmation over negation:

> The affirmative proposition is prior to and better known than the negative (since affirmation explains denial just as being is prior to not-being). (*Post. An.* 86b33–36)

> We say that he who knows that the thing is something has understanding to a higher degree than he who knows that it is not something. (*Metaph.* 996b14–16)

Perhaps, then, it is not that what is denied must first have been asserted, or that positive facts are more real or basic than negative ones, but simply that knowledge of a positive fact counts for more than knowledge of its negative counterpart.

The schism between the symmetricalists and asymmetricalists was quick to develop: the early commentator Alexander held affirmation to be crucially prior to negation, since denying 'lifts or destroys affirming', while the neo-Platonist (and proto-Fregean) Porphyry argued that since affirmation and denial are equally propositions with respect to truth and falsity, there is no significant asymmetry between them (Bosley 1975: 7).

The asymmetricalists found an early champion in Saint Thomas, who distinguished the linguistic, psychological, and ontological grounds of affirmative priority in his commentary on *De Interpretatione:*

> The affirmative enunciation is prior to the negative for three reasons. . . . With respect to vocal sound, affirmative enunciation is prior to negative because it is simpler, for the negative enunciation adds a negative particle to the affirmative. With respect to thought, the affirmative enunciation, which signifies composition by the intellect, is prior to the negative, which signifies division. . . . With respect to the thing, the affirmative enunciation, which signifies to be, is prior to the negative, which signifies not to be, as the having of something is naturally prior to the privation of it. (Saint Thomas Aquinas, book 1, lesson 13, in Oesterle 1962: 64)

Aquinas thus endorses theses (38a, b, d, e), while Aristotle himself directly supports only (38c) and perhaps (38i). Aristotle, however, is clearly an asymmetricalist with respect to negative terms; the in(de)finite nouns and verbs *not-man, not-white, not recovers,* and so forth, are explicitly set off as suspect, sharply distinguished from honest-to-goodness nouns and verbs proper (*De Int.* 16a30, 16b14, 19b8–11). Predicate denial, however, is less clearly tainted or, in modern jargon, marked with respect to affirmation.

It may be significant that Aristotle points out the one-to-one correspondence between affirmative and negative propositions: 'Every affirmation has a corresponding negation' (*Prior An.* 51b35). This observation, which has often been echoed:

> Everything which can be affirmed can also be denied.
> (Bosanquet 1888:294)
>
> For every thought there is a contradictory thought.
> (Frege 1919:131)
>
> There is one negative sentence corresponding to every positive sentence and vice versa. (Kraak 1966:89)

seems to align the Stagirite with the symmetricalists. The asymmetricalists would no doubt counter that Aristotle didn't put it the other way around, as 'Every negation (or denial) has a corresponding affirmation'.

It should be noted that this principle of one-to-one correspondence can only apply to propositions, not (at least literally) to sentences, pace Kraak. It has been pointed out (by, inter alia, Zimmer [1964:87], Kissin [1969:77], and R. Lakoff [1969:144]) that performative utterances cannot be directly negated, as seen in (40).

(40) I (*do not) now turn to the examination of . . .
I (*do not) hereby inform you that . . .

Similarly, the quasi-performative of *I guess it'll rain* allows a parenthetical sense or use which is lost under negation for those speakers who can't neg-raise over *guess* (Horn 1978b; Horn and Bayer 1984; cf. §5.3 below). The wide-scope quantifiers and adverbs of (40′a) also exclude any syntactically transparent contradictory negation, as of course do the classic instances in (40′b) of POSITIVE (or AFFIRMATIVE) POLARITY ITEMS, expressions which cannot normally appear in the direct scope of negation:

(40′) a. Many years ago I lived there.
Even an idiot can solve that problem. (cf. §2.5 below on *even* and negation)
{Some / Several} of my friends showed up.
b. I would(*n't) rather be in Montpelier.
She's (*not) {pretty tall / far taller than he is}.

On the other side of the coin are NEGATIVE POLARITY ITEMS, expressions which can only appear felicitously within the scope of negation (or a semantically related operator):

(41) He {isn't / *is} eating any meat tonight.
I {can't / *can} ever seem to make any progress.
She {hasn't / *has} been to Casablanca yet.

(See Baker 1970; Fauconnier 1975a, 1975b; Horn 1978a: §2; Ladusaw 1979; Linebarger 1981, 1987 for more on positive and negative polarity.)

Thus, some affirmative sentences have no directly corresponding negation, while some negative sentences lack any affirmative counterpart. Still other sentences—*He's hardly a linguist; Few students came to the party*—can be taken as either matchless affirmatives or matchless negatives, depending on the criteria for analysis (cf. Jespersen 1917; Klima 1964; Jackendoff 1969).

The Paradox of Negative Judgment

If an affirmative statement or judgment (*The cat is on the mat*) is about (corresponds to, refers to) some positive fact in the world, what on earth can a negative statement or judgment be about? To what negative fact, if any, does it correspond or refer? Are there negative facts in the first place? (Or in the second place?) These questions lead us directly to the PARADOX OF NEGATIVE JUDGMENT:

> A negative judgment declares what it is not, and how can this express it as it is? (Joseph 1916: 171)

> The paradox consists in this—that in negation the work of positive knowledge appears to be performed by ignorance. (Bosanquet 1888: 277)

> [The negative judgment] cannot be true if there is nothing in reality corresponding to its Not, and if there is something corresponding to its Not the judgment cannot be negative. (Raju 1941: 585)

> We cannot specify the fact of non-existence except in terms of the thing itself which does not exist. Thus the existing thing seems to be involved in this fact, as a constituent of it. Thus the thing has to exist as a condition that it should not exist. It seems that non-existence is a logical impossibility, which is surely absurd. (Toms 1972: 7)

This paradox clearly recalls its ontological analogue, expounded by Parmenides and formulated succinctly by Apostel (1972a:211): How can there exist something which does not exist?

Attempted resolutions of this paradox have proceeded by consigning negative facts either to oblivion (§1.2.1) or to a logical and/or ontological ghetto (§1.2.2). Negative judgments or statements in turn have been taken to signify indirectly, misleadingly, subjectively, or not at all. More recently, the asymmetricalist view has been formulated within the theory of markedness, and psycholinguistic evidence has been marshaled to buttress the earlier philosophical arguments for the priority of affirmation over negation (see chapter 3). In the remainder of the present section, §1.2, I shall take the Paradox of Negative Judgment as a device for framing my view of the battle of the symmetricalists and asymmetricalists; in doing so, I shall be more concerned with following the changing picture of negation that emerges than in resolving the paradox itself.

1.2.1 Eliminating the Negative

The world can be described without the use of the word 'not'.
(Russell 1948:520)

Negation as exclusion and difference

The first (Western) attempt to explain negation away is due to Parmenides and to the Megarians who sought to defend and extend his account: a negative sentence cannot refer to how things are, and so is necessarily indeterminate (cf. Wheeler 1983:289). There are no negative states of affairs or properties; *not-red* is no more a property than *not-Odysseus* is an individual. (It will be recalled that even for Aristotle indefinite, i.e., negative, names are not properly names as such.) Since the False is that which is not true, falsehood does not exist either. Armed with these doctrines, it is little surprise that Parmenides sought to banish all negative thought.

A different means to the same end—eliminating the negative—is offered by another of our earlier acquaintances, the Stranger in Plato's *Sophist,* who assures us that the Parmenidean swamp of nonexistence can be skirted if we take negation to represent not nonbeing, not oppositeness (contrariety), not contradiction, but simply OTHERNESS or DIFFERENCE.

The problem with Plato's ploy is that it seems to be either inadequate or circular. If difference is a positive entity, it is incapable of accommodating all the uses to which logical and linguistic negation is put. If it is intrinsically negative, we have not eliminated negation but merely relabeled it.[45]

In particular, if in stating:

(42) My hat is not red.

I am simply observing that my hat is (of some color) other than or different from red, I must be saying either (1) that all of its properties are nonidentical to redness (a hat can be wide-brimmed, which is different from being red, and still be red), or (2) that it is of some color incompatible with red, such as green or blue (a hat may be crimson or scarlet, which are colors nonidentical to red, but my hat's being crimson or scarlet does not make (42) true). Yet incompatibility is even more clearly than otherness or difference a negative relation.

Thus, for Plato's Stranger (as for Russell's dissimilaritarian below), the putative negative fact that A is not B is unpacked into the corresponding positive fact that A is other than, different from, or dissimilar to B. But 'unfortunately', as Toms (1972: 8) observes, 'for such a theory to be genuinely a theory of negation, the fact taking the place of the negative fact has still to exclude from existence the opposite positive fact'. And if there is no such positive fact, we have been redeposited into the nightmare realm of Parmenides—and dropped off within a deeper circle of that inferno: 'Instead of shaking off the paradox of non-existence, the standard theories of negation [i.e., those just cited and their successors, to be explored below] have the effect of showing that the paradox extends to the whole field of negation' (Toms 1972: 8). Briefly, the "otherness" criterion, in any of its various guises, either does or does not provide a relation which satisfies the Law of Contradiction (see §1.1.1); in the former case, it is circular, in the latter, inadequate.

The problems encountered by otherness and incompatibility analyses of negation (that is, problems with taking such analyses to be reductive or eliminative) have long been acknowledged (cf. Bradley 1883), especially by those aware of the difficulty in determining what counts as a negative judgment or proposition (see §1.1.4 above). Yet, as untenable analyses go, the Otherness thesis has proved surprisingly resilient, particularly in the form in which we shall encounter it in §1.2.2: some negative judgments (SIGNIFICANT negations, with a 'positive ground') reduce to assertions of otherness; the others (INSIGNIFICANT negations) aren't worth worrying about.

Negative facts

The battle over the nature and very existence of so-called negative facts has also raged on, fiercely but inconclusively. Demos's argument (1917: 189) against the existence of the negative fact is perhaps the most ingenious. Intoning the ritual exorcism—'Strictly negative facts are nowhere to be met with in experience; . . . any knowledge of a negative nature seems to be derived from perception of a positive kind'—he supports his contra position by appealing to the judgment of 'intelligent acquaintances' of his, who acknowledged 'that they had never encountered a negative fact'. While we may quibble with Demos's experimental design (especially since his write-up comes immediately after a well-taken warning about negative wolves in positive clothing), it is in fact hard to imagine any compelling empirical evidence for or against the existence of negative facts.

Russell apparently gave up his early (and well-founded) skepticism about the eliminability of negation to become an enthusiastic warrior for the cause. At the time of his *Philosophy of Logical Atomism* (1918:211), he not only valiantly overcame his self-acknowledged 'repugnance to negative facts', but was even willing to fight in their defense. He reports having earlier triggered 'a near-riot' at Harvard by arguing that there are indeed negative facts and finds himself 'still inclined to think that there are'. But he takes Demos's (1917) diatribe seriously enough to direct a thoughtful counterattack against it (Russell 1918:211–14):

> If I say 'There is not a hippopotamus in the room', it is quite clear there is some way of interpreting that statement according to which there is a corresponding fact, and that fact cannot be merely that every part of this room is filled up with something that is not a hippopotamus. . . . It is simpler to take negative facts as facts . . . otherwise you will find it difficult to say what it is that corresponds to a proposition. (Russell 1918:213–14)

And since there is in any case no formal test for establishing whether a given proposition is negative or positive (a point Russell was later to ignore), the whole exercise of negative bashing is doomed to futility (p. 215).

Thirty years later, Russell appears willing to go to any lengths and adopt any means to eliminate negation. He begins Platonically by identifying negation with difference:

> When . . . I say 'This is not blue', I may be interpreted as meaning 'This is a color differing from blue', where 'differing' is the positive relation [*sic*] that might be called 'dissimilarity'. . . . When I say truly 'This is not blue' there is, on the subjective side, consideration of 'This is blue', followed by rejection, while on the objective side there is some color differing from blue.[46]
> (Russell 1948:122)

But even if we ignore the Aristotelian objection that a sufficient condition for the truth of a negative like *This is not blue* would be an instance on which the referent of the subject phrase (the *this*) is not the sort of thing that can be blue—or any other color, either (e.g., justice or the square root of 2)—Russell's evidence for the positive nature of the 'dissimilarity' relation is not compelling. *This is red* can serve as a ground or basis for the assertion of *This is not blue* only by virtue of the mutual incompatibility of the properties expressed by *red* and *blue*, the fact that for all *x*, if *x* is red, *x* is not blue.

Russell next adopts the very tactic from Demos (1917) he had earlier (1918:221) rejected: negation can be eliminated through the notion of a 'true disbelief'. On this account, *This is not blue* expresses, by definition, a disbelief in what is expressed by the words *This is blue* (1948:124–26). But the key assertion that disbelief is 'a state just as positive as belief' is entirely unsupported, as is Russell's apparent leap to a subjective theory of negation. Affirmative and negative propositions can both be true even if there is nobody alive to believe or disbelieve them.

Russell's observation that the world can be described without the word *not* is of course trivially correct; we can, after all, use French or Kikuyu to describe it, or for that matter a periphrastic variety of English. Whether it can be described without negation is less clear, and the question is not resolved by a Russellian appeal (1948:520ff.) to an unspecified theory of 'impulses' and their inhibition, or by the limitation of the domain of inquiry to reports of perception (cf. also Russell 1940).

The repugnance to negative facts acknowledged by Russell, and his (temporary) adoption of the Platonic negation-as-positive-difference line, both have their echoes today. One may be found in the attempt by Katz (1964; 1972:157–71) to define negation in terms of an ANTONYMY primitive, derived from the earlier Aristotelian relation of contrariety. Patton (1968:230) points out that 'De Morganian theories of negation' like Katz's 'may be seen as stemming from a refusal to acknowledge the semantic primitiveness of negation'. While we may follow Katz in accepting (43b) as the antonymically motivated equivalent of (43a),

(43) a. My cook is not a woman.
b. My cook is a man.

what, Patton asks, might the positive counterpart of (43′) be?

(43′) The dog is not beside the man.

(This is not to suggest that there is any semantic equivalence between (43a) and (43b), but the problems encountered with such pairs are magnified when we look at terms that do not belong to a set of two.)

For Apostel (1972a:209), 'there are no negative facts; no non-existent things'. We can translate apparent references to nonexistents into positive statements about absences. Thus *There is no noise* is 'not a negative observation; it is the positive observation of silence', just as *There is no money in my wallet* offers positive information about the state of my wallet. But this translation procedure would seem inadequate, circular, or just plain silly when extended to other examples, for example, *There is no spaghetti left* (= the positive observation of spaghettilessness?), *There are no unicorns, There is no even prime >2,* and so forth. The negation in Apostel's example *The dress I see is not black* can be eliminated without undue difficulty, but what of the one in *I didn't eat an apple?* And even if we could successfully explain away negative facts, negations in nondeclarative moods would seem to resist elimination more fiercely. Even God (for whom there can—ex hypothesi—be no negative facts) might have found it difficult to avoid recourse to negation in issuing certain directives: *Thou shalt not kill, Of this tree do not eat.*

Gale (1976) provides a careful and evenhanded survey of the prospects of each side in the campaign for and against negative facts and events; he concludes that the former, but not the latter, must be admitted. But there may be certain advantages to allowing even negative events, ontologically unwelcome guests though they may be, into linguistic semantics if not logical representation. As Stockwell, Schachter, and Partee (1973 [SSP]: 250–51) note, generalizing from an earlier observation by G. Lakoff (1965: appendix F), 'There are certain cases where the negation of an event may, loosely speaking, itself be an event, e.g., *not paying taxes, not getting up early, not going to church* . . . (semantically, the "event" seems to be the breaking of a habitual or expected pattern of activity)'. Such negative "events"—note the scare quotes—may be modified by reason, frequency, and location adverbials, although they do not take instrumentals (SSP 251). Thus, we have

(44) a. I don't beat my wife because I love her. (G. Lakoff 1965: F-6-3)
b. He often hasn't paid taxes. (= SSP, chapter 5, (49b))
c. I don't get up early at home. (= SSP, chapter 5, (53a))
d. I don't cut my salami with a hacksaw (≠ It's with a hacksaw that I don't cut my salami.)

If such apparent exceptions to the ban on negative events cannot be explained, they must at least be explained away.

Interestingly, the distribution of adverbials also provides an argument against the existence of negative events. As Thomason and Stalnaker (1973: 218–19) observe, there are **IV**-phrase (**VP**) adverbs like *slowly* which can modify events or actions but not states, as seen in (45a, b):

(45) a. John slowly walks. (John walks slowly.)
b. *John is slowly tall.
c. *John slowly does not walk.
d. John does not walk slowly. [OK, but not as negation of (45a)]

As indicated by the ungrammaticality (or semantic deviance) of (45c), the negative counterpart of an event is evidently a negative state; what blocks (45c) on Thomason and Stalnaker's account is the principle that 'the function denoted by *slowly* will be defined only for events and actions', and *not-walking*, unlike *walking*, does not constitute an event. (Only predicate adverbs are restricted in this way; by Thomason and Stalnaker's criteria, the adverbials in (44a–c) are sentential, while the instrumental in (44d), which does indeed pattern like *slowly*, is a predicate modifier on events or actions.)

An analogous argument was extant in generative semantics circles in the late 1960s: adverbs like *until Sunday* and *for a week* show up in negative contexts and as modifiers of states and durative processes, but not with simple "punctative" events:

(46) John { *got here
didn't get here
is here
stayed
didn't stay } {until Sunday / for a week}.

If negation turns an action into a state or process, then the "negative polarity" *until* and *for* phrases are reduced to special cases of a more general phenomenon, namely, durative (interval-associated) adverbials.

The question of whether there are negative events cannot be answered directly, by invoking the evidence of natural language, especially in the absence of a consensus as to what counts as an event. Thomason and Stalnaker's verdict that negative predicates can only denote states is less convincing when we consider sequences like the following:

(47) <u>What happened next</u> was that the consulate {held up / denied / didn't give us} our visa (<u>for six months</u>).

In this token, the distributional evidence (cf. the underlined diagnostics) and the intuitive semantics converge to suggest that a negative predicate can be functionally equivalent to a (morphological) positive in denoting a simple event.

Whatever the ultimate verdict on the existence of negative events, it would appear that negative facts we shall always have with us. The ontologist's pen, like the poet's (cf. *Midsummer Night's Dream* 5.1.7), must 'give to airy nothing / a local habitation and a name'.

Eliminating the negative: Other tactics

If Plato's ploy—negation as positive difference—is not adequate to the task of eliminating all instances of superficial negation, what other devices can be tried? We have already come upon one such device in Hobbes's attempt to incorporate all contradictory negatives into the predicate, aptly criticized by Mill (see §1.1.4) and Bradley (1883). This approach is also dismissed by Joseph (1916: 172), who notes that negation (*A is not B*) cannot be eliminated by an incorporation into the predicate (*A is not-B*) unless *not-B* is positive, and this can occur only if all other alternatives can be ruled out. *Fido is not male* can be analyzed as *Fido is not-male,* that is, *Fido is female,* but *Fido is not a dog* cannot be reduced to a positive in the same fashion.[47] (Patton's critique of Katz's reductionist program, summarized in §1.1.4 above, hinges on a similar point.)

Even Demos, scourge of negative facts, points out that negation cannot simply be absorbed into the predicate, by 'defining a proposition like "X is not white" to be really "X is not-white"', since the negative element bears 'almost as often' [!] on the grammatical subject as on the grammatical predicate. An atheist can assert *God will not provide* on the grounds that there is no God. (This point does not originate with Demos, of course, but traces back to Aristotle.) Furthermore, the standard logical analysis of relations like *is to the right of* or *precedes* offers no grammatical predicate (no term which combines with a subject to form a sentence) for the negation in (48) to negate (Demos 1917: 190).

(48) X is not to the right of Y.
X does not precede Y.

While Demos objects to this particular collapse of contradictory and contrary negation, he ends up urging a different one, as we shall see below. (For another critique of the move from *A is not B* to *A is not-B,* see Wood 1933: 418–19.)

A more frequently encountered method for eradicating negation is that of identifying it with, and "reducing" it to, falsity. The scare quotes here are motivated by the question of whether such a "reduction," if it could be accomplished, would really accomplish anything. But there are in any case strong grounds for rejecting the proposed identification in the first place, without even considering its role within a reductionist program.

That negation and falsity might be identified, and eventually confused, with each other should not be surprising. Aristotle discusses 'being in the sense of true and non-being in the sense of false' (*Met.* 1027b18), and he seems to explicitly link the negated copula with falsity (as the affirmative copula is linked with truth): '"To be" and "is" mean that something is true, and "not to be" that it is not true but false. . . . For example, in "Socrates is musical", the "is" means that it is true that Socrates is musical, and in "Socrates is not-white", that this is true; but in "the diagonal is not commensurate with the side" the "is not" means that it is false that the diagonal is commensurate with the side' (*Met.* 1017a31ff.).

In Aristotle's simple correspondence theory of truth, framed within a two-valued logic, truth and falsity are interrelated as the two terms of a contradictory opposition. But contradictory negation does not reduce to falsity, since negation (i.e., predicate denial) and falsity are about different things and operate on different levels: 'A falsity is a statement of that which is that it is not, or of that which is not that it is; and a truth is a statement of that which is that it is, or of that which is not that it is not' (*Met.* 1011b25–27; cf. *De Int.* 18b2–4). Thus Aristotle resists Wood's attempt (1933:422) to recruit him to his own negation-as-falsity camp.

As we have seen, Leibniz sought to eliminate apparent propositional negation, 'by predicating falsehood of a term, rather than negating truth of it' (Castañeda 1976:484): 'If B is a proposition, not-B is the same as that B is false' (Leibniz 1966:58). But contradictory predicate denial (*A non est B*) and contrary privation (*A est non B*) are not directly affected by this identification.

The equation of negation—often specifically "logical" negation—and falsity is a frequent maneuver among the Idealists of the late nineteenth and early twentieth century:

> To say 'A is not B' is merely the same as to deny that 'A is B', or to assert that 'A is B' is false. (Bradley 1883:118)
>
> 'A is not B' means 'it is false, it must not be believed that A is B'. . . . Immediately and directly, the negation is a judgment concerning a positive judgment that has been essayed or passed. (Sigwart 1895:122)
>
> a is not b = that a is b is false. (Baldwin 1928:147)
>
> The pure negative judgment 'A is not B' is equivalent in every case to 'it is false that A is B' . . . 'Snow is not black' is a shorthand statement for 'snow is black is an erroneous judgment'. (Wood 1933:421)

As emerges clearly from the Sigwart and Wood citations, the identification of contradictory negation with falsity often goes hand in hand with a view of negation as a second-order comment on a first-order affirmation, and/or as a more subjective act than simple affirmation. I shall examine these tenets more closely in the following section, but it is relevant to cite once more the protean Russell, for whom every negation is a shorthand for some assertion of falsity, but at least some affirmatives are second-order as well: 'When you say "This is not cheese" you mean "the statement 'this is cheese' is false". . . . Just as the statement "Yes, this is cheese" really means "the statement 'this is cheese' is true"' (Russell 1940:74). 'It is unnecessary to have the two words "false" and "not", for, if *p* is a proposition, "*p* is false" and "not-*p*" are strictly synonymous' (Russell 1940:81).

Ayer, whom we shall later encounter in the cap of the symmetricalists, offers a variant of this position: *false* is logically superfluous, but so is *true*, both reducible to 'signs of affirmation or denial' (1936:17). Thus we have falsity-as-negation (and truth-as-affirmation), rather than vice versa.

Within the modern logical (and linguistic) tradition, the temptation to identify negation and falsity stems directly from the Fregean line that all negation is propositional and reducible to a suitably placed *it is not true that* . . . (cf. Prior 1967, Seuren 1969:159). In multivalued logics, there is one form of negation (internal, strong, choice) which does not display the logic of contradictory opposition, being governed by LC but not LEM (see chapter 2). Within such approaches, at least some negations cannot be reduced to assertions of falsity. Similarly, there may be illocutionary distinctions between the negation of a proposition and the statement that that proposition is false, as in Heinemann's differentiation (1944:143) of **not-p** ('$\bar{\mathbf{p}}$ is valid') from '**p** is not valid'. But even within classical two-valued logic itself, there are sufficient grounds for rejecting the identification of negation and falsity. Philosophers as diverse as Frege (1919), Austin (1950), Quine (1951), and Geach ([1972] 1980) have observed that the identification of *not* and *false* results from a confusion of language and metalanguage.

Here is Austin's symmetricalist manifesto, addressed to the view (represented by Ayer) that 'is true' and 'is false' are logically superfluous:

> An important point about this view is that it confuses falsity with negation: for according to it, it is the same thing to say 'He is not at home' as to say 'It is false that he is at home'. . . . Too many philosophers maintain, when anxious to explain away negation, that a negation is just a second order affirmation (to the effect that a certain first order affirmation is false), yet, when anxious to explain away falsity, maintain that to assert that a statement is false is just to assert its negation (contradictory). . . . Affirmation and negation are exactly on a level, in this sense, that no language can exist which does not contain conventions for both and that both refer to the world equally directly, not to statements about the world. (Austin [1950] 1970:128–29)

Quine (1951:27–28) is also at pains to distinguish the predicates 'is false' and 'is true', which are used to speak about statements, from the connective '$\sim$', which is used to make statements. Example (49a) is a statement about the statement *Jones is ill,* while (49b), read 'Jones is not ill', is a statement about Jones:

(49) a. 'Jones is ill' is false.[48]
b. ~Jones is ill.

Quine lays at the door of Russell and Whitehead the mistaken identification of '~' with falsehood (as well as the parallel misidentification of '⊃' with implication), but the underlying mistake both antedates and survives the *Principia*—as does its correction. The Stoics were careful to make the same distinction as Quine, that 'between the negation of a proposition and a (metalinguistic) statement that the proposition is false'; these two operations played different roles in the Stoics' account of syllogistic reasoning (Mates 1953:64–65).

In the same vein as Austin and Quine, Geach ([1972] 1980:76) inveighs against the 'widespread mistake' of assuming that 'the negation of a statement is a statement that that statement is false, and thus is a statement about the original statement and logically secondary to it'. The error of this approach emerges clearly when we look at nondeclaratives: ' "Do not open the door!" is a command on the same level as "Open the door!" and does not mean (say) "Let the statement that you open the door be false!" '

We have seen negation survive enough attempts at liquidation—negation as positive difference, negation as incompatibility, negation as dissimilarity, negation as true disbelief, negation as the affirmation of a negative predicate, negation as falsity—to qualify as the Rasputin of the propositional calculus. Perhaps the last word on negative elimination should be left to the multifaceted Russell.

In "The Metaphysician's Nightmare" (Russell 1954), Russell's old friend Andrei Bumblowski, ex-professor of philosophy at a nameless and now-defunct Central European university, relates a nightmare of Hell, in which he is invited to an audience with Satan himself, whom he recognizes at once from Goethe's description as 'der Geist der stets verneint', the Spirit of Perpetual Negation.[49] After an unfortunate slip on Russell's Paradox ('Henceforth I will not use the word "not" ', Professor Bumblowski declares, for which he is mercilessly derided by the Devil's metaphysicians), the Professor denounces Satan—and the negation he sponsors—as merely 'a bad linguistic habit', and vows to abjure both forever.[50] For 'This egg is not fresh', he would henceforth assert 'Chemical changes have occurred in this egg since it was laid'; for the tabooed 'I cannot find that book', he would substitute 'The books I have found are other than that book'; for 'Thou shalt not kill', 'Thou shalt cherish life'. This does the trick, and the 'murky air of Hell' dissolves into—or, rather, out of—nothingness, and Satan vanishes. The moral, according to Professor Bumblowski-Russell: 'Avoid the word "not" and His empire is at an end'. (Or, an idle *not* makes work for the Devil's hands.)

1.2.2 Accentuating the Positive

One single positive weighs more,
You know, than negatives a score.
(Matthew Prior, epistle to the Fleetwood Shepherd, 1689)

If negations cannot be legitimately evicted from the logical conveyance, perhaps they might still be restricted to the back of the bus, as second-order statements and second-class sentences. Advocates of the ontological priority of affirmation over negation include Plato: 'About each form there is much that it is, but an infinite amount that it is not' (*Sophist,* 256E). Or, in the words of an influential asymmetricalist of the Idealist school, 'Only a finite number of predicates can be affirmed of every subject, while an incalculable number can be denied' (Sigwart 1895: 119).

Since nonexistents cannot be individuated, every positive fact or event 'seems to carry on its back an infinite or indefinite number of negative fleas' (Gale 1976: 2)—we know (or can know) whether we are dealing with seven or seventy-seven bomb detonations or forest fires, but how do we begin to count nondetonations and non–forest fires?

As we have seen, Aristotle took the priority of affirmation over negation to be largely a matter of epistemology (hinging on such asymmetries as those just sketched), rather of logic or ontology. On the Aristotelian view, 'negation is not the rejection of a previous affirmation, negation is the rejection [and affirmation the acceptance] of a suggested connexion' (W. D. Ross 1923: 28; cf. *Met.* 1017a31–35). But commentators since the early medieval period have tended to endorse Avicenna's stronger position, that negation presupposes an affirmation against which it is directed and cannot be understood except through affirmation (Madkour 1934: 167).

The issue of a presuppositional asymmetry between negation and affirmation will be confronted below, but I shall turn first to another strain in the thought of early asymmetricalists, the view that the negative judgment or statement represents an unfortunately necessary way station on the long march from total ignorance to complete knowledge. In this perspective, negation is a useful tool for us humble and imperfect mortals, but would not be admitted by a perfect intelligence.

This idea is expressed first, and perhaps most eloquently, by Bacon, in the *Novum Organum* (book 2, aphorism 15):

> It is only for God (the bestower and creator of forms), and perhaps for angels and intelligences, at once to recognise forms affir-

> matively from the first glance of contemplation; Man, at least, is unable to do so, and is only allowed to proceed first by negatives, and then to conclude with affirmatives, after every species of exclusion. (Bacon [1620] 1853:474)

In the same vein, Kant observes that negative judgments 'are not held in very high esteem';

> They are regarded rather as the jealous enemies of our unceasing endeavour to extend our knowledge. . . . In respect to the content of our knowledge . . . , the task peculiar to negative judgments is that of rejecting error. (Kant [1787] 1964:574, A709/B737)

The Bacon-Kant view of negation as a necessary fiction, a tool for rejecting or warding off error which in an epistemically perfect state would simply wither away, resonates into our own century:

> [In negation] we take to task an interlocutor, real or possible, whom we find mistaken and whom we put on his guard. . . . [It] is of a pedagogical and social nature. It sets straight or rather warns, the person warned and set straight being possibly, by a kind of doubling, the very person that speaks.
> (Bergson 1911:289)

> In the complete grasp or experience of truth no negative judgment would remain. . . . The aim of negation is elimination. . . . In complete knowledge no trace of it survives. (Mabbott 1929:73)

> Negation is infected with error and ignorance. . . . 'This rose is not red' [necessarily involves] a fallible and partially ignorant mind erroneously attributing red to a rose of a definite hue other than red. . . . The occasion of negative judgment is one of intellectual frustration. . . . The negative judgment, thus pragmatically considered, is a memorandum of past failures to serve as a warning against similar attempts in the future. (Wood 1933:421)

> Negation is indispensable for a finite mind. (Heinemann 1944:152)

The pragmatic considerations leading Ledger Wood to accept negation as a temporary evil are also invoked by modern philosophers and linguists working within the tradition of the syntax/semantics/pragmatics trichotomy developed by Morris, Peirce, and Carnap, beginning with the man usually credited with introducing the term and field of pragmatics into the modern lexicon:

> The term 'not' is primarily of practical importance, since it allows reference to something other than what is specifically referred to

> without specifying what the other is. . . . The practical importance of the term is obvious, but it is not theoretically necessary, and certainly no existential "negative facts" need be invoked to correspond to it. (Morris 1938:28)

This view, similar to (one of) Russell's, has recently been echoed by Apostel: 'The universe can be given a complete but not pragmatically useful description without using negation' (Apostel 1972a:209).

For man, an imperfect creature mired in these nether swamps where pragmatics holds sway, positive facts may be accentuated and the affirmative (eventually) latched onto, but it is only in those higher realms where God and the angelic hosts work out Their pure syntax and semantics that negation can be fully eliminated.

Within this tradition, a negative sentence is not marked just with respect to the corresponding affirmative (see chapter 3), but in fact the negative morpheme represents a mark of Cain by which a sentence is irrevocably tainted. This 'taint of negation' (cf. Wood 1933:419) may result in the association of negation with ignorance and error in particular or with subjectivity in general.

The strongest statement of the thesis that negative judgments and sentences are inherently subjective comes from Bergson (1911:285ff.):

> While affirmation is a purely intellectual act, there enters into negation an element which is not intellectual. . . . Negation is only an attitude taken by the mind toward an eventual affirmation. (p. 287–88)

> From the point of formal logic, to affirm and to deny are indeed mutually two symmetric acts, of which the first establishes a relation of agreement and the second a relation of disagreement between a subject and an attribute. But how do we fail to see that the symmetry is altogether external and likeness superficial?[51] (p. 292)

For a 'passive intelligence', devoid of expectations, affirmative judgments would still be affirmed but negative ones would be impossible, since there can be no reason to deny what has not been affirmed. Negation inherently involves 'the disappointment of a real or possible expectation'. Once again, we have the view (later reiterated by Russell and others) that negative judgments are in fact second-order affirmations. *The table is black* is about the table, but *The table is not white* is about the (possible) judgment *The table is white:* '[It] implies that you might believe it white, that you did believe it such, or that I was going to believe it such. . . . An affirmative proposition expresses a judgment on an object; a negative proposition expresses a judgment on a judgment' (Bergson 1911:288).

Bergson's Subjectivity thesis is also endorsed by Joseph (1916:172), who argues that if no sentient being existed, *The wall is green* could be true, but not *The wall is not blue*, since the latter could be uttered only because someone might suppose or believe the wall to be blue. But this position, like the reductionist theses examined in section 1.2.1, is untenable. Bergson and Joseph would predict that all negative propositions—*2 + 2 ≠ 5, Water is not an element, The moon is not made of green cheese*—imply the existence of a judging intelligence, an implication which is counterintuitive in the extreme. Two and two would not equal five, nor would water be an element, even if nobody had ever been around to know it, and the timeless propositions expressing these "negative facts" should thus be true. As Matilal (1968:90) points out, the Subjectivist thesis reflects an equivocation between the statement as proposition or sentence type vs. the statement as utterance or sentence token. The critiques of negation-as-falsity and negation-as-second-order-affirmation by Austin, Quine, and Geach (cited in section 1.2.1) apply directly to Bergson's account; cf. Gale (1976:55–61) for additional commentary on the Subjectivity thesis.

The current incarnation of the Subjectivity thesis is in the view that negation is (always, necessarily) a speech act, reducible to speaker denial. The logical consequence of this position is of course that if there were no speakers (or only silent ones), there would be no negation (cf. §1.2.3 below). A related view is articulated by Apostel (1972a:277): 'Negation is a modality, in this sense that it expresses a propositional attitude of the subject [*sic*] towards the entity denied'.[52]

But it is hard to see how whatever insight is gained through this identification can compensate for the loss of the necessary distinctions blurred thereby. The true propositional attitudes—belief, knowledge, hope, fear, regret, desire, permission, obligation—crucially involve some creature whose psychology is characterized by the appropriate attitude or relation toward a given proposition. Speaker denial does constitute a propositional attitude in this sense, but predicate denial, propositional negation, and analogous operators with the semantics of contradictory opposition do not.[53]

Does negation presuppose affirmation?

Lurking behind the various asymmetricalist salvoes I have been, or will be, considering is a not entirely hidden agenda: the attempt to argue that every negative statement (judgment, proposition) PRESUPPOSES an affirmative, but not vice versa. This view, which can be traced from Śaṇkara in the eighth century to Givón in our own, collects a number of distinct subpositions, given the multiplicity of phenomena that have been subsumed under the term "presupposition".[54] Nor is it always clear just which affirmative a given writer takes a given negative to presuppose. I shall proceed to examine the purported presuppositional asymmetry in a more-or-less chronological fashion.

The earliest extant version of the doctrine that negatives necessarily presuppose affirmatives comes down to us from the eighth-century Indian idealist philosopher and logician Śaṇkara (Raju 1954:703). For Śaṇkara, every negation either has a positive basis or is insignificant. But the positive basis of a significant negation like *The pen is not red* is not the corresponding affirmative *The pen is red* but rather the distinct fact that the pen is black (or whatever). We must distinguish the POSITIVE BASIS of *A is not B*, that is, *A is C* (where C is incompatible with B), from its POSITIVE COUNTERPART, that is, *A is B*. (In this, and in his verdict that being and nonbeing are not on the same level, Śaṇkara fits squarely within the Parmenides-Plato tradition.)

Śaṇkara's distinction between significant and insignificant negation was (it would seem) independently redeveloped in the West, going back at least as far as Spinoza and his dichotomy of privation or particular negation (*privatio*) and absolute negation (*negatio*) reviewed above (§1.1.5).

Spinoza's dictum, *Determinatio est negatio,* is stood on its head by Hegel, who interprets it as the claim that every significant negation is a determination or limitation; significant negation must always occur on a POSITIVE GROUND (corresponding to Śaṇkara's 'basis'). For Hegel, a 'pure negative judgment' like *The rose is not red* suggests that a different predicate from the same semantic class applies to the subject: 'To say that the rose is not red implies that it is still coloured' (*Logic* §173, Hegel 1892: 306). 'If the rose is not red, it is assumed it has a colour—some other colour' (Hegel [1812–16] 1929:275). Such a 'simply negative' judgment does not constitute total negation; the judgment—that is, the relation of subject and predicate—is still 'essentially positive', and the subject is 'untouched by negation'.

On the other hand, the 'infinite' or 'negatively infinite' judgment, exemplified by such 'correct but absurd' propositions as *The rose is not an elephant, Understanding is not a table* are not really judgments at all, since their subject and predicate have no connection (Hegel [1812–16] 1929: 277ff.).

It will be noted that Hegel has performed a complete 180-degree turn in identifying the simple negative judgment with Aristotle's predicate term negation and the medievals' *negatio infinitans* (a restricted negative in which some other member of the predicate class is presupposed to apply to the subject) and the infinite judgment with Aristotle's predicate denial and the medievals' *negatio negans* (an unrestricted negative taken as simply denying the connection of subject and predicate). The discrepancy between the traditional and the Hegelian labels must be borne in mind when I extend the discussion of negative and infinite judgments below.

In any event, the Idealists of the late nineteenth and early twentieth century followed Hegel's lead on negation (as elsewhere), with Sigwart (1895: 126) glossing *Determinatio est negatio* as the proposition that 'a figure is determined insofar as it is not the space surrounding it, and thus can be thought of only by the aid of negation—as a limitation, i.e. negation of the infinite'. Contra Aristotle, the affirmative is logically prior to the negative: 'The object of a negation must always be either a completed or an attempted judgment, and for this reason we cannot regard the negative judgment as a species equally primitive with the positive judgment and coordinate with it. . . . The negative judgment presupposes the attempt, or at least the thought, or an affirmation' (Sigwart 1895: 119).

Sigwart later (p. 122) strengthens this notion of presupposition: 'Immediately and directly, the negation is a judgment concerning a positive judgment that has been essayed or passed'. We are back again to negation as second-order affirmation. (Elsewhere, Sigwart endorses Kant's characterization of negation as a means for averting error and Bergson's Subjectivity thesis.)

For Sigwart, it is only under an asymmetricalist banner that *Duplex negatio affirmat,* the classical Law of Double Negation, can be captured: 'As soon as we see that every negation presupposes a previous synthesis, its only object being to declare this synthesis invalid', we can improve on Aristotle (whose logic was blighted by his parallel treatment of affirmation and negation) and derive LDN (Sigwart 1895: 148). Since LDN functions as the crucial stepping-stone from LC to LEM, its derivability would be a major argument for the asymmetricalist thesis on negation—if Sigwart's claim is correct. But those symmetricalists who have worked within classic propositional logic, from the Stoics to Frege to Quine, have experienced no difficulty in taking LDN as a cornerstone of the logic of negation, albeit one which tends to be stipulated as an axiom rather than derived as a theorem.

Bradley (1883) adopts an intermediate position on the question of priority. On the one hand, he agrees with Sigwart that 'assertion and denial stand on different levels', and echoes the Śaṇkara-Hegel line on significant negation: 'Every negation must have a ground, and this ground is positive' (Bradley 1883: 112). 'Nothing in the world can ever be denied except on the strength of positive knowledge. . . . We can not deny without also affirming' (p. 120). 'We should never trust a negative judgment until we have seen its positive ground' (p. 200). But he cautions that denial cannot be reduced to or derived from affirmation, explicitly rejecting Sigwart's claim that negation presupposes an affirmative judgment: 'We must never say that negation is the denial of an existing judgment. For judgment . . . implies belief; and it is not the case that what we deny we must once have believed' (p. 110).

Another reluctant warrior in the Idealist battalion of the asymmetricalist army is Bosanquet ([1888] 1911, 1895), for whom the real question is not so much whether but what the negative judgment presupposes. Of the view that 'every negation presupposes an affirmation, so "A is not B" presupposes the affirmation "A is B"'—which he attributes to Sigwart—he declares:[55] 'I think it monstrous. I do not believe that you must find an affirmative standing before you can deny' (Bosanquet 1895: 132; cf. also [1888] 1911: 277). But the matter is more complicated: 'Negation is not, as such, the denial of affirmative judgment, and therefore does not presuppose the affirmation of that which is denied. . . . On the other hand, Negation does presuppose some affirmation' (Bosanquet [1888] 1911: 280). Bosanquet here refers to Bradley's notion (1883: 166) of 'a suggested affirmative relation', and endorses the latter's view that 'in the beginning, negation is a degree more remote from reality than is affirmation'. But while affirmation is epistemologically prior to negation, eventually 'affirmation and negation alike become double-edged, each involving the other' (Bosanquet [1888] 1911: 281).

For Bosanquet, 'all and only those negatives which presuppose an affirmative are significant' (Heinemann 1944). Any significant negative judgment of the form *A is not B* can always be analyzed as *A is not B but C*, or as *A is X, which excludes B* (as in 'a colour (undetermined) not red'). But insignificant or bare negations—*The lion is not an elephant, Virtue is not square*—do not posit a true contrary, since they do not limit the sphere of negation (Bosanquet [1888] 1911: 281–89, 1895: 130). We are back once again to Spinoza's *privatio* vs. *negatio*, particular vs. absolute negation.

Bosanquet distinguishes the positive ground for negation (identical with Śaṇkara's positive basis)—some contrary whose truth determines the truth of the negative proposition—from the positive CONSEQUENT, the indeterminate proposition which logically follows from the negative. Thus the positive ground of (50a) is (50b), while its positive consequent is (50c) ([1888] 1911: 289):

(50) a. This surface is not black.
b. This surface is (e.g.) green.
c. There is a color x, x ≠ black, such that this surface is x.

Notice that the relation between a negation and its ground is one of contrariety, while a negation and its consequent are in (essentially) contradictory opposition. Bosanquet's insightful remarks on the interaction of contradictory and contrary negation will be reviewed in chapter 5, where I shall use them as a launching pad for my consideration of contraries in contradictory clothing.

Joseph (1916: 172) echoes Bosanquet's line: 'We must accept the negative judgment as expressing the real limitation of things, but we must allow that it rests upon and presupposes the affirmative. . . . There is always a positive character as the ground of negation. Snow is not hot, because it is cold'. Plato's view of negation as difference or otherness is thus endorsed, but only as a means of explaining the function of (significant) negative statements in communication, not as a means of eliminating negation from logic, that is, the representation of thought.

A related but somewhat more unusual approach is pursued by Demos (1917). While all negation must be read as 'qualifying the entire content of the proposition', so that *X is not dead* is analyzable as **not (X is dead)**, the meaning of *not* can be given in terms of contrariety rather than contradiction. Thus (Demos 1917: 190), we have the equation

(51) John is not at home = not (John is at home) = an opposite of (John is at home) is true

If I believe that John is not at home, I believe an opposite, a contrary, of the proposition that John is at home. Negative assertions are therefore 'always positive in reference', but—since they refer ambiguously or indeterminately to their positive ground—'never positive in content' (p. 193).

Thus **not-p** is to be unpacked into 'some proposition is true which is a contrary of **p**', where the latter 'refers descriptively to that proposition which is true'. The claim that a negative proposition **not-p** denotes a proposition **q** which is a true contrary of **p** recalls a recent analysis of questions within an extended Montague grammar (Karttunen and Peters 1976) in which every question denotes the set of its true and complete answers. But just which contrary does the negation denote? If John is not at home, there are infinitely many true positive propositions which entail that he is not at home, each of which constitutes a contrary of *John is at home*.[56]

Demos joins Bergson and the Idealists in some of the asymmetricalist campaigns; while both negative and positive propositions refer to positive facts, 'in the first case the reference is indirect, and in the second direct. From this angle, a negative proposition may be defined as a referent to a referent or a description of a description' (Demos 1917: 194). But, contra Bergson and his Subjectivist troops, 'the negative proposition is an objective entity, not dependent on the mind's attitude towards it' (p. 195).

Two encyclopedia entries from this period on 'negation' and 'negative', prepared by Peirce and his colleagues, offer a measured state-of-the-art statement, not only summarizing the Idealists' view on the conditions for significant negation, but essentially prefiguring the position of Givón (1978, 1979) as well:

> In order that a negative statement may have any value, there must have been some reason to suppose that the affirmative statement of which it is the exact denial was true, either that it had been proposed for our acceptance by an interlocutor, that it had been part of our stored-up knowledge or purported knowledge, or that we had in mind what we took at the moment to be sufficient ground for its acceptance. . . . Negation is a secondary function of thought, which presupposes the existence of positive judgments.
>
> (Baldwin 1928: 146–48)

The site for the next major skirmish in the war of affirmative priority was the symposium on negation reproduced in the Aristotelian Society proceedings for 1929. The first speaker is Mabbott, who (following Hegel and Bosanquet) distinguishes significant negations of the type in (52a) from the 'impossible thoughts' of (52b, c) and from the 'pseudo-judgments' of (53):[57]

(52) a. Some politicians are not honest.
b. Some politicians are not $\sqrt{3}$.
c. Some politicians are not of.

(53) a. Virtue is not square.
b. The soul is not an elephant.
c. The soul is not a fire-shovel.

When the subject is not a 'possible candidate' for the predicate to apply to, and negation does not function to eliminate its candidacy and thereby narrow the field, the result is a TURKEY-CARPET JUDGMENT—so called because it bears the same relation to a real negative proposition (e.g., (52a)) that an oriental carpet bears to a painting (Mabbott 1929: 68ff.).

What remains are the cases of true ELIMINATIVE negation, constituting a pis aller which can be unpacked by situating the negated predicate within a disjunctive set: *My hat is not red = My hat is yellow or blue or . . .* Other instances of meaningful negation, involving 'neither doubt, ignorance, nor error' contain no implicit disjunction; members of this class, including *This is not a living wage* and *We cannot cross that field,* are branded TELEOLOGICAL negations and dismissed as 'damaged specimens' (Mabbott 1929: 72–76).

Price (1929) rejects Mabbott's Subjectivist approach to negation as well as the view that 'negation presupposes a positive suggestion'; at least this can't be a differentia of negation, Price notes, since affirmations may also presuppose positive suggestions. Further, Price defends *The soul is not a fire-shovel* as sensical and indeed true.

Ryle, in his contribution to the symposium, situates himself between Mabbott's hard-line asymmetricalist approach and Price's symmetricalist stance. While rejecting Mabbott's contention that we affirm out of knowledge and deny out of ignorance or error, as well as his repudiation of negative facts ('even negation is determination': Ryle 1929:96), Ryle does follow Mabbott in endorsing an 'elimination-within-a-disjunctive-set' analysis:[58] 'When I say "Mrs. Smith's hat is not green" I can equivalently say ". . . but some other colour". The "but some other . . ." is always there, sometimes explicitly, sometimes marked by tone of voice, or simply implied by the context' (Ryle 1929:85).

Without the *but* clause, 'negative sentences are elliptical', though still generally interpretable in context. Ryle identifies the missing step in this analysis: 'When I say "The hat is not green (but some other colour)", I am (not stating but) presupposing that the hat is coloured'. The predicate belongs to a contextually assumed set, some other member of which holds of the subject. Thus we can maintain, with Plato's Stranger and subsequent Otherness theorists, that 'the full explication of what is meant by a negative sentence necessarily takes the form of an assertion of otherness . . . otherness as specified or made determinate by mention of the particular disjunctive set to which the "others" belong as members' (Ryle 1929:89). Hence the oddness of Mabbott's Turkey-carpet judgments, like *Virtue is not square:* we can't continue '. . . *but some other shape*'.

A contemporary advocate of the disjunctive-set analysis is Apostel (1972b:396–97), who distinguishes two types of expansions for a negative sentence like *This stone is not red:* 'The logical negation designates the disjunction of all alternates, while the linguistic or psychological negation means [*sic*] only the disjunction of a few alternatives lying in some sense "close" to the negated sentence'. *Not* ***q*** is read as 'incompatible with, but in the neighborhood of, ***q***'.

But this entire line of argument, reasonable as it sounds when applied to carefully selected examples involving the colors of hats or stones, loses plausibility rapidly when it is extended to other predicates whose 'assumed disjunctive set' is much harder to fill in. Does *Sue didn't eat any pizza* really mean or presuppose that she ate (or did) something else? Is the disjunctive set invoked by *not red* a product of the meaning of *not* or of *red?* Does *John doesn't love Mary* presuppose that he loves another? When no obvious closed class of predicates can be reconstructed, Ryle's disjunctive-set analysis, as he seems to recognize (Ryle 1929:88ff.), works fairly well for instances of contrastive negation, where the focused element is stressed; it is far less convincing for more neutral cases.

Another difficulty for both the Mabbott-Ryle analysis of predicational sentences and the comparable line on negative existentials (as in Gale 1972:473–74) is that they seem to assume the appropriate disjunctive set will be manageable, or at least finite. But applying this style of analysis to (54a), not to mention (54b, c), proves extremely discouraging.

(54) a. There are no unicorns.
b. My favorite real number is not π.
c. The largest complex number is not $3i + 4$.

Mabbott would no doubt consign these recalcitrants to the dustbin of 'teleological negation', but it is hard to see what insight has been captured thereby.

On the hybrid asymmetricalist theory favored by Wood (1933), the negation-as-otherness line associated with Plato, Spinoza, and Hegel applies only to those equationals which reject a false identity claim, such as the 'pseudo-negatives' in (55):

(55) a. Black is not white.
b. 2 is not 3.
c. Mr. A is not Mr. B.

Predicationals like those in (56), on the other hand,

(56) a. Snow is not black.
b. 2 plus 2 does not equal 5.
c. Mr. A is not in the room.

are 'pure and genuine negative judgments', to be analyzed as second-order affirmations, 'affirmative judgments in disguise' (Wood 1933:421). (Given my earlier discussion of Hegel's, Sigwart's, and Mabbott's negative identities, it would seem that Wood's lumping of all such statements into the pseudo-negative bag occupied by the examples of (55) may be somewhat hasty, since the class of negative identity or equational sentences appears to be heterogeneous.)

In addition to disdaining, defending, and subsequently casting out negative facts, seeking to eliminate negation through the negation-as-true-disbelief and negation-as-falsity ploys, and codirecting the Bumblowskian campaign to spite Satan by banning negation outright (§1.2.1.), Russell—in his *Inquiry into Meaning and Truth*—ignores his earlier cogent argument from hippolessness and brings evidence from perception to bear on behalf of an analysis of negation as secondary to, and presupposing, affirmative judgment:

> Although they seem equally based upon sensible experience, the two statements 'there is butter [in the larder]' and 'there is not cheese' are really on a different level. There was a definite occurrence which was seeing butter . . . but there was no occurrence which could be described as 'not seeing cheese' or 'seeing the absence of cheese'. You must have looked at everything in the larder and judged, in each case, 'this is not cheese'. . . . To judge 'this

is not cheese', you must have the word 'cheese' . . . in your mind already. (Russell 1940:73)

It is for this reason that *This is not cheese* counts as a statement about a statement, equivalent to the statement that the statement *This is cheese* is false (1940:74; see §1.2.1 above). In general, then, 'while a positive basic proposition is caused only by a percept a negative one is caused by a percept plus a previous propositional attitude' (1940:163).

On this view, all negation is metalinguistic. This bears out the line taken elsewhere in the *Inquiry:* 'The word "not" is only significant when attached to a sentence, and therefore presupposes language. Consequently, if "**p**" is a sentence of the primary language, "not-**p**" is a sentence of the secondary language' (Russell 1940:64). But we are now back within the familiar Bergsonian encampment besieged so effectively by Frege, Austin, Quine, and Geach (cf. §1.2.1); the new reinforcement offered for this position by Russell is valid, if at all, only for that restricted class of sentences that report direct perception. As Gale notes (1976:60), what holds for the perception of negative facts and events does not necessarily hold for their existence, and it is the latter point which is at issue here. In the light of the questionable status of any generalization from perception reports to the entire range of affirmative and negative propositions, Russell's reassurance that 'we may safely treat "false" and "not" as synonyms' (1940:81) amounts to whistling in the dark.

As we have seen in this section, there are two primary ways in which the claim that every negative presupposes an affirmative can, and has been, understood. In the weaker sense, every "significant" contradictory negation "rests upon", or owes its truth to, some contrary (its "positive ground"). It is in this sense that *This hat is not red* presupposes that the hat in question is of some other specific (but perhaps unknown) color. In the stronger sense, every negative proposition of the form *not-**p*** (or *A is not B*) presupposes the corresponding affirmative ***p*** (or *A is B*), its "positive consequent". But this latter claim, as frequent today as it was in the days of Śaṇkara and Avicenna, relies on knowing what we mean by "presuppose".

Sigwart maintained (at times) that ***p*** must be affirmed before it can be denied, and Bosanquet thought him monstrous for doing so. Tesnière would have evidently stood with Sigwart, at least in terms of negation presupposing an earlier mental affirmation: 'Toute négation procède d'une affirmation. . . . Avant de nier le contenu d'une phrase l'esprit doit d'abord l'affirmer afin de pouvoir ensuite le nier' (Tesnière 1959:225).

The Presuppositionalist thesis most clearly presented in Peirce et al.'s encyclopedia entry (Baldwin 1928:147, cited above) is echoed in recent treatments of negation by those working within the perspectives of psycholinguistics (cf. the discussion of Wason's 'contexts of plausible denial' in chapter 3) or pragmatics. Thus, Ducrot (1973: 119) observes that

(57) Pierre n'est pas le cousin de Marie.

would be an odd thing to say if nobody had ever claimed that Pierre was Marie's cousin; hence the expected rejoinder when (57) is uttered in the absence of such a context,

(57′) Qui a jamais prétendu cela? (= Whoever claimed she was?)

In the same vein, Givón (1978: 79–81; 1979: 103–4) remarks on the oddness of a discourse-initial utterance of (58),

(58) Oh, my wife's not pregnant.

when the hearer cannot be expected to assume 'that there was some likelihood that my wife was pregnant, that the subject has been under discussion, that it had been considered as a probability, etc.' (emphasis in original). If the hearer cannot make this assumption, Givón notes (1979: 103), s/he is likely to respond accordingly:

(58′) Wait a minute—was she supposed to be pregnant?
Hold it—I didn't know she was supposed to be pregnant.

The affirmative counterpart of (58), like that of Ducrot's (57), is not comparably restricted.

Givón cites García's pragmatically fueled salvo against the purported 'objective' and 'grammatical' symmetry of affirmation and negation: 'Negative sentences communicate in terms of an implicit, but rejected, affirmation. . . . In terms of actual communication . . . , a negative sentence is, a priori, worth far less than an affirmative' (García 1975: 8–9; cited in Givón 1979: 111–12). And he translates this language into that of 'presuppositionality'. But Givón's attempt to show that negative sentences are 'more presuppositional' or 'presuppositionally richer' than affirmatives (1979: 108 and elsewhere) founders on an equivocation on the term "presupposition". He argues that 'from a strictly logical point of view, while the speaker asserts **~p** he presupposes **p**' (Givón 1978: 70; 1979: 92). Not surprisingly, this strikes Givón as 'a nonsensical conclusion', from which he deduces that formal logic is hopelessly flawed, or at least totally inadequate to describe the behavior of natural language negation.

Whether or not his conclusion is correct, Givón has hardly made a case for it, especially since no formal logic of presupposition (including the two Givón cites, those of Keenan [1971] and Herzberger [1971]) would accept the premise that **~p** logically presupposes **p**, that is, that the truth of the latter is a necessary condition for the former to be true or false. Nor would a Strawsonian be prepared to grant that **p** must be true in order for the question of the truth or falsity of ~**p** to arise. The Stalnaker-Karttunen notion of pragmatic presupposition (cf. e.g., Stalnaker 1974) fares no better here: **p** can hardly constitute a necessary condition for the appropriate or felicitous utterance of ~**p**. But a pragmatic notion of a different sort does seem to be involved here.

While Givón's original version of the Presuppositionality thesis, as reflected in the claim (1979:108) that negative sentences are used 'always in context where the speaker believes that the hearer holds a certain belief in the truth of the corresponding affirmative', is far too strong to uphold consistently, this untenable form of the thesis is later weakened (although the weakening is not acknowledged) to a more defensible, if vaguer, position: 'Negatives in general are uttered in a context where the corresponding affirmative has been discussed, or else where the speaker assumes that the hearer's bias toward or belief in—and thus familiarity with—the corresponding affirmative [*sic*]' (1979:139). But on this version, Givón's notion of presupposition is closer to the Praguean notion of GIVEN or OLD INFORMATION (cf. Firbas 1964, 1966; Kuno 1972; Prince 1981) than to either the logical/semantic or the formal pragmatic (Stalnaker-Thomason-Karttunen) approaches explicitly cited by Givón.[59]

Negation as a speech act

For Givón, the doctrine that negation is 'more presuppositional' than affirmation, that negative sentences involve, by their nature, 'denial of the hearer's belief' (1979:112), is built into his general depiction of the 'NEGATIVE SPEECH ACT'. The assumption that it makes sense to speak of the 'speech act of negation' is worth examining more closely.

Givón's characterization of negation as a speech act echoes the asymmetricalists of antiquity: 'Negative declarative sentences constitute a different speech act than the corresponding affirmatives. Affirmatives are used to convey new information on the background of assuming the hearer's ignorance. Negatives are used to correct misguided belief on the background of assuming the hearer's error' (Givón 1979:139). As in Kant, negatives function to ward off error; as in Bergson, they are crucially subjective.

Strawson (1952:7) seems to share this view of negation as a means for rendering explicit what he calls 'the function of exclusion', as a device used 'when we wish to contradict a previous assertion, or to correct a possible false impression, or to express the contrast between what had been expected, feared, suggested, or hoped, and the reality. . . . The standard and primary use of "not" is specifically to contradict or correct; to cancel a suggestion of one's own or another's'.[60] The equation implicitly assumed by Strawson and Givón is explicitly spelled out by Apostel (1972a:273):

(59) He is not poor = I deny that he is poor.[61]

But the doctrine that there is a special negative speech act, that negation is always or generally equivalent to speaker denial, has met with the expected resistance. Indeed, it was Frege who first stressed that negation cannot be reduced to denial (any more than it can be reduced to falsity). There is, to be sure, a truth-conditional relation between the two concepts or acts: 'When a sentence **p** may be truly asserted, the corresponding negative sentence not-**p** may be truly denied, and vice versa' (Frege 1919:129).[62]

Negation itself is a 'chimerical construction' for Frege, since all we really have is the fusion of ordinary judgment or assertion with a content that happens to contain *not.* In Frege's representation of assertion, ⊢**A**, the vertical stroke represents the JUDGMENT, the horizontal the CONTENT, the latter representing (when unasserted) a 'mere complex of ideas', which may or may not include negation (Frege 1919:130ff.). Crucially, Frege points out, negation may occur in unasserted contexts, such as the antecedent of a conditional; there is no notion of speaker denial which straightforwardly applies to the negative clause in sentences like those in (60).

(60) a. If Paris is not the capital of France, my itinerary is in trouble.
b. Either he isn't going to the opera tonight or he's going to miss the first act.

Similarly, we can entertain or report a negative proposition—for example, that Fermat's last theorem is not true—without rejecting the positive version of that proposition.

Within speech act theory, Searle defends exactly the move which Frege forswears: for Searle (1969:32), the ILLOCUTIONARY NEGATION ~**F**(**p**) contrasts with the PROPOSITIONAL NEGATION **F**(~**p**), where **F** is an illocutionary force indicator, and **p** the propositional content. Let **F** be the speech act of promising, **Pr**. Then ~**Pr**(**p**) can be read as, for example, *I do not promise to come,* that is, as 'a refusal to make a promise'. **Pr**(~**p**), for example, *I promise not to come,* is simply a promise with negative content.

Given that Searle's approach, unlike Givón's and perhaps Strawson's, does not seek to reduce all negatives to the speech act of denial, the questions remain: is illocutionary negation tenable at all, and—if so—how can it be characterized? These questions sparked a flurry of debate. For Sloman (1969:58ff.), one who performs the act corresponding to ~**F**(**p**) performs a speech act different from, but related to, **F**(**p**), one in which the commitments associated with **F** are at least temporarily rejected. Hare (1970: 12) argues that 'nearly all speech acts, including assertions, can be negated in these two ways': either 'externally' (via Searle's illocutionary negation) or 'internally' (via propositional negation). This extends to declaratives, where *The cat is on the mat* allows both internal negation (*The cat is not on the mat*) and external negation (*I don't say the cat is on the mat; The cat may not be on the mat*). Using Fregean notation, the former comes out $\vdash(\sim\mathbf{p})$ and the latter (in contravention of Frege) as $\sim\vdash(\mathbf{p})$.

Garner (1970) argues, contra Hare, that 'the very idea of external negation *of a speech act* is suspect'; consider the 'external negations' that would have to be provided for, for example, *I declare this bridge open, I bid you welcome,* and so forth. If external negation is motivated at all, it can only apply to sentences, not to speech acts. And, against Sloman, while *refusing* does indeed constitute a speech act, '*refusing to promise* is no more a modification or transformation of the act of promising than refusing to run is a modification of the act of running' (Garner 1970: 110).

Patton (1968: 231), in his characterization of negation as external to the propositional content, lines up against Frege and (apparently) alongside the advocates of 'illocutionary negation': 'Negations, like interrogatives and imperatives, are semantically distinguished from their sources not by content but as it were by differences in linguistic moves that can involve the same content'.

Apostel (1972a) adopts a related position: negative and positive propositions differ in content, à la Frege, but either type of proposition may be asserted or denied by a speaker. Asymmetry comes in at the point of the latter distinction, since whoever denies both denies an assertion and asserts a denial, while whoever asserts does not ipso facto deny anything. Thus, positive and negative (propositional attributes) are symmetrical, but affirmation and denial (speech act attributes) are not.

A quite different symmetricalist view of negation as a 'linguistic move' is suggested by Toms (1972): negation is not the result of performing an operation on a positive "source"; rather, both negative and affirmative propositions result from performing separate operations on a basic entity which is neither positive nor negative per se. Both the affirmative *There are unicorns* and the negative *There are no unicorns* descend from a common universal, rather than the latter deriving from the former in any direct way: 'A negative fact does not relate to the opposite positive fact, but . . . to a universal distinct from both positive and negative facts' (Toms 1972: 12). We can in fact trace this idea back to Hegelian dialectic (cf. Hegel [1812–16] 1929: 66), which insists on the existence of 'an **A** which is neither **+A** nor **−A**', a 'third term' indifferent to the polar opposition.

The symmetricalists strike back

I shall conclude this rather hyperextended section by reviewing some of the counterarguments that have been offered in our own century to the asymmetricalists' stance on negation-as-presupposing-affirmation and negation-as-speaker-denial. We have observed Frege's hostility to the translation of negative statements into instances of the speech act of denial or rejection; he also vigorously attacks the view that 'a negative thought is less useful than an affirmative one'; even Aristotle, as we have seen, might be held to have adopted the version of epistemological asymmetry Frege rejects here.

Following Frege's lead, the logical positivists launched an assault on these Idealist beachheads. For the Wittgenstein of the *Tractatus*, presuppositionality is a two-edged sword: 'The positive proposition necessarily presupposes the existence of the negative proposition and vice versa' (Wittgenstein 1922: §5.5151). And Ayer makes it clear that he would dissent from even the measured quasi-Praguean line on the putative discourse presuppositionality of negation, as endorsed by Peirce and the neo-Hegelians in their edict that negation is always directed against a 'suggested' or 'attempted' judgment, and by Strawson, Ducrot, and Givón in terms of the enriched context required for the felicity of the 'negative speech act'.

Ayer's manifesto ([1952] 1963: 39) is worth quoting in full:

> From the fact that someone asserts that it is not raining one is not entitled to infer that he has ever supposed, or that anyone has ever suggested, that it is, any more than from the fact that someone asserts that it is raining one is entitled to infer that he has ever supposed, or that anyone has ever suggested, that it is not. No doubt negative forms of expression are very frequently used to deny some previous suggestion; it may even be that this is their most common use. But whatever the interest of this fact it cannot be the ground of any viable distinction between different types of statement.

Ayer goes on to challenge the epistemological asymmetricalists on the move from negation-as-worth-less to negation-as-worthless: 'Why should it not be allowed that the statement that the Atlantic Ocean is not blue is as much a description of the Atlantic as the statement that the Mediterranean Sea is blue is a description of the Mediterranean?' Unlike Frege, Ayer is willing to grant that the negative might well be less informative but, he observes (in a passage which has since become quoted almost as often as Spinoza's one-liner), 'to say that a description is relatively uninformative is not to say that it is not a description at all' (Ayer [1952] 1963: 47).[63]

Just as Bosanquet was led to a recognition ([1888] 1911: 281) that affirmation and negation end up 'double-edged, each involving the other', Ayer notes ([1952] 1963: 48) that 'to say what things are not is itself a way of saying what they are. And conversely, to say what things are is itself a way of saying what they are not'. Negation is determination—but, then, so is affirmation.

Kissin (1969) is a moderate symmetricalist who fits neatly within neither camp. He argues, contra Frege (and Geach; cf. below), that not all negative statements can be reduced to 'an assertion-or-affirmation of a negation'. The denial or rejection of **p** cannot be represented simply as $\vdash(\sim\mathbf{p})$; rather, as Lukasiewicz maintained: 'we need another symbol of the same level as the assertion-or-affirmation symbol in order to symbolize denial-or-rejection, say $\dashv$. The point is that denial-or-rejection is an act with the same status as assertion-or-affirmation, and can't be reduced to assertion-or-affirmation' (Kissin 1969: 147).

At the same time, Kissin rejects the strong asymmetricalist position that 'a statement is a denial-or-rejection iff it is made by using a negative sentence straightforwardly'. The most we can say is that 'when a person uses a negative sentence straightforwardly, typically or standardly what he is doing is denying-or-rejecting something' (Kissin 1969: 149; emphasis mine). This weaker position is forced on Kissin by his recognition that a given negative (or positive) sentence may be used in different ways depending on the context, speaker intentions, and so on; he cites the parallel uses of *He's staying* and *He's not leaving* to issue the same denial, as well as the use of a given negative, for example, *There's nothing on the table,* either to deny an assertion or to assert a (negative) fact. Kissin thus alights next to Hare (1970): some but not all negative statements represent 'illocutionary negation'.

The pure Fregean line on behalf of propositional-content-internal negation is reiterated by Geach ([1972] 1980: 260), who argues that logic 'demands the use of a negation sign which is not polarly opposed to the assertion sign and does not express rejection of what is negated; and when a proposition is negated, we may equally well conceive this as asserting the negation of a proposition'. Dismissing Lukasiewicz's suggestion of $\dashv$ for negation as a 'futile complication', Geach contents himself with Frege's assertion sign and propositional negation, echoing Frege's caveat: 'Whatsoever is more than this cometh of evil' (Frege 1919: 125).

As we have seen, the identification of negative statements with the speech act (or mental act) of denial leads inevitably to the claim that (61a) and (61b) are logically equivalent and/or communicatively interchangeable.

(61) a. not-**p**. (It is not the case that **p**.)
b. I deny that **p**.

Apostel (1972a: 273), for example, suggests an equivalence between *He is not poor* and *I deny that he is poor,* and others (including Bergson and Givón) would seem committed to the same view. But Gale (1970: 201ff.;

1976:59) launches an effective counterassault. As Ayer observed, it is simply not true that the statement addressed by a negative must have been made—or envisioned as being made—by anyone. Further, since positive statements can also be used to deny another's assertion (as even Russell conceded), we can have no general equivalence between (61a) and (61b). If such an equivalence were to be maintained, then—given any proposition **p** and the Law of Excluded Middle, **p** ∨ ~**p**—(62b) should, like (62a), be necessarily true.

(62) a. Either it is not the case that **p** or [it is not the case that it is not the case that] **p**.
  b. Either I deny that **p** or I deny that $\left\{\begin{array}{l}\text{I deny that } \mathbf{p}.\\ \text{not-}\mathbf{p}.\end{array}\right\}$

Yet, as Gale points out, while (62a) is valid, (62b) is not. Nor does it follow from (61b) as it does from (61a) that **p** is false, since true statements can perfectly well be denied. To Gale's cogent arguments I can add one obvious codicil: the performative nature of (61b) is supported by the standard syntactic correlates, while (61a) is clearly nonperformative in nature:

(62′) a. *Hereby not-**p**. (*It is hereby not the case that **p**.)
  b. I hereby deny that **p**.

For the final salvo from the armory of the symmetricalists, I must by rights turn to their modern captain, Peter Geach. Having elsewhere attempted the demolition of the negation-as-falsity troops, he now aims his cannons at the remaining positions of the affirmative priority camp, beginning with the one just cited. Although Geach's remarks directly concern negative predicates rather than negative statements or sentences per se, his point is apposite:

> Verbally, the negation of a predicate is more complex than the predicate itself, since it contains an added 'not'; this has often led people to think that the understanding of the negative predicate includes something over and above the understanding of the affirmative predicate—viz. the understanding of negation. . . . But the understanding of 'not male' is no more complex than that of 'male': they go inseparably together. . . . Of course a predicate cannot really be any more definite than its negation is; the one is exactly as sharply defined as the other. (Geach 1972:78–79)

It will be noticed that Geach, in choosing the pair *male/not male* to make his point for symmetry, is stacking his deck as carefully as the Idealists stacked theirs in finding color predicates—*X is not black*—especially convenient in their defense of the asymmetricalist (elimination-within-a-disjunctive-set) line.

In any case, Geach goes on to blast the Presuppositionality thesis as well: 'We must *a fortiori* reject the view that a negative predication needs to be backed by an affirmative one—that we are not justified in predicating the negation of **P** unless we can predicate some **Q** which is positive and incompatible with **P**'.

As Geach recognizes in the above passage, one (often tacit) step in the asymmetricalists' argument that negation presupposes affirmation (but not vice versa) is provided by logical and linguistic syntax: standardly, a positive proposition is turned into a negative one by the addition of a logical constant symbol (~ and its variants), and generally, a positive sentence of natural language is turned into a negative one by the insertion of a negative word or morpheme. This has suggested to many (as it did to Saint Thomas) that positives or affirmatives are logically—and ontologically—more basic than negatives. I shall return to the asymmetry wars a bit later; in particular, the jump from the relative morphological complexity of negative statements to the alleged logical or psychological complexity of negation will be examined more closely in chapter 3. But first I shall endeavor to frame the issue within a broader philosophical, psychological, and cultural perspective.

1.3 Negation East and West

> Anything which could appropriately be called a 'world' must be such that one or other of every pair of contradictory propositions would apply to or be true of it, and such that all the propositions thus holding of it will be mutually consistent. (C. I. Lewis 1946:56)

> I contradict myself?
> Very well, then, I contradict myself.
> (I am large, I contain multitudes.) (Walt Whitman, "Song of Myself")

1.3.1 The Law of Contradiction and the "Eastern Mind"

While Hegelian philosophy has often been cited as dissenting from the 'pious acceptance' (Burtt 1955:211) traditionally accorded to the founding principles of Aristotelian logic—the Law of Contradiction (LC), the Law of Excluded Middle (LEM), and the Law of Identity—it is classical Indian logic which has been generally accused of (or credited with) having 'thrown these laws overboard' (Staal 1962:52): 'The Eastern mind is convinced that . . . it would be fatal to allow ourselves to be enslaved by these principles' (Burtt 1955:202).

The primary tool for removing the shackles of LC and LEM was historically the *caṭuṣkoti,* the PRINCIPLE OF FOUR-CORNERED (or FOURFOLD) NEGATION (Raju 1954; Burtt 1955), aka the TETRALEMMA (Robinson 1956).

This principle (henceforth L4CN) can be traced to the pre-Buddhist logician Sanjaya (< sixth century B.C.), but it was quickly adopted (at least for some purposes) by Buddha and his followers. The "four corners" are the four propositions purporting to exhaustively describe any subject **S** in relation to an entity or class **P**:

(63) a. **S** is **P**.
b. **S** is not-**P**.
c. **S** is both **P** and not-**P**.
d. **S** is neither **P** nor not-**P**.

It follows that there are four possible questions concerning the evaluation of anything: *Is it good? Is it not-good? Is it both good and not-good? Is it neither good nor not-good?* What L4CN amounts to is a rejection of all four propositions in (63), a negative answer to all four of the evaluative questions. Thus, all four of the propositions in (64) must be rejected (Burtt 1955:203):

(64) a. Nirvana is (some form of) being.
b. Nirvana is (some form of) nonbeing.
c. Nirvana is both being and nonbeing.
d. Nirvana is neither being nor nonbeing.

Similarly, for the Buddhists reality is neither being nor nonbeing nor both nor neither (Raju 1954:702).

What we have, in effect, is a metaphysical multiple-choice question on the qualifying exam for sagehood, taking the abstract form in (65):

(65) a. **S** is **P**
b. **S** is not-**P**
c. both (a) and (b)
d. neither (a) nor (b)

where the grading key stipulates that the successful candidate is to pencil in (e), 'none of the above'. (Or, more precisely, that he is to refuse to turn in his answer sheet altogether.)

One problem in pinning down the significance of the L4CN is determining exactly what each corner states. Crucially, no distinction between contradictory and contrary negation was regularly made within classical Indian logic. How then to represent the propositions in (63) and the form of their joint negation? One possibility (Robinson 1956:303) is that the four corners are built upon contradictory negation, as in (66):

(66) a. **Pa**
b. −**Pa**

c. $\mathbf{Pa} \wedge -\mathbf{Pa}$
d. $-\mathbf{Pa} \wedge -(-\mathbf{Pa})$

Under this interpretation, the third corner is a direct rejection of LC and the fourth a rejection of LEM. The joint denial of all four corners via L4CN then amounts to (66′):

(66′) $-\mathbf{Pa} \wedge -(-\mathbf{Pa}) \wedge -(\mathbf{Pa} \wedge -\mathbf{Pa}) \wedge -(-\mathbf{Pa} \wedge -(-\mathbf{Pa}))$

An alternative reading is probably more faithful, given the wide scope assumed for the subject term in the Indian texts. On this account, the form **not-P** figuring in the last three corners anticipates Aristotle's contrary (predicate term) negation. (Note that (63b) cannot be legitimately viewed as a contradictory of (63a), given the Nirvana example and its equivalents.) We then have the four alternatives in (67), where $\bar{\mathbf{P}}$ is the strong (logical) contrary of **P** (cf. §1.1.5):

(67) a. $\mathbf{Pa}$
b. $\bar{\mathbf{P}}\mathbf{a}$
c. $\mathbf{Pa} \wedge \bar{\mathbf{P}}\mathbf{a}$
d. $-\mathbf{Pa} \wedge -\bar{\mathbf{P}}\mathbf{a}$

Alternative (67d) does not constitute a rejection of LEM, any more than would, for example, *Socrates is neither well nor not-well* or *2 is neither red nor not-red* for Aristotle, since the terms of the correlative denial, (67a) and (67b), are contraries rather than contradictories. (The third corner, of course, still cannot be affirmed without rejecting LC, since this law applies to contrary as well as contradictory oppositions.)

Now the joint denial of all four propositions effected by L4CN amounts to the affirmation of (67′):

(67′) $-\mathbf{Pa} \wedge -\bar{\mathbf{P}}\mathbf{a} \wedge -(\mathbf{Pa} \wedge \bar{\mathbf{P}}\mathbf{a}) \wedge -(-\mathbf{Pa} \wedge -\bar{\mathbf{P}}\mathbf{a})$

But once again, the familiar landmarks of Western (and, as we shall see, Eastern) logic are lost in the mist: if **S** isn't **P** and it isn't not-**P**, it certainly isn't both **P** and not-**P** (which is just as well, given LC). But then how can it not be neither **P** nor not-**P**? Haven't we precisely just gotten through saying that it is neither **P** nor not-**P**? In accepting L4CN we seem to be simultaneously rejecting LEM and LC, since two contradictories, that is, $(-\mathbf{Pa} \wedge -\bar{\mathbf{P}}\mathbf{a})$ and $-(-\mathbf{Pa} \wedge -\bar{\mathbf{P}}\mathbf{a})$, are (simultaneously) simultaneously denied and simultaneously affirmed.

Doesn't the acceptance of fourfold negation (in the words of Burtt 1955: 205) 'encourage a disrespect for the requirements of consistent and rigorous thinking'? Doesn't it 'lead to a wholesale flouting of logic'? Well, yes and no. Just as Aristotle dismissed LC flouters as vegetables, the Buddhists tended to lampoon Sanjaya and his acolytes as 'eel-wrigglers', complaining that 'it was impossible to fix their position either for approval or rejection. They would criticize any view . . . but would not themselves hold any. And it was difficult for any person to enter into any controversy with them' (Raju 1954:695). Indeed, Sanjaya was notorious for his periodic lapses into the extended silence Aristotle described as the inevitable last refuge of archskeptics.

But the Buddhist tradition, as noted, accepts at least one application of L4CN, that which relates to transcendental propositions concerning the character of Nirvana or reality. So too, we find—under the heading 'Questions which lead not to edification'—in Sūtta 63 of the *Majjima-Nikāya,* the tale of a certain Malunkyaputta, who remonstrates with Gautama Buddha over the Latter's simultaneous rejection of the theories 'that the saint exists after death, that the saint does not exist after death, that the saint both exists and does not exist after death, [and] that the saint neither exists nor does not exist after death' (Warren 1896:117–22). If this paradox—a clear instance of L4CN—is not elucidated by the Blessed One, Malunkyaputta vows to abandon his religious training. The Buddha counters that a mortal lifetime does not suffice for such an elucidation.

Elsewhere, the Blessed One warns another disciple against the theory that the saint neither exists nor does not exist after death; this theory, he cautions, is 'a jungle, a wilderness, a puppet-show, a writhing, and a fetter, and is coupled with misery, ruin, despair, and agony, and does not tend to aversion, . . . quiescence, knowledge, supreme wisdom, and Nirvana' (Sūtta 72, in Warren 1896:124)—hardly a ringing cry for the overthrow of LEM. But then, to say that the saint is reborn, or is not reborn, or is both reborn and not reborn, or is neither reborn nor not reborn 'would not fit the case'. What, then, would fit the case?

One possibility worth considering is that L4CN might be taken, not as a proposition negating a four-term alternation or disjunction of possibilities, nor of course as the logically equivalent conjunction of four negations (as represented in (66′) or (67′)), but as an expression of the sage's unwillingness to commit himself to the truth of any of the four alternative corners. The Buddha refrains from 'elucidating' that $\mathbf{p}$, that $-\mathbf{p}$, that $-(\mathbf{p} \wedge -\mathbf{p})$, or that $-\mathbf{p} \wedge -(-\mathbf{p})$ 'because it profits not'; to say that none of these propositions 'fits the case' is not to say that each of them is false (or has a true contradictory). As Burtt (1955:203) and others have noted, the principles of logic, in classical Indian logic, are at the service of the theory of knowledge; a proposition is regarded as 'an epistemic act'. None of the alternatives can be fully grounded in what is known or knowable; hence none can be asserted or 'elucidated'.[64]

It is not surprising that Sanjaya, the reputed source of L4CN, is also known as the great skeptic of the Indian tradition. His Greek counterpart was Pyrrho, who was fond of declaring, 'I am not only not certain of the knowledge of any object, but also not certain that I am not certain of such knowledge' (cited in Raju 1954:695). For a true skeptic, neither Moore's paradox nor its epistemic counterpart (cf. Hintikka 1962) holds any terror.

The view that the Eastern mind has thrown LC and LEM to the four corners continues to hold its appeal (for the Western mind). Burtt offers a sociological explanation for the apparent Indian predilection for embracing contradiction: for the tolerant and sophisticated (if not inscrutable) Oriental, the dangers which forced Aristotle to invoke his (simplistic) principles had already been overcome. The Western logician must battle against dogmatic fluff and wishful thinking; the Eastern logician could relax and take an "I'm OK, you're OK" attitude. Indeed, such recent Western crusaders as Korzybski and Hayakawa and their General Semantics acolytes[65] or Haj Ross and his Elephant Theory of Linguistics strike a similar pose.[66]

But there are at least two other perspectives from which the apparent rejection of LC can be coherently viewed. Besides the epistemological consideration already invoked, it is clear that the context of interpretation plays an essential role in determining the application of the logical principles; if the context is not held constant, all bets are off. Aristotle himself was fully aware of this point; in his arguments against the 'Sophists' (whose number is apparently still growing), he explicitly embellishes his statement of LC with what he terms 'the customary qualifications', as in his standard disclaimer formula: 'It is not possible for the same thing both to belong and not to belong at the same time to the same thing and in the same respect (and let as many other qualifications as we might add against dialectical difficulties be added)' (*Met.* 1005b20–29, in the version of Dancy 1975:156).

The Jainists directly invoke a similar codicil when they observe that ***S*** *is* ***P*** and ***S*** *is not-****P*** can both be true from different standpoints (Raju 1954: 698–701; Burtt 1955:204–5). Within the Jainist theory of conditional truth, or truth relative to a standpoint, each of the fourfold alternatives (63a–d) can be affirmed and denied without however violating LC. For the same reason, we have no trouble declaring 'I'm happy and (yet) I'm not happy', 'It is and it isn't', 'Yes and no' (cf. Strawson 1952:7 on the exchange 'Were you pleased?' 'Well, I was and I wasn't'). As far as we can tell from reading Aristotle (and the Sophists against whom he addressed his remarks), this possibility is exactly what the classical definition of LC predicts when the context of evaluation is not held constant.

In this light, it is not surprising that the prominent "irrationalist" logician Nāgārjuna (second century A.D.), who joyfully endorsed L4CN (Robinson 1956:303), nevertheless explicitly invoked and/or tacitly assumed

LC (and LEM): 'For entity and negation of entity do not occur within a unity. He would be non-eternal and eternal, and that is not admissible' (Robinson 1956:295). Thus, LC (the incompatibility of A and not-A) is a basic principle at the empirical and logical level which may be suspended on the transcendental sphere (Sharma 1970:126). It has also been argued that the *paribhāṣā* (metalinguistic rule) of Panini assumes a version of LC for propositions, $-(\mathbf{p} \wedge -\mathbf{p})$ (Staal 1962:56).

Within the mainstream Nyāya school, predating the second century A.D., LC and LEM (along with the closely related Law of Double Negation, LDN) are periodically adopted and abandoned. Sridhāra explicitly refers to these principles in his argumentation (Randle 1930:212), and by the tenth century, LC was widely accepted within Indian logic and grammar (Staal 1962:58). Udayana's formulation of LC (tenth century) is very much in the spirit of Aristotle: 'When two are mutually opposed there is no occurrence (of both) within the same class' (Staal 1962:68).

On a deeper plane, Sharma (1970:59) points to the central role of double negation in the Indian metaphysics of the last three millenia. Nirvana, or emancipation, is absolute removal of pain, hence the negation of a negation. If violence *(hiṃsā)* is the absence of love, compassion, and peace, then *ahiṃsā*—the principle of active moral nonviolence guiding religious and social practice from the Upaniṣad (<700 B.C.) to Gandhi—is a 'double *not*', the absence of this absence.

In early Buddhist logic, every term or proposition can be defined as the negation of its negation; as Dharmakīrti observes, 'Affirmation is the denial of negation' (Sharma 1970:60, 111). In later Nyāya works, LDN is formally significant. As the law is typically instantiated, 'The absence of an absence of a pot is essentially identical with the presence of a pot' (Ingalls 1951:68). Similarly, a triple absence is identical to a (simple) absence.

This approach generally held sway despite the resistance of Śaṇkara (ca. 1600) and his followers (Raju 1954). An especially concise formulation of LDN appears in an anonymous early grammatical commentary (cited by Staal 1962:65): 'Two particles of negation give the meaning of the original'.[67]

The analysis of negation plays a central role in the evolution of the various schools of Indian logic. The general approach is akin to that of the moderate asymmetricalists of the Western schools. First of all, negation is basic and cannot be explained away, if only because of its essential role in defining change of state processes. As Gautama points out in the Nyāya Sūtra (second century A.D.), 'It is impossible to conceive of "becoming" without the notion of "not yet" (antecedent non-existence) and "no longer" (subsequent non-existence)' (cited in Randle 1930:330).

In the Nyāya-Vaiśeṣika tradition, *ABHĀVA*—the negative property of absence or nonexistence—is as real as presence. Both affirmation and negation may be realized directly in assertions or indirectly in inferences (Matilal 1968:90ff.). In the proto-Fregean Nyāya doctrine, the negative element is simply one component of the objective content; linguistic negation does not reduce to the psychological act of denial (Matilal 1968:93).

But there are, in Nyāya and Vaiśeṣika logic, two distinct species of *abhāva,* as discerned in the (familiar) contrast between *The lion is not an elephant* and *The pot is not blue*. The first is MUTUAL ABSENCE, representable as the negation of an identity claim, $\mathbf{x} \neq \mathbf{y}$. The second is RELATIONAL (or ATTRIBUTIVE) ABSENCE, $-\mathbf{Fa}$.

While the classical Western dichotomies between contradictory and contrary negation and between term and propositional negation find no parallel in the Indian schools, we have here a direct anticipation of the split between significant and insignificant negation later to be promulgated by Spinoza, Kant, Hegel, and the Idealists: a negation is real only if it has a positive counterpart (*PRATIYOGIN*). Mutual absence can be reduced (à la Plato) to pure difference or otherness, while relational absence, like the significant negation of the Idealists, necessarily involves 'suggestion and frustration, expectation and disappointment' (Raju 1941:600).

We are thus not surprised to read, in Sharma's summary (1970:118) of the Buddhist approach to relational negation, ca. A.D. 500, the view that 'the judgment "the book is not on the table" presupposes the judgment "the book is on the table"'—every absence presupposing the prior establishment of an expected presence (Sharma 1970:23). Once again, we are in the realm where all judgments are equally real, but some judgments are really more equal than others.

Indeed, the central questions posed by the Paradox of Negative Judgment: How is a negative judgment possible? How can we know, or assert, that something is not the case? What is the positive basis for negation? have the same resonance and trigger the same controversy for the Indian schools as they were to do (cf. §1.2) in the West. (A good summary of the parallels between the Eastern and Western traditions on negation is provided by Matilal 1968, esp. chapter 11.)

Of particular interest is Sūtra ii.2.8 of Gautama's *Nyāya-Sūtra,* which argues that the 'object of absence' is real, 'because when there are certain objects marked, the unmarked objects are characterized by the nonexistence of the mark' (Chattopadhyaya and Gangopadhyaya 1968:110). Vātsyāyana's commentary (the *Bhāṣya,* ca. A.D. 300; this is, incidentally, not the Vātsyāyana of the *Kama Sūtra*) renders this argument more concrete: Someone who is asked to fetch the unmarked cloths can identify them by virtue of their opposition with the marked ones. Thus the unmarked cloth is marked out and counts as a PRAMEYA, a valid object of cognition, because of its contrast with the marked cloth (Chattopadhyaya and Gangopadhyaya 1968:110; cf. Matilal 1968:106; Randle 1930:329–30).

This reflects what is for Vātsyāyana the true nature of opposition: 'That which is not is the means of apprehending that which is' (cited by Randle 1930:331). The same point is emphasized by Dharmakīrti: 'There can be no affirmation which does not exclude the other; nor can there be a negation of that which cannot be affirmed' (Sharma 1970:112). From here, Spinoza's *Determinatio est negatio* and Bosanquet's 'double-edged' affirmation and negation are not too far down the road.

While the Buddhist and Nyāya-Vaiśeṣika logicians (and grammarians) we have been looking at were, like their Western contemporaries, primarily concerned with the treatment of declarative statements, another tradition—focusing on the development of principles for interpreting the laws of Vedic ritual—is codified in the Mīmāṃsā (§§320–363 in Āpadeva's seventeenth century text; cf. Edgerton 1929).[68]

The most significant contribution of the Mīmāṃsākas to the theory of negation lies in the scope distinction they drew for negation in modal contexts. A Vedic injunction like *He shall not eat kalañja* (where *kalañja* probably denoted a kind of red garlic) is analyzed, in the default case, as an instance of *NIṢEDHA* (or *PRATISEDHA*), 'prohibition'. In the frame 'A should not do B', the speaker deters or prohibits A from doing B. The negation in this case is not construed with the verb root (= 'He shall not-eat kalañja') or with the object (= 'He shall eat non-kalañja'), but with the modal (optative) ending (Edgerton 1929:§320–§324).

But when the *niṣedha* reading is implausible or impossible in a given context, there emerges a second interpretation which the Mīmāṃsākas call *PARYUDĀSA* 'exclusion' (Edgerton 1929:§330ff.). Here we have not a negative injunction (or prohibition) against doing something, but a positive injunction to not do something. For example, following the phrase '*His vows are . . .*', where 'vow' *(vrata)* corresponds to a thing positively to be done, negation will be associated with the verb root. Similarly, the injunction *He shall not look on the morning sun* must be analyzed as 'He shall [not-look . . .]'. In other contexts, *paryudāsa* negation may focus on a nominal element instead of the verb root, while *niṣedha* always focuses on the verbal ending.

From their initial application to the optative sentences of the Vedic ritual, the notions of prohibition and exclusion are eventually generalized to indicatives as well. *Pratiṣedha* is extended from 'prohibition' to denote negation itself. *Prasajya-pratiṣedha* ('direct' or 'simple' negation) can refer to any negative expression where the negation is essential and the positive element secondary; *paryudāsa*, on the other hand, corresponds to a narrow-scope negative operator where the positive element is essential and the negation secondary or implied (Renou 1957:202, 230; Staal 1962:58; Matilal 1968:157; Sharma 1970:112–14).

Staal (1962:57ff.) attempts to show that LC fails to apply in the case of injunctions, whence the absence of any mention of the law in the Mīmāṃsā. Unfortunately, his argument is vitiated by an idiosyncratic, if not incoherent, logical notation. For *nekṣeta* 'he shall not-look', which must be assigned the *paryudāsa* (narrow-scope) reading following *Tasya vratam* (His vows are . . .), Staal assigns the representation **N**[∼**F**(**x**)], where **N** is the (logical? deontic?) necessity operator. So far so good. But *niṣedha* is interpreted as the negation of an injunction, and translated into ∼**N**[**F**(**x**)], as in 'he shall-not eat'. This notation suggests—wrongly—that the *niṣedha* or prohibition corresponding to a positive injunction is its contradictory opposite, equating in effect to 'he doesn't have to eat', 'the door needn't be locked', rather than to 'he shall not eat', 'the door should not be locked'.

Staal acknowledges (1962:58) that his rules require '∼**N**' to be analyzed as a 'prohibitive functor'; the problem is that no interpretation ever proposed for any system of modal or deontic logic yields a reading of prohibition for a negative operator outside the scope of a necessity or obligation operator.[69] Confusing the issue still further is the existence of a second (nominal) *paryudāsa* reading, for which Staal offers the formula **N**[**F**(∼**x**)], whatever that might mean.

Notwithstanding Staal's attempted argument, *niṣedha* and *paryudāsa* do not respectively constitute contradictory and contrary opposites of a given *vidhi* (injunction). In fact, *niṣedha* and *paryudāsa* represent two distinct contrary negations of the corresponding *vidhi*. The closest English equivalents to the two negatives of the Mīmāṃsā would involve **Aux** vs. **VP** negation (or in Aristotelian terms, predicate denial vs. predicate term negation), best distinguished in the corresponding pseudo-clefts:

(68) a. What A {must not (mustn't) do / should not (shouldn't) do / ought not do} is B. *(niṣedha)*

b. What A {must do / should do / ought to do} is not B. *(paryudāsa)*

Notice that in both cases, despite the clear meaning difference between (68a, b), the negation is contained within the scope of the modal operator; neither type corresponds to the (contradictory) negation of *What A must do is B*.

The question of how to capture this subtle distinction with formal deontic logic must at this point be left open, but a good place to start might be with Von Wright's two-negation deontic logic (1959: 27ff.), in which **O**(−**A**) 'it is obligatory to do not-**A**' is distinct from **O**(~**A**) 'it is obligatory not to do **A**', 'it is forbidden to do **A**'. This possibility will not be pursued here, although I shall return to Von Wright's logic of negation in a different connection in §2.4.

The Mīmāṃsākas speak to modern linguistic analysts not only on the issue of operator scope in nondeclarative sentences, but also on the importance of context in determining the interpretations of a statement or injunction. Thus a declaration like

(69) Not in the atmosphere, not in the sky shall he build the sacrificial fireplace.

does not constitute a prohibition, since prohibitions can only be given against actions that the addressee might have considered undertaking, and nobody would have thought of building an altar in the sky (Edgerton 1929: §342). By the same token, the negative injunction *He shall not kill* applies not to a man who spontaneously refrains from killing, but only to a man who is impelled to kill. The Mīmāṃsākas' general gloss on prohibition can be given as: 'What was regarded as a thing to be done, that is not to be done' (Edgerton 1929: §344).

While this observation may recall the Ducrot and Givón line on the 'discourse presuppositionality' of negative sentences (cf. the discussion of *My wife is not pregnant*), the Mīmāṃsākas are not opening an eastern front for the asymmetricalist legions. In fact, they go on to point out, positive injunctions are similarly constrained: we don't order a man to do something (e.g., to beat out the rice) if he was impelled to do it independently of our intercession (§344). What is involved here is what Searle calls a general preparatory condition on directive speech acts: S cannot felicitously order H to do A unless 'it is not obvious to both S and H that H will do A in the normal course of events of his own accord' (Searle 1969: 59–60, 66). This condition against pointless action is clearly connected with Grice's (1975) maxim of Relation and, as Searle suggests (Searle 1969; cf. also Searle 1982: 235), is ultimately derivable from Zipf's (1949) Principle of Least Effort (see Horn 1984b for related discussion).

Another instance of proto-Gricean exploitation of pragmatic inference appears in the grammatical commentary of Patañjali: 'By a restrictive condition (*niyama*) on what food is fit to be eaten is implied a prohibition (*pratiṣedha*) of what food is not fit to be eaten. . . . Or alternatively, by a prohibition of what is not fit to be eaten is implied a restrictive condition on what is fit to be eaten' (Chatterji trans., cited by Staal 1962: 64). Thus, if we are told that the domestic pig is not fit to be eaten, we can infer that the wild pig is fit to be eaten, and vice versa. In the same way, Patañjali comments, we can teach the correct grammatical forms directly (implying that alternate forms are incorrect) or indirectly (by teaching which incorrect forms are to be abjured).

Patañjali here seems to be alluding to the pragmatic principle of invited inference, and in particular to the rule of Conditional Perfection (Geis and Zwicky 1971). In order for the inference to go through, we must assume that the context can be filled in as suggested; it would not be a logical contradiction to extend the prohibition, as other religious traditions have done: *The domestic pig is not fit to be eaten, nor is the wild pig.* (Compare Geis and Zwicky on *If you mow the lawn I'll give you $10, . . . and in fact I'll give you $10 even if you don't.*)

Our tour of the proclivities of the Eastern mind concludes with a brief layover in China, where LC and LEM were both recognized in the Mohist canons (third century B.C.). Contradictory negation is characterized via a disputation over some unidentified object between 'one saying that it is an ox, the other that it is not': 'Their claims will not both fit, and if they do not both fit one necessarily does not fit. It is not like one's claim fitting a dog'. That is, the contrary opposition in which A claims that it is an ox and B that it is a dog, in which case they may both be wrong (Graham 1959:91). Note that as predicted by the Mohists' Stagirite contemporary, LC applies in both disputes, but LC only in the first.

LDN is alluded to in a sophisticated exposition of the Liar's paradox in the same text: one who considers all statements mistaken is mistaken. Evidence: his own statement. But not all statements can be accepted: one who rejects all denial is mistaken, since he must reject his own denial (Graham 1959:95).

I conclude this section by providing some comfort to the asymmetricalist forces. Wang Fu-Chih (seventeenth century) addresses himself to the Paradox of Negative Judgment, rejecting the Taoist doctrine of Nothing in favor of a (by-now-familiar) strategem wherein apparent negative propositions are really about a positive other. The concluding sentence of his argument must have soothed the restless spirit of Parmenides:

> One who says 'There is not' is provoked to denial by someone saying there is. He takes up what the other says and says there is no such thing. . . . If you say there is no hair on a tortoise, you are talking about (something on) a dog, not (nothing on) a tortoise. . . . A speaker must set something up before he can argue successfully. Now if he is to set a Nothing in front of us, he can search everywhere above and below, North, South, East, and

> West, in the past and the present, the surviving and the lost, without succeeding in getting to the end of it.
>
> (cited in Graham 1959:103–4)

Our tour of applied negation now leads us back to the West, where we shall touch down briefly in the domains of social and political theory, theology, and psychoanalysis.

1.3.2 Negation and Mind in the West: A Capsule View

> Traditional logic is based upon the law of contradiction, according to which **A** is not non-**A**. Hegel's entire *Logic* is built upon an ontological repudiation of this principle. (Friedrich 1953:xl)

While often taken to be an essential ingredient in the dialectic, the "repudiation" of LC by Hegel, like its jettisoning by the sages of Indian antiquity, is a sometime thing. Here is McTaggart (1922:8–9) on Hegel and the Law of Contradiction:

> It is sometime supposed that the Hegelian logic rests on a defiance of the law of contradiction. . . . Now if the law of contradiction is rejected, argument becomes impossible. . . . And indeed it is impossible, as Hegel himself has pointed out to us, even to assert anything without involving the law of contradiction, for every positive assertion has meaning only insofar as it is defined, and therefore negative. If the statement All men are mortal, for example, did not exclude the statement Some men are immortal, it would be meaningless. . . . If then the dialectic rejected the law of contradiction, it would reduce itself to an absurdity, by rendering all argument, and even all assertion, unmeaning.
>
> The dialectic, however, does not reject that law. An unresolved contradiction is, for Hegel as for every one else, a sign of error. The relation of the thesis and antithesis derives its whole meaning from the synthesis. . . . "Contradiction is not the end of the matter, but cancels itself". (Enz. Sec. 119, lect. note)

The Hegelian dialectic, far from simply rejecting LC, finds its motivation in the law.[70]

When two contradictories **p** and not-**p** cancel each other, the result in Hegel's system is not 'abstract identity'; the two contradictory opposites 'fall to the Ground', that Ground which contains 'identity as well as difference superseded' (Hegel 1892:223; cf. Kaufmann 1965:192). The key notion here is that of SUPERSEDING, *AUFHEBUNG.* The clash of positive thesis and negative antithesis produces an enriched synthesis in which the apparent contradiction is resolved at a higher level, *aufgehoben*—transcended, superseded, lifted up, or sublimated, according to the various imperfect English renderings of this term which Hegel wielded with joyful ambiguity:

> *Aufheben* . . . is one of the most important concepts of philosophy. . . . What sublimates itself does not thereby become nothing. *Aufheben* has . . . a double meaning in that it signifies conserving, preserving, and at the same time also making cease, making an end. . . . Thus what is *aufgehoben* is at the same time conserved and has merely lost its immediacy [like conserved fruit] but is not for that reason annihilated. . . . Something is *aufgehoben* only insofar as it has entered into a union with its opposite.[71]
> (Hegel's *Logic*, trans. and cited in Kaufmann 1965:192–93)

For Hegel, as we saw in §1.2.2, all significant negation occurs on a positive ground. But everything finite contains its own negation (*Determinatio est negatio*). The power of negativity is 'the life element of the Spirit and of Reason', 'the power to comprehend and alter the given facts in accordance with the developing potentialities by rejecting the "positive" once it had become a barrier to progress in freedom' (Marcuse 1954:433–34). The negation of this negation generates the forward movement of thought which constitutes historical progress.

Negation is central in the Marxist lexicon as well. The Hegelian dialectic, powered by negation and LC, is adopted—and adapted—by Marx and the Marxists. For Marx himself, the negativity of capitalist society resides in private property, that is, in the alienation of labor. The negation of this negation amounts to the abolition of alienated labor, the annihilation of private property (*Das Kapital* 1, chap. 24, sec. 7, cited in Bottomore 1983:352; cf. Marcuse 1954:282).

In Marxist as in Hegelian dialectics, double negation is always characterized as an *Aufhebung* in which the earlier contradictories are simultaneously preserved and superseded, rather than simply canceled out in the way LDN has operated in propositional logic since the Stoics. Engels views this dynamic conception of double negation as 'a general law of development of nature, history, and thought; a law which holds good in the animal and plant kingdoms; in geology, in mathematics, in history, and in philosophy' (*Anti-Dühring*, pt. 1, chap. 13, cited in Bottomore: 1983).

Similarly, for Lenin, negation must be taken 'as a moment of connection, as a moment of development, retaining the positive' (*Science of Logic* (1919), p. 226, cited in Bottomore: 1983). I can only conclude, with apologies to Goethe, that the dynamic force of negation has come to replace the *Ewig-Weibliche* as the spirit that *zieht uns hinan.*[72]

The knee-jerk response of many latter-day Hegelians and neo-Marxists against LC and indeed all laws of formal logic is probably best seen as an overreaction based on their disdain for classical historical explanation. Certainly, as Dancy (1975:24) points out, nothing in the dialectic or in Marxist thought hinges on the rejection of the standard axioms of formal logic—when these laws are applied to the domain for which they were intended. Similarly, Kierkegaard's instantiation of an apparent LC-defying conjunction—*Christ is a man and is not a man*—may also be seen as illustrating, not the defects of Aristotle's Law, but rather the importance of the 'customary qualifications' he explicitly builds into his statement of it (cf. my earlier discussion of Aristotle and the Jainists in §1.3.1). Like Superman, who both is and is not Clark Kent, but not at the same time in the same respect, Christ is a man in one respect and is not a man in another.

I have already touched (all too briefly) on the central role of negation in Buddhism and in Hindu 'negative theology'. In the West, mystics like Thomas à Kempis regarded God as the negation of the world (Royce 1917: 264). Divine intervention also plays a role in the asymmetry wars: for the true believer in the priority of affirmation over negation, only God can avoid negative judgments (recall the Bacon passage cited in §1.2.2). But even infinite minds must evidently resort to negation in instructing finite minds, as is clear from the canonical *Thou shalt not* form of the Commandments. With their negative character, observes Royce (1917:270), 'the Ten Commandments appear to make their appeal to an already more or less evil-minded, rebellious, or wayward people, whom the thunders of the law are to terrify into submission'. While Royce contrasts the negative nature of the Old Testament God of the Tablets with the more upbeat, positive tone of Jesus in the Sermon on the Mount, a significant rhetorical hallmark in the Sermon and throughout the synoptic Gospels is the recurring turn *not X but Y:*

(70) Do not store up your riches on earth, where moths and rust destroy them . . . but store up your riches in heaven. . . .
Judge not, lest ye be judged.
I tell you not to resist injury, but if anyone strikes you on your right cheek, turn the other to him too.
Do not think that I have come to bring peace to the earth. I have not come to bring peace but a sword.

It is not surprising that Jehovah and Christ, like the Vedic lawgivers, direct their negative commands to those who were prepared to act otherwise. As the Mīmāṃsākas recognized, there is no point in preaching nonviolence to a pacifist. The marked status of negative expressions makes them natural candidates for conveying guidance from above. Further, the linguistic polarity of positive and negative makes possible the expression of other dualities: 'Without negation, there would be no clearness with regard to values, no knowledge of heaven or hell, of good and evil' (Royce 1917:270).

But is there not an inner realm of the psyche in which duality has not yet emerged? Such, at any rate, is the claim of the theory of psychoanalysis (Freud 1910, 1925; cf. Buelens 1972). On the primary, infantile level, reflected in dreams and neuroses, there is no *not:* ' "No" seems not to exist as far as dreams are concerned. Anything in a dream can mean its contrary' (Freud 1910:155). Hearing the analysand insist of a dream character 'It's not my mother', the analyst immediately translates 'So it is his mother': 'In our interpretation we take the liberty of disregarding the negation and picking out the subject-matter alone of the association' (Freud 1925:235). On this level, it can truly be said that LC does not exist, for the simple reason that contradiction itself does not exist; ~**p** doesn't just presuppose **p** (as Givón's overcautious logic would have it), ~**p** asserts **p**!

Negation, on this theory, provides the means for the conscious mind to allow repressed material to penetrate. 'The content of a repressed image or idea can make its way into consciousness, on condition that it is negated. Negation is a way of taking cognizance of what is repressed' (Freud 1925: 235–36). Notice that negation thus viewed cannot be part of the objective content (or the subject matter) of what is said. Rather, we must line up with Kant and the Idealists and against Frege in maintaining the existence of a distinct 'negative judgment'. The task of the analyst is to address himself to 'conquering the negation', stripping away the veil, and revealing the repressed material beneath: 'A negative judgment is the intellectual substitute for repression; its "no" is the hall-mark of repression, a certificate of origin—like, let us say, "Made in Germany"'. Only this certificate is stamped 'Made in the Ego'.

On another level, the polarity of judgment corresponds to Freud's polarity of instinctual forces: 'Affirmation—as a substitute for uniting [the ego taking things into itself]—belongs to Eros; negation—the successor to expulsion [the ego rejecting the unwanted object]—belongs to the instinct of destruction' (Freud 1925:239).

A linguistic correlate of 'the dream-work's tendency to disregard negation' can be seen in 'the antithetical meaning of primal words' (Freud 1910). Freud adopts from the philologist Karl Abel (1882) the thesis that 'primitive languages' like ancient Egyptian, especially in the 'oldest' and most basic roots, contained a significant number of words which simultaneously denote two contrary notions (e.g., 'strong' / 'weak', 'light' / 'dark').[73] Examples adduced from Indo-European by Abel and Freud include Latin *clamare* 'to cry' vs. *clam* 'softly', *siccus* 'dry' vs. *succus* 'juice', Old English *bat* 'good' vs. Modern English *bad,* English *cleave* 'join together' or 'separate', German *stimme* 'dumb' vs. *Stimme* 'voice'. In addition to these instances of (self-acknowledged) *lucus a non lucendo* etymology, Freud cites the 'sound reversals' exhibited by pairs like German *Topf* 'pot' vs. *pot, Ruhe* 'rest' vs. *hurry, care* vs. *wreck,* and so on.

For Freud, this (putative) phenomenon serves to relate dream work and children's play, harkening back to that Elysian land of childhood before the reality of negation and contradiction rears its ugly head. But needless to say, no linguistic argumentation is offered to support these speculations, which are finally no more convincing than Freud's equally daring theory of the origin of language via the transfer of rhythmical utterances from the sexual act to work, investing the latter with some of the interest of the former (Freud 1924: 175).

It is significant that the antithetical word which has yielded the most philosophical mileage is one which hardly qualifies as "primal", in any sense of the word: Hegel's *aufheben* (see discussion above). For Hegel, as for Freud, 'the double usage of language, which gives to the same word a positive and negative meaning, is not an accident' (Hegel 1892: 180). The confluence of the two (or three?) meanings of this term may well constitute 'a joy for speculative thinking' (Kaufmann 1965: 192). But it is doubtful that either Hegel or Freud would argue for the inscription of *aufheben* or *Aufhebung* on the roster of *Urwörte*.

While Hegel's negation is a tool for generating a new and higher truth, Freud's negation is a tool for concealing truth (which the analyst reconstructs as a tool for revealing truth). But in each case, negation is secondary, something added on to an original positive or affirmative. One constant feature through all the various moral, psychological, historical, and social theories of negation I have touched on is this view of negation as a marked response to a basic positive. From the Vedic commentary of the Mīmāṃsā to the biblical commentary of Royce, from Hegel and Marx on the negative dialectic to Freud on the ego's negative censorship, the abstract logical symmetry of negation and affirmation—insisted on by Aristotle and Austin, by Frege and Quine, by Geach and Gale—gives way to a universe of discourse in which asymmetry is a given.

This basic asymmetry is spelled out by Kurrik (1979), who detects in the form and function of Chomsky's negation (presumably the Chomsky of *Syntactic Structures,* with its kernel sentences and optional meaning-changing transformations) an echo of Freud's:

> The simple negative transformation is a way to unsay what one is saying. Negation is addition because it has to include the positive statement it seeks to deny in its assertion. Negation superimposes itself on an assertion. . . . Negation is always tantalizing, pro-

> vocative, and ambiguous, a positive descriptive force which implies and promotes the very idea or thing that it seeks to deny. It is an absence yoked to a presence, or a presence-evoking absence. (Kurrik 1979:207)

Kurrik sees a significant difference in the grammatical role and psychological effect of negation as against deletion: 'Negation tries to "create" absence but fails, managing only to indicate it. Deletion, on the other hand, "creates" it'. Thus it is deletion which corresponds to repression, while negation is (as for Freud) merely a way to repackage the repressed material to render it more palatable: 'Consciousness appears only to be able to negate and remember, or to repress and be haunted by the repressed' (Kurrik 1979:208).

But this equation of psychic repression and grammatical deletion is only possible under a hopelessly naive conception of grammatical theory; it becomes evident, for example, that deletion is assumed here (Kurrik 1979:231) to be necessarily unrecoverable, hardly an orthodox view among syntacticians of any stripe. Kurrik's olla podrida of Freud, Hegel, Marx, and Chomsky, with a dash of the 'negative dialectics' of Nietzsche, Kierkegaard, Dostoevsky, and Beckett, manages in the end to be as 'tantalizing, provocative, and ambiguous' as negation itself.

Kurrik provides us here with one more instance of what we might call the Syntactic Fallacy (cf. Geach [1972] 1980:78–79): the argument from the greater grammatical complexity of the negative sentence to the greater complexity (and marginality) of the negative thought. Given its obvious appeal, it is worth trying to determine why this particular fallacy is so seductive. I return to this question in chapter 3.

I began my survey with an exposition of the major themes of the Aristotelian theory of negation, themes which will recur as the leitmotivs of my study. The Law of Contradiction and the Law of Excluded Middle, those first principles of both Aristotelian and Stoic logic, will be reencountered within the multivalued and truth-gap presuppositional logics touched on in chapter 2, where I shall expand the range of approaches to the issues of existential import, sortal incorrectness, and the scope of negation. The Asymmetry Thesis, with its forged (or forced) connections between the formal markedness of the negative statement and the allegedly superfluous, subjective, and second-class status of the negative judgment, will be reexamined in the light of psycholinguistic evidence in chapter 3; the pragmatic truce I propose there for the asymmetry conflicts reaffirms Aristotle's position that the inferiority of negation is a matter of epistemology, not of ontology or logic.

In chapter 4 I return to the Square of Opposition, focusing on the relation of subcontrariety; by treating the conversion between the 'verbally opposed' subcontraries of the **I** and **O** vertices of the Square in terms of Quantity-based conversational implicature, and retaining the classical semantic account of the subaltern inference from **A** to **I** and from **E** to **O**, I can capture the insights of Aristotle and his heirs while escaping the snares of analyses which recognize both inferences but fail to provide a theoretical mechanism for distinguishing them. The connection between Aristotle's two primary forms of opposition, contradiction and contrariety, lies at the heart of chapter 5, where I examine a variety of cases in which a certain range of formally contradictory negations are understood (through an independently motivated pragmatic strengthening rule) as acting contraries.

Finally, by chapters 6 and 7, the questions opened in this chapter can be reconsidered: Is negation ambiguous? If so, where does the ambiguity reside? Which readings should an adequate linguistic model distinguish, and how? Can all instances of negation in natural language be assimilated to logical operators, of whatever scope and character? What is the nature of semantically contradictory negation? Might there remain grounds for taking a page from Aristotle's term logic book to define wide-scope (sentential) negation as a mode of predication rather than, as in standard logical practice, as a one-place propositional connective?

In our pursuit of answers we must roam far beyond the Aristotelian purview, but it is doubtful whether the questions themselves could have been set in their current form had they not been broached by the Stagirite and subsequently explored by Avicenna, Spinoza, Frege, Russell, Lukasiewicz, Jespersen, Geach, Grice, Montague, and our other guides through the jungles of negation in this chapter and in the chapters to follow.

2 Negation, Presupposition, and the Excluded Middle

In my initial sorting of negation, we have seen the pretheoretical notion of opposition immediately splinter into two distinct categories, those of CONTRADICTORY VS. CONTRARY opposition, where the two categories are differentiated primarily by their interaction with the Aristotelian laws of contradiction and of the excluded middle. Contradictory opposition is defined by an adherence to both LC and LEM; contradictory statements are neither true together nor false together. Contrary opposition is defined by an adherence to LC; contrary statements (or terms) cannot simultaneously hold, but they may simultaneously fail to hold.

We observed that two singular expressions which appear to be contradictories may in fact be contraries, since the positive version (*Socrates is wise*) and its apparent negative counterpart may both be false, if the subject fails to refer and if the negation in question is read as the narrow-scope contrariety operator, Aristotle's predicate term negation (= *Socrates is not-wise*). Clearly in this case LEM does not obtain. Another member of the same class of constructions—those which seem to induce the assignment of LEM-violating contrary readings for apparent contradictory negations—is the category mistake, where a subject exists but its predicate cannot be 'naturally' predicated of it (*2 is red, 2 is not-red*).

As we shall see in this chapter, the vacuous subject case and the category mistake have both endured a bewildering variety of reanalyses over the twenty- three centuries since Aristotle took them on. These reanalyses typically abandon one or both of Aristotle's basic tenets—his premise that negation is ambiguous, and/or his insistence that the logic of narrow-scope (internal) negation, while different from that of contradictory negation, is nevertheless two-valued. Within one family of theories, the bivalence principle (often confused with LEM, although conceptually distinct from it), stipulating that every proposition is either true or false, is jettisoned in favor of the view that a proposition **p** may logically or semantically PRESUPPOSE a proposition **q**, such that in a context in which **q** does not obtain, the question of the truth or falsity of **p** does not arise. Whatever is presupposed by **p** will also be presupposed by the negation (or at least by <u>a</u> negation) of **p**.

We shall begin our journey by exploring neither vacuous subjects nor category mistakes, but yet-another context in which LEM and the principle of bivalence have been taken (by a group of scholars which may or may not include Aristotle) to fail.

2.1 Future Contingents: The Sea Battle and Other Skirmishes

Is every proposition either true or false? If so, is there a range of statements over which LEM must be suspended, so that a given statement **p** of this class and its apparent contradictory ~**p** may in certain circumstances both come out false? If not, what value is assignable when bivalence fails? I shall begin to address these questions by retracing the Aristotelian footprints once again to reach the still-controversial domain of future contingent propositions.

For Aristotle, truth-conditional logic is the study not of sentences but of propositions:

> Every sentence has meaning. . . . Yet every sentence is not a proposition; only such are propositions as have in them either truth or falsity. Thus a prayer is a sentence, but is neither true nor false. Let us therefore dismiss all other types of sentence but the proposition, for this last concerns our present inquiry, whereas the investigation of the others belongs rather to the study of rhetoric or poetry. (*De Int.* 17a1–8)

Thus, only declaratives may be true or false and hence express propositions. The rug under which Aristotle urged burying nondeclaratives has begun to fray only in the last few decades. But is it the case that all propositions (or declarative statements) must be true or false? In the long-notorious chapter 9 of *De Interpretatione,* Aristotle begins with the assumption that in general, one of a pair of contradictory propositions must be true and the other false in any state of affairs:

> For if all propositions whether positive or negative are either true or false, then any given predicate must either belong to the subject or not, so that if one man affirms that an event of a given character will take place and another denies it, it is plain that the statement of one will correspond with reality and that of the other will not. (*De Int.* 18a33)

Aristotle is arguing here for deriving LEM (the thesis that for any proposition **p**, either **p** or its contradictory ~**p** is true) from what I shall call the Law of Bivalence (the thesis that for any proposition **p**, either **p** is true or **p** is false). I shall return to the distinction between LEM and LBV below.

So far so good. But when we turn to predications relating to the future, 'the case is altered'. Take any two apparently contradictory future contingent statements, for example, (1) and (2) (19a30):

(1) There will be a sea battle tomorrow.

(2) There will not be a sea battle tomorrow.

Clearly, (1) and (2) cannot both be true; LC applies as vigorously to future contingents as to any pair of contradictories. But what of LEM? Can (1) and (2) be simultaneously false? Here is where the text becomes uncertain, and where the interpretations of *De Interpretatione* begin multiplying like a philosophical cancer. The difficulties culminate in this key passage with which Aristotle concludes and (apparently) summarizes his account:

> A sea-fight must either take place tomorrow or not, but it is not necessary that it should take place tomorrow, neither is it necessary that it should not take place, yet it is necessary that it either should or should not take place tomorrow. . . . One of the two propositions in such instances [e.g., (1) and (2)] must be true and the other false, but we cannot say determinately that this or that is false, but must leave the alternative undecided. One may indeed be more likely to be true than the other, but it cannot be actually true or actually false. It is therefore plain that it is not necessary that of an affirmation and a denial one should be true and other false. For in the case of that which exists potentially, but not actually, the rule [i.e., LEM] which applies to that which exists does not hold good. The case is rather as we have indicated. (19a30–b4)

It is unfortunately by no means clear just what has been indicated, nor is this made any clearer by variations in texts, systematic ambiguity in the Greek original, and the lack of formal devices for the essential scope disambiguation evidently needed here.

Lukasiewicz (1922, 1930, 1934) sees the argument in chapter 9 as involving not only a rejection of determinism (the thesis that if **p** is true tomorrow, **p** is necessarily true tomorrow and nothing we may do can alter the fact), but also a denial of what I am calling LBV, the Law (or Principle) of Bivalence: Every proposition is either true or false. Rather, according to Aristotle (according to Lukasiewicz), 'there are propositions which are neither true nor false but indeterminate'; as such, (1) and (2) 'are neither true nor false today' (Lukasiewicz 1922: 36–37; I shall follow Rescher 1963 in referring to this LBV-violating interpretation of chapter 9 as BOETHIAN, after one of its early adherents, the fifth-century commentator Boethius.) To such propositions, Lukasiewicz assigns a third truth value, **I** (for Indeterminate), distinct from the two classical (Aristotelian) values True and False.

But this analysis is strikingly less attractive for what we might term past contingents (or unknowables), that is, unverifiable and unfalsifiable statements about the past (e.g., *Aristotle ate no breakfast the day he died*). Nor is it any more appealing in the case of present unknowables, ignored by Lukasiewicz but quite familiar to the medievals, whose instantiation of choice was *The number of stars is odd (even)*. A more contemporary example is provided by Quine (1981:91), who is able and willing to shoulder—in the cause of the loyal order of the 'stalwarts of two-valued logic'—the burden of responsibility for the 'harboring of undecidables': 'We [stalwarts] declare that it is true or false that there was an odd number of blades of grass in Harvard Yard at the dawn of Commencement Day, 1903'. It is curious that the same man who vilified his contemporaries for infecting their philosophical logic with epistemology (Lukasiewicz 1934: 84) saw nothing amiss in his own advocacy of a hybrid system with two truth values and one 'indeterminacy' value.[1]

The view that Aristotle sought to consign future contingents to a 'truth-status limbo' (Rescher 1963:43), rejecting LEM and/or LBV, is not unique to Boethius and Lukasiewicz. Similar readings were standard among the Epicureans, who defended this position, and among the Stoics, who opposed it. The inviolability of LEM and LBV were especially dear to the Stoic Chrysippus, whence the claim that systems rejecting these laws are more accurately labeled 'non-Chrysippean', as in Lukasiewicz's practice, rather than 'non-Aristotelian', as in the slur of the General Semanticists (cf. Korzybski 1933; Hayakawa 1949). Aristotle's apparent derivation of determinism from LBV did not affright the Stoics, who were committed fatalists (on this and related issues, cf. Lukasiewicz's valuable history of LBV, appended to Lukasiewicz 1930).

From Ammonius and Boethius to Linsky and Prior, Aristotle has been seen as fitting neatly into the non-Chrysippean niche: every proposition is either true or false, except when it's a future contingent (or past unknowable, or present undecidable?). But a different interpretation of chapter 9 has been convincingly defended by Kneale and Kneale (1962:214) and Rescher (1963), and they too have centuries of precedent on their side. Here is Abelard's position (*Dialectica*, 210–22), as presented by the Kneales:

> No proposition *de contingenti futuro* can be determinately true or determinately false . . . , but this is not to say that no such proposition can be true or false. On the contrary, any such proposition is true if the outcome is to be true as it states, even though this is unknown to us.

This interpretive tradition may have originated with al-Fārābi (ca. 900) and was clearly accepted by Averroes, Saint Thomas (cf. Oesterle 1962, lectures 13–15), Duns Scotus, and Occam (Rescher 1969:45). As Rescher observes, these Muslim and Christian interpreters were faced with the task of reconciling the Philosopher's truth with God's truth, the latter encompassing both free will and divine foreknowledge.[2]

While it is ultimately impossible to know how Aristotle would have chosen to translate his argument into modern modal logic, Rescher makes a strong case for this latter, FARABIAN, interpretation, a case also supported by earlier writers, including Anscombe (1956) and Strang (1960). On this account, the exceptional nature of future contingents affects, not their truth status as such, but their necessary truth or falsity (cf. Ackrill 1963:140–41). Thus, the tenability of LBV (formalized as in (3a)) or its cognate LEM (given in (3b)) is never at issue.[3]

(3) a. $\mathbf{T(p)} \vee \mathbf{F(p)}$ (Every proposition must be true or false)
b. $\mathbf{T(p)} \vee \mathbf{T({\sim}p)}$ (Of a proposition and its contradictory, one must be true)

Rather, what is rejected (for instantiations of **p** as a future contingent) is the move from truth (falsity) to necessary truth (necessary falsity), as in (4):

(4) a. $\mathbf{T(p)} \rightarrow \Box\mathbf{T(p)}$ (or $\mathbf{T(p)} \rightarrow \Box\mathbf{p}$?)
b. $\mathbf{F(p)} \rightarrow \Box\mathbf{F(p)}$ (or $\mathbf{F(p)} \rightarrow \Box{\sim}\mathbf{p}$?)

Similarly, in the passage from *De Interpretatione* 19a30–b4 cited above, Aristotle can be read as rejecting the unqualified acceptability of the disjunction in (5), while endorsing the apodeictic proposition in (6):

(5) $\Box\mathbf{p} \vee \Box{\sim}\mathbf{p}$

(6) $\Box(\mathbf{p} \vee {\sim}\mathbf{p})$

Aristotle is indeed emphatic in opposing the view that contradictory future contingents can be simultaneously false (or not true): 'To say that neither the affirmation nor the denial is true, maintaining, let us say, that neither will take place nor will not take place, is to take up a position impossible to defend' (18b17–19). If (what is not certain) we can take Aristotle to be speaking with his own voice here, rather than as a representative of one of the several straw men of this chapter, he does seem to be claiming that either (1) or (2) is true today, but that whichever one is true is not determinately true and (without foreknowledge) cannot be known to be true. The Boethian reading, on which Aristotle assigns a third truth value (or no truth value) to future contingents, may thus stem from a misreading, although its philosophical interest may be none the less for this error, as

Rescher (1963:51) observes (and as the deconstructionist theory of creative misreading would predict).

From a Farabian perspective, those nonclassical logicians (e.g., Lukasiewicz 1930, 1934) who trace their own nonbivalence to Aristotle are as unjust to the Stagirite as is Quine, that two-valued classical heir of the Stoics, who dismisses as 'Aristotle's fantasy' the thesis 'that "It is true that p or q" is an insufficient condition for "It is true that p or it is true that q"' (Quine 1953a:65). What the Farabian attributes to Aristotle is the far-less-fantastic thesis that 'It is necessary that p or q' is an insufficient condition for 'It is necessary that p or it is necessary that q'.[4]

The contributions of the Scholastic participants in the battle over future contingents are evaluated by Michalsky (1937:285–301) and Baudry (1950). Rescher (1963:51–54) provides a comprehensive annotated bibliography on the Sea Battle, ranging from Cicero's defense of the Stoic determinists to a series of logical and historical analyses of the early 1960s. Another useful discussion of the literature and the issues is offered by Ackrill (1963:132–42), whose translations of the relevant passages from chapter 9 are significantly different from the Oxford version (by Edghill) cited above. While it may be a notorious truism of interpretation that the translator's hand must carry the baggage of its owner, Edghill's Farabian sympathies are particularly apparent. I shall explore some implications of the Farabian reading in §6.2 below.

2.2 Vacuous Singular Terms: From Socrates to the King of France

> By the law of the excluded middle, either 'A is B' or 'A is not B' must be true. Hence either 'the present king of France is bald' or 'the present king of France is not bald' must be true. Yet if we enumerated the things that are bald and the things that are not bald, we should not find the king of France on either list. Hegelians, who love a synthesis, will probably conclude that he wears a wig. (Russell 1905:485)

As we saw in chapter 1, the earliest disputes over the logic of nondenoting or vacuous singular terms involved, not the alopecia of the nonexistent French monarch, but the indisposition of an incorporeal Socrates. While Aristotle may or may not have argued for a truth-value gap in the case of future contingents, depending on whether one reads him with Boethian or Farabian lenses, his two-valued stance toward vacuous singular expressions is clear. It will be recalled (from §1.1.1) that if Socrates exists, affirmations concerning him, for example, (7a) or (7b):

(7) a. Socrates is sick.
 b. Socrates is well.

(8) a. Socrates is not sick.
  b. Socrates is not well.

(9) a. Socrates is not-sick.
  b. Socrates is not-well.

may be true or false, although mutually contrary affirmations like these may not be simultaneously true. If he does not exist, such affirmations are automatically false. The corresponding contradictory negations, that is, (8a) and (8b), are then automatically true; these 'predicate denials' are not to be confused with the predicate term negations (9a) and (9b), which are in fact affirmations (in which a negative term, *not-sick* or *not-well,* is affirmed of Socrates) and hence (like the positive affirmations in (7)) false in the same vacuous context. Thus affirmations, with either positive or negative predicate terms, entail the existence of their subjects, while negations (predicate denials) do not. For Boethius and other dissenters from this 'qualitative' approach to the problem of nondenoting singular terms, (7a) and (7b) are not contraries, but contradictories. (Cf. §1.1.3 for related discussion.)

Bradley (1883: 119ff.), who starts from an Aristotelian view of contradictory negation, seems to have been responsible for injecting royalty into the debate, with the usual troubling results:

> Sokrates may be not sick because he is well, or because there is now no such thing as Sokrates. . . . 'The King of Utopia did not die on Tuesday' may be safely contradicted. And yet the denial must remain ambiguous. The ground may be that there is no such place, or it never had a king, or he is still living; or, though he is dead, yet he died on Monday.

Thus, when the singular term fails to denote, both the denial and its denial come out true. But something is clearly amiss if the same positive knowledge (that is, that there is no king of Utopia) can serve as the ground for each member of a pair of (apparent) contradictories.

Bradley's fellow neo-Hegelian Idealist Sigwart rejects Aristotle's truth-functional, entailment-based analysis of sentences with empty singular terms in favor of an approach building on the suggestive, but somewhat inchoate, notion of presupposition; notice the proto-Strawsonian flavor of the concluding clause here: 'As a rule, the judgment A is not B presupposes the existence of A in all cases when it would be presupposed in the judgment A is B . . . 'Socrates is not ill' presupposes in the first place the existence of Socrates, because only on the presupposition of his existence can there be any question of his being ill' (Sigwart 1895: 122). Yet the Aristotelian asymmetry between affirmative and negative cases is retained, recast as an asymmetry of presupposition rather than of entailment: 'Since the negation only declares it to be false that Socrates is ill, the presupposition contained in it is certainly not so definite as in the affirmative judgment 'Socrates is ill'; for this may also be denied because Socrates is dead' (Sigwart 1895:124). Indeed, for Sigwart a prima facie contradiction like *The fire does not burn* can be true only vacuously, that is, when there is no fire.

Elsewhere, Sigwart (1895:152) offers an ordinary language argument for rejecting the Aristotelian position on which (7a) has (7b) (or the putatively equivalent (9a)) as its contrary and (8a) as its contradictory. He points out that (8a) is 'commonly understood' to signify 'Socrates does live but is ill'. Furthermore 'if we answer the question "Is Socrates ill?" by yes or no, then—according to our usual way of speaking—we accept the presupposition upon which alone the question is possible; and if we say of a dead man that he is not ill, we are guilty of using our words ambiguously'. Technically, however, Aristotle may be deemed correct: 'We must admit . . . that formally, the truth of the proposition [8a] is incontestable' if Socrates is not alive. Sigwart's conclusion that in a given context, a given statement may be true but misleading or inappropriate is very much in the spirit of somewhat later work on pragmatic presuppositions (cf. Stalnaker 1974; Bergmann 1977; Gazdar 1979a; Karttunen and Peters 1979).

Bosanquet ([1888] 1911:287–88) advocates a similar line on the status of negative judgments with empty subjects. Given the pair (10a, b),

(10) a. The house on the marsh is burnt down.
b. The house on the marsh is not burnt down.

he allows that (10b) must be considered true if there is no house on the marsh, although reality 'excludes the burning down of any such house'. In these circumstances, (10a) would be reckoned false, as in the Organon. Bosanquet confesses a 'strong sympathy' for the objection (straw or real) that an assertion like (10a) is 'not so much false as unmeaning' in such a case, so that its negation (10b) 'has meaning only if there is a house and presupposes or asserts that there is one'. But, he concludes, an unmeaning judgment is clearly not true'. (It will be observed that Bosanquet implicitly identifies the not true with the false, a move which would be seconded by Russell but not by the logical presuppositionalists of our own century.)

The first incorporation of a presuppositional account of singular terms into a formal semantic model is due to Frege (1892).[5] In his classic paper on sense and reference, Frege argues that both (11a) and its contradictory (11b) presuppose (*voraussetzen*) that the name *Kepler* denotes something.

(11) a. Kepler died in misery.
b. Kepler did not die in misery.

This presupposition is associated with periphrastic descriptions as well as names, so that (12) shares the presuppositional properties of (11).

(12) Whoever discovered the elliptic form of the planetary orbits {died / didn't die} in misery.

The presupposition of existence associated with singular terms is employed, along with the distinction between sense and reference and the doctrine of semantic compositionality, in the construction of an elegant argument for the counterintuitive conclusion that a sentence refers to its truth value. Let us grant, with Frege, that meaning (in particular, both sense and reference) must be compositional, in that the sense (reference) of an expression is a function of the sense (reference) of its parts. What then is the x such that a sentence, which expresses the proposition constituting its sense, refers to (denotes) x? Frege points out that x must be some entity which is elusive or unidentifiable just when the reference of the components of a sentence is elusive or unidentifiable. But a sentence like (13)

(13) Odysseus landed at Ithaca.

can be true or false (given Frege's line on (11)) only if the name *Odysseus* has a reference. Thus, 'we are driven' into concluding that x is the truth value of the sentence, and that (13)—along with all other declarative sentences, refers to one of the two truth values, the True or the False, or it refers to nothing at all (Frege 1892:62–63). Thus, if the presuppositions of any of the parts fail (e.g., if *Odysseus* does not refer), the presupposition of the whole will fail as well, and the sentence will induce a reference failure. Further, ignoring the complications brought in by intensional or opaque contexts (Frege 1892:66ff.), the substitution of a term with identical reference (*Penelope's husband* or *Polyphemus's slayer* for *Odysseus*) does not affect the reference (i.e., the truth value) of the sentence in which it appears.

Thus every sentence (affirmative or negative) with a singular subject (name or description) presupposes the existence of a (presumably unique) referent for that subject. But this presupposition is not part of the content of the expressions in question, and hence a sentence like (11a) does not entail the existence of Kepler—else the negation of (11a) would not be (11b), which preserves the presupposition, but rather (via De Morgan's Law) the disjunction in (14):

(14) Kepler did not die in misery, or the name *Kepler* has no reference.

By the same token, the corresponding periphrastic case of (12) would find its negation in the even clumsier presupposition-free disjunction of (14′):

(14′) Either whoever discovered the elliptic form of the planetary orbits did not die in misery or there was nobody who discovered the elliptic form of the planetary orbits.

While Frege (1892:68–70) seems to have viewed this option as a prima facie absurdity, its citation curiously prefigures the later emergence of a presupposition-canceling external negation operator with truth conditions precisely equivalent to those of disjunctions like (14) and (14′).

The dichotomy of internal vs. external negation, while harking back to the two negations of Aristotle (cf. §1.1 and chapter 7 below), emerges in its contemporary guise in the work of Russell (1905:490ff.). Seeking to unravel the puzzle with which I introduced this section, Russell urges the banishment of descriptions like *the king of France* from logical form. Once this exorcism is performed, sentences like (15) and (16) will no longer be analyzed as of subject-predicate form, their surface syntax notwithstanding.

(15) The king of France is bald.

(16) The king of France is not bald.

(15) emerges instead as the (false) proposition that there is one and only one entity which has the property of being king of France, and that this entity is bald; this is the existentially quantified conjunction we can represent as in (15′).

(15′) $\exists x(Kx \wedge \forall y(Ky \rightarrow y = x) \wedge Bx)$

But there are two different ways of unpacking the corresponding negative sentence, (16). If the description *the king of France* has a PRIMARY occurrence, we get the (false) proposition that there is one and only one entity which is king of France and is not bald, that is, (16′):

(16′) $\exists x(Kx \wedge \forall y(Ky \rightarrow y = x) \wedge \sim Bx)$

For Russell, (16′) is 'simply false' in the absence (or oversupply) of male French monarchs. But Russell admits a second reading of (16), equivalent to the proposition that it is false (or not true) that there is a unique entity which is king of France and is bald. Here, the description has a SECONDARY occurrence, within the scope of negation, as indicated in the logical form in (16″);

(16″) $\sim\exists x(Kx \wedge \forall y(Ky \rightarrow y = x) \wedge Bx)$

This results in a true proposition when France is a republic. The reading sought here is the one which is favored with the appropriate intonation contour and rectification:

(16!) The king of France isn't bald—there isn't any king of France!

I shall return to these linguistic correlates of Russellian external negation in chapter 6.

The negation in the logical form (16′) is internal to the basic proposition, while that in (16″) is external to it, whence the familiar labels for the two scopally distinct variants which have long since replaced Russell's terminology of primary vs. secondary occurrences. Notice that (16″), unlike (16′), fails to entail (17);

(17) There is a king of France.

Indeed, the falsity of (17) guarantees the truth of (16″).

What Russell has done here (and in later reworkings of his theory of descriptions) is to formally reconstruct Aristotle's truth-conditional theory of negation, in which contradictory predicate denial (as in (16″) = *The king of France [is not] bald*) is distinguished by scope from contrary predicate-term negation (as in (16′) = *The king of France is not-bald*). It may have been an accident that Russell apparently overlooked this precedent—although, given Russell's oft-voiced contempt for Aristotle as a logician, such an oversight was predictable. In any case, it was not accidental that Russell illustrated his analysis with an example based on a description rather than (as with Aristotle and Frege) with one based on a proper name.

Confronted with the familiar Aristotelian and Fregean negations (e.g., (8a), (11b)), Russell must treat what he dubs 'logically improper' names (Socrates, Kepler) as disguised definite descriptions (= 'the x such that . . .'), en route to translating each name into the description it 'abbreviates', and finally unpacking the resultant description into an existentially quantified conjunction as illustrated above.

If we accept the cogent arguments of Kripke (1972) against analyzing names as disguised descriptions (on any version of such theories), the only obvious alternative for preserving the Russellian line on the ambiguity of negative statements containing names (as well as descriptions) would require the conversion of such names into predicates. This move is vigorously supported, as it happens, by Quine (1948:7–12), who analyzes *Pegasus is winged* into 'The thing which pegasizes is winged'. On this approach, Aristotle's (8a) would be taken (on its 'primary' interpretation, with internal negation) as asserting that there is one and only one entity that socratizes, and that this entity is not sick.

Quine advocates this method for eliminating names by converting them into descriptions based on phantom predicates as a principled tool for ontological slum clearance: we can deny Pegasus's existence without presupposing that he exists and thus admitting nonexistent entities into our ontology. But there is—as our society has come to recognize—a price to pay whenever slums are cleared. In the present case, we are no longer capable of treating even the most obvious instances of subject-predicate sentences as such. Nor can we take much pride in barring Pegasus from the front door only to have the property of pegasizing fly in through the window.

Whatever we may decide as to the relative merits of propositional vs. term logic (cf. §7.2 for a reconsideration of this question), it is clear that the ambiguist line on negation strongly motivates the latter approach.[6] Henry (1972:74) forcefully sums up the difficulties for a Russellian theory of descriptions which grants the existence of two negations:

> The distinction which has evidently been desirable all along is introduced in a tortuous and *ad hoc* fashion under the misleading guise of the 'primary and secondary occurrence' of descriptions and all names have to be construed as disguised descriptions in order to be able to take advantage of this *ad hoc* distinction.

But the predicate term negation of Aristotelian term logic, unlike Russell's internal negation, can be truth-functionally distinguished from its (contradictory) counterpart without forcing the scopal analyst to cough and look the other way as proper names are transsubstantiated into descriptions based on otherwise nonexistent predicates.

Other commentators on the status of negative sentences with nondenoting singular subjects, especially those operating within the strictures of two-valued propositional logic, have tended to overlook or dismiss one or the other of the twin forks of Russell's ambiguity. Thus Wood (1933:421) simply reduces (16) to the statement that 'It is false that the king of France is bald'. On such an account, negation is necessarily external, and no existential commitment (as Fregean presupposition or Russellian entailment) can be inferred. Collinson (1937:89), on the other hand, places the subject outside the scope of negation, so that negative as well as affirmative singular statements entail existence: 'When we assert or deny (e.g.) redness of a subject, the subject remains unimpaired'. (It may be significant that Wood is writing as a philosopher and Collinson as a linguist. Aristotle and Russell were simultaneously philosophers and linguists, whence the insights and inconsistencies of their analyses.)

Reichenbach, on the other hand, accepts the Russellian package in toto, claiming for it 'the advantage that such statements as "The present king of France is forty years old" need not be regarded as meaningless, but are simply false; and that they can even be made true by the addition of a negation outside the scope' (Reichenbach 1947:263).

The 'simple' falsity of sentences like (15), as perceived by Russell,

Reichenbach, and of course Aristotle before them, was later to suffer an historic collision with the intuitions of Strawson (1950, 1952) and his Oxonian colleagues. Strawson agrees with Russell and Reichenbach that (15) is meaningful.[7] Meaningfulness and meaninglessness are for him, however, properties of sentences, while reference—and truth value—are properties of the statement the sentence may be used to make.

As with the analysis of Frege and the citation from Collinson above, Strawson's celebrated attack on Russell's theory of descriptions is premised on the assumption that negation, normally or invariably, leaves the subject 'unimpaired'. For Russell, (16)—on its 'primary' reading, the internal negation (16′)—comes out false in the absence of a French king; for Frege, the utterance of the analogous (11b) makes no assertion if there was no Kepler. For Strawson, someone who utters (16) does commit herself to the existence of a (unique) king of France but, contra Russell, she does not thereby assert (nor does her statement entail) the corresponding existential proposition (17). Rather, (16)—along with its positive counterpart (15)—PRESUPPOSES (17). If this presupposition is not satisfied, neither (15) nor (16) can be judged true or false. A statement is indeed made under these circumstances, pace Frege, but the question of its truth value 'fails to arise'.

Strawson tacitly lines up with Frege, and against Russell (and Aristotle), in regarding negative singular statements like (16) as essentially unambiguous. He does, however, recognize the marginal existence of a nonpresupposing negation, citing the exchange in (18),

(18) A: Does he care about it?
B. He neither cares nor doesn't care; he's dead.

where the fact that B's reply posits a nonexcluded middle testifies to the contrary nature of the opposition between *He cares* and *He doesn't care* (Strawson 1952: 18). (It will be recalled that contrary opposites, but not contradictory opposites, allow for the existence of a third proposition inconsistent with both, in this case the proposition that he's dead.) But in this passage, as in the seminal paper (Strawson 1950), the passing of the subject into the great beyond renders any statement about him not false (as it would for Aristotle or Russell) but rather immune to concerns of truth and falsity (barring, one presumes, resurrection).

If a statement is made which is meaningful, but which by virtue of reference failure in its subject term is neither true nor false, is this equivalent to the formal device of assigning a third truth value, distinct from the classical two values of the Aristotelian and Russellian programs? Is there simply a GAP, in effect a truth-conditional black hole, at the point where truth values are normally assigned? Are these instances of statements with vacuous subject terms collapsible in some sense with meaningless or ungrammatical sentences (and perhaps with future contingents), given that in these cases the question of truth or falsity also (arguably) fails to arise?

Strawson himself, ever skeptical that any system of formal logic could do justice to his intuitions about truth and meaning in ordinary language, would probably shrug off these questions. Yet, as we shall see in §2.4, his skepticism did not dissuade other philosophers and linguists from adapting Lukasiewiczian multivalued logics to represent a variety of presuppositional analyses of (inter alia) the vacuous subject term examples of (15) and (16). The first thirty years after the publication of Strawson's "On Referring" witnessed a rapid proliferation of three (or more)–valued logics in which truth-value gaps arise or nonclassical values are ascribed, that is, in which meaningful declarative statements can be made which in at least some contexts are assigned neither of the classical values **T** or **F**. Ironically, these neo-Strawsonian formal accounts of presupposition consistently assume an ambiguity for negation, a position advocated by the archclassicists Aristotle and Russell but never explicitly endorsed by Strawson himself.

2.3 Category Mistakes: The Significance of Insignificance

In the Organon, Aristotle repeatedly offers two reasons why certain apparently contradictory pairs of statements, such as those in (19a, b) or in (20a, b), are in fact opposed as immediate contraries rather than as contradictories.

(19)	a. α is healthy.	(19′)	a. α is not healthy.
	b. α is sick.		b. α is not sick.
(20)	a. α is even.	(20′)	a. α is not even.
	b. α is odd.		b. α is not odd.

The first reason is that α does not exist, in which case (as we have seen) both members of the pair are false. But even if α exists, it may be the sort of thing—the sort of subject—to which neither the (a) nor the (b) predicate 'naturally' applies. If α is a number, both (19a) and (19b) will be false (and their respective contradictories (19′a) and (19′b) true). If α is Socrates, (20a) and (20b) will both be false (on the relevant readings of the predicate terms), and (20′a, b) true. Similarly for predicate term negation: α is neither **P** nor not-**P** if α does not exist, <u>or</u> if **P** (and hence not-**P**) expresses a property which cannot be predicated of α (or equivalently, if α is not the sort of thing of which **P**, and hence not-**P**, can be predicated).

The class of negative propositions in which some property is denied of a subject not eligible to receive it has been encountered earlier in this study. In §1.2 I traced the notion of INSIGNIFICANT negation back to the *negatio* of Spinoza's *The wall does not see;* (21) and (22) are the insignificant negations of choice for the Idealists (cf. Sigwart 1895; Bosanquet [1888] 1911; Mabbott 1929).

(21) Virtue is not square.

(22) The soul is not a fire shovel.

These are Mabbott's Turkey-carpet judgments, negations which are insignificant because, as Hegel had observed, they do not 'narrow the field' (Mabbott 1929:68). We also recall Price (1929) defending these same judgments as 'perfectly sensible and indeed true'.

But what is the logical status of such insignificant negations? A quick tour of the recent literature on the topic might begin with the defense of Price's position by Ewing (1937:359–64) and Prior (1954). Ewing acknowledges true meaninglessness—incomplete and presumably ungrammatical sentences like *Cambridge is between York,* jabberwocky sentences whose components are semantically undefined, and word-salad sentences like *Are of fond not dogs cats*—but he argues that negative instances of type crossing or category mistakes (CMs) like those of (21) and (22) are not among them.[8] Sentence (23), for example, must express a proposition and not 'a meaningless set of words':

(23) Quadratic equations do not go to race-meetings.

since there are propositions which it entails (e.g., (24a)) and others which entail it (e.g., (24b)).

(24) a. Quadratic equations do not watch the Newmarket horse-races.
b. Quadratic equations do not move in space.

Besides, how could we know that quadratic equations do not attend horse races unless (23) were a proposition? Ewing considers and responds to a potential objection here: 'No doubt if I frequently made assertions such as [22] or [23] I should be in danger of being consigned to an asylum, and it may be asked why I should be regarded as a lunatic because what I say is true. The answer is that to qualify as a lunatic it is not necessary to say what is false or meaningless; it is sufficient to say what is true in an unsuitable context' (Ewing 1937:360–61).

The point is well taken. Yet we may be willing to grant Ewing's conclusion that (22) and (23) are (not meaningless but) true, and the corresponding affirmations (not meaningless but) false, without accepting his move (p. 361) of assimilating the oddness of these CMs to the oddness associated with the assertion, in a neutral context, of such indubitably true propositions as those in (25):

(25) a. 2 plus 2 equals 4. (uttered when totally irrelevant)
b. I did not commit more than six murders yesterday. (given that I murdered no one)
c. I worked an hour yesterday. (given that I worked eight hours)
d. He has not stopped beating his wife. (given that he never started)

Negative category mistakes (NCMs) may well be both true and pragmatically deviant, like the examples in (25) (cf. chapter 4), but they suffer from another defect which Ewing has not succeeded in isolating.

Prior (1954:159–60) offers a proof for (21) in support of Ewing's line on NCMs. In saying that virtue is not square, just as in saying that my left eye is not square, I am not saying that virtue (or my left eye) is of some other shape (Plato's Eleatic Stranger to the contrary notwithstanding); in both cases, the negation is true simply because the entity in question lacks the requisite property of squareness.

The Aristotle-Price-Ewing-Prior position that NCMs are not just meaningful but true (and their positive counterparts simply false) is also supported by Quine (1953b:449), who laments 'the recurrent notion among philosophers that a predicate can be significantly denied only of things that are somehow homogeneous in point of category with the things to which the predicate applies'.

Elsewhere, Quine (1960:229) addresses 'the concern among philosophers to declare meaningless, rather than trivially false' such classic positive category mistakes as (26a) and (26b), due to Carnap and Russell, respectively.

(26) a. This stone is now thinking about Vienna.
b. Quadruplicity drinks procrastination.

He finds this concern misplaced, attributing it largely to nothing more than 'a spontaneous revulsion against silly sentences'. Better to eschew any sorting among the class of statements which are unarguably not true: 'Tolerance of the don't cares . . . is a major source of simplicity in theory; and in the present instance it counts double, sparing us as it does both the settling of categories and the respecting of them'.

The class of philosophers under attack (or at least reproach) here includes Strawson, whose definitions of 'incompatibility-ranges' (1952:6) presuppose a theory in which NCMs are neither true nor false. For Strawson, a class expression like *green things* has both an extension (the class of things that are green) and an EXCLUSION (the class of things which are not green, or in Aristotle's terminology not-green, including cherries and crows, but not prime numbers or cardinal sins). The extension and exclusion of a class expression jointly determine its universe of discourse. Any entities about which 'it does not make literal sense to say that they are or are not green' fall outside the universe of discourse, beyond the pale, and into a truth-value gap.

But NCMs, as well as their affirmative counterparts, are not meaningless for Strawson; rather, as with the king-of-France statements of section 2.2, the question of their truth or falsity fails to arise. Lest one conclude that the critical remarks of Quine and other opponents of the meaninglessness analysis of category mistakes might have been directed, not at Strawson, but at Strawman, it should be noted that it was Russell (1908) who most clearly enunciated the true/false/meaningless trichotomy and who placed category mistakes—positive and negative alike—into the third class, through the application of type theory. Thus, *A is/is not B* is meaningless if A does not belong to the appropriate type defined by the predicate B. It is easy to share Quine's skeptical evaluation of the Russellian theory of types—or any similar notion of type or category—as a tool for making just the right cuts in predicting category clashes between subject and predicate. It would not require the surgical skills of a Dwight Bolinger or a Jim McCawley to demonstrate that a suitable type theory necessitates virtually as many categories as there are predicates.

The Russellian account of category mistakes, however, finds a champion in Pap, who adopts a Platonic perspective: 'In ordinary parlance a rejection of a statement of the form "x is blue" as false is equivalent to the assertion that x has some colour other than blue' (1960:41). Therefore, NCMs like (27a, b),

(27) a. The square root of 2 is not blue.
 b. The theory of relativity is not blue.

since they cannot be read as assertions of otherness (= 2 is green or red or yellow or . . .), are in fact (not true but) meaningless. Pap is thus a legitimate heir to the negation-as-otherness line established by Plato's Stranger, developed (via the notion of significant vs. bare or insignificant negation) by Spinoza, Hegel, and the Idealists, and reinterpreted (through the equation of significant negation to the assertion of an implicit positive disjunction) by Mabbott; recall the synopsis of this history in §1.2.2.

Pap's argument that NCMs are not true hinges on a controversial assumption: 'The negation of a meaningless sentence is surely itself meaningless: the relevant sense of "meaninglessness" here is "neither true nor false"'. By LC and LEM, any proposition (category-mistaken or not) and its negation share (non)bivalence; if *The theory of relativity is blue* is meaningless and hence (for Pap) neither true nor false, so is its negative counterpart (28b). Negation, represented as '**not-(S is P)**', is thus 'ordinarily construed as limited', equivalent to '**S is non-P**', where **non-P** is understood, à la Mabbott, as 'the disjunction of all the other predicates belonging to the same family as P': 'To deny that x is kind is to affirm that x is unkind'.[9]

Having drafted this notion of a PREDICATE FAMILY—'a set of predicates such that one and only one member of it must be true of anything of which some member of the set is true or false'—Pap (1960:48) enlists it in his campaign to recast Russell's type-theoretic account of category mistakes in terms of a Strawsonian presuppositionalist analysis. Pap acknowledges that the effect of presupposition failure in the CM cases (*Socrates is/is not a prime number*) differs from that in the standard cases of reference (*The king of Switzerland is/is not a pipe smoker*) and change-of-state verbs (*Mr. Miller has/has not stopped beating his wife*) in that, on Pap's own testimony, the first is meaningless while the others clearly are not. Pap attributes this difference to the necessary nature of the presupposition failure in the former case and its contingent nature in the other two; but this cannot be the whole story, as seen in the contrast between (28a, b), both of which involve a necessary instance of presupposition failure.

(28) a. The largest prime number is odd (divisible by 3, prime, . . .).
b. The largest prime number is blue (happy, unfair, . . .).

Whatever the resolution of these problems, Pap's presuppositional treatment of 'limited', type-internal negation provides only part of the NCM picture. Unlike Strawson, Pap reluctantly acknowledges the (marginal) existence of 'unlimited' (type-crossing) negation. In this case, '**not-(S is P)**' cannot be taken as equivalent to '**S is non-P**', but serves rather as a locution for warning against a category mistake (= S is not the sort of thing to which P can be ascribed). It is only in this specialized use of negation that an NCM—*Fire is not red, Socrates is not a prime number*—can be said to be true (1960:53–54).

Drange (1966:21) rejects Pap's attempts to co-opt 'ordinary parlance' into the cause of the meaninglessness analysis. Drange grants that laymen would regard NCMs like those of (27) as either meaningless or true, depending on whether they happen to be, in Drange's terms, 'inhibited' or 'uninhibited' thinkers. Essentially, an inhibited thinker is one who immediately sets up a limited universe of discourse (cf. Strawson 1952:112, cited above), while an uninhibited thinker does not. But the inhibited thinker's move from (29a) to (29b) is unwarranted as recognized by Ewing and Prior:

(29) a. X is not blue.
b. X is some color other than blue.

Even if it can be shown that 'in ordinary life no one who is speaking truthfully ever says of anything that it is not blue unless he believes it to have some color other than blue', this would not constitute support for Pap's view that (29a) is meaningless if (29b) does not hold.

Drange's own position is that positive category-mistaken sentences are meaningless, although not necessarily neither true nor false. With Strawson and against Pap, Drange rejects the identification of meaninglessness with truth-valuelessness; a sentence is meaningless only if it is neither true nor empirically false. Positive CMs are a priori false and hence meaningless. This asymmetrical theory yields the curious result that the negation of a meaningless sentence can itself be meaningful, a result which Drange (1966:23) defends: 'To say "Socrates is a color" makes no sense at all; to say "Socrates is not a color (but a person)" makes perfectly good sense'. Such NCMs, Drange notes, are appropriately used in enlightening a foreigner or a child (*No dear, you can't color your truck Socrates*).

Following Ewing and Prior, Drange offers a series of proofs for the thesis that NCMs are (necessarily) true statements. Drange's proof for (27b) (1966:24) can be given in abbreviated form (cf. Lambert 1968:83):

(27′) The theory of relativity is an abstract object.
<u>No abstract object is blue.</u>
∴ The theory of relativity is not blue.

If Drange's uninhibited negation does not preserve (in)significance, it does preserve (un)grammaticality: 'any sentence is grammatical if its negation is grammatical' (50). Since (27a, b) and their ancestors are grammatical (and indeed true), their positive counterparts, along with the Carnap and Russell classics of (26), are likewise grammatical (although meaningless). Drange thus lines up with Chomsky (1957), for whom (30) is grammatical but anomalous,

(30) Colorless green ideas sleep furiously.

and against Chomsky (1965), whose theory of selectional restrictions rules out (30) on syntactic grounds.

But Drange's identification of the negation of a declarative sentence **S** with the result of substitution into the standard formula 'it is not the case that **S**' yields the ill-starred prediction that the negations in (31) are grammatical.

(31) *It is not the case that here comes the bus.
*It is not the case that I now pronounce you husband and wife.
*It is not the case that I hereby sentence you to death.

This is, of course, the same point I raised in connection with Aristotle's position, echoed through the ages, that for every affirmation there is a corresponding negation (cf. §1.2.1); in the absence of a response from Drange, his argument for the (un)grammaticality-preserving character of negation is without force.

The observant reader will have noticed that most of the participants in the conflict over the truth, grammaticality, and meaningfulness of NCMs, including Quine, Strawson, and Drange, tend to share the presupposition that language has room for only one variety of negation, essentially the contradictory propositional operator (*apophatikon*) of the Stoics, or its Fregean counterpart, realized as either 'it is not the case that **NP VPs**' or as '**NP** {does/is} not **VP**'. Kissin (1969), accepting the Russellian ambiguity between internal and external negation, claims a distinction in his grammaticality judgments on (32a) and (32b).

(32) a. *These stones do not have cancer. (the asterisk is Kissin's)
b. It is not the case that these stones have cancer.

Kissin's account makes positive CMs and their internally negated counterparts ungrammatical, while external negation (as in (32b)) fails to preserve ungrammaticality. But given that we need to rule out other external negations as ungrammatical—**It is not the case that Bill arrived of Lucy a potato*—this approach too needs some refinement.

As we have seen, Drange's theory posits an asymmetry between positive CMs (which are meaningless and false) and negative CMs (meaningful and true). Indeed, what I have been calling NCMs are for Drange technically not CMs at all. Thus, his working definition of the type crossing as 'a sentence which ascribes to something of a certain type a property with which only a different type of thing is associated' (Drange 1966:93) must later be modified to assure that all such ascriptions of properties to things be done 'in a positive manner'.

The sentences (33a–c) are all used to deny rather than to ascribe a property to Socrates; Drange takes the three versions to be essentially equivalent.

(33) a. Socrates is not a person.
b. Socrates is a nonperson.
c. Socrates is something other than a person.

The distinction between ascribing and denying a property is intended to rule out of the category of CMs not only sentences of the form ***S** is not **P*** but also any sentence equivalent to a sentence of this form. But it is by no means obvious, contra Drange, that the sentences of (33) are equivalent; recall that for Aristotle only (33a) comes out true if Socrates does not exist. Further, given LDN, a doubly negated CM will involve positive ascription (Drange 1966: 96–97, n. 5). Thus, while it is meaningful to assert that $\sqrt{2}$ is not blue (or nonblue), both its positive congener (*$\sqrt{2}$ is blue*) and its negation (*$\sqrt{2}$ is not nonblue*) come out meaningless, a rather suspicious result. The crucial Drangean notions of ascription and denial of a property may well turn out to be either circular or incoherent. (Cf. §1.1.4 for related difficulties in the use of formal criteria to distinguish positive and negative propositions.)

One of Drange's more widely accepted views is the claim that the negatively affixed predicates of ordinary language (e.g., *is un* + ***Adj***) ascribe a property positively rather than denying it. In a paradigm like that of (34),

(34) a. The number 4 is tolerant of carelessness.
     b. The number 4 is not tolerant of carelessness.
     c. The number 4 is intolerant of carelessness.

(34c) 'certainly says more than merely [34b]', and indeed, like (34a) and unlike (34b), it may be said to 'have a positive content'. Contrary to the behavior of ordinary negatives and of the 'logicians' predicates' with initial *non-* (*nonblue,* etc.), predicate terms with *un-* and *iN-* affixes conform to the (revised) definition of type crossings; (34a) and (34c), but not (34b), come out meaningless.

The inherently problematic nature of Drange's attempt to divide predicates into positive and negative subclasses affects even this argument, however. As Routley points out (1969: 368), such predicates as *dislikes dancing, is unmagnetized, is irregular* do not exhibit the 'positive content' Drange attributes to *is intolerant;* the more contradictory the semantics of a given affixal negation, the more closely it approaches the behavior of the paradigmatic NCM cases.

Another weakness of Drange's exposition is his failure to recognize that (34b), unlike (34c), can be used in two different ways, and that in one of these understandings it is quite parallel to the affixal form. In effect, as we shall see in more detail in §6.4, the 'ambiguity' of (34b) is neutralized in (34c). This point, which may have originated with Aristotle, is not lost on Zimmer (1964: 23ff.), who distinguishes (35a), which may or may not be read as a CM, from (35b), which can only be a CM.

(35) a. Triangles are not intelligent.
     b. #Triangles are unintelligent.

The same observation is made by Bergmann (1977:65), who notes that (36b), but not (36c), follows from the fact that (36a) is not true.

(36) a. #The theory of relativity is interested in classical music.
 b. It is not the case that the theory of relativity is interested in classical music.
 c. #The theory of relativity is uninterested in classical music.

Sentence (36c) represents the external and (36c) the internal negation of (36a). As Zimmer predicts, the ordinary syntactic negation of (36a):

(36′) The theory of relativity is not interested in classical music.

can be read either as (36b) or as (36c)

The same distinction applies to vacuous subject examples as well (Zimmer 1964:23):

(37) The present king of France is {not intelligent / #unintelligent}—there isn't any.

For Zimmer, affixal negation in *un-*, *iN-*, and perhaps *non-*, yields contrary rather than contradictory negation, while ordinary particle negation allows both contradictory and contrary interpretations. For Drange, too, *un-* and *iN-* predicates produce contrariety, but both *not* and *non-* yield contradictory negation. I shall return to the semantic differences between particle and affixal negation, and between *non-* and less productive (or more lexicalized) negative prefixes, in chapters 5 and 6.

As we have seen, Russell, Pap, and their Idealist forerunners all adopt various versions of what I shall call (following Routley 1966) SIGNIFICANCE THEORY, in which for some sentential function, the significance or meaningfulness range of an argument place x, that is, the set of expressions that can be significantly substituted for x in, for example, *x is blue,* is a proper subset of the grammaticality range of that argument place. Ewing, Prior, and Quine, on the other hand, are practitioners of NO-TYPE THEORY, which regards all grammatical sentences as ipso facto significant. Strawson, for whom CMs are meaningful but necessarily nonbivalent, seems to fall between the theoretical cracks, while the hybrid program advocated by Drange casts him as a no-typer for negative sentences and a significance theorist for affirmatives.

The late 1960s witnessed the opening of an Australian front in this battle, with the troops for the significance theorists and the no-typers led by Routley (1966, 1969) and Lambert (1968), respectively. What is especially relevant here is the role played by negation (and by its purported ambiguity) in the history of this debate. Routley begins (1966) by attempting a reduction of Quinean no-type theories in which CMs like (38a) are simply false.

(38) a. The number 7 dislikes dancing.
b. The number 7 doesn't dislike dancing.

If (38a) is false, it follows (by LEM) that (38b) is true. But for Routley (as for Pap) (38b) is 'certainly not true' and indeed insignificant; hence the no-typer must either give up LEM altogether or accept the ambiguity of negation (with (38b) analyzed as an instance of non-LEM-preserving internal negation).

In his rebuttal, Lambert (1968) rejects the claim that (38b) is not true (a rejection in which he would be joined by Drange, himself a part-time significance theorist). Following van Fraassen (1966), Lambert argues that LEM does apply to (at least some) sentences which are neither true nor false, defends the postulation of two varieties of negation as independently motivated, and concocts a proof (in the style of Prior and Drange; cf. (27′) above) for the truth of the offending proposition:[10]

(39) The number 7 is an abstract object.
No abstract object dislikes dancing.
∴ The number 7 doesn't dislike dancing.

Routley (1969:368ff.), unconvinced, detects a crucial equivocation on the semantics of the negative operator corresponding to *no* and *not* (*n't*) in Lambert's proof. The second premise is presumably read as denying that abstract entities are the type of things that can be said (significantly) to dislike dancing. But this is postclassical, unrestricted negation, best represented not as $\sim\mathbf{p}$ (or as $\mathbf{F(p)}$ for '$\mathbf{p}$ is false'), but as $\sim\mathbf{T(p)}$, '$\mathbf{p}$ is not true'. The only conclusion warranted in (39) is then that (38a) is not true, leaving open the possibilities that it may be either false or nonsignificant. Thus we arrive at Routley's central thesis (1969:372) on 'the multiplicity of negations':

> The fact is that languages can, and do, contain more than one sort of negation. . . . In adequate sentential significance logics these three sentence negations are distinguished: an unlimited negation '$\neg$' ('$\sim\mathbf{T}$') read 'it is not true that', a restricted negation '$\sim$', and a falsity connective '$\mathbf{F}$' ('$\mathbf{T}\sim$') read 'it is false that'.

Crucially, Routley's basic negative operator, the 'restricted' negation (recall Pap's 'limited' negation and Drange's 'inhibited' negation), does not yield contradictory oppositions or obey LEM; nor does it equate to falsity. But both the contradictory (wide-scope, unrestricted) and falsity operators can be defined in terms of restricted negation in combination with the truth connective **T**. So much for formal significance logic, but what of natural language? 'That these different sorts of negation also occur in natural languages is not so decisively shown. However, they certainly occur in particular English idiolects, for example in the idiolects of speakers trained in significance logic and of philosophers like Pap'. And fortunately, Routley assures us, others can be trained to make the right distinctions.

While there is something classically absurdist about this line of argument (if you don't like the way my logic represents your language, I can teach you to change your language), this maneuver neither originates nor culminates with Routley. Nor are the alternatives without their own absurd touches, as we have seen; recall Drange's mixed theory in which certain false sentences are meaningless but have meaningful negations (but meaningless double negations), and negative predictions are always meaningful (except when the negation appears as an affix (except when the affix is *non-*)). I shall return to this dilemma, and attempt to sketch a possible escape from it, below.

Perhaps responding to the appeals of Pap, Drange, and Routley to the parlance of ordinary lay folk, Steinberg (1970) in fact constructed a psycholinguistic study designed to measure speakers' responses to category mistakes. Given a class of sentences and the choice of evaluations of them—synthetic (contingently true or false), analytic (true by virtue of meaning), contradictory (false by virtue of meaning), redundant, or amphigorous—subjects turned out to be highly consistent on which value to assign in most cases. Judging sentences of the form *The A is / is not a B*, the greatest disagreement emerged, not surprisingly, in the range of NCMs. While positive type crossings like those in (40) are almost unanimously judged amphigorous (meaningless), the corresponding negations in (41) split the respondents,

(40) The chair is a person.
The sheep is a man.

(41) The chair is not a person.
The sheep is not a man.

some subjects (presumably Papists and inhibited thinkers) finding them as amphigorous as their positive mates, others uninhibitedly branding them reduncant or analytic, and hence true.

Before gloating over the results of Steinberg's study, Drangeans should remind themselves that the experiment was hardly conclusive, especially insofar as it focused on reactions to identity statements rather than ordinary predications, while the latter class has provided virtually all the bones of contention for recent disputants in the CM wars.

For linguists, of course, the category mistake of the philosophical literature is more familiar under the *Aspects* rubric of SELECTIONAL VIOLATION. Classic examples include (in addition to (30) above) those in (42), from Chomsky 1965:149.

(42) Golf plays John.
The boy may frighten sincerity.

While the *Aspects* model (Chomsky 1965:148–60) tentatively took selectional restrictions and their violation to involve essentially syntactic properties of lexical items (typically, constraints stated as features on verbs governing the class of subjects or objects they accept), McCawley (1968) and Jackendoff (1972), approaching the question from rather different directions, converged on the argument that selection must be a matter of (at most) semantics. Unlike violations of category rules (**I potatoed the of*) or of strict subcategorization (**I slept the armadillo, *John found sad*), selectional violations produce anomaly but not ungrammaticality. McCawley noted that simple instances of selectional deviance may be constituents in well-formed complex sentences, especially when appearing in opaque contexts. Thus (43b), unlike (43a), is grammatical, significant, and—if the report is accurate—even true.

(43) a. *I dreamed that my toothbrush slept Mary an armadillo.
b. I dreamed that my toothbrush was pregnant.

Furthermore, selection can only be a relation between predicates and term phrases (**NP**s), not between predicates and individual lexical items (nouns). As McCawley has stressed, citing contrasts like those in (44),

(44) a. My {sister/#brother} is pregnant.
b. My {buxom neighbor/#virile neighbor} is pregnant.

the crucial factor is not what noun heads the **NP** which bears a given relation to the selecting predicate, but rather just what entity the **NP** in question refers to. Thus, as McCawley also observes, in languages with grammatical gender distinctions, it is invariably natural gender (e.g., the sex of the referent) which is relevant in determining selection and not the gender feature arbitrarily associated with a given noun. Further, as philosophers have long noted with respect to category mistakes, selectional restrictions may be freely violated in metaphorical or poetic contexts, while true syntactic restrictions are less vulnerable.

Given the arguments for treating selection as a matter of reference, it was natural for generative semanticists to subsume selectional rules (and their violation) within a more general theory of presuppositions (and their nonsatisfaction). In essence, every sentence can be taken to presuppose that its selectional restrictions are met; indeed, in the view of G. Lakoff (1971), we can speak of the well-formedness of a sentence only relative to the satisfaction of its presuppositions, including its selectional restrictions. To know whether a sentence like *My pet amœba believes that I'm a lousy cook* is grammatically well-formed, you have to know what my beliefs are, and hence whether a selectional rule on possible subjects of *believe* was violated.

Now it is true that given an arbitrary sentence **S**, we will tend to give the same answer to the three questions of (45):

(45) Is **S** syntactically well-formed (grammatical)?
Is **S** meaningful (significant)?
Is **S** (or better, Is this token of **S**) true-or-false (bivalent)?

The interdependence of these questions is undeniable; only grammatical sentences can be significant, and (pace Drange) only significant sentences can be used to make bivalent statements. But—and this caveat does not apply to Lakoff alone—the questions are not identical, and they need not receive the same answer. To subsume a mismatch between predicate and subject, that is, an instance of category error, type crossing, selectional violation, or sortal incorrectness, under the general heading of presupposition failure does not tell us more than what we already knew—or believed. What it does not tell us in particular is whether NCMs are or are not grammatical (cf. Kissin and Lakoff vs. Drange and Routley), meaningful (Russell, Pap, and Routley vs. Strawson and the no-typers), true (Drange and Lambert vs. Pap and Routley), or ambiguous (Pap and Lambert vs. Prior and Strawson). Nor, as we shall see, does it tell us whether the deviance of CMs—positive and negative—is to be situated within truth-conditional semantics, non-truth-conditional semantics, or pragmatics.

2.4 External Negation in Presuppositional and Nonpresuppositional Logics: New Solutions for Old Dilemmas

2.4.1 Negation(s) in Multivalued Logic

The thesis that any logical analysis of natural language sentences must countenance two varieties of negation, whose differences may be neutralized at surface structure, is (as we have seen) as old as Aristotle. The classical term-logic-based distinction of a narrow-scope, predicate term negation yielding contrary opposition (*A is not-B*) vs. a (relatively) wide-scope, predicate denial yielding contradictory opposition (*A is not B, A is-not B*) maps straightforwardly, as we saw in §2.2, onto Russell's ambiguity between narrow-scope, internal negation (where the subject term has a primary occurrence) and wide-scope, external negation (where the subject term has a secondary occurrence).

For Russell, working within a propositional logic, the ambiguity of negation requires treating descriptions—and names—as (in the Scholastics' term) EXPONIBLES, superficially simple expressions which can be unpacked to reveal a conjunction hidden within. Crucially, however, Russell shares with Aristotle the view that every meaningful sentence is either true or false (Russell 1957), as stipulated by LBV.

For both Aristotle and Russell, a vacuous subject term yields a proposition which is either meaningful and false (if it is affirmative or internally negated) or meaningful and true (if it is externally negated). LEM holds without exception for external negation (predicate denial): every meaningful indicative sentence expresses a proposition which is true or has a true denial (a true contradictory or external negation).

While Aristotle's analysis of future contingents leaves many questions unsettled, the evidence for a Boethian (LBV-violating) interpretation is not compelling. Category mistakes are either true or false for Aristotle (as for Quine); for Russell they are nonbivalent—but also meaningless. Thus neither Aristotle nor Russell can be said to endorse truth-value gaps in the relevant sense of the term.

One alternative tradition begins with the Stoics and extends through such recent opponents of logical presupposition as Kempson (1975), Boër and Lycan (1976), and Gazdar (1979a). According to this view, Aristotle and Russell are correct in spurning truth-value gaps and third or nonbivalent truth values, but misguided in finding a semantic ambiguity in simple negative sentences. Rather, negation is a semantically invariant truth-functional operator which takes any proposition **p** into its contradictory ~**p**. I shall return to this monoguist line on negation in chapter 6.

A third thesis can be extracted from the presuppositionalist approaches pursued by Frege (1892) and Strawson (1950, 1952): there are syntactically well-formed sentences which in some contexts or states of affairs cannot be used to make an assertion (Frege) or to make a statement which is true or false (Strawson), and such truth-value gaps affect positive and negative sentences alike (cf. §2.2). But these accounts cannot readily deal with the familiar observation that negation can be used to reject a presupposition, as in (46):

(46) Kepler didn't die in misery, because he never existed.
The king of France isn't bald—there <u>isn't</u> any king of France.
(= (16!))

The number 7 doesn't dislike dancing, because numbers have no feelings.

The natural step for a presuppositionalist to take when confronted with the acceptability of such examples is to preserve truth-value gaps by concealing the inherent ambiguity of negation, an ambiguity independently motivated by nonpresuppositionalists from Aristotle to Russell. Strawson in fact does seem implicitly to have taken this step (cf. Strawson 1952:18; 1964:95), although not in so many words. The third option, so amended, then slides into the fourth, that of modern multivalued logics.

As we have observed, Lukasiewicz introduced a third value into the philosopher's tool kit as a device for representing the indeterminist line on future contingents, that is, the position ascribed to Aristotle by the Boethian tradition to which he subscribed. He argues (1922:36) that his three-valued logic, in which sentences describing future contingent (or past unknowable) events are neither true nor false in the present, preserves LEM while rejecting LBV: a proposition may be neither true nor false, but every proposition is either true or has a true (external) negation.[11]

The formal programs developed by Lukasiewicz, Bochvar (1938), Kleene (1938, 1952), Smiley (1960), Herzberger (1970, 1971), and others have offered different solutions to the central question of compositionality within multivalued logics: how is the truth value of a complex expression determined by the truth value of its component parts?

Classical two-valued propositional logic had long since settled on the solution to the compositionality question: given propositions **p** and **q**, their truth functions are computed in accordance with the standard table in (47).

(47)

p	q	$\sim p$	$p \wedge q$	$p \vee q$	$p \rightarrow q$	$p \leftrightarrow q$
T	T	F	T	T	T	T
T	F	F	F	T	F	F
F	T	T	F	T	T	F
F	F	T	F	F	T	T

Any disputes arising over the adoption of the standard truth tables have tended to address the assumption that the connectives (in their linguistic guise) are truth functions. In particular, does the truth table for $p \rightarrow q$ accurately reflect the semantics of *if **p** then **q*** in natural language? Is ***p** or **q*** always semantically inclusive (as in the definition of $p \vee q$ above), or sometimes exclusive (and hence false when both **p** and **q** are true)? What are the non-truth-functional constraints on the appropriate use of sentences employing the connectives, and what is the logical import of these constraints? (Cf. Grice 1967, 1975; Cohen 1971; Barrett and Stenner 1971; Walker 1975; Gazdar and Pullum 1976; Gazdar 1979a; and chapters 4 and

6 below for some discussion of these issues.) For the most part, however, the assignment of particular values in these tables has not itself been a matter of dispute.

In multivalued logics, however, there has been little consensus on how to assign truth values to compound sentences in which one or more components lacks a (bivalent) truth value. (Recall the Mercer-Arlen warning against messing with Mr. In-Between.) In particular, a widely shared unease with the projection rules of Lukasiewicz's system led to the development of other projection tables, beginning with Bochvar 1938. Lukasiewicz and Bochvar agreed, however, in distinguishing an internal, presupposition-preserving negation from an external, presupposition-canceling negation.[12] Just in case a given affirmative proposition lacks a classical truth value, the internal negation of that proposition will lack one as well; the external negation is always true or false. This is shown in (48), where **N** denotes the neuter (nonbivalent, nonsense, neither-**T**-nor-**F**) value.

(48)

p	Internal Negation $\neg p$	External Negation $-p$
T	**F**	**F**
F	**T**	**T**
N	**N**	**T**

Within multivalued logic it is clear that *true* and *false* must be taken as mediate contraries rather than contradictories: a proposition is not true if—but not only if—it is false.

Bochvar's insight was to generalize this dichotomy between the negations of (48) to the binary connectives, distinguishing in each case a presupposition-preserving (truth-gap-inducing) internal connective from the corresponding two-valued external connective. The truth tables for the internal connectives assigned by Bochvar reflect his decision to treat nonbivalence as contagious; if one component is neither true nor false, so is the expression in which it is contained. For conjunction and disjunction, we obtain the tables in (49) and (50), respectively:

(49)

$p \wedge q$	q: **T**	**F**	**N**
p **T**	**T**	**F**	**N**
p **F**	**F**	**F**	**N**
p **N**	**N**	**N**	**N**

(internal conjunction)

$p \mathbin{\hat{\wedge}} q$	q: **T**	**F**	**N**
p **T**	**T**	**F**	**F**
p **F**	**F**	**F**	**F**
p **N**	**F**	**F**	**F**

(external conjunction)

(50)

$p \vee q$	q: T	F	N
p: T	T	T	N
F	T	F	N
N	N	N	N

(internal disjunction)

p ⩔ q	q: T	F	N
p: T	T	T	T
F	T	F	F
N	T	F	F

(external disjunction)

Furthermore, if we follow Bochvar in adopting a one-place truth connective or assertion operator **t(p)**, modeled on the Fregean horizontal, we can define external negation (as in (51)) and the other external connectives (including those in (52)) in terms of the corresponding internal connective and the truth operator (cf. Smiley 1960, Rescher 1969, Herzberger 1970, Donnellan 1970, and Bergmann 1977 for variations on this theme).

(51) $-\mathbf{p} =_{df} \neg \mathbf{t}(\mathbf{p})$

(52) **p** ⩓ **q** $=_{df} \mathbf{t}(\mathbf{p}) \wedge \mathbf{t}(\mathbf{q})$
p ⩔ **q** $=_{df} \mathbf{t}(\mathbf{p}) \vee \mathbf{t}(\mathbf{q})$

It is this truth connective—the 'Bochvar-Frege horizontal', as Herzberger and Bergmann dub it—that in effect filters out nonbivalence. The crucial tables for our purposes are those defining negation:

(53)

p	t(p)	¬t(p)	¬p	t(¬p)
T	T	F	F	F
F	F	T	T	T
N	F	T	N	F

Notice in particular that the third column of this table is identical to the column for external negation (**−p**) in (48); intuitively, we can agree that if a proposition is neither true nor false, it is clearly not true. As the fifth column suggests, we can also define a falsity connective based on the truth of the internal negation, such that $\mathbf{f}(\mathbf{p}) =_{df} \mathbf{t}(\neg\mathbf{p})$; cf. the citation from Routley.

What makes this approach more than merely a clever trick is the frequently made observation that the English sentences which (at least for some speakers) most closely correspond to the semantics of the external negation of (15) are not of the form of (16) but rather of (54a, b).

(15) The king of France is bald.

(16) The king of France is not bald.

(54) a. It is not true that the king of France is bald.
b. It is not the case that the king of France is bald.

Based on this intuition, it is tempting to conclude that (54a, b) are direct English translations of the formula ¬**t**[(15)]. Natural language negation, on this view, is not lexically ambiguous between the two readings signaled in (48), but scopally ambiguous as to its position in logical syntax (as it is, in a somewhat different way, within both Aristotelian and Russellian two-valued logic). The view that the formula *it is not {true / the case} that . . .* represents logical external negation will be examined more closely in chapter 6.

In any case, there are a number of serious problems that confront any attempt to provide a coherent and intuitively plausible semantic interpretation for three-valued logic; 'Beginning with Lukasiewicz's trivalent matrices', laments Herzberger (1970: 32), 'peculiar things happen'. A useful bestiary of these peculiar things is offered by Rescher (1969: 160ff.), who suggests that four-valued logic might score higher on the plausibility meter. I shall confine my remarks here to the representation of negation within multivalued logic (MVL) and to its interaction with the classical Aristotelian laws.

Given the multifarious forms taken by negation-like operators in MVL, it is not even clear a priori when a given MVL operator is a negation. Rescher offers a simple criterion: '**N** is a mode of negation iff it can never happen that **p** and **Np** are both true or both false'. Of course, it may be the case (as with the internal negation of the Lukasiewicz, Bochvar, and Kleene frameworks) that neither **p** nor **Np** is true, or that neither of them is false. Notice that the '¬' and '−' connectives as defined in (48) satisfy Rescher's criterion, and that both connectives are also NORMAL in the sense that they receive the same values as the negation operator of classical two-valued systems (cf. (47)) when only the classical values **T** and **F** are involved.

How do the classical laws—LC, LEM, LBV—fare in multivalued systems? As Rescher notes, everything hinges on just how these laws are formulated. Thus if LC is the thesis that '**p** ∧ **Np**' is logically false, it will govern the external negation of a Bochvar-like system, since '**p** ∧ −**p**' is self-contradictory and '−(**p** ∧ −**p**)' tautologous. But internal negation does not obey this formulation of LC, since the infectious nature of nonbivalence assures that whenever **p** is nonbivalent, both '**p** ∧ ¬**p**' and '¬(**p** ∧ ¬**p**)' will be nonbivalent as well.

But in MVL, as in classical logic, '**p** ∧ **Np**' can never be true for either internal or external negation. As Cresswell (1973: 41) observes, '**p** ∧ **Np**' can be true only if **N** is not a real negation, ∧ is not a real conjunction, or both. In other words, '**p** ∧ **Np**' can be true only when it is not a real contradiction, vindicating LC.

Similarly, LEM is standardly taken as stipulating that any proposition is true or has a true negation. But '**p** ∨ **Np**' can be true only when it has a classical truth value; LEM fails for Bochvar's internal negation (while holding, of course, for the neoclassical external variety). Indeed, as Rescher points out, if we were to read LEM as asserting that every proposition is either true or false (i.e., as equivalent to LBV), any self-respecting multivalued system must reject this principle if it is not to collapse with the classical two-valued system (cf. van Fraassen 1969:69). But the weaker principle that at least one of **p** and **Np** must be true can be maintained in MVL (although, as Bochvar demonstrates, it needn't be).

It is evident that in MVL, unlike classical logic, LEM (interpreted as '**p** ∨ **Np**') is not identical to LBV (interpreted as the thesis that every proposition is either true or false, or alternatively that one of **p**, **Np** is true and the other false). This emerges most clearly when we recall that **p** is not identical to **t(p)** (cf. the first two columns of (53)): for a speaker of MVL, *it is true that p* does not reduce to ***p***. For this reason, Kneale and Kneale's attempt (1962:46–48) to show that Aristotle errs in 'trying to assert the Law of Excluded Middle while denying the Principle of Bivalence' founders on their unwarranted assumption that ***p** or not-**p*** is 'plainly equivalent' to *it is true that **p** or it is false that **p***. If the Kneales' Boethian interpretation of *De Interpretatione* chapter 9 is correct (which is by no means obvious), Aristotle must be assuming a three-valued logic in which the falsity of ***p*** (e.g., *There will be a sea battle tomorrow*) cannot be inferred from the falsity of *it is true that **p***. (Cf. van Fraassen 1969:493 for related discussion.)

Various systems of MVL have chosen various approaches to the classical laws. In the intuitionistic logic of Brouwer (1908) and Heyting (1971), LC is preserved but LEM fails. Bochvar's model, with its external connectives, incorporates the classical bivalent system as a proper subpart. As illustrated in (49) and (50), when a proposition **p** contains only external connectives, no truth-value gap can arise; regardless of the value of **p**, '**p** ∧ −**p**' comes out false and '**p** ∨ −**p**' true. In such cases the classical laws are preserved. But the fact that both LEM and LC fall by the wayside when we consider internal negation has tended to depress the market for this system among potential consumers of MVL.

Advocates of two-valued logic have taken the failure of the classical laws as a step in the direction of reducing MVL to absurdity. Arguing from the assumed inviolability of LDN, Russell notes that 'when the law of excluded middle fails, the law of double negation also fails' (1940:271; cf. Sigwart 1895:148 for a similar argument). The observation is correct, but hardly telling against dual-negation logics—two-valued or multivalued—which explicitly or implicitly take LDN to apply just when LEM does, namely in the case of wide-scope or external negation only. Note that for

Russell himself, the king of France is neither bald nor not-bald (i.e., (15′) and (16′) are both true) when France is a republic.

In the same vein, Geach ([1972] 1980: 80–81) attempts to demonstrate that 'exceptions to the Law of Excluded Middle can . . . be allowed only if exceptions to the Law of Contradiction are also allowed—a much less popular concession'. But Geach's derivation of LEM from LC hinges on an equivocation over the distinction between internal and external negation; it remains possible to defend the position that MVLs can accept LC without a commitment to LEM. Such a defense is offered in appendix 1.

In fact, as van Fraassen's theory of supervaluations shows, a nonclassical model admitting value gaps may endorse LC, LEM, <u>and</u> LDN—if we are willing to abandon the truth-functional character of the connectives along with the Law of Bivalence; cf. van Fraassen 1966, 1968, 1969, for details.

Within MVL, a Strawsonian notion of presupposition can be defined in terms of (internal) negation and some version of an inference rule variously called SEMANTIC ENTAILMENT (Smiley 1960) or NECESSITATION (van Fraassen 1968). Van Fraassen (1968: 138) provides the definitions in (55):

(55) (i) A necessitates B if and only if whenever [if any situation or possible world in which] A is true, B is also true.
(ii) A presupposes B if and only if
(a) A necessitates B; and
(b) (not-A) necessitates B

Similarly, for Smiley (1960: 131),

(56) A presupposes B $=_{df}$ A ⊩ B and (~A) ⊩ B,

where '⊩' is read as 'semantically entails' and defined in the same manner as (55i). As Smiley points out, these definitions of logical presupposition embedded within a coherent version of MVL will capture the standard assumptions: any proposition and its (internal) negation share the same presupposition set, the presupposition of **p** is a necessary condition for **p** to be either true or false, and so on. It is an important consequence of these definitions that necessitation / semantic entailment within MVL, unlike either the material conditional or entailment within classical logic, does not allow contraposition: 'From the fact that A ⊩ B it does not follow that (~B) ⊩ (~A) (because A's having no [bivalent] truth value is compatible with A ⊩ B but not with (~B) ⊩ (~A))' (Smiley 1960: 129; cf. van Fraassen 1969: 81).

Similarly, while the analogues of the classical law of modus ponens will carry over (from A ⊩ B and A, we can infer B), modus tollens does not (from A ⊩ B and ~B, we can validly conclude only that A is not true, not that ~A is true). Let A be any presupposing sentence (e.g., (15) *The king of France is bald*) and B one of its presuppositions (e.g., (17) *There is a king of France*). Then A, and likewise ~A, necessitates, semantically entails, and indeed presupposes B, but all we can infer from ~B is that A, and likewise ~A, is not true.

Within a dual-negation MVL, '~' in the above discussion must of course be read as '¬', the internal presupposition-preserving connective. While internal negation maintains presuppositions and truth-value gaps, external negation—serving as a device for canceling all outstanding presuppositions—will be employed 'by someone who wishes not so much to contradict a particular assertion as to reject the ontology behind it' (Smiley 1960:131).

It is a curious fact that while Lukasiewicz may have developed his prototype MVL with the problem of future contingents in mind (cf. §2.1), other advocates of nonclassical systems from Bochvar to Smiley, Herzberger, and van Fraassen have tended (following Frege and Strawson) to focus on those (putative) truth-value gaps which stem from vacuous singular terms, nondenoting names, and descriptions (cf. §2.2). Bergmann (1977, 1981) applies her own version of MVL to sentences afflicted with sortal incorrectness, that is, category or type errors (cf. §2.3). As we have seen, category mistakes are like reference failure in that 'two senses of negation are distinguishable when the operation is applied to sortally incorrect statements' (Bergmann 1977:61). These are the restricted / limited vs. unrestricted / unlimited negations of Pap and Routley, mapping directly onto the internal / external dichotomy of Bochvar. External negation is, as Bergmann (1977:76) puts it, 'sortally opaque' with respect to type-based presuppositions.

Bergmann (1981) proposes a 'regimentation' of the intuitive concept of presupposition, covering nondenoting subject terms and factive predicates (cf. Kiparsky and Kiparsky 1971) as well as category mistakes. In her two-dimensional formal language, semantic presuppositions are admitted, but truth-value gaps are not; each sentence receives two separate binary valuations, one (its truth value) for what it expresses and the other (its SECURITY value) for what it presupposes. Thus, there are four distinct fully specified values assignable, since truth / falsity and security / insecurity are assessed independently. The internal negation of a proposition **p** will be true just in case the corresponding external negation is true and **p** is secure (nonanomalous). (I shall pay a brief return visit to Bergmann's two-dimensional logic in chapter 6.)

How closely do the presuppositions formally defined within the various multivalued systems fit the intuitive notion of presupposition associated with the Oxford school of analytic philosophy? Whether the principal modern champion of semantic presupposition and truth-value gaps would accept either the standard (Bochvarian) MVL line, the supervaluation model, or any formal hybrid as a means for capturing his intuitions on truth and meaning in ordinary English is a moot question, given his skepticism toward all varieties of formal logic: 'Neither Aristotelian nor Russellian rules [nor, we may presume, Bochvarian, van Fraassenian, or Bergmannian rules] give the exact logic for any expression of ordinary language; for ordinary language has no exact logic' (Strawson 1950: 344).

Kempson (1975: 86) has noted the irony of this concluding sentence from the presuppositionalist manifesto, containing as it does a definite description (*the exact logic . . .*) which evidently does not induce an existence presupposition. In fact, however, this apparent inconsistency is consistent with the revisionist position embraced in Strawson's later work (1964: 95ff.), where truth-value gaps arise only when a nondenoting singular term occurs in a referential position (typically as surface subject and/or topic).

Strawson observes that while we may 'feel squeamish' about assigning a truth value to (57a, b), we are more confident in assessing (58a, b) as false and true, respectively, given the monarch's nonexistence.

(57) a. The king of France visited the exhibition.
b. The king of France didn't visit the exhibition.

(58) a. The exhibition was visited by the king of France.
b. The exhibition wasn't visited by the king of France.

On this view, an existence presupposition will be triggered when a sentence purports to be ABOUT some entity when no such entity exists. Sentences (57a, b) are about the king of France, when the truth-value gap associated with the utterance of these sentences during a republican period, but (58a, b) are about the exhibition, so that the expression which is 'guilty of reference failure' is absorbed harmlessly into the predicate.

Even the classic truth-value-gap-inducing (15) may be simply false, Strawson concedes, if it is taken not as a description of the king of France (*Does M. le roi have need of a royal barber?—No, the king of France is bald*) but as a statement about the class of hairless entities (*What bald notables are there? Well, let's see, the king of France is bald*), in which case the subject is not functioning as the sentence topic, as the stress and intonation pattern show. (Cf. also Kuroda 1977: 84–85, McCawley 1979, J. D. Fodor 1979, and Horn 1986 for related discussion of presupposition and topichood; I return to this issue in §7.3.)

The other main innovation of Strawson 1964 is the observation that there is no knockdown argument either to prove or to disprove the need for truth-value gaps in an account of what (following Kuroda 1977) I shall nonprejudicially call presuppositional phenomena. Arguments can indeed be marshaled for and against the two-valued theory of Russell (1905, 1957), the informal truth-value-gap account of Strawson (1950, 1952), and the various formal nonclassical systems I have cited in this section, but for each argument there is a plausible (if not provably correct) counterargument. As Dahl (1981 : 197) points out, the more measured position defended in Strawson's later work is never acknowledged by those antipresuppositionalists (e.g., Wilson 1975; Kempson 1975) concerned with refuting Strawson 1950. And theoretical tolerance can be taken still further, as the live-and-let-live attitude of Strawson 1964 blossoms forth into the let-a-thousand [if not ∞]-logics-bloom pluralism of Rescher (1969:235): 'There can be no more question of the universally and generically "correct" logic than there can be of the universally and generically "correct" woodworking tool. There will be a range of legitimate choice, with borders delimited by functionally grounded regulative principles, but within which the alternatives are, from a purely conceptual standpoint, equally viable'.

2.4.2 Negation Weak and Strong

The logic of truth-value gaps, in its variegated multivalued flowering, is historically grounded in Lukasiewicz's Boethian interpretation of Aristotle's account of the future sea battle. As I have noted, this is not necessarily the most plausible reading (much less the only reading) available for chapter 9 of *De Interpretatione,* and it does not in any case generalize to the clearly two-valued approach Aristotle adopts elsewhere in the Organon, extending to cases of vacuous subject terms and category mistakes. But if the havoc-wreaking innovations of MVL are not sufficient to handle the full range of phenomena insightfully characterized by Aristotle, they may also be unnecessary.

The semiclassical propositional logic of Von Wright (1959), in which two negations are distinguished, but truth-value gaps do not arise, would therefore seem more faithful to the spirit of the Stagirite's intentions. In this system, STRONG negation ($\neg\mathbf{p}$) is an affirmation as well as a denial. As with the predicate term negation of Aristotle, LEM does not apply to a proposition and its contrarily opposed strong negation. Like the corresponding term logic predications *S is P* and *S is not-P,* **p** and $\neg\mathbf{p}$ may both be false, namely, when the subject doesn't exist or when it exists but the predicate cannot be naturally applied to it. WEAK negation ($-\mathbf{p}$), on the other hand, is a contradictory operator, corresponding to predicate denial, amounting to the proposition that it is not true that **p**.

Von Wright (1959 : 6–7) recognizes Aristotle's distinction between things that are not white and things that are not-white as more germane to a logic of natural language than Russell's distinction between meaningful and meaningless statements. Von Wright's ELP, the extended logic of propositions, is designed to reflect the Aristotelian judgment that (59a, c) are simply false and (59b) true, as against Russell's line that these sentences are meaningless.

(59) a. The number 7 is white. **[p]**
b. The number 7 is not white. [$-$**p**] ([~**p**] in Von Wright's notation)
c. The number 7 is not-white. [¬**p**]

On the Aristotle–Von Wright approach, but not the Russellian, negative sentences whose predicates are inappropriate for characterizing their subjects fall together with negative sentences whose subjects (extrinsically) happen not to denote (*the king of France*) or (intrinsically) cannot denote (*the largest prime number, the round square*). In each case the sentence expresses a proposition which is (automatically) true or false, depending on whether the negation is read weakly or strongly.

The logic of weak negation, like that of the external Bochvarian connectives, respects the laws of classical propositional logic. Reading $-$**p** as '**p** is not true' rather than '**p** is false' (Von Wright 1959:5ff.), we get these versions of the classical principles:

(60) LC: $-(\mathbf{p} \wedge -\mathbf{p})$ 'It is not true of any proposition that it is both true and not true'
LEM: $(\mathbf{p} \vee -\mathbf{p})$ 'Any given proposition is either true or not true'
LDN: $\mathbf{p} \leftrightarrow --\mathbf{p}$ '**p** if and only if it is not true that it is not true that **p**'

But the characterization of strong negation requires different laws. Starting with four axioms,

(61) (A1) $\neg(\mathbf{p} \wedge -\mathbf{p})$
(A2) $\neg\mathbf{p} \rightarrow -\mathbf{p}$
(A3) $\neg(\mathbf{p} \vee \mathbf{q}) \leftrightarrow \neg\mathbf{p} \wedge \neg\mathbf{q}$
(A4) $\neg\neg\mathbf{p} \rightarrow \mathbf{p}$

Von Wright explores the properties of EPL. The essential point is that while **p** and $-$**p** are opposed as contradictories, **p** and ¬**p** are contraries (misleadingly termed 'strong contradictories' by Von Wright). LC clearly holds for the latter opposition ($-(\mathbf{p} \wedge \neg\mathbf{p})$ is provable as a theorem), but LEM does not ($\mathbf{p} \vee \neg\mathbf{p}$ may be false, as in the disjunction of (59a) with

(59c)). Similarly, the law of double strong negation only holds in one direction, as stipulated by (A4); from **p** it cannot be derived that ¬¬**p**. And while (A3) is a biconditional, De Morgan's other law works only one way:

(62) $\neg(\mathbf{p} \wedge \mathbf{q}) \leftrightarrows \neg\mathbf{p} \vee \neg\mathbf{q}$

Von Wright proceeds (1959: 17ff.) to translate the dual-negation system of EPL into the language of modal logic, where ¬**p** corresponds to '**p** is impossible', −¬**p** to '**p** is possible (not impossible)', and ¬−**p** to '**p** is necessary (impossible . . . not)', and thence to that of deontic logic: 'That an act is prohibited [**O**(−**A**)] means that it is "positively" not permitted [¬**P**(**A**)] and not that it is "merely" not permitted [−**P**(**A**) = **O**(¬**A**)]' (Von Wright 1959: 27). Thus **O**(−**A**) entails −**P**(**A**), but not (contra Von Wright 1951) vice versa.

The two negations thus yield a distinction between weak and strong permission and between weak and strong prohibition: in each case the weak norm represents a mere absence of the contrary norm, while the corresponding strong norm is a positive statement of the rules of action. Just as 'to be not-A' is distinguished from 'not to be A' (cf. Aristotle's *Prior Analytics* 1, chap. 46, discussed in §1.1.1), so too can 'to do not-A' be distinguished from 'not to do A'.

There is evidence that this formal distinction, subtle as it may appear, does surface in natural language. In his classic grammar, Kruisinga (1931: §688) differentiates the weaker *may not* (no permission has been / will be given) from the stronger *must not* (there are facts, rules, circumstances prohibiting the action). The railway guard who begins at the first station by admonishing the passenger, 'You may not smoke here, sir', strengthens his warning by the second station to 'You must not smoke here, sir', and finally escalates by the third station to 'You shan't smoke here, sir', presumably uttered with grim determination, gritted teeth—and, in today's less civil world, probably a drawn weapon. As I mentioned in §1.3.2, the values of weak vs. strong prohibition may also serve to capture the analogous distinction in the *Mīmāṃsā* between the two forms of negative injunction, *niṣedha* and *paryudāsa*.

Von Wright's is not the only formal system which establishes a dichotomy between weak (contradictory) and strong (contrary) negation operators within a gap-free truth-conditional logic. The primary negation of the Intuitionists is essentially strong negation, or 'falsity "de jure"', but Heyting (1971: 18) also allows for the existence of a weaker 'falsity "de facto"' operator for nonmathematical discourse: 'Strictly speaking, we must distinguish the use of "not" in mathematics from that in explanations which are not mathematical but are expressed in ordinary language. In mathematical language no ambiguity can arise: "not" has always the strict meaning'. This 'strict meaning' of not-**p** is such that 'if we suppose the truth of **p**, we are led to a contradiction' (Heyting 1971 : 102).

Like Von Wright's strong negation, the negative operator employed in the Intuitionists' mathematical discourse is incompatible with LEM. Indeed, the rejection of LEM is so fundamental a component of the Intuitionist program for the reconstruction of mathematics that Brouwer felt compelled to seek ethical justification for the move: 'An incorrect theory, even if it cannot be inhibited by any contradiction that would refute it, is none the less incorrect, just as a criminal policy is none the less criminal even if it cannot be inhibited by any court that would curb it' (Brouwer 1923:336). The sympathetic hearing such appeals received prompted Brouwer's formalist adversary Hilbert to lament that 'the power of suggestion of a single man, however full of temperament and inventiveness, is capable of having the most improbable and eccentric effects', including the decision by 'a whole community of mathematicians' to reject LEM and the related ε-axiom. For Hilbert, giving up LEM is 'tantamount to relinquishing the science of mathematics altogether': 'Taking the principle of excluded middle from the mathematician would be the same, say, as proscribing the telescope to the astronomer or to the boxer the use of his fists' (Hilbert 1927:476).

Whatever the merits of the Intuitionists' overall program, it should be noted that their strict, 'de jure', LEM-violating negation is not equivalent to the strong negation of Von Wright. In particular, the Intuitionists accept the law $\mathbf{A} \rightarrow \bar{\bar{\mathbf{A}}}$ (i.e., $\mathbf{p} \rightarrow \neg\neg\mathbf{p}$) while rejecting its converse, precisely the opposite pattern from that laid out in EPL (cf. (A4) in (61) above). Noting this and other differences between the two systems, Von Wright argues convincingly (1959:27–30) that the 'intuitions' of Brouwer, Heyting, and others result in an 'unhappy hybrid' of classical and nonclassical approaches.

The theory proposed in McCall 1967a, in which a strong contrariety operator **R** is distinguished from the ordinary contradictory negation **N**, offers a closer parallel to Von Wright's EPL, although it was apparently developed independently. The crucial law **CRpNp** is essentially a Polish-notational variant of Von Wright's (A2), $\neg\mathbf{p} \rightarrow -\mathbf{p}$, although the basic observation goes back to Aristotle's argument at *Prior Analytics* 52a1 that *it is not-white* unilaterally entails *it is not white*. Von Wright's analogy between strong negation (contrariety) and impossibility in modal logic is also independently elaborated by McCall (1967a:124ff.). (I have previously noted the dissents by Geach ([1972] 1980) and Englebretsen (1974) from McCall's attempt to formulate a propositional, rather than term-based, logic of contrariety; the same objection automatically carries over to Von Wright's earlier efforts toward the same end.)

For another neo–Von Wrightian two-track two-valued semiclassical approach to negation, I turn to Kuroda's (1977) distinction between PROPER negation (defined in terms of truth-value reversal) vs. DENIAL negation (defined in terms of possible answers to yes-no questions). Kuroda characterizes semantic presupposition by reference to denial negation, differing from van Fraassen's and Smiley's definitions in (55) and (56) above in rejecting the MVL position that presupposition failure induces truth-value gaps. Rather, a presupposition for Kuroda is a kind of privileged or selected entailment, prefiguring the theory later elaborated by Wilson and Sperber (1979). Kuroda shares Strawson's intuition that a negative sentence like (16) is unambiguous, implying (as would its positive counterpart (15)) the unique existence of the referent of the subject term. While we may read (63) as a proper negation (sans entailment),

(63) It is {not true / not the case / false} that the king of France is bald. (cf. (54a, b))

this reading is 'contrived' (cf. Kuroda 1977: §5 for discussion). But while ordinary, auxiliary negation gives the canonical-denial negation of the corresponding affirmative sentence, syntactic form provides neither a necessary nor a sufficient criterion for the presence of denial negation, as sentences with quantified **NP**s make especially clear.

The mismatch between the syntax and semantics of negative sentences also emerges in Katz's rather differently conceived dual-negation system (1972: 329). For Katz, **S′** is the NEGATION of **S** if **S** and **S′** are sentences differing only in that one contains a *not* in its main **VP** where the other does not. Thus, negation is a purely syntactic relation, distinguished from the logical relation of DENIAL; **S′** is the denial of **S** if **S** and **S′** are incompatible. It will be noticed that Katz's denial relation, like Kuroda's, is a form of contrariety, but for Katz any two (weak) contraries are each other's denials, for example, *That's red* vs. *That's blue, That's alive* vs. *That's a rock*. Thus, the denial of a sentence need not be its negation, nor is a negation necessarily a denial.

Yet one more echo of EPL is sounded by L. Carlson (1983: 220ff.), who argues that for any sentence based on an emotive factive (cf. Kiparsky and Kiparsky 1971) like *odd,* we must distinguish '[its] DENIAL (true if and only if the sentence is false) and its REJECTION or categorical denial (true if and only if the sentence is not true)', where *not true* ≠ *false* (Carlson 1983: 220–21). Like Kuroda, Carlson determines the presuppositions of a sentence (and of its denial) based on what follows from the corresponding yes-no question; (64a, b) both presuppose that he won, since this is assumed by the corresponding interrogative, (64c).

(64) a. It is odd that he won.
b. It is not odd that he won.
c. Is it odd that he won?

(Note that the term 'denial', as employed by Kuroda, Katz, and Carlson for a contrary relation, does not correspond to the contradictory relation of predicate denial used in translations of Aristotle.)

But doesn't (65), whose existence is acknowledged by Carlson, show that (as argued by, e.g., Boër and Lycan [1976]) factivity (and more generally semantic presupposition) is a myth?

(65) It is not odd that he won, for he did not win.

Carlson's analysis of (65) recalls Strawson's discussion of (18) above or Pap's characterization of 'unlimited negation'.[13] Carlson admits to feeling 'slightly uncomfortable' about (65), which he views as comparable to a sentence in which negation is used to point out a category mistake, for example, (66):

(66) Friday is not in bed, it is a date.

Thus, (65) does not constitute a straightforward reading of the negative sentence in (64b), involving instead the 'marked option' of categorical denial.

LDN cannot apply across the two disparate varieties of negation: 'The categorical rejection of an ordinary negative sentence does not amount to dropping both denials' (Carlson 1983:221)—that is, in Von Wright's notation, $-\neg \mathbf{p} \not\rightarrow \mathbf{p}$. Thus, the joint categorical denial in (65′) does not result in inconsistency

(65′) It is neither odd nor not odd that he won; for he did not win.

any more than does Strawson's parallel example, *He neither cares nor doesn't care; he's dead.*

But it is never made clear whether Carlson's reduction of 'factivity errors' like those in (65), (65′) to category mistakes (as in (66)) is to be carried out within a gap-free, two-valued system (à la Von Wright or Kuroda), a multivalued truth-functional system (à la Lukasiewicz or Bochvar), a non-truth-functional system with two values, gaps, and supervaluations (à la van Fraassen), a two-dimensional gap-free system (à la Bergmann), or none of the above.

I have been implicitly treating a dual-negation propositional logic like Von Wright's as isomorphic to Aristotle's term logic, in which contradictory predicate denial (corresponding to weak propositional negation) is distinguished from contrary predicate-term negation (corresponding to strong propositional negation). In Aristotle's formulation, there are two possible 'modes' or 'manners' of predication: any predicate (positive or negative) may be either affirmed or denied of the subject (cf. Sommers 1970:5 and chapter 1 above). Within propositional logic, external or weak negation can only be represented as a *de dicto* unary truth-functional connective, taking one sentence or proposition, **p**, into another, **−p**. But internal or strong negation can be taken as a non-truth-functional sentential operator with the semantic properties ascribed to it by Von Wright (1959) and McCall (1967a), or as a *de re* operator taking one predicate **P** into another, representable as **not-P**, **non-P**, or $\bar{\mathbf{P}}$ (cf. R. Clark 1974). Thus when neo-term-logicians like Sommers (1963, 1965, 1970) or Englebretsen (1976, 1981a, 1981b) claim that Aristotle's term logic included no external negation and two kinds of internal negation, this claim is as misleading as it is literally accurate. While predicate denial and predicate term negation are indeed both syntactically internal to a proposition, the former—but not the latter—is semantically analogous to garden-variety external or weak negation in a modern propositional logic like Von Wright's. Predicate denial effectively takes wide scope over the subject and over the subject-predicate connection, so that (as we have come to see) we can truly deny a predicate of an empty subject or one to which the predicate fails to apply.

The term-logic revivalists Sommers and Englebretsen join with Aristotle, Leibniz, and, to an extent, Strawson (as against the received doctrine of the Stoics, Abelard, Frege, and Russell) in recognizing the centrality of the subject-predicate distinction: 'Subjects and predicates may both contain negative terms, but only predicates may be denied—or affirmed' (Englebretsen 1981a:46). In the dual-negation logic of Sommers (1965:273ff.), denial is distinguished from negation in what is essentially a term-logic translation of the method of Von Wright's EPL. NEGATION (**−S**) is the weak sister, applying (de dicto) to the whole sentence **S** and respecting LEM; **S** v **−S** is an axiom. The strong sister is DENIAL (**S′**), which applies directly (de re) to the predicate term and only indirectly to the sentence; like the corresponding strong or internal connectives of EPL and MVL, it is a contrary operator which allows a nonexcluded middle. LDN applies as a biconditional to two negations (Sommers specifies that **−(−S) = S**), and—contra Von Wright's strong operator—to two denials (**S″ = S**), but not to a mixed formula; negation does not cancel a denial (**−S′ ≠ S**).

When neither **S** nor **S′** is true (or when both are false), a category mistake is involved. Thus, either (67a) or its negation (67b) must be true, and in fact the latter is.

(67) a. The equator is clean.
 b. Not (The equator is clean)
 c. The equator is {not-clean / unclean}.

But neither (67a) nor (67c) is true, and the latter is the denial of the former, so (67a, c) are category mistakes. What is not clear on this account is just where empty-subject sentences fit in, given that—at least in the two-valued dual-negation systems of Aristotle and Russell—both *The king of France is bald* and its denial *The king of France is not-bald* come out false, where no category mistake is involved.

Another version of term logic does invoke an explicit truth-conditional distinction between sentences with empty subjects and those with category mistakes. Englebretsen (1976:538), following Sommers 1963, elaborates four interrelated 'levels of rectitude' for grading species of failure that may afflict a given sentence:

(68)

Level of Failure	Diagnosis	Laws Violated	Example
Level 3	Empirically false	Laws of physics	*Bill is now both inside and outside this room.*
Level 2	Inconsistent	Laws of logic	*All men are mortal and not mortal.*
Level 1	Category mistaken	Rules of sense	*2 is red.*
Level 0	Ungrammatical	Rules of grammar	*Of slept she up.*

In this schema, 'A sequence incorrect at some level must be correct at all lower levels and is neither correct nor incorrect at any higher level' (Sommers 1963:348), so that any category mistake, even an apparent contradiction (*His anger was triangular and not triangular*) can never be logically inconsistent or a fortiori empirically false. It may, however, be a priori false, as it would for Drange (1966)—and for Sommers 1965 (see above).

On Englebretsen's account, empty subject cases are merely instances of sentences which are empirically false and have an empirically false denial (predicate-term negation). This much is in the spirit of Aristotle (and of course Russell). But Englebretsen's innovation is to regard an NCM as false on both its **O**, predicate-denial reading (*2 is not red*) and its **E**, predicate-term negation reading (*2 is not-red*). No support is offered for this stipulation, which departs radically from the standard approaches of dual-negation logics in which the **O** reading is assessed as true and the **E** reading as either simply false, a priori false, neither true nor false, false and insecure, or meaningless, depending on whether the assessor is, respectively, an Aristotelian, a Drangean, a Bochvarian, a Bergmannian, or a Russellian.

A rather different approach is endorsed in Englebretsen 1981a. Here, LBV is explicitly repudiated and truth-value gaps allowed. Category mistakes are neither true nor false, à la Strawson, while instances of empty-subject vacuosity are still treated essentially à la Russell. But there is a new innovation here, in the form of an additional reading for negation now admitted by Englebretsen. Alongside predicate denial (a truth-functional contradictory operator) and predicate-term negation (a non-truth-functional contrary operator), he recognizes the use of negation as a METALINGUISTIC operator (glossed as 'it is untrue that . . .') affecting presuppositions as well as entailments. Whether three such distinct interpretations must be countenanced, and whether Englebretsen has successfully characterized the metalinguistic use of negation, are questions to which I return in chapter 6.

In our dizzying march through multivalued and allied nonclassical (more accurately, non-Fregean) systems of logic, we have repeatedly come upon a distinction between two varieties of negation, but that distinction has appeared to us in a discouraging array of guises. But within this array a pattern begins to emerge. If we include some of the analogous dichotomies drawn in the medieval and Idealist literature summarized in §1.2, we can draw up the chart in (69):

(69)

	COLUMN A	COLUMN B
Aristotle	Predicate denial [**S** is not **P**]	Predicate term negation [**S** is not-**P**]
Avicenna, Burleigh, William of Sherwood	Negative judgment, *negatio negans* [**S** non est **P**]	In(de)finite judgment, *negatio infinitans* [**S** est non **P**]
Spinoza	*negatio*	*privatio*
Śaṇkara, Hegel	Insignificant / bare negation	Significant negation
Hegel [1812–16] 1929	(Negatively) infinite judgment	(Simply) negative judgment
Sigwart 1895, Bosanquet [1888] 1911	Negative judgments without positive ground	Negative judgments with positive ground
Russell 1905	Negation with secondary occurrence of description [wide scope ~]	Negation with primary occurrence of description [narrow scope ~]
Mabbott 1929	Turkey-carpet judgments; teleological negation	Eliminative negation

	COLUMN A	COLUMN B
Bochvar 1938	External negation	Internal negation
Von Wright 1959	Weak negation [~**p**]	Strong negation [¬**p**]
Smiley 1960	Secondary negation [~**p**]	Primary negation [~**p**]
Pap 1960	Unlimited negation	Limited negation
Sommers 1965	Negation [−**S**]	Denial [**S**′]
Drange 1966	Uninhibited thinkers' negation	Inhibited thinkers' negation
Routley 1966, 1969	Unlimited negation [¬**p**, ~**T**(**p**)]	Restricted negation [~**p**]
McCall 1967a	Negation [**Np**]	Contrariety [**Rp**]
van Fraassen 1969	Exclusion negation	Choice negation
Keenan 1969	External negation [**S**]	Internal negation [~**S**]
Herzberger 1970	Complementation [**P**]	Choice negation [~**P**]
Heyting 1971	Falsity de facto	Falsity de jure
Henry 1972 [glossing Burleigh, Anselm, etc.]	Propositional negation [~(ϕ(**x**))]	Nominal negation [*N*(**a**)]
R. Clark 1974	de dicto (external) negation [−**Pa**]	de re (internal) [$\bar{\mathbf{P}}$**a**]
Kuroda 1977	Proper negation	denial negation [$\bar{\mathbf{S}}$]
Bergmann 1977	External negation [−**A**]	Internal negation [~**A**]
Bergmann 1981	External negation [¬**A**]	Internal negation [~**A**]
Englebretsen 1981a	Predicate denial [**S** is not **P**]	Predicate term negation [**S** is non**P**, **S** is not-**P**]
L. Carlson 1983	Rejection, categorical denial	Denial

To be sure, the distinctions among the pairings above are often significant. Among other relevant parameters, the interpretation of Column A and Column B negations may be affected by whether the operators are motivated semantically or syntactically, whether they operate within a term-based or propositional logic, whether they figure in a two-valued or multivalued system, whether the definer's concerns are linguistic, psychological, metaphysical, or strictly logical, and so on. In some theories, one of the negative operators is defined in terms of the other (e.g., the Column A negation in terms of the Column B negation together with a one-place truth connective); in other systems, both are logically primitive. Some approaches take Column A negation to be ontologically prior (especially those which are linguistically oriented or founded on a classically oriented logic), others Column B (especially those which are psychologically oriented or founded on a multivalued and/or truth-gapping logic), while still others place the two negations on equal footing. But several generalizations can be maintained across these various frameworks—beyond the obvious one, namely, that there seems to have been a remarkable *n*-plication of effort expended. Some of the more consistent regularities are sketched in (70), where I maintain the practice of employing '$-$' and '$\neg$' as metanotations to symbolize Column A and Column B negation, respectively, ignoring the actual practice of a given practitioner.

(70) a. Column A negation may be regarded as a propositional operator or as a mode of predication (affecting the way subject and predicate combine); Column B negation is basically an operation on or within the predicate, although it may be formulated in terms of a propositional function.

b. Column A negation is weaker than Column B negation in the sense that $\neg\mathbf{p}$ unilaterally entails $-\mathbf{p}$.

c. Column A negation is a truth function of the proposition it negates ($-\mathbf{p}$ is true iff **p** is not true, and false iff **p** is true); Column B negation is generally not a truth function ($\neg\mathbf{p}$ is false if **p** is true, but may be true, false, or neither true nor false, otherwise).

d. Column A negation is logically contradictory (**p** and $-\mathbf{p}$ are in contradictory opposition in that in any context one must be true and the other false); Column B negation is logically contrary (**p** and $\neg\mathbf{p}$ are opposed as 'strong' or 'absolute' contraries, in the sense of the terminology developed in chapter 1; they are mutually exclusive but need not be mutually exhaustive).

e. Column A negation conforms to LC and LEM; Column B negation usually obeys LC (under an appropriate formulation) but not LEM.

f. In systems with truth-value gaps, Column A negation is always consistent with LBV ($-\mathbf{p}$ is always true or false), while Column B negation cannot obey LBV without collapsing into its Column A counterpart.

g. Semantic presuppositions, in models admitting them, are typically blocked or canceled by Column A negation; Column B negation preserves presuppositions ($\neg\mathbf{p}$ does, but $-\mathbf{p}$ does not, presuppose whatever **p** does).

h. Column A negation is understood in an absolute, unrestricted sense; Column B negation is often understood against the assumption of a restricted universe of discourse.

These principles are not intended to be either mutually independent or conjointly exhaustive, but are meant to illustrate the sorts of claims typically made within various logical systems that—whatever their other differences may be—share one essential property, that of admitting (at least) two distinct negative operators.

In fact, the generalizations in (70a–h) pertain largely to the post-Russellian systems incorporating dual-negation analyses within a more global formal framework. These generalizations tend not to apply directly to the earlier nonformal approaches, but my intention here has been to point out the parallels that can nonetheless be drawn between systems of thought stipulating the existence of what are essentially two scope-differentiated negative operators. Among the missing in this roster are Kleene, whose delineation of strong vs. weak connectives (1938, 1952) could have been extended, and later was extended (by others), to negation (resulting in precisely the opposite pattern of labels from those provided by Von Wright), and Katz, whose 'denial' is a typical Column B operator opposed to a syntactically based notion of 'negation' which cannot easily be situated in either of the two columns.

The absence of two more familiar paired negations may be noticed. Why not assimilate the NEXAL VS. SPECIAL negations of Jespersen 1917 and the SENTENCE VS. CONSTITUENT negations of Klima 1964 under the headings for Column A and Column B? As we shall see in later chapters, a closer examination of these pairs from the linguistic literature shows that Jespersen's and Klima's syntactically motivated notions are in fact quite distinct from the largely semantically defined operators of (69).

Those theories that accept logical or semantic presuppositions, and the truth-value gaps that arise when these presuppositions are not satisfied, almost inevitably reach the conclusion that natural language negation is ambiguous. Presupposition is typically defined in terms of internal (Column B) negation; those presuppositionalists (notably Frege 1892 and Strawson 1950) who do not admit an external (Column A) negation which is situated outside the scope of the presupposition would seem to be whistling in the dark, given the apparent counterexamples of the type cited in (16!), (65), and (66) earlier in this chapter. Indeed, Frege and Strawson themselves both tacitly allowed for an external, presupposition-canceling use of negation, while not treating it as a separate reading of negative sentences; recall (14), (18), and the accompanying discussion in §2.2. (I shall argue in chapters 6 and 7 that the independent Fregean and Strawsonian reluctance to fully endorse external negation may in fact constitute not a vice but a virtue.)

Thus, a systematic ambiguity for negation—the choice of one from Column A and one from Column B—figures crucially in all theories admitting semantic presuppositions and truth-value gaps, but (as illustrated by Aristotle, Russell, Von Wright, et al.) not only in such theories. Unless negation is treated as semantically ambiguous, Thomason (1973) points out, 'it's not clear that the semantic notion of presupposition can be defended'.

I shall return in later chapters to the viability of dual-negation systems and of the semantic accounts of presupposition they facilitate. But before we leave the ambiguist and presuppositionalist theories entirely behind us, I shall touch on one influential approach which has attempted to combine an ambiguist treatment of negation with a nonsemantic (or at least non-truth-conditional) analysis of presuppositional phenomena.

2.5 Conventional Implicature and Contradiction Negation

Lauri Karttunen and Stanley Peters, in a series of individual and joint publications culminating in Karttunen and Peters 1979 (henceforth K&P), have sought to marry Grice's notion of conventional implicature to Montague's truth-conditional formal semantics and syntax for English. As in Montague's work, the offspring displays an intensional but classically two-valued logic. Within their framework, (71a) entails (72) and indeed, since the entailment is mutual, is truth-conditionally equivalent to it.

(71) a. John managed to solve the problem.
b. John didn't manage to solve the problem.
c. It was difficult for John to solve the problem.

(72) John solved the problem.

However, (71a) differs from (72) non-truth-conditionally, in that *manage to* contributes a CONVENTIONAL IMPLICATURE to (71a)—and to its ordinary negation, (71b)—both of which will thereby end up suggesting something like (71c), their conventional IMPLICATUM. The conventional implicature associated with (71a, b) is part of the meaning of these sentences and is thus distinct from the notion of conversational implicature, also due to Grice (1967, 1975).

Some of the essential differences between conventional and conversational implicatures are spelled out in (73); cf. also Grice 1967, 1975, 1978; Walker 1975; Sadock 1978; Gazdar 1979a; Karttunen and Peters 1979; Hirschberg 1985.

(73)

<table>
<tr><th></th><th>Conventional Implicata</th><th>Conversational Implicata</th></tr>
<tr><td>a.</td><td colspan="2">Make no contribution to TRUTH CONDITIONS, but constrain APPROPRIATENESS of expressions with which they are associated.</td></tr>
<tr><td>b.</td><td>UNPREDICTABLE, arbitrary part of meaning; must be learned ad hoc.</td><td>NATURAL concomitant of what is said or how it is said; NONCONVENTIONAL by definition.</td></tr>
<tr><td>c.</td><td>NONCANCELABLE; apply in all contexts of utterance.</td><td>CANCELABLE, either explicitly (by linguistic context) or implicitly (by extra-linguistic context).</td></tr>
<tr><td>d.</td><td>DETACHABLE: two synonyms may have different conventional implicatures.</td><td>NONDETACHABLE if arising via one of the content maxims (Quality, Quantity, Relation). DETACHABLE if arising via the Maxim of Manner.</td></tr>
<tr><td>e.</td><td>NOT CALCULABLE through any procedure; must be stipulated.</td><td>CALCULABLE through the Co-operative Principle and the Maxims of Conversation.</td></tr>
<tr><td>f.</td><td>Akin to pragmatic presuppositions (noncontroversial propositions speaker posits as part of common ground); cf. Karttunen 1974 and Stalnaker 1974.</td><td>Conceptually related to Mill's (1867) 'sous-entendu of common conversation' (see below) and to Ducrot's (1972) 'sous-entendu' as discourse or rhetorical notion.</td></tr>
<tr><td>g.</td><td>Exhibit a well-defined set of PROJECTION PROPERTIES enabling the implicata of larger expressions to be computed from those of their subparts (K&P).</td><td>Projection properties unclear, since conversational implicatures 'may be indeterminate' (Grice); but cf. Gazdar 1979a, 1979b; Hirschberg 1985 on the determination of implicata.</td></tr>
</table>

As the table in (73) indicates, both conventional and conversational implicata are part of what is meant or conveyed by a given speaker in a given utterance at a given place and time of utterance, without being part of what is said or literally expressed by that speaker in that utterance. The general Gricean framework in which these notions are embedded can be given schematically as in (74):

(74) WHAT IS CONVEYED
- WHAT IS SAID (truth-conditional aspects of meaning)
- WHAT IS IMPLICATED
  - CONVENTIONALLY
  - nonconventionally
    - CONVERSATIONALLY
      - generalized conversational implicature
      - particularized conversational implicature
    - nonconversationally

[*SEMANTICS*]: WHAT IS SAID, CONVENTIONALLY

[*PRAGMATICS*]: CONVENTIONALLY, nonconventionally

The notion of conversational implicature and its relation to negation will be the focus of several of my later chapters, in particular chapter 4.

It should be acknowledged that neither the original Gricean notion of conventional implicature nor its (re)working out by K&P is uncontroversial. In particular, the properties in (73c, g) may be mutually incompatible: cf. inter alia Wilson 1975; Gazdar 1979a, 1979b; Soames 1979; Horn 1979, 1981a. Further, the entire Gricean apparatus, with its key distinction between what is said and what is implicated, has been challenged by recent work within Relevance theory (cf. Carston 1985a, 1985b; Kempson 1986; Sperber and Wilson 1986). Since I am concerned here with the interaction of conventional implicature and negation, I shall pass over the more general issues concerning the nature and behavior of implicata.

The importance of the role played by conventional implicature in the schema delineated by (73) and (74) is apparent: as part of the meaning of an expression and yet not part of its literal meaning (that aspect of meaning which affects truth and satisfaction), conventional implicata are located simultaneously within semantics (construed as indicated above) and pragmatics. Under these assumptions, the truth conditions for (71a, b) cannot be affected by the falsity of their conventional implicatum, (71c); the positive version is true iff (72) is true, the negative iff (72) is false. But (71c) represents an appropriateness condition on the normal, felicitous utterance of both (71a) and (71b).

But is it really the case that (71b) always implicates (71c) and is thus always infelicitous if (71c) cannot be assumed within the context of utterance? Here is where the now-familiar ambiguity of negation resurfaces in yet another guise. K&P concede that in fact (71b) can be uttered felicitously in a context in which (71c) is not only not implicated or taken for granted as part of the common ground, but represents the very component of meaning being denied or negated. On this interpretation, negation 'seems to block off the implicatures of the sentence it has scope over' (K&P, p. 46).

This reading of (71b), which K&P dub CONTRADICTION negation and link up with the familiar external negation operator of MVL discussed in section 2.4, emerges more clearly with the right intonation contour (cf. Liberman and Sag 1974; Ladd 1980) and an appropriate continuation, as in (75) (cf. (16!) above):

(75) John didn't manage to solve the problem—it was quite easy for him.

K&P formalize this marked sense of negation by assigning it wide scope with respect to that material which is conventionally implicated, as in (76b); as seen in (76a), implicata appear outside the scope of ordinary negation.

(76) a. ORDINARY NEGATION OF ϕ: $\langle \neg\phi^e; \phi^i \rangle$
b. CONTRADICTION NEGATION OF ϕ: $\langle \neg[\phi^e \wedge \phi^i]; [\phi^i \vee \neg\phi^i] \rangle$
(where ϕ^e represents the truth-conditional meaning of ϕ and ϕ^i its conventional implicata; the members of each ordered pair denote, respectively, the 'extension expression' and the 'implicature expression' for the form specified)

Substituting into these formulas from the example under consideration here, we determine that the ordinary negation of (71a) amounts to conveying the conjunction in (77), although the first conjunct is implicated and the second entailed. The contradiction negation (as in 75) amounts to the negated conjunction in (77b):

(77) a. It was difficult for John to **VP** $\wedge$ ~(John **VP**'d)
b. ~(It was difficult for John to **VP** $\wedge$ John **VP**'d)

As K&P note (p. 47), the contradiction negation defined in (76b) and exemplified in (77b) is 'by itself non-specific (in the absence of contrastive intonation) in regard to what it is that the speaker is objecting to'; note that both entailment (72) and implicatum (71c) are within the scope of this marked negation operator.

In the language of Karttunen's earlier work (e.g., Karttunen 1974), ordinary negation is a HOLE to presuppositions, aka. conventional implicata, where contradiction negation is a PLUG. (The plug nature of contradiction negation is represented in (76b) by the assignment of a tautological implicatum to this reading; an analysis in Ducrot 1972 prefiguring that of K & P in this and other respects will be discussed in chapter 6.)

This approach echoes Russell's scope-differentiated dual-negation analysis for singular propositions: just one selected conjunct is negated in (16′) and in (77a), while the entire conjunction is negated in (16″) and in (77b). The crucial question is how to assure, without stipulation, that it is just the appropriate conjunct (the one corresponding to the existential proposition and to the 'difficulty' implicature, respectively) which gets selected for special treatment. Two devices proposed for accomplishing this task in work postdating Russell and predating K & P were a kind of translucent bracketing in Grice (1967, 1981) and the 'heavy parentheses' notation for enclosing presupposed material in Katz (1972: 167–78). (Katz differs from K & P in accepting semantic presupposition, while Grice has wavered on this question.)

The presuppositional phenomena on which K & P exercise the force of their argument tend to involve not the traditional bones of philosophical contention I have picked over earlier in this chapter—singular terms, category mistakes, future contingents—but rather such "small" lexical items as *manage to* (as in (71)), *even*, and *too*. Similarly, the locus classicus of Grice's conventional implicature is the range of adverbs and conjunctions (e.g., *but, therefore*) which seem to contribute some non-truth-conditional aspect of meaning to the sentences in which they occur.

Thus, Grice argues that (78) and (79) are assigned the same truth conditions as the corresponding simple conjunctions.

(78) Mr. X is a politician but he is honest.

(79) Henry is an Englishman; he is, therefore, brave.

Sentence (78) is true iff Mr. X is both a politician and honest, (79) iff Henry is a brave Englishman. What *but* and *therefore* contribute are conventional implicatures involving contrast with expectation in the former case and causal connection in the latter. If these conditions are not satisfied in the context in which (78) and (79) are uttered, the utterance is inappropriate, but the proposition thereby expressed (i.e., the simple conjunction) remains true so long as each conjunct is true.[14]

Similarly, on K & P's theory, (80)—like (71a)—has the same truth conditions (is assigned the same extension expression) as (72),

(80) Even John solved the problem.

(72) John solved the problem.

differing from (72) in conventionally implicating that others solved the problem and that John was the least likely (of a contextually designated set) to have done so (cf. K & P, p. 23–32 for details).

More graphically, the K&P analysis for a sentence involving a simple NP-focus instance of *even* like that in (81) proceeds as indicated in (81′):

(81) Even BILL likes Mary

(81′) **Focus (even)**: 'Bill'
Scope (even): 'x likes Mary'
Existential implicature: 'There are other x under consideration besides Bill such that x likes Mary'
Scalar implicature: 'For all x under consideration besides Bill, the likelihood that x likes Mary is greater than the likelihood that Bill likes Mary'

As I just noted, (81) will come out truth-conditionally identical to the corresponding *even*-less statement, that is, *Bill likes Mary.*

While there is much to be said both for and against the K&P approach to *even,* I shall concentrate my fire on the interaction of *even* and negation.[15] The first problem here is that the implicata induced by *even* are too strong, to the point of apparent invulnerability to 'contradiction' (external, plug) negation. Thus the oddness, for many speakers, of the examples in (82), alongside the parallel cases of well-behaved plug negation in the *manage* paradigm (cf. (75) above):[16]

(82) a. # It's not the case that even Bill likes Mary—he likes her, but {he's the only one who does / that's to be expected}.
b. # Susan didn't solve even the last problem—she solved it, but {it was the only one she solved / it was certain she'd solve that one}.

The impossibility of these cancelations is especially striking, given the fact that the implicata in question can be removed without the use of negation. Thus we have the following attested examples:

(83) a. Living in such glass houses, even a President—perhaps particularly a President—should hesitate to throw campaign stones. (Tom Wicker's column, *New York Times,* 12 September 1980)
b. Heterosexuality, like motherhood, needs to be recognized and studied as a political institution—even, or especially, by those individuals who feel they are . . . the precursors of a new social relation between the sexes. (Adrienne Rich, "Compulsory Heterosexuality and Lesbian Experience," *Signs* 5 [1980])
c. —Surely even the Japanese understand quid pro quo.
—On the contrary, they've made an art of it. (Conversation reported by Bob Ladd.)

d. Nobody can invent a new prayer from the heart five days a week. Not even a Congressman. Even especially [*sic*] a Congressman. (Arthur Miller, on prayer in public schools, *New York Times,* 12 March 1984)

Particularly favored as a means for detaching the implicata of (81′) is the *even . . .* {*particularly / especially*} gambit, often accompanied by a free-standing *no* of correction or midcourse adjustment (see DuBois 1974 and chapter 6):

(84) a. This has not always been the view even of grammarians, or perhaps, more correctly, particularly of grammarians. (Fromkin and Rodman, *Introduction to Linguistics,* p. 10, on the putative equality of all languages and dialects)
b. There was always something sad about him, even—no, especially—in his smile. (Ruth Prawer Jhabvala, *In Search of Love and Beauty*)
c. Even Mark Langston—no, especially Mark Langston—is impressed with what Dwight Gooden has done this season. (opener of *New York Times* piece, 14 September 1984, on the two rookie strikeout phenoms of 1984)

Then there is the all-purpose metalinguistic implicatum canceler illustrated by David Lewis (1979:339) in his imaginary exchange in (85):

(85) —Even George Lakoff could win.
—Whadda ya mean, 'even George'?

Why then are the simpler cancelations of (82) virtually unimaginable? Notice that to negate a presupposition or conventional implicature (frequently) is to turn that aspect of non-truth-conditional meaning into an assertion, as in (75) or in (86), where the normal assertion / presupposition relation of meaning components is reversed by the application of the metalinguistic use of negation (cf. Horn 1985 and chapter 6):

(86) He hasn't stopped beating his wife, he never started.
I'm not here again, I'm still here.

But the counter-to-expectation aspect of meaning associated with *even* and related adverbs evidently cannot be focused on by negation. Thus, compare the behavior of the adverbs in the frame of (87):

(87) He didn't resign {because he had tenure. / #although he had tenure. / #despite having tenure.}

Nor can counter-to-expectation clauses be directly focused on in positive assertions:

(88) It was {because / only because / #although / *even though} he had tenure that he resigned.

But this fact, whatever its ultimate motivation, cannot be the whole story, since (as (82) indicates) the existential implicature of (81) is no more permeable to wide-scope contradiction negation than is the scalar implicature. Notice in this connection the peculiarity of (89), where the existential implicature contributed by *too* (i.e., that Bill likes some x ≠ Mary) cannot be contradicted:

(89) #Bill doesn't like Mary too—she's the only one he likes.

Once again, other metalinguistic devices for removing the existential implicatum seem to fare better (*Whaddaya mean, he likes Mary too?!*) The non-truth-conditional semantics of *too* are discussed by K&P, pp. 32–43; cf. also Horn 1972: §1.12 for a tentative preliminary stab at the impermeability of the existential implicature associated with *too* and *even*.

An even more basic problem about *even* sentences and their interpretations, for the K&P model or anyone else's, is that no ordinary negation of a sentence like (81)—preserving its presuppositions or implicata while denying its entailment—is available either. Thus, in the paradigm of (90),

(90) a. Even Bill likes Mary.
b. Even Bill doesn't like Mary.
c. Not even Bill likes Mary.

the K&P analysis correctly predicts that *even* in (90b) takes wide scope with respect to the predicate negation, so that this sentence differs from (90a) both in asserting that Bill does not like Mary and in implicating that someone other than Bill does not like Mary and that Bill is the least likely member of a contextually designated set to not like Mary. But there is no way for this or any other straightforward compositional approach to account for the fact that (90c) is neither the ordinary (hole) negation of (90a) nor its contradiction (plug) negation. Instead, (90c) must be assigned the same truth-conditional meaning and implicata as the syntactically dissimilar (90b).[17]

In fact, (90a) has no negation. Similarly, Kuroda (1977:70–71) observes that there is no 'denial' available for (91a), maintaining its presuppositions but denying its entailments; he notes that (91b) cannot be read as the negative statement paired with the positive (91a) as possible answers to the question *Does John even love Mary?*

(91) a. John even loves Mary.
b. John does not even love Mary.

In (91b), as in (90c), *even* must be assigned wide scope with respect to negation; the sense is approximately that of 'Even Mary is such that John doesn't love her'.

It is the type of pattern reflected here that motivated my earlier treatment of *even* as a 'neg-raising predicate', allowing the coderivation of (90b) and (90c); cf. Horn 1969:105; 1971:128–32. But even if we quite properly reject such a scheme out of hand, we must be able to deal with the semantic equivalence between (90b, c).[18] One of the virtues of Karttunen and Peters's theory of presuppositional phenomena is its falsifiability, deriving from its formal explicitness and, in particular, from the strong constraint on possible interpretations forced by the rule-to-rule hypothesis and well-formedness constraint built into their Montague-based format. In the case of the syntax-semantics pairings in negative *even* sentences (as I argued in more detail in Horn 1979), the theory is not only falsifiable but falsified.

I shall return in chapter 6 to the task of situating the K&P theory of implicature and contradiction negation with respect to other dual-negation frameworks, bivalent and multivalued, and with respect to those frameworks in which the ambiguity thesis for negation is rejected. The principal rival to K&P in accounting for the projection properties of conventional implicata (aka. pragmatic presuppositions) is Gazdar (1979a, 1979b), who falls within the camp of the radical monoguists on negation. I shall argue that neither his position nor K&P's, nor in fact those of the unrepentant ambiguists (classical or multivalued), permits us to capture the whole picture of the unity and diversity of natural language negation.

In chapter 1, I looked in some detail at the properties of one particular dual-negation theory, that of Aristotle. In this chapter, I have explored the relation between negation and the classical (Aristotelian) laws of contradiction and of the excluded middle. We have seen that there is no royal road to negation (even under the protection of the king of France), although we have followed our quarry down some rather picturesque (if poorly maintained) logical byways.

I began by focusing on the behavior of negation within a variety of long-studied but still controversial constructional types—future contingent statements (§2.1), singular expressions with nondenoting subject terms (§2.2), and category mistakes (§2.3)—which some have seen as the death knell for the classical principle of bivalence, the view that every proposition must be either true or false. I explored a range of analyses of these constructions and entertained the possibility of modifying Aristotle's dual-negation two-valued logic to handle the data presented. But nothing in what we observed forced us to abandon the essential Aristotelian premises; in fact, as we saw in (§2.4), the problems encountered in the formal elaboration of multivalued logic suggest that a refinement of Aristotle's own approach to the central questions (often echoed, without acknowledgment, by Russell) might be more rewarding. One such approach is that developed by Von Wright (1959), although this involves the somewhat uncomfortable translation of the term-logic notion of contrariety into a proposition-based logic where it is less at home. In the last two sections, I surveyed a variety of frameworks predicating an ambiguity for negation, with the readings differentiated either lexically or by the scope of operators in logical form. This survey concluded in (§2.5) with the Karttunen and Peters (1979) theory of conventional implicature and non-truth-conditional semantics, in which negation is not always contradictory but the logic is always bivalent.

The thesis that negation is ambiguous, assumed or defended in several of the theories under consideration in these first two chapters, will reclaim our attention in the last two. But before returning to that fray, I have other conflicts to take up. The first of these involves the so-called markedness of negation and its relevance to the asymmetry wars whose history I began chronicling in §1.2. It is to this question that I now turn.

3 Markedness and the Psychology of Negation

> With respect to vocal sound, affirmative enunciation is prior to negative because it is simpler, for the negative enunciation adds a negative particle to the affirmative.
>
> (Saint Thomas Aquinas, commentary on Aristotle, *De Interpretatione*, in Oesterle 1962:64)

> The fact that affirmatives are unmarked and negatives are linguistically marked is completely correlated with the finding in this study and previous studies that affirmatives are psychologically less complex than negatives. (Just and Carpenter 1971:248–49)

As we have seen, Saint Thomas's attempt to correlate the morphosyntactic marking of negation (as against the absence of marking for affirmation) with the purported priority of affirmation over negation was echoed by a millenium of asymmetricalist scholarship (§1.2). But, as we have also seen, this correlation has been challenged, most directly by Geach ([1972] 1980:78–79). Geach's remarks, however, are confined to the relative complexity of of affirmative and negative predicates, and it might be maintained by proponents of two-negation systems (cf. chapter 2) that while negative predicates might indeed be ontologically no more complex than their affirmative counterparts, notwithstanding the morphosyntactic asymmetry, negative propositions are inherently more complex than their positive mates.

Another would-be exorciser of the Thomist argument for affirmative priority is Sommers (1970:6), who inveighs against the 'nominalistic error' of thinking that negative terms are necessarily derived from neutral (or positive) terms. It is, Sommers suggests, an unfortunate 'accident' that there is no term **plus-wise* alongside and symmetrical to the occurring *unwise*. Sommers's approach is to follow the 'classical' tradition in which terms are given in contrary pairs, and it is arbitrary which member of the pair we take as positive and which as negative. As I mentioned in my discussion of Geach's argument, however, this line is most convincing when we are dealing with exhaustive sets of two contrary terms where the term negation can be expressed affirmatively (e.g., *male* vs. *non-male* or *female; odd* vs. *not-odd* or *even*). The argument loses force when we turn to pairs of mediate, especially polar, contraries (cf. §1.1.5): is it really an accident that we have a simple lexical expression for 'white' but not for 'not-white'? Is it any more accidental that the realization of 'wise' or 'happy' tends to be simple while that of their immediate or logical contraries does not?[1]

Within modern linguistics, such a verdict of "accidental" for a systematic correlation between formal (morphosyntactic) and functional (semantic) asymmetries would be overturned on appeal. Universalists of every stripe—Praguean (e.g., Jakobson, Firbas), empirical-comparativist (e.g., Greenberg), generativist (e.g., Chomsky, Bickerton), neofunctionalist (e.g., Halliday, Kuno), pragmatic (e.g., Grice), or Californian (e.g., Bolinger, Givón)—have sought to explain surface regularities of natural language by invoking deeper and/or more general principles governing the innate linguistic mechanism (aka. UG), the exigencies of rational communication, or the packaging of information. For such approaches, the accident invoked by Geach and Sommers is simply a null hypothesis devoid of independent interest.

The Thomist premise instantiates a pattern ranging far beyond negation and affirmation. The notion of MARKED OPPOSITION in language was developed by Jakobson, Trubetzkoy, and other members of the Prague linguistic circle, building on earlier work by Saussure and Bally (cf. Jakobson 1939). Markedness involves the correlation of a formal asymmetry with a functional or semantic asymmetry. Formally the key notion is that of Jakobson's *SIGNE-ZERO:* one member of an opposed pair is literally MARKED (overtly signaled) while the other is UNMARKED (signaled via the absence of an overt signal).[2] Semantically, the marked category is characterized by the presence of some property **P**, while the corresponding unmarked category entails nothing about the presence or absence of **P** but is used chiefly (although not exclusively) to indicate the absence of **P** (Jakobson 1939).

In the clearest cases of opposition, the formal and semantic criteria coincide. Thus *lion* is formally unmarked with respect to *lioness,* given the absence vs. presence of the suffix, and this suffix 'states the presence of a certain property A', (+ *female*), while the corresponding unmarked category may or may not be understood as lacking that property (i.e., *lion* may denote either a male *Felis leo* or simply a *Felis leo* undifferentiated for sex). Similarly, within the human realm, *woman* is treated formally and functionally as a marked *man;* cf. Bodine (1975), Miller and Swift (1976), and Martyna (1983) on the history, extent, and significance of what Martyna has dubbed '*he/man* language'.

For Jakobson and for Greenberg (1966), it is the unmarked term of a binary opposition which tends to be formally less complex (often with zero realization, e.g., *lion* vs. *lioness*), morphologically more irregular, distributionally less restricted (and more frequent in text tokens), semantically more neutral (*How tall is Chris?* vs. *How short is Chris?*), syntactically more differentiated, less likely to undergo syncretization or neutralization, and more likely to govern *a potiori* agreement.

Now by the formal criterion, the marked member of the positive/negative opposition is clearly negation (Greenberg 1966:26). On the basis of his characteristically careful and extensive survey of the world's languages, Greenberg (p. 50) concludes that 'the negative always receives overt expression while the positive usually has zero expression'. While Vietnamese contains a formal marking for the positive category, this marker is not (unlike its negative counterpart) obligatory, and the far more frequent pattern is for the positive or affirmative category to be indicated simply by the absence of a specific negative marker, either a particle (*John is happy* vs. *John is not happy*) or an affix (*John is happy* vs. *John is unhappy*).[3]

This formal pattern is reflected even within the austere confines of mathematical and logical symbolism, as Greenberg (1966:25) notes. Thus, negative numbers must be so designated (e.g., -5), while their positive counterparts may be indicated by lack of overt marking (5 or $+5$). The absolute value (where the polarity distinction is neutralized) is given in the unmarked, positive form, as $|5|$ rather than $|-5|$. And the inherently positive number of the *true/false* dichotomy gives its name to the category as a whole: we speak of *truth* (not *falsity*) value and *truth* (not *falsity*) conditions.

But negation is marked semantically, as well as formally.[4] Greenberg (1960:50) cites a variety of syntactic patterns in a cross section of natural languages as 'evidence for the marked character of the negative as opposed to the positive'. One such pattern is the tendency for verbal categories which are distinguished in positive clauses to collapse in the negative; thus, the Shilluk present and future inflections are syncretized under negation. Ancient Greek would seem to present a counterinstance to Greenberg's generalization, in that the indicative and subjunctive fall together in the affirmative, negation often keeping them separate (cf. Mirambel 1946). But the preponderance of the evidence tilts overwhelmingly in Greenberg's direction. Bhatia (1977:§3.1.2.2), for example, shows that in Indian languages (both Indo-Aryan and Dravidian), verbal distinctions maintained in the positive are characteristically neutralized in the negative, subject to other rules and strategies of the language in question.

An extensive list of such neutralizations toward the unmarked positive values is offered by Givón (1978, 1979), who seeks to link the formal and presuppositional markedness of negation (cf. §1.2.2 above) to the distributional restrictions placed on negative clauses. For Givón, negation—like other marked constructions—is allowed both 'less freedom in the distribution of elements or structures embedded in it' and 'less freedom in embedding itself in other structures or contexts' than the corresponding unmarked (i.e., affirmative) constructions (1978:91). Givón's evidence for the restricted distribution of—and under—negation ranges from the scope of English adverbials and modals to the patterns of verb-focus and tense-aspect constructions in Bantu languages (1978:81–87, 92–101); some of his English-based contrasts (from 1978:95–96) are reproduced below:

(1) a. When John {comes / ?doesn't come}, I'll leave.
b. When did John {arrive / ?not arrive}?
c. How did he {do it / ?not do it}?
d. With what {did he / ?didn't he} cut the meat?
e. I had the doctor {examine / ?not examine} Mary.
f. I {want to / ?want not to / don't want to} work.
g. She was as fast as he {was / ?was not}.
h. And then {came / ?didn't come} John.
i. There {stood / ?didn't stand} a man in front of the mirror.

While aspects of Givón's data, argumentation, and conclusions may well be challenged, the essential point is unarguable: negative structures are cross-linguistically barred from a wide range of environments in which the corresponding affirmatives are well-formed. (A closer look at the universal "conspiracy" to avoid overt negation in embedded nonfinite clauses is offered in Horn 1978a: §5.)

The converse claim, that negative structures embed a more restricted class of syntactic sequences than positive structures do, runs up (as Givón acknowledges: 1978:96–97) against the fact that many languages contain a large—often fully productive—class of NEGATIVE POLARITY ITEMS, expressions whose distribution is limited to environments containing a commanding overt or incorporated negative (or related "affective" element, e.g., question or conditional). Positive or affirmative polarity items do exist, but their number, productivity, and strength are less impressive, presumably because their trigger can be specified only negatively, in terms of the absence of a negative or affective element.[5]

A related issue—the restricted distribution of the marked and inherently negative member of contrary adjective pairs (e.g., *short* vs. *tall, bad* vs. *good;* cf. Givón 1978:104–5) will be discussed in more detail below. But it is worth noting here that while the marked member of an opposition is standardly viewed as conveying more information than its unmarked counterpart (*How short is she?* is more informative than *How tall is she?* since the former conveys that [the speaker believes that] she is short, while the latter is neutral), the marked member of the positive / negative asymmetry, negation, has been recognized by commentators from Plato to Givón as contributing less information to the discourse than the corresponding unmarked affirmative (*My hat is not red* vs. *My hat is red*).

Furthermore, the positive / negative opposition, in predicates, propositions, or sentences, does not directly reflect one of the fundamental properties of marked oppositions, that property cited in the passage from Jakobson above, or in Greenberg's discussion (1966:25) of 'the pervasive nature in human thinking of [the] tendency to take one of the members of an oppositional category as unmarked so that it represents either the entire category or *par excellence* the opposite member to the marked category'. That is, positive sentences and predicates do not normally exhibit an extended use which is neutral as between the unmarked (positive) and marked (negative) members of the entire category. *Pat is a man* cannot be neutral as between 'Pat is a man' and 'Pat is not a man' in the way that *man* (representing *Homo sapiens*) may be neutral between '(male) man, *vir*', and 'human, *homo*', that *lion* includes both '(male) lion' and 'lioness', and that 5 stands for −5 as well as +5. (I may note in passing, however, that a neutral, nonconducive yes-no question—e.g., *Is Pat a man?*—is invariably formed from the positive member of the opposition; cf. Bolinger 1957.)

In any case, assuming that these problems with the notion of markedness, as it applies to negation vs. affirmation, can be successfully addressed, the question remains: what are the implications of the formal and functional markedness of negation for the grammar and logic of negative statements? Kissin (1969:28) defends the strong view: 'The basic fact which the grammar of negation must contend with is that negative sentences contain an overt element that nonnegative sentences do not contain . . . but there is no such nonnegative element to be pronounced in a nonnegative sentence.' Kissin sees this asymmetry in overt marking as the connecting link between the syntax and semantics of negation, although he stipulates, rather than demonstrates, that the asymmetry described by Greenberg must be reflected by a syntactic and semantic asymmetry in the analysis of negation in a synchronic grammar of (e.g.) English.

Kissin's contention that negative sentences must be treated as 'parasitic' on their nonnegative congeners, in the manner of Klima (1964), as against either the contrary (straw) theory in which negatives are basic or the view that each sentence type is equally basic, provides a straightforward means for capturing the marking asymmetry, but we are never told how universal grammar might be set up to force our metatheoretical hand. Indeed, Kissin's point might itself be circular if we were to define the negative member of a contradictory pair $\langle S_{pos}, S_{neg} \rangle$ as that member whose deep structure differs from that of its counterpart by containing an overt element in a specified position. Negation would then be the marked (or parasitic) member of the opposition by fiat.

Perhaps the most direct reflex of the priority of unmarked over marked terms in general, and affirmation over negation in particular, is provided by the linear order of conjuncts (and disjuncts) in what have been variously called FIXED BINOMIALS (in Malkiel 1959) and FREEZES (in Cooper and Ross 1975). A freeze is a conjunction in which 'the ordering of the two conjuncts is rigidly fixed in normal speech' (Cooper and Ross 1975:63), for example, *cat and mouse* vs. *?mouse and cat, bigger and better* vs. *?better and bigger, fore and aft* vs. *?aft and fore*. Linguists since Pāṇini have investigated the phonological and semantic constraints on order within freezes, and there seems to be a clear consensus on the role of a variety of factors, if not on the relative weighting to be assigned to each factor. The central semantic principle is that, ceteris paribus, the unmarked or "easier" member of an opposition will precede the marked or "harder" member (Clark, Carpenter, and Just 1973), for example, *big and small, this and that, man and woman*. In particular, the affirmative member of an opposed pair of contrary terms always takes linear precedence over its negative counterpart. Cooper and Ross (1975:65, domain (13)) include such examples as

(2) positive or negative   many or few
all or none   more or less
plus or minus   win or lose

The negative term relegated to second place may be overt, as in *yes and no* or the first two examples in (2); it may be covert or "inherent", as in the other examples above, or in

(3) pro and con
tall and short
good and bad

(Cf. H. Clark 1974 and Givón 1978 for arguments that the marked member of contrary adjective pairs is inherently negative.) The connective may be *and* or *or*, depending on the context (*plus or minus / pluses and minuses; rich {and / or} poor*). In some cases, the first (affirmative) entry into the freeze may be both formally and semantically unmarked with respect to the second; such instances comprise Cooper and Ross's category of 'A or Neg-A', for example, *happy or unhappy, like or dislike*. In one instance, however, these two criteria clash, yielding a paradox of self-reference:

(4) marked {and / or} unmarked (cf. ?unmarked {and / or} marked)

Here the formal parameter wins out, suggesting that the first conjunct may be (at least literally) the *marked* member of the opposition.

Another counterexample to the general unmarked / marked, positive / negative pattern in freezes comes from Chinese, but its delineation requires some additional background. Osgood and his associates (cf. Boucher and Osgood 1969, Osgood and Richards 1973, Hoosain and Osgood 1975) see in the opposition between an unmarked positive and a marked negative an everlasting war-of-the-words between two cosmic principles, YIN and YANG. Affirmation and negation are simply the cognitive reflex of 'a polarity between two global forces which can only be termed THE POSITIVE AND THE NEGATIVE' (Osgood and Richards 1973:380; capitals in original).

This characterization derives from the *I Ching*, the ancient (4,000-year-old) Chinese Book of Changes, and its Confucian commentary. In the metaphysics of the Confucians (and the Taoists, e.g., Givón 1978:109), the universe divides into two apparently infinite sets of opposed pairs:

(5)	**Yang**	**Yin**
	positive	negative
	light	dark
	heaven	earth
	high	low
	creative (active)	receptive (passive)
	male	female
	gods	ghosts
	large	small
	hard	soft

Osgood et al. stress the correlation of these ancient oppositions with markedness, both formal and functional, as defined by Greenberg (1966), arguing that the positively evaluated (*yang*) member of a bipolar cognitive opposition tends to be more frequent and overwhelmingly more likely to accept a negative affix if either member does (e.g., *unhappy* / **unsad;* cf. Zimmer 1964 and §5.1 below). Osgood and Richards show that subjects in a study of cognitive interaction are much more likely to link two adjectives with *and* when the adjectives are both *yang* (*sweet and* / **but kind*) or both *yin* (*ugly and* / **but cruel*), while *but* is the connective of choice for mixed adjectival marriages (*sweet but* / **and cruel, ugly but* / **and brave*).

But the question remains: 'What determines cognitive polarity—the Yang and Yin of things?' (Osgood and Richards 1973:409). Frequency alone (a language user's greater exposure to the unmarked *yang* term) is not sufficient to make this determination; cf. Greenberg (1966:100). The POLLYANNA HYPOTHESIS (Boucher and Osgood 1969; cf. also Leech 1983) posits a socially adaptive value to communication about positive rather than negative aspects of life ('Accentuate the positive, eliminate the negative', once more with feeling)—but this can't be the whole story either: in what sense does width (and not narrowness) or thickness (and not thinness) represent a 'positive aspect of life'? Osgood and Richards's phylogenetic speculations (1973:410) are no more compelling. Givón (1978:103ff.), who is entirely sympathetic to this approach, focuses on the perceptual salience of the unmarked term, which seems plausible enough in the case of *wide* vs. *narrow* or *big* vs. *small*, but somewhat less convincing for *happy* vs. *unhappy*, *believe* vs. *doubt*, or *accept* vs. *refuse*.

Whatever the ultimate determining factor(s) for yang- and yinhood, Osgood and Richards are confident in depicting a universal, perhaps innate, preference for yang over yin, plus over minus, with evaluation (rather than potency or activity) the central parameter of opposition, reflecting 'different reinforcing mechanisms' for the two polar opposites:

> It would appear that from time immemorial humans have been differently reinforced for strength (rather than weakness), for activity (rather than passivity), for height (rather than shortness), and . . . for maleness (rather than femaleness); that humans have found believing more reinforcing than doubting, certainty more than uncertainty, plenitude more than scarcity, asserting more than denying. (Osgood and Richards 1973:411)

Whether a Cro-Magnon, confronted with a tail-wagging saber-tooth or woolly mammoth, would really have found the yang qualities of belief, trust, and hope more 'reinforcing' than the opposed yin qualities of doubt, distrust, and fear is a question best left to the cognitive and behavioral psychologists of prehistory. But one surprising empirical finding is that, contrary to our a priori expectation and to the observed result of that expectation reflected in the practice of Osgood and Richards (where we obtain over twenty sentence-internal sequences of the sequence yang . . . and . . . yin from the title to the final sentence, as against no occurrences of the form $_s$[yin . . . and . . . yang]), the actual form of the freeze in ancient Chinese was invariably Yin and Yang. Why this should be is unclear, although Hugh Stimson has suggested to me that the freeze order might reflect the fact that the ontological priority of yang is overridden by the cosmological priority of yin: Yin precedes yang in that as the physical universe was created out of nothingness, so was yang created out of yin.

3.1 Markedness and the Acquisition of Negation

Marked categories tend to be harder, and take longer, for children to acquire, and negation does not disappoint this expectation. To be sure, there is some question as to just what constitutes the task we call the acquisition of negation. At the same time, negation has provided an attractive target for psycholinguistic research, precisely because of its linguistic complexity as against its universality, its early appearance in the child's repertoire (by 18–24 months), and its presumed logical simplicity (at least within the two-valued, one-negation propositional model). Let us survey some of the central findings—and speculations—in recent (and not-so-recent) research on the acquisition of negation.

Based on evidence from the acquisition of English, French, Russian, and Japanese, Bellugi concludes that children's earliest (STAGE 1) negative utterances contain an initial or (less frequently) final negative marker (Klima and Bellugi 1966: 191–93; Bellugi 1967; cf. also McNeill and McNeill 1968). The examples of peripheral Stage 1 negation cited for English (e.g., *no drop mitten, no the sun shining, wear mitten no*) conflate a number of different negative functions. In later stages, the separate functions and distinct placements of negation are gradually differentiated.

McNeill and McNeill (1968: 72–73) distinguish four values of negation for children learning Japanese:

(6) *nai* (aux), attached to verbs and adjectives = FALSITY
nai (adj), with verbal force = NONEXISTENCE
iya, 'I do not want' = INTERNAL DESIRE
iiya, 'What was just said was wrong; something else is right' = ENTAILMENT

McNeill and McNeill claim that the earlier uses of negatives in children's Japanese involves the general sense 'not here', only later developing into the more abstract meaning 'false'; negation first marks relations involving internal states and the external world, before extending to mark relations about language (as is implicit in declarations of falsity).

The more influential taxonomy posited by Bloom (1970: chapter 7) contains three developmental categories of negation, with a built-in temporal ordering. The categories are

(7) NONEXISTENCE > REJECTION > DENIAL (where > is read as 'precedes in acquisition')

Bloom's first two categories correspond to McNeill and McNeill's categories of nonexistence and internal desire, respectively, but her third category merges the McNeills' entailment and falsity. The following features, examples, and glosses are offered by Bloom (1970: 172–73):

(8) NONEXISTENCE (existence vs. nonexistence)
no pocket (in Mommy's shirt)
gloss: Referent not manifest in context, where it was expected.
REJECTION (internal vs. external)
no dirty soap (i.e., 'I don't want . . .')
gloss: Referent existed or was imminent, but was rejected or opposed by child.

DENIAL (entailment vs. nonentailment)
no truck (after being given a car: *There's the truck*
—*No truck,* i.e., 'This isn't a truck')
gloss: An actual or supposed predication was not the case.

Bloom's first and third categories both can be read as involving an assertion of the absence of something expected; in the former case, the expectation is set up by the child's belief world, in the latter by the prior linguistic context. But the second and third categories are similar in that both involve a comment on a present referent (rejecting it in the former case and rejecting a characterization of it in the latter). It is for this reason, Bloom suggests (p. 219), that nonexistence must be (and is) expressed first: 'The child needed to express nonexistence syntactically in order to transmit information, whereas syntactic expression of rejection and denial was less necessary'.

Rejection need not be verbalized, since it can be adequately signaled nonverbally; when it is verbalized, its expression can be in the form of a simple *no,* while nonexistence requires a more complex syntactic realization. The function of denial is acquired last, since it involves the most abstract relation, with a symbolic referent. Contrastive denials (cf. the McNeills' category of entailment) involve 'holding two propositions in the mind at once', as in Bloom's citation *That's not mines, that's dolly's,* representing a still-higher level of complexity.

Bloom's speculations on why the proposed temporal ordering should exist are not obviously incorrect, but they are not totally convincing either. It is not always the case that we know how to classify a given instance of a child's negative utterance, nor do the classifications we do arrive at invariably support her ordering hypothesis. Bowerman (1973), for instance, in her intensive study of the acquisition of Finnish, finds that the data are often hard to interpret within Bloom's model; to the extent that we can interpret them, they seem to point to denial as the most basic (earliest and most frequently instantiated) category for Finnish negatives, contrary to the predictions of the taxonomy in (7).

Bloom's rejection category corresponds to what philosophers—at least since Peirce—have long identified as the SUBJECTIVE or PRELOGICAL negative. Heinemann (1944 : 138) glosses this 'prelogical use of negation' as 'I do not wish (will, desire, etc.) that' or 'It is not in my interest that', alongside the 'logical' negation of 'It is not true that'. On this view, the rejection category should antedate both nonexistence and denial; that it does not (at least in Bloom's data) may reflect the difference between possessing a concept and expressing it syntactically (a point made by Bloom herself in the passage cited above).

Whether we follow Plato's identification of negation with difference or a

Freudian model emphasizing the traumatic nature of the infant's realization of otherness (Mother ≠ me), we may be tempted to agree with Heinemann's distinction (1944:140) between a 'lower (prelogical) level' on which negation can signify 'to separate, exclude, reject, or eliminate' and a 'higher (logical) level' on which only the eliminative function remains. But Heinemann's argument that 'the function of negation' cannot be 'limited to the sphere of logic' because negation can appear 'in refusal, prohibition, imperatives, and wishes' (p. 137) is only valid if we restrict ourselves (as Von Wright, Rescher, Lewis, and others show we need not do) to a logic of declaratives. It is precisely to avoid sweeping negation and similar operations under the 'nonlogical' or 'prelogical' rug that deontic and imperative logics have been devised.[6]

The importance and cognitive complexity of the *no* of rejection are emphasized by Royce (1917:265), who observes that 'a conscious voluntary action is possible only to a being who understands the meaning of "not"'. The child's *no* / *not* of rejection is 'a fundamental tool used to express, not necessarily disobedient refusal, but objecting or unwillingness, or a preference and desire standing in some sort of negative contrast to the modes of actions which the questions and proposals of his elders or playmates suggested'. At this point, of course, we have an instance of 'holding two propositions in the mind at once', and the line between Bloom's categories of rejection and denial begins to blur.

Russell (1948:500ff.) offers an interesting, although necessarily speculative, theory of the 'ontogenesis' of negation. The logical *not* derives from a more basic *no,* learned early on with the association of 'unpleasant feelings'. This *no* is not the child's sign of rejection but the parent's sign meaning 'disadvantageous to act upon': '"Yes" means "pleasure this way" and "no" means "pain that way"', either direct pain via reality or indirect pain via social sanctions. How do we get from this *no* as a warning signal to the *not* of propositional logic? 'We may say that "not" means something like "You do right to reject the belief that . . .". And "rejection" means, primarily, a moment of aversion. A belief is an impulse toward some action, and the word "not" inhibits this impulse'.

While this account has some plausibility, and indeed the view of negation as a generalized "inhibiting" operation seems to appeal to psychologists of our own era (cf. Vandamme 1972:59–64), Russell must execute a number of prodigious leaps of faith over the apparent holes in his argument. Does *not* really mean—in childhood, adulthood, or anywhere else—'You do right to reject the belief that . . .'? (Note, incidentally, the neo-Bergsonian subjectivist view of negation this approach seems to build in.) Is the aversion (if any) involved in disbelief related (ontogenetically or ontologically) to the infant's aversion to painful stimuli, partially conditioned by the parent's *no?* Are we willing to accept Russell's identification of belief as 'an impulse toward some action'? A negative answer to any of these questions entails a negative verdict on Russell's negative ontogenesis.

Recent empirical work in developmental psycholinguistics has focused on the pragmatic and discourse properties of negation that the child must learn along with negative syntax and semantics. Volterra and Antinucci (1979) ground their 'pragmatic study' of negation in child language in the classical asymmetricalist theses that negatives are necessarily second-order statements and less informative than the corresponding affirmatives; they cite Kant, Bergson, and Russell (cf. §1.2) approvingly to this effect. On Volterra and Antinucci's view (1979:283), 'the core of negation is . . . to deny a corresponding (implicit or explicit) affirmative statement'. In uttering a negative sentence, the speaker (invariably?) 'denies a corresponding positive presupposition attributed to the listener'. The problems I have already surveyed in pinning down precisely what constitutes an 'implicit affirmative statement' or in assimilating all instances of negation (including embedded negation; cf. Frege 1919) to a speech act of denial or presupposition cancelation will not be rehearsed here.

Volterra and Antinucci posit a taxonomy for child (and possibly adult) language in which four types of negation are distinguished, each corresponding to a different species of "presupposition":

(9)

Presupposition	Effect
TYPE A: S believes H is doing / about to do P	Command: S doesn't want H to do A (= negative imperative)
TYPE B: S believes H believes P	Assertion: S doesn't want H to believe P (= negative information or correction)
TYPE C: S believes H wants S to do A	Assertion: S doesn't want H to believe S will do P (= refusal)
TYPE D: S believes H wants S to (dis)confirm P	Assertion: S (dis)confirms P (= negative reply to question)

For the four Italian- and English-speaking children in Volterra and Antinucci's longitudinal study, all four types of negation were found to be present from the earliest sessions (at fifteen months) and the four types are sufficient to exhaustively classify instances of negation spontaneously produced (up to thirty-six months). In the earlier examples, the presupposed information (Hegel et al.'s 'positive ground') is present in the context as an ongoing event or an immediately preceding utterance (p. 290). Later the child must 'reconstruct it [the presupposition] internally'.

While Volterra and Antinucci's classification is defined by different criteria (essentially pragmatic rather than semantic), there is some overlap with the taxonomies of McNeill and McNeill (1968) and Bloom (1970). Indeed, the earlier taxonomies sometimes seem to fare better with Volterra and Antinucci's own data. Consider the following example (pp. 292, 296) uttered by Francesco at age 2:0:

(10) Hai itto no c'è a panta!? 'Look, there is no bell'

'Apparently', Volterra and Antinucci explain, 'Francesco's experience with bells has been restricted to church towers, and this hospital tower violates the norm that all towers have bells' (p. 292). But this example, and Volterra and Antinucci's exegesis of it, would appear to fit more neatly into Bloom's category of nonexistence ('Referent not manifest in context, where it was expected') than into Volterra and Antinucci's Type B negation, where they in fact place it ('The speaker does not want the listener to believe P'). While Volterra and Antinucci's analysis and classification of negation are expressed within the language of speech act theory, the classic treatise on speech acts—Searle (1969)—expressly warns against defining assertion in particular, and speech acts in general, in terms of their actual or intended perlocutionary effect (i.e., the effect they may or do have on other interlocutors in the speech situation). Yet that is precisely what Volterra and Antinucci seek to do here.

Keller-Cohen et al. (1979) investigate the somewhat different notion of DISCOURSE NEGATION, 'the logical negation or rejection of a proposition or presupposition in a prior speaker's utterance' (p. 305). Overt negative morphemes are neither a sufficient criterion for membership in this category (Francesco's utterance of (10) in the context described would not qualify as an instance of discourse negation), nor a necessary criterion (as Keller-Cohen, Chalmer, and Remler note, a child who responds to an adult utterance *This truck is red* by shaking his head and saying *Blue* has issued a discourse negation).

To be sure, there is a functional kinship between the act of saying *It isn't red, it's blue* (or *It isn't red but blue*) and the act of shaking one's head and saying *Blue* in the above context. Indeed, I can reject the 'proposition or presupposition' in your utterance of, say, *You stink* by replying *No I don't,* by responding *So do you,* by sneering *Yo' mama!,* by thumbing my nose, by leaving the room, or by throttling you, but it's altogether unclear what insight we have gained by assimilating all these acts to the general category of discourse negation. In particular, there is something odd about a category of negation—whatever we call it—which would include all the above acts while excluding Bill's utterance in the sequence below:

(11) Mary: John thinks it's gonna rain.
Bill: {No / Don't worry} it's not gonna rain.

Nevertheless, there are many useful observations in Keller-Cohen, Chalmer, and Remler's study, beyond the (not surprising) result that discourse negation, as defined above, shows up early, the conclusion that its English realization takes one of the canonical forms *no*, [*no* + {*N*,*V*,*VP*}], or [*no* + affirmative sentence], and the speculation that a 'happy' discourse negation is 'one which both negates and introduces new information' (p. 320). This last category is the authors' ELABORATED discourse negation, apparently corresponding to the McNeills' category of entailment negation and to Bloom's subcategory of contrastive denial. But while such dual-function negations are undeniably more complex, at least as a production, than their "unelaborated" counterparts, it is not clear that they are always "happier." A simple response of *No* (or *No, I'm not*) to the query *Are you tired?* or to the assertion *You're an idiot* would seem to be at least as natural, proper, and mature as any elaborated alternative introducing new information.

Pea (1980a, b) offers a detailed review of previous studies on the acquisition of negation. While providing a more systematic taxonomy of negative utterances than the four-category analysis of the McNeills and the three-category analyses of Bloom and Volterra and Antinucci, Pea invokes a Wittgensteinian 'family resemblance' theory to collect the various negative types. He argues convincingly (1980b: 31) that Volterra and Antinucci's emphasis on the role of the addressee and his or her 'presupposed' beliefs is untenable, since there is no evidence that a negation-wielding pre-two-year-old can infer specific beliefs of others.

Pea (1980b) posits a developmental sequence of child negation moving from REJECTION—typically nonverbal—to DISAPPEARANCE to TRUTH-FUNCTIONAL NEGATION or FALSITY. The six children in his longitudinal study differed individually, but affective negation—expressing either unfulfilled expectation or self-prohibition—invariably preceded the development of a truth-functional negation (which Pea views as metalinguistic).[7] Pea also establishes a clear gestural priority for the expression of rejection; head shakes (exhibited as early as one year) are followed one to nine months later by a verbal negation usually, but not always, signaling rejection.

Pea echoes Russell (1948) in seeking to derive the *not* of logical negation from the *no* (or *don't*) of parental prohibition. The early signaling of rejection in the child's gestures and speech would then reflect the predominant value of negation in comprehension during the first year. External prohibition then develops into internal (self-) prohibition or rejection, although the child's signaling of self-prohibition may not serve as a deterrent.[8]

By age two, the child's use of negation to deny propositions earlier expressed by another (Keller-Cohen, Chalmer, and Remler's discourse negation) or herself is typically well established. The two-year-old is particularly partial to 'antithetical phrases' marking minimal contrast. Pea's examples (1980a) include

(12) a. Not that boat hot, that boat hot. (pointing at two different boats)

b. That light. No, is vacuum cleaner.

One constant in the work of various researchers on child language is the early association of negative utterances with what Pea calls 'unfulfilled expectation', a departure from a habitual norm. As Pea notes, the norm may be local or idiosyncratic (as with the *No bell* example from Volterra and Antinucci) or general. The same correlation exists, of course, in adult speech. Barring any overriding idiosyncratic norm, a sentence like (13a) is felt to be more natural (as a discourse opener) than those in (13b), while (13c) is more unnatural still:

(13) a. There's no beer in the fridge.
b. There's no halvah in the fridge.
c. There's no corpse in the fridge.

Ducrot's *Pierre n'est pas le cousin de Marie* and Givón's *My wife's not pregnant* (cf. §1.2)—not to mention Nixon's celebrated *I am not a crook*—serve to make essentially the same point. And, as we are about to see, the related notion of 'plausible denial' has played a major role in contemporary psycholinguistic work on the relative complexity of negative sentences.

3.2 Markedness and the Processing of Negation

While the 1960s and early 1970s may be more widely associated with other social and political developments, this era was also marked by an intense empirical investigation of the psychology of negative sentences, as measured in comprehension and verification tests. The central finding of these studies, notes Clark (1974), is that 'negation is more difficult to comprehend than affirmation'. All things being equal, a negative sentence takes longer to process and is less accurately recalled and evaluated relative to a fixed state of affairs than the corresponding positive sentence. But all things are not always equal, and the more illuminating results emerge when the context is varied along different parameters.

Why are negative statements responded to more slowly than their corresponding affirmatives, even when the same information is conveyed (cf. Wason 1961)? The standard assumption by psycholinguists in the immediate post–*Syntactic Structures* era was that negatives—like interrogatives, passives, and other 'nonkernel' sentence types—are relatively difficult to produce and to comprehend because of their relative transformational complexity. The extra rule(s) of negative placement would presumably require additional processing time. Yet this suggestion, like other predictions of the so-called Derivational Theory of Complexity, has never been directly confirmed.

Nor is length (the fact that negative sentences are typically longer than their affirmative counterparts) clearly a factor; when both transformational history and length are controlled for, negative sentences are apparently still harder than affirmatives (Fodor and Garrett 1966; Just and Carpenter 1971). As Fodor and Garrett conclude (1966: 148), 'Perhaps negatives are more complicated than affirmatives simply because they are negatives'.[9]

Earlier psychological studies focused on demonstrating the superiority of positive to negative information in concept attainment and in problem solving (cf. Wason 1959, Cornish and Wason 1970, and Jacobsson 1970: 18–19 for summaries of the literature). The point at issue here was one familiar to Plato and Aristotle: positively presented information is intrinsically more valuable or, in modern terminology, more adaptive than negatively presented information: 'It is no good knowing what something is not unless that helps to eliminate possibilities about what it is' (Wason 1959: 103).

One significant and puzzling result from the earlier studies (Wason 1959, 1961) was the ranking of different types of sentence-situation pairings by reaction time in verification tasks. The four types of sentences evaluated and an example of each are given in (14):

(14) **TA** (true affirmative) *24 is an even number.*
FA (false affirmative) *25 is an even number.*
TN (true negative) *27 is not an even number.*
FN (false negative) *26 is not an even number.*

It is not surprising that the affirmative sentences are easier than the negatives; subjects are faster and more accurate in labeling a **TA** sentence as true and an **FA** as false than they are with **TN**s and **FN**s. What is surprising is the consistent finding (cf. Wason 1961, 1965, 1972; Wason and Jones 1963; Clark 1974; Pea 1980a) that while false affirmatives take longer to verify than true ones, the longest verification time of all is required for true negatives, rather than false ones. Thus, the ranking by response latencies obtained in these studies is not (15a) but (15b):

(15) a. **TA** < **FA** < **TN** < **FN**
b. **TA** < **FA** < **FN** < **TN**

Whence this asymmetry between affirmation and negation?

One clue is revealed in Wason's subjects' introspection on what they are doing while performing their verification tasks. Affirmatives are processed directly, but negatives tend to be converted into affirmatives in one of two ways. Either the negation is "mentally deleted" at the first stage of calculation (and later restored, in effect, by "flipping" the truth value eventually assigned) or it is removed by translation, the subject (often consciously) converting a negative (*X is not even*) into an ostensibly equivalent affirmative (*X is odd*).

These two approaches for dealing with (or more correctly for not dealing with) negation in experimental situations correspond essentially to Clark's TRUE model and CONVERSION model, respectively. The true model—in which a negative proposition (*A is not B*) is mentally represented as the proposition that its positive counterpart (*A is B*) is false—is so called because it predictably yields the true or correct result, whatever the predicate involved; no information is lost or damaged in processing. The conversion model, on the other hand, works perfectly only for what Clark calls contradictory pairs, for example, Wason's *even / odd* example, the immediate contraries of §1.1.5. In cases of mediate (simple or polar) contraries, the conversion model involves 'cheating', since, for example, *A isn't above B* is not equivalent to *B is below A*, nor is *X isn't white* synonymous with *X is black.*[10] Of course if the context (i.e., the experimental design) serves to exclude the middle, so that *A* must be either above or below *B*, and *X* can only be either black or white, the cheating is rendered legitimate; I shall return in chapter 5 to the effect of context in rendering mediate contraries immediate.

Given the reasonable assumption that the translation of negatives—by whichever method is employed—takes some time and effort, the greater response latencies for negative vis-à-vis affirmative statements makes sense (as does the finding that negative sentences are processed less accurately). But Wason and Jones (1963:307) seek to isolate not just one but two factors for why negatives are harder, only the first of which reflects translation or conversion time. These factors are:

(16) a. The assumed tendency to translate negative statements into an affirmative form because of the role denials play in language, and

b. An inhibition of response specifically associated with the prohibitive connotations of the word 'not' (emphasis mine).

The effect of (16b) is difficult to pin down, but Wason and Jones point to data showing that subjects respond faster to an uninterpreted nonsense syllable assigned the logical value of negation than to explicit negation itself. Two earlier studies speak to the importance of factor (b). Wason (1959: 105) stresses the emotional side of negation: 'Words which imply positive actions or states-of-affairs have, in general, a more pleasant or creditable connotation than their opposites. "Yes" is permissive; "no" prohibitive'.

Of course the emotive connotations in question are eloquently conveyed by the extended use of the adjectives *positive* and *negative* themselves.

In support of the importance of this 'unpleasant hedonic value' associated with negation, Wason cites introspections of (British) subjects in his reaction-time study:

> ' "Not" gave me a sort of tremor half-way through'.
> 'I don't like "not"—it's a horrid word'.

Wason posits an inhibiting effect of negation on these and other experimental subjects who confessed to being frightened by what we might think of as a *not* in the throat.[11]

In the same spirit, Eifermann (1961) shows that the Hebrew negators *lo* and *eyño* differ in that the former, which corresponds to English *not* in its distribution and meaning, requires a greater mean response time than the latter, which appears directly before the predicate and cannot be used to express prohibition. Eifermann's conclusion, endorsed by Wason and Jones, is that the prohibitive value of negation, established in early childhood (as suggested by Russell and Pea), has an 'inhibiting' effect on responses. Hebrew *eyño,* as a nonprohibiting and hence noninhibiting negative particle, is affected only by factor (a) and is thus "easier" than *lo,* which is affected by both factors.

Also supporting the putative effect of factor (b) is the consistent finding that overt, explicit instances of negation (*A is not present, B is not happy*) take longer to verify than implicit negation (*A is absent, B is sad*); cf. Wason (1972); Clark (1974); and Fodor, Fodor, and Garrett (1975).

But the role of factor (b)—the implicit emotive content of negativity—in contributing to the psychological complexity of negation remains largely a matter of speculation, and a number of the speculators have expressed their doubts. Wales and Grieve (1969: 330) offer an alternative account of the greater complexity of *lo* over *eyño* negation in Eifermann's study, one which focuses on the greater ambiguity of the former rather than its association with contexts of prohibition. Clark (1974) concludes that the 'prohibitive connotation' factor in the processing of negation alleged by Wason and his colleagues remains an unproved hypothesis in the absence of any valid evidence directly confirming it. It is in any case factor (a) which has received the lion's share of psycholinguistic attention in empirical work over the last two decades.

If negation acts as a kind of dam to the stream of understanding, an interesting body of data shows that the strength of this dam varies with the character of the stream. Consider the classic minimal pair from the work of Wason and Clark:

(17) a. The whale isn't a fish.
b. The whale isn't a bird.

Both statements are of course true, but the latter is distinctly peculiar, and turns out to require more processing time in (at least some) experimental contexts.[12] What we are dealing with here is the long- (and often-) observed association between the appropriateness of a negative utterance and the plausibility or accessibility of its positive counterpart, an empirical reflex of the thousand-year-old tradition of analyzing negative statements as presupposing (in a Praguean rather than Fregean sense; cf. §1.2.2) the corresponding positive proposition.

While this association is thus hardly novel, it was Wason who recommended it to the attention of the psycholinguistic community, under the rubric of CONTEXTS OF PLAUSIBLE DENIAL. Very simply, the function of negative sentences is 'generally to emphasize that a fact is contrary to an expectation' (1965:7). Negative statements by their nature 'assume and depend on a prior state of affairs, either existent or supposed. . . . It is unlikely that the sentence "It is not x" would be uttered unless there were good reason to suppose that it might have been "x" or that someone thought it might' (Cornish and Wason 1970:113). Thus, *5 is not even* is harder to process and takes longer to verify than *5 is odd,* but the difficulty is mitigated if we set up a 'context of denial': *4 is even {and / but} 5 is not even* (cf. Greene 1970a:18; Wason 1972:28). Psychologically, if not ontologically, negation seems to require—or at least to strongly prefer—an affirmative context against which to operate.[13]

Wason (1965) seeks to support a corollary of plausible denial, the EXCEPTIONALITY HYPOTHESIS: negation is most natural when it is associated with a dissimilar item (the 'figure'; cf. Sigwart 1895; Givón 1978) set off against the rest of a class of similar items (the 'ground'). When eight circles are presented, with one—circle 7—colored blue and the rest red,

(18)
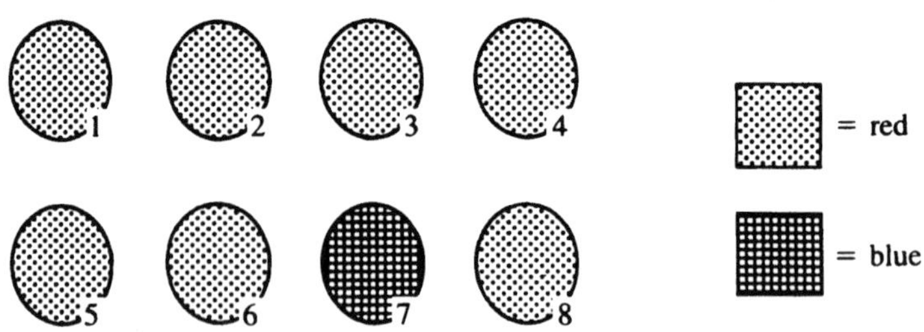

subjects have little trouble verifying affirmative sentences (*Circle 4 is red, Circle 7 is blue*) or the negative sentence about the exceptional case (*Circle 7 is not red*), but a negative sentence about one of the unexceptional cases (e.g., *Circle 4 is not blue*) proves far more troublesome. These results are consistent with the exceptionality hypothesis and with the notion of plausible denial: given the proportion of red and blue circles, we would (a priori) expect circle 7 to be red, but there is no reason to expect circle 4 to be blue.[14]

The Exceptionality Hypothesis is independently supported by a recent artificial-intelligence-based empirical study focusing on the pragmatic differentiation of negation and affirmation vis-à-vis their logical symmetry. Shanon (1981:42) begins with the by now familiar premise that 'in uttering a negative statement a speaker not only states that a certain state of affairs does not hold but that there is something special about the fact that it does not'. In the language of Minsky's frame-oriented approach to artificial intelligence, a negative sentence evokes a frame in which the negated value is present: the utterance 'A-G and I-Z are <u>not</u> in the room' is natural only when the relevant individuals 'have disappeared, are expected, or are supposed to be present', while the corresponding positive statement that H is in the room builds in no such background assumption. Whence the contrasts experimentally established by Shanon:

(19) a. —Why did you pick up the food by yourself?
—Because I saw that there was no {waiter / #airplane}.
b. We don't have any {furniture / #diamonds}.
c. The ceiling has no {overhead light / #carpeting}.

As in the analogous contrasts cited by Wason, Ducrot, Givón, and Volterra and Antinucci reviewed earlier in this chapter and in §1.2, the plausibility of a given negative utterance depends on the accessibility of the corresponding positive proposition in the context.

A decade after Wason's initial circle study, its findings were confirmed in an adaptation for children by de Villiers and Flusberg (1975). Given a stimulus set of seven cars and one bottle, a group of 2½-4½-year-old subjects consistently took significantly longer to correctly complete, and made more errors in processing, the implausible negative statement *This is not a bottle* as compared with the more plausible *This is not a car.* As Pea concludes from this and related studies (1980a:33), 'Apparently even 2½-year-olds are aware of the social, pragmatic conditions for negation'. Pea (1980a, 1980b) offers a useful summary of the literature on children's awareness of the plausibility requirement on negation.[15]

Greene provides evidence that a negation is processed more easily when it relates two sentences, either implicit or explicit—that is, when it is used to deny a proposition present in the discourse context—rather than, as with affirmations, when it relates a sentence directly to a state of affairs. The natural function of negation as a means 'to signal a change in value' is exercised 'when a statement is being contradicted, a request refused, a misconception corrected or a difference pointed out' (Greene 1970a: 17).

Greene's conclusion that subjects have little difficulty with negation when it performs its natural function of denial is supported in a study by Johnson-Laird and Tridgell (1972), who show that, unlike in tasks of interpretation (i.e., matching tasks) where affirmatives prove consistently easier than negatives, negatives win out in tasks involving the establishment of opposition: 'In denying a statement negatives are easier than affirmatives. . . . It is easy to grasp that a negative denies an affirmative; but exceedingly difficult to grasp that an affirmative denies a negative' (Johnson-Laird and Tridgell 1972:90).

Cornish and Wason (1970: 109) find that recall tasks not only illustrate the usual superiority (outside denial contexts) of affirmative over negative information, but more particularly demonstrate that the difficulty associated with recalling negative clues depends on 'their inappropriateness in the situation, namely in the absence of any prior expectations'. When the positive ground of a negation is directly accessible, the latency for understanding the negative sentence is correspondingly reduced.

A related study of the pragmatic aspects of negation by Cornish shows that when partially red and partially blue circles are presented to subjects, the completion time or evaluation time for a sentence like

(20) The circle is not all red.

is largest when the circle contains only a small proportion of red, gradually decreasing as the subject is presented with circles containing more and more red. Similarly, (20) is increasingly likely to be given as a description of a circle as the proportion of redness in the circle is increased. More generally, 'The sentence "X is not all y" applies with increasing appropriateness as X increases in y" (Cornish 1971:510)—up to the point when X is entirely y, for example, the circle is all red, in which context *X is not all y* (e.g., (20)) becomes noticeably inappropriate.[16] Cornish's conclusion is a restatement of Wason's contexts of plausible denial: 'The closer the negative is to the presupposition (without being identical to it, rendering it false), the more appropriate it is'.

Given the principle of plausible denial, which he explicitly links to Strawson's remark (1952: 18; cf. §1.2.2 above) that 'the standard and primary use of "not" is specifically to contradict or correct; to cancel a suggestion of one's own or another's', Wason offers an account of the puzzling result mentioned above, the **TA** < **FA** < **FN** < **TN** ordering of complexity as measured by processing time on verification tasks:[17] 'The "false negative" is analogous to the negative which corrects a preconception rather than one which maintains a truth. . . . The true negative does not conform to the ordinary usage of a natural language: it maintains truth rather than denies what is supposed to be true' (Wason 1972: 17, 35).

As observed by Clark (1974), the complexity of **TN**s can be illustrated by evidence from introspective reports: some subjects 'report changing their answer from *true* to *false* and back again on True Negatives'. Reflecting this point, Wason (1972: 24) offers a new characterization of the four types of sentence-assessment values:

(21)
TA = 'a fact'
FA = 'a falsehood'
FN = 'denial of a fact'
TN = 'denial of a falsehood'

Significantly, it is the true and not the false negative statement which thus emerges as an empirical double negation.

3.2.1 Issues and Problems in the Psycholinguistics of Negation

While the results from the studies summarized here are suggestive of the nature and extent of an apparent pragmatic constraint on the processing of a certain class of negative statements, there are some problems with the design of these studies and especially with some of the interpretations they have received.

The validity of the studies on 'plausible denial', based as they are on isolated sentences and situations, is challenged by De Mey (1972: 149), who offers a different psychological explanation for the correlation between processing time and the correcting function of negation:

> 'Natural' negation only involves objects or elements a speaker or listener is attending to. . . . It makes no sense to instruct a listener to suppress a thought he is not considering or an idea he is not having. The time needed for conversion of negative statements into affirmative ones could well be the time needed for focussing attention on the context of the statement, a process which is, most of the time, unnecessary in 'natural' conversation.

The Wasonians' basic thesis that 'negation is typically difficult to process' is itself disputed by Wales and Grieve (1969: 327), who correctly point out that the interpretation of the data in earlier studies often 'relies heavily on the notion that negation is some sort of operation on [affirmative] statements'. They argue that this view is compatible only with cases of contradictory, 'mutually exhaustive and exclusive' pairs (e.g., *odd / even*, where *This number is not even* will automatically be equivalent to *The number is odd*) and does not extend automatically to nonexhaustive (mediate) contrary oppositions (e.g., *hot / cold*). On the other hand, Wales and Grieve (p. 330) also see problems of plausibility in studies like that of Wason 1961, which apply negation to predications with *odd* and *even:* 'It is eccentric to use a negative with either member of an antonym pair whose range is logically exhaustive'.

However, as Wason (1965) argues, this eccentricity is reduced in a context where the negative can appear as a direct denial (*4 is even but 5 is not even*).[18] Furthermore, as Wales and Grieve themselves acknowledge, some contradictory pairs—typically 'gradable' opposites like *open / closed*—allow negation more easily than others; *The door is not open* is clearly less eccentric out of context than *5 is not even.*

In their reply, Greene and Wason (1970) point out that Wales and Grieve's own data confirm that ceteris paribus negatives do take longer to process, and they reiterate the position of Wason 1965 and later papers, namely, that exceptionality is more of a factor in processing negative sentences than in affirmatives. (In the circle study cited above, the 'unexceptional' affirmative *Circle 4 is red* and the 'exceptional' affirmative *Circle 7 is blue* are equally easy to verify against the context of (18); the discrepancy shows up, as noted, only under negation—*Circle 4 is not blue* proving consistently more difficult than *Circle 7 is not red.*) But while Greene and Wason do not acknowledge it, the observation by Wales and Grieve on the differential behavior of contradictory vs. contrary opposites was taken into account in later work, stimulating the distinction between the two negation-eliminating strategies formulated in different ways by research teams headed by Wason, Clark, and Trabasso (cf. Wason 1972:25).

Wales and Grieve share one flaw with the researchers they criticize. They take a negative sentence like (22)

(22) The drink is not hot.

to be ambiguous as between the two senses in (22′a, b):

(22′) a. The drink is cold.
b. The drink is lukewarm.

Since we do not know which of these senses is intended by a given utterance of (22), we cannot take the negation to result in the simple affirmative *The drink is hot,* since the negation of (22′a) is compatible with the truth of (22′b), and vice versa (Wales and Grieve 1969:327–28).

This discussion reflects a lack of appreciation for the distinction between ambiguity and vagueness (or lack of specificity) and for the criteria that serve to distinguish them. Sentence (22) is not ambiguous, but vague (general, unspecified) as between the two UNDERSTANDINGS (to adopt Sadock's neutral term) depicted in (22′a, b)—and others.[19] Seuren (1967: 348–49) and Kempson (1977: chapters 7 and 8) provide useful expositions of the view that a negative sentence is (generally) general, rather than ambiguous, as among the various circumstances under which it is true. Note that the standard linguistic criteria for ambiguity (cf. Zwicky and Sadock 1975) fail to respect the two understandings of (22) distinguished in (22′a, b); consider, for example, the "crossed understandings" available for (22″a), as opposed to the impossibility of such a crossed reading in the case of the true ambiguity in (22″b):

(22″) a. My drink is not hot [since it's lukewarm] and neither is yours [since it's cold].
b. #My tostada is not hot [since it's lukewarm] and neither is yours [since you didn't get the salsa].

(This test and others apply with greater difficulty to negative statements with vacuous subjects, a point to which I return in chapter 6.)

The same critique applies to members of Wason's research team. Cornish (1971: 510) claims an ambiguity for *X is not all y* parallel to Wales and Grieve's for (22); the two putative readings are 'X is predominantly y' and 'X is not y at all'. Again, there is no evidence that such an ambiguity exists. Rather, *X is not all y* is true under the two sets of circumstances described by Cornish, as well as others (e.g., when X is half or two-thirds y); it is simply unspecified as to which of these circumstances obtains when it is true. This is not to say that such a negation is not more appropriately uttered when it is known that one context obtains rather than another, as predicted by the plausible denial thesis Cornish invokes.

This leads us to another point which is worth raising against some of the studies under review here. As is typical, if not inevitable, in the type of psycholinguistic experiment conducted by Wason, Clark, and others, subjects are usually provided with full information about the data with which they are confronted. For example, in the circle studies of Wason (1965) and Cornish (1971), all circles can be perceived in their entirety, whether each circle is either all red or all blue (as in Wason's) or some mixture of red and blue (as in Cornish's). Furthermore, Wason's monochrome circles are all presented to the subjects in a fully perceptible set.

It is clear that Wason's subjects prefer to deal with negatives like *Circle X is not y* when most of the circles are y and are known to be y, and equally clear that Cornish's subjects prefer negative universals like *Circle X is not all y* when X is—and is known to be—mostly (but not all) y. But in real life, such epistemically perfect situations are not always so thoughtfully provided. What happens, for example, when one of Cornish's graduate student subjects, let's say Oscar, leaves the lab and is handed the message in (23)?

(23) The students in your 10:30 Psych 101 section are not all psych majors.

Oscar may be willing to infer that the registrar (or whoever is the source of the message) doesn't know for a fact that two students are majors and the rest nonmajors, given the misleading nature of (23) in such a context. But what if the registrar only knows—or cares—about the confirmed nonmajors: does this contingency render (23) implausible? What (23) in fact suggests is not that only one—or a couple—of the students are nonmajors, but rather that no more than a few are known to be nonmajors.

The situation is thus exactly analogous to that of (other) cases of Gricean conversational implicature, especially those deriving from the maxim of quantity (see Horn 1972, 1973, and chapter 4): in saying that some of the students are majors (or that ten of the students are majors), I implicate but do not say that for all I know not all (no more than ten) of them are majors.[20] Cornish does acknowledge the pragmatic (context-dependent) nature of the inference from *X is not all y* to *X is mostly y,* but she fails to bring out its scalar nature, the fact that the inference correlates with the conversational constraint on speakers to give the strongest information available to them consistent with truth and relevance (Grice 1961, 1975; Horn 1972; Gazdar 1979a). If I know that circle 3 is entirely blue and that this information is relevant to you, it is misleading for me to tell you that it's not all red, when I could have expressed the stronger proposition, namely, *Circle 3 is not at all red* (or . . . *is all blue*) directly.

Note in particular how the interaction of relevance and quantity applies to Cornish's case as to other instances of scalar implicature: if all you care about is finding a circle which is entirely red, the information that *Circle 3 is not all red* would eliminate circle 3 from contention without suggesting, implicating, or (in Cornish's terms, presupposing) that it is mostly red—in the same way that *Max has three children* does not convey that he has at most three if all that is relevant in the context of utterance is whether he has at least three.

Another criticism that might be leveled against these studies is more terminological than substantive. A number of claims that seem to be being made are in fact not being made (or at least not being supported), the problem being one of truth in labeling. Wason proclaims in the title of his overview article that "In Real Life Negatives are False." Does this proclamation signal that '*2 plus 2 does not equal 5*' cannot be uttered truthfully (in real life), that in fact the class of real-life **TN** statements is empty? Hardly. Instead, what is being claimed (after discounting Wason's intention to *épater les bourgeois, i.e., les philosophes*) is simply that negative statements, as ordinarily used in natural language, 'are usually false rather than true' in the (rather unlikely) sense that 'they correct a false preconception rather than simply being true relative to a state of affairs' (Wason 1972: 32). But the claim that negatives are false certainly sounds as though it is saying something different (and stronger) than the claim that negation is used for falsifying or denying one's own or another's earlier statements or beliefs.

Wason goes on to observe that 'the real life corollary of the true negative is not, of course, a lie but a vacuous statement, e.g., "*The train wasn't late this morning*", when nobody expected it would be late'. But this sense of VACUOUS, applied to Wason's punctual train (or to Ducrot's noncousin, or Givón's nonpregnant wife in my earlier discussion), is clearly distinct from the standard sense in which sentences like (24a, b) have been described as vacuous (or vacuously true).

(24) a. The train wasn't late this morning, {it didn't come at all/there was no train}.
b. #The train wasn't {divisible by 3/pregnant} this morning.

A less blatant but equally misleading instance of misterminology occurs in Clark's useful summary article. Clark (1974: 1325) concludes that problems involving the unmarked, inherently positive comparative adjectives (e.g., *better, more, faster, taller*) are solved faster and more accurately than problems based on their marked, implicitly negative counterparts (*worse, less, slower, shorter*). But for some (insufficient) reason, he chooses to label the latter class QUANTIFIER NEGATIVES. Marked or inherently negative prepositions such as *under* (vs. *over*), *below* (vs. *above*), *behind* (vs. *ahead of*), and *in back of* (vs. *in front of*), which also predictably take longer to verify than their (parenthesized) positive counterparts, are also 'quantifier negatives'.

In their psycholinguistics text, Clark and Clark (1977: 452–57) are more circumspect. Longer processing time is associated here with INHERENT negatives, whether morphologically marked (*unhappy/happy*) or not (*sad/happy*), whether the category is verbs (*forget/remember, lose/find*), prepositions (*out of/into, from/to*), conjunctions (*but/and, unless/if*), contrary adjectives (*short/tall, low/high*), or 'contradictory' adjectives (*absent/present, dead/alive*).

The extra processing time is represented (but not explained) via the invocation of a 'semantic procedure' *not (x)* linked to inherently negative expressions, analogous to, but distinct from, the procedure *false (x)* associated with overt negation. The analogy between overt and inherent negation is seen as clearest in the case of the so-called contradictories, where, for example, '*absent* is merely the phrase *not present* compressed into a single word' (Clark and Clark 1977:457). This equivalence glosses over Aristotle's insight, representable as the observation that while Socrates and the king of France were not present at the first inauguration of Ronald Reagan, neither were they absent from it. Further, as Clark and Clark (1977) recognize elsewhere, *not present* and *absent* function entirely differently with respect to Klima's syntactic criteria for (sentential) negation. Thus, the "compression" of *not present* into *absent* is neither semantically nor syntactically straightforward. (I shall return to the issue posed by the extra processing time associated with inherent negation in appendix 2).

Given that negatives—or at least one class of negatives—are normally used to correct (reject, deny) some preconception (misconception, earlier proposition), two questions remain to be answered: (1) who is responsible for (associated with the assertion of) the rejected proposition? and (2) what is the theoretical status of the relation between the negative statement and the proposition it corrects? There is an easy—and wrong—answer to the first question: it is the listener (hearer, addressee) who is responsible for the proposition. Here is Wason's version of this answer: 'If I were to say, "The train wasn't late this morning", then in one sense the statement could count as a true negative, but that overlooks the reason for saying it. What the sentence does is to falsify the preconception of my listeners ("His train is always late"). And in this sense the statement is a false negative' (Wason 1972:32).

Leaving aside the question of whether the statement is a false negative in <u>any</u> sense (and thus assimilable to Wason's earlier examples of false negatives, for example, *92 is not an even number*), must my listeners really believe that the train is always late for me to be able to say that this morning it wasn't?

Or take Clark's defense of this thesis, on which I can say *Helen isn't at home* appropriately if and only if 'I thought that you had expected Helen to be home, or had said so, or had implied so in what you had just said' (Clark 1974:1312). This appropriateness condition derives from the premise that the speaker 'normally supposes that the listener does or could well believe in the truth of what is being denied' (p. 1313).

Operating on this shared assumption, Wason and Clark would apparently predict that sentences like those in (25) and (26) could never be uttered appropriately:

(25) {As you know/You were right}, the train wasn't late this morning.

(26) {Just as you predicted/You win the bet}, Helen isn't at home.

But of course there is nothing remotely odd or even mildly deviant about these sentences. Nor must the speaker have believed the denied proposition if the addressee didn't; cf.

(27) a. We were right: the train wasn't late this morning.
b. Just as we guessed, Helen isn't at home.

What seems to be going on is that to the extent that a negation is normally taken to deny the corresponding affirmative proposition, that proposition need only be in the common ground or discourse model, however it got there—from the beliefs or claims of the speaker, the hearer, some third party, or some more nebulous source like the collective mind of the speech community; cf. the minimal pair *The whale isn't a fish / ?bird* ((17) above).

The nonspecificity of the source of the understood proposition has in fact long been recognized, as a brief review of the evidence (reprised from §1.2.2, emphasis added) reveals:

> The negative judgment presupposes the attempt, or the thought, of an affirmation. (Sigwart 1895:119)
>
> [In negation] we take to task an interlocutor, real or possible . . . (Bergson 1911:289)
>
> . . . the chief use of a negative sentence being to contradict and to point a contrast . . . (Jespersen 1917:4–5)
>
> There must have been some reason to suppose that the affirmative statement of which [the negation] is the exact denial was true, either that it had been proposed for our acceptance by an interlocutor, that it had been part of our stored-up knowledge or purported knowledge, or that we had in mind what we took at the moment to be sufficient ground for its acceptance. (Baldwin 1928:146)
>
> 'This rose is not red' [involves] a fallible and partially ignorant mind erroneously attributing red . . . (Wood 1933:421)
>
> The standard and primary use of 'not' is . . . to cancel a suggestion of one's own or another's . . . (Strawson 1952:7)

What is crucial in this 'standard and primary use' of negation is evidently not that it necessarily cancels one of the hearer's or the speaker's earlier beliefs or assertions, but that it is a second-order statement seeking to rectify some proposition that was directly inserted into the discourse model (by someone's—anyone's—previous assertion) or that can be indirectly placed into the model by a reasonable inference as to the hearer's likely beliefs, or the beliefs of anyone relevant to the discourse context.[21]

Note that the first-order proposition need not be reasonable or plausible in any global sense. In discourses like those in (29) or (30),

(29) A: Pigs can fly.
B: No, you idiot, pigs can't fly!

(30) A: The robin and the whale are my two favorite birds.
B: The whale isn't a bird.

neither of B's utterances is at all peculiar (much less 'vacuous') in context, despite the a priori unreasonableness, unexpectedness, and implausibility of the misconceptions they deny.[22]

Thus, an expectation may be sufficiently plausible in a given context to motivate a denial without ever being directly subscribed to by anyone in the discourse context. This emerges especially clearly in the exception-to-a-generalization negatives of the Wason's (1965) circle study: nobody need have actually said or believed that circle 7 in (18) was red in order for that proposition to be reasonably contradicted; all that is required is that the positive proposition be somehow accessible as a good or natural guess.

One final empirical study of negation is worth citing here by way of a conclusion to this section on the processing and interpretation of negation in ordinary discourse. Based on her examination of spoken and written instances of negative sentences in corpora of fifty thousand words each, Tottie (1982) concludes that negation is twice as frequent in speech as in writing. She seeks to explain this discrepancy by invoking a revised version of the taxonomies of negation I explored earlier in this chapter.

Tottie offers the overall schema in (31):

```
(31)        NEGATION
              /   \
     REJECTION     DENIAL
                   /    \
           EXPLICIT      IMPLICIT
```

Tottie's category of rejection, subsuming the McNeills' category of INTERNAL DESIRE (or refusal), corresponds essentially to Bloom's; it is canonically realized in spoken language by a free-standing *No* followed by a negative utterance, as in (32a, b). For Tottie, as for Bloom, denials may be explicit, as in (32c, d).

(32) a. —Would you care for some scotch?
—No thanks, I don't drink. (REJECTION)

b. —Come and play ball with me.
   —No, I don't want to. (REFUSAL [subtype of Rejection])

c. —That dress must have been pretty expensive.
   —It wasn't (expensive), in fact I bought it on sale.

d. —What a hypocrite you are!
   —I am not (a hypocrite)—I'm being perfectly honest with you.

(c–d: DENIALS)

Bloom's (and the McNeills') category of NONEXISTENCE is (correctly) viewed by Tottie as a subinstance of denial: *There isn't any (more) soup* denies the proposition that there is (more) soup, just as *It isn't raining out* denies that it's raining.

But not all denials are as explicit (or immediately motivated) as those in (32c, d). As we have repeatedly seen in this chapter and in my historical overview in chapter 1, a felicitous instance of negation in a given discourse may deny, not an asserted, but a presupposed proposition. If you tell me that John's wife is a teacher, I may deny your claim explicitly (*No, she's a doctor*) or I may, as Tottie notes, deny it implicitly (*John isn't even married*). Tottie associates her notion of EXPLICIT DENIAL with Bolinger's EXTERNAL NEGATION, in which 'the speaker denies something that has supposedly been affirmed' (Bolinger 1977:44; note that Bolinger is not taking external negation in the logical sense discussed in §2.4 above). Explicit denials tend to be elliptical when actually uttered in discourse, since their 'positive ground' is directly recoverable in the linguistic context. In implicit negation, on the other hand, the speaker can be characterized as 'rebutting what he assumes that the other speaker might be thinking'. The earlier assertion he denies in such cases is itself implicit, one he views as 'not actually claimed but as in the mind of his interlocutor'.

While rejections and refusals, like the 'discourse negations' of Keller-Cohen I touched on above, are not ipso facto either necessarily linguistic or even human—Tottie observes that a dog can refuse food or to come—denials constitute the 'linguistic category of negation par excellence' (Tottie 1982:96). Lyons (1977:777) to the contrary notwithstanding, denials cannot be assimilated to rejections without committing what we might think of as Bergson's fallacy: treating all negation as a rejection of an earlier overt assertion. Tottie properly rejects this radical asymmetricalist stance, especially in the light of the predominance in the written language of negative sentences used to express implicit denials.

If we assume that rejection and explicit denial are both unlikely interpretations of negation in the written medium (except, of course, in the representation of spoken dialogue and perhaps in rhetorical discourse), we would seem to have—as Tottie observes—an account of the preponderance of negatives in the spoken over the written corpus. But, as pointed out to me by a reader of an earlier version of this chapter, we still lack an explanation for why we simply don't find more implicit denials in written discourse, to take up the slack. Neither Tottie nor I have any nonspeculative response to this point.

Tottie's depiction of implicit denial as implying the (global or local) existence of some 'unfulfilled expectation' is reminiscent of Wason's notion of plausible denial (a notion which has its own philosophical predecessors, as we have seen). One central question remains to be addressed: what is the nature of the relation between the proposition expressing this unfulfilled expectation, that is, the positive counterpart of negation and the negative statement used to deny it? This issue is fundamental enough to deserve its own section, one in which I shall return to the even more basic question first broached in §1.2: given that negation and affirmation are in some sense linguistically (if not logically) asymmetric, how is that asymmetry best captured within a general theory of the form and function of negation in natural language?

3.3 Markedness and the Asymmetry Thesis

The validity of the psycholinguistic studies on negation is affected by one variable which I have not yet touched on: the morphosyntactic form of the negative element in the test sentences. Clark (1974:1312) brings this out by defining his version of Wason's contexts of plausible denial as a condition on (what he calls) explicit denial or S(entential) negation.[23] The construct of sentence negation employed by Clark and most other contemporary laborers in the field of negation is standardly defined by reference to the diagnostics provided by Klima (1964), whose test frames include those in (33), where only the (a) examples pass the test for SENTENTIAL (S-) negation; the (b) sentences contain CONSTITUENT negation:

(33) (i) *either* (vs. *too*) tags:
a. Mary isn't happy and John isn't happy either.
b. Mary is unhappy and John is unhappy {*either/too}.
(ii) *neither* (vs. *so*) tags:
a. Mary isn't happy and neither is John.
b. Mary is unhappy and {*neither/so} is John.

(iii) negative appositive (e.g., *not even*) tags:
 a. The attacks weren't successful, not even the last one.
 b. *The attacks were unsuccessful, not even the last one.
(iv) positive (vs. negative) confirmatory tag questions:
 a. It isn't possible to solve that problem, is it?
 b. It is impossible to solve that problem, {#is it / isn't it}?

Similarly, it is only those fronted adverbials expressing S-negation that trigger subject-auxiliary inversion, as seen in the pair in (v) (Klima 1964: 300) and in its classic adaptation (by Charles Bird via Jackendoff 1972: 364) in (vi):

(v)	a. Not even two years ago could you swim there.	(S-negation)
	b. Not even two years ago you could swim there.	(constituent negation)
(vi)	a. With no clothes is Sue attractive.	(S-negation)
	b. With no clothes Sue is attractive.	(constituent negation)

Additional diagnostic environments for S-negation have been proposed since Klima, for example (cf. Ross 1973a; Culicover 1981), the possibility of inserting a postnegation negative parenthetical:

(vii) a. It isn't possible, I don't think, to solve that problem.
 b. *It is impossible, I don't think, to solve that problem.

But it has been argued that all these tests prove to be insufficient for deciding the crucial cases, in that they often give conflicting results; cf. Jackendoff (1969, 1972), Attal (1971), and Ross (1973a) for discussion. Ross, for example, cites contrasts like those in (viii) and (ix):

(viii) He hardly damaged the car, { ??did he? / ?and neither did you? / not even by filling it with gravy. }
 [judgments are Ross's]
(ix) a. Nobody saw John, did(*n't) they?
 b. John saw nobody, did*(n't) he?

The variables determining negative strength cited by Ross include the degree of overtness of the negative element (*not* > other *n*-initial morphemes (e.g., *no, never*) > others) and the grammatical relation of the nominal containing the incorporated negation (the more accessible the nominal on the Keenan and Comrie (1977) hierarchy—Subject > Direct Object > Indirect Object > . . .—the more sentential the behavior of the incorporated negation).

Klima's distinction between sentence and constituent negation fits within a long tradition of scholarship, beginning apparently with the distinction between QUALITATIVE and QUANTITATIVE negation drawn by Gebauer (1885) in his analysis of Old Bohemian. Qualitative negation is realized as a negated finite verb, resulting in sentential negation, while quantitative negation focuses on some other (nonverbal) constituent, which may or may not result in the semantics of S-negation.[24]

The utility of this dichotomy has been challenged, notably by Delbrück (1910:36ff.) and Jespersen (1917:69–71); cf. also Coombs 1976 for discussion. But Jespersen himself adopts a similar distinction, that between NEXAL and SPECIAL negation. Nexal negation is clause-based, marked in the auxiliary, while special negation has clause-internal scope and is typically marked by a negative (lexically incorporated or not) which immediately precedes or is part of the element on which it focuses. While Jespersen's distinction does not map directly onto Klima's, both are crucially syntactic in definition and spirit and thus distinguishable from the semantic approaches to S-negation favored by Jackendoff (1969) and Seuren (1969).

To take one classic pair of examples, (34a) is a case of special (constituent) negation for Jespersen (1917:44), while the auxiliary negation in (34b) counts as nexal:

(34) a. Not many of us wanted the war.
  b. Many of us didn't want the war.

But for Jackendoff, (34a) is a more likely candidate for sentence negation than (34b), since only the latter can be paraphrased by (34′)

(34′) It is not so that many of us wanted the war.

Klima's diagnostics are indecisive in determining whether or not either of these two negatives is to be assigned to the class of sentential negations; cf. Attal (1971); Ross (1973a).

Jackendoff's semantic criterion for S-negation, essentially equivalent to the standard truth-value-reversal criterion for contradictory negation, is not only less "squishy" (cf. Ross) but also more easily universalizable than the English-based syntactic criteria utilizing position and distributional diagnostics. But it too leads to some odd results. As Attal (1971:106) points out, *I don't want to leave* (like its French counterpart, *Je ne veux pas sortir*) has an interpretation (the so-called neg-raising reading: see §5.2 below) on which it is not a contradictory but a contrary of *I want to leave (Je veux sortir)*. When *I don't want to leave* is not equivalent to 'It is not so that

I want to leave', but rather to 'I want to not-leave', Jackendoff's criteria banish it from the ranks of S-negations; yet on both intuitive and syntactic grounds it ranks with undoubted sentential negations like *I don't have to leave,* as Jespersen's and Klima's syntactically oriented criteria predict.

Similarly, we might note that *You shouldn't go* passes the syntactic tests for S-negation, but does not qualify as a Jackendoffian S-negation in that it is (generally) not paraphrasable by 'It is not so that you should leave'. As with Attal's 'want' case, a syntactic S-negation may count semantically as a contrary rather than a contradictory negation (cf. chapter 5 for related instances of this phenomenon, all of which are potentially hazardous for any semantically based definition of S-negation which is intended to have syntactic consequences).

On the other hand, a morphological negation like *That's impossible* is opposed as a contradictory to the corresponding affirmative *That's possible,* thus passing Jackendoff's criteria for S-negation (*That's impossible* = 'It's not so that that's possible'). This ignores the essential structural difference between *That's impossible* and a true S-negation, *That isn't possible* (cf. (33iv, vii)).[25]

Additional problems with, and implications of, Klima's and Jackendoff's approaches to S-negation are discussed by Kraak (1966:101–3), Seuren (1967:336–37), Attal (1971), Bald (1971:3–7), Stockwell, Schachter, and Partee (1973:257ff.), and Culicover (1981). It is clear that both the definiens and the definiendum for S-negation present unresolved problems. Unfortunately, these are problems whose resolution directly affects the value of the studies on the psycholinguistics of negation I reviewed in the last subsection.

Virtually alone among the researchers on the topic, Clark (1974) recognizes that the relevant studies on the processing of negation are directed almost entirely at the sentential subspecies of negation, however (syntactically and/or semantically) that notion is ultimately to be defined. The literature on contexts of plausible denial can be read seriously only if we bear in mind that negative morphology is neither a necessary nor a sufficient condition for an expression to count as a (sentential) negation.

In Clark's terms, Klima's diagnostics for S-negation are actually tests for denial, picking out overt negations (*no, nobody, not, never*) and other explicitly negative quantifiers implying partial rather than full negation (*scarcely, hardly, few, seldom, little*). What is excluded is the class of IMPLICIT negatives concealed within words like *except* (but not), *without* (not with), and *absent* (not present).

In studies cited by Clark, implicit negatives (*except, absent*), while they take longer to process than the corresponding affirmative, are significantly easier than the explicit negatives (putatively) synonymous with them (*but not, not present*). Another class of implicit negatives that consistently correlates with longer reaction time in a variety of studies is the group of marked adjectives—both simple and comparative—with unmarked antonymous counterparts, for example, *small(er), short(er), slow(er),* as against *big(ger), tall(er), fast(er).*

In their related study of overt and covert negation, Fodor, Fodor, Garrett (1975) (FFG) distinguish four classes of negative morphemes:

(35) Class 1: explicitly negative free morphemes, e.g., *not.*
Class 2: explicitly negative bound morphemes (morphological negatives), e.g., *un-, iN-, never.*
Class 3: implicitly negative morphemes, e.g., *doubt, deny, fail.*
Class 4: pure definitional negatives (PDNs), e.g., *kill* (cause to become not alive), *bachelor* (man who has never married)

As FFG note, only class 1 (and some class 2) negatives satisfy Klima's criteria for sentence negation, but both class 2 and class 3 negatives trigger negative polarity items ({*I doubt/It's unlikely*} *he's ever eaten sea cucumber au gratin*) and so may be said to exhibit syntactic (or distributional) as well as semantic negativity. PDNs, on the other hand, contain no negative morpheme, overt or incorporated, and trigger no S-negation diagnostics or polarity items; their negativity, FFG argue, is spurious, an artifact of a particular theoretical commitment (i.e., that of generative semanticists).

The results of the reaction time experiment conducted by FFG (1975: 552) are as follows:

(36) a. Class 2 and class 3 negatives are somewhat easier than their explicit counterparts (as shown in the studies cited by Clark (1974) but not mentioned by FFG).
b. There is no significant processing difference between morphological and implicit negatives.
c. Arguments containing class 4 'negatives' are significantly easier; PDNs 'do not act as though they contain a negative element in their linguistic representation'.

While the finding in (36c)—unsurprising as it may now seem, with the generative/interpretive wars behind us (cf. Newmeyer 1980)—represents the main thrust of their study, FFG acknowledge that morphological (class 2) and implicit (class 3) negatives 'must somehow be classed with negative words by the grammar', either by decomposition or stipulation.

Unfortunately, as Herb Clark has pointed out to me, the strongest candidates for psycholinguistically complex PDNs are not examples like *kill* or *bachelor,* but the set of marked scalar adjectives which Clark and others had convincingly demonstrated to require more processing time: *small, short, low, narrow*. The point of the earlier studies on negation was to show that while these adjectives are indeed easier than their overtly negative synonyms (*not big, not tall*), they are clearly harder (pace FFG) than their unmarked antonyms (*big, tall*). Since FFG are arguing from null effects (a notoriously difficult argument to establish in empirical work), and since the experiment whose results they cite has never been directly reported in the literature, it is hard to determine the validity of their results. In any event, their primary conclusion, (36a) above, is consistent with the earlier findings on negation reported in Clark 1974.

Clark's identification of denial as (Klima's) S-negation (1974:1312ff.; cf. also Kissin 1969:86) suggests the Aristotelian notion of predicate denial, with its semantics defined in terms of contradictory opposition, that is, truth-value reversal. The identification of logical denial with linguists' sentential negation would be more convincing if the specimen linguist were not Klima (for whom S-negation is crucially a syntactic, rather than semantic, construct), but Seuren (1969) or Jackendoff (1969), for whom S′ is the negation of S just in case it can be paraphrased as 'it is not so that S'. But when Clark shifts to a discussion of the pragmatic criteria for denial or (in Wason's terms) plausible denial—as in the claim that 'a denial is specifically a sentence that asserts that something is false, where that something is presupposed to be possible' (Clark 1974:1315)—we sense an equivocation. Here, denial is not a logical or syntactic operation, but evidently a speech act.

As I observed earlier, however, not every instance of negation—or even of predicate denial, that is, contradictory negation—can be treated as an instance of the speech act of denial; in particular, as Frege points out, an embedded negation asserts nothing. Moreover, if we take Keller-Cohen, Chalmer, and Remler's category of 'discourse negation', the 'rejection of a proposition or presupposition in a prior speaker's utterance', to constitute the core case of the speech act of denial, not every instance of denial can be an instance of contradictory or sentential negation, since discourse negation may not involve any speech act or any overt negation at all, as Keller-Cohen, Chalmer, and Remler (1979) point out (§3.1 above).

Instead we are brought back to a key phrase in the much-cited but often misread Strawsonian definition: it is not negation itself but 'the standard and primary use' of negation that can be identified with Givón's 'negative speech act', Searle's 'illocutionary negation', Keller-Cohen's 'discourse negation', or Wason's 'plausible denial'. The most illuminating way to view the correlation between negation as discourse denial and negation as logical denial is one in which the former notion represents the functional core or PROTOTYPE of the latter, in the sense that 'focal red' is the prototype of the catetory *red* (Rosch 1977; cf. Putnam 1973 on STEREOTYPES).

I shall return below to a defense of this prototype view of negation, but I shall focus here on another aspect of the last citation from Clark, namely, the claim that the 'something' asserted to be false in a denial is the 'something' which is simultaneously 'presupposed to be possible'. This presupposition is of course intended to capture the relation of plausibility between a negation (*A is not B*) and its positive counterpart (*A is B*). But is the term PRESUPPOSITION, tentatively adopted here by Clark and vigorously promoted by Givón (1978, 1979) in the same connection, really an appropriate label for this ancient (cf. §1.2) connection? Clark himself tries to assimilate the negative cases to the classic instances of logical presupposition (he cites such examples as *You should stop beating your wife*) and quickly recognizes that the two relations don't directly line up. In the end, Clark judiciously drops the *pre-*fix and treats *John isn't present* as supposing rather than presupposing the proposition that John is present. Hence the language of passage cited earlier, 'A negative sentence will be easy if the supposition of that negative is plausible in that context' (Clark 1974: 1333; emphasis mine).

Clark concludes (p. 1337) that a negative sentence is mentally represented as an embedding of its positive supposition within a frame involving the falsity operator. *Helen isn't at home* is represented as (*false (Helen at home)*), while the corresponding positive is represented simply as (*Helen at home*). It will be noticed that this thesis on the mental representation of negation calibrates with the asymmetry theses on negation surveyed in §1.2, especially those of negation-as-falsity and negation-as-second-order-affirmation.

While an explicit negative (e.g., *Helen isn't present*) denies a positive supposition (*Helen is present*), an implicit negative (e.g., *Helen is absent*) on Clark's account affirms a negative supposition (*Helen isn't present*). But this negative supposition is itself psychologically complex, *ex hypothesi*, predicting (incorrectly) that implicit negatives should actually require more rather than less processing time vis-à-vis their paraphrases with explicit negation. Having rejected Wason's theory in which the greater latency for explicit negation is due not so much to the complexity of overt negations as to their inhibiting emotive associations, Clark is left without a convincing alternative hypothesis to explain the asymmetry. (Cf. appendix 2 for an alternative hypothesis.)

What does it mean, exactly, to say that explicit negatives suppose rather than presuppose their positive counterparts? What rough semantic or pragmatic beast is this supposition? One natural, and I would argue correct, guess is that it is an instance of conversational implicature (Grice 1961, 1967, 1975), the relation that obtains (cf. Horn 1972, 1973, and chapter 4) between an utterance of (37a) and the proposition in (37b):

(37) a. Chris has three children.
b. Chris has no more than three children.

Notice that unlike semantic or pragmatic presuppositions or conventional implicatures (cf. §2.5), but like conversational implicatures, the Clarkian supposition can be freely canceled:

(38) a. #I don't still beat my wife, but in fact I never did.
#Only I can do it, and in fact I can't.
#John managed to solve the problem, but in fact it was easy.
#Even I can do it, but nobody else can.
b. Chris has three children, and in fact he has four.
Susan was able to solve the problem, but she didn't solve it.
c. The whale isn't a bird, but then nobody ever thought it was.
Circle 7 isn't red, but whoever thought it was?

(Cf. Grice 1967, 1975; Horn 1972; Sadock 1978; Levinson 1983; and Hirschberg 1985 on the cancelability of implicatures.)

Like (other) conversational implicatures, our supposition is also extremely sensitive to context, coming and going (as we have witnessed) far more freely than a well-behaved presupposition or conventional implicature. In addition, as noted in Horn (1978c:204), the relation between a negative statement and its positive counterpart shares another feature with conversational, as opposed to conventional, implicata: it is NONDETACHABLE, adhering to any expression conveying denial within a given discourse context. Compare (39a, b):

(39) a. I want to marry you.
b. I'll be damned if I'll marry you.

Even if the subject of matrimony had never been broached between them, (39a) would be (linguistically, at least) an appropriate remark for Mr. X to make to Ms. Y, and might in fact constitute an indirect proposal, context permitting (ring proffered, on bended knee). But (39b), with its conveyed negative force, is pragmatically restricted (like the overt negation of *I won't marry you* or *I don't want to marry you*) to contexts in which the question of marriage has been entertained, if it is not under active deliberation.

For the supposition associated with negation to qualify as a conversational implicature, however, it must be not only cancelable and nondetachable, but also CALCULABLE, that is, derivable from Grice's Cooperative Principle via one of the maxims of conversation. I will suggest a derivation of the Clarkian supposition as an implicatum once I have introduced some additional descriptive apparatus.

3.3.1 Negative Uninformativeness within a Dualistic Model of Inference

The neo-Gricean model of nonlogical inference I shall assume here and in later chapters is taken from the exposition in Horn 1984b, itself prefigured in Atlas and Levinson 1981 and further developed (along somewhat different lines) in Levinson 1987a, b. The essential (and by no means novel) idea is that there is in language a systematic interaction between two antinomic forces identified by George Kingsley Zipf (1949:20ff.). The Force of Unification, or Speaker's Economy, is a correlate of Zipf's Principle of Least Effort, a drive toward simplification or minimization which, operating unchecked, would result in total homonymy or lexical versatility, yielding 'a vocabulary of one word which will refers to all the m distinct meanings' the speaker might want to express. The antithetical Force of Diversification, or Auditor's Economy, would expand the inventory to guarantee 'a vocabulary of m different words with one distinct meaning for each word'. More generally, the Speaker's Economy places an upper bound on the form of the message, while the Hearer's Economy places a lower bound on its informational content.

These two mutually constraining mirror-image forces are periodically invoked in the linguistic literature. Here are some sample citations:

> The more economical or more abundant use of linguistic means of expressing a thought is determined by the need. . . . Everywhere we find modes of expression forced into existence which contain only just so much as is requisite to their being understood. The amount of linguistic material employed varies in each case with the situation, with the previous conversation, with the relative approximation of the speakers to a common state of mind.
>
> (Paul 1898:351)

> In order to understand how and why a language changes, the linguist must keep in mind two ever-present and antinomic factors: first, the requirements of communication, the need for the speaker to convey his message, and second, the principle of least effort, which makes him restrict his output of energy, both mental and physical, to the minimum compatible with achieving his ends.
>
> (Martinet 1962:139)

> The speaker always tries to optimally minimize the surface complexity of his utterances while maximizing the amount of information he effectively communicates to the listener.
>
> (Carroll and Tanenhaus [1975:51], defining their MINIMAX PRINCIPLE)

The evolution of language can be seen as resulting from the dynamic tension between these two functional principles. In the phonological sphere, the speaker-oriented least-effort principle tends toward maximization of sensorimotor discriminability and the minimization of movement from rest, while the hearer-oriented counterforce tends toward maximization of salience and of perceptual discriminability. The goal of the linguistic sound pattern can be seen as the achievement of the greatest perceptual benefit at the least articulatory cost.[26]

In the lexical and semantic sphere, the speaker's force can be identified with the Law of Differentiation (Paul 1898; Bréal 1900), the principle of Preemption by Synonymy (Clark and Clark 1979), or the Avoid Synonymy principle (Kiparsky 1983; E. Clark [forthcoming]). The essential idea here is that languages tend not to allow a given semantic slot to be filled by two distinct lexical expressions; more precisely, a lexicalized item tends to preempt the filling of its slot by a less-lexicalized form that would have precisely the same meaning. The hearer's economy is realized as the corresponding principle which we might label Avoid Homonymy; cf. Bloomfield 1933, Zipf 1935, Menner 1936, E. R. Williams 1944, and Bolinger 1961 for the appropriate formulation of the principle of homonymic clash and illustrations of its predictive force.

I focus here on the application of the two countervailing Zipfian forces to a program for nonlogical inference. Grice (1967, 1975) attempts to show how participants in a conversational exchange can compute what was meant (by a speaker's utterance at a given point in the interaction) from what was said. The governing dictum is the Cooperative Principle (Grice 1975:45): 'Make your conversational contribution such as is required, at the stage at which it occurs'. The CP in turn is analyzed into four specific subprinciples, the general and presumably universal maxims of conversation on which all rational interchange is putatively grounded.

(40) The Maxims of Conversation (Grice 1975:45–46):

QUALITY: Try to make your contribution one that is true.
1. Do not say what you believe to be false.
2. Do not say that for which you lack evidence.

QUANTITY:
1. Make your contribution as informative as is required (for the current purposes of the exchange).

2. Do not make your contribution more informative than is required.

RELATION: Be relevant.

MANNER: Be perspicuous.

1. Avoid obscurity of expression.
2. Avoid ambiguity.
3. Be brief. (Avoid unnecessary [*sic*] prolixity.)
4. Be orderly.

There is, a priori, no privileged status to this fourfold classification of maxims (except perhaps for its echo of the similarly labeled Kantian categories), nor to the effective total of nine distinct subprinciples, and much of neo- and post-Gricean pragmatics has been devoted to a variety of reductionist efforts. If I assume (with Horn 1984b and most other work in this area, and against Sperber and Wilson 1986) that Quality (or what Lewis 1969 has called a Convention of Truthfulness) is primary and essentially unreducible, I can attempt to boil the remaining maxims and submaxims down to two fundamental principles responding to the two basic forces identified by Zipf and others. I use **Q** to evoke Quantity (i.e., Quantity$_1$) and **R** Relation, with no commitment to an exact mapping between my principles and Grice's maxims.

(41) Minding our **Q**s and **R**s (slightly revised from Horn 1984b):

THE **Q** PRINCIPLE (Hearer-oriented)	THE **R** PRINCIPLE (Speaker-oriented)
Make your contribution SUFFICIENT:	Make your contribution NECESSARY:
Say as much as you can (given both QUALITY and **R**).	Say no more than you must (given **Q**).
LOWER-BOUNDING principle, inducing UPPER-BOUNDING implicata	UPPER-BOUNDING principle, inducing LOWER-BOUNDING implicata
Collects Grice's QUANTITY$_1$ maxim and MANNER$_{1,2}$	Collects Grice's RELATION maxim, QUANTITY$_2$, and MANNER$_{3,4}$

The functional tension between these two fundamental pragmatic principles motivates and governs a wide range of linguistic phenomena, synchronic and diachronic, lexical and syntactic, ranging from implicature and politeness strategies to the interpretation of pronouns and gaps, from lexical change to indirect speech acts, from the interpretation of case marking in so-called split ergative languages to the analysis of recorded conversational interaction, from the pragmatic strengthening of apparent contradictory negation to the weakening effect of "logical" double negation (cf. Horn 1984b; Levinson 1987a, b; and the chapters below).

Crucially, my two antinomic principles are not in simple opposition, but interact (in the classical Hegelian manner) in a dialectic process in which each inevitably appeals to and constrains the other. Notice that Grice is forced to build in the **R** Principle in defining the primary **Q**-based maxim (Make your contribution as informative as is required' [emphasis added]), while Quantity$_1$ is similarly built into the definition of Quantity$_2$.[27] Further, the second Quantity maxim essentially incorporates Relation: what would make a contribution more informative than is required except the inclusion of material not strictly relevant to the stage of the exchange at which it occurred?

The **Q** principle is a lower-bounding law in terms of information structure which may be (and systematically is) exploited to generate upper-bounding implicata. **Q**-based implicature is essentially negative in character, proceeding from a speaker's nonuse of a stronger or more informative form to the inference that the speaker was not in an epistemic position to have employed the stronger form. The locus classicus is scalar implicature: a speaker in saying '. . . P_i . . .' implicates that for all s/he knows '. . . at most P_i . . .', that is, that it is not the case that '. . . P_j . . .' for any P_j stronger than P_i in some relevant sense of 'stronger than' to be made (somewhat) more precise in chapter 4. Thus '*some* . . .' implicates '. . . not all . . .' and '. . . *warm* . . .' '. . . not hot . . .', '*I believe that S*' implicates that I don't know that S, and so on.[28]

The **R** principle is an upper-bounding law which may be (and systematically is) exploited to generate lower-bounding implicata: a speaker in saying '. . . P_i . . .' implicates '. . . P_j . . .' for some P_j stronger than P_i and/or representing a salient subcase of P_i. The locus classicus here may be indirect speech acts and/or euphemisms, where a speaker's use of a weaker form may be filled in by an addressee who recognizes that some particular stronger or more informative meaning may have been intended. Because there is no essential reference here to what a speaker might have said but did not say, **R** inferences are essentially positive in character. As we shall see in some detail in chapter 5, the motivation for **R**-based implicata is not linguistic (as with **Q** inference) but typically social or cultural.

Schematically, we can distinguish the two patterns of inference as follows:

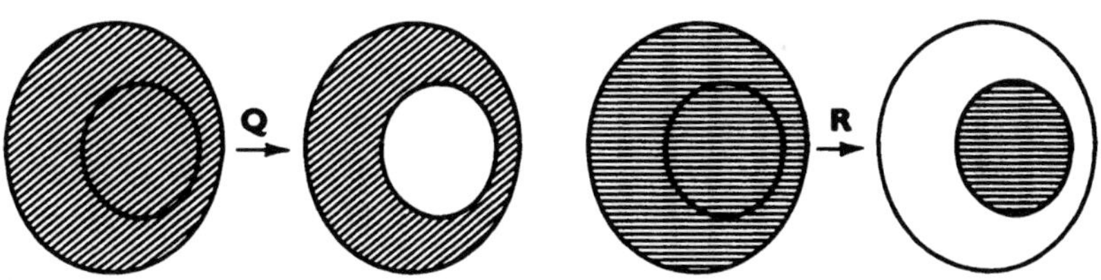

The class of **R** implicata (the informativeness or **I** implicata of Atlas and Levinson 1981) may appear heterogeneous, but it involves in each instance what Atlas and Levinson call an inference to the best interpretation, as determined by their PRINCIPLE OF INFORMATIVENESS (Levinson 1983:146–47): 'Read as much into an utterance as is consistent with what you know about the world'. Examples include the 'conditional perfection' of *if p then q* into 'if and only if p then q' (Geis and Zwicky 1971), the strengthening of *p and q* to 'p and then q' and thence to 'p and therefore q' (Grice 1975; Schmerling 1975; cf. also chapter 6), the inference from *Lee and Kim moved the piano* to 'Lee and Kim moved the piano together', and that from *Chris ate the cake* to 'Chris ate the whole cake' (Harnish 1976), and the frame-based or bridging inference from *I have a new car but the carburetor is clogged* to the assumption that the clogged carburetor is the one in my new car (Charniak 1972; Clark and Haviland 1977). Other specimens of **R**-based implicature will be collected in the course of our negative travels in later chapters.

As in any model of nonlogical (i.e., cancelable or defeasible) inference, we must produce an algorithm for computing which of the two opposed principles and inference strategies prevails in a given discourse context; this issue has been addressed (cf. Grice 1975; Harnish 1976; Atlas and Levinson 1981; Horn 1984b; and especially Levinson 1987a, b), but not definitively solved. Maxim clash, for example, arises notoriously readily in indefinite contexts; thus, an utterance of *I slept in a car yesterday* licenses the **Q**-based inference that it was not my car I slept in (or I should have said so), while an utterance of *I broke a finger yesterday* licenses the **R**-based inference that is was my finger I broke (unless I know that you know that I am an enforcer for the mob, in which case the opposite, **R**-based implicatum is derived). When the application of Quantity tends to contradict the stereotyped 'conventions of noncontroversiality' assumed within a given culture, the rival **R** principle takes precedence (Atlas and Levinson 1981).

Let us consider a more complex case. Grice (1975:51–52) cites the exchange in (42) between A and B; the former is planning an itinerary on which he would like to visit C if it does not take him too far out of his way:

(42) A: Where does C live?
B: Somewhere in the south of France.

Here it clearly would have been relevant to A's needs to know just where in the south of France C lives. Grice's gloss is that B's apparent infringement of Quantity is explainable within the Cooperative Principle only if we suppose that B couldn't have been more informative without violating Quality (Don't say what you lack evidence for); thus, B's reply **Q**-implicates that he doesn't know where in the south of France C lives.

But it is also possible that B is opting out of the Cooperative Principle for some principled reason taking precedence over Quantity. Indeed, Collinson (1937:47) eerily foreshadows Grice's example by recalling that a soldier's letter sent from the front in World War I would bear the return address *Somewhere in France*—not because the writer didn't know just where he was writing from or because the information was deemed of no interest to the recipient, but rather because of a military prohibition against revealing his more precise location. Collinson also comments on the widespread use of euphemisms of the form *to go somewhere* for 'a place it is not considered seemly to mention'; the informativeness requirement is similarly overridden in the use of expressions like *a drop of something, the dog has done something, someone asked after you this morning* (Collinson 1937:62). As we shall see in §5.3, the euphemistic flavor of these examples epitomizes the **R**-based inference to a specific understanding.

The opposition of the two Zipfo-Gricean forces may result, not simply in maxim clash, but in a resolution of the conflict through what I have called the DIVISION OF PRAGMATIC LABOR (Horn 1984b:22ff.). This principle is inspired by the Elsewhere Condition in morphology and by the program for lexical pragmatics suggested in McCawley 1978: given two coextensive expressions, the briefer and/or more lexicalized form will tend to become associated through **R**-based implicature with some unmarked, stereotypical meaning, use, or situation, and the marked, more complex or prolix, less lexicalized expression tends to **Q**-implicate a marked message, one which the unmarked form could not or would not have conveyed. Schematically,

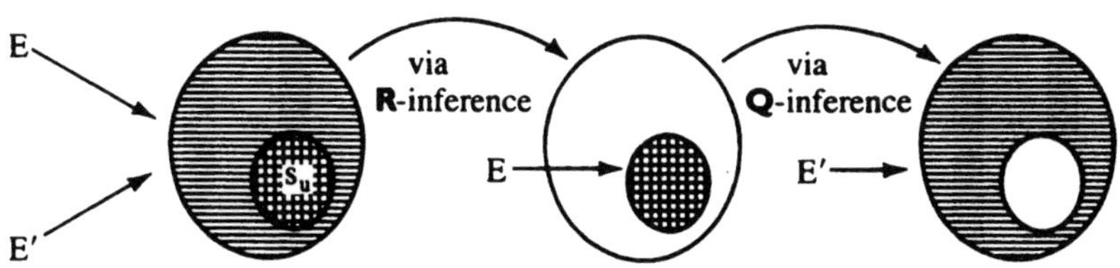

Thus, where the use of a modal question (*Can you pass the hot sauce?*) **R**-implicates a request (Pass the hot sauce), the use of the periphrastic alternative (*Do you have the ability to pass the hot sauce?*) will license the **Q**-based inference that only the literal question understanding was intended. Or if I chose to say not *My brother went to jail,* with its (conventionalized) **R**-based stereotyped understanding, but rather *My brother went to the jail,* you will infer that he was only there for a visit, not an incarceration. Or, to borrow an example from McCawley (1978), if I go out of my way to tell you not that Amanda killed the sheriff, but that she caused the sheriff to die, you will presumably infer that the causation was not of the stereotypic (direct, unmediated) variety **R**-associated with *kill,* but was in some sense marked—perhaps Amanda strewed plastic shrimp in the sheriff's bed, knowing of his weak heart and pathological fear of crustaceans. I return to the Division of Pragmatic Labor, and in particular to an apparent exception to this pattern, in appendix 2.

Armed with this reanalysis of the mechanism of conversational implicature, how might we account for the Clarkian supposition (or the Wasonian plausibility of denial), the tendency for the use of a (sentential) negation to involve the expectation or plausibility of the corresponding positive state of affairs within the discourse frame? As it happens, there are already at least two (woefully sketchy) proposals on the books for forging this connection within a theory of conversational implicature.

Horn (1978c: 203) tentatively suggests that our supposition—there called the MARKEDNESS IMPLICATURE for negation—arises through Grice's Maxim of Relation (Be relevant): 'There should be a reason to utter a sentence and, for a negative sentence, that reason . . . is generally the earlier consideration of its contained affirmative counterpart.' Thus, the argument goes, if Talmy Givón tells me that his wife is not pregnant, I would conclude either that he is violating the maxim and so opting out of the Cooperative Principle, or—more likely—that his utterance is intended to exploit the maxim by generating the implicatum that it was expected, likely, or plausible that his wife should be pregnant, that his wife's pregnancy was under consideration, and so forth. But what remains unclear on this account is just why the reason for uttering a negative sentence is 'generally the earlier consideration of its . . . affirmative counterpart'.

Leech (1981: 431; see also Leech 1983: 100–102, 165ff.) offers a different derivation of the markedness implicature. His starting points are (1) Grice's first submaxim of Quantity (i.e., my **Q** principle) and (2) Leech's PRINCIPLE OF NEGATIVE UNINFORMATIVENESS:[29]

> Negative propositions are generally far less informative than positive ones, simply because the population of negative facts in the world is far greater than that of positive facts. Consider the sentences
>
> **a**. Bogota isn't the capital of Peru.
>
> **b**. Bogota is the capital of Colombia.
>
> Both statements are true, but assuming a current United Nations membership of 132, **a**. is 131 times less informative than **b**. Hence to reconcile such a negative proposition with the first Maxim of Quantity, we must assume a context in which the negation of X is precisely as informative as is required.

Such a context is one in which I can assume that you (or someone else relevant to the discourse) believed or stated that the positive supposition of (a)—*Bogota is the capital of Peru*—was (possibly) the case. The Principle of Negative Uninformativeness can thus be used to satisfy our quest, that is, 'to provide an explanation of why negative propositions are, in pragmatic terms, denials of positive propositions which are in some sense "present in the context"' (Leech 1983: 101).

Curiously, Leech leaves unmentioned both the psycholinguistic support for Negative Uninformativeness (although his Bogota examples translate directly into the blue and red circles of the seminal Wason [1965] study) and the much longer lineage of the same notion within the philosophical literature. Both his principle and the demographic imbalance of positive and negative facts on which it is based, which Leech associates with Givón (1978), have in fact been part of the asymmetricalist canon from Plato's *Sophist* and Aristotle's *Metaphysics* ('He who knows that thing is something has understanding to a higher degree than he who knows that it is not something': *Met.* 996b14–16) through Bacon and Kant to the neo-Hegelians and to Ayer's Specificity thesis (see chapter 1 for discussion).

The link between (lack of) informativeness and discourse markedness emerges more clearly when we expand the data base. Grice observes (1967: lecture 1: 17ff.) that a negative sentence like (43),

(43) The man at the next table is not lighting his cigarette with a $20 bill.

discussed elsewhere by Austin and Searle, while undeniably true if the referent is employing the conventional lucifer, is nevertheless inappropriate in the absence of any special context. He endorses a 'condition on assertibility' adapted from Searle: 'There should be, or it should be supposed that there is, some chance that the asserted proposition is false'. If this condition is unsatisfied, as it normally is in a random true utterance of (43), the utterance is otiose or pointless.

But this condition, familiar to us from the *Mīmāṃsā* (§1.3.1 above), cannot be directly pinned to the presence of negation in sentences like (43). Indeed, Grice invokes the identical condition to explain the normal inappropriateness of a variety of true positive utterances, as in the examples in (44), the first three of which are taken from Grice (1967).

(44) I went to the meeting of my own free will.
I remember my own name.
Your wife is faithful.
The 1988 presidential election will be held.
The dean is breathing.

When a positive sentence is less informative, more otiose than its corresponding negation, it is the positive sentence which is 'presuppositionally richer' or odder in the absence of a special context. (See here Givón 1978 on the reversal of figure and ground.)

The extra supposition normally (although not invariably) associated with negation is generated as a **Q**-based implicatum. As I have noted, the **Q** principle is bound by both **R** and Quality. If I suppose that you are only concerned with whether or not Bogota is the capital of Peru, my utterance of the negative proposition (Leech's **a**) licenses no upper-bounding implicature. On the other hand, in a context where I assume it would be relevant for you to know not just that Bogota isn't the capital of Peru, but what country it is the capital of, I may still utter **a**. What I then implicate is that there is no stronger, more informative proposition that I could have uttered (consistent with the Maxim of Quality, the requirement that I say only what I believe and have evidence for). Just as my assertion that Chris has three children, in a context where I assume that you care whether or not he has more than three, **Q**-implicates that for all I know he has only three, so too my saying that Bogota isn't the capital of Peru in the described context implicates that for all I know it may not be the capital of Colombia, that is, that I don't know for a fact that it is. Only when you can assume that I do have full knowledge, and that I believe this knowledge to be relevant to your conversational goals, would such a statement be unnecessarily weak and thus unhelpful, misleading, or implausible.

One finding which goes unexplained on the epistemic/informational **Q**-based account of the markedness or (pre)suppositional richness of negative statements is the persistence of the affirmative/negative response-time differential in cases of contradictory (or immediate-contrary) opposition.[30] We may assume that (unlike Leech's capital cities or Wason's colored circles) a given integer must be either odd or even and is no more likely to be one or the other. Sentences (45a) and (45b) will then impart the same information and reflect the same degree of knowledge.

(45) a. The number 5 is not even.
b. The number 5 is odd.

Why then is the decontextualized occurrence of (45a) both intuitively and experimentally harder or less motivated than that of (45b), as we saw in the previous section? Recall that the use of (45a), but not that of (45b), implicates that someone might have believed or said that 5 was even (or at least that other integers under discussion were; cf. my earlier discussion of *4 is even {and/but} 5 is not even.*

Leech (1983:101) considers an analogous pair,

(46) a. Our cat is not male.
b. Our cat is female.

in which (46a), though informationally equivalent to (46b), 'still strikes one as being "marked", and as requiring special interpretation as a denial of what someone else has asserted'.[31] To explain the persistence of the markedness asymmetry, Leech invokes the difficulty in processing negation, as confirmed by the psycholinguistic studies reviewed above. But this deposits us back at square one, sadder but no wiser: why are negative sentences harder to process in the first place (when indeed they are)? The explicandum has been taken for the explicans.

Perhaps it is partly a case of the negative sentences in (45a) and (46a) being longer, by one morpheme or one word, than their contradictorily opposed positives. But again we would like to explain the formal markedness of negation by virtue of some aspect of its meaning, rather than treating this Thomistic correlation as an accident.

There seems to me to be another possible line worth pursuing. The extra implicature associated with negation does indeed derive from the prototype situation in which a negative statement (e.g., *A is not B*) is less informative—often infinitely less informative, as in *My favorite number is not 5*—than its affirmative basis or positive ground (*A is B′* where $B' \neq B$). But the markedness of negation, born in the pure pragmatics of conversational implicature, may tend to become partly conventionalized, with the result that all negative statements (or, more properly, all main clause predicate denials), even those which (like (45a) or (46a)) convey the same information as their positive bases, are affected (or, à la Wood 1933, <u>in</u>fected) by this implicature.

Thus it should not be surprising that the same implicature should be variously analyzed as deriving from Relation (Horn 1978c), from Quantity (Leech 1981), and from Manner (Leech 1983). In fact, the marked status of negation results from the interaction of the **Q**-based requirement that speakers be as informative as possible—where positive statements are prototypically (although not invariably) more informative than negative statements—with the **R**-based principle directing the speaker to omit anything irrelevant to the concerns of his interlocutor which might increase processing effort.

If my account of the markedness of negation is even generally correct, the asymmetry thesis I have been exploring off and on since §1.2 applies at the level of pragmatics. Negative propositions are typically, but not necessarily, less specific and less informative than positive propositions. As noted by Apostel (1972b), however, the real asymmetry is located, not in the relation of negative to positive propositions, but in the relation of (speaker) denials to assertions.

As Strawson stresses, the 'standard and primary <u>use</u>' of negatives is 'to correct and contradict' (emphasis mine)—but, pace Wittgenstein and

Givón, use is not meaning. As a formal philosopher primarily concerned with meaning and truth, Frege (1919) focuses on the logical symmetry between positive and negative propositional content and correctly rejects the identification of negation with the speech act of denial; for one thing, as Frege teaches us, an embedded negative clause cannot be analyzed as asserting or denying anything. Linguists, psychologists, and psychologically oriented philosophers (including Hegel and the neo-Hegelians, Bergson, the later Wittgenstein, Wason, Clark, García, Givón, and Shanon) have tended, understandably enough, to focus on the asymmetry in the use of negatives and affirmatives. Indeed, for at least one linguist, Lörcher (1900), any negative judgment which is not used to correct or contradict, and hence has no affirmative supposition, must instantiate (in the words of the title of his treatise on Old Saxon negatives) *die unechte Negation:* false or spurious negation.

The disparity between the logical symmetry and functional asymmetry of affirmation and negation is nicely brought out in an insightful encyclopedia entry by Josiah Royce (1917). Royce begins with the putative one-to-one relation between positive and negative propositions and the fact that for any proposition (or action), exactly one member of the pair ⟨*x, not-x*⟩ must be taken to be true (or to be executed). But whence, then, the 'common sense' feeling that negative propositions and predicates are not on all fours with affirmatives?

Royce (1917:269) determines that it is the interactions of the '*not*-relation' with other relations which leads to the (pragmatic) asymmetry: 'In a limited universe of discourse, one of two terms each of which is the negation of the other may have a value superior to that possessed by the other'. These 'extrinsic considerations' lead us to take one of the two contradictories as 'the positive, the required, the superior member' of the opposed pair. In some cases, there may be 'a definable or empirically obvious distinction in value, dignity, or desirableness' between the two terms in opposition, so that one may be considered the 'privation' of the other (e.g., *prime* vs. *nonprime* integers, *rational* vs. *irrational* numbers, *elements* vs. *nonelemental substances, winning* vs. *losing, acceptance* vs. *refusal*).

In other cases, although this is not noted by Royce, the positive, unmarked member of the opposed pair will be the one which is perceptually salient. As Greenberg, Clark, Givón, and others have observed, the functionally unmarked member of a pair of adjectival opposites is often more easily perceived—as well as emotively more positive and easier to process—than its marked counterpart (*tall/short, big/small, wide/narrow, high/low*): cf. Givón 1978:105 for additional examples and discussion. Nor does Royce discuss, as Ayer (1952) does, the epistemological factor in the determination of which predicate or proposition counts as negative (but cf. Axinn 1964 and Gale 1976:21–32 on problems with Ayer's specificity criterion). But in the end, Royce and Ayer dissociate themselves from both the radical asymmetricalists and the single-minded logical symmetricalists, coming down on the side of a more complex position wherein all propositions—negative and positive alike—are inherently equal, but some speech acts (affirmations) are pragmatically and functionally more equal than others (denials).

When Bergson (1911:289) tells us that negation is necessarily 'of a pedagogical and social nature', when Wood (1933:421) brands negation as 'infected with error and ignorance', when Wittgenstein (1953:§447) remarks that 'the feeling is as if the negation of a proposition had to make it true in a certain sense in order to negate it' (and when Givón [1978:70] legitimizes that feeling by taking **~p** to logically presuppose **p**), when Apostel (1972a:277) analyzes negation as a modality or propositional attitude, they are yielding to the temptation of placing the pragmatic cart before the semantic horse.[32]

Negatives (pace Wason 1972) are by nature no more false than affirmatives, but prototypically they are psychologically harder and more loaded, epistemologically less specific and hence less valuable, emotively more inhibiting (or at least less highly valued), and pragmatically more difficult to use appropriately within an arbitrary discourse context. Not every negation is a speaker denial, nor is every speaker denial a linguistic negation, but the prototypic use (or—following Volterra and Antinucci 1979:203—the core) of negation is indeed as a denial of a proposition previously asserted, or subscribed to, or held as plausible by, or at least mentioned by, someone relevant in the discourse context.

Thus, while affirmation not only can but standardly does function to introduce a proposition into the discourse model, negation—in its 'chief use' (Jespersen), its 'most common use' (Ayer), its 'primary and standard use' (Strawson), its 'straightforward use' (Kissin)—is directed at a proposition already in the discourse model. Further, as we shall see in chapter 6, we can isolate a METALINGUISTIC use of negation in English and other languages, specialized precisely for just this function of negative statements. But, as we shall also observe, not all instances of negation can be characterized in this way.

Like Freud's dreamer in §1.3.2 above, for whom *It is not my mother* really means 'It is my mother', the strong Asymmetricalist thesis is literally false but psychologically true.

4 Negation and Quantity

> *Not* means 'less than', or in other words 'between the terms qualified and nothing'. Thus *not good* means 'inferior', but does not comprise 'excellent'. . . . This is especially obvious if we consider the ordinary meaning of negatived numerals: He does not read three books in a year / the hill is not two hundred feet high / his income is not £200 a year . . .—all these expressions mean less than three, etc.
>
> But the same expression may also exceptionally mean 'more than', only the word following *not* then has to be strongly expressed . . . , and then the whole combination has generally to be followed by a more exact indication: his income is not two hundred a year, but at least three hundred / not once, but two or three times, etc.
>
> (Jespersen 1924:325–26)

As I mentioned in chapter 3, the locus classicus of Quantity (or **Q**-based) conversational implicature in natural language is provided by the phenomenon of scalar predication. In this chapter, I shall explore the structure of quantitative scales and investigate their interaction with negation. I begin with the range of cases touched on in the epigraph from Jespersen.

For Jespersen, nexal (sentential) negation ordinarily yields contradictory opposition (*John is coming / John is not coming*), while special (constituent) negation yields contrary negation (*John is happy / John is unhappy*). But applied to certain predicates, those denoting scalar or gradable values (cf. Sapir 1944; Bolinger 1972; Horn 1972, 1973; Ducrot 1972, 1973; Fauconnier 1975a, 1975b, 1976; Gazdar 1979a; Hirschberg 1985), negation appears to take on a special value.

Jespersen's first observation in the passage just cited is that, given a gradable predicate P, the corresponding negative predication is understood as predicating the value 'less than P' of its subject. Translated into truth-conditional terms, Jespersen's position results in the claim that (1a) and (1b) may both be judged false if Tolstoy's masterpiece is excellent, hence excluding both *good* and *not good* (inferior).

(1) a. *War and Peace* is a good book.
    b. *War and Peace* is not a good book.

Similarly, (2a, b) may both fail to be true in a case where the subject in question reads four or forty books in a year.

(2) a. He reads three books in a year.
b. He does not read three books in a year.

We are, of course, familiar with affirmative/negative pairs in contrary opposition; some of these pairs of mediate contraries go back to Aristotle:

(3) a. She is happy.
b. She is unhappy.

(4) a. All men are just.
b. No men are just.

(5) a. I want to leave.
b. I don't want to leave. (on neg-raised understanding; cf. §3.3, §5.2)

But in each of these cases, the failure of the Law of Excluded Middle is indeed linked to an unexcluded middle, a context in which the opposed terms permit mediation by a possible third term falling in between the opposites. In the oppositions of (1) and (2), however, we seem to be dealing instead with an excluded <u>non</u>middle.

Does the negation of a scalar term really yield neither contradictory nor contrary opposition, as maintained by Jespersen, Tasmowski-De Ryck (1972:172), and others? What is the structure of opposition defined by positive and negative scalar predications? What problems are presented by the existence of the exceptional and linguistically marked variety of negation Jespersen refers to in the second paragraph of the cited passage, and how are these problems to be dealt with in a unified (or nonunified) theory of natural language negation? While I shall begin to address these questions in the present chapter, the search for answers will carry me through chapters 5 and 6 as well.

4.1 Scalar Predication and Subcontrariety I

Any investigation of how negation interacts with scalar predicates (as represented by such predicate terms as the *good, three,* or *two hundred* of the Jespersen epigraph) presupposes an analysis of scalar predication itself. We cannot determine whether (6a, b)

(6) a. **a** is good.
b. **a** is not good.

are contradictory opposites, contrary opposites, or neither or the two, until we decide what sort of meaning—or at least what sort of truth conditions—we are to assign to (6a) and (6b).

Let us begin by positing the following three assumptions:

(7) A1: **a** is a singular term whose referent exists and is in the domain of the evaluative predicate (i.e., (6a) and (6b) are not category mistakes or otherwise vacuous).

A2: We know the necessary and sufficient conditions for situating **a** on the evaluative scale (i.e., we know when **a** is excellent, when **a** is mediocre, and so on).

A3: We know which context of evaluation (situation, state of affairs) obtains (i.e., we know whether in the actual context of utterance **a** is or is not excellent, mediocre, and so on).

Against the background of these assumptions, we can distinguish at least four distinct contexts or situations in which (6a, b) might be uttered, C_1–C_4; by (A3) we can agree on which context we are in.

(8) C_1: **a** is in fact excellent.

C_2: **a** is in fact good but not excellent.

C_3: **a** is in fact mediocre (not as good as 'good' but better than 'bad').

C_4: **a** is in fact bad ('inferior').

It is, I trust, noncontroversial that (6a) is true if we are in C_2 and false if we are in either C_3 or C_4. By the same token, (6b) is clearly true in C_3 or C_4 and false in C_2.[1] But what happens when we carry our two statements into the world of C_1? Is (6a) true? False? True but misleading? Neither true nor false? And what of (6b)? Let me delimit the range of possible answers to these questions:

(9) a. (6a) is false in C_1, and (6b) true.

b. (6a) and (6b) are both false in C_1.

b′. (6a) is false in C_1, and (6b) is false as well on its ordinary interpretation, but (6b) allows a second, extraordinary reading, on which it is true in C_1.

c. (6a) is true (or true but misleading) in C_1, and (6b) false.

c′. (6a) is true but misleading in C_1; (6b) as normally understood is false in C_1. If (6b) is assigned the appropriate intonation contour, appropriately grounded in the discourse context, and followed by 'a more exact indication' (à la Jespersen), it may be understood as indirectly conveying a true proposition; this understanding is associated with the negative operator, not with the scalar predicate per se.

d. (6a) is lexically ambiguous; on reading (1) *good* means 'good but not excellent', and (6a) is false in C_1, while on reading (2) *good* means 'at least good', 'good if not excellent', and (6a) is true in C_1. (6b) is ambiguous in the same way, and yields the corresponding (i.e., opposite) truth values.

e. (6a) and (6b) are not lexically or semantically ambiguous, but—because of the effect of the scalar predicate *good*—each is logically or propositionally ambiguous, capable of expressing one proposition which is true and another which is false.

Position (9a), on which *good* is invariably read as excluding *excellent*, represents an expositionally useful but (I believe) historically straw proposal. Position (9b), in the form of its refined variant (9b′), is adopted by Jespersen (1917, 1924); it is on this view that (6a, b) do not (or do not ordinarily) qualify as contradictory or contrary opposites. Position (9c) is derivable from observations by De Morgan (1847), Mill (1867), and especially Grice (1961, 1975); it is defended by Horn (1972, 1973), Fauconnier (1975a, 1975b), Gazdar (1979a, 1979b), Cornulier (1984), and Hirschberg (1985). The closely related (9c′) is supported in Horn 1984a, 1985, and in the following pages (see especially chapter 6). Position (9d) is the view maintained in Lehrer and Lehrer 1982, to which I shall return below. A parallel ambiguist line on other scalar operators is adopted by Aristotle for *possible* but not *some*, by Hamilton (1860) for *some*, by Smith (1970) and Löbner (1985) for the cardinals, and by Klein (1980) and Anscombre and Ducrot (1978) for the equative (*as X as*) construction. A view assimilable to (9d) is also supported in the earlier papers by practitioners of what I have elsewhere (Horn 1984a) called the London School of Parsimony: cf. Kempson 1979, 1980; Cormack 1980; Burton-Roberts 1984. The London School has since moved into the position defined by (9e); see especially Carston 1985a, 1985b, and Kempson 1986.

One essential feature shared by all these approaches except that of (9b, b′) is that once we assign a truth-conditional meaning to (6a)—one which may (by (9a, c, c′)) or may not (by (9d, e)) constitute its sole truth-conditional meaning—we can take (6b) to (normally) express the contradictory opposite of (6a). But the differences among the positions outweigh this one similarity. It should be borne in mind that all these incompatible approaches to positive and negative scalar predications are designed to account for the same set of intuitions: Under certain conditions (6a) conveys that context C_2 is known to hold, while under other conditions it conveys that either C_2 or C_1 holds. Under certain conditions (6b) denies that either C_2 or C_1 holds, while under other conditions (6b)—or a longer variant of it—simply denies that C_2 holds. But how are these intuitions to be accommodated in a theory of natural language semantics and pragmatics? What are the truth (or satisfaction) conditions, implicatures (conventional or conversational), and/or presuppositions (logical or pragmatic) we must adopt if we are to regiment these intuitions into a coherent, plausible, natural, and maximally simple account of (6a, b) and similar oppositions? In this chapter, I shall concentrate on developing the program suggested by position (9c), while touching on its various competitors along the way.

4.1.1 Merry Merry Subcontrary, How Does Your Logic Go?

The relation of *good* to *excellent* is parallel to that of *some* to *all,* or of *possible* to *necessary;* in each case, one member of the pair—the weaker element (*good, some, possible*)—appears to be simultaneously compatible and incompatible with its stronger counterpart (*excellent, all, necessary*). In recognizing this parallel and its linguistic correlates to be explored below, we can affirm that the problem of scalar operators has its historical roots in the logic of subcontrariety, to which I now turn.

It will be recalled that Aristotle establishes four basic logical types for general statements,

(10) **A**: All men are mortal.
I: Some men are white.
E: No men are omnipotent.
O: Not all men are just. (Some men are not just.)

which are interrelated by several distinct types of opposition. If the four statement types are realized with the same subject and predicate terms, for example, in that canonical example of the ancients, {*All / Some / No / Not all*} *pleasure is good,* these oppositions can be mapped out via the traditional Square of (11), based on the commentaries of Apuleius and Boethius (see chapter 1; classes are assumed to be nonempty, to avoid the complications addressed in §1.1.3).

(11)

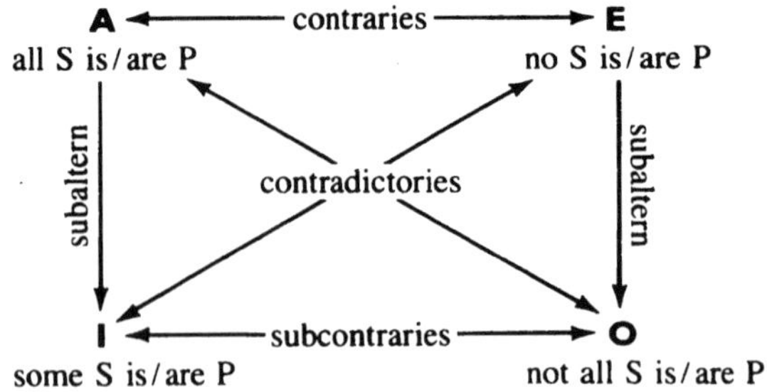

The key terms are understood as indicated in (12); cf. §1.1 for details.

(12) a. Corresponding **A** and **E** statements are CONTRARIES and cannot be simultaneously true (though they may be simultaneously false).
b. Corresponding **A** and **O**, and **I** and **E**, statements are CONTRADICTORIES; members of each pair cannot be simultaneously true or simultaneously false.
c. An **I** statement is the SUBALTERN of its corresponding **A** statement (and **O** of **E**); the subaltern is unilaterally entailed by its corresponding superaltern.
d. Corresponding **I** and **O** statements are SUBCONTRARIES and cannot be simultaneously false (though they may be simultaneously true).

The relation I am concerned with in this chapter is (12d), subcontrariety. As we saw in the earlier discussion, the members of an assertoric subcontrary opposition—the particular affirmative (*Some pleasure is good*) and particular negative (*Some pleasure is not good*)—Aristotle holds to be 'only verbally opposed' (*Pr. An.* 63b27). But while these two statement types are not strictly speaking in opposition (since they may both be true in a given context and hence, unlike Aristotle's true oppositions, do not obey the Law of Contradiction), they are nevertheless logically distinct and rhetorically contrastive within the classical system, as noted by Joseph (1916: 229–30).[2] Crucially, the **I** statement is compatible with its superaltern **A**, but not with its contradictory **E**, while the **O** statement is compatible with its superaltern **E**, but not with its contradictory **A**. More graphically, *Some men are mortal* is true, given that all men are mortal, and *Some men are not omnipotent* (*Not all men are omnipotent*) is true, given that no men are omnipotent.

But when the scene shifts to the modals, the analysis shifts as well. As we have seen, Aristotle posits an ambiguity for *possible* (*endekhomenon, dunaton*), the modal operator corresponding to the particular affirmative (**I**-type) assertoric of the southwest corner of the Square. One-sided possibility (at least possible), like *Some S is P,* is compatible with its superaltern, that is, necessity. But two-sided possibility (possible but not necessary, neither necessary nor impossible) is incompatible with necessity and logically equivalent (via complementary conversion) to its **O** subcontrary alternant.

Translated into modern terms, Aristotle tacitly advocates an asymmetry between the quantificational values (*all, some, none*) and their modal analogues (*necessary, possible, impossible*): *Some S is P* is unambiguous and true in a C_1 context where all S is in fact P, while *S is possibly P* (*S may be P, It is possible for S to be P*) is ambiguous between its one-sided and two-sided readings, these readings being, respectively, true and false in a C_1 context (in which S is necessarily P).

Essentially, then, Aristotle adopts Position (9c) for *some* and (9d) for *possible*. But this mixed approach ignores the important logical and linguistic parallels between the quantificational and modal notions, the most central of which is definitional: Aristotle himself recognizes an intimate connection between the necessary and that which is always the case, on the one hand, and between the possible and that which is sometimes the case, on the other (cf. Hintikka 1973, Waterlow 1982, for discussion). It was apparently Leibniz who first identified necessary truth with truth in all possible worlds; the possible is that which is true in some possible worlds, and the impossible in none. This thread was picked up by De Morgan (see below), Russell, Carnap, and proponents of possible-world semantics (cf. von Wright 1951 : 19). We might even say that a proposition is possible but not necessary—bilaterally possible—if it obtains in some but not all possible worlds. This semantic identification is a springboard for a wide range of other logical and linguistic signals of the kinship between *some* and *possible* (and between *all* and *necessary*); cf. Horn 1972: §2.3 for an inexhaustive inventory. By severing the treatment of *possible* from that of *some*, Aristotle makes it (somewhat) impossible to capture this parallel.

On the standard logical account of the subcontraries, particularity and possibility are treated as parallel and unambiguous, but only at the cost of ignoring the intuition that led Aristotle to the formulation of complementary conversion (*possible* [*p*] <--> *possible* [~*p*]). Just as *Some S {is/are} P* has been regarded (since Aristotle) as true so long as at least one S is P, so too *S may be P* or *It is possible for S to be P* has been taken (since the early commentator Theophrastus) to be true provided it is at least possible for S to be P; *some* is COMPOSSIBLE, mutually consistent, with *all*, and *possible* with *necessary*. The 'one-sided' versions of both operators have thus won the day, while their 'two-sided' competitors (*some but not all, possible but not necessary*) have been relegated to the role of secondary, composite operators, when they are mentioned at all. This approach has proved to be especially compelling for the general assertoric statements, where a millennium of logicians have followed Avicenna's lead: 'If you say "some men are so-and-so", it is not necessary that some others are not so-and-so. If the proposition is about all, it is also about some' (Avicenna 1971 : 24).

In the mid-nineteenth century, Sir William Hamilton of Edinburgh revived the debate over the proper treatment of the subcontraries. Distinguishing the INDEFINITE (one-sided) *some* from the SEMIDEFINITE (two-sided) *some*, Hamilton regarded the latter as basic: '*some*, if not otherwise qualified, means *some only*—this by presumption' (Hamilton 1860 : 254).[3] On this reading of the particular, the two statements *Some men are learned* and

Some men are not learned are not only (as for Aristotle) compossible, given that their conjunction is consistent, but logically indistinct. The purported opposition between the two subcontraries, charged Hamilton (1860: 261), was 'only laid down from a love of symmetry, in order to make out the opposition of all the corners in the square of opposition'.

Unfortunately, the Edinburgh Aristotle was as inconsistent in wielding his two *some*s as was his Greek counterpart with his two *possible*s. This inconsistency results in the ultimate incoherence of Hamilton's entire logical system, as his archrival De Morgan was quick to observe. While acknowledging the existence (at least in 'common language') of Hamilton's 'presumption' whereby *some* conveys *not all* (*some not*), De Morgan (1847) defends the standard practice of relegating this inference to an extralogical domain:

> In common conversation the affirmation of a part is meant to imply the denial of the remainder. Thus, by 'some of the apples are ripe', it is always intended to signify that some are not ripe.
> (De Morgan 1847:4)

> *Some*, in logic, means *one or more, it may be all*. He who says that *some are*, is not to be held to mean *the rest are not*. 'Some men breathe' . . . would be held false in common language [which] usually adopts the complex particular proposition and implies that some are not in saying that some are. (p. 56)

> Common language makes a certain conventional approach to definiteness, which has been thrown away in works of logic. 'Some' usually means a rather small fraction of the whole; a larger fraction would be expressed by 'a good many'; and somewhat more than half by 'most'; while a still larger proportion would be 'a great majority' or 'nearly all'. (p. 58)

In earlier studies (Horn 1972, 1973), I have reviewed the post-Aristotelian fate of the subcontraries and argued that *some, possible,* and related operators are unilateral or lower-bounded by their logical form, but may become bilateral in conveyed meaning through the accretion of an upper-bounding conversational implicatum. The relevant principle determining this implicatum is Grice's first submaxim of Quantity, my **Q** principle: the dictum that the speaker is to provide the addressee with the strongest relevant information available.

If I know that all men are mortal, and this information is relevant to you, I mislead you in saying (only) that some are, although (13a) expresses a true proposition.

(13) a. Some men are mortal.
b. It is possible that 2 + 2 = 4.

Similarly, (13b) is not ambiguous between a unilateral sense on which it is true and a bilateral sense on which it is false, contra Aristotle and Burton-Roberts (1984). Rather it too is simply a true statement whose utterance implicates something false, that is, that (for all I know) it's not necessarily true (or, for all I know, true at all) that 2 + 2 = 4.

The same principle that renders a relatively weak assertion like (13a, b) misleading when the speaker knows that a stronger proposition obtains can also be EXPLOITED, in Grice's sense, to generate conversational implicata. If I tell you that some of the dinner guests are smokers, and if it's relevant to you whether all of them are smokers, and if I can be expected to know, for each guest, whether or not s/he is a smoker, then our mutual awareness of the Maxim of Quantity licenses the inference that not all of them are smokers. In the same way, my informing you that I may finish my book by 2001 yields the nonlogical inference that (for all I know) I may not finish it by 2001. Thus, *S may be P* does mean that *S is P* is at least possible, as the unilateralists maintain, but it is standardly used to convey that *S is P* is only possible (i.e., not known to be necessary or actual), as the bilateralists point out. Nor does *some* mean 'some only', 'some but not all' (pace Hamilton), although it may be—and generally is—used in natural discourse to convey this two-sided or semidefinite understanding. The symmetrical inference between the southwest and southeast corners of the Square is thus valid, not as a logical or semantic principle, but as a context-dependent, generalized conversational implicature. (Cf. Grice 1975; Horn 1972, 1973, for details.)

The Gricean line on the subcontraries traces back (at least) to Mill:

> If I say to any one, 'I saw some of your children to-day', he might be justified in inferring that I did not see them all, not because the words mean it, but because, if I had seen them all, it is most likely that I should have said so: even though this cannot be presumed unless it is presupposed that I must have known whether the children I saw were all or not. (Mill 1867:501)

In this proto-Gricean argument, Mill is careful to provide an epistemic rider on quantity-based inferences: the use of a weaker predicate suggests (implicates) that for all the speaker knows, the stronger predicate on the same scale could not have been substituted *salva veritate*.

The **Q** principle responsible for Mill's inference is first explicitly formulated by Strawson (crediting the essential idea to 'Mr H. P. Grice') as the 'general rule of linguistic conduct' that 'one should not make the (logically) lesser, when one could truthfully (and with equal or greater clarity) make the greater claim' (Strawson 1952 : 178–79). Grice's own 'first shot' at this principle, offered within his program for a causal theory of perception (1961 : 132), is the dictum that 'one should not make a weaker statement rather than a stronger one unless there is a good reason for so doing'. Six years later, in his William James lectures, Grice reformulated this rule as the first submaxim of Quantity—'Make your contribution as informative as is required (for the current purposes of the exchange)' (1975 : 45)—and situated it within a general program for deriving nonlogical inferences within a conversational context.

Given this independently motivated principle of linguistic (and, as Grice notes, nonlinguistic) interchange, no special logical treatment of the inference from *some* to *not all* (*some not*) is required—which is just as well, since the context dependence and epistemic qualification associated with this inference would vitiate a logical treatment in any case. But this fact has not deterred others from 'adopting into logic', as Mill (1867) charged Hamilton with seeking to do, 'a mere *sous-entendu* of common conversation in its most unprecise form'.[4] Thus, Kuroda (1977 : 97–98) posits an 'every-day reading' of (14a), which is 'assumed to entail' (14b), so that (14a, b) come out 'logically equivalent'.

(14) a. Some animals are white.
b. Some animals are not white.

Kuroda is not dissuaded from this 'logical equivalence' by his recognition that on its everyday reading (14a) cannot serve as the contradictory of *No animals are white,* since both these propositions would be false if all animals are white.

A similar semantic account of both assertoric and modal subcontrary pairs is offered by Morpurgo-Tagliabue (1981 : 502):

> The ∃x, the 'possible', may, to some extent, come nearer and nearer to the 'all (x)', the 'necessary', without ever reaching it, like Achilles and the tortoise. . . . It is excluded that while saying 'not-all' (O) one could mean 'nobody' and saying 'not-nobody' (I) one could mean 'all'. . . . If I say 'not all people are clever' (O), this means that there are some who are stupid.

The Hamiltonian identification of *some* with *not all* is reiterated by Jespersen (1917, 1924) and Collinson (1937), as we shall see below.

My pragmatic account of the subcontrary relation generalizes to all relatively weak scalar operators, including cardinal numbers and evaluative or gradable adjectives like *good* (cf. (6a)). The Gricean story proceeds in the same way in each case. Let us take as an instance my assertion of the sentence in (15a): why does this statement normally convey the proposition in (15b) if this proposition is not part of its meaning? And why does it seem to exclude (15c)?

(15) a. Pat has three children.
b. Pat has exactly three children.
c. Pat has four children.

The argument from (15a) to (15b) proceeds as follows:

(16) i. Cardinals like *3* are lower-bounded by their literal or conventional meaning; hence (15a) means (is true iff) Pat has at least three children.
ii. There is a stronger statement than (15a), that is, (15c), such that the latter unilaterally entails the former but not vice versa. (Actually, there are in fact infinitely many such stronger statements.)
iii. Given **Q**, if I know or believe that Pat has (at least) four children, and that it would be relevant to you to know this fact, it would be misleading for me to tell you that he has three children.
iv. You are prepared to assume that I am abiding by the Cooperative Principle and its component maxims, including **Q**; I know this, as well, and you know I know it. Thus, you take me to be observing **Q** unless I indicate otherwise.
v. Therefore, you infer that the reason I chose not to express the stronger proposition in (15c) is that I didn't know for a fact that it was true.
vi. You infer that for all I know (15c) is false, that is, that Pat has fewer than four children.
vii. If, in addition, you assume that I know how many children Pat has, you can infer that I know (and am informing you) that Pat has fewer than four children; hence (given (i)) that (15b) is in fact the case, Pat has exactly three children.

Instantiations of this argument schema appear in Grice 1975, 1978; Gazdar 1979a; Levinson 1983; and Hirschberg 1985, although Gazdar and Levinson seek (incorrectly, in my view) to assimilate steps (vi) and (vii) of the schema in their account of scalar implicature.[5]

Since the upper bounding of the original statement in (15a) to the conveyed message in (15b) is mediated by principles whose applicability is context-dependent, I can block, cancel, or suspend the implicature (by uttering (15a) in a linguistic or extralinguistic context that removes the implicatum), or I can reinforce it (and thereby assert or entail, rather than merely implicate, the upper bound). Examples of linguistic devices for canceling and reinforcing the implicatum are illustrated in (17) and (18), respectively (see Horn 1972: chapter 1 for details and examples).

(17) Pat has { at least three children. / three children and possibly four. / three children and for all I know four. / three children if not four. / three or more children. / three or even four children. / three, indeed four children. / not just three but (in fact) four children. }

(18) Pat has { exactly three children. / three and only three children. / three children but not four. }

These constructions surfaced earlier in the debate between the arch-bilateralist Hamilton and his unilateralist foes. Hamilton himself seems to have recognized the contextual, pragmatic nature of his upper-bounding 'presumption', without, however, realizing the import of this fact: 'We ought to be able to say *some at least* when we do not know, and cannot, therefore, say determinately either that *some only* or that *all* is true' (Hamilton 1860:254). If *some* really meant, rather than implicating, 'some only', *some at least* would be a contradiction. Furthermore, the radical bilateralist would predict that expressions like *some but not all,* and indeed *some only* or *only some* (along with the examples of (18)), would be semantically redundant, which they clearly are not.

In his 1906 *Logic,* Keynes defends the standard one-sided *some.* His evidence includes the argument that in a situation in which the speaker's knowledge is incomplete, if all the Ss s/he knows about are P, s/he can't use either 'all Ss are P' or—with Hamilton's bilateral (*only*) *some*—'some Ss are P': 'The only solution . . . is to say that all or some S's are P's. The complexity that this would introduce is obvious' (Keynes 1906:202–3). Of course speakers who find themselves in such an epistemically imperfect state often do resort precisely to this 'obvious complexity': *Some or all of the dinner guests are nonsmokers, Some, if not all, men are irrational animals.*

Despite his unilateralist stance, Keynes (p. 200) concedes that many logicians 'have not recognized the pitfalls surrounding the use of *some.* Many passages might be quoted in which they distinctly adopt the meaning—*some, but not all*'. To which Jespersen (1924:324) retorts, 'in the name of common sense', with the rhetorical question: 'Why do logicians dig such pitfalls for their fellow-logicians to tumble into by using ordinary words in abnormal meanings?' Like his forbears Aristotle, Hamilton, and the pitfallen logicians chastised by Keynes, Jespersen fails to recognize that meaning, in the broad sense, may be determined not only by rules of logical form (meaning "proper") but also by pragmatic rules which determine the use of an expression within a context of utterance. We can rescue the baby of Aristotelian and Hamiltonian insight into the function of the subcontraries in ordinary language from the bathwater of logical inconsistency by accepting the logical status of the subaltern inference (*all* → *some, necessary* → *possible*), while demoting the subcontrary conversions (*some* <--> *some not/not all, possible* <--> *possible not/not necessary*) into a sous-entendu or context-dependent generalized conversational implicature. On the Mill-Grice program, the oddness of the assertions in (19),

(19) Some men are mortal.
It is possible that 2 + 2 = 4.
There are seven planets.
War and Peace is a good book.

as well as the systematic "ambiguity" of *good, some, possible,* the cardinals, and other weak-scalar values, will receive a pragmatic rather than strictly semantic explanation.

Before expanding on this explanation, which involves an elaboration of the notion of the quantitative scale defined by logical entailment and of the Q(uantity)-based implicatures defined off such scales, I shall offer a brief history of the relations between negation and the logical operators of the Square of Opposition.

4.2 Negation and the Three-cornered Square

Jespersen (1917: chapter 8; 1924: 324–25) regiments the logical operators into a TRIPARTITION defined by a set of equivalence rules. He begins (1917: 86) with the quantificational values:

(21)					
A: all	everything	everybody	always	everywhere
B: some/a	something	somebody	sometimes	somewhere
C: none/no	nothing	nobody	never	nowhere

and later (pp. 92–93) extends the tripartition to the modal and deontic (obligation- and permission-based) categories, which he takes to constitute 'special instances' of the original values:

(22)			
A: necessity	must/need	command	must
B: possibility	can/may	permission	may
C: impossibility	cannot	prohibition	must not/may not

Other candidates for the tripartition, not dealt with by Jespersen, are the epistemic adjectives (*certain, possible, impossible*) and the deontic causatives (*require, permit/allow, forbid/bar*).

For each case in (21) and (22), the opposed values are interdefinable in accordance with a set of proposed equivalences which appear (along with some of Jespersen's instantiations) in (23):

(23) i. ~**A** = **B** (*not all* = *some, something; not always* = *sometimes; not necessary* = *possible*)
ii. ~**C** = **B** (*not for nothing* = 'to good purpose'; Latin *non-nemo, non-nulli, non numquam* [lit., 'not nobody/none/never'] = 'somebody, some, sometimes'; *not impossible* = *possible*)
iii. **A** . . . ~ = **C** (*everybody was unkind* = 'nobody was kind'; *necessarily not* = *impossible*) [the *all* . . . *not* construction is discussed in §4.3 below]
iv. **C** . . . ~ = **A** (*nobody was unkind* = 'everybody was kind'; Latin *nemo non vidit* [lit., 'nobody doesn't see'] = 'everybody sees'; *impossible not* = *necessary; cannot but* = 'must'; Latin *non potest non amare* [lit., 'can't not love'] = 'must love')

Based on these laws, Jespersen (1917:91–92) proposes the following generalization: 'When the absolute notion (**A** or **C**) is mentioned first, the absolute element prevails, and the result is the contrary notion [cf. (23iii, iv)]. If on the other hand, *not* comes first, it negatives the absolute element, and the result is the intermediate [**B**] relative [cf. (23i, ii)]'.

From the perspective of the standard predicate calculus, the interdefinability of **B** and **C** stipulated in (23ii) reflects the Law of Double Negation. Analyzing the morphologically negative **C** forms (e.g., *nobody, impossible*) as the negation of the corresponding **B** forms (*not somebody, not possible*), we are left with the truism that ~(~**B**), the negation of the negation of a **B** operator, yields the **B** operator itself.[6] Furthermore, the decomposition of **C** into ~**B** reduces Jespersen's twin equivalences in (23iii, iv) reproduced in (24), to the familiar logical laws of quantificational and modal logic given in (25) and (26), respectively:

(24)	(25)	(26)
$\mathbf{A} = {\sim}\mathbf{B} \ldots {\sim}$	$\forall x\phi \leftrightarrow {\sim}\exists x{\sim}\phi$	$\Box\phi \leftrightarrow {\sim}\Diamond{\sim}\phi$
$\mathbf{B} = {\sim}\mathbf{A} \ldots {\sim}$	$\exists x\phi \leftrightarrow {\sim}\forall x{\sim}\phi$	$\Diamond\phi \leftrightarrow {\sim}\Box{\sim}\phi$
${\sim}\mathbf{A} = \mathbf{B} \ldots {\sim}$	${\sim}\forall x\phi \leftrightarrow \exists x{\sim}\phi$	${\sim}\Box\phi \leftrightarrow \Diamond{\sim}\phi$
${\sim}\mathbf{B} = \mathbf{A} \ldots {\sim}$	${\sim}\exists x\phi \leftrightarrow \forall x{\sim}\phi$	${\sim}\Diamond\phi \leftrightarrow \Box{\sim}\phi$

For any operator **P**, let us (following Löbner 1985) call ~**P** its OUTER NEGATION and **P**~ its INNER NEGATION. Then the interdefinability of a corresponding pair of **A** and **B** operators, as reflected in (24)–(26), amounts to the principle that either member of such a pair may be defined as the outer negation of the other's inner negation; equivalently, the outer negation of either member (the **A** or **B** item from any of the columns of Jespersen's extended tripartition) is identical to the inner negation of the other.

Any two operators **A** and **B** which are interdefinable in this way, and which differ in that any simple, unembedded sentence containing **A** unilaterally entails the corresponding sentence containing **B**—*All Ss are P* $\not\rightleftarrows$ *Some Ss are P; Necessarily* (**p**) $\not\rightleftarrows$ *Possibly* (**p**)—are known as DUALS. Thus, on a standard account of the quantificational and modal operators, ⟨all, some⟩ is a pair of duals, as is ⟨necessary, possible⟩. But for Jespersen, no such dualism can exist.

The key, once again, lies in the analysis of the subcontraries. For *all* . . . to entail *some* . . . (in non-null sets) or for *necessary* . . . to entail *possible* . . . , the unilateral definition of the **B** terms as subalterns must be adopted. This step Jespersen is clearly unwilling to take; on the contrary, he explicitly defines *some*—along with the other 'intermediate' **B** values—bilaterally, 'in the ordinary meaning it has in natural speech [some but not all] and not in the meaning logicians sometimes give it, in which it is the positive counterpart of *no* (*nothing*) and thus includes the possibility of all' (1924:324). Whence the "equivalences" of (23i): *not all = something, not necessary = possible*. Where Aristotle (sometimes) interprets *possible* bilaterally, and Hamilton (usually) takes *some* bilaterally, Jespersen consistently treats both operators as instances of Category **B**, which is (it would appear) always two-sided.[7]

One striking result of Jespersen's adoption of the bilateral **B**s, as defined by (23i), inheres in the tripartition itself. In the classical Square of Opposition (cf. (11) above), the four opposed vertices are defined by the laws in (12); note especially (12c), the principle of subalternation. The paired subcontraries (*some / some not, possible / possible not*) remain distinct through the relation of each to its own superaltern (*all / none, necessary / impossible*). But Jespersen's bilateral subcontraries, with no subalternation (entailment) relation to support them, collapse into each other. The four corners of the post-Aristotelian Square map onto the three points of the tripartition, which must evidently be assigned a triangular figure whose vertices are all linked by contrariety:

(27)
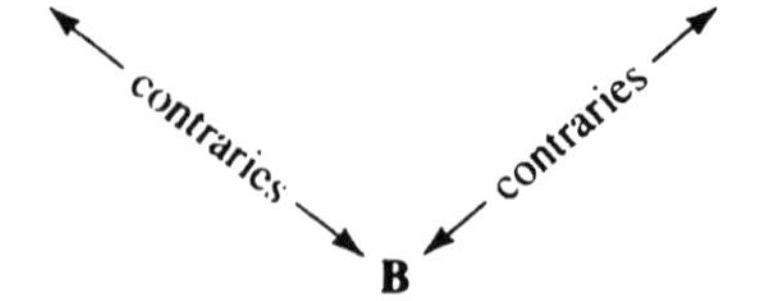

Crucially, the apex (or nadir) of the Triangle of Opposition, Jespersen's **B**, corresponds neither to the **I** nor the **O** vertex of the traditional Square, but rather to their conjunction.

This point is clearly recognized by De Morgan. Writing sixty years before Jespersen, he warns that the TRICHOTOMY apparently possible to one with complete knowledge must yield, in a logic based on the imperfectly epistemic human condition, to the classical four-way opposition (with qualitative and quantitative axes) mapped out in the Square:

> There are three ways in which one extent may be related to another . . . : they are, complete inclusion, partial inclusion with partial exclusion, and complete exclusion. This trichotomy would have ruled the forms of logic, if human knowledge had been more definite. . . . As it is, we know well the grounds on which predication is not a trichotomy, but two separate dichotomies. . . . Must be, may be, cannot be, are the great distinctions of ontology: necessity, contingency, impossibility. This was clearly seen by the logicians. But it was not so clearly seen that this mode of predication tallies, not with the four ordinary forms *A, E, I, O,* but with the three forms *A, (OI), E.* As in the following:—Every *X* is *Y,* which is the consequence of necessity; Some *X*s are *Y*s and some are not, which is the consequence of contingency; and No *X* is *Y,* which is the consequence of impossibility.
>
> (De Morgan 1858 : 121)

Thus, as De Morgan saw, when contingency or possibility is bilateral, it represents (like the *some* of Hamilton and Jespersen) a complex rather than simple operator. The simple notion of possibility, corresponding to the classical *some,* has no negative component in logical form.

Leaving aside the controversial and ultimately untenable bilateralist principle (23i), the interdefinability of the operators predicted by Jespersen's other three equivalences (23ii–iv) is freely exploited by natural languages, as even a quick survey of the literature amply attests. Pott (1859: 359–60) offers some of the same Latin examples as Jespersen—*nonnulli* = *aliquot* (~**C** = **B**) vs. *nulli non* = *omnes* (**C** . . . ~ = **A**)—and provides a table which directly anticipates Jespersen's Tripartition.[8]

The interdefinability of the modals, in particular, is reflected in a wide variety of doubly negated periphrastic expressions, as well as the scope distinctions predicted by Jespersen's generalization. In Mandarin Chinese, for example (Chao 1955), negation may precede the modal (*bu idinq* 'not necessarily', *bu neng lai* 'not able to come'), follow the modal (*idinq bu* 'certainly not', *neng bu lai* 'able not to come'), or both precede and follow the modal (*bu neng bu lai* 'cannot but come').[9] In Yoruba (Banjọ 1974) the same facts obtain but the outer and inner negations differ morphologically:

(28) a.	Ade lè korin.	'Ade may/can sing'	$[\Diamond \mathbf{p}]$
b.	Ade kò lè korin.	'Ade cannot sing'	$[\sim\Diamond \mathbf{p}]$
c.	Ade lè mà korin.	'Ade may [not sing]'	$[\Diamond\sim \mathbf{p}] \leftrightarrow [\sim\Box \mathbf{p}]$
d.	Ade kò lé má korin.	'Ade can't not sing'	$[\sim\Diamond\sim \mathbf{p}] \leftrightarrow [\Box \mathbf{p}]$

Similar instances of 'can't not' for 'must' are cited by Harries (1973):

(28′) **German**: Hans kann *nicht* den Mann *nicht* hassen.	'Hans can't not (must) hate the man'
Hungarian: John *nem* tudta *nem* szeretni öt.	'John couldn't not (had to) love her'
Latin: *Non* possum *non* amare.	'I can't not (must) love'
Russian: Ja *ne* mog *ne* dat' emu nagrádu.	'I couldn't not (had to) reward him'

Nor is Yoruba the only language to exploit the interdefinability of the modal duals to eliminate one of the two operators, generally the **A** (necessity) operator. The standard expression for 'must' in Malagasy is *tsy maintsy* (lit., 'not able not'), while that in Basque is *ezin bertze* (lit., 'impossible not').[10]

As we have seen, the four-way opposition embodied in the Square and the three-way contrast reflected in the Tripartition apply to the notions of the quantificational, modal, and deontic systems. But not to these alone; Peter of Spain offers a medieval version of the table of equivalences which includes the suppletive forms for quantifiers restricted to domains of two. Peter's equivalences (from the *Summulae Logicales,* Tractatus 1; cf. Mullally 1945:lxxi) can be mapped onto the traditional Square as in (29):

(29)

no one not : every
not any one not : all
neither not : both

no one : every not
not any one : no one
not one of : neither

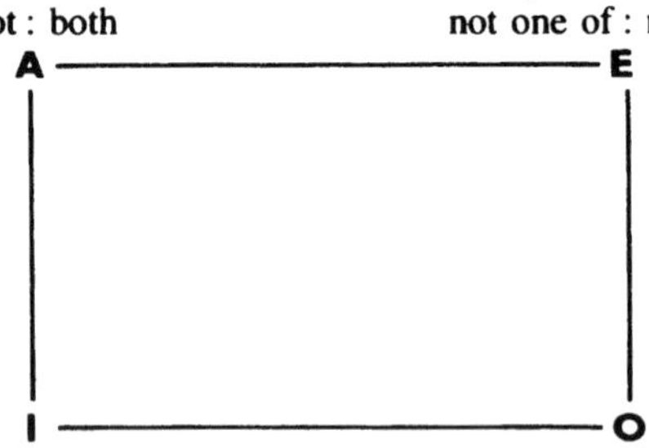

not no one : some

not every : some not

These equivalences presuppose a treatment of quantifiers over two-object domains in which *both* and *neither* are the contrary universals, with *one* (*of the two*) as the subaltern of the former and the contradictory of the latter. As Collinson (1937:94) puts it, 'both is the all of two'. Within the tripartite model, *both, one* (*of the two*), and *neither* fall under Jespersen's **A**, **B**, and **C** categories, respectively. The fourth corner of the Square, orphaned by Jespersen, would be satisfied by *not both* (corresponding to the general-domain *not every, not all*).

This very identification is in fact made in Sapir's quadripartite version of Jespersen's table. With characteristic insight, Sapir (1930:21) opts for a solution midway between the classical Square and the Jespersenian Triangle. His particular subcontraries are logically neither strictly bilateral nor absolutely unilateral:

> '*Not everybody came*' does not mean '*some* came', which is implied, but '*some* did *not* come'. Logically, the negated totalizer [*not every*] should include the totalized negative, i.e., opposite or contrary [*none*], as a possibility, but ordinarily this interpretation is excluded and the totalized negative (contrary) is expressed by negating the corresponding unitizer or non-specifying selective [*not* {*one*/*any*/*a*} . . .].

Sapir adopts neither the orthodox logical approach (in which *some* = *at least one*) nor the revisionist Hamiltonian-Jespersenian line (in which *some* = *not every*), foreshadowing instead the implicature-based theory outlined above; note especially his use of *is implied* (vs. *means*) and of the qualifier *ordinarily*, suggesting the essential role of the context in licensing the implication (or implicature) in question.

Sapir's own program for the quantity expressions (1930:22, reproduced in (30)) builds in the two-object-domain operators, as well as the standard quantifiers and quantificational adverbs; he does not, however, extend his account to the modals.

(30)

Positive Totalizer	Negated Totalizer	=	Partial	Negated Unitizer	Totalized Negative (contrary)
all the men	not all the men		some of the men	not one man not a man not any of the men	no men none of the men
all of it	not all of it		some of it	not {any/ one bit} of it	none of it

Positive Totalizer	Negated Totalizer	=	Partial	Negated Unitizer	Totalized Negative (contrary)
everybody, everyone	not every-body, not everyone		somebody, someone	not anybody, not (a) one	none, nobody
both of them	not both of them		one of (the two of) them	not either of (the two of) them	neither of them
always	not always		sometimes	not {ever/at any time}	never, at no time

Despite Sapir's identity or equality sign linking the second and third columns (i.e., the two subcontraries), it is clear from the passage quoted above that this is not intended as logical equivalence so much as discourse commensurability, mediated by contextual considerations.

Like Pott and Jespersen, Sapir provides an account of the different forms of double negation: a 'negated totalized negative' (*not none*, Lat. *nonnulli, non numquam*) reduces to a 'partial' ('some', 'sometimes'), à la (23ii), while a 'totalized double negation' (*There was none but was present, There was none who was not present*) cancels out to a 'positive totalizer' (*Everyone was present*), à la (23iv).[11]

As signaled by Sapir, *both* and *one of* (*the two of*) are dual operators over two-member sets, just as *all* and *some* (= *at least one*) *of* are duals over multiple-member sets. But the members of a set can be enumerated individually rather than quantified over, and the standard logical devices for doing so are intrinsically binary. The classical constants here are of course *and* and (inclusive) *or*, which constitute one more pair of duals. Given the standard truth tables for these connectives (cf. chapter 2, (51)), the two connectives are interdefinable via negation, and an unembedded expression containing *and* unilaterally entails the corresponding expression containing *or*. The principles stipulating the interdefinability of the binary connectives, the equivalences in (31),

(31) a. $\sim(\mathbf{p} \wedge \mathbf{q}) \leftrightarrow \sim\mathbf{p} \vee \sim\mathbf{q}$
b. $\sim(\mathbf{p} \vee \mathbf{q}) \leftrightarrow \sim\mathbf{p} \wedge \sim\mathbf{q}$

are familiarly known as DE MORGAN'S LAWS, but their history in both Western and Eastern logic long antedates that great nineteenth-century formal logician.

For the Stoics, the logical constant corresponding to (*either*) . . . *or* was viewed as 'true' or 'proper' only when one disjunct is true and the other false; whence the validity of the Stoics' fourth and fifth 'indemonstrable syllogisms', given in (32a, b), respectively:[12]

(32) a. Fourth indemonstrable syllogism

p or **q**
p
∴ not-**q**

b. Fifth indemonstrable syllogism

p or **q**
not-**p**
∴ **q**

Alongside their exclusive (or 'proper') disjunction, some Stoics and early medieval logicians admitted an inclusive ('improper') disjunction, usually disparaged as a 'paradisjunction' or 'pseudodisjunction'. These two operators in question, with their assumed truth conditions, are represented in the last two columns of (33); conjunction is included to fill out the tables.

(33)

p	q	p ∧ q	p ∨ q (inclusive)	p w q (exclusive)
T	T	T	T	F
T	F	F	T	T
F	T	F	T	T
F	F	F	F	F

As Peter of Spain observed, only the syllogism in (32b), not that in (32a), can be retained under the inclusive definition of *or*, while exclusive *or* validates both syllogisms; on the other hand, ADDITION (the inference from ***p*** to ***p*** ∨ ***q***) is compatible only with the inclusive definition.

By the late thirteenth and early fourteenth centuries, in the work of Albert of Saxony, William of Ockham, Peter of Spain, and their colleagues, the inclusive reading of *or* came to predominate. With it came some strikingly terse formulations of the so-called De Morgan's Laws, the equivalences in (31) which go through only for inclusive disjunction:

> The contradictory of the affirmative disjunctive is a conjunctive composed of parts which are contradictories of the parts of the disjunctive.
>
> (= (31b), from Albert of Saxony's *Logica 3*, chapter 5; cited in Moody 1953:41)

> A conjunction and a disjunction with mutually contradictory members contradict one another.
>
> (= (31a, b), from Peter of Spain's *Summulae Logicales;* cited in Lukasiewicz 1934:81 and Mullally 1945)

In fact, an incipient version of (31b) can be seen in the work of the tenth-century commentator Avicenna (ibn-Sīnā), for whom a negative disjunctive judgment is equivalent to the joint denial of the terms of the disjunction (Madkour 1934: 168).

Earlier still, the practitioners of the Navya-Nyāya school of Indian logic (cf. §1.3.2), for whom *and* and *or* are (as for the Stoics) truth functions which are (as against the Stoics) assigned the values in the third and fourth columns of (33), respectively, exploited the interdefinability of duals to represent *or* in terms of *and* and negation, that is;

(34) $\mathbf{p} \vee \mathbf{q} =_{df} \sim(\sim\mathbf{p} \wedge \sim\mathbf{q})$

While De Morgan's Law is not explicitly stated as such, it is applied in logical argumentation (Ingalls 1951:63–64, 142; Sharma 1970:70–71). In particular, the form of negation known as *ubhayābhavā* (the negation of both) presupposes the relation formalized in (31b).

Whatever their provenance, De Morgan's Laws can be viewed as special instances of the laws of quantifier negation; (31a, b) correspond directly to the last two equivalences of (25). The key principle here is the (ancient) insight that every universally quantified proposition is essentially identical to a (perhaps infinite) conjunction, and every existentially quantified (or particular) proposition to a disjunction. (Cf. Horn 1972: §2.13 for some linguistic correlates of this logical principle.) Thus, beyond the fact that *and* and (inclusive) *or* are logical duals, there is a direct correlation of the form *and : all :: or : some*. This correlation is reflected in the shared function of *both* and *neither*—as quantifiers over sets of two, suppletive with *all* and *no(ne)*, as noted above, and as 'correlative conjunctions', paired with *and* and *or*. To our ever-expanding tripartition, we can now add the two columns in (35):

(35) **A**: both (of them) — (both) α and β
B: one (of them) — (either) α or β
C: neither (of them) — (neither) α nor β

In the light of this parallel, it is essential to recognize that exclusive disjunction, *α or β but not both*—as defined in accordance with the rightmost column of (33)—has precisely the same theoretical status as the bilateral particular of Hamilton and Jespersen, that is, *some but not all*. Yet this is not always acknowledged. Thus, Strawson (1952:91) rejects truth-functional inclusive disjunction as a model for English *or*, along with the entailment from $\boldsymbol{p} \wedge \boldsymbol{q}$ to $\boldsymbol{p} \vee \boldsymbol{q}$. But, as Geach (1972:68) points out, he has no similar qualms in identifying *some* with the standard (unilateral) existential quantifier or in accepting the corresponding subaltern inference from *all* to *some*.

Others, including Quine (1952:5) and Geach himself, recognize both an inclusive and an exclusive truth-functional disjunction, labeled *vel* and *aut*, respectively:[13] ' "***p*** *vel* ***q***" is true iff one of the propositions "**p**" and "**q**" is true, and "***p*** *aut* ***q***" is true iff one of the propositions is true and the other false. (I use Latin words as connectives in order to dodge the idiotic but seemingly perennial dispute as to the "proper" meaning of "or" in ordinary language.)' (Geach 1972: 15).

The unspoken motivation for this terminology, involving the behavior of the Latin conjunctions, is spelled out by Collinson: 'Latin clearly distinguishes a disjunctive alternative of mutually exclusive terms expressed by aut from an alternative not excluding both expressed by vel. Finnish uses tai and vai and Welsh ynte and neu respectively to make the same distinctions' (Collinson 1937: 95). But on closer inspection, the claim that ***p*** *aut* ***q*** and ***p*** *vel* ***q*** represent prototype instances of exclusive (***p*** ∨ ***q***) and inclusive (***p*** w ***q***) disjunction, respectively, tacitly subscribed to by Quine, Geach, and other logicians, is untenable. There is clear evidence from texts, dictionary citations, and (in the case of Finnish and modern Welsh) speakers' intuitive judgments that whatever distinguishes the two Latin disjunctions (and their putative cross-linguistic counterparts), it is not that the former is invariably an exclusive, and the latter an inclusive, truth-functional disjunctive operator. Indeed, this fact is implicit in Collinson's wording: to say that ***p*** *aut* ***q*** is used when **p** and **q** are mutually exclusive terms (*Is he alive or dead? Kim or Chris will win the race*) is to admit that we cannot tell (or do not care) whether the disjunction in question is in fact inclusive or exclusive, since the one situation which distinguishes the two truth functions ***p*** ∨ ***q*** and ***p*** w ***q***, as the truth table in (33) illustrates, is that in which both disjuncts are true.

This point is forcefully made by Barrett and Stenner (1971), in their attempt to relegate the 'purely truth-functional exclusive *or*' to the mythical realm of unicorns, mermaids, and centaurs. Other arguments against a logically exclusive disjunction for natural language are given by McCawley (1972, 1981), Gazdar and Pullum (1976), Gazdar (1977), and Pelletier (1977), while the purported ambiguity is defended by Lang (1977); I shall return to this issue in chapter 6.

If *or* corresponds logically to *some*, we should expect to find ***p or q*** standardly used to convey that the stronger expression (***p and q***) does not hold, or at least that it is not known to hold. Our logically inclusive ***p*** ∨ ***q*** (***p or q or both, p and/or q***) conversationally implicates (but does not entail) that (for all the speaker knows) ~(***p*** ∧ ***q***), thus conveying the upper-bounded disjunction '**p** or **q** but not both', just as *some* implicates *not all*, thus conveying 'some but not all'. *Or* is no more lexically ambiguous than *some*— or *one*. Ironically, Geach's own truth-functional definition of *vel*, cited above, builds in the assumption that *one* means 'at least one', yet in exactly the same way, *or* means 'at least or'. Just as *one* standardly implicates 'no more than one', 'not two', 'not all', so too the use of *or* standardly implicates 'not and'.

Thus, while there is a tendency to take disjunctions exclusively—in Mill's words, 'when we say A is either B or C, we imply that it cannot be both'—this implicature is not a logical inference: 'If we assert that a man who has acted in a particular way must be either a knave or a fool, we by no means assert, or intend to assert, that he cannot be both' (Mill 1867:512). Certain nineteenth-century politicians may have come to Mill's mind here, as certain of their twentieth-century counterparts come to ours.[14]

Before returning to address the question of formal representation for the subcontraries and related scalar operators, I shall pause to undertake a brief reconsideration of the third of Jespersen's equivalences in (23) and of a set of apparent counterexamples to it.

4.3 *All that Glitters:* Universals and the Scope of Negation

Jespersen's third equivalence, (23iii), correlates **A**~ expressions (*everybody . . . not, both . . . not, necessary . . . not*) with the corresponding **C** (~**B**) expressions (*nobody, neither, impossible*). But, he observes (1917: 86ff.), there is a tendency—often disparaged as 'illogical'—for an apparent universal negation ({*all/every*} . . . *not*) to be read as a negated universal (*not* {*all/every*}). Given the collapsing of the subcontraries within Jespersen's tripartition, an **A**~ quantifier (*all . . . not*) equates to the corresponding **B** value (*some*). In the alternative four-cornered picture modeled by the Square of Opposition, the result of the inner negation of **A** is not the predicted contrary (**E**) term but apparently the contradictory (**O**); in either case, we have some explaining to do.

The phenomenon in question is as widespread as it is long-lived. In addition to the locus classicus of (36),

(36) All that glisters is not gold.

Jespersen cites a wide range of attested examples, dating back to Chaucer, where negation takes wide scope over a preceding universal, so that *all . . . not* must be read as *not all:*

(36′)	All things are lawful unto me, but all things are not expedient.	(1 Cor. 6:12)
	Every one cannot make music.	(Walton)
	Tout le monde n'est past fait pour l'art.	(Rolland)
	Thank heaven, all scholars are not like this.	(Richardson)
	All is not lost.	(Milton, Shelley)

Each man kills the thing he loves / Yet each
man does not die. (Wilde)

As Jespersen points out, if the negation is lexically incorporated into the predicate it cannot take scope over the universal; the ~**A** reading of *all things are not expedient*, for example, disappears if the negative is imported into the adjective (*all things are inexpedient*). This observation is also made by Wagenaar (1930), in his discussion of the Old Spanish version of the *all that glitters* phenomenon, while Collinson (1937:91) adds that incorporation is standardly employed precisely to disambiguate a universal denial: 'For everyone did not smile (= 'No one smiled'), we get Everyone refrained from smiling.'

Similarly in French, the ordinary interpretation of *tout* (*tout le monde, chaque, chacun, n'importe qui*) + *ne . . . pas* is ~**A** rather than **A**~ (cf., e.g., Grevisse 1969:1164, citing Diderot's plaint that *Chaque âge n'a pas son Homère* and analogous examples). The standard example is the equivalent of (36) above:

(36″) Tout ce qui reluit n'est pas or.

Tobler's paper of the same name, now over a century old (Tobler 1882a), still offers us an insightful commentary on this construction. For Tobler, the "illogical" reading of *tout . . . ne V pas, all . . . not* as 'not all' is a perfectly logical reading, and French and English are no more to be scorned for admitting it than is German (which permits only *Nicht alles, was glänzt, ist Gold*) to be valued for excluding it.[15] If we think of the negation in (36)–(36″) as saying that it cannot be predicated of the subject (*all that glisters, tout ce qui reluit*) that it is gold, then the ~**A** interpretation is precisely what we should expect. Notice that the observation of Jespersen and Wagenaar on the narrow-scope (**A**~) reading of incorporated negation (cf. *Everyone was unkind*, as against *Everyone was not kind*) bears out Tobler's view, since a predicate-internal negation—as Aristotle stressed—is not a predicate denial.

But every *tout . . . ne V pas* construction with unincorporated negation is not assigned the "illogical" ~**A** reading; when La Bruyère discusses a 'maxime usée et triviale que tout le monde sait, et que tout le monde ne pratique pas', the sense is clearly that the maxim is neglected by all, and not just by some. The variables determining the appropriate reading—~**A** or **A**~ (NEG-Q or NEG-V, in the terminology of Carden 1970)—include context and (in English) intonation. In the former (~**A**) case, Tobler depicts the speaker as rejecting a universal judgment, in the latter (**A**~) as universalizing his own negative judgment. This distinction is represented in the dialogue sequences of (37) and (37′):

(37) A: Everybody came.
B: (No,) Everybody didn't come.

(37′) A: {Somebody / The Schwartzes} didn't come.
B: (Yes, in fact) Everybody didn't come.

Notice that respondent B in (37) can be taken only as negating or rejecting a universal, in Tobler's terms, while his counterpart in (37′) is universalizing a negative—although, contra Tobler, it is not his own, but A's, negative that B chooses to strengthen. While the latter sequence type is indeed possible, it is somewhat marked in both English and French, which, as Tobler recognizes, may be attributed to the existence of an alternative, less-marked device for signaling universal negation by denying a particular, as in *nobody came* or—in place of the A~ (NEG-V) reading of *Tous ne savent pas*—*{Nul / Aucun / Personne} ne sait.* I return to this factor, and its place in a principled explanation of quantifier-negation scope ambiguities, in §7.3.

The unavailability of a parallel construction in French (**Pas tout ce qui reluit est or*) would seem to put the ~**A** / NEG-Q reading on sounder prescriptive footing in that language. Yet the standard grammars persist in vilifying (36″) as a notorious 'paragrammaticalism', reflecting nothing more significant than a 'mauvaise placement de *tout*' (Le Bidois and Le Bidois 1968). If Tobler is correct, no misplacement of *tout*—or of negation—is in fact involved.

Something crucial is left unexplained in both Tobler's and Jespersen's accounts of *all that glitters.* Why is a nexal negation following a particular or existent subject (one containing a **B**, rather than **A**, operator, to use Jespersen's labels) always assigned narrow scope relative to that of the **B** operator? That is, why is NEG-Q available only when Q is in the **A** category? Why is there (apparently) no interpretation of (38a) on which it is equivalent to (38c) rather than to (38b)?

(38) a. Somebody didn't come. — Something that glitters is not gold.
b. Not everybody came. — Not everything that glitters is gold.
c. Nobody came. (~[somebody came]) — Nothing that glitters is gold.

If it is sometimes—in fact, usually—the case that '**A** . . . ~' = '~**A**', why is it (apparently) never the case that '**B** . . . ~' = '~**B**'? These questions will be revisited, if not disposed of, in chapter 7.

More recent treatments of *all . . . not* in the generative literature do not address the fundamental semantic questions, largely focusing instead on the empirical issue of which readings are available for which speakers in which linguistic contexts. Carden (1970, 1973) posits the existence of three distinct dialects, NEG-V (whose speakers accept only that "logical" reading), NEG-Q (whose speakers accept only that "illogical" reading), and AMB (whose speakers comprise the 'relatively uncommon group' accepting both readings for (36), (36′), and parallel sentences). His work, correlating subjects' grammaticality judgments on the basic data with their reactions to transformationally related sentences, has been challenged by Heringer (1970), Stokes (1974), Labov (in unpublished work), and Baltin (1977), with Labov and Baltin arguing that the idiosyncratic three-way dialect split attested by Carden in fact represents a differential ability to contextualize the two possible readings. By adjusting the linguistic and extralinguistic parameters appropriately, on this argument we can assimilate every speaker into the AMB group. This conclusion is supported by comparing the NEG-Q-forcing (37) with the NEG-V-forcing (37′), as I did above. Even without a prior discourse context, the a priori less accessible NEG-V reading is selected in a frame like that of the classic advertising jingle *Everybody doesn't like something / But nobody doesn't like Sara Lee,* where the first line cannot be interpreted as 'Not everybody likes something'.[16]

Unsurprisingly, the **A**~/~**A** ambiguity triggered by the universal quantifier surfaces as well for its suppletive variant *both* and for the corresponding binary connective *and,* as emerges from an examination of the paradigm in (39):[17]

(39)	a. All of them didn't come.	(all . . . not / not all)
	b. Both of them didn't come.	(both . . . not / not both)
	c. (Both) Lee and Kim didn't come.	(both Lee and Kim . . . not / not both Lee and Kim)

In each case, the subaltern or weak scalar dual (i.e., the **B** category operator) excludes the wide-scope reading for negation in the corresponding examples:

(40)	a. Some of them didn't come.	(= some . . . not / ≠ not some)
	b. One of them didn't come.	(= one . . . not / ≠ not one)
	c. (Either) Lee or Kim didn't come.	(= Lee or Kim . . . not / ≠ neither Lee nor Kim)

Furthermore, the three examples of (39) can be disambiguated intonationally in the same way (Horn 1972: 94–96; Jackendoff 1972: §8.6): high stress on the **A** word (*all, both, and*) and a final rise combine to yield the NEG-Q/~**A** reading; normal stress and a final fall are associated with the NEG-V/**A**~ interpretation. The most comprehensive and revealing study of the effect of intonation on the assignment of NEG-Q vs. NEG-V readings in particular, and on the scope of negation in general, is that of Ladd (1980: 145–62). After finding Jackendoff's account of fall-rise suggestive but ultimately insufficient and/or ad hoc, and Liberman and Sag's (1974) theory fundamentally mistaken in its collapsing of two distinct patterns (the fall-rise contour and the so-called TILDE or contradiction contour, to which I shall return in chapter 6), Ladd seeks to extend Jackendoff's program for focus and presupposition into a general and coherent semantico-pragmatic analysis of the fall-rise pattern.

Among the examples Ladd considers are the pairs in (41), where fall-rise and simple falling contours are annotated by ˇ and ˋ, respectively.

(41) a. ˇAll the men didn't go. (NEG-Q, 'not all')
b. ˋAll the men didn't go. (NEG-V; 'all . . . not')

(42) [Did you see anyone you haven't met?]
a. I haven't met ˇone of them. (there is one I haven't met)
b. I haven't met ˋone of them. (I haven't met any)

(43) a. I don't want to talk to ˇanyone (not to just anyone)
b. I don't want to talk to ˋanyone (to no one)

(44) a. John doesn't drink because he's unˇhappy (It's not because he's unhappy that he drinks)[18]
b. John doesn't drink because he's unˋhappy (It's because he's unhappy that he doesn't drink)

In each case, Ladd observes, a fall-rise (ˇ) contour on the focused element signals a SUBSET or HYPONYM relation: the focused element represents a proper subset or member of a contextually accessible set. Thus the fall-rise conveys a partial denial, while the straight fall (ˋ) contour signals a full or simple denial. This emerges especially clearly in exchanges like those in (45) and (46), from Ladd's discussion:

(45) A: You have a VW, don't you?
B: I've got an ˇOpel. (Well, not exactly, but . . .)

(46) A: That new military base is going to be as big as Texas.
B: As big as ˇNew Jersey, maybe. (Well, not quite)

The responses triggering a fall-rise function as qualifiers; they convey, politely, something like 'What you said is not quite true, but a related [often entailed] proposition _is_ true'. A straight fall, on the other hand, may be understood as a direct denial (*I've got an `Opel*) or as an invocation of a superordinate (*As big as Alaska!*).

Similarly, in the negative examples of (41)–(44), the fall-rise triggers a subset interpretation. In (41)—and in related examples with *both* and *and* (cf. (39))—the subset relation associated with the fall-rise is semantically incompatible with the **A**-category operator (*all* can't pick out a proper subset, nor can *both* or *and*), and the sentence is essentially reprocessed with the tacit caveat '*All* can't be a subset, so it must mean *not all*' (Ladd 1980: 161). Ladd offers related, but more straightforward, accounts for the disambiguating effect of intonation in those examples where subset readings are possible. In (44), for example, the fall-rise is interpreted as focusing on one among a set of possible reasons for John's not drinking; the falling contour presupposes no set of possible reasons and is thus fully compatible with a narrow-scope adverbial. (This analysis must assume a separate account of why it is that the wide-scope reading can be assigned to *because*, but not to *since* or *although* clauses, as I observed in §2.5.)

While Ladd's general program of taking scope differences to follow from inferences based on the lexical meaning of ˇ is promising (but cf. Ward and Hirschberg 1985), it seems counterintuitive to view the NEG-Q interpretation associated with (36), (36′), (41a), and related sentences as the result of a conscious reasoning-out process based on the meaning of fall-rise; some degree of conventionalization is likely to be involved.[19] Given the possibility of similar pairings of "illogical" ~**A** semantics with **A** . . . ~ syntactic form in languages with considerably different intonational means at their disposal, any general account of the *all-that-glitters* phenomenon must extend well beyond the ups and downs of the fall-rise contour.

4.4 Scalar Predication and Subcontrariety II

CAN SEX ONCE A WEEK HELP FEMALE INFERTILITY?
(Southfield, Michigan, *Northwest Newsday*)
It wouldn't hurt any. (*New Yorker*, 27 September 1987)

In the previous sections, I have taken subalternation—the relation between dual pairs like *all*/*some*, *necessary* (*certain*)/*possible*, *and*/*or*—as special cases of the more general phenomenon of scalar predication. It is now time to explore the nature of these relations and of their possible representations.

In Horn 1972, quantitative scales are defined by entailment; $\mathbf{P}_j$ outranks $\mathbf{P}_i$ on a given scale iff a statement containing an instance of the former unilaterally entails the corresponding statement containing the latter. As examples of such scales, I can cite those in (47), where $\langle \ldots, \mathbf{P}_j, \mathbf{P}_i, \ldots \rangle$ indicates that $\mathbf{P}_j > \mathbf{P}_i$, that is, that $\mathbf{P}_j$ outranks (is stronger than) $\mathbf{P}_i$ on the relevant scale:

(47) ⟨all, most, many, some⟩ ⟨always, usually, often, sometimes⟩
⟨and, or⟩ ⟨. . . , 6, 5, 4, 3, 2, 1⟩
⟨must, should, may⟩ ⟨necessary, (logically) possible⟩
⟨certain, {probable/likely}, possible⟩ ⟨obligatory, permitted⟩
⟨boiling, hot, warm⟩ ⟨freezing, cold, cool, (lukewarm)⟩
⟨beautiful, pretty, attractive⟩ ⟨hideous, ugly, unattractive, plain⟩
⟨adore, love, like⟩ ⟨loathe, hate, dislike⟩
⟨excellent, good, OK⟩ ⟨{terrible/awful}, bad, mediocre⟩

In my first stab at schematizing the generation of quantity-based implicata (Horn 1972:ex. (2.69)), I offered the following rule:

(48) Given a quantitative scale of n elements, $\langle \mathbf{P}_n, \mathbf{P}_{n-1}, \ldots, \mathbf{P}_2, \mathbf{P}_1 \rangle$, and a speaker uttering a statement **S** which contains an element $\mathbf{P}_i$ on this scale, then
 (i) the listener can infer $\sim\mathbf{S}(\mathbf{P}_i/\mathbf{P}_j)$ for all $\mathbf{P}_j > \mathbf{P}_i$ ($j \neq n$) [where $\phi(\mathbf{P}_i/\mathbf{P}_j)$ denotes the result of substituting $\mathbf{P}_j$ for $\mathbf{P}_i$ within ϕ]
 (ii) the listener must infer $\sim\mathbf{S}(\mathbf{P}_i/\mathbf{P}_n)$
 (iii) if $\mathbf{P}_k > \mathbf{P}_j > \mathbf{P}_i$, then $\sim\mathbf{S}(\mathbf{P}_i/\mathbf{P}_j) \rightarrow \sim\mathbf{S}(\mathbf{P}_i/\mathbf{P}_k)$

Various objections have been raised to this first attempt at a definition of scalar implicature, most of which are justified. First of all, the distinction between what the listener can infer and what the listener must infer—though I would claim it is supported by a real intuition which has not been otherwise captured—is not very well elucidated anywhere in Horn 1972 or in subsequent work and seems worth abandoning until it can be (see Hirschberg 1985 for a good critique).

Second, as Hirschberg points out, the very decision to define implicature in terms of what the listener can or must infer is fundamentally misguided, or at least unfaithful to the spirit of Grice, who characterizes implicature as part of nonnatural meaning (MEANING$_{nn}$), defined in turn as an aspect of the speaker's intentions. Implicature is best defined directly as an attribute of the speaker meaning, and only indirectly in terms of what the addressee is entitled to infer. (Elsewhere in Horn 1972 it is informally stated that a speaker uttering a given (weak-scalar) predication implicates that for all s/he knows, no stronger statement could have been employed; it is this approach which was echoed in the earlier discussion in §4.2).

The third objection is acknowledged in Horn 1972 and formulated more systematically in Fauconnier 1975a, 1975b, and Gazdar 1979a: chapter 3: the definition of quantitative scales by logical or semantic entailment is too narrow, since nonlogical inference also supports the construction of scales and the associated implicatures. I return to this point below.

Fourth, as Gazdar (1979a: 56) observes, not just any sentence or statement containing a weak scalar operator ($\mathbf{P}_i$ in (48i) above) licenses an upper-bounding implicature; if I tell you that *Paul ate some of the eggs,* you are entitled (ceteris paribus) to draw the inference that he did not eat all of them, but the statement that *It is not the case that Paul ate some of the eggs* does not license the implicatum that [it is not the case that it is not the case that] he ate all of them. The problem, Gazdar argues, is that the implicature mechanism only operates on scalar items in a logically SIMPLE position, that is, not embedded within the scope of another logical functor—here, sentential negation realized as *it is not the case that.* (A related problem is the spelling out of the conditions under which the potential implicatum gets realized as an actual implicatum; cf. Gazdar 1979a, Levinson 1983, and Hirschberg 1985 for discussion.)

Gazdar's own formulation of the function f_s for generating potential scalar quantity implicatures (1979a: 58–59) is given in (48′); his prose elucidation (edited here) follows.

(48′) $f_s\,(\Psi) = \{\chi : \chi = K\neg\phi(\alpha_i)\}$
for all $\phi(\alpha_i)$ such that for some quantitative scale $\mathbf{Q}$, α_i, $\alpha_{i+1} \in \mathbf{Q}$,
(i) $\Psi = \mathbf{x}^\frown\phi(\alpha_{i+1})^\frown\mathbf{y}$ where **x**, **y** are any expressions, possibly null
(ii) $[\Psi] \subseteq [\phi(\alpha_{i+1})]$,
where $\phi(\alpha_i)$ and $\phi(\alpha_{i+1})$ are simple expression alternatives with respect to α_i and α_{i+1}.

> This says that Ψ [potentially] scalar-quantity-implicates that the speaker knows that it is not the case that ϕ if and only if there is some sentence ϕ', just like ϕ except that it contains a 'weaker' scalar expression, and which is entailed by ϕ and is either identical to Ψ or forms a part of it, subject to the constraint that the scalar expressions are not within the scope of any logical functors in ϕ or ϕ'.

While it is both more explicit and more accurate than (48), Gazdar's function does not represent the final solution to the characterization problem either. Two objections are worth noting here. First of all, as I have already noted (in §4.2), the epistemic condition **K¬** (a Hintikka 1962–style abbreviation for 'speaker knows that it is not the case that') is in general too strong; it is preferable to derive the weaker conclusion that **¬K**

(~**K**), that is, 'it is not the case that speaker knows that . . .', or—utilizing Hintikka's epistemic possibility operator **POSS**, the dual of **K**—the equivalent **POSS**~ (for all speaker knows, it is not the case that).[20] In just those contexts where the addressee assumes that the speaker possesses all the relevant information, **POSS**~ can be strengthened to **K**~.

In addition, where my original formula was not restricted enough (in generating implicata from positions where these implicata are not licensed), Gazdar's version is too restricted. For the implicatum to go through, the scalar expression in question need not be logically unembedded, so long as it is not embedded under certain operators. Which operators are these? Hirschberg (1985:73) notes that while we may want to follow Gazdar in blocking the move from (49a) to (49a′), the comparable moves in (49b–d) are legitimate, although the weak-scalar predicate does not have a simple, unembedded occurrence in these examples any more than in (49a):

(49) a. It is not the case that Paul ate some of the eggs.
a′. **POSS**~(It is not the case that Paul ate all of the eggs)
(= **POSS**[Paul ate all the eggs])
b. It is possible that Paul ate some of the eggs.
b′. **POSS**~(It is possible that Paul ate all of the eggs)
c. Paul ate some of the eggs or Paul is a liar.
c′. **POSS**~(Paul ate all of the eggs or Paul is a liar)
d. Some people think Paul ate some of the eggs.
d′. **POSS**~(Some people think Paul ate all of the eggs)

But in place of Gazdar's overly strong restriction on (48′), Hirschberg offers her own overly weak restriction (albeit one not as overly weak as the nonrestriction in Horn 1972). For Hirschberg, only overt negation (as in (49a)) blocks the implicature function. As demonstrated by the substitution of *impossible* for *possible* in (49b, b′), or that of *doubt* for *think* in (49d, d′), this will not do. Rather, we seem to need to restrict the set of logical functors in question to those which include negation and other scale-reversing operators (cf. Fauconnier 1976), that is, to the set of downward-entailing operators in Ladusaw 1979. Rather than attempting to re-revise the formulation of the implicature function in the light of these factors (and others raised by Harnish [1976] and Hirschberg [1985]), I shall turn to some of the linguistic analogues of this phenomenon.

In Horn 1972, quantitative scales are correlated with syntactic frames like those in (50), where $\mathbf{P}_j > \mathbf{P}_i$ throughout.

(50) a. (at least) $\mathbf{P}_i$, if not (downright) $\mathbf{P}_j$
$\mathbf{P}_i$, {or/and possibly} even $\mathbf{P}_j$
$\mathbf{P}_j$, or at least $\mathbf{P}_i$

b. $\mathbf{P}_i$, {indeed/in fact/and what's more} $\mathbf{P}_j$

not even $\mathbf{P}_i$, {let alone/much less} $\mathbf{P}_j$	not only $\mathbf{P}_i$ but $\mathbf{P}_j$

The environments in (50a) are SUSPENDERS: the speaker is explicitly leaving the possibility open that a higher value on the relevant scale obtains, with the suggestion that his or her knowledge of the actual state of affairs is incomplete. The environments in (50b) do not just suspend, but CANCEL or BLOCK, the upper-bounding implicatum, with the assertion that a higher value on the scale is in fact known to obtain.[21]

In each case, the order of the scalar predicates cannot be reversed without incoherence or anomaly; thus compare:

(51)	a.	at least some if not all	#at least all if not some
		at least possible if not necessary	#at least necessary if not possible
	b.	not only three but four	#not only four but three
		not only warm but (downright) hot	#not only hot but (downright) warm
	c.	Pat or Lee, indeed Pat <u>and</u> Lee	#Pat and Lee, indeed Pat <u>or</u> Lee
		one of them, indeed <u>both</u> of them	#<u>both</u> of them, indeed <u>one</u> of them

These diagnostics, along with the distribution of scale-sensitive modifiers like *almost, barely, not quite,* and so on, lead to the assumption that positive and negative quantifiers, modals, and related operators must be represented on distinct, though related, scales. There can be no single scale on which operators like *some* and *not all,* or *possible* and *unlikely,* can be plotted. Rather, there is one scale defined by the positive operators and one by their negative counterparts. Only in this way can we predict the distribution of scalar elements in the test frames of (50). For example, the weak and strong elements on the positive quantifier scale define permissible sequences like *some or even all* (vs. *#all or even some*); the corresponding values on the negative scale similarly yield *not all or even none* (*#none or even not all*). But values from distinct scales cannot be combined in either direction: *#some or even {none/not all}*; *#not all or even {some/all}*. In traditional terminology, the two values in the scalar diagnostic constructions must differ in quantity but not quality.

Similarly, the epistemic modals pattern as in (51′)(cf. Horn 1978b: 194):

(51′)	a.	possible if not likely	#likely if not possible
		likely if not certain	#certain if not likely
		unlikely if not impossible	#impossible if not unlikely
		uncertain if not unlikely	#unlikely if not uncertain

b. #{possible / likely / certain} if not {uncertain / unlikely / impossible}
#{uncertain / unlikely / impossible} if not {possible / likely / certain}

Thus, alongside the positive epistemic scale ⟨certain, likely (probable), possible⟩ from (47), we can assume a parallel negative epistemic scale ⟨impossible, unlikely (improbable), uncertain⟩ and in like fashion a negative quantifier scale ⟨none, few, not all⟩ alongside its positive counterpart ⟨all, most, many, some⟩. But just what is the relationship between these parallel positive and negative scales? How can this parallelism be represented?

The structure of the scales can be overlaid directly onto the traditional Square of Opposition, with the relation between strong and weak values (indicated by the head and tail of the arrow, respectively) corresponding to the relation between a subaltern and its superordinate:

(52)

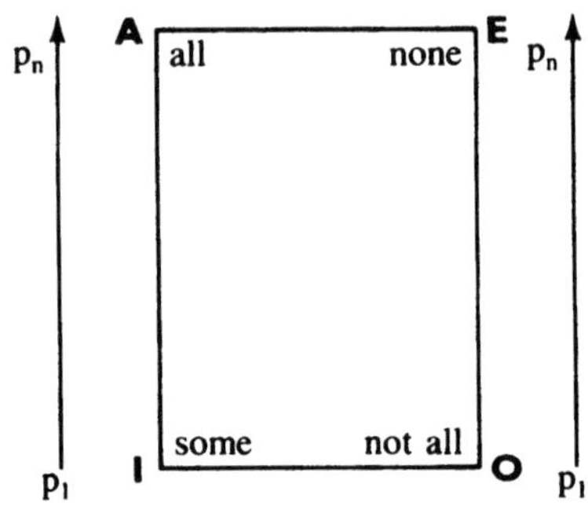

As always, the horizontally opposed (**A** and **E**, **I** and **O**) terms differ in quality, while the vertically opposed (**A** and **I**, **E** and **O**) terms differ in quantity. More generally, two terms in a quantitative opposition will occupy different positions on a single scale, while two terms in a qualitative opposition will occupy analogous positions (weak, intermediate, or strong) on corresponding (parallel) scales.

The notion of correspondence between matched positive and negative scales can be elucidated by arithmetizing the Square. Let the values on the positive scale in question range from 0 to +1 and those on the corresponding negative scale from 0 to −1. Each operator is ranked in accordance with its lower bound (recall that, e.g., *some* says 'at least some' and implicates 'at most some'), such that a simple proposition containing a scalar operator **P** will be true at all positions at or above the position assigned to **P**. Schematically, the result we obtain in the case of the quantificational determiners can be given (approximately) as in (53):

(53)
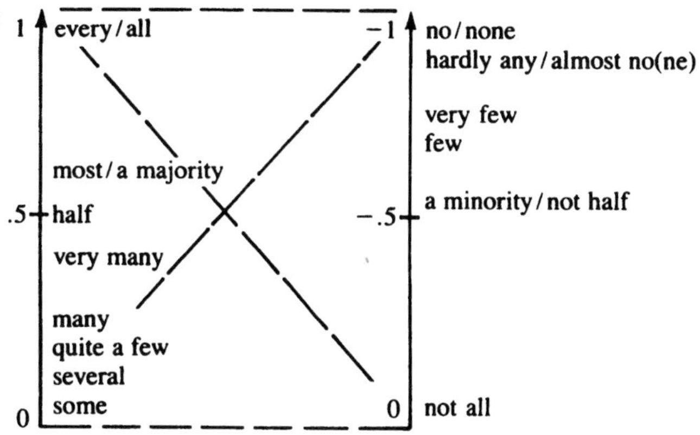

The duals (*all / some; none / not all*) represent the weakest and strongest values on their respective scale, the weakest situated just above ±0 and the strongest at ±1. Only the strongest value can be modified by *absolute(ly)* and its synonyms (Horn 1972: §2.34): cf. *Absolutely {all / none / #many / #few} of them can make it; It's absolutely {certain / impossible / #possible / #unlikely} that he'll win.*

Further, we observe that for any positive operator **P**, the sum of its scalar value and that of its inner negation **P**~ directly across from it will always be 0. These two operators, **P** and **P**~, will be contraries if the value of **P** is greater than .5 and subcontraries if it is equal to or less than .5. The crucial parameter here is whether the conjunction (marked with either *and* or *but*) of an operator with its inner negation is logically consistent. Thus, compare the consistent statements in (54a), where **P** and **P**~ are relatively weak (≤.5), with the inconsistent statements in (54b), where **P** and **P**~ are relatively strong (>.5) scalar values.

(54) a. Many of my friends are linguists and many of them aren't.
Some of my friends are linguists and some of them aren't.
It's possible that she'll win and possible that she won't.
It's fifty-fifty that she'll win and fifty-fifty that she won't.
b. #All of my friends are linguists and all of them aren't.
#Most of my friends are linguists and most of them aren't.
#It's certain that she'll win and certain that she won't.
#It's likely that she'll win and likely that she won't.

Adopting the terminology of Löbner 1985, I shall say that an operator **P** is TOLERANT if the conjunction [**P**(**p**) ∧ **P**(~**p**)] is logically consistent (as in (54a)) and INTOLERANT if [**P**(**p**) ∧ **P**(~**p**)] is logically inconsistent (as in (54b)). A tolerant determiner (*some, many, half*) is situated at or below the midpoint (.5) of its scale and is compatible with its inner negation (i.e., its subcontrary); an intolerant determiner (*most, almost all, all*) is situated above the midpoint and is incompatible with its inner negation (i.e., its contrary).

In this model, (sub)contrary pairs are parallel on the vertical (quantitative) axis; the terms of this opposition (**P**, **P~**) are linked by a horizontal line. In accordance with the classical definition, I take contradictories to differ from each other both quantitatively and qualitatively; the terms of a contradictory opposition (**P**, **~P**) are linked in the model of (53) by a diagonal line.[22] While the sum of the values of **P** and **P~** is always zero, the sum of the values of **P** and **~P** (e.g., of *some* and *none*, of *all* and *not all*) is always just over |1|, the absolute value of 1.

Since every positive quantified expression **P** is interpreted as 'at least **P**', picking out the interval above (the lower bound of) **P**, its negation **not-P** signals that no value within this interval obtains. Thus, *not many* means 'less than many', *not half* 'less than half'. Not surprisingly, these equivalences are not simply facts about English. Bhatia (1977: chapter 4) points out that the same *not* ***P*** = 'less than **P**' correspondences obtain in Hindi (and other Indo-Aryan and Dravidian languages of the subcontinent). He cites (1977: 59) such examples as (55):

(55) Usne ādhī kitāb nahī̃ parhī — 'He did not read half the books'
he half book NEG read — (i.e., he read {less / *more} than half of them)

The relative positions of the elements intermediate between the duals can be fixed by applying the diagnostic tests in (49). Thus, *most* outranks (is stronger than) *many* on the positive scale because of examples like *many if not most* / *#most if not many; many or even most* / *#most or even many*. That *many* is below and *most* above the midpoint is confirmed by the tolerance test illustrated in (54): *Many Americans smoke and many don't* / *#Most Americans smoke and most don't*.

In the same way, we can see that the negative scale for quantificational frequency adverbs has the form ⟨never, hardly ever, rarely/seldom, not always⟩, contrary to the claim in Stockwell, Schachter, and Partee (1973: 292) that there is a 'low-level rule' converting *hardly ever* into *seldom*. In fact, there is a direct mapping between the quantificational adverbs and their determiner counterparts: *always* corresponds to *all* (at all times, on all occasions), *almost always* to *almost all, usually* to *most, often* to *many*, and *sometimes* to *some;* on the negative side, *never* (at no time, on no occasions) maps onto *no (none), hardly ever (almost never)* onto *hardly any (almost none), rarely* or *seldom* onto *few*, and *not always* onto *not all*.

That the purported equation *hardly ever* = *seldom* is incorrect can be shown by the usual scalar tests:

(56) a. I've seldom—indeed, hardly ever—smoked cigars.
#I've hardly ever—indeed, seldom—smoked cigars.
(cf. *Few—indeed, hardly any—of my friends smoke cigars* vs. *#Hardly any—indeed, few—of my friends smoke cigars*)
b. I've seldom been to the Midwest and hardly ever to Chicago.
#I've hardly ever been to the Midwest and seldom to Chicago.

The scalar representation of quantificational and modal operators correctly predicts the correlation of these values with a wide range of other—not obviously logical—predicates. The same criteria that determine the positioning of *all* and *some, always* and *sometimes, and* and *or, necessary* and *possible* as strong and weak values on their respective scales apply equally to predicate *n*-tuples like ⟨adore, love, like⟩, ⟨beautiful, pretty, attractive⟩, ⟨fascinating, interesting⟩, ⟨boiling, hot, warm⟩, and ⟨excellent, good, OK⟩. Thus, if *I love you* entails *I (at least) like you,* then my acknowledging that I like you, in a context where my loving you would be relevant, will (ceteris paribus) **Q**-implicate that I don't in fact love you. As expected, we get patterns like those in (57) and (58):[23]

(57) He likes you, and he may even love you.
#He loves you, and he may even like you. [but see note 23]
He absolutely {adores / ?loves / #likes} you.

(58) It's warm if not hot out today.
#It's hot if not warm out today.
#It's absolutely {boiling / ??hot / #warm} out today.

As with the logical constants, our familiar diagnostics require the construction of separate but parallel scales for contrary values, for example, ⟨boiling, hot, warm⟩ vs. ⟨freezing, cold, cool, lukewarm⟩, given the distribution in frames like (59).

(59) It's cool if not {cold / freezing / #lukewarm / #warm / #hot / #boiling}
It's warm if not {hot / boiling / #lukewarm / #cool / #cold / #freezing}

Comparatives of scalar adjectives can be understood only against the assumption of paired scales. Just as *not-P* is interpreted as 'below **P** on **P**'s scale', so too is ***P-er*** interpreted as 'more in the direction of the endpoint of **P**'s scale', and not as 'nearer [my God] to **P**'. Thus *warmer* and *hotter* are both read as 'higher on the ⟨. . . , hot, warm, . . .⟩ scale', *cooler* and *colder* as 'higher on the ⟨. . . , cold, cool, . . .⟩ scale'. It is only for children who have not yet acquired the subtleties of full scalar competence that *warmer* can denote 'less hot', 'closer to (exactly) warm'.

As stressed by Fauconnier (1975a, 1975b, 1976) and Hirschberg (1985), scales may be defined not only universally by virtue of the meaning (entailment) relations definable on the elements involved, as in (47), but also locally, where a given context establishes the pragmatic implications on which the scale is based. Quantitative scales may thus be taken as the limiting case wherein every pragmatic model or context assumes the scale in question, while other sets of predicators are less consistent across models. This position is justified by the existence of scales established between pairs of expressions which do not participate in a logical entailment relation of the type I considered above.

Thus, to borrow one of Fauconnier's examples, there may well be contexts or situations in which Hercules can lift only extremely heavy rocks; his lifting a heavy one will thus not entail that he can lift a lighter one. Yet the normal pragmatic implication is that (60b) does indeed follow from (60a).

(60) a. Hercules can lift a rock weighing **n** pounds.
b. Hercules can lift a rock weighing **n** – **k** pounds.
(where **n**, **k** > 0)

Given this implication, the superlative (61) will receive a 'quantified' reading, and is thus paraphrasable by other expressions utilizing universals or diagnostics for absolute scalar elements:

(61) Hercules can lift the heaviest rock.

(61′) Hercules can lift {even / absolutely} the heaviest rock.
Hercules can lift any rock, {even the heaviest / however heavy}.

Similarly, on the assumption that anyone who laughs at a given joke will laugh at a funnier one, the three versions of (62) are pragmatically equivalent.

(62) The funniest jokes don't make Alexander laugh.
Even the funniest jokes don't make Alexander laugh.
No jokes (however funny) make Alexander laugh.

As Fauconnier also observes, it is a characteristic property of negation and other polarity triggers that they reverse the ordering of elements on a scale. Thus, alongside the ordinary (positive) scale defined in (63a) we get the negative scale in (63b); notice that while the former scale has no maximum or absolute value, given the nature of the cardinals, the latter one does.

(63) a. . . . → Odette has three children → Odette has two children → Odette has one child
b. Odette doesn't have a child → Odette doesn't have two children → Odette doesn't have three children → . . .

Under negation—as on the great Day of Judgment (cf. *Mark* 10:31)—those that are first shall be last, and the last first.

The importance of Fauconnier's insight in generalizing the notion of scale to the pragmatically as well as semantically defined cases is brought out when we begin to collect examples of the $\mathbf{p}_i$-if-not-$\mathbf{p}_j$ construction and its analogues.[24] To the citations collected in Horn 1972:ex. (1.90), I can now provide some additional instances:

(64) Overt antifeminism, if not homosexuality, may be the result of such experience in the male. (from *The Parenting Advisor,* on the failure to shift gender identification from mother to father)
In the Netherlands the crowds [for the Pope] were small, the welcome lukewarm if not cold. (*New York Times,* 19 May 1985)
Most photographers were inarticulate if not subhuman. (from a novel by Elliott Chaze)
The picture of Chiang Kai-Shek that emerges is one that rivals Mussolini, if not Hitler, as the very model of a modern major dictator. (from a review of Sterling Seagrave's *The Soong Dynasty*)
Of course, not all our teachers made *the* or even *a* difference in our lives. (from a *New York Times* Op-Ed pieces; italics in original, underlining mine)

While it might be maintained that, for example, *cold* and *lukewarm* are situated on a semantically defined scale here (perhaps by extension from their literal, climatological meanings), no such semantic criteria are presumably available for proper names; yet we clearly draw the implication from the penultimate example that there does indeed exist a scale on which dictators can be ranked, and furthermore that the Führer clearly outranks (outgrosses?) il Duce on this scale.[25]

A rival account of scalar operators is offered by Ducrot and his colleagues (cf. especially Ducrot 1973; Anscombre and Ducrot 1976, 1978, 1983). On this view, scales are essentially not quantitative (in the sense of Horn 1972) or pragmatic (in the sense of Fauconnier 1975a, 1975b, 1976), but argumentative. Ducrot's *échelles argumentatives* share many of the properties I have described, including (1) the plotting of elements by their relative strength (as defined, however, by their argumentative power rather than by entailment or pragmatic implication); (2) the rhetorical suggestion (*sous-entendu*) by the use of a weaker expression that—for all the speaker knows—the stronger expression does not apply (recall my discussion of Ducrot's *Loi d'exhaustivité* in §4.1 above); and (3) the motivation of paired scales for contrary terms, as in the thermometric scales of (65) (glosses are mine):[26]

(65)	↑ il fait glacial	'it's freezing'	↑ il fait brûlant	'it's boiling'
	il fait froid	'it's cold'	il fait chaud	'it's hot'
	il fait frais	'it's cool'	il fait assez chaud	'it's warm'

Moreover, as for Fauconnier, negation produces a *renversement des échelles;* from the positive scale (66a) we derive the negative (66b).

(66) a. il fait glacial → il fait froid → il fait frais
b. il ne fait pas frais → il ne fait pas froid → il ne fait pas glacial

Two expressions may denote the same objective reality but differ in terms of the conclusions they can be used to argue for; cf. *The glass is half empty* (--> we should fill it, or buy another) vs. *The glass is half full* (--> we should, or can, empty it). This difference (cf. Ducrot 1973:236–37) is brought out under scale reversal: a glass which is not half empty is more full than one which is not half full, since *not* here (as for Jespersen) equates to 'less than'. But for Ducrot, the 'less than' interpretation associated with descriptive (ordinary) negation must be stipulated, via his *Loi d'abaissement*.

This is because, as Fauconnier (1976) points out, the effect of negation can be predicted by the argumentation model only if there is what Ducrot calls a *graduation objective homologue* to appeal to. Since scalar predications are not in principle objectively measurable, there remains no way for Ducrot to avoid an ad hoc account of the less-than interpretation in non-objective scalar contexts—*This soup isn't good* (is less than good), *Your statue isn't beautiful, My story isn't interesting*—where the effect is as real and general as in the objective *It's not warm* (It's less than warm) *out, It's not half* (It's less than half) *full.*

Since we need recourse to logical entailment, pragmatic implication, and **Q**-based implicature independently of scalar phenomena, the argumentation-based scale favored by Ducrot and Anscombre is best viewed as dependent on, rather than prior to, a pragmatically generalized quantitative model of the type(s) depicted in Sapir 1944; Horn 1972; Fauconnier 1975a, 1975b, 1976; Harnish 1976; Gazdar 1979a, 1979b; Atlas and Levinson 1981; and Hirschberg 1985.[27]

It is easy to see how the 'less than' interpretation of ordinary negation is explained within the scalar model I have been assuming. A proposition $\mathbf{p}_i$ containing a scalar operator $\mathbf{P}_i$ (in an appropriately simple occurrence) will be true whenever any proposition $\mathbf{p}_j$ is true, where $\mathbf{p}_j$ is just like $\mathbf{p}_i$ except for the substitution of the stronger $\mathbf{P}_j$ for the weaker $\mathbf{P}_i$. But the reverse, of course, is not the case. Thus, the weaker the scalar item, the wider the conditions under which its containing proposition is true, as exemplified in (67):

(67) ↑ it's boiling } **True** (it is hot)
it's hot }
it's very warm } **True** (it is very warm)
it's warm } **True** (it is warm)

The negation of a scalar predication, $\sim\mathbf{p}_i$, is true if its contradictory $\mathbf{p}_i$ is false; this amounts to the claim that no proposition of this type containing $\mathbf{p}_i$ or any stronger value on its scale is true. Just as *it is warm* means that (is true iff) it is at least warm, its negation *it is not warm* means 'it is not at least warm', that is, 'it is less than warm'. In the same way, *not pretty* is interpreted as 'less than pretty', *not happy* as 'less than happy', and—as Jespersen points out in the epigraph to this chapter—*not good* as 'less than good', and *not three* as 'less than three'.

Constituent negation, depending on its form, may or may not receive a scalar interpretation. Stockwell, Schachter, and Partee (1973) note that while *not* means 'less than' in *not three hundred feet away,* no such interpretation is assigned to *not quite three hundred feet away.* But here *not quite* is clearly a constituent (special) negation; *not quite 300* does not equate to **less than quite 300,* because it is not the negation of *?quite 300* (cf. Bolinger 1972, Sadock 1981, and Atlas 1984 for more on *not quite X*). Jespersen (1949:435) distinguishes *no more than three* (surprisingly, only three) from *not more than three* (three at most), and likewise (1917:83) *no less than thirty* (surprisingly, exactly thirty) from *not less than thirty* (at least 30).[28]

If *not warm* in principle designates a position below *warm* on the ⟨hot, warm⟩ scale, why does it seem so often to designate a temperature for which a value on the parallel ⟨cold, cool⟩ scale would be appropriate? We can represent this fact by recognizing that whenever we are dealing with paired contrary gradables, for example, *warm / cool, attractive / unattractive, like / dislike,* and the paired scales they define, the tail of each scale automatically extends through the excluded middle between the contraries into the rival scale. Graphically, the EXTENDED positive and negative temperature scales will take the form in (68).

(68)

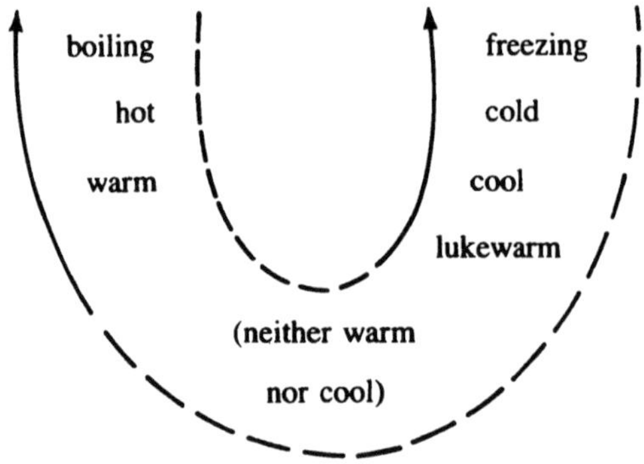

Thus, *not warm* (less than warm) will apply to temperatures which are neither warm nor cool and, a fortiori, to those which are located at or below *cool* on the extended *warm* scale; *not cold* will exclude freezing and cold temperatures while admitting all other values on the extended scale.

In the same way, if a temperature can be described by a value on the ⟨hot, warm⟩ scale, no value on the ⟨cold, cool⟩ scale can felicitously apply to it; a day which is not cool may be moderate (neither cool nor warm), warm, or a real scorcher, that is, less than cool on the extended ⟨. . . , cold, cool, . . .⟩ scale. Similarly, *not good* comprises any value below *good* on the extended ⟨excellent, good, OK, . . .⟩ scale, including the neutral (*average*) and negative (*mediocre, bad, inferior, terrible*) values.

The first comprehensive treatment of gradable (scalar) predicates is that in Sapir 1944. Although he does not provide a formal representation, Sapir seems to conceive of a scale not as two paired, systematically related lines, but as two distinct directions of orientation associated with a single line. Both my quantificational scales (cf. (53)) and my scales for antonymic predicate pairs like *hot* and *cold* (cf. (68)) can be mapped into Sapir's unitary, bidirectional scales, which I shall label GRADATIONS. Let us represent the two gradation types as in (69a, b):

(69) a.

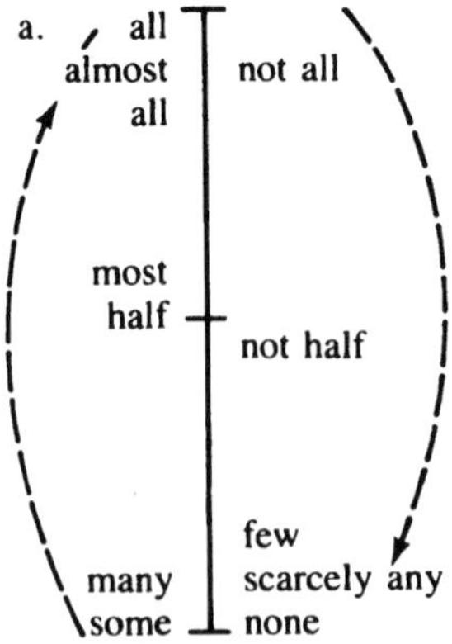

b.

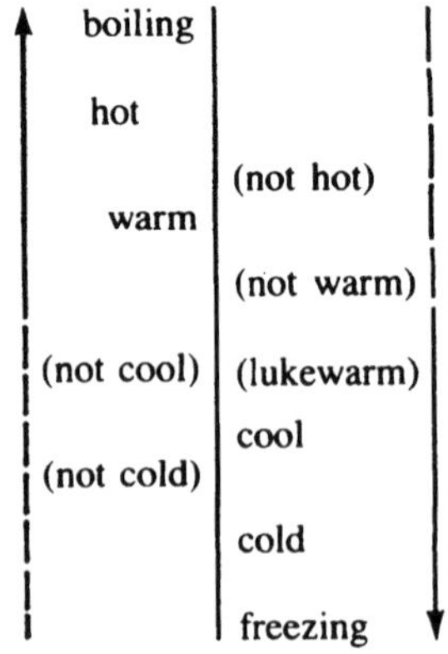

There is a significant structural difference between these two gradations. This difference, which applies equally to the paired unidirectional scales described above, results from the fact that while quantificational (and analogous) operators apply to open or closed sentences (or propositions) and define their (sub)contraries in terms of inner negation (**P**~ is the (sub)contrary of **P**, as I have noted), simple first-order predicates like *hot / cold, good / bad,* and *love / hate* have no inner negations or duals. Thus, the analogy with the Square of Opposition which I employed in my discussion of the annotated Square in (53), and which applies, mutatis mutandis, to the Sapirian gradation in (69a), has no direct counterpart in the *hot / cold* example represented in (69b).[29]

Notice also that the partitive gradation (69a) has two absolute endpoints—*(absolutely) all,* representing 100 percent of the members of the set quantified over, and *(absolutely) no / none,* representing 0 percent—while the temperature gradation, as we normally conceive of it, has no literal endpoint. When we declare that it's absolutely boiling (freezing) out, we are speaking metaphorically.

The diagnostic evidence (cf. (50), (51)) can be accommodated within Sapir's gradation analysis if we include the caveat that any two terms can appear in the relevant frames ($\mathbf{P}_i$ if not $\mathbf{P}_j$, $\mathbf{P}_i$ or even $\mathbf{P}_j$, etc.) only if they have the same direction of orientation (if $\mathbf{P}_i$ and $\mathbf{P}_j$ are on the same arrow shaft) and, of course, if $\mathbf{P}_j$ is a stronger or more extreme value than $\mathbf{P}_i$ with respect to that orientation (if $\mathbf{P}_j$ is closer to the arrowhead than $\mathbf{P}_i$). We can motivate the positioning of the negative values on the paired scales as well: *Not only is it not warm, it's downright {cool / cold / freezing / #warm / #hot}; Not only is it not cool, it's downright {warm / hot / #cool / #cold / #freezing}.*

Sapir notes that we often use a 'notational kinesthesia' corresponding to the direction of the arrow; increase is marked by 'upward grading' and decrease by 'downward grading'. This point is well taken, and may argue for the psychological superiority of Sapir's gradations over my scales. On the other hand, the gradation in (69b) seems to suggest that *warm* should be regarded as psychologically close to *cool,* despite their difference in orientation; but, as Sapir acknowledges (1944: 133), this predicted closeness does not in fact obtain. In any case, as stressed here, the similarities between my paired unidirectional scales (cf. also Ducrot 1972) and the unitary bidirectional gradations employed by Sapir (as I represent him here) outweigh any notational differences between the two models.

A more rigorous definition of the positive and negative quantifiers, one which is compatible with, but rather differently conceived from, the scalar model I have presented in this chapter, is offered by Barwise and Cooper (1981), henceforth B&C. Within this approach and in formal work associated with it, the QUANTIFIER in a sentence like *All men are mortal* is the entire NP *all men,* while *all* is a DETERMINER. Syntactically, quantifiers are NPs; semantically, they are sets. The crucial notion for our purposes is MONOTONICITY, defined as follows (B & C: 184ff.)

(70) a. A quantifier **Q** is MONOTONE INCREASING (**mon** ↑) if for any set $\mathbf{x} \in \mathbf{Q}$, **Q** also contains all the supersets of **x**.
b. A quantifier **Q** is MONOTONE DECREASING (**mon** ↓) if for any set $\mathbf{x} \in \mathbf{Q}$, **Q** also contains all the subsets of **x**.
c. A determiner **D** is monotone increasing (decreasing) if it always gives rise to monotone increasing (decreasing) quantifiers.

B & C employ the following linguistic diagnostic (among others) for determining whether a quantifier is **mon** ↑, **mon** ↓, or neither:

(71) Let $\mathbf{VP_1}$ and $\mathbf{VP_2}$ be two verb phrases such that the denotation of $\mathbf{VP_1}$ is a subset of the denotation of $\mathbf{VP_2}$. Then **NP** is **mon** ↑ if (i) holds, and **mon** ↓ if (ii) holds:
i. If **NP** $\mathbf{VP_1}$, then **NP** $\mathbf{VP_2}$.
ii. If **NP** $\mathbf{VP_2}$, then **NP** $\mathbf{VP_1}$.
If neither (i) nor (ii) holds, **NP** is nonmonotone.

In other words, a **mon** ↑ quantifier allows the predicate to be weakened *salva veritate,* while a **mon** ↓ quantifier allows the predicate to be strengthened *salva veritate* (cf. Löbner 1985). Where $\mathbf{VP_1}$ is *entered the race early* and $\mathbf{VP_2}$ *entered the race* (the denotation of the former constituting a proper subset of the denotation of the latter), we get results like the following:

(72)

a. Some men entered the race early ⇄ Some men entered the race (→ holds, ← fails)
b. No men entered the race early ⇄ No men entered the race (← holds, → fails)
c. Exactly five men entered the race early ⇄ Exactly five men entered the race (neither direction holds)

Thus, *some men* is **mon** ↑ (as are other **NPs**/quantifiers that behave in the same way, e.g., *he, Chris, the girl; many men, most women, all linguists, everybody, at least five frogs*), *no men* is **mon** ↓ (along with *nobody, at most five frogs*), and *exactly five men* is nonmonotone, since the biconditional in (72c) goes through in neither direction.

The determiners which are **mon** ↑ by B & C's criteria correspond to the operators which figure on some positive scale (*all, most, many, some, (at least) n, n or more*), the **mon** ↓ determiners to those which figure on some negative scale (*no, not n, few, at most n, fewer than n*). Nonmonotone determiners (in particular, *exactly n*) are not scalar operators at all.

Corresponding to the dictum that negation reverses scales (Horn 1972; Ducrot 1973; Fauconnier 1975a, 1975b, 1976) we have the rule (B&C: 186) that negation reverses monotonicity: if **Q** is **mon** ↑, **~Q** and **Q~** are both **mon** ↓, and if **Q** is **mon** ↓, **~Q** and **Q~** are both **mon** ↑. In B&C's example, *One man ran* contains a **mon** ↑ quantifier whose outer negation (*Not one man ran*) and inner negation (*One man didn't run*) are both **mon** ↓ quantifiers. If **Q** is **mon** ↑, its dual **Q** (= ~[**Q**~] = [~**Q**]~)—for example, *Not one man didn't run*—is also **mon** ↑.

One area of particular interest emerging from the behavior of **mon** ↑ and **mon** ↓ quantifiers and determiners involves the question of conjoinability (B&C: 193ff.). The pattern $\mathbf{Q}_1$ **CONJ** $\mathbf{Q}_2$, where **CONJ** ∈ {*and, or*}, is possible when $\mathbf{Q}_1$ and $\mathbf{Q}_2$ are both **mon** ↑ or both **mon** ↓.

(73) a man and three women
some students and every professor
most men and any women

(74) few violins and no violas
no men and few women
none of the dogs and hardly any of the cats

But when $\mathbf{Q}_1$ and $\mathbf{Q}_2$ differ in their monotonicity properties, ordinary conjunction is impossible:[30]

(75) *three men and no women
*few women and those men
*most of the dogs and none of the cats

As B&C point out, these mixed conjunctions cannot be ruled out pragmatically, in the light of such contrasts as that in the minimal pair in (76):

(76) John was invited and no woman was.
*John and no woman was invited.

Thus, conjunction, like the diagnostics of (50), (51), and (51′), respects the integrity of a scale, of Sapirian orientation, or of monotonicity type. (Cf. Brown 1984 and Löbner 1985 on the relations between the generalized quantifiers of B&C and the Square of Opposition.)

I will cite one more parallel between the **mon** ↑ and **mon** ↓ quantifiers of Barwise and Cooper, the positive and negative scalar operators of Horn, Ducrot, and Fauconnier, and the upward- and downward-oriented gradations of Sapir (cf. also Ladusaw 1979, 1980 on upward- and downward-entailing operators). Among the quantifiers, what sort of account are we to assign to expressions like *a few* α and *only* α, which seem to be analyzable either dynamically, as scalar (monotone) values, or statically, as conjunctions of two values differing in orientation? B&C point out that if *a few* corresponds to 'some but not many', it is a nonmonotone determiner, but if it means simply 'at least a few' (with the upper bound brought in by a **Q**-based implicatum, as for the cardinals), it is **mon** ↑. But which is it to be?

The evidence from conjunction (*most women and a few men, no women* {**and* / *√but*} *a few men*) and from the suspension diagnostics (*a few if not* {*more* / **less*}, *a few or even* {*a lot* / **none*}, **a few if any* (vs. *√few if any*)) reveal that *a few*—unlike *few*—can never be **mon** ↓ and that it can (at least sometimes) be **mon** ↑. Yet the upper bound associated with *a few* seems to be much more strongly built in than the upper bound normally implicated by the use of a relatively weak positive scalar (**mon** ↑) operator. Not surprisingly, the same question—and essentially the same answer—arises in Sapir's framework. Unlike *few,* which he characterizes as 'grading downward from something more', Sapir (1944: 134–35) takes *a few* to be 'static' or 'essentially noncommittal' with respect to directionality, while conceding that it can be given 'an upward trend' with the appropriate (fall-rise?) contour. *Quite a few,* as Sapir notes, is clearly 'upward-tending', that is, positive scalar or **mon** ↑.

While *a few* may hover uneasily between **mon** ↑ and nonmonotone status, *only* seems to hover between **mon** ↓ and nonmonotone. The facts here are rather complex; perhaps it is not surprising that *only* constructions are avoided as scrupulously by B & C as by Karttunen and Peters (who limit themselves to noting [1979: 32] that 'the case of *only* is more complicated [than that of *even*], although it also involves a distinction between focus and scope'). The essential issue is one debated in earlier treatments of *only* by analysts ranging from Saint Thomas, Peter of Spain, and William of Sherwood (cf. Mullally 1945 and Kretzmann 1968) to Geach (1962), Horn (1969, 1972, 1979), and Levergood (1984): is *only* α negative in meaning and positive only by presupposition or implicature, or does it abbreviate a conjunction (*only* α = 'α and nothing {other / more} than α')? In the former event, *only* α should pattern with the **mon** ↓ quantifiers; in the latter event, it should pattern with the nonmonotones, themselves typically analyzable into conjunctions (e.g., *exactly 5* = '[at least] 5 and at most 5', *some but not all,* etc.).

The monotonicity question for *only* α in fact arose in the medieval period, in the work of Saint Thomas Aquinas and William of Sherwood. Saint Thomas observes that the addition of *solus* 'only, alone' to an NP affects the entailments associated with the statement containing that NP: 'An exclusive expression [e.g., *solus*] so fixes the term to which it is joined that what is said exclusively of that term cannot be said exclusively of an individual contained in that term: for instance, from the proposition *Man alone is a mortal rational animal,* we cannot conclude, *therefore Socrates alone is such*' (*Summa Theologica*, Q31, art. 3: Aquinas 1945:311–12). Without the addition of *solus*, the entailment in question goes through, providing of course that we accept the hidden premise that Socrates is a man.

It was William of Sherwood who broadened this observation by applying the monotonicity test formalized seven centuries later by Barwise and Cooper (see (71) above). As William points out (*Syncategoremata* 11:7, in Kretzmann 1968:73), the ordinarily straightforward entailment in (77a) is blocked when *solus* is attached to the subject, as in (77b).

(77) a. Socrates is running ⊩ Socrates is moving
  b. Only Socrates is running ⊩ Only Socrates is moving

In B&C's terms, *Socrates* is **mon** ↑, while *only Socrates* is not. What remains to be determined is whether the opposite entailment, that is, (77′),

(77′) Only Socrates is moving $\Vdash^{?}$ Only Socrates is running

is valid; if so, *only* α is indeed **mon** ↓ rather than nonmonotone. This question is hard to decide, except by fiat: Peter of Spain (whose conjunction analyses of *only* phrases are cited and discussed in Mullally 1945:106ff.) would vote nay, along with Saint Thomas, while William of Sherwood, along with Geach (1962:§108), would presumably have begged to differ. The latter position strikes me as more theoretically coherent and arguably more faithful to natural language intuitions and distributional evidence, but I shall not seek to defend this claim here.

A somewhat confusing (if not confused) protomonotonicity classification in Horn 1969:103–5, utilizing an entailment test similar to B&C's, concludes that while *a few* α is an '**M**-CLASS' (essentially **mon** ↑) quantifier, *only a few* α, like *few* α, is in the '**L**-CLASS' (**mon** ↓) group.[31] This claim, supported by such (prefeminist) examples as those in (78),

(78) a. **M** girls are both clever and seductive → **M** girls are clever and **M** girls are seductive (where **M** = {some, many, at least n, a few, . . .})
  b. **L** girls are clever and **L** girls are seductive → **L** girls are both clever and seductive (where **L** = {no, at most n, few, only a few, . . .})

accords with the position that *only* is negative by assertion (or logical form) and affirmative by presupposition (or implicature), given the premise that 'entailment relations are determined by assertions alone' (Horn 1969:105).

Thus *only*, like simple negation, reverses scales or monotonicity (cf. also Ducrot 1973:237 on the reversing effect of *seulement*)—at least sometimes. In the case of NPs of the form *only n CN* (CN = common noun), the context helps determine whether *only n* is to be read as a **mon** ↓ expression (= 'at most *n*') or as a nonmonotonic conjunction (≅ 'exactly *n*, surprisingly'). As William put it (Kretzmann 1968:95), *only three* can exclude a greater number only, as in *Only three are running*, or a greater and smaller number, as in *Only three are hauling the boat*. But in the default context, where no jointly performed activity is expressed by the predicate, *only n*—like *at most n*—is **mon** ↓: 'If one says "only three", one cannot infer "therefore not two", but instead "therefore not four or five"' (Kretzmann 1968:82).

In this chapter, I have been expounding a view of the subcontraries and related scalar operators on which these expressions are unambiguous in semantic representation or logical form but open to two different uses according to whether or not the Mill-Grice Maxim of Quantity applies to a given utterance token involving that operator. Thus, each of these values—*some, a, possible, warm, cool,* and so forth—is lower-bounded (with respect to the relevant scale or direction of orientation) by its literal meaning and potentially upper-bounded by implicature.

Some recent accounts of the same phenomena have challenged this picture; for some of these developments, cf. Cormack 1980; Sadock 1981; Burton-Roberts 1984; Sadock 1984; Atlas 1984; Carston 1985a, 1985b; Kempson 1986; and Sperber and Wilson 1986 (cf. also Horn 1984a, 1985). The division of labor between logico-linguistic and pragmatic rules I have argued for has been particularly called into question for the cardinal numbers. It is doubtful that anyone would now seek to defend the original position of Smith 1970, on which *1, 2, 3,* and the infinitely many values in this series are all semantically or lexically ambiguous (as between 'at least n' and 'exactly n' readings). What Kempson and her colleagues have argued for is a position on which cardinals are semantically univocal but propositionally or logically ambiguous, where the semantic interpretation of an expression in general radically underdetermines its contribution to filling out propositions and hence to determining truth conditions as well as utterance meaning (cf. §6.5 below). Hans Kamp has suggested, on the other hand, that determiners of the form *n + CN* are truth-conditionally identical to those of the form *exactly n + CN*. Kamp's position, worked out within his theory of discourse representations, is challenged by Kadmon (1984), who adduces additional evidence from anaphora in favor of a discourse representation version of a mixed semantico-pragmatic theory of cardinals akin to that of Horn 1972, where the 'exactly' reading is generated indirectly from the 'at least' reading.

It is undeniable that the upper-bounding implicature mechanism behaves differently with respect to cardinals than it does with other scalar values (e.g., the quantificational determiners, modals, and first-order gradable contraries). Kadmon points out that the quantity-based implicature is stronger for *Myers has three kids*—and especially for *Myers has one kid*—than for *Myers has a kid* and offers a straightforward Gricean explanation for the difference.[32] She also observes (1984: 30ff.; cf. Horn 1972: 45) that rounded numbers (*200 Hondas are defective*) are easier to get under the non-upper-bounded 'at least' reading than are unrounded numbers (*278 Hondas are defective*).[33]

Sadock (1984) discusses the problem posed for a minimalist theory of the cardinals in giving the truth conditions of such mathematical statements as *2 + 2 = 3*, *The square root of 9 is 2*, each of which would seem to have a true reading on the 'at least' understanding of the cardinals in question (*2 + 2 is not only 3—it's 4!*). As Sadock points out, the cardinals differ from the 'inexact quantifiers' (*some, many, none*) and from other scalar terms in the contextual reversibility of the direction of their scale; while the normal numeric scale is easily reversed, 'it does not seem possible to use *some*, for example, in such a way as to implicate "at most some"' (Sadock 1984: 143).[34] Whether other factors may be involved or not, the exactness built into the cardinals as against the noncardinal determiners and the greater knowledge or information consequently presupposed in their felicitous use certainly seem to add an additional reinforcement to the scalar implicata they induce.[35]

Addressing a related issue, Campbell (1981: 97–99) distinguishes CRYPTIC (unconscious) from PHENIC (conscious) processes in pragmatic inference. He notes that the upper-bounding quantity-derived implicatum derived in a context like that of B's response in (79)

(79) A: How many children do you have?
B: Two.

is cryptic or automatic, requiring no 'real cognitive effort' or deliberation by A (of the form *Does B mean exactly two or at least two?*). In a context like that of (79′), however,

(79′) A: Do you have two children?
(B: {No, I have three / Yes, in fact I have three})

the addressee may indeed consciously wonder whether A meant exactly two or at least two, perhaps resolving the question in the former direction by a phenic application of Quantity in a context where the maxim can be appropriately invoked.

Thus the inference whose application in effect pragmatically disambiguates an equivocal scalar operator may involve a greater-or-lesser degree of conscious effort—that is, less or more conventionalization or 'short-circuiting' (Morgan 1978). That a given pragmatic inferencing mechanism may operate with varying degrees of directness (or unconsciousness) is by now clear; this question is allied to the broader issue of the 'conventional' vs. 'natural' status of implicature in particular and pragmatic rules in general, an issue to which I shall return in chapter 5. But first I shall explore an important linguistic conspiracy located at the crossroads of negation, scalar implicature, and lexicalization.

4.5 The Story of O: Quantity and Negative Incorporation

O me no O's (Ben Jonson, *The Case Is Altered*)

Why do some logical squares—those depicted in Jespersen's Tripartitions (§4.2)—have only three corners? As we have seen in our earlier incursions into logical geometry, Jespersen's **A** and **C** categories map onto the **A** and **E** vertices of the traditional Square of Opposition. The **B** category has the lexical membership of the **I** vertex (*some, possible*), but its semantics suggests a neutralization or conjunction of the two (**I** and **O**) subcontraries (*some but not all, possible but not necessary*), as illustrated in (80):

(80)

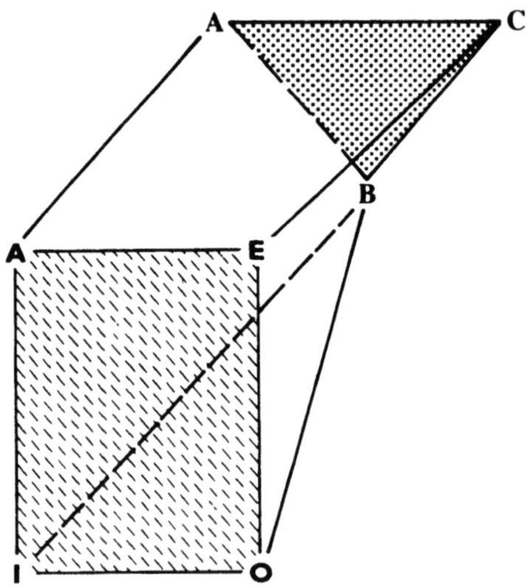

The result is either logically incoherent or radically counterintuitive. If **B** terms are really bilateral, they are incompatible with, rather than simply weaker than, their **A** counterparts. Further, as I noted earlier, terms from the **B** and **C** categories are not (pace Jespersen) true contradictories if the former is defined bilaterally: *some but not all α VP* and *no α VP* can both be false in the same context. Yet Jespersen (like Hamilton before him) is correct in recognizing a warp in the logical square.

If we follow Jespersen in regarding **B** as basically positive, and if we assume, contra Jespersen, that the members of the **B** category are given a unilateral representation (lower-bounded by meaning, upper-bounded by implicature, as laid out in this chapter), we obtain the three-cornered square in (81), a logical amputee of the traditional figure:

(81)
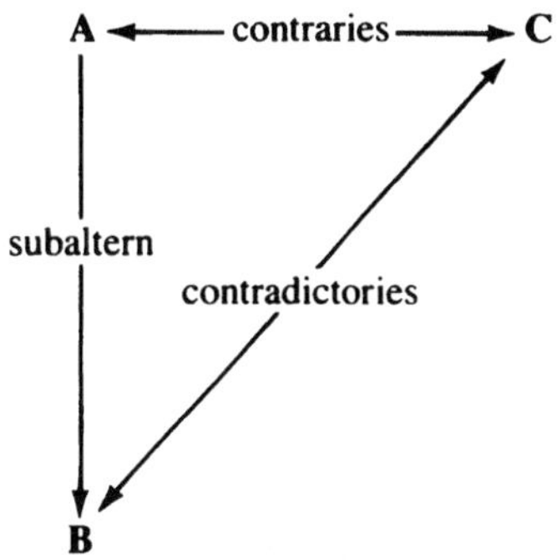

For every triad of **A**, **B**, and **C** terms, now mapping directly onto the **A**, **I**, and **E** vertices of the square, symmetry would demand a **D** term mapping onto the (missing) fourth, **O** vertex. But in each case, it will be observed, this **D** value cannot be expressed straightforwardly as a single lexical item. Instead of corresponding to one of the items in the paradigm, the anticipated **D** (**O**) term corresponds to a paradigm gap: we have *all, some, no,* but *not all* (*some . . . not*); *everybody, somebody, nobody,* but *not everybody* (*somebody . . . not*); *always, sometimes, never,* but *not always* (*sometimes . . . not*).

That the **O** vertex, unlike its three companions, allows no simple representation was recognized by Saint Thomas, who observed that whereas in the case of the universal negative (**A**) 'the word "no" [*nullus*] has been devised [*sic*] to signify that the predicate is removed from the universal subject according to the whole of what is contained under it', when it comes to the particular negative (**O**), we find that 'there is no designated word, but "not all" [*non omnis*] can be used. Just as "no" removes universally, for it signifies the same thing as if we were to say "not any" [i.e., 'not some'], so also "not all" removes particularly inasmuch as it excludes universal affirmation' (Aquinas, *In Arist. De Int.*, lesson 10, in Oesterle 1962:82–83).

In Greek, Latin, and English, negative (**mon** ↓) quantifiers and quantificational adverbs are available (whether or not we view them as having 'been devised' for the purpose) to express universal negation—for example, *no, none, nothing, nobody, nowhere, never*—although the same notions may be expressed equally by lexically complex constructions of various types, as Jespersen, Sapir, and others have noted—*nobody* = *not* *anybody* = *everybody* . . . *not* (on its NEG-V reading). But no negative quantifiers, no one-word lexical items with or without negative morphology, are available to express particular negation (the negation of a universal).

Nor is this asymmetry restricted to the Indo-European family. In unrelated languages throughout the world, we often find a portmanteau lexicalization translatable as *all not* (*not some*) or *none*. Thus, for example, Malagasy *tsy* 'not' and *misy* [miš] combine to form [tsiš] 'no', literally 'not some'; there is no corresponding lexicalization for 'not all' or 'some not'. In other languages, no negative-incorporated quantifiers are available at all, but the **E** form tends to be expressed more simply or directly than the **O** form. Thus, in Japanese, we find 'not all' distinguished from 'no' by the insertion of the *wa* topic marker:

(82) a. Minna ikana katta yo. 'Nobody went'
b. Minna wa ikana katta yo. 'Not everybody went'

The same asymmetry extends to the binary suppletive forms of the quantifiers (or quantificational determiners) and to the binary connectives as well. Thus, alongside the lexicalized **A**, **B**, and **C** forms in the Jespersenian tripartite table in (83), the corresponding **D** row is lexically unrealized.

(83)

	Ordinary Quantifiers	Quantificational Adverbs	Binary Quantifiers	Correlative Conjunctions	Binary Connectives
A	all α, everybody	always	both (of them)	both . . . and	and
B	some α, somebody	sometimes	one (of them)	either . . . or	or
C	no α, nobody	never	neither (of them)	neither . . . nor	nor
	(= all ~ / ~some)	(= always ~)	(= both ~ / ~either)	(= [both . . . and]~)	(= and ~)
D	*nall α, neverybody	*nalways	*noth (of them)	*noth . . . nand	*nand
	(= some ~ / ~all)	(= ~always)	(= either ~ / ~both)	(= [either . . . or]~)	(= and ~ / ~or)

I have argued that the conjunction (*both* . . .) *and* corresponds logically, pragmatically, distributionally, and intonationally to the universal *all*, the disjunction (*either* . . .) *or* to the existential or particular *some*, and the joint negation (*neither* . . .) *nor* to the universal negation *no, none*. This fills in three of the four corners of the Square of Opposition, but the fourth, southeast corner, the **O** position, remains lexically fallow. Why should this asymmetry exist? Whatever became of the missing **O**?

A solution to the puzzle is proposed in Horn 1972: chapter 4 (cf. also Blanché 1969 for a similar idea). Given the existence of the generalized **Q**-based implicature, the use of either of the subcontraries (the **I** or the

O value) tends to implicate the other; as Mill observed (see §4.1 above), if I say (in a neutral context) that *some are*, you will infer that *some are not* (= *not all are*), and vice versa. Thus, in terms of what is communicated or conveyed rather than in terms of what is literally said, the two subcontraries are informationally interchangeable. That they are nonetheless not interchangeable in an arbitrary discourse context (*salva felicitate*), as (84) indicates:

(84) A: All of your friends showed up.
B: No, {some of them didn't. / not all of them did. / {just / only} some of them did. / ?#some of them did.}

A: None of your friends showed up.
B: No, {some of them did. / ?#just some of them didn't. / ?#(just) not all of them did.}

reflects the fact that the two subcontraries occupy corresponding (weak) positions on different scales: it is **O** which provides the contradictory of an **A** statement, and **I** of **E**. But the fact remains that speaker A, if he accepts B's rebuttal, is left assuming the same proposition at the end of the two dialogues in (84): some of B's friends showed up and some did not.

If **I** and **O** statements tend to result in the same conveyed information, with the choice between them partly determined by the context, a language conforming to the account of the subcontraries I have drawn here does not essentially need separate lexicalizations for both subcontraries.[36] And this is what we find: cross-linguistically, at least one universal, strong scalar quantifier, or determiner is lexicalized (corresponding to the **A** position), and often both (filling the **A** and **E** slots). But at most one of the particular, weak scalars is lexicalized, invariably the subaltern of **A** (and contradictory of **E**). The asymmetric pattern of lexicalization represented by the English state of affairs:

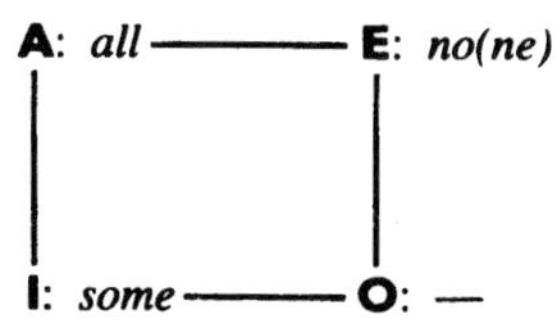

is typical rather than exceptional.

The alternative, even more frequently attested, pattern is found in languages with greater restrictions on the lexical incorporation of negation:

neither of the negative positions (**E** or **O**) is lexicalized. But the generalization remains sound: **A**, **I**, and often **E** values may lexicalize, **O** values may not. And this suffices: in any situation in which I possess (and am known to possess) complete knowledge, and in which that knowledge is (and is known to be) relevant to you, I can convey that information to you via a proposition containing one of the three values *all, some* (implicating *some not, not all*), or *none*. The fourth value (**nall* = *not all,* implicating *some*) is functionally (although not logically) expendable.

Precisely the same pattern is found with the binary connectives. The two values standardly lexicalized occupy the **A** and **I** slots, represented in English by *and* and *or,* respectively. If a third position is represented, typically through negative incorporation, it maps onto the **E** vertex. In fact, there are two distinct semantic candidates for lexicalizing a negative conjunction, both historically linked to English *nor.* Jespersen (1917: 108) cites Old English (and Old Norse *nē*) as a negative phrasal conjunction 'looking before and after', so that *sūð nē nōrð* in *Beowulf* stands for 'neither south nor north'.[37] German offers a (rarely invoked) modern equivalent: *in Wasser noch in Luft* 'neither in the water nor in the air'. Technically, these constitute the only instances of a true **E**-valued binary connective.

The standard (and later the only) value taken by *nor* is simply 'and not', so that ***p nor q*** affirms **p** while denying **q**. And the connective has acquired a further restriction to overtly or tacitly negative contexts. While Tennyson could still write *Ida stood nor spoke* (for 'stood and did not speak'; cf. Jespersen 1917: 114), far more typical today is the use of *nor* as a noncolloquial connective paraphrasable by 'and also not', 'and not . . . either' (rather than by 'but not'), following a clause or sentence with negative form and/or force. Thus, compare

(85) I {*could / √couldn't} do it then nor can I do it now. (Kruisinga 1931: §2395)
#He is rich nor (is he) handsome. / √He is rather poor, nor is he exactly handsome.
He was upset about it. Nor was she {totally thrilled / #unhappy} herself.

While stylistically and functionally restricted, then, naked *nor* survives with the value 'and not'.

Another candidate for this slot, as argued by Dieterich and Napoli (1982), is comparative *rather.* They gloss ***x*** *rather than* **y**, where ***y*** contains a tensed verb as in (86):

(86) Harry walked to work rather than drove.

as '**x** and not **y**', accompanied by the conventional implicature (see §2.5 above) that someone has previously asserted or assumed that Harry drove to work.[38]

As Dieterich and Napoli note (1982:163), their analysis of *rather than* as 'a viable candidate for the logical connective "and not"' stands as a counterexample to the claim of Gazdar and Pullum (1976:224) that no natural language connective can be noncommutative; (86) is obviously not equivalent to *Harry drove rather than walked to work.* Indeed, the same can be said of *nor* in examples like Tennyson's. More generally, Gazdar and Pullum predict no difference in the status of connectives representing the complexes (*neither* . . .) *nor* and *not both,* yet such a difference clearly exists. In no language do we find any lexical candidate to occupy the southeast corner of the Square, corresponding to 'or not', 'not both (and possibly neither)'.

Since the work of Peirce, Nicod, and Sheffer, it has been appreciated that all the truth-functional connectives of propositional logic, including negation, can be generated from one basic truth function, and that there are two equally viable candidates for this truth function: joint denial (**p** ↓ **q**, read as 'neither **p** nor **q**') and the Sheffer (or Nicod) stroke (**p** | **q**, read as 'not both **p** and **q**'). The values associated with these connectives are given in (87):

(87)

		Joint Denial	Sheffer Stroke
p	**q**	**p ↓ q**	**p | q**
T	T	F	F
T	F	F	T
F	T	F	T
F	F	T	T

Depending on which of the two connectives is chosen as the generator, propositional negation (~**p**) can be defined directly (as **p** ↓ **p** or **p** | **p**), and the binary truth functions (*and, or, if then, if and only if*) receive equally straightforward and somewhat less counterintuitive definitions; it will be observed that **p** ↓ **q** is the contradictory of **p** ∨ **q**, and **p** | **q** of **p** ∧ **q**. But this solution is as unnatural as it is elegant, as emerges clearly from the attempt to think intuitively of *not p* as shorthand for 'neither **p** nor **p**' or for 'not both **p** and **p**'.

While the argument that the Sheffer stroke is not conceptually primitive (Gale 1976:6) applies to the joint denial operator as well, there is nevertheless an essential difference between the two cases. As we have seen, some languages (albeit a minority) provide a simple lexicalization for the **E** value of joint denial.[39] While the naked *ne, nor* was once employed to this end in English, the discontinuous correlative (itself dating back to Beowulf: *nē lēof nē lāð* 'neither dear nor loathsome', cited by Coombs 1976:177) is now virtually *de rigueur,* in the form of *neither . . . nor;* Tennyson's Ida must now clothe her naked *nor* as in (88a) to avoid offending linguistic sensibilities (altering both meter and sense, but so goes progress).

(88)	a. Ida neither stood nor spoke.	(Ida both did not stand and did not speak)
	b. *Ida (noth) stood nand spoke.	'Ida did not both stand and speak'; 'Ida didn't stand or didn't speak (and possibly did neither)'

Correlatives of the (88a) type, often of the form **X** . . . **X** . . . , as in French *ni . . . ni . . .* , are legion. But neither English nor any other language allows a simple or correlative form corresponding to the **O** value, as hypostatized in (88b).

More generally, as against the lexicalized **E** forms in (89a), we get no lexicalized **O** forms of the type represented in (89b):

(89) a. Lee can't come { and Kim can't either. / nor can Kim. / and neither can Kim. }
Neither Lee nor Kim can come.
b. Pat can't come { or (else) Sandy can't. / *nand can Sandy. / *or noth can Sandy. }
*Noth Pat nand Sandy can come.

This universal seems to have been detected independently by Horn (1972:§4.23) and Zwicky (1973:477), the latter commenting that despite the logical self-sufficiency of the Sheffer stroke, 'no language seems to have a conjunctive root <u>nub</u>, with the property that <u>A nub B</u> means "not both A and B"'. On the present account, this gap is attributable to the existence of a 'conjunctive root' which conveys (although it does not literally express) this meaning, that is, *or.*

As with all weak scalars, disjunction is lower-bounded (inclusive) in meaning and (unless the context stipulates or presumes otherwise) upper-bounded (exclusive) in ordinary use, so that ***p or q*** says ($\mathbf{p} \vee \mathbf{q}$), implicates (ceteris paribus) ($\mathbf{p} \mid \mathbf{q}$)($= \sim[\mathbf{p} \wedge \mathbf{q}]$), and ends up conveying the conjunction of the two, that is, the exclusive disjunction ($\mathbf{p}$ w $\mathbf{q}$). Ordinary language can make do with either of the two subcontraries and, as with the quantifiers, chooses to latch on to the affirmative (*or*) and eliminate (from the lexicon) the negative (*nub*, aka *noth, nand*).

The next installment in the story of **O** features a familiar cast of characters, the modals. In the case of the modal auxiliaries, lexical items formed by negative incorporation take the form of contractions: *can't, couldn't, shouldn't, mustn't,* and so forth. Consider now the ambiguity of (90):

(90) A priest could not marry. (~◇ or ◇~)

On the former (Catholic) reading, this statement asserts that it is not possible (or not allowed) for a priest to marry; on the latter (Episcopalian) reading, that it is possible (or allowed) for him (or her) not to marry. Parenthetical material inserted after the modal forces the latter reading:

(90′) A priest could {always / if he wishes / of course}
not marry. (only ◇~)

But contraction of the modal-negative complex disambiguates (90) in the opposite (Catholic) direction:

(90″) A priest couldn't marry. (only ~◇)

The same results obtain with can, which also allows 'orthographic' contraction:

(91) a. You can not work hard and still pass. (~◇ or ◇~)
b. You {cannot / can't} work hard
and still pass. (only ~◇)

Whether the modal is understood as denoting logical possibility, physical possibility (ability), or permission, only an outer negation can incorporate into the contracted form.

What of the other modals? The generalization seems to be that ~◇ and □~ sequences contract, while ◇~ and ~□ sequences do not. Thus, alongside *can't* and *couldn't* (~◇), we get

(92) a. You mustn't go. (you must [not go], □~)
b. You shouldn't go. (you should [not go], □~)

For most American speakers, *mightn't* and *mayn't* are virtually nonoccurring forms. *Mightn't,* if it shows up at all, does so in the one environment where a negative associated with *might* must take wide scope over the modal:

(93) He might go, {might he not / mightn't he}? (~◇[he go]?)
?*It mightn't rain today. (◇~[it rain])

When *mayn't* occurs, it also involves outer negation (and hence the deontic reading, given the impossibility of associating wide-scope negation with epistemic *may*):

(93′)	You mayn't go out.	(only ~◇, or rather ~**Perm**)
	?*It mayn't rain today.	(out on ◇~ reading)

In all these cases, modal + negative sequences expressing **E** values tend to lexicalize, while those expressing **O** values do not (cf. Horn 1972: §4.1 for a more thorough discussion.)

The existence of one clear counterexample to this tendency, *needn't* (= ~□, ≠ □~) is partly explained by the status of modal *need* as a negative polarity item; we get *He need not (needn't) go* and (in some registers) *Need he go?*, but not **He need go*. While quantificational **O** values never lexicalize, modal **O** values are relatively free to do so, it would seem, provided there is no possibility of misinterpreting the resultant form. Whenever an **E** reading is possible, it in effect preempts the potential **O** reading. In the case of *needn't*, the polarity status of *need* precludes any **E** interpretation (with outer or wide-scope negation over □), and the isolated **O** is free to undergo contraction.[40]

Other **E** / **O** asymmetries in the modal system obtain in the adjectival and verbal lexicon. Alongside the **E** value *impossible*, we would seem to have the **O** values *unnecessary* and *uncertain*. But there is a difference: a state of affairs can be *logically impossible*, but not *logically unnecessary; unnecessary* seems to be limited to the deontic spheres. Similarly, it can be *impossible* (not possible) that the Indians will win the pennant, but (for many speakers) it cannot be *uncertain* (not certain) that the Yankees will win. *Uncertain* is also restricted syntactically vis-à-vis *not certain*, in that it fails to allow subject raising: *The Yankees are* {*not certain* / **uncertain*} *to win*.

In other languages (e.g., Latin and French) the cognate of the **E** adjective (*impossibile, impossible*) has no lexical **O** counterpart (**innecessarius*, **innécessaire*). In general, we find that any language containing a lexicalization of ~□ also contains a lexicalization of ~◇, but not vice versa; furthermore, if either of the resultant forms is more fully lexicalized (in terms of lack of productivity of the affix, as with *iN*- vs. *un*- in English; cf. §5.1 below), and / or freer in its distributional and semantic potential, it will always be the **E** value, ~◇, not the **O** value, ~□.[41]

Another reflex of the **E** / **O** asymmetry is the existence of negative deverbal adjectives, for example, English *unreadable, unsolvable*, with the logical form *un*-[[V]*able*], that is, ~◇V (incapable of being read, solved, etc.). It is rare that a lexical item—in English or other languages—will incorporate affixes for possibility and negation in such a way that the former takes wide scope over the latter. And when this does occur, as in the Turkish verb forms in (94) (from Payne 1985 : 227), the **E** value will tend to be less transparent or more lexicalized:

(94) a. gel- emiy-ecek
come- IMP- FUT
'He will not be able to come', where *emE* represents an unanalyzed impossibility marker

b. gel- miy- ebil- ecek
come-NEG-POSS- FUT
'He will be able to not come', 'He may not come'

Alongside the vast array of causative verbs lexicalizing the semantic complex CAUSE $\sim\Diamond$ in English, we can find only a couple of candidates for the corresponding logical form CAUSE $\sim\Box$ (CAUSE $\Diamond\sim$):

(95) a. 'make, cause to become not {possible / legal / moral}'

ban	enjoin	interdict	proscribe
bar	exclude	preclude	refuse
deter	forbid	prevent	veto
disallow	inhibit	prohibit	withhold

b. 'make, cause to become not {necessary / obligatory}' ({possible / legal / moral} not)
excuse exempt

Confirming the universal lexicalization preference for **E** over **O** values, a wide range of modal and quantificational complexes which appear (by morphosyntactic criteria) to represent **O** forms are in fact assigned **E** semantics. Russian *nel'zja* 'impossible, forbidden' ought on etymological grounds to denote, not an **E**, but an **O** value, since it derives from the negation of a (now-archaic) root *l'zja* 'good, useful'.

Similarly, the frozen English adverbial *not at all,* like its French cousin *pas du tout,* appears to have strengthened **E**-wards from its original source as the negation of a universal (positive *at all* survives in Irish dialect with the meaning 'wholly, altogether' [OED **all**, 9b]; cf. Van Dongen 1918). The evidence is even clearer in Old English, where the lexical item *nalles, nealles*—representing at first glance a counterexample to the constraint blocking quantifiers of the form **nall*—in fact is attested only with the value 'no, not, not at all', never 'not all'. (Other OE quantificational expressions included *næfre* 'never', *næðor* 'neither, nor', *náht* 'nothing', *nán* 'no one, none', and *náhwær* 'nowhere', all occupying the kosher **E** rather than the treyf **O** slot.)

The conspiracy toward simpler expressibility of **E** over **O** values is also supported by a well-attested **O** --› **E** semantic drift. The outer negation associated with a necessity predicate often seems to develop an inner negation reading, the contradictory of the **A** form turning into its contrary. This development, to which I return in chapter 5, is perhaps seen most clearly in the French construction (96):

(96) Il ne faut pas que tu meures. 'You must not die' ($\Box{\sim}$**p**), lit. 'It is not the case that you must die' (${\sim}\Box$**p**)

first systematically investigated in the paper of the same name by Tobler (1882b). *Falloir* 'must', *devoir* 'must, should', and their cross-linguistic correspondents often take a preceding negative within their scope, either optionally or obligatorily. In the case of (96), as Tobler demonstrates, the *unlogisch* **E** reading coexisted alongside the (expected) **O** sense as early as the fourteenth century and eventually evicted it entirely.

The tendency for **O** forms to strengthen until they take on **E** values is often subsumed under the general phenomenon of so-called neg-raising (cf. Horn 1978b and §5.2 below), but the same strengthening or drift can be seen in another class of cases where the neg-raising analysis is more problematic. In many languages, a negation outside the syntactic scope of a causative element (verb or bound morpheme) is interpreted as inside its semantic scope, resulting in the reading 'cause not' or 'not let' (cf. Tobler 1882b; Spitzer 1927; Cornulier 1973). Thus, the Italian example (97b), as Cornulier notes,

(97) a. Il caffè mi fa dormire. '(The) coffee {makes / is making} me sleep'

b. Il caffè non mi fa dormire.

is normally read as the contrary ('Coffee doesn't let me sleep', 'The coffee is causing me not to sleep') rather than as the contradictory ('Coffee doesn't make me sleep') of its strong causative base, (97a).

Similarly, McGloin (1982: 11) observes that in Japanese, 'Negative causative sentences . . . predominantly give a CAUSE-NOT (i.e., not let) reading. A NOT-CAUSE (i.e., not make) reading is marginal in many cases'. She illustrates this contrast by citing the (marginal) ambiguity of (98):

(98) Watashi wa otooto ni hon-o yom- ase- nakat- ta.
I TOP *younger brother-my book*-ACC *read*-CAUSE- NEG- PAST
'I {didn't let / ?didn't make} my younger brother read the book'

As discussed in Horn 1978a: §5.3 and Horn 1978b: 213–14, Biblical Hebrew, Turkish, Amharic, Czech, and Jacaltec are among those languages in which a negation of a strong causative (lit., 'not make', 'not cause') yields a complex which may, and in some cases must, denote 'not let', 'make not', 'prevent'. Even in English, some lexicalized negative causatives are subject to the **O** --> **E** drift:

(99) a. That approach didn't please me. ('didn't cause me to become pleased' / 'caused me to become not pleased')

b. That approach displeased me. (only = 'caused me to become not pleased')

This diachronic drift has its analogues in performance as well. One curious (and unexplained) result of the experiments on the processing of negation was the finding (in Wason 1959; cf. Wales and Grieve 1969 and discussion in §3.2 above) that subjects had difficulty in dealing with conjunctions of the form *not both . . . and . . .* , often treating it as though it were instead *both . . . not* (*neither . . . nor*). Like the diachronic shifts represented by the *il ne faut pas* construction and by causative negation, this process of spontaneous translation reflects the instability of the **O** category.

The general explanation for the **E** / **O** asymmetry assumed here (and given in more detail in Horn 1972: chapter 4) leans on the observation that the two subcontraries tend to result in the communication of the same information, despite a significant difference in how that information is packaged. This can be illustrated graphically in the table in (100), where the appropriate quantificational, modal, and deontic values are plotted against the speaker's knowledge of the situation. Let **n** range over possible epistemic states of the speaker; **n** will vary from 0 (indicating total negative certainty, knowledge that . . . ~ . . .) to 1 (total positive certainty). The context of utterance is as neutral as can be mentally arranged. Simple lexical items for scalar values are indicated in boldface.

(100) (adapted from Horn 1972: ex. (2.109))

Vertex	Quantifier	Modal	Deontic Value	Knowledge
A	**all** α	**necessary**	**obligatory**	$\mathbf{n} = 1$
I v **A**	at least some α; some if not all α	at least possible; possible if not necessary	at least permitted; permitted if not obligatory	$1 \geq \mathbf{n} > 0$
I	**some** α	**possible**	**permitted**	$1 > \mathbf{n} > 0$
O	not all α; some α not	not necessary; possible not	not obligatory; permitted not	$1 > \mathbf{n} > 0$

Vertex	Quantifier	Modal	Deontic Value	Knowledge
O v **E**	not all α, if any	not necessary, if (even) possible	not obligatory, if (even) permitted	$1 > \mathbf{n} \geq 0$
E	not any α; **no** α; **none**	**impossible**	not permitted; **forbidden**	$\mathbf{n} = 0$

Notice that the information represented in the third and fourth rows of the table is identical, as follows from the symmetrical conversion properties of the subcontraries. The two alternatives differ in logical force, since what is asserted by each is implicated by the other; they also differ, as we have seen, in their privileges of occurrence in well-formed discourse. But their informational parity results in the three-cornered square I have depicted, in which only one of the subcontraries need be directly realized in lexical forms.

But why is it always the positive value that is lexicalized? Given the symmetry between the **I** and **O** values represented in (100), why don't we ever find a quantificational system in natural language with lexicalizations for *all, none,* and *not all,* but not for *some?* Why are the three lexical binary connectives never (*both*) *and,* (*neither*) *nor,* and *not both*? Why are there no modal systems based on the obligatory, the forbidden, and the possible not?

The three-cornered square for natural language results from the vertical asymmetry of the traditional Square, the fact that the subaltern relation (*all* → *some, none* → *not all*) is a one-way street, and that the subcontraries (*some, not all*), unlike the contraries, are compossible (logically compatible). Given the horizontal symmetry of the Square, nothing tells us why, if we are to eliminate one of the subcontraries, or reduce it to second-class (lexically complex) status, it should always be the negative one that is shunted aside.

Nothing, that is, within the pure logic of opposition symbolized in the Square or within the Maxim of Quantity which licenses the paradigm gaps I have explored here.[42] The key, of course, is the priority of the affirmative which I have dealt with at length (in §1.2 and chapter 3), where I ultimately argued for a pragmatic treatment of the positive-negative asymmetry. It is the markedness of negative statements and the formal marking of negation itself which are responsible for the fact that the three-cornered squares defined by natural language scalar operators have the form of (101) rather than of (102).

(101) 

(102) 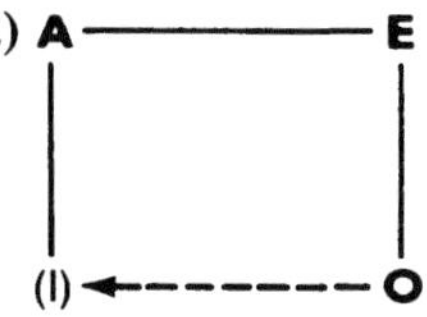

As we have seen (and cf. Horn 1972: §4.2–4.3), strong scalars incorporate an inner negation and weak scalars an outer negation. What of the intermediate scalar operators? What we find is that any operator whose lower bound is above the midpoint on its scale will incorporate only an inner negation; any operator at or below the midpoint will incorporate only an outer negation. In the terms of Löbner 1985, **P**~ is a possible lexical operator only when **P** is INTOLERANT, ~**P** only when **P** is TOLERANT.

Thus we get negative operators like *few,* which can be viewed either as the outer negation of a tolerant determiner (= 'not many') or as the inner negation of an intolerant one (perhaps = 'a significant majority . . . not . . .'). In the same way, the negative (**mon**↓) quantificational adverb *rarely* (along with *seldom* and *infrequently*) equates to 'not often' or 'usually not'. The complexes denoted by *many . . . not, often . . . not,* on the other hand, do not lexicalize.

Since ~**P** will be intolerant just in case **P** is tolerant, I predict that a lexicalized negative quantifier or quantificational adverb should itself always be intolerant, and that is precisely what we find:

(103) a. He often goes to church {and / but} he often doesn't go.
(similarly, *sometimes, not always*)
b. #He rarely goes to church {and / but} he rarely doesn't go.
(similarly, *never, always, usually*)

(104) a. Many of my friends are linguists, {and / but} many of them aren't.
(similarly, *some, not all*)
b. #Few of my friends are linguists, {and / but} few of them aren't.
(similarly, *none, all, most*)

Essentially the same predictions are made by the MONOTONICITY CORRESPONDENCE UNIVERSAL of Barwise and Cooper (1981: 186): 'There is a simple [NP] which expresses the **mon**↓ quantifier ~**Q** iff there is a simple NP with a weak non-cardinal determiner which expresses the **mon**↑ quantifier **Q**'. Since a weak noncardinal **mon**↑ determiner will be one of our tolerant positive scalar values (*a, some, a few, many*), B&C's proposed universal successfully rules out 'not every', 'not both', 'not all', 'not most', and so forth, as possible values for simple determiners, while admitting 'neither', 'no', 'few' (not many), and so on. But without a grounding in the pragmatics of scalar operators, B & C's approach to the constraints on quantifier lexicalization is ad hoc; worse still, it fails to generalize to modals and other nonquantificational values.

As I would predict, the paradigm of (103) and (104) obtains for the modals as well. The negative epistemic scalars *not probable* and *not likely* allow—and indeed favor—a neg-raising interpretation under which they are taken to signify 'probable not', 'likely not'; this instantiates my **O** --› **E** drift. Their lexicalized counterparts, however, can only be read in this stronger way, their internal morphology notwithstanding:

(105) a. It's {not probable / not likely} that a fair coin will land heads.
(ambiguous; true on outer [contradictory] reading of negation)
b. It's {improbable / unlikely} that a fair coin will land heads.
(for most speakers, unambiguously inner [contrary] negation; hence false)

(106) a. It's not likely that the Yankees will win and not likely that they'll lose.
(allows 'tolerant' reading)
b. # It's unlikely that the Yankees will win and unlikely that they'll lose.
(allows only 'intolerant', hence anomalous, reading)

This pattern extends to the deontic analogues of *improbable* and *unlikely: undesirable* and *inadvisable* are unambiguously inner negations ({desirable / advisable} . . . not . . .), while their periphrastic counterparts *not desirable* and *not advisable* are ambiguous. In both epistemic and deontic cases, then, the **O** --› **E** drift affecting the use of the unlexicalized forms *not* α, rendering the "illogical", contrary (inner) negation readings more natural than the "logical", contradictory (outer) readings, is fossilized as an aspect of literal meaning in the lexicalized *iN* + α, *un* + α forms. We shall encounter other instances of such fossilization in chapter 5.

I began this chapter with Jespersen's observation that negation, when associated with numerals and related (gradable) notions, means 'less than'. I have argued that this meaning is not a matter of negation mysteriously taking on a reading which is neither contradictory nor contrary in such contexts, but results instead from the nature of scalar operators. Quantifiers and quantificational determiners and adverbs, modals, and other scalar values are lower-bounded by their literal meaning and upper-bounded (in default and certain other contexts) by quantity-based conversational implicature. Since the implicature relation is context-dependent, we systematically obtain two understandings for each scalar value **P** ('at least **P**', 'exactly **P**') without needing to posit a semantic ambiguity for each operator. Negation, in contradicting the literal meaning contributed by the operator, applies to the pre-upper-bounded value, yielding the value 'not (at least) **P**', that is, 'less than **P**'. (An obvious, but only apparent, counterexample to this claim will be the focus of chapter 6.)

Jespersen's Tripartition model of the quantificational and modal notions, incorporating upper bounding as a logical rather than pragmatic rule, suffers the same ultimate incoherence as Aristotle's modal logic and Hamilton's quantificational logic. But each of these systems suggests the appropriate pragmatic amendment we need to tack onto the traditional Square of Opposition (or to more contemporary formal accounts of the semantics of the logical operators). As we have seen, not all four corners of the Square are equal under the law of natural language. The pragmatic mechanism relating the positive and negative subcontraries, **I** and **O**, results in the superfluity of one of these subcontraries for lexical realization; the functional markedness of negation assures that the superfluous, unlexicalized subcontrary will always be **O**.

5 The Pragmatics of Contra(dicto)ry Negation

Things fall apart,
The center cannot hold. (W. B. Yeats)

Keiner oder alle. Alles oder nichts.
Einer kann sich da nicht retten.
Gewehre oder Ketten.
Keiner oder alle. Alles oder nichts. (Bertolt Brecht)

If you're not part of the solution, you're part of the problem.
(Eldridge Cleaver)

That negation cannot in general be reduced to contrariety or antonymy has been recognized at least since Plato's mouthpiece, the Stranger, pointed out in the *Sophist* that the not-great cannot be identified with the small, any more than with the medium-sized, but embraces both (cf. §1.1.1). The received position for both Aristotle (and his Peripatetic followers) and the Stoics was that ordinary negation or predicate denial (the Stoics' *apophatikon*) is semantically contradictory. At the same time, an affirmation may also have a contrary opposite of either positive or negative form (Aristotle's predicate term negation or privative opposition, the Stoics' *sterētikon*).

As we have seen, any two mutually inconsistent terms are contraries in the broad sense; two sentences are in contrary opposition if they can be simultaneously false but not simultaneously true. Sentences (1a, b) are thus contraries, since x may be neither black nor white.

(1) a. x is black.
  b. x is white.

(1′) a. y is odd.
  b. y is even.

But for Aristotle, the sentences in (1′) are also in contrary opposition, although their middle is excluded. If y denotes an integer either (1′a) or (1′b) will indeed be true and the other false. But if y happens to be Socrates, both (1′a, b) are false. Similarly, when two sentences differ in that one affirms a positive predicate term of its subject and the other affirms the corresponding negative term of the same subject, as in (1″):

(1″) a. *z* is bald.
  b. *z* is nonbald. (or, *z* is not-bald)

the result is again a contrary, not a contradictory, opposition: if *z* is something of which baldness (and hence nonbaldness) cannot be naturally predicated (e.g., if *z* = the number 17), or if *z* fails to denote an actual existent (e.g., if *z* = the present king of France), (1″a) and (1″b) both come out false.

Indeed, by this line of argument there are no contradictory terms; the contradictory of (1′a) can only be its predicate denial *y is not odd,* which consists of precisely the same terms as does (1′a). In the same way, the contradictory of (1″a) is not (1″b) but the denial *z is not bald,* in which baldness is denied rather than nonbaldness affirmed, of *z* (cf. chapters 1 and 2 for discussion). Since I shall not be concerned in the subsequent discussion with vacuous subjects or category mistakes, I shall take the liberty (as I did in chapter 3) of restricting contrariety in this chapter to the mediate or weak contraries, those which allow an unexcluded middle (as in (1)).[1] Strong or immediate contraries will be assimilated to contradictories; the sentences in (1′a, b) and (1″a, b) will be taken to constitute contradictory oppositions, contra Aristotle, and the constitutive terms (*odd/even, bald/non-bald*) will also count as contradictories. I thus adapt the Tree of Contrariety (§1.1.5, (34)) as shown.

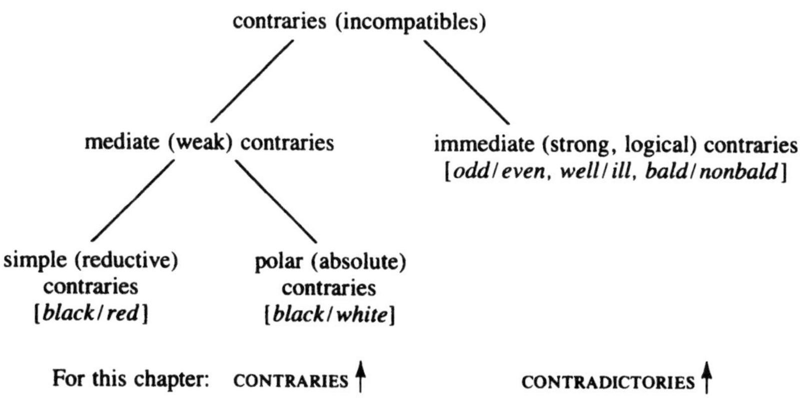

More graphically, we have the distinction illustrated in the diagrams in (2) and (3); the governing laws (defined in chapter 1) are repeated here in (2′) and (3′).

(2)

Contradictory Opposition

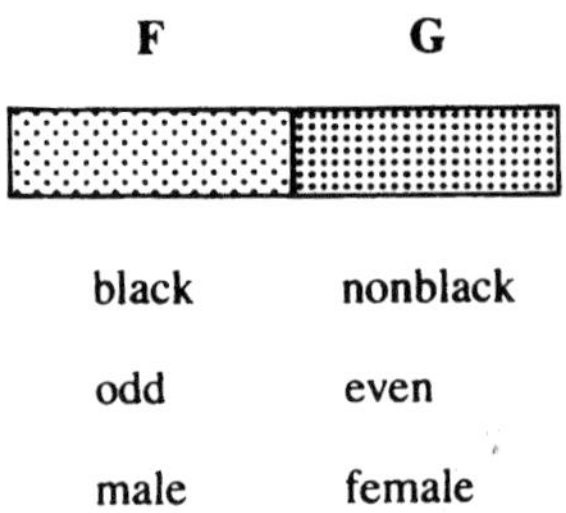

(3)

Contrary Opposition

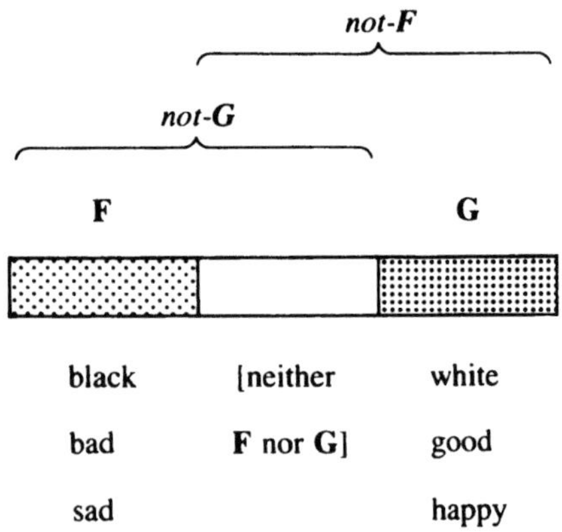

(2′) Contradictory opposition is governed by the Law of Contradiction (LC) and the Law of Excluded Middle (LEM): if two terms **F** and **G** are contradictories, then
(i) by LC, for any α in the relevant domain, $\sim(\mathbf{F}\alpha \wedge \mathbf{G}\alpha)$.
(ii) by LEM, for any α in the relevant domain, $(\mathbf{F}\alpha \vee \mathbf{G}\alpha)$.

(3′) Contrary opposition is governed by the LC but not LEM.

Contradictory terms (*black/nonblack, odd/even, male/female*) exclude any middle term, an entity satisfying the range of the two opposed terms but falling under neither of them: a shirt which is neither black nor not-black, an integer which is neither odd nor even. Mediate contraries, by definition, allow a middle: my shirt may be neither black nor white, my friend neither happy nor sad. Yet it may occur that the gap between the contraries narrows and even disappears, so that the middle is effectively excluded or swallowed up.

The centrifugal politics and/or theology of polarization, as reflected or warned of in the epigraphs from Yeats, Brecht, and Cleaver, tends to force every entity within the range of the polar contraries to choose one of the two terms to fall under. In this setting, evoked by revolutionists of every stripe, everything is black or white; there are no shades of gray.[2] Hitler, for one, was described by a contemporary as operating under 'a two-valued classification scheme where everything was Black (dark, evil, Jewish) or White (pure, good, Aryan)' (Richter 1944: 194). But Jesus, too, endorsed a binary taxonomy:

(4) a. He that is not with me is against me.
(*Matthew* 12:30; *Luke* 11:23)

b. He that is not against us is for us.
(*Mark* 9:40; *Luke* 9:50)

Notice that which polar contrary was to absorb the middle seems to have been answered differently depending on whether that middle was occupied by an outsider to be rejected (4a) or by a potential convert to be assimilated (4b).[3]

Whence this polarizing tendency, this drift of middle-allowing lexical contraries into middle-excluding acting contradictories? In his seminal investigation of gradable terms, Sapir (1944:133) points to the existence of a 'psychological excluded middle': 'Three-term sets [*superior/average/inferior, good/moderate/bad, big/medium/small, warm/lukewarm/cool*] do not easily maintain themselves because psychology, with its tendency to simple contrast, contradicts exact knowledge, with its insistence on the norm, the "neither nor"'. It is because of this psychological preference for simple, either-or contrast that the 'normed' or middle term, occupying a ZONE OF INDIFFERENCE, tends to be 'quasi-scientific rather than popular in character' and that it is itself typically ungradable (*?more average, ?more lukewarm*). Nor is it an accident—as Sapir and Aristotle have both noted—that the zone of indifference must often be characterized negatively, as '*neither X nor Y*'.[4]

A speculative mind might attribute the polarizing tendency to the presumed survival value for the primitive language wielder in perceiving and classifying the universe into various series of either-or, black-or-white, or (following Osgood and Richards 1973; cf. chapter 3) yin-or-yang pairs, ignoring the quasi-scientific niceties of the zone of indifference.

In any case, we have what appears to be a productive, context-dependent process: polar contraries are treated as mutually exhaustive as well as mutually inconsistent, contradictories in contrary clothing. When all values but *p* and *q* have been discarded, we obtain the disjunction in (5a), functioning as an instance of (5b), that is, the Law of Excluded Middle.

(5) a. $\mathbf{p} \vee \mathbf{q}$
b. $\mathbf{p} \vee {\sim}\mathbf{p}$

As the neo-Hegelian Sigwart observed, the efficacy of LEM derives in fact from the possibility of establishing just such pragmatic disjunctions between semantic contraries: 'We are able, on the ground of our knowledge and of the particular contents of our subjects and predicates, to frame two positive statements, of which we know [as with] contradictory judgments that while both cannot be true together, neither can both be false; and in this case we gain, by denial of either member of the disjunction, a definite, unambiguous affirmation' (Sigwart 1895:155).

The assumption of **p** ∨ **q**, together with the negation of either disjunct, licenses the inference of the other disjunct. This principle has been known since the Stoics established their 'fifth indemonstrable syllogism' (§4.2, (32b)), and comes down to us as the law of MODUS TOLLENDO PONENS (MTP):

(6) $\mathbf{p} \vee \mathbf{q}$
$\underline{\sim\mathbf{p}}$
$\therefore \mathbf{q}$

The crucial step, of course, is the initial establishment, in context, of the disjunction: 'If we could solve all difficult questions by starting right off with "it is either so or so"—"he is either mentally healthy or diseased in mind", "the number is either odd or even"—then indeed the principle of excluded middle would be an invincible weapon' (ibid.). No disjunction of contraries, no LEM; no LEM, no MTP.

But when, precisely, can we extend this strategy of "divide and assert" to semantic contraries? One class of cases has already been mentioned, those resulting from the polarizing tendency of Manichaean credos. When the center does not hold, there are only two possibilities: if everything is either good or evil, and something isn't good, what else can it be? But if *evil* expands to cover the ground of 'not good', *not good* is essentially reduced to 'evil'. Thus a formal contradictory (*not good* vis-à-vis *good*) is strengthened in terms of the relevant scale to yield the assertion of a contrary (*bad, evil*). This is what Sigwart has in mind in warning that 'The opposition of predicates [e.g., *good* vs. *evil, white* vs. *black*] has substituted itself unnoticed for the mere negation, and the negative statement [*x is not good, y is not white*] seems to tell us more than it really does; it is understood as if it applied to the truth of the proposition with the opposite predicate' (Sigwart 1895: 195).

How general is this sleight of mind? For Sigwart's contemporary and fellow Idealist Bosanquet, it is essentially universal: 'The essence of formal negation is to invest the contrary with the character of the contradictory' (Bosanquet [1888] 1911: 281). To support his position that 'negation always involves contradiction between contraries', rather than simple contradictory opposition on the one hand or pure contrariety on the other, Bosanquet cites the apparent 'mere contradiction' between *He is good* and *He is not good,* where the latter is semantically a relatively weak, 'non-informative form' which in practice is 'filled in', 'so that from "he is not good" we may be able to infer something more than that "it is not true that he is good"' (p. 293).

As a related illustration of the same tendency to 'fill in' a literal contradictory, Bosanquet recognizes 'the habitual use of phrases such as [*I do not believe it*], which refer grammatically to a fact of my intellectual state, but actually serve as negations of something ascribed to reality. . . . Compare [Greek] *οὐ φημι* [lit., 'I do not say'], which means 'I deny', or our common phrase 'I don't think that'—which is really equivalent to 'I think that—not'. (Bosanquet [1888] 1911:319)

Thus, both simple first-order contradictory negation and second-order (displaced, anticipated, or transported) negation associated with an embedded clause must be understood—in at least some of their instances—by virtue of a 'filling-in' process: just as the middle between polar contraries may be excluded, or filled in, by a context-dependent polarizing process, so too the 'mere contradictory' (*not good, not believe*) may be filled in to express the corresponding contrary (*bad, disbelieve*). In each case, contrary meaning is invested in contradictory form.

The basic principle identified by Bosanquet and Sigwart can be expressed straightforwardly in the language of modern pragmatics: In a context licensing the pragmatic assumption **p** ∨ **q**, to assert **not-p** is to implicate **q**. Under the right conditions, then, a formally contradictory negation **not-p** will convey a contrary assertion **q**—but just what are the right conditions? When is the assumption of **p** ∨ **q** (for polar contraries **p**, **q**) pragmatically warranted? What is the nature of this pragmatic implicature, if such it is, and when does it become strengthened or "institutionalized" into a convention of usage or meaning? With these questions, which were never directly addressed by the neo-Hegelians (or anyone else), I launch my inquiry.

5.1 Affixal Negation

5.1.1 Contrariety and E-Negativity: The Case of *iN-* and *un-*

I pick up the spoor of contrary-negation-in-contradictory-clothing, not in either of Bosanquet's suggested lairs, but in a third haunt. It has long been recognized (cf. inter alia Sigwart [1895:138], who cites *unhappy, unwise, unfeeling,* and *speechless,* and more recently Kruisinga 1931:§1620; Zimmer 1964; and Funk 1971) that negative affixation, especially when it involves the English prefixes *un-* and *iN-* and their cross-linguistic analogues, admits or tends to develop a contrary, rather than merely contradictory, interpretation. Here is Jespersen's statement of the generalization:

> The modification in sense brought about by the addition of the prefix [*un-*] is generally that of a simple negation: *unworthy* =

> 'not worthy', etc. . . . The two terms [*X*, *unX*] are thus contradictory terms. But very often the prefix produces a 'contrary' term or at any rate what approaches one: *unjust* generally implies the opposite of *just; unwise* means more than *not wise* and approaches *foolish, unhappy* is not far from *miserable,* etc.
> (Jespersen 1917:144; cf. Jespersen 1942:466–67)

What is less well understood is (1) when the contrary readings tend to arise—for which senses of which stems, and for which applications of which affixes; and (2) how the emergence of these readings correlates with the equally often observed fact that the same prefixes tend to yield derived forms which are associated with a depreciatory or pejorative (i.e., emotively negative) content. This latter point is also recognized by Jespersen (1917:144): 'The same general rule obtains in English as in other languages, that most adjectives with *un-* or *in-* have a depreciatory sense: we have *unworthy, undue, imperfect,* etc., but it is not possible to form similar adjectives from *wicked, foolish,* or *terrible*'.

Jespersen and Zimmer (1964:10ff.) review a number of early discussions of negative affixation, including those of Jhering ([1883] 1923), Wundt (1886), Noreen (1904), and van Ginneken (1907), focusing on what Wundt labels the *Unlustaffecte* of negatively affixed forms in particular and negation in general. In German, Swedish, French, and English, as these lexical studies demonstrate, disproportionately many negatively affixed adjectives are depreciatory, derogatory, or evaluatively negative in denotation or connotation. Following Cruse (1980), I shall adopt E-POS and E-NEG as shorthand for evaluatively/emotively positive and negative, respectively. Then the formula expressing the relevant generalization can be given as in (7):

(7) negative affix + e-pos base → e-neg derived output
[*un-, iN-*] [*happy, due*] [*unhappy, undue*]

For these late-nineteenth- and early-twentieth-century scholars, the affective e-neg coloration of negatively affixed forms is simply a special case of the emotive content inextricably linked to all varieties of negation, akin to Russell's 'moment of aversion' or Wason's 'prohibitive' factor. But the fact remains that this correlation of negative form and e-neg content is far stronger in the case of partially productive negative affixes.

Thus, in English we have *unhappy* but not *unsad, unwise* but not *unfoolish,* and a variety of other asymmetrical pairings illustrated in the table in (8):

(8) uncivilized *unboorish, *unbarbarian
unclean *undirty

unfriendly	*unhostile, *unantagonistic
unhappy	*unsad
unhealthy, unwell	*unsick, *unill
uninteresting	*unboring, *undull
unjust, unfair	*unwrongful
unkind, uncivil	*unrude
unsympathetic	*unantipathetic
untrue	*unfalse
unwise	*unfoolish
infertile	*unbarren

In these examples, only an e-pos or neutral stem can serve as a natural base for a negatively affixed adjective, and only e-neg derived adjectives are therefore possible.

Elsewhere, as in (9), we find an (e-neg) *un-* or *iN*-prefixed adjective based on an e-pos stem alongside which there is no obvious e-neg stem to serve as a source for an e-pos derived form:

(9)	impossible	inconsistent	unfit
	inappropriate	irrelevant	unfortunate/unlucky
	incoherent	unapt/inept	unsound
	inconsequential	unbecoming	unsuitable

There are also a number of *un-* adjectives with no—or no extant—relevant positive counterpart whatsoever: *unabashed, unassuming, unbending, uncouth, unflagging, unheard-of, unkempt, unparalleled, unprecedented, unruly, unscathed, untouched, untoward,* and so on. Significantly, the majority of such forms (many of which are evaluatively positive) represent the productive *un* + *V* + PTCPL pattern discussed below; those that fail to reflect this pattern do seem to share the e-neg quality of the adjectives in (8) and (9).

As Cruse (1980) notes, the category of e-pos adjectives is semantically heterogeneous. In the case of ANTONYMS (gradable contraries), where there is a midinterval between the unmarked and marked qualities, the unmarked term (*happy, wise, interesting, tall, big*) denotes a positive attribute or property which can be present in varying degrees and there are no literal endpoints of the relevant scales (*#absolutely* {*happy/sad/tall/short*}). In the case of GRADABLE COMPLEMENTARIES (contradictories), where there is no midinterval, the unmarked term (as determined by the usual distributional tests; see chapter 3) denotes the absence of some negative or undesirable property (cf. *clean* vs. *dirty, safe* vs. *dangerous*) and the unmarked term does contain a scalar endpoint (*absolutely* {*clean/#dirty, safe/#dangerous*}). In the former case, the e-pos term is also Q-

POS (quantitatively positive, denoting a salient property); in the latter case, it is the e-neg term that is q-pos: 'All gradable complementaries denote degrees of some undesirable property, like dirtiness, or danger; antonyms always indicate degrees of either a neutral property, like length, or weight, or a desirable one, like beauty, merit, or intelligence' (Cruse 1980:21). But, as shown by pairs like *unclean* / **undirty*, *unsafe* / **undangerous*, and the existence of unpaired neg-prefixed adjectives from q-neg but e-pos bases—*unfaithful, dishonest, imperfect, impure*—it is e-polarity rather than q-polarity which determines the availability of an adjective for negative prefixation. (Cf. Sapir 1944; Givón 1970; Ljung 1974; Lehrer 1974, 1985; Lyons 1977; and Lehrer and Lehrer 1982, for further discussion of antonymy, markedness, and gradable adjectives.)

How is the asymmetry pointed to in (8) and (9) to be explained? We cannot simply invoke Wundt's *Unlustaffecte* of negation, or other correlates of the widely held position (encountered frequently in earlier chapters) that all negation is inextricably tied up with the unpleasant or forbidden. Whatever the justification for such views, they do not distinguish the morphologically restricted cases of negative prefixation from the unrestricted syntactic potential of ordinary sentential or phrasal negatives; however marked might be a discourse occurrence of *She isn't sad, You're not foolish, That movie wasn't boring*, and so on, these sentences are undeniably grammatical, while the corresponding morphological negatives (**She's unsad*, **You're unfoolish*) are equally clearly ungrammatical.

The essential insight on the road to explanation is provided by Zimmer (1964): the less productive the affixation process, the more likely it is that the result will be interpreted as a contrary (rather than contradictory) of its base, and the stronger will be the restriction to e-pos bases (and, correspondingly, to e-neg resultant meanings for the derived negative adjective). This correlation emerges especially clearly when we turn to more productive subrules for *un-*. At least since the OED, it has been noticed that *un-* attaches freely to stems with deverbal suffixes, that is, *-able* and the participial *-ed* and *-ing*. Indeed, *un-* prefixation is virtually unrestricted in these cases, constrained only by the existence of lexicalized *iN-* forms occupying the same slot.[5] But in just these same contexts, the affixation rule produces derived forms which are contradictory and emotively either neutral (as in (10a)) or positive (as in (10b)).

(10) a.		b.	
	imperceptible	unbeaten	untarnished
	irreducible	unbigoted	unblamable
	undecidable	unblemished	unconquerable
	uneat{able / en}	undaunted	incorruptible
	unexpired	undefeated	indomitable

unprefix{able/ed}	undeterred	unimpeachable
unxerox{able/ed}	unharmed	unobjectionable
un-cross-examined	unscathed	irreproachable
un-mouse-eaten	unsullied	invulnerable
untidewashed		

Thus, 'the negative content of simplex words differs from the negative content of forms derived according to some synchronically productive and frequently encountered pattern' (Zimmer 1964:38).[6] It is for this reason that Jespersen, in discussing the tendency for *un-* negatives to be read as contraries, explicitly exempts 'words in *-able* (*-ible*) and participles'; his examples (1917:144) include *unabsorbable, unadaptable, unabbreviated, unadapted, unavailing,* and *unbefitting.*[7]

If the *un-* and *iN-* attachment rules can only peek at the next-most-internal material in the stem, these formations do not constitute counterexamples to the claim that prefixation only occurs with a positive base. What we need is the ADJACENCY PRINCIPLE of Siegel (1977): 'No word-formation rule can involve X and Y, where X is an affix, unless Y is uniquely contained in the cycle adjacent to X'. (Cf. Zimmer 1964:38 and Allen 1978 for related discussion.)

The constraint corresponding to (7) can now be given more explicitly: the BASE of *un-*, the material uniquely contained in the adjacent cycle when the rule applies, cannot contain negation (and in particular, the prefix *dis-*).[8] Siegel's formulation of this principle allows her to distinguish the ill-formed e-pos *un*-words in (11a) from the acceptable formations in (11b):

(11) a. *un[#dis[#courteous]], *un[#dis[#honest]]
b. un[#[dispute#]ed], un[#[distinguish#]ed], un[#[dis[#hearten]#]ed]

The words in (11b), along with the parallel formations in (10b), do not violate Siegel's constraint: at the time *un-* is attached, the base (*-ed, -ing, -able*) is not negative (although the stem containing that base may well be, as in Siegel's last example and those of (10b)).

But something else seems to be going on in these cases. Notice that productivity alone does not explain the cases in (10a, b), in particular those involving the nonproductive *iN-* prefix. As Jespersen (1917:140) points out, 'While most of the *in-* words are settled once and for all, and have to be learned by children as wholes, there is always a possibility of forming new words on the spur of the moment with the prefix *un-*.' Although there are isolated examples of new formations with *iN-* (Zimmer cites an acknowledgment by Kennedy to Khrushchev in October 1962 'that you and I were aware that developments [in the Cuban missile crisis] were approaching a point where events could have become immanageable'), Jespersen's observation is essentially correct and is reflected in the different behavior of *iN-* and *un-* with respect to phonological and morphological properties of the corresponding affixation rules (cf. Siegel 1974; Allen 1978).[9] The e-pos *in-V-able* forms in (10b) thus remain problematic, as do their emotively neutral counterparts in (10a). The Level I formation *inedible*, for example, is exactly as contradictory in its semantics as *uneatable*.

It will be observed that many of the examples in (10b) involve passive formations based on an active transitive verb with the general meaning of 'defeat'; while there are well over fifty verbs in this semantic class (e.g., *beat, conquer, deter, outdo, overcome, subdue, vanquish;* cf. Roget's *Thesaurus*, 731.8–10), there exists no simple lexicalization for the converse relation, that is, a transitive verb **glarf* meaning 'lose to', such that *The Pats glarfed the Bears* would signify 'the Pats lost to the Bears', 'the Bears squelched the Pats'.

Dowty (1982:111–12), following observations by Fillmore (1968), draws analogous correlations between lexical semantics of given verb classes and what we might call "direction of lexicalization". Dowty's principles, held to be valid cross-linguistically (with the possible systematic exception of "deep ergative" languages; cf. Dixon 1979), are given in (12):

(12) (i) If for any $\langle x, y\rangle \in R$, x is an entity that causes something to happen to y, R (and not R^{-1}) is lexicalized (*build, kill, ignite, move*).
(ii) If for any $\langle x, y\rangle \in R$, x is a sentient being that perceives something about y or has an emotion or attitude toward y, R (and not R^{-1}) is lexicalized (*see, love, believe*).
(iii) If for any $\langle x, y\rangle \in R$, x is moving and y is stationary, R (and not R^{-1}) is lexicalized (*enter, overtake, pierce, collide with*).

My principle can be given an analogous Dowty-style formulation:

(12′) If for any $\langle x, y\rangle \in R$, x defeats or prevails over y, R (and not R^{-1}) is lexicalized.

One indication of the strength of this principle, at least in English, is that when the verb *to best* entered the lexicon—the first OED citation is 1863—it did not fill the vacant slot, the lexical gap corresponding to R^{-1} in (12′). The OED observes (under '**best**, v. colloq.') that 'the form is hardly in accord with the sense, which is nearly equivalent to the existing vb. *to worst*', whose meaning (in line with (12′)) was well established over two hundred years earlier. A tendency must be strong indeed to result in the synonymy of *worst* and *best*. (The neologism *underwhelm* also fails to refute the principle in (12′), since it clearly does not mean 'to be overwhelmed by'.)

Note further that complex **V** + **Prep** forms of the *lose to* type do not act as syntactic or semantic units: cf. *Chrissie was* {*spoken to*/?*said goodbye to*/**lost to*} *by Martina*. Since there are predictably no prefixal adjectives to fill the e-neg semantic slots of **unglarfed* 'winless' or **unglarfable* 'incapable of winning', the corresponding e-pos slots are liberated to be filled as in (10b).

Additional confirmation of the asymmetry of e-pos vs. e-neg *un-* formations is provided by the history of English. A number of potentially occurring *un-X* adjectives are now unavailable, presumably blocked or preempted by a previously existing and more lexicalized simple contrary with the same meaning; but this was not always the case. Here again is the OED (**un-**[1], 7):

> There is . . . considerable restriction in the use of *un-* with short simple adjectives of native origin, the negative of these being naturally supplied by another simple word of an opposite signification. There is thus little or no tendency now to employ such forms as *unbroad, undeep, unwide, unbold, unglad, ungood, unstrong, unwhole,* [etc.] which freely occur in the older language.

It will be noticed that these cited examples, like those of (8) and (9), are e-neg forms with e-pos bases; no *unshallow, unnarrow, unsad, unbad,* or *unweak* seem to have occurred 'in the older language' (cf. Zimmer 1964:41).

The 'restriction against the use of *un-* with adjectives that have obvious simplex antonyms', cited and experimentally confirmed in Zimmer 1964, instantiates the more general Avoid Synonymy principle ('The output of a lexical rule may not be synonymous with an existing lexical item'; Kiparsky 1983). The same principle—a subcase of my Division of Pragmatic Labor, summarized in §3.3.1—predicts the narrowing of the semantic domain of such derived adjectives as *unhappy* and *unintelligent* in such a way that their meanings are palpably different from (and, in particular, weaker than) those of the corresponding underived *sad* and *stupid* (cf. Zimmer 1964; Lehrer 1985). Despite this difference in strength between derived and simple e-neg adjectives, *unhappy* and *unintelligent* still constitute contrary, rather than contradictory, opposites of their bases *happy* and *intelligent:* someone who is neither happy nor sad may, but need not, be unhappy. (An alternative explanation for the nonexistence of *ungood* et al. is offered by Marchand [1960:151], but it is flawed by an unsupported assumption that *un-* forms are in general contradictories rather than contraries of their stems.)

The privative denominal adjective-forming *-less* suffix is, like deadjectival *un-*, partially and variably productive, producing lexicalized forms which often drift away from compositionality. For this reason, there is no necessary semantic correspondence between a negative adjective of the form *N-less* and the related positive *N-ful* derivative, if indeed the latter even exists (cf. *careless/careful, helpless/helpful, boundless/*boundful, reckless/*reckful*). Thus, the way is often open (i.e., unblocked) for the formation of an *un-N-ful* derivative which, as once again predicted by Avoid Synonymy, will fail to be equivalent in its senses or uses to the *N-less* derivative if one exists: *uncareful ≠ careless, unhelpful ≠ helpless, unrestful ≠ restless*. Indeed, the fourteenth century witnessed such forms as *unmindful* (quite distinct from the already-existing *mindless*) and *unfruitful* (distinct, although not quite so obviously so, from *fruitless*). (Additional examples and related commentary are offered by Marchand 1960: 231, 261–63; and Funk 1971).

Also cited in the OED, however, and reproduced by Marchand (1960) and others, are a plethora of *un-X-less* forms, in apparent violation of the principle in (7) (and of the more specific Zimmer-Siegel-Allen ban on affixing *un-* to formally negative bases), culled from seventeenth- and eighteenth-century texts. But on closer inspection, these forms, exemplified in (13):

(13)	unboundless	unguiltless	unnumberless	unshameless
	undauntless	unhelpless	unquestionless	unshapeless
	uneffectless	unmatchless	unremorseless	untimeless
	unfathomless	unmerciless	unrestless	unwitless

do not constitute actual counterexamples to these claims. Crucially, the meaning of *unmatchless,* for example, was 'unmatched' or 'matchless', rather than (as its form would suggest) 'not matchless'; *unmerciless* corresponded not to *merciful* but to *merciless* (or *unmerciful*). Thus, these examples, like Modern English *irregardless* (or Ger. *unzweifellos* 'doubtless'), must be interpreted as containing redundant or pleonastic (rather than mutually annihilating) double negation (cf. Horn 1978a: §3.2): an e-pos base is transformed into a morphologically peculiar but semantically regular e-neg derived adjective.

As noted by Jespersen, Marchand, and other descriptive morphologists, *un-* and (especially) *iN-* derivatives tend to negate the emotive senses of the stems to which they attach, while *non-* and (to a lesser degree) *un-* prefixes negate objective or descriptive content. Among the minimal pairs that have been cited to illustrate this contrast are the derived forms in (14):

(14)

immoral : nonmoral	unprofessional : nonprofessional
irrational : nonrational	unprofitable : nonprofit
un-American : non-American	unremunerative : nonremunerative
un-Christian : non-Christian	unrhythmical : nonrhythmical
unnatural : nonnatural	unscientific : nonscientific

In each case, the *iN-* or *un-* form is understood pejoratively and is in contrary opposition with the corresponding positive stem, while the *non-* derivative is a simple, evaluatively neutral contradictory.[10] Algeo (1971: 90), echoing Jespersen and Marchand, describes the contrast: 'A Moslem is a *non-Christian,* but only a Christian can be *un-Christian* in behavior. A *nonrealistic* novel is one whose goal is other than a realistic view of the world, but an *unrealistic* novel is likely to be one that aims at, and fails to achieve, realism'.

A parallel contrast often obtains between the synchronically unproductive *iN-* prefix and its semiproductive *un-* counterpart. Jespersen (1942: §26.3) draws attention to the distinction between *inartistic* (outraging the canons of art) and *unartistic* (not concerned with art), and to the contrast between *immoral* and *unmoral* ('Children are naturally neither moral nor immoral, but merely unmoral'). In these cases, the post-Jespersenian ear might detect a more fully neutral and contradictory negative adjective formed with the fully productive *non-*, that is, *nonartistic* and *nonmoral* (or *amoral*).

We occasionally even get a three-way contrast, as in the examples of (14′), where the derived forms become gradually more descriptive and contradictory as we move from left to right:

(14′) inhuman : unhuman : nonhuman
irreligious : unreligious : nonreligious
impious : unpious : nonpious

In each case, the contrary reading is possible to the extent that an adjectival stem can be regarded as a GRADABLE or SCALAR value (cf. Sapir 1944; Horn 1972; Ducrot 1973; cf. chapter 4 above). Applying the standard tests for gradability, we get the correct prediction that *iN-* and (usually) *un-* adjectives can be inserted into the scalar frame *X was {somewhat/rather/extremely/very/awfully/downright} ADJ,* while the nonscalar *non-* forms cannot occur in that frame.

(15) downright {un-American/#non-American}
very {un-Christian/#non-Christian}
extremely {unnatural/#nonnatural}
somewhat {immoral/#nonmoral}

awfully {irrational / #nonrational}
rather {unscientific / #nonscientific}

Hence also the impossibility of *un-* prefixation for binary ungradables (**unmale,* **unfemale,* **unodd*) and the semantic restriction of other derived forms to scalar senses or contexts: the surface of my table can be *uneven,* but the number 7 cannot be; your decision may be *unfair,* but not your complexion.[11]

Given the evaluative vs. descriptive parameter, we obtain some striking lexical gaps. Alongside *unmaternal, nonmaternal, unmotherly,* we have no adjective **nonmotherly:* while the stem *maternal* may be construed either as a descriptive or an evaluative adjective, *motherly* can only be evaluative (Zimmer 1964:33). Similarly, we have *nonmale* but not **nonmanly* (cf. *unmanly* / **unmale*). The same consideration rules out **nondecent* and **nonrespectable.*

The other side of the coin, as pointed out by Funk (1971), is that adjectives based on *iN-* and (semiproductive) *un-*, even when they originate as evaluatively neutral and semantically contradictory senses, tend to develop a contrary, affective, and typically depreciatory meaning or connotation. Funk's examples of this process include *inadequate, inappropriate, inconvenient, incorrigible, infertile, irrelevant, uninteresting,* and *unsatisfactory.* And only a failed comedy may be *unfunny,* not a successful tragedy.

In some cases involving technical vocabulary, however, the predicted distinction between a contrary form in *iN-* (or the equally unproductive *a-*) and a contradictory in *non-* does not correlate with any depreciatory nature—or gradability—for the former terms: a two-place relation which is not *transitive* (*symmetric, reflexive*) is necessarily nontransitive (*nonsymmetric, nonreflexive*), but not necessarily *intransitive* (*asymmetric, irreflexive*). Yet none of the terms involved is scalar or depreciatory (cf. Lyons 1977:154 or Allwood, Andersson, and Dahl 1977:88–90 for definitions).

In summary, then, we have the following correlations:

(16) *iN-* tends to combine only with scalar predicates on their evaluative readings; the resultant derived forms are lexicalized, semantically and phonologically opaque, and tend to be assigned a contrary and generally depreciatory (e-neg) sense or connotation, often involving an opposition to some expected or established norm.

non- is much freer in its combinations (cf. §5.1.2 below); the resultant derived forms are in general unlexicalized, semantically and phonologically transparent, and involve the formal and/or descriptive (rather than emotive or evaluative) dimensions of meaning.

un- forms are situated between the *iN-* and *non-* forms with respect to these criteria, depending on how productively or freely the prefix combines with a given base: the less productive, the more like *iN-;* the more productive, the more like *non-*.

The correlation between productivity and transparency is too well known to need detailed elaboration here; it is treated extensively both in the traditional sources on word formation and by Siegel ([1974] 1979), Aronoff (1976), and Allen (1978) within level-ordered generative morphology. The application of this correlation to the negative prefixes is specifically addressed by Siegel and Allen. I need only point to the phonological opacity of nonproductive formations like *impious* and *infamous,* citing the retraction of stress, the vocalic reduction, and the nasal assimilation, as contrasted with the virtual compoundlike stress pattern manifested in *nonpious;* this contrast correlates exactly with the semantics—opaque (noncompositional) in the former case but transparent (compositional) in the latter.[12] As usual, the *un-* forms occupy an intermediate position, causing some problems for a two-level system like Siegel's or Allen's; if *iN-* is a Level I prefix and *non-* Level II, *un-* seems to need to hover between (in Class I½?), to flit back and forth, or to apply at both levels, none of these solutions being entirely straightforward.[13]

The principal difficulty resides in a level-ordering paradox diagnosed and treated in various ways by Allen (1978), E. Williams (1981), Lieber (1981), Selkirk (1982), Kiparsky (1982, 1983), and Pesetsky (1985). The paradox can be reconstructed as follows: First, given its phonological properties and the fact that (unlike Level I *iN-*) it attaches only to full words, we must assume that *un-* is a neutral, Level II, #-boundary affix. Second, I adopt the core generalization of level-ordered morphology, due originally to Siegel (1974: 163) and adopted by Aronoff (1976), Allen (1978), and others, that Level II affixes may appear outside Level I affixes but not vice versa. Third, I accept the evidence of Siegel and Allen that negative *un-* is essentially a non-category-changing adjectival prefix. Fourth, I follow Aronoff in treating the nominalizer *-ity* as a Level I, +-boundary suffix, as its phonology, morphology, and semantics demand. Then it clearly follows that alongside the acceptable *impartiality, inability,* and *instability,* we should get no derived *un-X-ity* words of the form of *unclarity, ungrammaticality, unoriginality, unreality,* and *unworkability.* Unfortunately, we do.

It is clear that adjective-yielding *un-* is category preserving. Even Marchand (1960: 152), with no theoretical ax to grind, maintains that *un-X-ity* forms 'must . . . be analyzed as derivatives from negative adjs'. There is strong systemic pressure to preserve the generalization that negative *un-* applies only to adjectives; this pressure was in fact the principal motivation for the position that 'unpassive' participles (*uninhabited, uncollected, uncalled-for*) must be de-adjectival rather than deverbal in nature (cf. Siegel 1973; Hust 1977; Wasow 1977). It is true that both Marchand and the OED attest a variety of other nominal *un-* formations in which no such derivation is possible: *unbelief, unbeliever, uncandor, uncrime, unluck, unphilosophy, unpromise, unreason, unrest, unsuccess, unvalue,* and so forth. But most of these nominals strike us as nonce forms, as back formations, and/or as—in the OED's delicate phrase—'not intended seriously'; cf. Alice's *unbirthday present,* or (a century later) the *uncola* and McCawley's (1980) theory of *unsyntax.* Other *un-* nominals—*unintelligence, untruth, unwisdom*—involve the apparent attachment of *un-* inside a nonproductive Level I nominalizing suffix. These formations will presumably be handled by whatever mechanism is ultimately capable of providing an escape from the *ungrammaticality* paradox.

But just what mechanism is that? Williams (1981 : §3) and Lieber look for a solution in the observation that 'lexical semantics is in principle autonomous from the structural aspects of generative morphology' (Lieber 1981 : 70). But, as Selkirk and Kiparsky point out, this is a red herring, since the problem involves morphological as well as semantic compositionality. How do we attach a Level I suffix onto an adjective which has already acquired a Level II negative prefix? Alternatively, how do we attach a de-adjectival negative prefix onto a form which (at the output of Level I and the input to Level II) is no longer an adjective?

For Selkirk, the answer is that when *-ity* attaches "outside" *un-*, it is because *un-* in these forms is a Level I, not a Level II, prefix. That is, there are two negative *un-* prefixes in English, just as—on Aronoff's (1976: §6.2) argument—there are two *-able* suffixes, a Level I + *abl* and a Level II *#abl.* The problem with this line, besides the unparsimonious duplication of *un-* prefixes (and of other prefixes with the same properties), is that, as Kiparsky (1983 : §5) points out, Selkirk's Level I *un-* never brings off the kind of phonological legerdemain associated with self-respecting Level I (or root) affixes. In particular, it fails to assimilate in such forms as *unpopularity, unreality* (**umpopularity,* **urreality*). Nor (unlike Aronoff's + *abl*) does it attach to roots as well as full words. There is, in short, little reason (outside of the level-ordering paradox itself) to believe that *un-* is ever Level I, and good reason to believe it is not.

Kiparsky (1982, 1983) considers an analysis in which a form constructed in Level II, such as *ungrammatical,* can somehow loop back and undergo the rules of Level I, such as *-ity* suffixation. The problem here, however, is that not just any Level I affix can apply to a word of the form *un-X.* Kiparsky's ultimate solution, which strikes me as a good starting point, is to suppose that *un-X-ity* words undergo a MORPHOLOGICAL RE-

ANALYSIS: the noun [$_N$[$_A$grammatical]ity] is a 'marked exception' to Bracketing Erasure (Kiparsky's version of the Siegel-Allen Adjacency Condition), and the Level II prefix *un-* can effectively peek inside to attach to the adjectival component. When it does so, the resultant derived form [$_N$un [$_N$[$_A$grammatical]ity]] is reanalyzed as [$_N$[$_A$un[$_A$grammatical]]ity].

This analysis essentially treats *ungrammaticality* as a marked blend of two impeccable forms, the Level I noun *grammaticality* and the Level II adjective *ungrammatical.* It correctly predicts that *un-X-ity* formations will be possible only when the corresponding positive *X-ity* noun is possible and, since the admissibility of *un-X-ity* is a fact about *-ity* rather than a fact about *un-*, that other Level II prefixes will automatically trigger the same reanalysis (cf. *semigrammaticality, noncompositionality, extrametricality, bilaterality*). Other instances of reanalysis include compounds incorporating Level I affixes taking semantic scope over the entire compound (cf. Levi 1978): *three-dimensional, set-theoretical,* and so on.

Of course, the Level I suffixes *-th, -ce, -dom,* which are even less productive than *-ity,* must also (as seen above) be marked as exceptions to Bracketing Erasure, a somewhat disconcerting result. Other cases which might prove problematic for Kiparsky's approach are *unemployment, unfulfillment, uninvolvement* (Selkirk 1982: 130). Whether the deverbal nominalizing suffix *-ment* is only Level I (as Aronoff assumes) or both Level I and Level II (as Selkirk argues), the blend involved here seems to be one of, for example, [$_A$un[$_A$[$_V$employ]ed]] and [$_N$[$_V$employ]ment], where something more than the erasure or transparency of brackets seems to have to be assumed. We might simply relax the restriction of negative *un-* to adjectives, but these forms don't share the nonce or jocular flavor of simple *un* + *N* forms (*uncola, uncrime, unsyntax, unword*). We begin to appreciate why Chomsky, from his earliest writings on syntactic theory, has always discussed the property of ungrammatical<u>ness</u>.

Whatever the final resolution, if any, of the level-ordering paradox, there is no gainsaying Zimmer's basic finding that *un-* varies in productivity and compositionality, and that the variation along these two parameters is strongly correlated. While lexicalized *un-* forms tend to drift to contrary and e-neg interpretations, the morphophonemic and suprasegmental behavior of the prefix is uniformly that of a word-level, neutral, Level II affix. But it is not clear how deep a fact this is: the parallel French negative prefix *iN-*, which must effectively occupy the space of both *iN-* and *un-* in English, varies between Level I and Level II phonological behavior, depending on its productivity and compositionality (cf. Zimmer 1964: 50–51; Aronoff 1976: 125).

The generalizations in (16) seem secure, then, however they are to be reconciled with the facts just outlined. But it is important to recognize that not just any positive scalar value can serve as the base for less than fully productive negative affixation. As Zimmer observes, *iN-* and *un-* forms tend not to accept superlative, or in my terms strong scalar, bases. Thus, we have no derived adjectives of the type illustrated by **undelicious*, **unexcellent*, **unsuperb* (Zimmer 1964:44). The general principle can be stated as in (17):

(17) The stem to which a relatively nonproductive negative affix can attach tends to be an UNMARKED, WEAK POSITIVE scalar value.

Thus, alongside *unhappy* we have no **unecstatic* (from the strong positive *ecstatic*), **unsad* (from the weak negative *sad*), or **unmiserable* (from the strong negative *miserable*). Verbal prefixes behave the same way: we have *dislike*, but not **dislove*, **disadore*, **dishate*, or of course **disdislike*. In each case, the permissible derived forms incorporate stems which represent unmarked scalar values, as can be confirmed by applying the standard diagnostics: *How happy is he?* vs. *How {ecstatic/sad/miserable} is he? How much do you like me?* vs. *How much do you {love/adore/hate/dislike} me?* In §5.2 and §5.3, I shall investigate other contexts in which these same unmarked, weak positive scalar values tend to trigger contrary readings for a formally contradictory negation outside their scope.

5.1.2 Other Prefixes, Other Problems

The negative prefixes *iN-* and *un-*, along with other nonproductive affixes I did not consider in §5.1.1 (e.g., the adjectival *dis-* of *disadvantageous, discourteous, dishonest, disreputable*, and the *-less* suffix of *breathless, careless, faithless, lifeless*) correspond to the privatives of Aristotle (§1.1.1) and the Stoics (§1.1.2), as of course does the linear descendant of the Greek privative, the *a(n)-* of *ahistorical, amoral, anesthetic, anorexic*. As we saw in chapter 1, the privatives represent for Aristotle the clearest instance of the contrariety operator he calls predicate term negation; I shall return in chapter 7 to the formal characterization of word-internal negation and its relation to contrariety.

Whether or not a given negative adjective counts as a contrary in the sense of this chapter—as a mediate or weak contrary—seems to depend on the sense of the stem and the degree of productivity of the relevant affixation pattern with the base in question. As Zimmer observes (1964:87), ungenerated (lexically listed) forms tend to have contrary interpretations, and generated forms (those resulting from a productive rule), contradictory. Intuitively, if a derived form is produced by a general, predictable, quasi-syntactic process, its meaning must be predictable (compositional) as well, or speaker and hearer would fail to communicate. For a negative adjective, this will generally determine a contradictory rather than contrary meaning, since (weak) contrariety is not a function (cf. §1.1.5).

Other negative affixes seem to involve a different dimension altogether. Alongside negative *un-*, which attaches (almost) only to adjective stems (cf. OED under **un-**^{1}; Kruisinga 1931: §1620; Marchand 1960; Zimmer 1964; Siegel [1974] 1979; Allen 1978), the homophonous (but historically separate) REVERSATIVE *un-* prefix (the OED's **un-**^{2}, Jespersen's PRIVATIVE *un-*) forms verbs from either verbs or nouns: *unbend, uncage, uncover, undo, unfasten, unfold, unlock.*

In the presence of a deverbal, adjective-producing suffix, we get structurally ambiguous lexical items—*unbending, uncoiled, unfoldable, unlockable, unwrapped*—each of which may be taken either as $[_{ADJ}\mathbf{un}^{1}[_{ADJ}[V]X]]$ or as $[_{ADJ}[_{V}\mathbf{un}^{2}[V]]X]$.[14] The stress pattern may disambiguate the resultant homonym (Jespersen 1917: 149; Kruisinga 1931: §1623), but then again it may not. In keeping with the conspiracy against lexicalization of semantic complexes corresponding to the **O** (southeast) vertex of the logical square (§4.5), any adjective of the form *un-V-able,* where *un-* is the ordinary negative prefix, will always be analyzable as the **E** value $[_{ADJ}un[_{ADJ}[V]able]]$ (incapable of being Ved), with negation taking wide scope over the possibility modal (Horn 1972: §4.13). The modal suffix can take wide scope only in the structure $[_{ADJ}[_{V}un[V]]able]$, where *un-* is the reversative verbal prefix.

Are the negative and reversative *un*-s synchronically, as well as historically, unrelated? The standard answer is given by Covington (1981: 34): the two prefixes are 'of course distinct'. But is this in fact the case? Marchand (1960: 153) argues that as early as the OE period, the prefix *on(d)-* (source of reversative *un-;* cf. the cognate German *ent-*) 'had come to be felt connected with the negative prf *un*', with which it merged orthographically and, presumably, phonologically. Given that what so-called reversative *un-* actually reverses is not the action denoted by the verbal base but rather the result of that action, the semantic relation between the two sets of derived forms may be closer than first appears, as well: 'What distinguishes *unbound* 'not bound' from *unbound* 'loosened' is only the additional idea of an action preceding the state of being loosened, but the state itself is the same' (Marchand 1960).

In the second edition of his *Word Formation,* Marchand (1969: 205) lays out this relationship explicitly and seeks to capture it in the (deep) syntax:

> The reversative type **untie** has the meaning 'undo the result of a verbal action', more precisely 'cause the object of the verb to be no longer -ed'. *Tie a package* means 'cause the package to be tied', *untie a package* means 'cause the package to be tied no longer'.

> The 'tiedness' (passive state) of the package is undone. . . . At the level of the underlying syntactic structure, the analysis is . . . 'cause to be *un-* (= not) *-tied*'.

Marchand's proto-generative-semanticist account, cited and discussed by Dowty (1979:258), generalizes naturally to semantically related verbal prefixes, including *dis-* and *de-* (Marchand 1972; Dowty 1979:§5.7).

Working in ignorance of Marchand, Horn (1978c) in fact offers a Lakoff-McCawley-style decomposition of *dis-* verbs. In his rebuttal of G. Lakoff's (1969) decompositional unpacking of *dissuade* into *persuade not,* Hust (1975) cites as potential counterexamples to this approach the *dis-* forms in (18), which contain no overt complement structure and are not paraphrasable by their counterparts in (19).

(18) a. I have disarmed the prisoners.
b. I have disassembled the mechanism.

(19) a. I have not armed the prisoners. [≠ (18a)]
b. I have not assembled the mechanism. [≠ (18b)]

But (18a, b) can be decomposed instead into (20a, b), respectively,

(20) a. I have caused the prisoners to come not to have arms.
b. I have caused the mechanism to come to be {not/no longer} assembled.

where the negation is internal to the causative element. In these paraphrases, 'the inchoative element builds in the markedness implicature that the prisoners had previously been armed and the mechanism assembled'; the internal position of the negation predicts correctly that, for example, (19a) does not entail that I—or anyone—had previously armed the prisoners (Horn 1978c:205; see Dowty 1979:chapter 5 for insightful discussion of such abstract internal negations and appendix 2 for more on *dis-* verbs and the markedness implicature for negation).

I cannot make the case here for the connection between the negative (adjectival) **un-**[1] and the reversative or privative (verbal) **un-**[2], nor can I take the space to explore the semantic and pragmatic intricacies of the latter construction and of its rivals *de-* and *dis-*, all of which are used to 'express the undoing of a previous state' (Marchand 1972:636; cf. Covington 1981; Horn and Covington 1987). It may be worth noting, however, that the phonological item *dis-* is, like *un-*, morphologically polymorphous.

First of all, some *dis-* verbs denote, not activities or accomplishments, but states: *disbelieve, dislike, distrust.* Nor is the *dis-* prefix limited to verb formation; it also surfaces as a negative prefix on Romance-based adjectives (*disadvantageous, discourteous, dishonest, disloyal, disobedient, disreputable*). Furthermore, as Marchand (1960: 112–13) points out, there are two semantically distinguishable classes of *dis*-prefixed nouns, those interpreted privatively as denoting 'absence, lack of ——' (*discomfort, disease* [originally 'lack of ease'], *disharmony, disregard, disunity*) and those interpreted reversively or contrarily as 'opposite of ——' (*dishonor, dislike, disobedience, disorder, displeasure*). The appearance of *dis*- forms in all three major categories, and in different subcategories within those categories, supports the plausibility of defining a synchronic connection among the adjectival, verbal, and (marginally) nominal *un*-s. The versatility of both *dis*- and *un*- would thus call into question the Unitary Base Hypothesis (Aronoff 1976)—the proposal that the input to a morphological rule can be specified in terms of a single syntactic category.

Like other forms incorporating negative prefixes of limited productivity, the *dis*- derivations, regardless of category, tend (as Jespersen (1917: 146) recognizes) to receive mediate contrary interpretations—and privative or e-neg readings—whenever possible. This applies clearly in the case of stative *dis*- verbs: to *disbelieve* someone or something is not simply to fail to believe someone or something, nor is to *dislike* or *distrust* simply to fail to like or to trust. It should be pointed out that under some circumstances we do seem to be able to equate such *dis*- predicates with simple negatives:

(21)		(21′)	
a.	I disbelieve your story.	a.	I don't believe your story.
b.	I dislike chocolate oysters.	b.	I don't like chocolate oysters.
c.	I distrust such analyses.	c.	I don't trust such analyses.
d.	I'm disinclined to accept that.	d.	I'm not inclined to accept that.

But, as I shall argue in §5.3, the problem here is not that the complex negatives of (21) are interpreted as contradictories, but that the (apparently) simple negatives of (21′) are interpreted as contraries.

The one negative prefix universally regarded as producing contradictory negatives is of course *non*-. This feature of the prefix, which we have already encountered, correlates with the fact that *non*- is morphologically opaque and extremely productive—especially, as Jespersen notes (1917: 147), with those stems for which no *in*- or *un*- prefixed lexical negation is available. As we saw in chapter 2, it has been claimed (by, inter alia, Zimmer [1964] and Drange [1966]) that a predication of the form *S is non-P,* like *S is not P* but unlike *S is un-P,* tends to be interpreted as a weak negation of *S is P,* one which is vacuously true if *S* is not the sort of thing of which *P* can be predicated—whence the minimal pairs of the *un-American*/*non-American* genre exemplified in (14) above. The same point is made by Funk (1971: 31), who associates the (purported) invariant contradictory semantics of the *non-* prefix with its interpretation as not just 'that is not **Adj**', but 'that is other than **Adj**'.[15]

But even in this paradigm case of a negative prefix whose semantics is stereotypically neutral, objective, and category-external and whose freedom approaches that of a syntactic operation, there are rumblings of latent contrary/e-neg tendencies. First of all, we find a substantial set of lexicalized *non-* prefixed adjectives and nouns which have no obvious synchronic source: *nonchalant, nondescript, nonentity, nonplussed,* etc. Other derived *non-* words do have a positive source but do not constitute a contradictory negation of that source: a *nonsporting* dog or breed of dog is not just any dog/breed that is not a sporting dog—terriers and hounds are neither sporting nor nonsporting dogs—while a *nonpaper* is a particular kind of unofficial bureaucratic document. A *nonconformist* is not simply someone who is not a conformist, nor is *nonviolence* simply the absence or lack of violence.

Then there are the 'voguish' uses of *non-* described by Algeo (1971) and Bauer (1983:279–85), such as the PEJORATIVE *non-*, attaching to nouns, glossed by Algeo as 'possessing the superficial form but not the values of': *nonactor* (on one meaning), *nonbook, noncandidate, nonevent,* etc. And, as with its more contrary counterparts, *non-* often attaches only to the unmarked member of an opposition: in our demographic taxonomy, we have *nonwhites,* but not usually *nonblacks* or *noncoloreds,* etc.

As I have noted, negative prefixes exhibit a general correlation between degree of productivity, tendency toward contradictory interpretation, and morphological and phonological transparency. *Non-* not only fails to trigger segmental sandhi phenomena and stress shifts (like *un-* and unlike *iN-*), it also retains a considerable portion of its own stress. Indeed, as Allen (1978) has observed, a word of the form *non-X* (where *X* may be an adjective or noun) often behaves morphologically and phonologically more like a compound than a simple derived form. The unpredictable semantics of some of the lexicalized *non-* formations supports this perspective, since compounds themselves are notoriously capable of noncompositionality.

Furthermore, while it is true that *S is non-P,* like *S is not P,* ordinarily yields contradictory negation, we can isolate another compound symptom in the former construction. Zimmer (1964:44) notes that a *nonbarking dog* 'would appear to designate a dog which does not usually bark, rather than a dog which is not barking at some particular moment'. This sense of characteristic or habitual association, which strikes me as far weaker (if not simply absent) when we refer to a *not-barking dog,* is (as Zimmer [1971] and others have noted) a general property of compounds.

In their (usual) status as predicate terms with the semantics of contradictory opposition, *non-* adjectives evidently occupy a slot intermediate between sentential negation, on the one hand, and ordinary, colloquial, contrary-tending affixal negation, on the other. Whence the innovation of the pseudopredicates in such classc citations as those in (22):

(22) a. The king of France is nonbald (—there is no king of France).
b. 2 is nonblue.

No ordinary affixal negation of limited productivity can co-occur with these stems—*bald* is too e-negative and *blue* has no polar contrary—so *non-* forms must be stipulated. While *un-* and *iN-* adjectives may represent contradictories, strong contraries, or polar contraries of their stems, they never apply within sets of multiply opposed terms, so *non-* must volunteer to leap into the breach. For Zimmer, Drange, and Funk, the contradictory semantics associated with *non-* renders sentences like those in (22) true (as external negations), where an *un-* or *iN-* adjective would result in falsity, meaninglessness, or truth-value gaps ({*The king of France*/*The number 2*} *is unhappy*).

But then why don't the sentences of (22) strike us as the least bit plausible, much less true? The problem resides in an equivocation on the notion of contradictory opposition. In the terms of my discussion in chapters 1 and 2, both ***non-P*** and the even more artificial turn ***not-P*** (which I have employed for rendering Aristotle's predicate term negation) are in fact properly viewed as yielding, not contradictories, but strong or immediate contraries of their base ***P***. The distinction between contradiction and strong contrariety, although I have been ignoring it for the purposes of the present chapter, becomes crucial in instances of vacuous subject terms and category mistakes. These are precisely the cases illustrated in (22a, b), where—given the ontology and semantics of the model—the predicate cannot hold of its subject. But while it does not yield sentential negation (predicate denial), ***non-P*** will predicate something different from ***un-P*** in just those cases where the latter has or acquires a mediate contrary interpretation.

One feature which *non-* shares with its semantically less transparent, less productive prefixal cousins is the power to alter the category of its base. As *un-*, *de-*, and *dis-* may convert nouns into verbs (*unearth, unnerve; deflea, deplane; disbar, dismember*), so too does *nonskid* turn a verb into an adjective. Other *non-* representatives of what Marchand (1960: 130) disparages as 'commercial jargon' include *nonstick* skillets and *nonstop* and/or *nonsked* (*nonscheduled*) flights. Then there is the apparently related commercial (and nonprofit) use of *no-* as an adjective-forming prefix applying to verbs, nouns, or adjectives, as in *no-good* rascals, *no-fault* insurance, *no-iron* fabrics, *no-strings* relationships, *no-wax* floors, and *no-win* (or *no-lose*) situations. What makes these examples significant is the comparative rarity of category-changing prefixes in the language.

If we view the syntactic role of affixes as that of functions from sets (categories) into sets, then inflectional affixes define operations, since they take a word of a given category into an inflected word of the same category. Derivational suffixes, on the other hand, are in principle not operators, since they are capable of converting a root, stem, or word of one category into a stem or word of a different category. But prefixes (which in English can only be derivational) are in general operators, leaving the category of their stem unaffected.

This asymmetry between prefixes and suffixes is captured in Edwin Williams's RIGHTHAND HEAD RULE (RHR), the principle which stipulates that the head of a morphologically complex word is invariably the right-hand member of that word, and that it is that head which determines the category of the resultant word (Williams 1981:248). Williams acknowledges (p. 250) the 'systematic exceptions' to the RHR posed by the essentially unproductive prefix *eN-*, which forms verbs, often with a causative interpretation, from adjectives (*embitter, enable, enlarge, enrich*) or nouns (*empower, enchain, encode, encircle, enslave*). He does not observe the equally "exceptional" behavior of de-adjectival and denominal *be-* verbs (*becalm, befoul, belittle; becloud, befog, behead, bewitch*), cited by Lieber (1981:57) along with the parallel German prefixes *ver-* and *be-*, which form verbs from nouns, adjectives, verbs, and even adverbs. These classes too, however, are nonproductive in the modern languages.

It is the thriving class of denominal verb-forming negative prefixes (especially *de-*) and the marginal but clearly productive classes of deverbal and denominal adjective-forming *non-* and *no-* that pose the most serious threat to the strong version of Williams's RHR. Given these clear counterexamples, we are compelled to abandon the RHR in favor of an approach like Lieber's, in which features always percolate up from the affix regardless of the handedness of that affix, or like Selkirk's (1982:87–88), in which a weakened version of the RHR marks, but does not rule out, prefixes as heads.

But why do prefixes involving negation constitute such a large subset of the counterexamples to the RHR? Jespersen (1917, 1933) views the predominance of prefixal over suffixal negatives as a reflex of the general tendency for negation to be attracted leftward, and generally to precede the material over which it has scope. While there are clear counterexamples to this tendency in both the syntactic and morphological spheres (cf. older English *I know not . . .* , colloquial French *Elle vient pas*, and the English and German privative suffixes *-less*/*-los* and *-free*/*-frei*), the fact remains that both affixal negation (as also noted in Zimmer 1964) and sentential negation (as shown by Dahl 1979) display a strong leftist bent, both where we expect it—on the basis of general typological correlations and the semantics of individual constructions—and where we do not. This tendency 'to put the negative word or element as early as possible, so as to leave no doubt in the mind of the hearer as to the purport of what is said' (Jespersen 1933:297) is strong indeed; it may even, in the words of Poldauf (1964:369), 'assert itself at the cost of clear expression', as in the so-called neg-raising phenomenon (see §5.2) and in the evolution of "illogical" expressions like *Don't let's go just yet* and *I didn't go because I was afraid.* Historically, if not synchronically, the existence of negative category-changing prefixes (where we might have expected suffixes) can be linked to this tendency, which I shall dub NEG FIRST; return to this principle in chapter 7.[16]

How do affixal negations behave in other languages with respect to the e-neg restriction and the contrary-contradictory dichotomy? The study of negative adjectival affixes reveals that, in the words of Zimmer (1964:82), 'for any given language, negative affixes that are distinct from the particle(s) used in sentence negation are likely to have a greater affinity for evaluatively positive adjective stems than for evaluatively negative ones'. It is this same class of affixes which tend to maintain contradictory interpretations rather than developing toward contrariety.

Thus, for example, Zimmer shows that French *iN-* shares many of the attributes I have associated with English *un-* (it selects contrary, e-neg readings, especially in lexicalized forms where it completely assimilates; in more productive combinations, especially with *-able* bases, it may be interpreted as a contradictory, it often produces e-pos or neutral adjectives, and it may not assimilate completely in combination with vowel-initial stems), while the *non-* prefix, homophonous with the free particle ('no'), essentially mirrors its English homograph.

German *un-*, like its English cognate, tends to form contrary negatives with e-neg meanings and is restricted to neutral or e-pos bases: we get *ungesund* 'unhealthy', *unklug* 'unintelligent', but no *unkrank* 'unsick', *undumm* 'unstupid'. As in English (and French), when the base of *un-* is itself a productive adjectival suffix, the result (*un-X-bar, un-X-ig, un-X-isch, un-X-lich*) is semantically transparent, contradictory, and evaluatively unrestricted. The *nicht-* prefix, identical (as with French *non-*) to a free negative particle, is consistently contradictory and tends toward the technical register (although this tendency is not as pronounced as with French and English *non-*). And, as we would predict, we get the same minimal pairs as in English; Zimmer (1964:82) cites *unchristlich* 'un-Christian' vs. *nicht-christlich* 'non-Christian'.

Other languages in which some partially nonproductive negative affix manifests the restriction against e-neg bases which emerge from Zimmer's survey include Finnish (*epä*), Yoruba (*ai-*), Chinese (*fēi*), and Japanese (*hi-*, *fu-*). But when a language contains no affixal form distinct from free sentential negation, the affixal use of the negative particle is much less subject to these constraints, as predicted by Zimmer's generalization. Russian *ne-*, for example, seems to act like a neutralization of the two prefix types. Like English and German *un-* (and unlike English and French *non-*), *ne-* is not felt to belong specifically to the learned or technical register; nor is it barred from combining with e-neg stems: *neploxoj* (lit. 'nonbad/unbad') is just as natural as its e-neg counterpart *nexorošij*. Similarly, a form like *nemuzykal'nyj* corresponds semantically to both of its English counterparts, the e-neg *unmusical* and the neutral, exclusive *nonmusical*.

But even here, there are signs of a preference for the unmarked form. In the case of paired adjectival antonyms differing in perceptual saliency, *ne-* combines naturally only with the unmarked, more salient member: we have *nevysokij* ('lowish', lit. 'nonhigh') but not **nenizkij* ('highish', lit. 'nonlow'). And, as Zimmer stresses, it is significant that *nexorošij* is felt to be closer to *ploxoj* 'bad' than *neploxoj* is to *xorošij* 'good'. I shall return to these contrasts and their implications in §5.3.[17]

When we are working in an area of natural language characterized by systematic and partly conventionalized schemata of nonlogical inference, it is often instructive to survey the behavior of artificial languages to gain a perspective for comparison. Mathematical and computer languages, which are designed to minimize ambiguity and context-dependent vagaries of interpretation, eschew affixal negation (Jespersen's 'special' negation) entirely. But (would-be) international auxiliary languages, aspiring as they do to a greater degree of naturalness, cannot be so circumspect. What choices have tended to be made about the form and meaning of negative affixes and what can we learn from them?

Zamenhof and his followers in the Esperanto movement distinguish a simple negative *ne-* which yields contradictory negations (*ne'ebla* 'impossible', *ne'kutina* 'unaccustomed', *ne'sci'anta* 'unaware'), a privative prefix *sen-* which forms adjectives interpreted as 'without ——' or '——less' (*sen'forma* 'formless', *sen'honta* 'shameless, unashamed', *sen'pova* 'unable'), and a direct opposite or contrary-producing prefix *mal-* (*mal'alta* 'low', *mal'bona* 'bad', *mal'varmo* 'cold'). Ido employs the same prefixes in essentially the same ways, as does Romanal (except that here the contrariety operator is *dis-* rather than *mal-*). Novial, Jespersen's nominee for an auxiliary language, builds in a two-way distinction: the simple negative *non-* (*nonposibli, nonreal*) contrasts with the contrary prefix *des-* (*desfaçil* 'difficult', *desquieti* 'anxious', *desagrabli*, etc.), producing such minimal pairs as *nonutili* 'not useful, useless' vs. *desutili* 'hurtful' (Jespersen 1928:123).

While it is true that international auxiliary languages are expressly designed to be morphologically analytic and semantically transparent, these treatments of negative prefixes shows that it is possible to carry a good thing too far. One result, as aptly noted by Guérard (1922:254), is that 'many of the MAL words in Esperanto strike us as childish or inaccurate'. Inaccurate, because *mal-*, especially in combination with verb stems, is too evocative of the Latin prefix for 'bad(ly)': *mal'fermi,* the Esperanto rendering of 'to open', sounds a lot more like it ought to mean 'to close badly, to misclose'. But why childish? Guérard's intuition, which I share, seems to stem from the linguistic perception that only children regularize and generalize morphology and lexical semantics in the manner and to the extent demanded by Esperantists. Adult speakers of natural languages are prepared to tolerate, and indeed to exploit, precisely the kind of duality of function these artificial languages rule out.

If Esperanto, Ido, et al. are "unnatural" in their general insistence on the correlation of one sound with one meaning and their specific decision to build in a consistent (and obligatory) differentiation of contradictory vs. contrary negation, can an artificial language arrange to be at worst "nonnatural" in its expression of affixal negation? The unfortunately yclept Volapük mirrors Russian (and other Slavic tongues) in assigning *ne-* for both contradictory and contrary negation. In Lancelot Hogben's Interglossa, a single prefix with phonologically conditioned allomorphs *no-/non-* yields such derived forms as *no-batho* 'shallow' (nondeep), *no-puro* 'dirty', *non-holo* 'incomplete', *non-iso* 'unequal', *no-volo* 'unwilling'. Hogben's statement of purpose (1943:94) is brief but eloquent: 'Admittedly, a negative is not necessarily an opposite; but it is the idiom of *Interglossa* to leave as much as possible to context'.

Finally, there is the grim example of Newspeak. Orwell, admittedly influenced in his linguistic views by Hogben, shows us the dark side of the linguistic utopians' lunatic vision. Newspeak is intentionally evocative of the international languages current in its day (Esperanto, Ido, Interglossa, Basic English, et al.): a hyperanalytic, transparent idiom favoring economy over nuance and contextualization. Yet of course it is used by its speakers precisely to conceal the very truth it seems to express so lucidly and economically.

Affixal negation in Newspeak consists of the prefix *un-*, which serves to create both contradictory and contrary negation, and (by a sort of backwards blocking) serves—like other derivational affixes in the language—to facilitate a purging of the inventory of stems. As stated in the Appendix ('The Principles of Newspeak') to *1984* (Orwell 1949:305), 'any word . . . could be negatived by adding the affix *un-*', so that 'given, for instance, the word *good,* there was no need for such a word as *bad,* since the required meaning was equally well—indeed, better—expressed by *ungood*'. But lexical reductivism can move in either direction: '*dark,* for example, could be replaced by *unlight,* or *light* by *undark,* according to preference'. Given these options, it is not so clear that affixal negation in Newspeak really does result in a reduction of the stem inventory; what does seem clear is that in a world where love is hate and peace is eternal war, the unraveling of e-pos vs. e-neg meanings is an exercise in futility.

Having come full circle, from the universe of Hitler, Jesus, and Eldridge Cleaver, in which everything is either good or evil, to the universe of Orwell's Ministry of Love, in which the bad is good and the good is ungood, it is time to move back one space and consider the circumstances under which the not ungood might be almost, but not quite, good.

5.1.3 A Not Insignificant Problem: The Logic of Double Negation

But grammar's force with sweet success confirm:
For grammar says (O this dear Stella weigh,)
For grammar says (to grammar who says nay)
That in one speech two negatives affirm.
(Sir Philip Sidney, "To His Mistress Who Has Said 'No, No' ")
from *Astrophel and Stella*

Nothing shows why
At this unique distance from isolation,
It becomes still more difficult to find
Words at once true and kind,
Or not untrue and not unkind. (Philip Larkin, "Talking in Bed")

Language has a logic of its own, and in this case its logic has something to recommend it. Whenever two negatives really refer to the same idea or word the result is invariably positive; this is true of all languages. . . . The two negatives, however, do not exactly cancel one another so that the result [of *not uncommon, not infrequent*] is identical with the simple *common, frequent;* the longer expression is always weaker: 'this is not unknown to me' or 'I am not ignorant of this' means 'I am to some extent aware of it', etc. The psychological reason for this is that the *détour* through the two mutually destructive negatives weakens the mental energy of the listener and implies . . . a hesitation which is absent from the blunt, outspoken *common* or *known.* In the same way *I don't deny that he was angry* is weaker than *I assert* . . .
(Jespersen 1924:332)

As the title and epigraphs indicate, I shall be concerned in this subsection with the variety of double negation in which each negative marker retains its semantic identity, thus (essentially, if not exactly) tending to annul rather than reinforce each other (see Horn 1978a: §3 for a typology of multiple negation). I am also excluding from consideration here those instances in which two negatives do not, in Jespersen's terms, 'refer to the same idea or word', but nevertheless result in conveying some affirmative, often the dual of the operator sandwiched by the two negations; cf. chapter 4 and Horn (1978a: §3.1) for discussion.[18]

There are two distinct environments in which negation is directly associated with an already-negated constituent: the prenominal position ({*a* / *the*} *not un-adj N*), where both negative elements function as TERM, SPECIAL, or CONSTITUENT negations (in the terminology of Aristotle, Jespersen, and Klima, respectively), and the predicate position, where at least one available reading for the outside negator is as PREDICATE DENIAL, NEXAL, or SENTENCE negation (***NP** is not un-**Adj***). While there are important syntactic and semantic differences between these two categories of double negation (cf. Langendoen and Bever 1973), neither environment involves the simple cancelation effect of double negation.

The principle that two negations with the same focus ought to cancel each other stems from the logical Law of Double Negation; as we saw in chapter 1, LDN (*Duplex negatio affirmat*) has a respectable lineage within both Western (Stoic) and Eastern (Buddhist and Nyāya) logical traditions. Prescriptive grammarians have long cited this law in defense of the position that, as the influential prescriptivist Bishop Lowth (1762: 126) put it, 'Two negatives in English destroy one another, or are equivalent to an affirmative'.[19] But this stance has always been directed primarily against the reinforcing or pleonastic negation which has always flourished in nonstandard dialects (and in certain constructions within the standard dialect; cf. Jespersen 1917; Labov 1972; Horn 1978a: §3.2), rather than against the observation of Jespersen and others that linguistic double negations don't quite cancel out. (Even Bishop Lowth doesn't commit himself to just which affirmative two negations are equivalent to.)

When *not un-X* fails to reduce to *X*, as in (23),

(23) a. A not unhappy person entered the room.
b. He's a not unhappy person.

What affirmative does this construction equate to, and when and why is it used? Jespersen's 'detour' through the double negatives has often been assumed to be available only in elevated registers; Marchand, in fact, ascribes the absence of positive *un-* formations (§5.1.1) to the unnatural sophistication required to make this detour: 'No un- words are formed from such adjs as in themselves denote the absence of something, as bad, evil, . . . Natural linguistic instinct would not make the sophisticated detour of negativing a negative to obtain a positive'[20] (Marchand 1960: 151–52). For Marchand, *not uncommon, not unhappy,* and (less convincingly) *not bad,* while possible collocations, fall outside what is permitted by the 'natural linguistic instinct'.

Seright (1966: 123) too takes logical double negation to be 'limited to the speech of the educated', but he echoes Jespersen in observing that doubly negated adjectives (*That is not {unlikely/unnatural/inconceivable/impossible}*) do not simply reduce to the corresponding affirmative adjective. But there are really two different issues involved in the nonredundancy of these forms. As recognized by Zimmer (1964: 22) (and as predicted by Aristotle), when the negative-prefixed adjective constitutes a contrary of its stem, a contradictory negation will not simply 'destroy' it: **contradictory (contrary (Adj$_i$))** $\neq$ **Adj$_i$**. This explains the lack of cancelation in the first of Seright's examples: if something is *not unlikely,* it may be likely, but it may also fall within Sapir's zone of indifference, that which is neither likely nor unlikely. Similarly, a man may be *not unhappy* because he is happy or because he is situated in the nonexcluded middle between the two contrarily opposed terms. But if something is *not inconceivable* or *not impossible,* what else can it be but *conceivable* or *possible?* Where is the zone of indifference, the unexcluded middle, in these cases? Why don't these doubly negated forms, amounting presumably to the contradictory of a contradictory, result in complete redundancy?

If the 'sophisticated detour' embodied in the *not un-X* collocation is, as Zimmer (1964) notes, 'logically quite justified' when *un-X* constitutes the contrary of *X,* it must be justified elsewhere rhetorically—if at all. When a Federal district judge in New Mexico, reopening a lawsuit filed by Pueblo Indians against a local newspaper charged with having 'demeaned and ridiculed' the tribe, finds that the tribe's complaint was 'not unjustified' (as reported in the *New York Times,* 7 October 1984), do we find his circumlocution equally not unjustified? Orwell, in his prescriptivist mode, questions whether any such justification is ever possible, singling out for his particular calumny the *not unjustifiable assumption:* 'Banal statements are given an appearance of profundity by means of the *not un-* formation. . . . It should be possible to laugh the *not un* formation out of existence. One can cure oneself of the *not un* formation by memorizing this sentence: *A not unblack dog was chasing a not unsmall rabbit across a not ungreen field*' (Orwell 1946: 357, 365).

Orwell's argument is, of course, somewhat tendentious in that it exploits a variety of constraints on negated prenominal adjectives of the type figuring in his laughing cure; it is to these constraints that I now turn. The first restriction to be noted (cf. Zimmer 1964: 91; Klima 1964: 310; and especially Langendoen and Bever 1973, henceforth L&B) is that no simple adjective, whether inherently (e-) positive or negative, may appear in the frame [**Det** not **Adj N**]:

(24) a. He sent me a not {unfriendly / *friendly} letter. (Zimmer 1964)
b. A not {unhappy / *happy / *sad} person entered the room. (Klima, 1964; L&B)

Note that this restriction applies only to prenominal or attributive adjectives:

(24′) a. She is a not {unattractive / *attractive} woman.
b. She is not an {unattractive / attractive} woman.

Second, the morphologically negative adjectives which do occur in this frame must be available independently in the lexicon; since we do not find Orwell's *unblack, unsmall, ungreen* as independently existing items, their negations will not be well formed either. Furthermore, the doubly negated attributive adjective must allow a contrary and gradable (scalar) interpretation; hence we do not get **Sheila wants to meet a not unmarried man* (as noted by Ross, personal communication, cited by L&B). Sheila may, however, locate a *not ineligible* bachelor, and secure him via *not illegitimate* ploys or manœuvers. But we will not reckon any lawful progeny resulting therefrom as **not illegitimate children.*

Another constraint on doubly negated attributive adjectives (henceforth DNAAs) is that the *not un-X* construction is semantically parasitic on, although not synonymous with, the corresponding simple adjective *X*. A DNAA of the form *a not unhappy person* is normally interpreted as 'a slightly-to-moderately happy person', in L&B's gloss. This parasitic relation is especially salient in cases where the negatively prefixed adjective stem is assigned a reading not available to it outside the DNAA form. Kruisinga (1931: §1246) notes that we may speak of *a not inconsiderable number,* with the sense of 'a rather considerable [i.e., large] number', although *inconsiderable* does not normally allow the sense 'small'. L&B illustrate the same point with the sentence *The president fled to Venezuela with a not unhealthy* ['rather sizable'] *share of the profits,* where *unhealthy* cannot ordinarily serve as an antonym of 'sizable'. (The possibility of such DNAAs is not unrestricted, however: we cannot refer to 'a somewhat-to-moderately smooth operator' as *a not unsmooth operator.*)

On the other hand, if an *un-* adjective bears a meaning unpredictable from its positive base, this meaning will be lost in the DNAA form, while it is of course retained in the corresponding straightforward sentence negation (predicate denial). Thus compare (25a, b):

(25) a. Kevin isn't an unworthy creep. (~[Kevin is an unworthy creep])
b. #Kevin is a not unworthy creep. (#Kevin is a rather worthy creep)

These facts—like the contrasts in (24)—demand an explanation, particularly within a lexicalist theory on which we cannot "peek inside" a lexical item to determine whether its morphology allows it to participate in a given syntactic construction. L & B suggest (p. 405–6) that an acceptable DNAA like *a not unhappy man* is misanalyzed perceptually as if it were of the same form as *a not very happy man,* with the negative prefix being 'treated as a negative intensifying adverb that modifies the adjective *happy*', just as in the case of the positive intensifying prefixes in (26).

(26) a. his not overdeveloped muscles
b. a not all-powerful deity
c. a not supersaturated solution
d. a not underdeveloped tribe

But of course *un-* is not really an intensifying adverb, so the DNAAs of (24), unlike the examples in (26), are ungrammatical; that they are nonetheless acceptable (to non-Orwellians) is accounted for, on L & B's modular theory, by an appeal to this misanalysis. The *not* of *not un-*, like that of *not very, not too, not overly,* and so forth, is not read as a contradictory negation applying to the following sequence of "adverb" + unmarked adjective, but essentially forms a constituent with the following "adverb". Just as a *[not very] happy* man is one who is 'slightly-to-moderately unhappy' (rather than simply one who is other than very happy), a *[not un]happy* man is one who is 'slightly-to-moderately happy'. (L & B offer a Gricean explanation for why just these particular readings should emerge in such cases; I shall return in §5.3 to the intricacies of the *not very* ***Adj*** construction.)

But, beyond any methodological qualms one might have in embracing the notion of "acceptable ungrammaticality" (cf. Otero 1972 in defense of this theoretical construct and Bolinger 1980 for a more skeptical view), the L & B account encounters certain empirical difficulties. Notice first that the *un-* of DNAA constructions does not pattern consistently with the intensifying adverbs purportedly parallel to it:

(27) That was a not {very / exactly / *un}, shall we say, intelligent thing to say.

Further, it should be noted that (contra L & B) there are cases where a DNAA is possible even when the corresponding simple positive adjective is not entirely (or at all) well formed. Bolinger (1980) finds one of L & B's unacceptable "ungrammatical" DNAAs—*a not inordinate amount of money*—perfectly acceptable; compare the parallel *not un-* examples in (28).

(28) a. %a not unsavory character (cf. ?*a savory character)
b. √a not unprecedented result (cf. ??a precedented result)

c. √a not unheard-of development (cf. ?*a heard-of development)
d. √a not unfounded rumor (cf. a {?*founded/√well-founded} rumor)

Confronted with the critiques of Aitchison and Bailey (1979) and Bolinger (1980), Langendoen retreats to a more conservative account of DNAAs, on which the ungrammatical-but-acceptable phrases of the 'misanalysis' analysis of L&B are reanalyzed as the 'not ungrammatical' constructions predicted by morphological rules and lexical features in Langendoen 1982.[21]

Whatever the correct analysis of DNAAs, it is undeniable that prefixally negated attributive adjectives figuring felicitously in these constructions tend to be interpreted as contrary opposites of their stems; note that we do not get, for example, **a not intransitive verb*, **a not unprefixable stem*, **a not nontoxic solution*. But, as we have seen, this restriction does not apply in absolute form to negated predicate adjectives, that is, predicate denials (sentence negations) of the form ***NP** is not un-X*. Jespersen himself happens to note (1917:70) that Kant's table of categories is 'not unobjectionable', and a majority of the *not un-* and *not iN-* forms (e.g., *not unuseful* [from 1657], *not inconsiderable [damage]*) cited in the OED under **not**, 10c, strike me as negations of logical contradictories. If someone is ascribed '*a certain air of dignity, not unmingled with insolence*', it is not implied that dignity might be neither mingled nor unmingled with insolence, nor is it clear how finding a suggestion *not unuseful* differs from finding it *useful*.

But, as I noted above with respect to *not impossible* and *not inconceivable*, there seems to be a rhetorical or pragmatic, if not a logical, difference between the doubly negated forms and their simple positive counterparts. Contra Orwell (and Tesnière [1959:233], who excoriates *nec non dixit* as *une des fausses élégances du latin*), there could well be a rationale for at least some logically superfluous instances of the *not un-* formation, although those *not unblack* dogs and *not unsmall* rabbits may remain beyond the pale. In his *Colloquia* ([1519] 1650:87), Erasmus—a rather different sort of stylist—indeed recommends the use of the double negative as 'graceful' (see below). Similarly, the best-selling American prescriptive authority Lindley Murray, who had once echoed Bishop Lowth's line that double negatives are (at best) equivalent to an affirmative and to be abjured therefore (1803:136–37, rule 26), rethinks the matter in his second ('improved') edition, adding the postcolonic *but* amendment to the last sentence of his grammar entry on this construction:

> Two negatives, in English, destroy one another, or are equivalent to an affirmative: as, '*Nor* did they *not* perceive him', that is,

> 'they did perceive him'. 'His language, though inelegant, is *not ungrammatical*'; that is, 'it is grammatical'. It is better to express an affirmation, by a regular affirmative, than by two separate negatives, as in the former sentence: but when one of the negatives is joined to another word, as in the latter sentence, the two negatives form a pleasing and delicate variety of expression.
>
> (Murray 1814:1:187)

More recently, Sharma (1970:60) lauds double negation as 'often extremely useful and by no means superfluous', as when *not impolite* is used 'to convey the fact that the person in question was not polite either'. But what renders a given construction simultaneously laughable, graceful, pseudoelegant, pleasing and delicate, and extremely useful, depending on the context and the evaluator? What, precisely, is the difference between *X* and *not un-X,* when *un-X* does not represent a contrary negation of *X?* What motivates a sophisticated detour, when the through road is (not im)passable?

The standard position is Jespersen's (in the epigraph to this subsection): a doubly negated adjective in either attributive or predicate position is somehow weaker, more hesitantly expressed than the corresponding simple positive. This weakness may be identifiable in the semantics (as in Zimmer's 'justifiable' negations-of-contraries—*not unhappy, not unintelligent, not impolite*) or it may be only pragmatic or rhetorical. For Seright (1966:124), the use of double negation 'results from a basic desire to leave one's self a loophole: certainly it is much easier to get out of a situation, to equivocate, if one has said "it is not unlikely" instead of "it is not likely" or "it is likely"'

And indeed, many of the citations we can observe of this construction do seem to involve the conscious or tacit goal of loophole procurement: the speaker describes something as *not un-X* in a context in which it would be unfair, unwise, or impolitic to describe that entity as *X*. Such attention to avoiding direct commitment is so notoriously characteristic of the political and governmental domains that we might suggest, after the fashion of Sapir's 'psychological excluded middle', a complementary 'bureaucratic unexcluded middle'. But this tendency is broader than such a label would imply, as I illustrate here with some attested examples:

(29) a. I do not pretend to be a "pure" bachelor. I was married for five years, and it was, to use a cowardly double negative, not an unhappy experience. (Philip Lopate, introducing his collection of autobiographical essays, *Bachelorhood*)

b. *New Yorker* cartoon (6 February 1971): couple standing before a mat inscribed *NOT UNWELCOME;* wife to husband: 'See what I mean? You're never sure just where you stand with them.'

c. *New York Times* editorial title (2 August 1985) on the exoneration of New York City Medical Examiner Elliot Gross of criminal wrongdoing: *Dr. Gross: Not 'Not Guilty'*.

As with Sharma's *not impolite*, the implication in each case is clear: a not unhappy marriage is not precisely a happy one, the guests are made to feel not quite welcome, and Dr. Gross is judged to be (morally if not legally) not entirely innocent.

There is often a sense here that the positive evaluation associated with the doubly negated adjective represents a concession, wrung out reluctantly from the source of the lefthanded *not un-* compliment. This understanding is especially salient in the frequently attested *not un-X . . . but . . .* construction. Here is Naomi Bliven, reviewing Kevin Starr's book on the California Progressive movement (*New Yorker*, 12 August 1985):

(29′) I wish Mr. Starr had been kinder to California's Progressives. He is certainly *not unjust* in pointing out their limitations, *but* it seems to me that we do not recognize all that the pre-WWI reform movements in this country accomplished. (italics mine)

It might also be noted that Bliven's syntax tends to suggest that one might have considered Mr. Starr to have been unjust (or to have thought that Bliven so considered him), a suggestion the *not un-* phrase explicitly dispels. Similarly, when a different reviewer writes that '*[Anais] Nin's life was not uninteresting*', we contextualize the remark either as a concession ('Contrary to what might have been expected from the foregoing') or as anticipating one ('but perhaps not as interesting as she suggests'). To say that Seright's analysis is *not unlike* Jespersen's, or that a sneeze is physiologically *not unlike* an orgasm, is to suggest that one might have expected more of a dissimilarity between the terms of comparison. In the same way, such standard rhetorical turns as *not without* (*The life of a millionaire is not without its compensations*) and *nothing if not* (*He's nothing if not earnest*), play off the actual or implicit suggestion that one might otherwise have thought the contrary (i.e., that such a life was without compensations, that he is not earnest). In each case a concessive *but* clause seems to be tacitly understood.

The rhetorical figure instantiated in all these examples is LITOTES, a form of understatement in which an affirmative is expressed by the negative of the contrary.[22] Like the *not un-* formation in particular, the superordinate category has been much maligned over the ages, most notably in the virulent attack by one (actually more than one) "Martinus Scriblerus"—in fact

Alexander Pope, Jonathan Swift, et al.—in "his" *Art of Sinking*. Just as hyperbole (according to Aristotle's *Rhetoric*) is a figure fit for Young Men of Quality, and ellipsis the favored figure for politicians, Scriblerus sees in litotes 'the peculiar Talent . . . of Ladies, Whisperers, and Backbiters' (Scriblerus [1727] 1952:115). Of course, it is just these classes of individuals (along with some others whose identity the Scriblerians delicately refrain from disclosing) who might have reason, through choice or necessity, to conceal their true feelings, avoid direct commitment, and leave themselves loopholes.

But rather than appealing, with Jespersen, to the metaphysical (and somewhat neo-Victorian) image of double negation sapping the listener's mental energy, I would assimilate the weakening effect to the Division of Pragmatic Labor (cf. Horn 1984b and §3.3.1 above): the use of a longer, marked expression in lieu of a shorter expression involving less effort on the part of the speaker signals that the speaker was not in a position to employ the simpler version felicitously. As we saw in the case of indirect speech acts and the use of lexical vs. phrasal causatives, there is a correlation between the stylistic naturalness of a given form, its relative brevity and simplicity, and its use in stereotypic situations; this reflects the operation of the **R** Principle. The corresponding periphrastic forms, stylistically less natural, longer, and more complex, are restricted, via **Q**-based implicature, to those situations outside the stereotype, for which the unmarked expression could not have been used appropriately. Litotes, and in particular "logical" double negation, where the two negatives do not cancel out functionally or rhetorically even when they do logically, illustrates the same equilibrium.

From this perspective, we would expect to find a variety of contexts in which the use of the double negative might be motivated by a desire to avoid the simple positive description, including contexts in which *not un-X* would come out not weaker but stronger than *X*, conveying not 'slightly-to-moderately *X*' (as in the DNAA construction) but 'extremely *X*'.[23] Under the suggestion *Commuta in Negationem* (Change it into a Negative) in his *Colloquia*, Erasmus (1650) urges the use of litotic formulæ ('Your letter was no small joy', 'Wine pleases me not a little') for 'modesty's sake'. The Latin equivalent of the *not un-* formation, rather than deprecated, is similarly recommended as a discreet means for conveying a strong positive, for example, *non ineloquens* for (the blatant and overdirect) *eloquentissimus*. Erasmus is clearly a different breed of rhetorician from Orwell: *Non ingratum, pro valde grato: non vulgariter, pro singulariter, recte & venuste dicimus*. (= For we say correctly and elegantly *not ungrateful* for 'very grateful', *not vulgarly* for 'singularly') (Erasmus 1650:89).

It has been argued more recently that double negation is characteristically employed to reinforce rather than qualify a description. Like Jespersen, Sigwart (1895:149) observes that even when the doubly negated adjective is not 'richer in meaning' than the simple positive alternative, 'it is not altogether to no purpose [!] that this circuitous route is taken', but the purpose he attributes to its employers is precisely the opposite of Jespersen's: 'Resisted attacks increase the psychological firmness of conviction; the affirmation which has fought through a negation seems to stand firmer and to be more certain'.

While Sigwart's line may strike us as at odds with, and much less plausible than, the standard Jespersen-Seright-Sharma story on double negation, especially when we bring the negated-contrary (*not unhappy, not unintelligent*) class into the picture, the two views are not as incompatible as they may appear. When a prefixal negative is itself negated so as to yield a positive, any one of a number of motivations may be at work, not all of which are subsumable under one simple rubric or metaphor, be it Sigwart's doubly-negated-affirmation-as-good-soldier, Jespersen's weakened mental energy, Marchand's sophisticated detour, or Seright's loophole. When a simple positive is abjured, and a double negation substituted, there is always (given the Division of Labor principle) a sufficient reason for so doing, but it is not always the same reason.

The same perspective informs my analysis of the stylistic disagreement over the rhetorical validity of litotes and double negation. While Pope and the Scriblerians, Orwell, Tesnière, and stylistic purists cut from the same cloth may scorn the use of this form of indirection, given its built-in prolixity and obscurantism, the same expression is praised by those who, like Erasmus (or Henry James), are concerned with the subtle niceties concerning what is communicated and how, and with preserving standards of politeness and delicacy. The clash between these two orientations thus parallels other instances in which rival pragmatic principles are weighted differently depending on one's communicative goals and constraints; cf. Tannen 1975, 1983; E. O. Keenan 1976; and Horn 1984b for discussion.

Before I close my inspection of logical double negation, it would be useful to try to determine what the logic of double negation is. The standard modern view is given by Geach, who acknowledges that a double negation, *not (not (**P**))*, 'looks like an added piece of meaning' and so might well be thought to involve a different sense from that of the basic element ***P***. But citing Frege (to whom I shall return), Geach ([1972] 1980:80) concludes that 'the right rejoinder is just to deny that the doubly negated predicate has got a different sense'. We are left with no explanation for the perceptible weakening or attenuation associated with double negation by Jespersen and others.

For Hintikka (1968:47), while a proposition or predicate may be logically equivalent to its double negation, the doubly negated form tends to convey one of three 'residual meanings':

(30) (i) to indicate hesitation or uncertainty
(ii) to signal diffidence
(iii) to express irony.

Hintikka argues that 'no one of these residual meanings helps to understand the others, which makes a paradigmatic analysis of the meaning of a double negative completely useless'. Nor is it obvious how to derive the difference(s) among (i), (ii), and (iii) as a 'mere difference in use', since the pragmatic rule needed for such a derivation would have to operate on some invariant meaning.

But Hintikka's pessimism about the possibility of reaching a generalization may be unwarranted. Hesitation, uncertainty, and diffidence are all subcases of the same general motivation for the use of the apparently brevity- (or **R**-) violating prolix form: the desire to avoid the direct expression of the positive attribute in order to leave oneself a loophole (à la Seright). Given the straightforward instances in which the double negative (*not un-X*) is semantically weaker than the corresponding positive (*X*), that is, where *un-X* represents a contrary of *X*, we can see how that feature of the use conditions on double negation might have become partially conventionalized to the form, extending to those instances in which *un-X* is a contradictory. The clash between this non-truth-conditional aspect of the 'residual meaning' or conventional use of an expression and the logical value of that expression leads to our intuition that the *not un-* form, when it represents a contradiction of a contradictory, often possesses an ironic quality. In any case, it is clear that the negations of both contradictory and contrary negations are generally weaker than their simple affirmative counterparts, as indicated by the standard scalar diagnostics (cf. Horn 1972 and chapter 4):

(31) a. She's happy, or at least not unhappy. (*not unhappy or at least happy)
b. It's possible he can do it, or at least not impossible. (*not impossible or at least possible)
c. Not only is it not <u>un</u>true, it's true! (*not only true but not <u>un</u>true)

Finally, the ironic feel of many instances of logical double negation derives from the appearance of diffidence or hesitancy when no diffidency or hesitancy is really felt (and when the speaker, exploiting the Gricean paradigm, assumes the addressee will recognize this). The third 'residual meaning' of double negation is thus parasitic on the first two (which are in fact one).[24]

I have been assuming that when *un-P* is a contradictory of ***P***, *not un-**P*** reduces logically (although not rhetorically) to ***P***, just as two contradictory negations applied to a single proposition cancel out via LDN. As we have seen, Geach and Hintikka share the assumption that doubly negated propositions are logically indiscriminable from the simple affirmative that results from the cancelation of the two negative operators via LDN. But this assumption is not unimpeachable.

As Lewis (1970:31–32) points out, Carnap's notion of intensional isomorphism yields the prediction that $\sim(\sim\alpha)$—for any expression α—cannot be identical to α, since the former, with its more complex logical form, must have a correspondingly more complex meaning. The two logical forms may have the same intension, and presumably (given LDN) they do. But meaning is a more fine-grained notion that intension. This point is classically demonstrated with belief contexts; it is precisely because such intensionally identical but nonsynonymous expressions as *bachelor* and *unmarried man*, or *3* and *the square root of 9*, behave differently in belief contexts that we need a subtler tool than intension.

While propositional identity may be defined in terms of intensions (functions from possible worlds to extensions), synonymy must be defined in terms of the more discriminating notion of meanings. Thus, for Lewis and for Cresswell (1973:44–47), the double negation of α expresses the same proposition as α without being synonymous with it.[25] Cresswell cites the behavior of these two expressions within opaque contexts to buttress this conclusion: just as the truth of (32a) is compatible with the falsity of (32b), even though a bachelor is necessarily an unmarried man, so too the intensional identity between *possible* and *not impossible* fails to license the inference of the sentences (32′b) on the basis of the corresponding examples in (32′a).

(32) a. John maintains that bachelors are not unmarried men.
  b. John maintains that bachelors are not bachelors.

(32′) a. Sue claims that it is not {impossible/inconceivable/untrue} that Bill is lying.
  b. Sue claims that it is {possible/conceivable/true} that Bill is lying.

As we have seen, Fregean semantics allows for but one negative operator, the contradictory-forming propositional operator. Not unexpectedly, Frege (1919:130) proclaims the logical superfluity of double negation: 'Wrapping up a thought in double negation does not alter its truth value'. Within this metaphor, ($\sim\sim$**p**) and $\sim$($\sim$**p**) are simply different ways of wrapping up the thought or proposition **p**. But, we may observe, there are times and places where a man cannot go naked. Wrapping him in clothes serves an ecological and/or social function, while leaving the inner man unaltered.

So too with naked thoughts—especially unappealing ones, which profit most from the protective layers of double negation.

If implicatures are read off logical form, we can then account for why there may be an implicature carried by double negation which is absent from the logically equivalent unnegated proposition. In any case, as we are about to observe, the conventionalized use of negation as a device for attenuating an assertion, or for qualifying the speaker's commitment to the truth of the expressed proposition, is a widespread and systematic phenomenon, one which demands—and will presently receive—an investigation of its own.

5.2 Neg-raising and Contrariety

In §5.1.1, I observed that Bosanquet's 'invest[ment] of the contrary with the character of the contradictory' is systematically exemplified by lexicalized affixal negation of the type of *unhappy, dislike*. I am now ready to turn to a second instance of contraries-in-contradictory clothing, that which is instantiated in 'our common phrase "I don't think that"—which is really equivalent to "I think that—not"' (Bosanquet [1888] 1911:319).

The NEG-RAISING phenomenon (NRP)—the availability of a lower-clause reading or understanding for a higher-clause negation—has a rich history, both outside and within the generative paradigm.[26] Writing in the eleventh century, Saint Anselm observed in his argument against Paulus, a Roman jurist six centuries his senior, that *non . . . omnis qui facit quod non debet peccat, si proprie consideretur*—not everyone who does what he *non debet* (not-should) sins, if the matter is considered strictly (i.e., with the contradictory reading of negation as suggested by the surface structure); the problem is that we tend to use the form *non debere peccare* to convey *debere non peccare,* rather than its literal contradictory meaning (it is not a duty to sin). A man who does what is not his duty does not necessarily sin thereby, but (because of the interference of the NRP) it is hard to stipulate something like (33a)—the proposition that a man need not marry—without seeming to commit oneself to the stronger (33b), an injunction to celibacy (Henry 1967: 193ff.; cf. C. J. F. Williams 1964; Horn 1978b:200).

(33) a. non debet ducere uxorem — lit., 'NEG [he should take a wife]'
b. debet non ducere uxorem — lit., 'he should NEG [take a wife]'

From Anselm to Quine, who notes (1960: 145–46) as 'an incidental idiomatic complication' for the logic of belief the 'familiar quirk of English' whereby (34a) is taken to convey (34b) rather than, or as well as, (34c):

(34) a. **x** does not believe that **p**.
b. **x** believes that not-**p**.
c. It is not the case that **x** does believe that **p**.

philosophers and logicians have typically sought to bury, rather than explain (much less praise) the phenomenon in question.[27] Hintikka (1962:15) notes that 'the phrase "a does not believe that p" has a peculiarity . . . in that it is often used as if it were equivalent to "a believes that -p"', while Deutscher (1965:55) regrets that '"I do not believe that p" can be unfortunately ambiguous between *disbelief* [$\mathbf{B_a}$-**p**, in his and Hintikka's representation] and *not belief* [-$\mathbf{B_a}$**p**]' (emphases mine).

Barnes (1969:304–5) remarks less disparagingly on the same phenomenon: 'The everyday negative "doesn't believe that" is ambiguous. We can say of both the atheist and the agnostic that he doesn't believe that God exists, but when we say it of the former we mean that it is the case that he believes that God does not exist, and when we say it of the latter we mean that it is not the case that he believes that God does exist.' On the atheist reading, Barnes notes, the proposition that **a** doesn't believe that **p** is incompatible with the proposition that **a** doesn't believe that not-**p**; on the agnostic sense the two propositions are compatible. While he does not condemn this 'everyday' reading as such, Barnes evidently assumes, with his philosopher and linguist colleagues, that the phenomenon it reflects is restricted to belief contexts.

I have argued elsewhere that far from representing an incidental complication, quirk, or peculiarity afflicting the processing of a verb or two in English (or Latin), the tendency pointed to by Anselm, Quine, and others in fact comprises a fundamental grammatical, semantic, and pragmatic process manifested across distinct, but systematically related, classes of predicates in genetically and typologically diverse families of languages.

The first systematic treatment of the NRP is provided by Tobler, in his description of the 'logically unwarranted position' (*logisch ungerechtfertigte Stellung*) of negation in French *Il ne faut pas que tu meures* (Tobler 1882b:205, cited in §4.5 above). Tobler cites a variety of French and German verbs reflecting this same pattern, including German *wollen* 'want' in (35a) and, recapitulating Anselm, *sollen* 'should, ought to' in (35b):

(35) a. Ich will nicht, daß man mir dergleichen hinterbringe.	'I do not want to be told about such things'
b. Du sollst nicht stehlen.	'Thou shalt not steal' (lit., 'NEG [thou shalt steal]')

But it is Tobler's acolyte Kalepky (1891) who first offers a semantic catalogue of the verb classes permitting the NR understanding (the association of a finite matrix negation with a dependent infinitive).

In his thesis on negation in Old Provençal, Kalepky comments on the *Voranstellung der Negation*, noting that the apparent main clause (modal) negation of (36), can be read—like its German equivalent (cf. (35b))—either as a contradictory or a contrary of its affirmative counterpart.

(36) Ja non degra dir ver esquern. 'He should [not scorn]' (lit. 'NEG [he should scorn]')

Along the same lines, the distinction between French *je n'espère pas* (I do not hope) and *j'espère que non* (I hope not) is neutralized in the German *Ich hoffe nicht;* it cannot be a coincidence, Kalepky points out (1891:22), that it is just these verbs that allow the collapsing (*Zusammenfall*) of the distinction between contrary and contradictory readings of negation.[28]

But when does this *Zusammenfall* take place? To Old Provençal *dever* in (36)—cf. also Latin *debēre,* French *devoir*—and German *sollen* in (35b), Kalepky (1891:23–24) adds the predicate classes exemplified by the following German verbs (my glosses give the closest English NR counterparts):

(37)
i. '*meinen* (*denken, glauben*)' [be of the opinion, think, believe]
ii. '*wollen*' (cf. Lat. *nolle* < *non velle*) [want]
iii. '*sich ziemen*', '*sich schicken*' [be proper, fit]
iv. '*den Schein erwecken*' [seem, appear]
v. '*sagen* (*im futurem*)' [would say]

In several of these cases, Kalepky notes, the distinction between the literal sense (the contradictory) and the usurper (contrary) is not too great, at least pragmatically (*für die Praxis*). I shall return to the relevance of this point below.

Our regular guide through the negative jungles is not unaware of 'the strong tendency in many languages to attract to the main verb a negative which should logically belong to the dependent nexus' (Jespersen 1917: 53). For Jespersen, (38a) is ambiguous and has a sense in which it is synonymous with (38b).

(38) a. I don't think he has come.
b. I think he has not come.

This 'attracted', anticipated, or adherescent negation, as it is variously known in traditional grammars of Ancient Greek, Latin, French, and English, is treated by Jespersen (and others) as a special case of contrary negation, an approach which I have argued (Horn 1978b: 210) is ultimately insufficient (cf. also §5.3.1 below).

While (38a) does seem to allow two readings, other instances of the NRP, particularly with infinitive or subjunctive dependent clauses, virtually force the "illogical" contrary reading. Kalepky (1891: 22) borrows a metaphor from real estate law to represent the contrary sense as an interloper which has taken over the rightful (but underutilized) property of the contradictory, this tendency being especially clear in the case of *falloir* (cf. Spitzer 1927: 69 and my discussion in §4.5).

In some cases, the "logical", embedded placement of the negation is syntactically unavailable, which may prompt the illicit seizure declaimed by Kalepky. Martinon (1927: 536) observes that (39a) has taken over (*accaparé*) the sense of (39b), *qui ne se dit pas.*

(39) a. Je ne veux pas que vous sortiez. 'I don't want you to leave'
b. ??Je veux que vous ne sortiez pas. 'I want you not to leave'

This same process has occurred in the cases of *vouloir* + infinitive (*Je ne veux pas sortir* for *?Je veux ne pas sortir*), *falloir,* and *devoir:* with these verbs, 'le transfert de la négation est devenu indispensable' (Martinon 1927: 537).

The same functional explanation has been independently appealed to for English; Epstein (1976: 91) maintains that 'the use of a sentence like [(40a)] may be motivated by the unavailability of [(40b)] as an acceptable alternative sentence'.

(40) a. John doesn't want to kiss Mary.
b. ?John wants not to kiss Mary.

But, as noted in Horn 1978b: 180–81, this line of argument, while it has some explanatory power, fails to generalize to similar cases (including the English gloss of the unlike-subject case (39a)) in which embedded nonfinite negation is acceptable and the illogical understanding of higher negation is still available (if not preferred); cf. (40′a, b):

(40′) a. John doesn't want me to kiss Mary.
b. John wants me not to kiss Mary.

Another stab at motivating the NRP begins with Jespersen's generalization which I have touched on earlier and labeled Neg First: negation tends to be signaled as early as possible within the sentence, for ease in processing. It is this principle which is responsible for a number of the diachronic shifts in the morphological realization of negation which have come to be known as Jespersen's cycle (cf. §7.1). In the present context, negative clarity leads to a predisposition to force the negative ahead to its illogical position within the main clause, even when it "belongs" in the subordinate.

To this tendency, Spitzer (1927:70–72) adds his own proto-radical-pragmatic analysis, in which the AFFECTIVITY of negative sentences plays a key role. For Spitzer, the lower-clause reading of higher-clause negation is not (pace Tobler) perverted and senseless (*verkehrt und sinnlos*), or even *unlogisch*—it is, rather, *affektvoll*. Spitzer's negative affect crucially involves the imputation to humanity of a fuzzy sort of 'negative predisposition' which leads us to prefer denial (*ich will nicht*) to affirmation (*ich will*).

But Spitzer's passionate conclusion notwithstanding—'Ist das alles Unlogik? Oder Logik des Gefühls? Le cœur aussi a ses raisons que la raison ne connaît pas' (Spitzer 1927:73)—we are unswayed by his emotional appeal.[29] If the NRP really existed as a way of emphasizing, rather than of attenuating, negation, the higher-neg versions of pairs like (38) would be felt to be stronger than their embedded-neg counterparts; *je ne {veux/pense} pas* would similarly serve to reinforce, rather than weaken, the negation of *je {veux/pense} que non*. Yet, as has been frequently observed, the opposite is in fact the case. Further, if we were really inclined to "accentuate the negative, eliminate the positive", as Spitzer in effect maintains, we would expect positive sentences to be more marked morphologically and functionally and to be harder to process, than their negative counterparts. If there is any agreement from those who have toiled in the negative fields over the last 2,500 years, it is that the reverse is true (whatever we end up choosing to make of this fact).

Those traditional grammarians who took on the NRP, whatever their theoretical predilection (if any), tended to describe the relation between members of pairs like (38)–(40′) in terms of a dynamic metaphor: attraction of a negative (Jespersen 1917), *transfert de négation* (Nyrop 1930), *déplacement de négation* (Grevisse 1969), and even 'the shifting of *not*' (Poutsma 1928:105). With the coming of age of transformational grammar, and in particular the appearance of Fillmore 1963, this metaphor is fleshed out.[30]

The transformation variously dubbed 'Transposition of NOT' (Fillmore 1963), *not*-hopping, Negative Transportation (e.g., in R. Lakoff 1969), or (e.g., in Kiparsky and Kiparsky 1971 and here) Neg-Raising (NR), is generally taken to be a lexically governed, cyclic, structure-preserving rule which extracts a negative element from a lower clause (presumably from a pre-neg-placed position within that clause) and raises it one clause up (into a similar pre-neg-placed position) over a predicate marked to allow the rule's application. The operation can apply iteratively, provided all intervening verbs are neg-raising predicates, as in Fillmore's example (41).

(41) a. I don't believe that he wants me to think that he did it.
b. $_S$[______I believe $_S$[______he wants $_S$[______I think $_S$[**not** he did it]]]]

NR is without question—in the words of Lerner and Sternefeld (1984)—'eine Transformation aus dem goldenen Zeitalter', but must it necessarily be scorned therefore?

Evidence for coderiving (38a, b) has focused on the interaction of the NRP with negative polarity: strict polarity items like *until midnight, in weeks,* or *for some time* in English, normally requiring a tautoclausal negative trigger (or durative predicate), are acceptable in the frame of (38a) although this requirement is apparently violated at surface structure. Thus compare the pattern in (42) with that in (42′):

(42) a. I don't {think/suppose/imagine} (that) he has come here in weeks.
b. I {think/suppose/imagine} (that) he has not come here in weeks.

(42′) a. *I don't {regret/claim} (that) he has come here in weeks.
b. I {regret/claim} (that) he has not come here in weeks.

If (42a) originates in a deep structure directly underlying (42b), and if the trigger requirements on *in weeks* can be satisfied at that level, we can predict its acceptability. *Regret* and *claim* are ruled out in the same structure because these verbs do not permit NR; there is no reading on which (42′a, b) are coderived.

This argument, which seems to have been first suggested by Kajita and first published in R. Lakoff 1969, is analyzed in some detail in Horn 1978b: 136–50, and I will not review its strengths and weaknesses here. Nor will I deal with the related grammatical arguments for and against a rule of NR based on phenomena involving rule interaction, mood, complementizer type, opposite-polarity tags, object case marking, sentence pronominalization, anaphoric destressing, sluicing, subject-aux inversion, queclaratives, or the syntactic reflex of De Morgan's Law (Horn 1978b: 151–77). In the end, I see no reason to repeal my conclusion in the 1978 paper: the strongest positive arguments on behalf of a syntactic rule of NR prove to be untenable, indecisive, or dependent on additional (often tacit) assumptions which are at best theoretically and/or empirically dubious.

Of course, within the extended standard theory (or any of its hyperextensions), the very question hardly gets to arise. Whether it would be formulated as in the more traditional (43),

(43) $[_S \mathbf{X}\ \mathbf{V}\ \mathbf{Y}\ [_{\bar{S}} \mathbf{COMP}\ [_S \text{not } \mathbf{W}]_S]_{\bar{S}}]_S$
$\quad\quad 1\ \ 2\ \ 3\ \quad 4 \quad\ 5\ \ 6$
$\quad\quad \Rightarrow 5 + 1, 2, 3, 4, \emptyset$ (or **t**?), 6

or simply treated as a mutant subcase of Move α, NR would be in arrant violation of at least the tensed S Condition (as noted in Wasow 1972), or any of its parametrized offspring, as well as of the æsthetic sensibilities of the theory's practitioners.

In addition, even most of today's non–fully extended standard theoreticians would be suspicious of, if not downright horrified at, a rule playing such havoc with derivational morphology. Any coherent transformational program for the NRP must countenance a syntax in which Neg-Raising feeds those incorporation rules which result in the formation of the lexical items in (44) (cf. Horn 1978b: 170–71):

(44)

nobody	-er than
neither . . . nor	few
{neither/none} of	scarcely (any)
only	doubt

Thus, the sentences of (44′) are understood with the (italicized) incorporated negative taken in each case as semantically within the scope of the (boldface) neg-raising trigger to its right:

(44′) *Nobody* **supposes** that nuclear war is winnable.
Neither Mutt *nor* Jeff **think(s)** that Chris has been here in weeks.
{*Neither*/*None*} *of* them is (are) **likely** to marry you.
Only Kim **intends** to seek reelection.
I spent *more than* I **should** have.
{*Few*/*Scarcely any*} of my friends **believe** you'd lift a finger for me.
I *doubt* that he {**wants**/**plans**} to resign just yet.

Thus, the first example is taken as suggesting that everybody supposes nuclear war is not winnable, the second that both Mutt and Jeff think that Chris hasn't been here in weeks, and so on.

Indeed, the existence of this pattern, if not its theoretical implications, was recognized by the discoverer of the NRP himself. For Saint Anselm, a sentence of the form *Solus homo debet facere* . . . (Only (a) man ought to do . . .) tends to be read as asserting that *Quisquis non est homo debet non facere* . . . (Any nonman ought not to do . . .): 'The negative force implicit in the applicative "only" can extend, so to speak, beyond the main verb of the clause in which it appears to infinitives governed by the main verb' (C. J. F. Williams 1964: 137). The assertion that 'only a man ought to make the offering for sin' will be true for Anselm 'only if anyone who is not a man has an obligation not to make the offering for sin' (C. J. F. Williams 1968:611).

For those who subscribe to the currently received view that rules of word formation (as opposed to rules of inflectional morphology, on some accounts) are not to be interspersed with rules of syntax, the thesis that a syntactic rule of NR applies to the output of the lexical formative processes involved in the creation of Anselm's *solus* and of the operators in (44) amounts to a reductio of the syntactic program for the NRP.

What alternatives can we find? As it happens, there are ample grounds to doubt both the feasibility and the desirability of a grammatical treatment of the NRP. To begin with, there is the question of semantics. Fillmore (1963), like the traditional grammarians before him, offered no syntactic evidence for his 'Transposition of NOT', relying solely on the alleged paraphrase relation holding between the lower-neg (45b) and the relevant (transported) reading of its putative congener (45a).

(45) a. I do not think that he will come.
b. I think that he will not come.
c. It is not the case that I think that he will come.

As has been frequently remarked, a higher-neg sentence like (41a) typically allows a ('literal') interpretation (à la (45c)) on which the negative has not been transported, the not-belief interpretation cited by Deutscher (see above); while less salient than the transported reading, it may be brought out by contrastive stress on *I* or *not*.

But does (45a), on its relevant understanding, really constitute a paraphase of (45b)? Can we assume, with Bierwisch (1971:425–56), that these two statements are fully synonymous? From Poutsma's remark (1928: 105) that 'the shifting of *not* often has the effect of softening down the negativing of a sentence' to Dwight Bolinger's observation (cited as a personal communication in R. Lakoff 1969 and G. Lakoff 1970) that the negative force in (45a) is perceptibly weaker than that in its nontransported partner (45b), there have been effective challenges leveled against this purported paraphrase relation. As noted in Horn 1978b:131–32, this same difference in negative force can be observed in other languages manifesting the NRP, including Swahili and Turkish: while a negation syntactically outside the scope of a verb denoting mental activity (think, believe) or desire (want) can be—and standardly is—understood as inside its semantic scope, the result is a somewhat weakened or attenuated negation.

In fact, this softening process appears to be an epiphenomenon of all rules and processes affecting the placement and surface realization of negation. In a paradigm like that in (46):

(46) I think she's sad.
I think she's unhappy.
I think she's not happy.
I think she isn't happy.
I do not think she's happy.
I don't think she's happy.

each version strikes us as a bit weaker or milder in e-negativity or negative force, or in the degree to which the speaker is committed to the embedded subject's unhappiness, than the one immediately above it.

Such paradigms reflect the interaction of two general principles: (*a*) negative force weakens with the distance of the negative element from the constituent with which it is logically associated, and (*b*) negative force strengthens with the degree of incorporation (or morphological absorption) of the negative element. The former principle is at work within the NRP, while in principle the latter is exemplified in affixal negation (cf. §5.1) and in Negative Incorporation (cf. Klima 1964; Sheintuch and Wise 1976). Thus, while the neutral (47a) is weakened in its transported counterpart (47b), it is strengthened in the neg-attracted (47c) (cf. Sheintuch and Wise 1976:548 for discussion).[31]

(47) a. I think I didn't see anyone in the room.
b. I don't think I saw anyone in the room.
c. I think I saw no one in the room.

The Poutsma-Bolinger Uncertainty Principle may thus be viewed as a special case of a more general interaction involving negation and morphosyntactic space, which may in turn be seen as a corollary of the law of gravitation, wherein the force between two objects is seen to vary inversely with the square of the distance between them. Coming back to earth, I conclude that a syntactic account of the NRP can be salvaged only if we are willing to give up the thesis that transformations preserve meaning (in which case the perceived strength difference between (45a) and (45b) may be read off the posttransportation surface structure), or if (with G. Lakoff 1970) we are willing to make the application of the rule of NR contingent on the speaker's uncertainty.

If the Uncertainty Principle casts doubt on the putative synonymy between (45b) and the transported reading of (45a), the putative ambiguity of the latter is no less controversial. In a brief discussion of the NRP in a footnote to her paper on belief sentences, Partee (1970:335–36) argues that—given a pair of sentences like (48a, b):

(48) a. A doesn't believe that **S**.
b. A believes that not-**S**.

—the claim that (48a) is ambiguous between a higher- or wide-scope-neg reading mapping directly onto its surface form and a lower-neg reading derived from the structure directly underlying (48b) is 'semantically indistinguishable from the claim that [(48b)] entails [(48a)] and not conversely'.

If A's belief world is consistent, (48a) will indeed be true in any context in which (48b) is true, but not vice versa; this much is clear. The two understandings of (48a), or of any other putative instance of NR, thus constitute privative opposites in the sense of Trubetzkoy and of Zwicky and Sadock: 'Understandings $\mathbf{U_1}$ and $\mathbf{U_2}$ are PRIVATIVE OPPOSITES with respect to some semantic feature **F** if $\mathbf{U_1}$ can be represented as identical to $\mathbf{U_2}$ except that $\mathbf{U_1}$ includes some specification for **F** that is lacking in $\mathbf{U_2}$' (Zwicky and Sadock 1975:6).

Zwicky and Sadock point out that, while the logic of privative opposition 'makes it difficult to distinguish ambiguity from lack of specificity [aka generality or vagueness]'—in particular, the classic identity-of-sense tests for diagnosing ambiguity are useless in these cases, since 'the existence of the more general understanding guarantees that we will get all possible understandings'—there are a substantial number of privative oppositions which nevertheless represent (at least arguably) true ambiguities.

Thus Partee's a priori assumption that the privative relationship between the two understandings of (48a) precludes or obviates an NR-type analysis positing a semantic ambiguity for such sentences, like Kempson's broader rejection of all privative ambiguity, is only as valid as the latter's demand (Kempson 1980:16) that 'no sentence be assigned two distinct semantic representations if one interpretation is logically dependent on the other'. I have argued elsewhere (Horn 1984a, c) that a wide variety of privative ambiguities must in fact be countenanced within a maximally simple semantics for natural language, both on the lexical level (where a substantial number of AUTOHYPONYMOUS lexical items allow two readings, one of which extensionally includes, and intensionally is included by, the other, such as *dog, man, Yankee,* German *Frau*) and on the constructional level as well.

A rich lode of privative oppositions can be mined in the exploration of scope ambiguity. In examples like *At least two languages are spoken by everyone in this room* (Chomsky 1965:224) and *Jones believes that someone is a spy* (Quine 1960), the specific or transparent reading in each case unilaterally entails the corresponding nonspecific or opaque reading, yet semantic ambiguity is not ruled out a priori on that account. Closer to home, and to the NRP, is the ambiguity of apparent universal negation discussed in §4.3 above and §7.3 below: a sentence like (49a) is generally acknowledged to be ambiguous between a 'NEG-V' reading equivalent to (49b) and its 'NEG-Q' reading equivalent to (49c).

(49) a. All the men didn't leave.
b. None of the men left. (All of the men stayed)
c. It is not the case that all the men left. (Not all the men left)

Just as in the putative examples of NR—(34a), (38a), (45a), (48a)—the ambiguity of (49a) hinges on the scope assigned to negation; in both cases, the inner-neg reading (= (45b)/(49b)) unilaterally entails the outer-neg reading (= (45c)/(49c)).[32]

While it may be difficult to determine whether a given pair of privatively opposed expressions constitutes a true lexical or semantic ambiguity, this is finally not an epistemological but an ontological question, one which cannot be answered a priori in the absence of a comprehensive theory of the phenomenon in question. In any case, returning to Partee's original hypothesis, it will be noted that the unidirectional entailment relation obtaining (modulo doxastic inconsistency) between the belief sentences of (48) is mirrored in the sentences in (50).

(50) a. A {didn't claim/isn't certain} that **S**.
b. A {claimed/is certain} that not-**S**.

Assuming again that Mr. A is consistent in his statements and beliefs, the truth of (50b) requires the truth of (50a), but not vice versa. And yet no question of ambiguity, or of an NR-type coderivation, arises for these sentences. An explanation must still be sought for the fact that while (48a) can be, and standardly is, used to convey (48b), (50a) simply cannot be used to convey (50b).[33] But just what sort of explanation might that be?

In her attempt to explain away the NRP, Partee observes that the scope of this phenomenon is largely restricted to, or at least favored by, first-person contexts. In his own rejection of the transformational solution to the NRP, Jackendoff (1971) concedes an ambiguity between 'non-committal' and 'committal' (or 'active disbelief') readings for (51a), for which he proposes a semantic-interpretive account (later fleshed out in Pollack 1974),

(51) a. I don't think that Bill went.
b. John doesn't think that Bill went.

but he finds the parallel third-person example (51b) unambiguous and 'vaguely non-committal': 'The synonymy between *John thinks that Bill didn't go* and one reading of [(51b)] is inferential in character and has nothing to do with the syntactic component—it may even have nothing to do with the semantic component' (Jackendoff 1971:291).

Later developments in the analysis of the NRP have proved Jackendoff's once-radical assertion overly conservative. The dichotomy between first-person and third-person subjects drawn by Partee and Jackendoff does not seem particularly compelling, especially when we move away from the *believe, think* class of neg-raisers: Are the examples of (44′) any less ambiguous with their third-person subjects than the corresponding first-person examples would be? Are the NEG + *want* examples of (40a) and (40′a) any more 'vaguely non-committal' than their first-person counterparts (*I don't want (you) to kiss Mary*)? If so, isn't it likely that the explanation for this subtle distinction might itself 'have nothing to do with the semantic component' per se? Not too surprisingly, Epstein (1976) judges even first-person examples like (51a) 'unambiguous both syntactically and semantically', offering a promissory note that 'the apparent ambiguity is best explained in terms of general principles of semantics and pragmatics'.

The first detailed exposition of such principles within a pragmatic account of the NRP is given by Bartsch (1973); her analysis is later independently echoed by Halpern (1976). For Bartsch, as for Partee, (52a) is not ambiguous between the understandings in (52b) and (52c), as it would be on an NR analysis:

(52) a. Peter glaubt nicht, daß Hans kommt. — 'Peter doesn't believe that Hans is coming'
b. Peter glaubt, daß Hans nicht kommt. — 'Peter believes that Hans is coming'
c. Es ist nicht so, daß Peter glaubt, daß Hans kommt. — 'It is not the case that Peter believes that Hans is coming'

rather, it has only one meaning, identical to (52c) which (again, we must assume, given consistent belief systems) is unilaterally entailed by the so-called neg-raising reading.

Under certain pragmatic conditions of applicability (*pragmatische Verwendungsbedingungen*), the weaker proposition (52c) can be used to convey the stronger (52b). Sentences (52a) and (52b) will then share their truth value and end up equivalent; they need not—should not, Bartsch argues—be assigned the same semantic representation, but will express the same information relative to a given context of utterance (*Sprechsituation*). But what are the relevant pragmatic conditions?

Mirroring the entailment (*semantische Implikation*) from (53b) to (53a):

(53) a. **a** glaubt nicht, daß **p**. — '**a** doesn't believe that **p**'
b. **a** glaubt, daß nicht **p**. — '**a** believes that not-**p**'

there is a *pragmatische Implikation* from (53a) to (53b), given the assumption that the subject will be taken to have given some thought to the truth of

p and to have come to some conclusion about it; Bartsch formulates this assumption as the disjunction **F(a, p)** v **F(a, ~p)**. In the unmarked discourse situation, (53a) will normally be used when the speaker assumes that (53b) is in fact the case, rather than that **a** hasn't thought about **p** at all or is neutral as to whether **p** or **~p** is in fact the case.

This schema of pragmatic inference can be spelled out as in (54):

(54)		
	i. **F(a, p)** v **F(a, ~p)**	[context-dependent 'assumed disjunction']
	ii. **~F(a, p)**	[utterance of, e.g., (53a) by speaker]
	iii. ∴ **F(a, ~p)**	[i.e., (53b); from (i), (ii) by modus tollendo ponens]

It is striking how closely Bartsch's approach to the NRP mirrors the neo-Hegelian line on contextually derived contrary negation (see the citations from Sigwart and Bosanquet earlier in this chapter). The cornerstone, and stumbling block, of Bartsch's inference chain is the middle-excluding assumed disjunction, the initial premise (54i); the mystery is to determine exactly when the assumption is valid. Bartsch stipulates that the disjunction, and hence the NRP, will be triggered whenever **F** is instantiated by a propositional-attitude verb, and the majority of NR predicates (*believe, think, suppose, want*) do indeed fall within this class. But there are potential counterexamples in both directions to this claim: (1) propositional attitude predicates like *hope, realize, know, regret,* and *be certain,* whose negations do not permit a lower-clause understanding or trigger strict polarity items; and (2) neg-raising triggers like *probable, likely, should, ought to,* and *advise,* as well as certain quantifiers and quantificational adverbs and possibly (on some analyses, although not mine) *true* and a variety of tense-aspect markers, which are not obviously propositional-attitude verbs in Bartsch's sense.

How can Bartsch's definition be narrowed and widened to take in just the right class(es) of predicates? As Epstein notes (1976: 158) in discussing a related pragmatic account of the NRP, the semantics of the upstairs predicate should not, strictly speaking, enter into the determination of when an assumed disjunction or bivalent situation obtains: 'If **P(S)** v **P(~S)** is among the shared beliefs of participants in a discourse, then no matter what verb **P** is formed from, a communication of **~P(S)** automatically communicates **P(~S)**.' The burden of proof on pragmatic theories like those of Bartsch (1973), Halpern (1976), and L. Carlson (1983), all of which invoke a version of the assumed disjunction, lies in demonstrating why it is that when the context permits eliminating (or assimilating) the middle, so that '**~P(S)** automatically communicates **P(~S)**', it nevertheless does not follow that ~**P(S)** automatically shares the properties of **P(~S)** with respect to the meaning imparted or to the standard syntactic correlates of the NRP.

As Epstein points out (1976: 158), if everyone in the discourse context 'has extremely strong feelings about the prospect of Wallace's nomination, so that everyone either fears that Wallace will be nominated or fears that he will not be nominated', an utterance of *John doesn't fear that Wallace will be nominated* should (via the schema outlined in (54)) 'yield' as its conveyed meaning *John fears that Wallace won't be nominated,* and yet it does not: 'This sort of "derived negative transportation" seldom if ever occurs in speech'.

When a non-NR trigger has its middle excluded by the context, it still fails to be transparent to negation in the manner of actual NR predicates, as signaled by the distribution of strict polarity items:

(55) a. I asked him whether or not Mary's written, and he told me in no uncertain words:
{ he said she hasn't written in weeks. / *he didn't say she's written in weeks. }
b. I have pretty strong feelings about this trip he's planning—
{ I hope he isn't going there until July. / *I don't hope he goes there until July. }

Even without the (ungrammatical) polarity item in the higher-neg versions, a negation over *say* or *hope* cannot be associated with the complement clause, even though the disjunction of (54i) can be assumed, ex hypothesi.

Moreover, the arguments first leveled by G. Lakoff (1970) against a purely semantic treatment of the NRP still militate against any of the variety of nonsyntactic analyses proposed since. As is well known, the availability of NR understandings is subject to semantically unmotivated lexical exceptions. In English, *suppose* neg-raises on its parenthetical reading for all speakers, but *guess* does so only for some (*I don't {suppose / %guess} Lee will arrive until midnight*). *Want* neg-raises freely, *wish* somewhat less so, and *desire* only with difficulty; the same pattern obtains for *expect* and *anticipate*. It is hard to detect any relevant non–ad hoc semantic or pragmatic distinction between *want* and *desire,* between *expect* and *anticipate,* or between parenthetical uses of *suppose* and *guess,* which could account for this distinction.

If NR were a grammatical rule, as Lakoff notes, we could invoke the independently necessary exception mechanisms motivated elsewhere in the grammar to account, inter alia, for the fact that *probable* is an exception to subject raising, while its synonym *likely* is not (*Sandy is {likely / *probable} to win*); cf. Green 1974: 18–20 for related discussion. But if a semantic principle of interpretation or, a fortiori, a general pragmatic process is invoked to account for the phenomenon of lower-clause understandings of higher-clause negation, it is not clear how this variation is to be treated. In particular, the notion of a lexical exception to a pragmatic principle requires some explication.

The plot thickens when the action shifts overseas. While similarities in the semantic classes of NR triggers outnumber the differences, cross-linguistic disparities arise as to just which semantically qualified predicates allow NR understandings in a given language. Among verbs of opinion, Hebrew *xošev* 'think' is an NR trigger, but *maamin* 'believe' is not; the opposite pattern obtains in Malagasy. NR in Hindi applies to complements of *lagnaa* 'seem', but not of *soocnaa* 'think' or of X-*koo khvaal hoonaa* 'have the opinion', and to *caahnaa* 'want' in Equi contexts (à la (40a) above) but not with unlike subjects (à la (40′a)), and so on. (See Horn 1978b: 183–93 for further examples and references.)

In particular, we find that negation over *hope* does not in general limit lower-clause understandings or trigger strict polarity items (as seen in (55b); cf., however, *I never saw a purple cow / I never hope to see one*) but its cross-Germanic counterparts do, including German *hoffen* and the cognate verbs in Danish and Dutch. While the Latin verb *sperare* was a neg-raiser (and *non spero* standardly translates as 'I hope not'), its French daughter *espérer* is transparent to negation only with subjunctive complements, and for some speakers not even then (*Je n'espère pas* = 'I don't hope so, ≠ 'I hope not', as noted above). *Souhaiter* is an NR trigger, but its semantics seem to correspond more closely to 'wish' than to 'hope'; cf. Horn 1978b: 183–87 for more on this and other aspects of the *hope* class.

Given the extent of the intra- and cross-linguistic variation on the membership of the set of NR predicates, we must apparently conclude with Epstein (1976: 160) that even if the NRP is pragmatically motivated, the process within a given language 'may be conventionalized, so that its end results are automatic'. Carlson (1983: 120–21) reaches a similar conclusion. After detailing his own version of the Bartschean assumed disjunction hypothesis, which combines with a preference for the 'guarded', 'polite' quality of the syntactically external negation to license 'the general tendency of any expression of doubt or indecision to suggest disbelief or disinclination', Carlson concedes that 'the lexical selectivity of the negative transportation phenomenon does suggest that whatever process creates it has become a conventional rule in the clear cases'.

The essential task is to determine the nature of this conventionalization process and the whereabouts of the resultant rule within a linguistic description. While I will not be able to complete this task until §5.3, I shall begin by returning to the issue first confronted by Kalepky (1891): the semantic characterization of the classes of predicates which qualify for status as potential NR triggers within a given language. These classes can be labeled and instantiated in English as follows:

(56) **a**. OPINION: *think, believe, suppose, imagine, expect, reckon, feel, (%guess, %anticipate)*
a′. PERCEPTION: *seem, appear, look like, sound like, feel like*
b. PROBABILITY: *be probable, be likely, figure to*
c. INTENTION/VOLITION: *want, intend, choose, plan*
d. JUDGMENT/(WEAK) OBLIGATION: *be supposed to, ought, should, be desirable, advise, suggest*

Cross-linguistic classification of NR triggers within these classes is fairly straightforward, although (as I noted above) whether a lexical item within one of these classes will trigger NR within a given language, dialect, or idiolect may not be predictable. Russian neg-raisers, for example, include (**a**) *dumat'* 'think' and *ožidat'* 'expect', (**a′**) *pokazat'sja* 'seem' and *vidno* 'appear, look like', (**b**) *verojatno* 'probable', (**c**) *xotet'* 'want', *zamyšlat'* 'plan', and *namerevat'sja* 'intend', (**d**) *dolžen* 'should' and *sovetovat'* 'advise', and so on (cf. Luborsky 1972). Similar categorizations can be made in other languages exhibiting the NRP, including Basque, Estonian, French, German, Hebrew, Hindi, Korean, Japanese, Latin, Malagasy, Old (and new) Spanish, Swahili, and Turkish; cf. Horn 1978b for examples and discussion.[34]

While there have been a variety of predicates cited as participants in the NRP which do not seem subsumable under any of these categories, closer inspection often indicates that such subsumption is indeed possible.[35] For example, one French NR trigger cited by Tobler (1882b:202) and others (cf. Nyrop 1930:41; Le Bidois 1968:§985; Grevisse 1969:884) is *faire semblant (de)*, apparently corresponding to 'pretend'. But as Tobler observes, *Il ne fit pas semblant de comprendre* is taken to convey not that he didn't pretend (i.e., pretended not) to understand, but that he didn't act as though he understood (*tat als verstünde er nicht*), that is, acted as though he didn't understand. On this reading, *faire semblant* fits naturally with the perception verbs of Category (**a′**).[36]

But assuming the essential universality of the classes in (56), why is it that just these predicates allow a lower-clause understanding of upper-clause negation, while others (*know, regret, claim, say, manage, try, be able, be possible, be certain, be odd*) do not? The first relevant observation on this score is the Kiparskys' recognition that factives are universally excluded from the ranks of potential neg-raisers. The Kiparskys are correct in claiming that no factives are neg-raisers, later critics to the contrary notwithstanding.[37] But their syntactic account of this correlation (which is flawed on independent grounds) would have a chance at validity only if all nonfactives were NR triggers, which is clearly not the case.

G. Lakoff (1970: 158) ties the absence of neg-raising factives to the Poutsma-Bolinger Uncertainty Principle identified above: 'Since it is presupposed that the subject of a factive verb [better, the speaker of a sentence containing a factive] knows that the complement of a verb is true, he cannot be uncertain about it'. This suggestion extends naturally to the implicatives (Karttunen 1971), predicates whose complements are entailed rather than presupposed, and which, when negated, entail the negation of those complements; entailment is as incompatible with speaker uncertainty as is presupposition. Thus, *be likely* is a candidate for the NRP, but (factive) *be odd* is not; *want to* and *figure to* are, but (implicative) *manage to, venture to, forget to,* and *bother to* are not.

But some basic questions remain: Why are such predicates as *able, possible,* and *allow,* whose complements are neither presupposed nor entailed, never found among the NR triggers of English or any other languages? What work can the Uncertainty Principle do for us in separating the wheat of (56c, d) from the related chaff (nontriggers *insist, have to, order*), when none of these root or deontic values bear any immediately evident connection to either certainty or uncertainty? Where do we go from here?

The answer, it is suggested in Horn 1975, 1978b, lies in the quantitative and/or pragmatic scales I explored in chapter 4. Following the principles outlined in my earlier discussion, let us assume the metascale in (57), incorporating both epistemic and deontic operators since, as it turns out, the same lexical modal or verb may serve both functions, often distinguished by complementizer type. (For typographic convenience and ease of interpretation, this scale is a ninety-degree rotation from those in §4.4.)

(57)

be able	believe, suppose, think	know, realize
be possible	be likely, probable	be certain
	figure to	be clear, evident, sure
	seem, appear, look like	be odd, significant, tragic
0 <--WEAKER--	.5	1
	--STRONGER-->	
may, might	be supposed to	must, have to
can, could	should, ought to, better	need, be necessary
allow, permit, let	be desirable, advisable	be obligatory, mandatory

be allowed	be a good idea (to)	make, cause, force, insist
be legal, ethical	want, choose, intend, plan {to/on}	order, demand, require

Parallel to this scale, again as in §4.4, I could (but will not) construct a metascale for the negative counterparts of these modal values, yielding the following correspondences:

(58) i. The (contradictory) negation of a weak scalar value (e.g., *possible, allow*) will be a strong value on the corresponding negative scale (*impossible, forbid*).
ii. The negation of a strong scalar value (e.g., *certain, have to*) will be a weak value on the corresponding negative scale (*not certain, do(es)n't have to*).
iii. The negation of an intermediate scalar value (e.g., *likely, advisable*) will be an intermediate value on the corresponding negative scale (*not likely, not advisable*).

I shall illustrate by extracting the epistemic adjectives, yielding the gradation in (59) or, alternatively, the annotated Square in (59′) (see §4.4 for more on these models):

(59)

+	−
certain (**p**)	impossible (**p**)
likely (**p**)	[unlikely (**p**)] not likely (**p**)
possible (**p**)	not certain (**p**)

(59′)

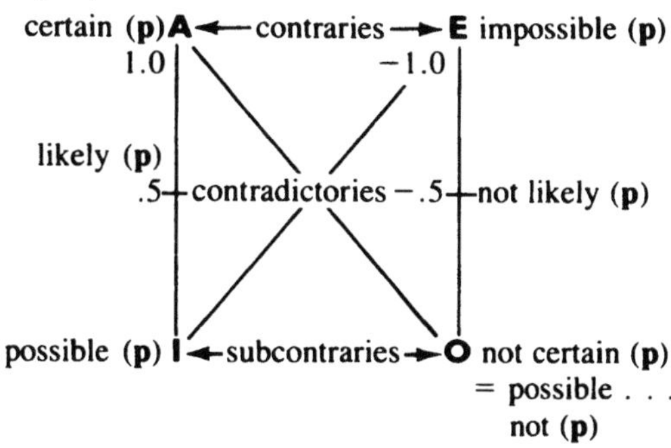

As we have seen (chapter 4, (54)), *likely* and *certain* are INTOLERANT predicates (à la Löbner 1985), since they cannot simultaneously be predicated of a proposition and its negation (*#It's likely she'll go and likely she won't*), while *possible* is TOLERANT, since the corresponding conjunction is consistent (*It's possible she'll go and possible she won't*). The same reasoning that leads to the placement of (the lower bound of) *likely* just above the midpoint of the positive epistemic scale, given the epistemic inconsistency of (60a), leads to the same move for the other 'midscalar' values in the metascale of (57). Since (60b) acknowledges an inconsistent belief set and (60c) reports contradictory desires, *believe* and *want* must be above the midpoints of their respective scales.

(60) a. #It's likely she'll go and likely she won't go.
b. #I believe she'll go and I believe she won't go.
c. #I want her to go and I want her not to go.

It is this class of midscalar or weakly intolerant (henceforth **WI**) predicates which harbors virtually all NR triggers; weaker (i.e., tolerant) and—with an exception to be noted—stronger predicates typically fail to participate in the NRP.

We have already seen why strong scalar epistemics exclude NR: their use presupposes or entails their complement, thus precluding the operation of the Uncertainty Principle. While *It's likely* (or *I think*) *Kim won* leaves open the possibility that she lost, *It's certain (I know) Kim won* explicitly forecloses this possibility. Thus, not only factives and implicatives but the semiimplicative IF verbs of Karttunen 1971, those (e.g., *certain, clear, sure, evident*) from which an entailment is deducible only from positive instances, are excluded from the ranks of potential NR triggers.

With the tolerant weak epistemic scalars, corresponding to Karttunen's ONLY-IF predicates, no entailment follows from a positive occurrence, but negated *able* and *possible* sentences do entail the negation of their complements. We must evidently inspect each pair of the form ⟨**P(p)**, ~**P(p)**⟩ and determine if an entailment is derivable from either member; if so, **P** is scratched from the roll of prospective neg-raisers.

Intuitively, what is common to all NR triggers is the relative slenderness of the functional difference between the preraised form with lower negation and the logical form with the upstairs negative taking wide scope. It is the closeness of the external (contradictory) readings of *not likely, not believe, not advisable* to *likely not, believe not, advisable not,* respectively, which renders the negated predicates potential neg-raisers, and the relative distance of *not possible, not realize, not obligatory* from *possible not, realize not, obligatory not* which removes these from that category.

This position is reminiscent of (part of) the account of the NRP by Stockwell, Schachter, and Partee (1973:253–56):

> Non-factives express 'propositional attitudes'; in some cases it happens that a negative attitude toward a positive sentence may be very nearly or perhaps perfectly equivalent to a positive atittude toward a negative sentence; this seems true when either (i) the attitude is a moderate one, such as <u>think</u>, <u>believe</u>, <u>seem</u>, or (ii) the attitude is dichotomous, such as <u>true</u> or <u>false</u>. When the attitude is a strong one such as <u>claim</u> or <u>sure</u>, however, the equivalence fails.

But Category (ii)—which would in fact counterexemplify my scalar hypothesis—does not represent a true instance of the NRP. Notice that while (61a, b):

(61) a. It isn't true that Chris will get here (*until midnight).
b. It's true that Chris will not get here (until midnight).

may indeed be 'very nearly or perhaps perfectly equivalent' to each other, given the transparency of *true* to negation and to everything else (at least within a bivalent semantics), so that these sentences vacuously satisfy the Bartschean disjunction, there is no argument for coderivation here, as confirmed by the ungrammaticality of strict polarity items in the context of (61a).

Category (i) does come closer to describing the set of 'moderate' NR triggers, but the definition of this class in terms of propositional attitudes is untenable. Exceptions to this restriction include 'moderate' **WI** quantificational determiners (*most (of the) α, a majority of (the) α*) and quantificational adverbs (*usually, most of the time*):

(62)	a. I don't think that most of my friends would approve.	[◇ = I think that most wouldn't]
	b. She doesn't usually attend church.	[◇ = She usually doesn't]

Weaker (tolerant) quantificational values (*some* and *many, sometimes* and *often*) and stronger values (*all, always*) rule out any lower-clause reading for upstairs negation in the same contexts. (See Horn 1978b:203–4 for elaboration.)

Indeed, if the 'moderate attitudes' of Category (i) are to be characterized as inducing a near, but imperfect, equivalence between higher-neg and lower-neg versions, the best representative of this class may be the epistemic **WI** predicate *likely,* which does not denote an attitude per se. The only circumstance in which *It is not likely that* **p** and *It is likely that not-***p** would differ in truth value is presumably when **p** has a fifty-fifty chance of occurring or coming true, for example, in the proposition that a fair coin will land heads. (Cf. Horn 1978b:196–97 for a defense of the status of *likely* and (for most speakers) *probable* as NR triggers.)

Of course no entailments are derivable with deontics—strong, intermediate, or weak, affirmative or negative. The deontic analogue of certainty is presumably obligation, but if something is obligatory (or forbidden), it does not follow that it will (or won't) occur, only that it had better (or better not). In the case of **WI** deontics no "absolute" obligation is derivable from either positive or negative use; both *not* {*advisable* / *desirable*} and {*advisable* / *desirable*} . . . *not* express a mild suggestion or weak obligation. Thus, the functional principle may be extended: NR understandings will not be available where they would systematically result in the emergence of pernicious ambiguities, that is, when the higher-S and lower-S readings of main clause negation would carry a high functional load, leading to a possible breakdown in communication.

But the very fact that absolute obligation is harder to pin down than absolute certainty (or, for that matter, logical necessity) evidently licenses the tendency cited in §4.5, for a number of languages to permit apparent neg-raising with strong scalar deontics (including French *falloir, Russian velet'*, Basque *behar*). The epistemic distance from likelihood to certainty is psychologically greater than the corresponding deontic distance from weak intolerant *devoir* to strong intolerant *falloir.*

While strong intolerant predicates may or may not license NR understandings, tolerant (weak scalar) predicates never do. Why is it that the negations of *able, possible, allow, be permitted,* and so forth, consistently exclude lower-clause interpretations? Why does a predicate **P** qualify for participation in the NRP only when (although not always when) it is intolerant, that is, when it cannot be conjoined with its inner negation **P**~ without inconsistency?

I submit that this fact results from the logical properties of these scalar values, as depicted on the annotated logical square in (59′). As seen in §4.4 and exemplified in the paradigm in (63),

(63) a. *It's not likely that the Yankees will win.*
outer-neg reading = ~(LIKELY . . .) [contradictory]
inner-neg reading = LIKELY . . . (~ . . .) [contrary]
b. *It's not possible that the Yankees will win.*
outer-neg reading = ~(POSSIBLE . . .) [contradictory]
*inner-neg reading = POSSIBLE . . . (~ . . .) [subcontrary]

the inner-neg reading of a (weakly or strongly) intolerant predicate will always yield a contrary negation, while the inner-neg reading of a tolerant predicate will always result in a subcontrary of the corresponding affirmative. In the latter case, no NR reading will be available. Whenever an outer negation does allow an NR (inner-neg, lower-clause) understanding, this understanding must represent a strengthening of the contradictory to a contrary, rather than a weakening to a subcontrary. The appropriate generalization is given in (64):

(64) The NR understanding is always stronger than the contradictory (outer) negation, in that it applies to a proper subset of the sit-

> uations to which the contradictory applies (is true in a proper subset of the worlds in which the contradictory is true). As with (other) indirect speech acts, the literal interpretation of the outer negation is true but too weak, and the addressee applies a (short-circuited) conversational implicature to 'fill in' the stronger proposition (cf. Horn and Bayer 1984 and §5.3 below on NR, polarity, and short-circuited implicature).

This principle, combined with the observation that a lower-clause understanding for higher-clause negation tends to be possible only when the two readings for the outer-neg sentence that would result are almost, but not quite, truth-conditionally identical (see discussion above), generates the table in (65) with its twin parameters for NR triggerhood,

(65)

	low functional difference between **P** . . . ~/~**P**?	**P** . . . ~ a contrary of **P**?
strong scalars	–	+
mid-scalars (**WI**)	+	+
weak scalars	–	–

and the correct prediction that mid-scalar (**WI**) predicates are the most trigger-happy, followed by strong scalars, with weak scalars ruled out entirely. As in the case of affixal negation (§5.1.1), inherently negative predicates—*doubt, disbelieve, deny, dislike, forbid, prevent*—universally rule out NR understandings.

We have seen (in §4.5) that the negative subcontraries of the southeast, **O**, corner of the Square—*some not/not all, possible not, not obligatory*—tend not to become lexicalized in natural language. Similarly, *disbelieve* lexicalizes *believe not,* the contrary of *believe,* while *disallow* lexicalizes *not allow,* the contradictory of *allow* (rather than *allow not,* its subcontrary).

I also noted earlier (in §5.1.1) that while negative affixes tend to result in the contrary of their positive bases, there are a number of exceptions to this general tendency:

(66) a. unabsorbable, unadaptable, unconquerable, . . .
b. impossible, incapable, inedible, unfeasible, . . .
c. unable, unapt, impractical, . . .

The contradictory reading assigned to the verb-based *un-X-able* forms of (66a) might be attributed to the oft-observed correlation between contradictory opposition and productive affixation. But the fully lexicalized root-based entries of (66b), as well as the unsuffixed forms of the same semantic class given in (66c), must equally be understood as contradictories of the corresponding positive adjectives.

It is now clear that these negative adjectives are contradictories because that is the strongest reading they can receive. *Unhappy* can be, and hence is, read as a contrary, but an adjective of the form *un-X*—where *X* is a weak scalar adjective, whether productively derived as in (66a), unproductively derived as in (66b), or underived as in (66c)—can only be read as a contradictory or a subcontrary. In such circumstances, given the absence of a weakening (contradictory --› subcontrary) rule alongside the strengthening (contradictory --› contrary) process I have attested in this chapter and chapter 4, only the contradictory interpretation can be assigned to the adjectives of (66).

In each of these cases, it is the reading closest to the **E** and farthest from the **O** pole of the negative scale which tends to be facilitated or selected, whether that reading amounts to a contrary or a contradictory negation of its positive base or counterpart. Data from lexicalization, semantic drift, affixation, and the neg-raising phenomenon converge into the generalization in (67):

(67) Contrary negation tends to be maximized in natural language.
Subcontrary negation tends to be minimized in natural language.

In referring to the conditions under which an upstairs negation can be associated with the lower clause, rather than the conditions under which a downstairs negative can be raised, I have tacitly endorsed the semantic or pragmatic approach to the NRP, without having dispelled the problems posed by the intra- and cross-linguistic variation in the selection of trigger predicates. This promissory note will be partially redeemed in §5.3.1, where I consider the neg-raising phenomenon against other contexts in which a superficial contradictory negation is pragmatically strengthened to a contrary.

5.3 Contrariety and Understatement: From Rhetoric to Rule

The negative words not, no, &c., have two kinds of meaning which must be carefully distinguished. Sometimes they deny, and nothing more; sometimes they are used to affirm the direct contrary. In cases which offer but two alternatives, one of which is necessary, these amount to the same thing, since the denial of one and the affirmation of the other are obviously equivalent propositions. In many idioms of conversation, the negative implies affirmation of the contrary in cases which offer not only alternatives, but degrees of alternatives. Thus, to the question, 'Is he tall', the simple answer, 'No', most frequently means that he is the contrary of tall, or considerably under the average. But it must be re-

> membered that in all logical reasoning, the negation is simply negation, and nothing more, never implying affirmation of the contrary.
> (De Morgan 1847:3)

> I've been living in the dark too long
> When something's not right, it's wrong.
> (Bob Dylan, "You're Gonna Make Me Lonesome When You Go")

My third instance of contraries in contradictory clothing is Bosanquet's first: the tendency to fill in a relatively uninformative sentence negation, converting it into an assertion of the contrary. I begin by noting Jespersen's comment (1917:43) that 'there is scarcely any difference' between (68a) and (68b).

(68) a. She isn't happy.
b. She is unhappy.

Jespersen's remark seems to have perplexed his admirer Poldauf, who correctly observes that the two forms of negation are clearly nonequivalent, given the possibility of an explicitly allowed middle: '*She isn't happy but she can't be said to be unhappy either—she is bored and indifferent*' (Poldauf 1964:369).

While Poldauf's rejoinder is well taken, is it really credible that someone as cognizant as Jespersen was of the contrary nature of the *happy/unhappy* opposition could be guilty of overlooking such a fundamental aspect of affixal negation? I submit that Jespersen's point is not that the affixal or special negative of (68b) is contradictory, like the nexal of (68a), but that the nexal negation may be contrary, like the affixal.

The same semantic lapse Poldauf attributes to Jespersen clearly afflicts Englebretsen's argument (1981b:46) that Aristotle (along with his modern interpreter Fred Sommers) takes the denial of a predicate to be equivalent to the affirmation of its logical contrary when the subject is a singular term, 'for Socrates is unhappy if and only if he isn't happy'.

Given its source, this claim is surprising on at least two counts: (*a*) *unhappy* does not constitute the logical, immediate contrary of *happy*, but its mediate contrary, since Socrates was presumably often neither happy nor unhappy, but somewhere in between; and (*b*) for Aristotle, even the true logical contrary of *happy*, which I have translated as *not-happy*, does not reduce to a predicate denial, since *Socrates is not-happy* (as well as *Socrates is unhappy*) can come out false when the contradictory-denoting predicate denial *Socrates is not happy* is true, namely, when *Socrates* names something that doesn't exist, or something that neither *happy* nor *not-happy* can be naturally predicated of—a mountain, an amoeba, or (for Aristotle) an infant. What is especially curious is that both (*a*) and (*b*) are amply documented in Englebretsen 1981a, as we observed in chapter 1.

Like his fellow Dane, Englebretsen may have been seduced into his historical faux pas by the pragmatic tendency to read an attribution of *not happy* to a subject term as if it were the corresponding attribution of *unhappy*. But why should this tendency exist? Where is it manifested? What sort of descriptive account shall we give it?

As I noted in the discussion of prefixal negation (cf. (22), (22′) above), the contrary reading associated with *dis-* verbs (*I {disbelieve / dislike / distrust} you*) is sometimes shared by their counterparts with unincorporated negation (*I don't {believe / like / trust} you*). This problem is often formulated by invoking the sentential vs. constituent negation dichotomy of Klima (1964) and Jackendoff (1969), yielding the dual analyses of sentences like (69).

(69) John [(doesn't like) mushrooms]. (= (104) of Chomsky 1970)

Here is Chomsky (1970:71) on this phenomenon: 'In [(69)] we can take the negative element to be associated with the verb, so that it means *John dislikes mushrooms*, or with the verb phrase, in which case it means: *it is not so that John likes mushrooms*. In other words, either the parentheses or the brackets express a possible interpretation'. Only on the former (constituent negation) reading is there 'a presupposition that he has the relevant experience with mushrooms': 'Thus if John has never tasted mushrooms and it is asserted that he likes them, I can deny the assertion by stating [(69)], interpreting the negation as associated with the predicate phrase, but not by stating *John dislikes mushrooms*, or [(69)] with the negation interpreted as associated with the verb'.

Under certain circumstances, then, to deny that John likes mushrooms counts as an assertion that he actively dislikes them, just as (for Sigwart and Bosanquet) to deny that someone is good may count as an assertion that s/he is bad, and just as (for Bosanquet and Bartsch) to deny that one thinks that **p** (or wants **q**) may count as an assertion that one thinks that not-**p** (or wants not-**q**). In the same way, denying that Socrates is happy (or good) may amount to an assertion that he is unhappy (or bad).

We seem to have stumbled upon one more source of contraries in contradictory clothing, triggered by (among other factors) the psychological preference for simple binary classification, as stressed by Sapir (1944:101): 'To the naive, every person is either good or bad'. And, we might add, we are all naive at least some of the time. The result is the assumed disjunction and the modus tollendo ponens–driven inference schema we have already encountered. Here is Lyons's version of the tendency cited by De Morgan and Sapir: 'Gradable antonyms are frequently employed in everyday language-behavior as contradictories rather than contraries. If we are asked *Is X a good chess-player* and we reply *No*, we may well be held by the questioner to have committed ourselves implicitly to the proposition that X is a bad chess player' (Lyons 1977:278). While there is no general entailment of the form *X is not good* ⊩ *X is bad*, the context permits such inferences for a particular token (cf. Bartsch's *pragmatische Implikation* in (53)). Unless otherwise stipulated, this middle-excluding inference will go through: 'If the speaker did not wish to be committed to the implication, he could have been expected to make it clear that a first approximation was insufficiently precise, by saying, for example, *X is not good, but he's not bad either: he's fair/pretty good/just about average*' (Lyons 1977:278).

There are, then, 'two kinds of propositional negation: one of which converts the proposition into its contradictory and the other into its contrary'; for Lyons, as for Sapir, the emergence of the second, contrary-yielding operator is favored—if not forced—in scalar contexts: 'It seems to be the case that the application of propositional negation to a gradable expression (e.g., '*like*') will always tend to produce a contrary, rather than a contradictory, whether the language-system lexicalizes the contrary (e.g., 'dislike') or not' (Lyons 1977:773). This tendency, which Lyons views as 'seem[ing] to depend upon the negation being more closely associated with the predicate, than with the subject-predicate link or nexus', extends beyond the simple first-order negation of *I don't like modern music* and its incorporated affixal counterpart (*I dislike modern music*) to the neg-raising ('transfer of negation') phenomenon I explored in §5.2; it afflicts 'so-be-it' imperatives (*Don't trust him*) along with 'it-is-so' indicatives (*I don't trust him*) (Lyons 1977:774).

But there is another important factor triggering the contrary readings of contradictory negations not touched on by the establishment of De Morgan, Bosanquet, Sigwart, Sapir, Chomsky, Bartsch, and Lyons. The choice of the weaker-seeming contradictory negation over the stronger affirmation of the contrary is characteristically prompted by a desire to avoid direct expression, to present one's contribution in a manner often labeled (e.g., by Jespersen 1933:285 or L. Carlson 1983:120) 'polite' or 'guarded'. Leech (1983:101–2) brings this point out clearly in his description of the use of the negative sentences like *I don't like Kenneth, We don't agree, He doesn't believe in marriage* 'as a form of understatement': 'Negation here is apparently a hedging or mitigating device, the motivation for which may be politeness or simply euphemistic reticence in the expression of opinion and attitudes'.

I shall return to the 'politeness or euphemistic reticence' of contradictory negation later in this section. But it is worth noting here that if the choice of the semantically weaker form is partly motivated by the avoidance of highly charged judgments, the relevant strengthening inference will tend to be favored in contexts like Leech's (and Lyons's), where there is some plausible reason to mask the speaker's true opinion (*I dislike Kenneth, We disagree, He is a bad chess player*). These contexts characteristically involve not only gradable predications, as Sapir and Lyons observe, but more specifically those gradable predications involving desirable properties, those whose denial would reflect undesirably on the subject, speaker, and/or addressee.

There does indeed appear to be a palpable asymmetry in the availability of contrary readings for contradictory negations. Alongside Leech's understating negations reproduced in (70), the negations in (70′) are far more likely to be taken as simple contradictories of the corresponding affirmatives:

(70) a. I don't like Kenneth.
b. We don't agree.
c. He doesn't believe in marriage.

(70′) a. I don't {dislike/object to} Kenneth.
b. We don't disagree.
c. He doesn't {disbelieve in/reject} marriage.

If I tell you that I don't approve of your behavior, you may infer that (presumably to spare your feelings) I am concealing my active disapproval—although I could have added, with perfect consistency, that I don't disapprove of it either, in which case this pragmatic inference will be blocked. But if I acknowledge that I don't disapprove of your behavior, you are less likely to conclude (however strongly you may wish to) that I actively approve of it.

This asymmetry extends to other evaluations in the moral sphere as well. If something is *not right*, it is (assumed to be) wrong, but if something is *not wrong*, it is not ipso facto (assumed to be) *right*—only, more weakly, that it is *all right*. *Not nice* may amount to *nasty* or *naughty*, but *not nasty* and *not naughty* never quite make it up to *nice*. The analogous asymmetry in French is represented graphically by Ducrot (1973: 123) in the following schema:

(71)

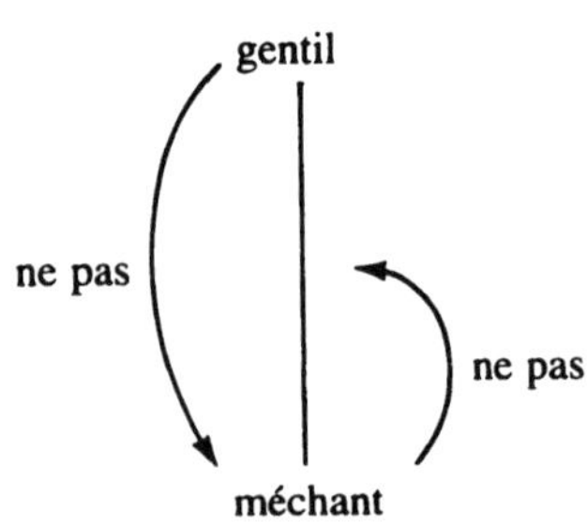

That is, (72a) is closer to (72b) than (72′a) is to (72′b):

(72) a. Pierre n'est pas gentil. 'Pierre isn't nice'
     b. Pierre est méchant. 'Pierre is {naughty / nasty}'

(72′) a. Pierre n'est pas méchant. 'Pierre isn't {naughty / nasty}'
      b. Pierre est gentil. 'Pierre is nice'

As before, the inference from (72a) to (72b) can be blocked in context by reinforcing the unexcluded middle between the contraries—*Non, Pierre n'est pas gentil, mais il n'est pas non plus méchant* ('but he isn't *méchant* either')—but in the absence of such a continuation, the pragmatic move will be made in (72) and not in (72′). Ducrot concludes that only when ordinary negation is applied to the favorable, unmarked term of a marked / unmarked opposition is it 'quasi-equivalent' to the contrary affirmation.

Markedness is similarly appealed to in Cornulier's observations on the asymmetry between 'not rich' and 'not poor':

> On peut être ni *riche*, ni *pauvre;* mais la négation peut produire avec *riche* une ambiguité: l'expression *un homme (qui n'est) pas riche* peut signifier *un homme (qui est) pauvre,* aussi bien que la simple contradiction de *riche;* au contraire, *un homme (qui n'est) pas pauvre* ne signifie jamais grammaticalement *un homme (qui est) riche:* on ne peut obtenir ce sens qu'au niveau stylistique (litote). (Cornulier 1974:55)

Nor is the asymmetry restricted to the moral sphere, as Cornulier's *riche / pauvre* example demonstrates; what is crucial is the opposition of an unmarked vis-à-vis a marked term, preferably buttressed by some sense (however elusive) that the unmarked term is evaluatively positive (or at least neutral). Sentence (73a) is felt to be a rather pessimistic assessment, while (73b) is not particularly optimistic; if the speaker hasn't formed an opinion either way, only the latter is a likely utterance.

(73) a. I'm not optimistic about his chances.
     b. I'm not pessimistic about his chances.

In the same vein, consider the following stance toward a tax bill announced by Senator Dave Durenberger (R-Minn.):

(74) I'm not endorsing it or not endorsing it.

If the senator's candid opinion is interpretable at all, it can only be taken as in (74′):

(74′) I'm neither endorsing it nor not-endorsing it.

That is, the second negation in (74) must induce a contrary reading, with the senator clinging to the reinforced middle ground he sees looming between endorsement and counterendorsement (opposition?). In any event, as odd as (74) may strike us, its marked counterpart (74″) must be still less conceivable.

(74″) #I'm not opposing it or not opposing it.

Just how seriously are we to take the hyphen in the gloss in (74′) or Chomsky's parentheses in (69)? Both Chomsky (re *doesn't like*) and Ducrot (re *ne pas gentil*) seek to assimilate the contrary reading of superficially contradictory negations to the phenomenon of constituent negation. But there is little evidence that the contrary-producing negations in (69), (72a), (74), and similar sentences (e.g., (68a), (70a–c), and (73a)) are (or must be) analyzed as syntactically distinct in any way from the corresponding but nonimplicating negations in (72′a), (70′), (73b), or (74″). In particular, the strengthening inference applying to the former set of examples is fully compatible with the diagnostics for sentential negation proposed in Klima 1964, Kraak 1966, and Attal 1971 (cf. §3.3 above):

(75) a. John doesn't like [--› dislikes] mushrooms, not even golden shiitakes.
b. I don't like [--› dislike] Kenneth, and neither does Chris.
c. He doesn't believe in [--› rejects] marriage, does he?
d. She isn't happy [--› is unhappy], and he isn't happy either.

Notice that even when the negative assertions in (75) are taken in context to convey the corresponding stronger negations in the brackets, the latter cannot substitute for the former in these S-negation frames *salva grammaticalitate*. Nor (pace Chomsky, Cornulier, and Lyons) is there any compelling evidence that the sentences in question are semantically or lexically ambiguous. I conclude that the apparent sentential and contradictory negations of (75) are in fact sentential and contradictory on the grammatical level, although I must still account for the strong pragmatic tendency to strengthen them to contraries.[38]

An extreme form of asymmetry in the lexicalization of contrary negation is realized in languages in which the only way to form the antonym of an unmarked, positive adjective is by negating that adjective; this tendency never operates in the opposite direction, as Greenberg (1966:52) observes: 'A considerable number of languages, African, Amerind, and Oceanic, have no separate term for "bad", which is expressed by "not good". On the other hand, there is as far as is known to me, no language which lacks a separate term for "good" and expresses it normally by "not bad"'. Greenberg extends the asymmetry to unmarked/marked oppositions differing in relative saliency (rather than e-polarity per se), for example, those designating long/short, wide/narrow, deep/shallow. Thus, the Hausa equivalent for 'narrow' or 'shallow' is literally 'lacking width/depth', while Spanish employs the periphrastic *poco profondo* (little deep) for 'shallow'. But again, width and depth are never defined negatively via the absence of narrowness or shallowness.

Along the same lines, Zimmer (1964:78) cites the South American Indian language Tukano in which an e-neg adjective can be obtained only by negating the corresponding positive, unmarked form: the designators for 'dirty', 'narrow', and 'short' are formed by negating the terms for 'clean', 'wide', and 'long', respectively, although both 'good' and 'bad' are assigned simplex lexemes. In Mayan, on the other hand, 'good' can only be designated indirectly as *ma'alob* (lit., 'not bad'). This counterexample to Greenberg's universal hypothesis reflects a countertendency to which I shall return later in this section.

5.3.1 Irony, Conventionalization, and the Inferential Short Circuit

As we have seen, a contrary reading is available for the negation associated with certain positive scalar values. But this is not the whole story. Just as in the other two environments for the strengthening (contradictory --› contrary) inference we have already explored, that is, partially nonproductive affixal negation and the NRP, no other scalar value triggers this inference. Thus compare the behavior of the unincorporated negatives in (76a, b) with their lexicalized analogues:

(76) a. She is not happy *vs.* She is not {ecstatic/sad/unhappy/miserable}.
(cf. unhappy *vs.* {*unecstatic/*unsad/*unmiserable})
b. She doesn't like him *vs.* She doesn't {love/adore/dislike/hate/loathe} him.
(cf. dislike *vs.* {*dislove/*disadore/*dishate/*disloathe})

The parallel is clear: a negation of the form . . . *not **P*** . . . can be understood as yielding the contrary of ***P*** under essentially the same circumstances as those in which a contrary-producing affixal negation can be incorporated onto an adjectival or verbal stem ***P***, namely when ***P*** represents the unmarked positive value on its scale.[39] As argued in Horn 1978b and §5.2 above, the NRP is similarly conditioned. My findings are summarized in (77):

(77) a. A contrary reading is available for a sentential negation just in case the negated predicate is positive and relatively weak on its scale.

b. A contrary (NR) reading is available for a negated proposition–embedding higher predicate only when the negated predicate is positive and relatively weak (i.e., just above the midpoint on its scale).

c. A contrary reading is available, typically preferred, and often required for a derived word of the form $[\alpha[\beta]]$, where the negative prefix α is relatively nonproductive (with respect to the base of β); in such cases, $[\alpha[\beta]]$ will be lexicalized only when β is positive and relatively weak on its scale.

Typical instances of each pattern include *I don't believe you* for (a), *I don't believe you came* for (b), and *I disbelieve your claim* for (c).

But isn't a generalization being missed here? The parallel among the three patterns in (77) is striking enough at first glance, and we can easily find ways to reinforce it. I noted, with Leech (1983), the 'hedging' or 'mitigating' effect associated with the pattern in (77a) and the resultant 'polite' or 'euphemistic' flavor of the resultant strengthened contradictory negations, compared with other (typically affixal) expressions of contrary negation. But affixal negations too have standardly been taken as euphemistic alternatives to the corresponding simple, unanalyzable e-neg adjectives.

Thus, as noted by Stern (1937 : 332), we may choose to substitute *unclean* for *dirty*, *untruthful* for *lying*, *intemperate* for *drunken*, *unwise* for *foolish*, *unsafe* for *dangerous*, *impolite* for *rude*, and so on. These substitutions are cited as parallel to *a lady of a certain age*, *a woman no longer young*, or *leaves much to be desired*, and we could add to Stern's list such contemporary favorites as *you have been misinformed* or *the President misspoke himself*. In each case, a more precise expression referring to an unpleasant state of affairs is rendered more palatable by the substitution of a vaguer or more general expression. In each case, the addressee may decode the euphemism by applying an **R**-based strengthening inference, yielding a more specific—and in practice more negative—understanding. And in each case, the more direct negative expression is avoided precisely because it is more direct; cf. Brown and Levinson (1978 : §5.4 and passim) on avoiding negative face.

But it is precisely this same factor, the avoidance of negative face and the desire to leave open different options of interpretation, which is repeatedly cited as the major motivation for the NRP, that is, the higher-clause appearance of a negation which in some sense ought to be located in the dependent. While the politeness or hedging nature of the NRP is often implicit in the treatments of the phenomenon by Jespersen, Bolinger, and others, it is with Prince 1976 that we arrive at an explicit analysis of raised negatives as 'metastatement hedges', motivated by the desire to express negative opinions as tentatively as possible. Prince's correlation of the NRP and tentativeness in English and French is extended by Shnukal (1980) and independently remarked on by L. Carlson (1983:120–21), who cites the 'guarded', 'polite' quality associated with the 'milder form of doubt' in the higher-neg version of pairs like ⟨*I don't suppose he's right*/*I suppose he's not right*⟩. Since the same association of raised negs with politeness, hesitancy, and/or uncertainty has been observed (cf. Horn 1978b) in languages as diverse as Hindi, Japanese, Swahili, and Turkish, it appears to be inherent in the very nature of the NRP.

Given the correlations we have observed, it is especially tempting to assimilate the patterns in (77a, b), and—as we have also seen—this practice has a rich history. Jespersen (1917:52–53) follows tradition in seeking to reduce the NRP to a special case of litotes or rhetorical understatement. On this view, the fact that Greek *phēmi* attracts a negative which is interpreted as logically attaching to the infinitive is a simple instance of litotes, just as when verbs denoting 'love' or 'like' attract a negative to form a single-clause contrary or antonym in Old Church Slavic (*nenaviděti*) and Greek (*ou stérgō*). Indeed, where Bosanquet (1888), Kalepky (1891), and Jespersen explicitly assimilate the NRP to the general figure of litotes, others collapse the two processes as a matter of course. Barnes (1969:303) does so within the space of a single sentence: 'It does appear that "a doesn't believe that p" at least sometimes means the same as "a disbelieves that p"—I can reject a story as well by saying "I don't believe you" as by the more ponderous "I disbelieve you"'.

More recently, Cornulier (1973, 1974) has pointed out that the apparently parallel ambiguity between contradictory vs. contrary readings in the two sentences in (78) seems to demand a parallel treatment within a grammar,

(78) a. Je n'aime pas l'ail. 'I don't like garlic'
b. Je n'aime pas étudier. 'I don't like to study'

yet obviously no NR analysis is possible in the former case. Rather than adopting a movement analysis in the latter case (deriving the *J'aime ne pas étudier* reading of (78b) by extracting the negative from the embedded clause), Cornulier urges a direct capture of the parallelism in (78) and analogous examples by opting for a uniform treatment of both cases as instances of lexical ambiguity: there is a 'strong' and a 'weak' *aimer*, a 'strong' and a 'weak' *vouloir*, and so on.

Cornulier eloquently serves notice that the defender of a syntactic approach to the NRP who ignores the parallel between (78a, b) can but 'tranche au couteau dans un nuage'. Difficult as it may be to locate the appropriate knife for cloud slicing, it is no more appealing to provide two senses for each NR trigger, as Cornulier's account demands. Nor can we find comfort in acknowledging that *hope* will lack a weak sense in English, while *hoffen* will not, or that the weak sense of affirmative *guess* available to all speakers (in *I guess he'll go*) mysteriously vanishes under negation for just those speaker who block the NRP in *I don't guess he'll go*. In any case, as we shall soon observe, the NRP resists this takeover attempt by ordinary litotes.

To collapse the NRP of (77b) with the simple litotes of (77a) is to reaffirm not only the judgment of Bosanquet, Kalepky, Jespersen, and Cornulier, but also the pragmatic line on so-called neg-raising endorsed by Bartsch and Halpern (cf. §5.2). In Bartsch's schema, repeated here for convenience,

(54) i. **F(a, p) ∨ F(a, ~p)**
ii. **~F(a, p)**
iii. ∴ **F(a, ~p)**

a context permitting the establishment of the disjunction in (i) (*Either I believe that **p** or I believe that not-**p***) will license the inference from the main-clause negation of the positive disjunct in (ii) (*I don't believe that **p***) to its lower-clause counterpart in (iii) (*I believe that not-**p***).

This approach, which essentially recapitulates the neo-Hegelian line on contrary negation in general and the NRP in particular, is echoed by Lauri Carlson (1983: 120) in his discussion of 'the general tendency of any expression of doubt or indecision to suggest disbelief or disinclination': 'It often seems as if Christ's famous law of excluded middle in questions of faith were valid: who is not with me is against me. . . . Given such a law of excluded middle, the scopes of propositional attitudes and negation become interchangeable, so that [*I don't suppose he is right*] comes to mean the same as [*I suppose he is not right*]'. (I might note here that it is only Christ's exclusive formulation of the law—cf. *Luke* 11: 23 and (4a) above—that yields a rule for litotic inference in natural language; the inclusive version of *Luke* 9: 50 and (4b) does not generate a parallel rule.)

But is the NRP in fact merely a special case of litotes or understatement affecting all (or a semantically coherent subset of) apparent contradictory negations, strengthening them (pragmatically, if not grammatically) into contraries? The first point to observe is that the NRP cannot be directly reduced to litotic strengthening in multiclause sentences, since not every contrary negation of **a** Fs that **p** is equivalent to the assertion that **a** Fs that not-**p**. Thus, for example, when *not good* embeds a complement, it conveys a contrary (by (77a)) without, however, reducing to a predication of *good that . . . not . . .* (by (77b)); (79a) is not merely a contradictory of the corresponding affirmative, but neither does it share a reading with (79b).

(79) a. It isn't good that he was reelected. [--› it's rather bad that he was reelected]
b. It's good that he wasn't reelected.

Similarly, *I don't like it that he was reelected* can convey (by litotes) a strong negative reaction to his reelection (= *I dislike it . . .*), but it cannot convey satisfaction with his defeat.

This is not surprising, given the factive nature of these predicates with indicative complements and the Uncertainty Principle in §5.2; nor is it surprising that the same predicates when appearing in hypothetical, nonfactive contexts act like good mid-scalar (**WI**) deontics with respect to triggering the NRP:

(79′) a. It wouldn't be {good / a good idea} for us to elect him. (≅ It would be {good / a good idea} for us not to elect him)
b. I wouldn't like his being reelected. (≅ I would like his not being reelected)

But this discrepancy makes it clear that the NRP, while rooted in the litotic pragmatics of contra(dicto)ry negation, has taken on a life of its own.

The same point is stressed by Tobler (1882b), who entertains the possibility of analyzing the 'logically unwarranted position of negation' in *Il ne faut pas que tu meures (--› Il faut que tu ne meures pas)* as an instance of litotes, citing the examples in (80) as clear instances of the latter process (the translations from the Old French and German are mine):

(80) the holy saints by whom God was not hated (= was loved)
he will do you little good (= all the harm he can)
we weren't in the best possible situation (= in the worst situation)
he didn't show that he had forgotten their kindness (in context, = he showed that he hadn't forgotten)

But Tobler correctly points out that modern examples of the NRP, operating across verbs like French *falloir, devoir, vouloir,* or German *sollen, wollen, meinen,* do not involve the conscious sense of irony associated with the figure of litotes which emerges in my reconstruction of the intended meaning of the examples of (80). Curiously, Tobler's disciple Kalepky seems to have overlooked this key insight, to the point of attributing his own NRP-as-litotes position to Tobler (Kalepky 1891:26). But it is Tobler's view which is correct and which anticipates my own treatment (below) of NRP as a pragmatically based but partially conventionalized process.

I might begin my reinforcement of Tobler's observation by noting that while the neg-raised, lower-clause understanding of higher-clause negation always represents a contrary of the corresponding affirmative, it does not necessarily constitute the (unique) contrary of that proposition. To say that it's not likely that a fair coin will land heads may amount to an (indirect) assertion of the proposition that it's likely that a fair coin won't land heads. This constitutes an instance of both litotic strengthening to a contrary (as in (77a)) and the NRP (as in (77b)). But *likely not* is not the only contrary negation of *likely; certain not (= impossible)* fits the bill just as well. The apparent contradictory negation in (81a) may indeed be pragmatically strengthened into either of the corresponding contraries in (81b, c), yet it is only the latter move which retains the ironic flavor characteristic of non-conventionalized litotes.

(81) a. It's not likely she'll accept your invitation.
b. It's likely she won't accept your invitation.
c. It's certain she won't accept your invitation.

Notice also that predicates outside the positive **WI** class, including *hate, be in the best possible situation,* and *show* (from Tobler's examples in (80)), may qualify for litotic interpretations even when they are barred as NR triggers. In the same way, the strong negative epistemic in (81c) can be indirectly conveyed not only by denying a **WI**-class neg-raiser as in (81a)/(82a), but also by denying the strong positive as in (82b), but this cannot be analyzed as an instance of the NRP, as the polarity evidence makes clear:

(82) a. It's not likely she'll accept your invitation (until you apologize).
b. It's not certain she'll accept your invitation (*until you apologize).

For Tobler, as for Epstein (1976:160) and L. Carlson (1983:120) a century later, the NRP is linked with—but not reducible to—ordinary litotes, the strengthening of a contradictory to a contrary. The difference lies in the partly conventionalized status of the former process, resulting in the absence of conscious irony (Tobler), the 'automatic' nature of the results (Epstein), and the 'lexical selectivity' (Carlson) associated with the process of so-called neg-raising. But what sort of conventionalization is involved here?

As we saw in §5.2, neither the grammatical nor the pragmatic approach to the NRP is without significant empirical and/or metatheoretical difficulties. On balance, a nonsyntactic account is clearly desirable, but not obviously possible. What is needed is a way of treating the lower-clause understanding of certain higher-clause negatives, and the syntactic correlates of that understanding, as a pragmatic association made possible by a certain semantic configuration, arguably involving the assumed disjunction of Bosanquet and Bartsch, together with the positive mid-scalar/**WI** criterion outlined in §5.2 and spelled out in more detail in Horn 1978b—an association which is, however, not freely available to every predicate in every language with the proper semantic qualifications. That is, we need a device for allowing at least some pragmatic rules to admit lexical exceptions.

Fortunately for us, just such a device is now on the descriptive market, having been designed to handle a superficially quite distinct phenomenon: the Indirect Speech Act. It has long been recognized that apparently synonymous expressions may differ in their indirect illocutionary act potential. Thus, the request in (83a) can be, and standardly is, indirectly conveyed by asking the question in (83b), but not (at least not without several additional degrees of indirectness) by asking the questions in (83c, d).

(83) a. Close the window.
b. Can you close the window?
c. Are you able to close the window?
d. Do you have the ability to close the window?

If (83b) conversationally implicates (83a) through the exploitation of Grice's Relation maxim, as Searle and others have supposed, this implicature should be nondetachable; its disappearance when we shift from (83b) to the basically synonymous (83c, d) is thus a mystery. It is just this mystery which motivates Sadock (1972) to reject the conversationalist line on indirect speech acts in favor of a theory which posits a semantic ambiguity for (83b), with the request reading constituting a SPEECH ACT IDIOM. Searle (1975:76ff.) counters with the thesis that 'there can be conventions of usage which are not meaning conventions', and that by these conventions 'certain forms will tend to become conversationally established as the standard idiomatic forms for indirect speech acts', for example, (83b) but not (83c, d) for indirect requests.

Carrying Searle's idea one step further, Morgan (1978) analyzes the inference from (83b) to (83a) as an instance of SHORT-CIRCUITED IMPLICATURE (SCI); such implicatures are in principle calculable (as are all conversational implicatures, by definition) but are not in fact calculated by speakers operating with the relevant usage conventions. Like Searle, Morgan finds

Sadock's speech act idiom thesis unparsimonious, arguing that it is, however, conventional to use (83b), with its literal interrogative meaning, to convey (83a). He proposes a variety of additional candidates for the status of SCIs, including the conventional use of the formula in (84a), but not the semantically akin (84b), to wish a performer good luck prior to a theatrical performance, and the use of (85a), but not (85b), as an emphatic affirmative.

(84) a. Break a leg!
b. Fracture a tibia!
Break your leg!

(85) a. You can say that again!
b. You can repeat that!
You're {able / permitted} to say that again!
It's possible for you to say that again!

Morgan notes that while these conventions of usage do not themselves constitute idioms (*John really broke a leg last night* ≠ 'John really performed well'), such a development—from metaphor to idiom, from usage convention to meaning convention—is quite plausible, and indeed frequently attested. Grice himself (1978 : 58) acknowledges that it may well be possible 'for what starts life . . . as a conversational implicature to become conventionalized', and the SCI is a natural halfway house along this route. As instances in which this shift has become complete, Morgan offers the valedictory *Goodbye* (from earlier *May God be with you,* but now sincerely utterable even by atheists) and the euphemism *go to the bathroom* (whose shift in literal meaning is demonstrated by the possibility of someone's dog doing so on the living room rug).[40]

The Searle-Morgan approach explicitly recognizes the existence of speakers' PRAGMATIC CONVENTIONS, alongside the better-understood conventions of syntax and semantics. Since the short-circuiting of implicatures is a matter of convention, we expect to find differences between speakers and between languages as to just which conventions of usage are operative. And indeed, as demonstrated by Searle (1975) and Green (1975), there is considerable cross-linguistic variation as to which questions can be used to convey which requests. But, as Green also points out, some indirect speech acts are more indirect than others, and the more indirect requests—like (86a, b) as hints for getting someone to close the window:

(86) a. The rain that's coming in the window is ruining the rug.
b. It's freezing in here.

do not exhibit the same variation. This is precisely to be expected if, as is plausible, hints involve non-short-circuited, and hence nondetachable, implicatures: only the literal meaning and extralinguistic context can be relevant for determining what is being hinted at, not the choice or form of expression used to convey that meaning.

One important morphosyntactic correlate of the short-circuiting of the request implicatum is the much-discussed preverbal *please;* as noted by Gordon and Lakoff, Sadock, and others, this item appears readily in (83b), as in the literal request (83a), but not in (83c, d). The fact that preverbal *please* is also ruled out in the hints of (86):

(87) a. Please close the window.
b. Can you please close the window?
c. Are you able to (*please) close the window?
d. It's (*please) freezing in here.

is correctly predicted if the distribution of this item is assumed to be sensitive to the usage convention posited by Searle and Morgan.

But the conveyed meaning of so-called neg-raising constructions can now be subsumed under the treatment prescribed for (other) indirect speech acts.[41] While questioning the hearer-based preparatory condition on requests is a natural way to convey that request indirectly (Searle 1969, 1975), only certain forms of those questions are conventionalized for that purpose. So too, while we have seen that predicates of a given semantico-pragmatic nature, that is, (more or less) positive **WI** propositional attitude predicates (a class including *think, believe, want, hope,* and *guess,* but not *know, try,* or *claim*), are candidates for the strengthening implicature permitting downstairs understandings of upstairs negation, whether a predicate within the appropriate class does in fact trigger the NR understanding depends on whether this implicature has been short-circuited into a usage convention.

This explains why we get exactly the same sort of variation across languages, dialects, and idiolects in neg-raisability that Searle and Green document for indirect speech act potential. Thus, the notion of short-circuited implicature captures just what is pragmatic (natural) and what is arbitrary (conventional) about the NRP.

An even stronger parallel can be drawn between indirect speech acts and the NRP. The Bartschean (and neo-Hegelian) pragmatic inference schema for generating NR understandings as strengthened forms of contradictory negation involves **R**-based implicature (Horn 1984b; cf. chapter 3). But this is (mutatis mutandis) just how Searle motivates indirect speech acts; in both cases, the additional conveyed meaning serves to strengthen and narrow the literal meaning in the appropriate context, via Atlas and Levinson's (1981) 'inference to the best interpretation'. Indeed, if scalar implicature constitutes the locus classicus for **Q**-based implicature, the stereotype of

R-based inference is the indirect speech act.[42] The detachability of the relevance-based neg-raising implicatum is simply an aspect of the general detachability of indirect illocutionary force, and both are to be treated through the admission of pragmatic conventions applying to a proper subset of linguistic expressions appearing in the relevant frame.

The distribution of preverbal *please* cited above is just one instance of a diagnostic item co-occurring with both direct and short-circuited indirect speech acts of a given type; other instances are cited in Horn and Bayer 1984:406–8. To take just one example, WH-queclaratives (cf. Sadock 1974) like (88a) are (pace Sadock) literal questions which are conventionally used to convey the corresponding universally quantified negative assertion (equivalent, of course, to a negative existential), here (88b):

(88) a. Who (but a total idiot) would have said a thing like that (?)
b. Nobody (but a total idiot) would have said a thing like that.

The parenthesized *but* phrase in these examples is normally a diagnostic for universally quantified assertions, as illustrated in (89):

(89) Everyone but Mary — Nobody but John
Anyone but Carter — *Somebody but Kim
Anywhere but here — *Somewhere but here
{All/*Most/*Many/*Three/*Some/None} of my friends but Chris
Everything but the kitchen sink
None but the brave deserves the fair. (Dryden)
No man but a blockhead ever wrote except for money. (Dr. Johnson)

Notice in particular that non-queclarative questions, which may implicate declaratives in a given context, but do not do so by convention, exclude this diagnostic:

(90) ?*Who but Leslie is coming to the party?

Seen in this light, the extension of the distribution of strict negative polarity items (NPIs) like *until, in weeks, for a year* from environments containing a tautoclausal negative to those in which the triggering negation is separated from the polarity item by a neg-raising predicate correlates directly with the presence of the SCI associated with such predicates, in the same way that preverbal *please* extends from direct to short-circuited indirect requests. Given that all English speakers conventionally use verbs like *think* and *suppose* so as to transmit their negation downstairs, while only some speakers operate with the analogous convention for *guess*, and none do for *hope*, the distribution of polarity *until* follows accordingly:

(91) a. I don't think they'll hire you until you shave off your beard.
b. %I don't guess they'll hire you until you shave off your beard.
c. *I don't hope they'll hire you until you shave off your beard.

In effect, whether a given predicate will be transparent to negation depends on the presence of the SCI; the acceptability of the NPI in contexts like (91) depends in turn on this transparency.

On the present analysis, (91a) is not rendered acceptable by being co-derived with (91′),

(91′) I think they won't hire you until you shave off your beard.

nor by sharing any semantic representation with (91′), but rather because it is conventionally used to implicate (91′). Since no such usage convention exists for any speaker with *hope*, the polarity item in (91c) is ruled out, even when a negative proposition is indirectly conveyed in a given context (see (55b) in §5.2).

This approach might illuminate a recurring perception that at first glance seems simultaneously real and ineffable. Tobler (1882b:204) sees the NRP as involving not the conveying of a contrary *tout court* so much as a fusion of the contrary and the contradictory understandings. Along the same lines, Bolinger (1968:23–24) suggests that 'the idea that in *I don't think he's coming* we have a negative element that belongs truly to the subordinate verb and that can be transferred, like a syntactic ping-pong ball, to another position without altering its logical connections, I think is not quite true. . . . It does not merely hop from one [clause] to another but belongs semantically to both'. And Cattell (1973:636) concurs that 'in the "negative transportation" sentences . . . there is a requirement that each clause be under the umbrella of a negation'. These three similar but independent observations might all be subsumed under the umbrella of the SCI approach to so-called neg-raising, on which the main clause negative belongs to one clause—its own—both syntactically and semantically, while being conventionally associated with another.

While conversational implicature is basically a matter of *parole*, the short-circuiting of implicatures into usage conventions takes place on the boundary between *parole* and *langue;* it is this feature which enables SCIs to trigger various linguistic correlates, including the strict NPI in (91a) and in the relevant dialect (91b). Indeed, for a significant range of their distribution, strict NPIs seem to be licensed by the conventionally signaled presence of negation in the clause or proposition in which they occur, whether the convention in question involves meaning or usage. Thus consider, alongside the paradigm in (91), the contrasts in (92) and (93):

(92) a. I'll be damned if I'll hire you until you shave off your beard.
b. *I'll be surprised if he hires you until you shave off your beard.

(93) a. Why get married until you absolutely have to?
b. *Why are you getting married until you absolutely have to?

In each pair, only the (a) construction is conventionally used to express the negative proposition which licenses *until,* that is:

(92′) I won't hire you until you shave off your beard.

(93′) You shouldn't get married until you absolutely have to.

The non-short-circuited negative implicata which may be associated with the (b) examples are insufficient to trigger such strict NPIs, although weaker (more lenient) polarity items may be acceptable in the same environments (cf. Linebarger 1981, 1987):

(94) I'll be surprised if he {hires anybody for that position/lifts a finger to help you}.

This pattern extends to other languages as well. Consider the distribution of strict NPIs in Japanese, as presented by McGloin (1976; 1982: chapter 4). These expressions, including adverbs like *kesshite* 'never', *dare no* 'nobody', and *mettani* 'rarely', normally 'require the presence of an overt negative morpheme in the same simplex sentence' (McGloin 1982:88), but this requirement is systematically relaxed in two sets of circumstances. The NPI can be separated from its negative trigger by a neg-raising predicate, as in (95):

(95) a. Konna ii hanashi wa mettani aru to wa omowanai.
'I think such an offer will rarely come along' (lit., 'I rarely think . . .')
b. Konna koro wa mettani atte hoshikunai.
'I want that such a thing would rarely happen' (lit., 'I rarely want . . .')

Here, *mettani* is taken to modify the embedded verb, yet its negative trigger *nai* is attached to the main verb.

But strict NPIs can also occur with no overt negation at all, provided that a negative proposition is conventionally signaled by the syntax (à la (92a), (93a)):

(96) a. Kesshite iku monoka. 'I will never go' ('No way I will ever go')
b. Dare mo iku monoka. 'Nobody will go'
c. Konna ii hanashi wa mettani aru monoka. 'A good deal like this will rarely come along'

Like *be damned if* and *why* suggestions in English, *monoka* 'no way' is negative not in form but in force; these are sentences in which the spirit of negation, like the ideal novelist of Flaubert's dictum, is everywhere present and nowhere visible. It is this covert but conventional negative force which allows the embedding of strict NPIs. Another such expression is *(-te) tamaru ka,* which McGloin glosses 'I'll be damned if ———':

(97) Anna yatsu ni wa kesshite makete tamaru ka.
'I'll be damned if I'll lose to that kind of person'

The same items which co-occur freely with *monoka* and *-te tamaru ka* cannot appear within the scope of constructions which may suggest a negative proposition, but are not conventionally used to do so, for example, *hatashite . . . daroo ka* 'I wonder if ———' (cf. English (92b), (93b)):

(98) *Hatashite dare-mo kunu daroo ka. 'I wonder if anybody is coming' (cf. (96b))

Just as in English (see (94)), more lenient NPIs—*amari, sonnani* '(not) too'—are freely triggered by implied negation, where the stricter items are blocked from occurring.

For McGloin, these facts suggest that the tautoclausal requirement normally associated with strict NPIs is systematically relaxed for expressions which 'conventionally imply a negative assumption of the speaker'; it is not clear whether she intends this characterization to extend to the neg-raising contexts of (95), but it certainly could—and I would argue, should—be so extended. But where McGloin alludes only to conventional implicature, the generalization covering both sets of cases requires us to adopt the same statement as for English: strict NPIs appear in clauses which are conventionally used to signal a negative proposition, whether the negation is syntactically overt, conventionally implicated, or conversationally implicated via an SCI. This approach to strict polarity is consistent with the comprehensive theories of negative polarity recently proposed by Ladusaw (1979, 1980) and Linebarger (1981, 1987), without directly following from either of them.

In the examples I have been considering, the critical environments for determining the acceptability of a given item are those where an expression not only can be used, but is conventionally used (within the relevant dialect) to convey something not literally expressed. It would appear that Searle's nonmeaning conventions, Morgan's short-circuited conversational implicatures, or some more precise and explanatory sharpening of these notions which has not yet emerged, will play an essential role in any successful account of the NRP and of other pragmatically or functionally based phenomena which are less than fully productive across a given syntactic construction type or semantic class.

But an important methodological issue remains to be addressed. Given our current state of knowledge, it must be conceded that ascribing some phenomenon to the presence of an SCI may amount more to labeling than explaining that phenomenon. By pushing the problem of variation in indirect speech act potential back to the pragmatics, we (along with Searle and Morgan) have in some sense reconstructed Sadock's speech act idiom analysis in different garb, rather than replacing it with a new, improved theory. By the same token, my treatment of the strengthening inference responsible for the NRP as a short-circuited implicature—like Tobler's rejection of the reductionist argument—tells us more about what so-called neg-raising isn't than about what it is.

Our goal must be to discover just why the variation in usage conventions should exist, and why it should exist just where it does. One means of constraining the application of the incredibly powerful device represented by SCIs, as Alice Davison has suggested to me, might be derived from the nature of the implicatures subject to short-circuiting. The two principal cases of short-circuiting I have explored here, indirect speech acts and the NRP, both involve an **R**-based implicature, as noted above, and it might be maintained that this is not a coincidence.

More specifically, we can note a functional kinship between the indirect speech act example in (99a) and the so-called neg-raising example in (99′a):

(99) a. I believe your answer is not wholly satisfactory.
b. Your answer is not wholly satisfactory.

(99′) a. I don't think your jumpsuit is entirely appropriate.
b. I think your jumpsuit is not entirely appropriate.

In both cases, the effect of the SCI is to express a given proposition (i.e., (99b), (99′b)) in a qualified, weakened, or hedged way. The short-circuiting in each case (as well as in the related nondeclarative indirect speech acts in (87b) and (93a)) seems to serve the aims of politeness or face-saving. This feature has often been acknowledged in connection with these constructions; cf. Searle 1975:64, Fraser 1975, and various papers by R. Lakoff on indirect speech acts as hedged assertions, requests, and so forth, and

Prince, Shnukal, and L. Carlson on the politeness or attenuation effect of so-called neg-raising.[43]

It seems plausible to derive the politeness effect from the fact that in each case the weaker (a) version is in effect pragmatically ambiguous between two understandings which stand in a privative relation (cf. Zwicky and Sadock 1975; Horn 1984a), with the stronger (b) version available to the addressee but not forced on him.[44] The analysis presented here, on which this stronger understanding is derived as a short-circuited implicature, correlates with the intuition that for most contemporary speakers the politeness associated with examples like (99a) and (99′a) is often felt as conventional or pro forma only and not really heartfelt.

This point is brought out nicely in the following excerpt from the remarks of *New York Times* executive editor A. M. Rosenthal, addressing the detention by Chinese authorities of *Times* reporter John Burns: '[The detention] is not only bad for John Burns, but it is not good for relations between China and the United States. . . . The idea that China is opening has taken hold in the United States and abroad and I don't think this will do it any good at all, to put it politely' (*New York Times* 22 July 1986). While hedging his remarks with four weakening devices, including a *not good* for *bad* in the first sentence and a triple hedge in the second (involving one indirect speech act, one instance of neg-raising, and one simple contrary-in-contradictory-clothing), Mr. Rosenthal effectively reveals the true intended force of his remarks in the coda. To put it somewhat less politely, for *I don't think this will do it any good at all*, read *This will do it a good deal of harm*.

My observation on the pro forma nature of negative politeness is not new. In his account of the NRP in French, Martinon (1927 : 536) comments on the gradual eclipse of the 'logical' embedded negation '*Je veux que vous ne sortiez pas*' by the 'illogical' (NR) reading of '*Je ne veux pas que vous sortiez*': 'Il est assez probable que cette forme illogique a été employée d'abord dans le dessein d'atténuer la rigueur de la défense; mais la défense est devenu tout ainsi rigoreuse dans cette nouvelle forme, dont elle a fait disparaître le sens propre'. The identical process, as traditional grammarians were well aware, characterizes the development of (other) euphemistic forms, as each successive proxy becomes in turn infected by the nature of the object being indirectly (and eventually directly) described. The diachrony of reference to human elimination affords the most graphic, but hardly the sole, illustration of this process; cf. *undertaker / mortician / funeral director; poor country / underdeveloped country / emerging nation.*

I am arguing that both indirect speech acts—in particular, those in which an act is indirectly performed via the assertion of the sincerity condition on that act, as in (99a)—and the NRP are essentially euphemistic. In each case the extension of a given expression is **R**-narrowed to a particularly (and negatively) charged subdomain of that extension. The speaker triggers this pragmatic narrowing by appealing implicitly to the addressee's ability to recognize that the speaker has sufficient social motivation to avoid the more direct (but more face-threatening) utterance. This socio-pragmatic procedure is carefully limned by Brown and Levinson (1978), whose account of understatement (pp. 268–70) predicts the pragmatic asymmetry between *not moral* and *not immoral,* or between *not good* and *not evil* (cf. our discussion of Ducrot's minimal pair in (71) above).

If, as Urmson (1952:484) notes, the 'parenthetical use' of verbs like *believe* constitutes a 'warning device' serving 'to modify or weaken the claim to truth which would be implied by a simple assertion', then the 'softening down' of negation effected by 'the shifting of *not*' (as perceived by Poutsma [1928:105]) is a corollary of this tendency to modify or weaken. Notice that the speaker's weakening (of a given assertion), by attenuating it through one form or another of indirection, is merely the other side of the illocutionary coin from the hearer's strengthening inference; if you utter (99a) or (99′a) to me, in the appropriate context, I must fill in what you left out, arriving at (99b) or (99′b), respectively.

I have maintained that the phenomenon of so-called neg-raising, while sharing the restriction to unmarked (positive weak intolerant) scalar values governing other instances of **R**-based strengthening of a formal contradictory negative into a functional contrary (cf. (77) above), differs from simple negative understatement (i.e., (77a)) in its degree of conventionalization. Thus, Tobler's argument against collapsing the NRP with ordinary litotes is vindicated. At the same time, I should stress that the strengthening inference responsible for the NR effect remains under this analysis a conversational implicature (albeit one which is short-circuited), an aspect of usage (albeit of conventionalized usage) rather than of literal meaning. But lexicalization through a negative affix literalizes the strengthened, contrary understanding which the NRP pragmatically sanctions. Thus, to reprise an example from §4.5, we get the contrast in (100), where only the contrary (inner neg, lower-clause) understanding can be lexicalized:

	outer-neg reading	inner-neg reading
(100) a. It's not {likely/probable} that a fair coin will land heads.	**T**	**F**
b. It's {unlikely/improbable} that a fair coin will land heads.	—	**F**

Disbelieve differs from *not believe* in the same way.

We thus distinguish three degrees of conventionalization of the negative strengthening inference, corresponding to the three environments distinguished in (77). But even in the case of simple understatement, the process depicted in (77a), there are pockets of conventionalization. Consider, for example, the use of negated intensifiers, as in the collocations of (101),

(101) Pat is not {especially / overly / particularly / so / terribly / that / too / very} bright.

where a speaker denies the location of an argument at a high point on a positive scale to convey its actual placement on the corresponding negative (unfavorable) scale.

This use of negative understatement has been the focus of not a little attention, most notably on the part of Bolinger (1972: 115–25), who points to the heterogeneity of the class of qualifiers appearing in the frame of (101). While *not very* ***Adj*** has functioned for some time as a gentle equivalent of 'rather un-', to adopt the OED's gloss, the attenuation produced thereby has come to be felt as too conventional (as with the process affecting euphemism and the NRP noted above), with the result that newer forms have been innovated. Bolinger cites examples like those of (102),

(102) He's not overly bright. (rather underly bright, rather stupid)
She isn't too much of a housewife.
You weren't too careful that time, were you?

where the affirmative evaluation being (literally) negated is itself evaluatively negative (*He's overly bright, She's too much of a housewife, You were too careful*). 'Presumably', Bolinger comments, 'it is safer to point out that something is not excessively good than to say it is not very good'.

Notice that the corresponding e-neg affirmative in such cases is not the proposition really being negated here—*not too* ***Adj*** ≠ 'not [too **Adj**]'—a fact which predictably arouses the ire of prescriptivists. Warner (1946: 302–3) detects 'an aura of depreciation or stoicism' in the use of this illogical 'absolute negative comparative':

> 'How are you?' draws the response, 'Not too good'. Plays are 'not too clean', race horses run 'not too fast', the world outlook is 'not too happy', and statesmen are 'not too optimistic'. . . . In all such usages there is no comparison to be completed. The discussions do not really concern too clean plays or too fast horses or over-optimistic statesmen.

But this nonequivalence is hardly limited to the *not too* construction: to be *not quite satisfied* is not simply to fail to be *quite satisfied* (as noted by Bolinger 1972: 101, the two expressions differ in register as well, with *not quite* ***Adj*** occupying a more colloquial slot than *quite* ***Adj***); to be *not very* *tall* or *not exactly happy* is something altogether different from the failure to be *very tall* or *exactly happy* (if the latter collocation even occurs: cf. *?#He's exactly happy*).

The strongest condemnation, not too surprisingly, is reserved, not for those negated intensifiers which merely fail to correspond to their positive counterpart, but for those to which there is no positive counterpart at all. An irate editorialist for the *Boston Herald-Examiner* (1 August 1972), evidently unfamiliar with Buyssens 1959 and Baker 1970 on negative polarity, chose to fulminate as follows against the *not all that* ***Adj*** construction:[45]

> 'All that' to introduce an adverbial [*sic*] phrase must, to be used properly, refer to something. In a common usage today it doesn't. It is used to provide emphasis, without any reference whatever, and when it is used that way it is misused. For instance, in a discussion of almost anything one suddenly reads that it 'really is not all that hard'. . . . All what hard, for heaven's sake? Or a cohort will tell us he had a good time at the theatre but 'the show really wasn't all that good'. Eh? How good is all that? And by what grammatical contortion does 'all that' pre-empt 'very', 'extremely', or a number of adverbs that would make more sense?

Ironically, of course, even if it 'would make more sense' to say *The show is not very good,* the sense it would make is normally not that of the contradictory negation of *The show is very good.*

The *Herald*'s second proffered option, *extremely* (*The show was not extremely good*), does not (at least for me) allow the strengthened understanding available for its apparent synonyms in (101). Similarly, I find it hard to assign indirect force to certain other negated intensifiers, including *thoroughly* and perhaps *absolutely,* although there seems to be no principled synchronic reason for this discrepancy; thus compare *I'm not thoroughly pleased* (which supports only a literal understanding) with *I'm not {altogether / totally} pleased* (which can convey 'I'm rather displeased'). Still other intensifiers occur only affirmatively (Bolinger [1977:26] lists *awfully* in this category). In this respect, as in the diachronic shift (i.e., the innovation of *not too*) signaled by Bolinger and in its spread to intensifiers which do not occur affirmatively, the strengthening inference associated with the NEG-intensifier-**Adj** construction seems to involve some degree of nondetachability, that is, of conventionalization of usage.

The strengthening rule in question, which can be characterized as in (103),

(103) ***(not + intensifier) + Adj***$_i$ --> *rather un-****Adj***$_i$
(or --> *rather* ***Adj***$_j$, where ***Adj***$_j$ is the antonym of ***Adj***$_i$)

largely shares the restriction to unmarked (e-pos) adjectives of the patterns in (77), as seen in (104); cf. Langendoen and Bever (1973) for related examples.[46]

(104) not particularly {friendly / †unfriendly}
not too {happy / †sad}
not overly {bright / †stupid}
not all that {sympathetic / †unsympathetic}
not especially {optimistic / †pessimistic}
(where † marks unavailability or marginality of litotic reading)

This is to be expected: the mutual knowledge that one should seek to avoid the direct categorization of someone or something e-negatively as *unfriendly, sad,* or *stupid* motivates the indirect (litotic) means of casting the evaluation, which is then unraveled by the addressee via the process in (103), while the usual absence of any reason to avoid categorizing someone or something e-positively as *friendly, happy,* or *bright* precludes the natural use of the more complex indirect form to convey this meaning.

But unlike the patterns in (77), the rule in (103) is not limited to relatively weak positive scalar adjectives. The effect of litotes or rhetorical understatement can be found in the examples of both (105) and (105′):

(105) not too happy
not especially bright
not exactly pleased

(105′) not too ecstatic
not especially brilliant
not exactly thrilled

Even e-neg predicates allow negative litotes in context (recall Tobler's examples in (80) above): cf. *We weren't exactly on the worst possible terms, He's not exactly stupid, She doesn't exactly hate you.* This freedom is precisely what we should expect if the strengthening involved in these examples operates through a conscious application of ironic interpretation, as Tobler suggests (cf. Grice 1975, Wilson and Sperber 1981 for suggestions on how such an interpretation might work). In fact, these same ironic readings are available when the intensifier is absent, especially when the scalar predicate in the focus of negation is assigned heavy stress—*He isn't* **brilliant**, *He's not* **stupid**, *I'm not* **thrilled** *with that proposal*—but once again, as with Green's hints, the effect is more indirect and more subject to the vagaries of the linguistic and extralinguistic context than in the more conventionalized patterns.[47]

There is one strong candidate for the status of counterexample to the claim that only unmarked, positive midscalar or weak intolerant predicates allow the (automatic) strengthening of their apparent contradictory negations to contraries. This candidate emerges from the most basic e-pos/ e-neg antonymic pair of them all. While *right* differs from *wrong* as predicted—to say that something (just) isn't right is often tantamount to saying that it's wrong, but not vice versa—we find that *bad* seems to allow negative strengthening as easily and fully as *good*. The occurrence of *not bad* and *not half bad* as positive descriptors equivalent to 'pretty good' or even 'very good' is remarked on by Stoffel (1901:126), who sees in this usage a reaction against exaggeration, a 'studied modesty of expression', a 'shirking of the least semblance of hyperbole'.[48]

Indeed, it is this particular inference which represents the core instance of the category:

> **litotes**: understatement in which an affirmative is expressed by the negative of the contrary (as in 'He's not a bad ball-player')
> (from *Webster's Third New International Dictionary*)

It is only by an extended usage that the term has come to be identified with rhetorical understatement in general.

It has long been recognized that the litotic understanding of *not bad* is contextually—and intonationally—determined: '*Not bad*, taken literally, leaves a large latitude, from *indifferent* to *excellent*, and may mean [*sic*] either, depending on the intonation used and the circumstances' (Stern 1937:312). For Bolinger (1972:115), *not bad* with 'a terminal fall rise' damns with faint praise, suggesting that this is the best or most positive evaluation the speaker can muster, while *not bad!* with 'an intonation of surprise' conveys 'very good'—praising, as it were, with faint damn.

Cutler (1977) draws the same distinction: in dialogue (106B_1) the effect of the fall-rise is 'to negate the literal reading of the utterance and convey instead the speaker's opinion that the color scheme is not good' (especially if accompanied by a wrinkled nose?), but in (106B_2) the speaker conveys that it's 'pretty damn good':

(106) A: How do you like my new color scheme?

B_1: Not bad. B_2: Not bad.

Cutler compares the effect produced by the fall-rise in (106B_1) to the application of 'ironic intonation' on, for example, *Sue's real smart*, but it is not clear that the analogy is motivated. As Ladd notes (1980:218), the fall-rise in Cutler's *not ˇbad* and in related midscalar evaluations (*OˇK, all ˇright, I ˇlike her*) is fully compatible with the 'focus within a given set' meaning he assigns to this contour in general. The fall-rise here 'sets up a hierarchy of possible interpretations and explicitly puts the evaluation given somewhere between the possible extremes'. Hence the 'nuance of mediocrity' (cf. Sapir's 'zone of indifference'), absent from the parallel examples with simple falling tone, for example, *not `bad* ($106B_2$), *O`K, all `right.*[49]

The two conflicting inference schemata associated with *not bad* should come as no surprise. As I noted earlier, instances of 'maxim clash' signaled by Grice and others can in general be attributed to the dialectic tension between the lower-bounding, hearer-based **Q** Principle, which induces upper-bounding implicata (cf. chapter 4), and the upper-bounding, speaker-based **R** Principle, which generates lower-bounding implicata (those discussed in this chapter). This tension is clearly operative in the case of negated scalar values, particularly when neither inference has become institutionalized (via incorporation or the NRP).

Thus, Ducrot notes (1972: 132) that an addressee, on hearing (107a), may infer that the speaker intended to convey (107b), through the exploitation of the (**R**-related) 'Loi d'informativité', that is, the assumption that the hearer does not already know the information the speaker is conveying.

(107)	a. La situation n'est pas excellente.	'The situation isn't excellent'
	b. Elle est franchement mauvaise.	'It's pretty bad'

But since the speaker has not said (107b), he can always retreat to the literal meaning of what he has said, that is, (107a), with its 'large latitude' of denotative meaning, in Stern's words. On the other hand, given the (**Q**-based) 'loi d'exhaustivité' (the principle which demands that the speaker provide the strongest possible information he possesses which may interest the hearer; cf. Ducrot 1972: 134 and §4.2 above), someone who utters (107a) may, in the appropriate context, implicate that the situation is pretty good.

While these same two interpretations are present as well in the case of *not bad,* as noted by Stern, Bolinger, and Cutler, there does seem to be a greater degree of conventionalization built into the **R**-implicated interpretation here. In Tobler's terms, there is no longer any conscious sense of irony accompanying this inference, either in the English phrase or in its French adverbial equivalent. The characteristic intonation contours assigned to the two understandings, whatever their precise characterizations, represent one clue that no on-line inferencing is necessarily practiced here.

A second clue, at least for the French version, is syntactic. Not only can we detect, at least since Molière, a functional correspondence in familiar speech, signaled by Le Bidois and Le Bidois ([1935] 1968: §1697) inter alia, between *pas mal* ('pris d'une façon ironique') and *assez* ('rather, quite'), but *pas mal de* serves as an element on the positive quantificational scale, located (by Le Bidois) somewhere between *assez* and *beaucoup de*. We also find a conventionalization within a conventionalization; *Elle n'est pas mal* (lit., 'She isn't bad') has taken on the secondary idiomatic reading 'She's quite good-looking'.

But to what do we owe the development of this conventionalized litotic use or meaning of *not bad/pas mal*, given that it is normally the unmarked, positive scale adjectives (e.g., *good*) whose negations tend to be interpreted via **R**-based strengthening as conveying values on the corresponding contrary (e-neg) scale? The answer may be found in the function of **R**-based meaning shift, as touched on above.

Both **Q**-based and **R**-based lexical narrowing are amply attested in diachrony (cf. Horn 1984a–c for examples). The former variety is linguistically motivated, in the sense that an already existing lexical item (often more basic or more fully integrated into the linguistic system) serves to limit or restrict the use—and sometimes eventually the meaning—of a more productively formed lexical item or expression (see references on BLOCKING cited in §5.1.1). Thus the existence of *thumb* tends to restrict the domain of *finger* to nonthumbs (even though a thumb is a finger); in the same way, *rectangle* tends to denote nonsquares, given the existence of the more specific, more informative item *square*.

But **R**-based narrowing, the shift of a lexical item originally denoting a category or set to one denoting a salient subset or stereotypic member of that set, is not linguistically, but culturally or socially, motivated. The clearest instance, as we have seen, is euphemism, which prompts the avoidance of more specific information (and hence the overriding of the **Q** Principle). When *drink* and *smell* take on narrowed readings denoting a particular type of drinking and smelling (*He drinks too much, Something smells around here*), with the resultant autohyponymy of the lexical item (cf. Horn 1984a), it is because we can count on an addressee who shares our culture to be able to figure out just which salient, highly charged member of the extension the speaker would have sufficient reason to avoid naming directly. The same reasoning applies to standard examples of euphemism in which the meaning shift has progressed even further (*go to the bathroom, sleep with, pass away, disease, accident*).

In the case of the euphemisms I have been exploring in this chapter, it is normally some e-neg attribution that a speaker would be motivated to refrain from expressing directly, leading her to employ litotes (the negation of the contrary) to convey the e-neg attribution while avoiding negative face and preserving delicacy (and deniability). In the case of *not bad*, however, a direct e-pos attribution is avoided and the denial of the e-neg form is employed litotically to designate this positive value indirectly.

We can see in the substitution of *not bad* for *pretty good* another form of euphemism or taboo avoidance, one prompted, not by politeness, but by the same factors that led the Greeks to call the Furies (Erinnyes) the Blessed Ones (Eumenides), or the Russians to call the bear 'honeyeater'. In various cultures, the real names of gods, demons, ancestors, or other feared and/or potent objects are avoided, and a more indirect means of reference is conventionally settled on. Within certain Western cultures, there seems to be a similar covert taboo against direct reference to positive evaluation, particularly when emotions or assessment of self-worth might be involved. Good feelings are not directly acknowledged, lest one jinx the source of those feelings (compare the nonverbal practice of knocking on wood); nor, within our sophisticated—if not cynical—culture, do we want to appear too positive or enthusiastic (recall Stoffel 1901 on the shirking of hyperbole). When emotive or subjective evaluations are not involved, the markedness built into (77a) takes over and the asymmetry between *not good* and *not bad* reemerges:[50]

(108) A: How are you feeling?
B_1: Not good. [--› fairly bad]
B_2: Not bad. [--› fairly good]

(108′) A: How's the battery?
B_1: Not good. [--› fairly bad]
B_2: Not bad. [-/-› fairly good]

When the litotic understanding of *not bad* does arise (as in (108) but not (108′)), there is the sense that the evaluation is an admission, reluctantly conceded by the speaker, often with a kind of grudging admiration (*'Not bad', he admitted*).

Under certain circumstances and in certain subcultures, both positive and negative direct evaluations are regularly to be avoided. One case in point is the Minnesotan dialect of English, as described in the Minnesota Language Systems self-teaching cassettes advertised on the Prairie Home Companion; see Mohr 1987. Here are some relevant excerpts from "The Power of the Negative" (lesson 2 from Mohr 1987):

> Minnesotans prefer to express their positive feelings through the use of negatives, because it naturally levels things out. . . . If you just got married or bought a late-model pickup under book price with low mileage and hardly any rust, or it's dawn on opening day of the duck season, a Minnesotan would say
> —'I wouldn't want you to think I'm not happy'.
> That's a strong statement here.
> *Not too good* and *not so good* are worse than *not too bad* and *not*

> *so bad.* . . . When somebody asks you how you slept on the guest bed with the bar that cuts across your back and gives you shooting pains down your legs, you will say 'Not too bad' because you don't want to hurt their feelings, but how you actually slept was *not too good.* (Mohr 1987:6–9)

In some contexts, then, second-order euphemisms are the order of the day.[51]

Nor is this linguistic behavior restricted to speakers of Minnesotan. In a *New Yorker* 'Talk of the Town' squib (4 February 1985, p. 32), a student of the late Robert Fitzgerald describes her participation in a versification class at Harvard, in which the renowned non-Minnesotan classicist and poet would assign grades based on the following scale (given in descending order):

(109) NB
NTB
NTG
NG

'There was also a rare NAAB—Not At All Bad', the student recalls, 'and anyone whose paper was so graced was jealously fêted'. With Fitzgerald's system of evaluation, the practice of attenuation has perhaps reached its modern pinnacle.

We have seen in the strengthening of contradictory to contrary negation a classic **R**-based inference pattern, functionally related to euphemism, motivated by the goal of avoiding the direct assertion of some negative proposition in a context in which it would tend to offend the addressee, overcommit the speaker, or otherwise count as inappropriate. Far from constituting a 'peculiar Talent . . . of Ladies, Whisperers, and Backbiters', as Martinus Scriblerus ([1727] 1952:115) would have it, litotes is a fundamental means for conveying a strong negative proposition while observing the amenities of civilized social interchange.

(At least) one question remains to be addressed: Why is it not just any positive value, but (almost) always the unmarked (positive weak intolerant) scalar value whose negation is pragmatically strengthened? While any answer would be necessarily speculative, I would begin by observing that it is harder to contextualize the negation of a marked term—*I'm not ecstatic, I'm not sad, I'm not miserable,* as opposed to *I'm not happy*—without constructing a discourse frame in which the negated term had itself been previously applied. In such discourse frames, there is no functional motivation for moving beyond the straightforward (contradictory) interpretation assigned by the syntax. Indeed, when the negation of an unmarked term appears in a context of this type, it too is interpreted as a contradictory of the corresponding affirmative:

(110) A: Are you happy?
B: No, I'm not happy.

(110′) A: So you believe the fog will lift?
B: No, I don't believe it will lift.

But unlike the marked cases, the negation of an unmarked value may felicitously initiate an exchange (*I'm not happy, I don't believe the fog will lift*), in which case the addressee—given Negative Uninformativeness and the Division of Pragmatic Labor (cf. §3.3)—will tend to strengthen the negative statement into an informationally sufficient proposition affirming the contrary.

The correlation of the processes discussed in this chapter involves one repeated premise: **in a context licensing the pragmatic assumption *p* v *q*, to assert *not-p* is to implicate *q***. Thus, a formally contradictory negation **not p** will tacitly convey a contrary assertion—but only when **p** is a relatively weak positive scalar predicate, representing the unmarked term in its contrast set. Conversational (Gricean) and discourse (Praguean) principles converge to predict that it is just in such cases that a contradictory negation **not P** will tacitly convey a contrary assertion—but only when **P** is a relatively weak positive scalar predicate, representing the unmarked term in its contrast set. Conversational (Gricean) and discourse (Praguean) principles converge to predict that it is just in such cases that a contradictory negation may be strengthened (or 'filled in', à la Bosanquet) to yield a tially fossilized as a short-circuited implicature or convention of usage in the NR cases of §5.2 (whence the unmediated nature of the inference and the lexical exceptions associated with the NRP), and is partially or fully conventionalized in the lexicalized affixal negations of §5.1.

6 Metalinguistic Negation

> With quantitative terms *not* nearly always means 'less than' . . . but exceptionally these combinations [*not once, not much, not three, not half full*] may convey another meaning; this is the case if we stress the word following *not* and give it the peculiar intonation indicative of contradiction, and especially, if the negation is followed by a more exact indication: not lukewarm, but really hot; not once but two or three times, etc. (Jespersen 1933:300–301)

> It's not a car, it's a Volkswagen. (VW commercial and advertisement)

I must now set out to redeem one of my promissory notes. As Jespersen observes in the epigraph, there is an 'exceptional' or marked reading available for scalar negation, one which seems to be incompatible both with the 'less than' gloss he provides for ordinary negation and with the Gricean theory of scalar predication I outlined in chapter 4. In this chapter, essentially a revised and expanded version of Horn 1985, I shall entertain the possibility that Jespersen's exceptional reading can be considered, along with at least some instance of so-called semantic external negation (cf. chapter 2), as a metalinguistic use of the negative operator.

As we saw in chapter 2, so-called EXTERNAL negation is standardly exemplified by the reading of (1) which is forced by the continuation in (1′)

(1) The king of France is not bald.

(1′) The king of France is not ˇbald—(because) there is no king of France.

and which is true if France is a republic; the INTERNAL reading, by contrast, is either false or truth-valueless in the same circumstances. The semantic ambiguity of negative sentences is adopted, within otherwise different theories, by Aristotle, by Russell (1905), by Karttunen and Peters (1979), and by proponents of three-valued logics. But the major recent trend among philosophers and linguists, represented by Atlas (1974, 1977, 1979, 1981), Kempson (1975, 1986), Gazdar (1979a), and Carston (1985a, 1985b), has been to reject this putative ambiguity, along with the purported existence of truth-value gaps and semantic presuppositions, and to assimilate all instances of natural language negation to a single truth-functional and/or semantically general operator.

Both views contain much insight and some truth, yet both are incomplete. While two distinct uses of sentential negation must indeed be admitted, the marked, nondescriptive variety is not a truth-functional or semantic operator on propositions, but rather an instance of the phenomenon of METALINGUISTIC NEGATION—a device for objecting to a previous utterance on any grounds whatever, including the conventional or conversational implicata it potentially induces, its morphology, its style or register, or its phonetic realization.

The chapter is structured as follows. First, in §6.1, I review the earlier discussion from chapter 2 on the alleged semantic ambiguity of negation and consider the counterproposal of the monoguists in which all instances of negation are assimilated to a single operator. In §6.2 I present evidence to support the view that negation must be taken as pragmatically ambiguous, with marked negation constituting an extended metalinguistic use of the ordinary descriptive (object-language) operator. At least some tokens of external negation can be identified with a more general phenomenon, a speaker's use of negation to signal his or her unwillingness to assert, or accept another's assertion of, a given proposition in a given way; metalinguistic negation focuses, not on the truth or falsity of a proposition, but on the assertability of an utterance. We shall also see that other logical operators (*and, or, if-then,* and *WH*-binding) display extended metalinguistic uses of their own. In §6.3 the interaction of metalinguistic negation with scalar and nonscalar implicature is considered, with special attention devoted to the use of negation to cancel the upper-bounding **Q**-based implicatum induced by scalar predications and to the immunity of **R**-based implicata to metalinguistic negation. Section §6.4 is devoted to the investigation of three morphosyntactic correlates of the descriptive-metalinguistic dichotomy: the inability of metalinguistic negation to incorporate prefixally, to trigger negative polarity items, and to co-occur with one particular variety of *but* clause. We shall see that the two uses of negation interact in an interesting way with the two *but*s of English and other languages. In §6.5 I examine other recent approaches to the unity or duality of natural language negation in the light of my own account and in §6.6 I offer some concluding remarks, along with a brief cross-linguistic excursus into some of the surface manifestations of metalinguistic negation.

6.1 On the "Ambiguity" of Negation

As we have seen (§2.2), the origin of semantic external negation can be traced to the (re)discovery by Russell (1905) of the apparent ambiguity of sentences like (1), in which a subject description may be taken alternately as inside (1a) or outside (1b) the scope of negation.

(1) The king of France is not bald.
a. INTERNAL: $\exists x(Kx \wedge \forall y(Ky \rightarrow y = x) \wedge \sim Bx)$
[The king of France is not-bald]
b. EXTERNAL: $\sim\exists x(Kx \wedge \forall y(Ky \rightarrow y = x) \wedge Bx)$
[not (the king of France is bald)]

The former is false and the latter true if there is no king (or more than one king) of France.

While the modern parents of logical presupposition, Frege (1892) and Strawson (1950), disregarded the marked reading (corresponding to (1′) and (1b)) on which the existential commitment normally associated with descriptions or names (cf. Frege on *Kepler {died/did not die} in misery*) is canceled or suspended, the development of three-valued logic—from Lukasiewicz and Bochvar to Smiley and Herzberger (see §2.4)—was largely prompted by the need to accommodate Russell's intuition that (1) and analogous sentences containing both negation and presupposition-inducing terms are semantically ambiguous. On the normal reading, with internal (choice) negation, presuppositions are preserved and truth-value gaps arise when one or more of the presuppositions fail; on the marked reading, with external (exclusion) negation, presuppositions are potentially removed or transformed into simple entailments and the sentence is bivalent (see §2.4 for details).

Different theories of presuppositional logic propose different characterizations of the two negations; the ambiguity may be lexical (with the presupposition- and truth-value-gap-preserving operator in $\neg\mathbf{p}$ opposed to the presupposition-canceling and bivalence-preserving operator in $-\mathbf{p}$) or scopal. Under the latter option, an abstract truth connective is typically introduced, with external negation defined in terms of ordinary negation applied externally to the proposition formed by this connective. The external reading of (1) is thus identified with the English sentence (2):

(2) It is not true that the king of France is bald.

This identification is often supported by an appeal to the purported intuition that it is indeed the syntax of (2), rather than that of (1), which is normally used to express the external, presupposition-free reading of negation, and that (2) allows only this interpretation.

However the analysis is to be worked out, the existence of marked negative statements which are clearly true when their affirmative counterparts—singular expressions, factive predications, category mistakes—are (ex hypothesi) neither true nor false has led most proponents of logical presupposition (but cf. Burton-Roberts (1987) for a dissenting view) to conclude that natural language negation must be treated as ambiguous, either by allowing dual interpretations for a single surface operator or by providing dual scope possibilities for negation in logical form.

The ambiguist position on negation has been supported recently by Martin (1979:43): 'Methodologically, of course, the multiplication of senses beyond necessity is undesirable . . . but a few such ambiguities, especially that of negation, seem perfectly reasonable'. Perfectly reasonable to some, perhaps, but not to those sharing the perceptions of Lycan (1984:97) concerning 'Martin's repellent proliferation of lexical ambiguities'.

For those in Lycan's camp, the principal difficulty for the ambiguist view of negation sketched above (and in more detail in chapter 2) is that it is by no means obvious that negative sentences like (1) are semantically ambiguous. Furthermore, as stressed by Atlas (1974), neither the periphrastic construction in (2) nor its counterpart in (2′)

(2′) It is not the case that the king of France is bald.

clearly disambiguates (1) in favor of the presupposition-free external reading (1b), except within the suspect dialect of a readily definable subset of philosophers and linguists. This point is independently reiterated by Kuroda (1977:105): 'It seems to me that the expression *it is not the case that* is a ready-made all-purpose form of negation which can in most, if not all, cases submit itself to logical convenience where the occasion demands'. For Kuroda, as for Atlas, (2) and (2′) are just as compatible (or as incompatible) with the internal, presuppositional understanding of negation (Kuroda's DENIAL negation) as with the external, presupposition-free understanding (his PROPER negation).

Within the last decade, a countervailing consensus in opposition to the ambiguists has emerged from the work of Allwood (1972), Atlas (1974), Kempson (1975), Boër and Lycan (1976), Gazdar (1979a), and Lycan (1984). Where Russell struggled with diligence and ingenuity to untie the Gordian knot constructed by the king of France, these monoguists—wielding Occam's razor as a samurai sword—seek to sever it with one blow; for them, negation is simply not ambiguous, in either meaning or scope.

The burden of proof is clearly on the ambiguist, as the defensive tone in Martin's passage indirectly concedes; like other abstract *entia*, senses *praeter necessitatem non multiplicanda sunt*. This is Grice's 'Modified Occam's Razor principle' (Grice 1978:118–19), and Ziff (1960:44) advances the same doctrine as 'Occam's eraser'.

It is, moreover, exceptionally difficult to prove that the presupposition-laden internal understanding and the presupposition-bare external understanding of (1) are semantically distinct, given that the former unilaterally entails the latter: if the existent king of France is not-bald (or nonbald), it is certainly not the case that the present king of France is bald (where the latter negation is taken externally). What the ambiguist must demonstrate between internal and external readings is a privative ambiguity, of the sort claimed to hold for such examples as (3).

(3) I just bought a dog.
  a. *canis familiaris,* male
  b. *canis familiaris*

Yet as I noted in §5.2, it is just such ambiguities which are the hardest to substantiate by the standard linguistic tests.[1]

For the linguist, a particularly telling argument against the ambiguist position is the fact (noted by Gazdar 1979a:65–66) that no natural language seems to employ two distinct negative operators which correspond directly to internal and external negation. This is especially striking when we consider the many languages which do contain two or more negative markers. In particular, as exemplified in (4), a number of languages draw an opposition between an ordinary declarative negation and some marked form (often labeled 'emphatic') which is restricted to embedded nonfinite and/or nonindicative contexts, frequently co-occurring with subjunctive mood (cf. Horn 1978a: §5 on one motivation for this pattern).[2]

(4)

	Unmarked Declarative Negation	Marked (Emphatic) Negation
Cantonese	m	m hai
Old English	ne	no, na
Estonian	ei	mitte
Ancient Greek	ou	mē
Modern Greek	ðen (< ouðen)	mi
Hungarian	nem	ne
Latin	nōn	nē
Modern Irish	nach	gan
Sanskrit	na	mā
Tagalog	hindi	huwag
Yoruba	kò	má

In other languages, including French, negation can be marked in a variety of morphologically distinct or overlapping ways, depending on the syntactic environment and the semantic context (cf. Gaatone 1971; Heldner 1981):

(5) (ne) . . . {pas/point/aucun/personne/rien/jamais}
  {aucun/nul/personne/rien . . .} ne . . .

non (pas) + **Adv**
in-, non- + **stem**

Similarly, Swahili sports several types of suppletive and redundant marking for negation:

(6) a. ni- na- ku- pend- a vs. si-ku-pend-i
I -PRES-*you*-*love*- IND ⌊ NEG ⌋
'I love you' 'I don't love you'
b. tu- na- ku- pend- a vs. ha-tu-ku-pend-i
we-PRES-*you*- *love*- IND ⌊ NEG ⌋
'we love you' 'we don't love you'

In some languages the marker for sentence negation may be determined by the tense / aspect of a given sentence, in others by the grammatical category of the predicator (cf. Payne 1985 for examples and §7.1 below for additional discussion). Evidently, languages freely utilize morphologically differentiated negative forms for syntactic, semantic, or synchronically arbitrary reasons, but never, significantly, for marking the one distinction which Russell and the three-valued ambiguists would lead us to expect.

It might be suggested that the external variety of negation need not display a separate morphological coding, since it is associated with the occurrence of *true* (or *the case*) within its scope, as suggested in the earlier discussion. Thus, Karttunen and Peters (1979:47) propose that 'the "external" negation of ϕ . . . might be rendered into English as "It is not true that ϕ"'. But this approach is vitiated by the fact, noted above, that the occurrence of the formula *It is not {true / the case} that* is neither a necessary nor a sufficient condition for the emergence of a nonpresuppositional understanding of a negative sentence. A semantic theory which invokes an abstract truth predicate (TRUE, as in Linebarger 1981) at either the object-language or metalanguage level, as a kind of *animus ex machina* for just those sentences whose negation seems to behave externally, is as unconvincing as a syntactic theory which invokes phonologically and semantically null inaudibilia without providing solid independent motivation for the existence of such constructs.

Another fundamental problem for the utilization of a truth predicate in the representation of external negation is the fact (to be explored in more detail in §6.5 below) that regardless of the analysis I provide for the relation of truth to negation, the function of TRUE as a metalinguistic operator in a truth-conditional semantic theory cannot be directly assimilated to the behavior of *true* (or of its cross-linguistic counterparts) in ordinary language.

As mentioned above, the monoguist thesis on negation is usually taken to preclude an effective defense of semantic presupposition. Indeed, monoguists like Atlas, Kempson, Boër and Lycan, and Gazdar have been more eager than reluctant to jettison any remnant of semantic presupposition. But at least one major approach has attempted to combine an ambiguist line on negation with a nonsemantic—or at least non-truth-conditional—analysis of presuppositional phenomena. This is, of course, the formal theory of conventional implicature represented in Karttunen and Peters 1979 and earlier works (cf. also Grice 1975).

Briefly reviewing the discussion in §2.5, we recall that for K&P, (7a) and (7b) both conventionally implicate (7′), provided the latter is read as an ordinary negation.

(7) a. Chris managed to solve the problem.
b. Chris didn't manage to solve the problem.

(7′) It was difficult for Chris to solve the problem.

But in some contexts, especially with the right intonation contour (cf. Liberman and Sag 1974 on the contradiction contour and Ladd 1980 on fall-rise) and an appropriate continuation, (7b) is realized as (8), with a CONTRADICTION negation assigned wide scope to the potential implicatum (7′).

(8) Chris didn't ˇmanage to solve the problem—it was quite easy for him.

Karttunen and Peters point out (1979:46–47) that contradiction negation—unlike ordinary, conventional-implicature-preserving negation—is incapable of triggering negative polarity items. Thus the ordinary negation of (9a) is (9b), where the existential shows up as *any* in the scope of negation. But with contradiction negation, as in (9c), no *some/any* suppletion is possible.[3]

(9) a. Chris managed to solve some problems.
b. Chris didn't manage to solve any problems.
c. Chris didn't manage to solve {some/*any} problems—he solved them easily.

Similarly, we find *already* rather than NPI *yet* in (10) (= K&P's (77b)), where contradiction negation removes the conventional implicatum associated with factive *forget*.

(10) Bill hasn't already forgotten that today is Friday, because today is Thursday.

K&P's contradiction negation is a plug to conventional implicata, while ordinary negation is a hole. (As we saw in §2.5, certain implicata—in particular, those associated with *even*—do not seem cancelable by contradiction negation.) The formal representations for ordinary and contradiction negation are repeated here as (11a, b) respectively (see earlier discussion for key to notational conventions).

(11) a. ORDINARY NEGATION OF ϕ: $\langle \neg\phi^{e}; \phi^{i} \rangle$
b. CONTRADICTION NEGATION OF ϕ: $\langle \neg[\phi^{e} \wedge \phi^{i}]; [\phi^{i} \vee \neg\phi^{i}] \rangle$

What the first part of (8) amounts to is thus a negated conjunction, with one conjunct corresponding to the entailment and the other to the implicatum of the corresponding affirmative:

(11′) a. (7b) as ORDINARY NEGATION:
implicature: It was difficult for Chris to solve the problem.
entailment: ¬(Chris solved the problem)
b. (7b) as CONTRADICTION NEGATION (à la (8)):
implicature: (vacuous)
entailment: ¬(It was difficult for Chris to solve the problem and Chris solved the problem)

We have arrived by now at the situation in (12). (Notice that for Strawson negation is unambiguously internal—by default, since he does not directly tackle sentences like (1′) or (8).)[4]

(12)

	Do truth-value gaps exist?	Do semantic presuppositions exist?	Is negation semantically ambiguous?	
Strawson:	yes	yes	no	
Aristotle, Russell:	no	no	yes	AMBIGUISTS
Lukasiewicz, Smiley, Herzberger, Katz:	yes	yes	yes	AMBIGUISTS
Karttunen and Peters:	no	yes (as conventional implicatures)	yes	AMBIGUISTS
Atlas[5], Kempson, Boër and Lycan, Gazdar:	no	no	no	MONOGUISTS

On the one hand, we have the original ambiguist thesis on negation in both its classical (Aristotelian-Russellian) and revisionist (three-valued) versions—as well as the K&P compromise position, which some would suggest has all the vices of each without the virtues of either. On the other hand, we have the monoguist antithesis which Occamistically denies any ambiguities in natural language negation, but offers no ready explanation for the intuition shared by ambiguists of all camps that negative sentences like (1) and (7b) may be used in two radically different ways, with (as K&P observe) distinct linguistic correlates in each case. Like Russell, we need a Hegelian synthesis—one, we must hope, more explanatory than the proposition that the king of France wears a wig.

6.2 Metalinguistic Negation and Pragmatic Ambiguity

They weren't people, Sir, they were the enemy.
(Lt. William Calley, on My Lai massacre victims, cited in Lang 1984:279)

In the synthesis I shall advocate here, negation is effectively ambiguous, contra Atlas, Kempson, and Gazdar. But contra Russell, Karttunen and Peters, and the three-valued logicians, it is not semantically ambiguous. Rather, we are dealing with a pragmatic ambiguity, a built-in duality of use. If I am correct, we must reject the classical (Fregean) view—cited by Prior (1967:459) and endorsed in various ways by virtually all previous analysts—that 'all forms of negation are reducible to a suitably placed "It is not the case that"'.

That we must reckon with a special or marked use of negation which is irreducible to the ordinary truth-functional operator is best seen, not in examples like Russell's (1)–(1′) or K&P's (7b)–(8), but in environments like (13), where what is negated is a conversational implicatum (viz. the **Q**-based inference of *not all* from the use of *some;* cf. chapter 4).

(13) a. Some men aren't chauvinists—all men are chauvinists.
b. Chris didn't manage to solve some of the problems—he managed to solve all of them.

Such examples cannot be collapsed with (8) under K&P's approach without incorporating conversational implicata (like the conventional implicatum in (11b)) into the logical form of these sentences.[6] But conversational implicata by definition are not part of logical form (cf. Grice 1975, 1978; Karttunen and Peters 1979). Crucially, however, these examples share with (1′) and (8) the linguistic correlates cited by K&P—and by Jespersen (in the epigraph to this chapter): the assignment of a 'peculiar intonation indicative of contradiction' and the proposed substitution of 'a more exact indication' for the focus of negation.

The cases below are even more devastating for any generalized semantic account of marked negation, which would presumably be driven to import phonetic representation and inflectional morphology into logical form, within the scope of negation.

(14) a. (So, you [mɨ́ʸəṇɨǰd] to solve the problem.)
No, I didn't [mɨ́ʸənɨǰ] to solve the problem—I [mǽnɨǰd] to solve the problem.
b. He didn't call the [pólis], he called the [polís]. (courtesy of Andy Rogers)
c. I didn't manage to trap two mongeese—I managed to trap two mongooses.

A related use of negation is found in the French example below, where the grammatical gender assignment and the woeful English accent are somehow brought within the scope of negation:

(15) (Esker too ah coo-pay luh vee-and?)
Non, je n'ai pas 'coo-pay luh vee-and'—j'ai coupé la viande.[7]

Analogously, we see in (16) that one speaker may employ negation to reject the pragmatics associated with the register or stylistic level chosen by another speaker in the discourse context, typically because of insufficient or oversufficient delicacy.

(16) a. Now, Cindy, dear, Grandma would like you to remember that you're a young lady: Phydeaux didn't shit the rug, he {soiled / had an accident on} the carpet.
b. Grandpa isn't feeling lousy, Johnny, he's just a tad indisposed.
c. We didn't {have intercourse / make love}—we fucked.
d. It's not stewed bunny, honey, it's civet de lapin.

In (17), one description is jettisoned in favor of another whose contributions to truth-conditional meaning are (arguably) identical to it in the relevant context, but which differs in focus or connotation:

(17) a. Ben Ward is not a black Police Commissioner but a Police Commissioner who is black. (*New York Times* editorial, 8 January 1983)
b. I'm not his daughter—he's my father. (after Wilson 1975: 152)
c. I'm not his brother—he's my brother.
d. She is not Lizzy, if you please—she's Her Imperial Majesty.

e. For a pessimist like you, the glass isn't half full—it's half empty.
f. I'm not a Trotskyite, I'm a Trotskyist.
g. They're not the best at what they do—they're the only ones who do what they do. (critic evaluating the Grateful Dead)
h. Winning isn't everything—it's the only thing.[8] (attributed to football coach Vince Lombardi)

Closely related to the examples in (17) is the use of negation to focus on and object to a previous speaker's racist or sexist vocabulary. The truth conditions of sentences like (18)

(18) {Niggers / Broads} will benefit from improvements in medicine.

have received a good deal of recent attention (cf. Grim 1981; Stenner 1981; Taylor 1981) as philosophers have debated whether an objection to the world view licensing the use of loaded words like *nigger* and *broad* is sufficient to render the statements these offensive descriptions are used to make automatically false or devoid of truth value. For Grim, if (18) is bivalent (whether it is true or false), it commits us to the excluded-middle disjunction in (18′):

(18′) It is either true that {niggers / broads} will benefit from improvements in medicine or false that {niggers / broads} will benefit from improvements in medicine.

Yet a commitment to (18′) seems to entail a commitment to {racism / sexism}.

However we may deal with Grim's quandary (cf. discussion below), it is relevant that a speaker can employ negation metalinguistically to reject the bigoted or chauvinistic point of view embodied in an earlier utterance within the discourse context:

(19) I beg your pardon: Lee isn't an uppity {nigger / broad / kike / wop}—(s)he's a strong, vibrant {black / woman / Jew / Italian}!

For someone who utters (19), as with (16) and (17), the denotative meaning of the statement under attack (what was said) may well have corresponded exactly to that of the rectified statement; the connotative meaning (what was implicated) cannot be allowed to stand unchallenged. In the light of the observation (by R. Lakoff [1975: 19–27], inter alia) that euphemisms are generated in the same referential contexts where we also have slurs and obscenities, it is significant that we find both euphemisms and their metalinguistic rejections in the Grim contexts:

(20) I'm not a 'colored lady'—I'm a black woman!
I'm not a 'gentleman of the Israelite persuasion'—I'm a Jew!

The use of negation explored in (13)–(17) and (19)–(20) may be marked, but it is by no means marginal or inconsequential in communication. Indeed, the example in (17e) has attained the status of a cliché, second perhaps only to that in (21), where the play between ordinary and marked uses of negation has entered immortality by way of the vaudeville stage:

(21) (Who was that lady I saw you with last night?)
That was no lady, that was my wife!

Note that the deliverer of the punch line in this routine does not intend to suggest that his wife is not a lady; rather, the negation attaches to the implicature associated with the set-up line. The pragmatic mechanism here is akin to that in an example from Grice:

(22) X is meeting a woman this evening.

The utterer of (22) 'would normally implicate that the person to be met was someone other than X's wife, mother, sister, or perhaps even close platonic friend' (Grice 1975:56). While not all speakers may agree with Grice in finding this implicature 'noncontroversial', it appears that—to the extent that it is felt to be present in (22)—it can be removed through negation:

(22′) No, he's not (meeting a woman this evening)—he's meeting his wife!

While the relevant implicata being denied or forestalled in (21) and (22′) result from the exploitation of the content maxim of Quantity (Make your contribution as informative as is required), manner-generated implicata may also be rejected by negation:[9]

(23) Miss X didn't 'produce a series of sounds that corresponded closely with the score of "Home Sweet Home", dammit, she <u>sang</u> "Home Sweet Home", and a lovely rendition it was too!

Here, what is denied is the reviewer's implicatum, devolving from an exploitation of the Brevity submaxim, that is, that 'Miss X's performance suffered from some hideous defect' (Grice 1975:55–56).

A similar analysis can be provided for the conjunction examples of (24):

(24) a. They didn't have a baby and get married, they got married and had a baby. (cf. Wilson 1975; McCawley 1981)
b. Mozart's sonatas weren't for violin and piano, they were for piano and violin.

Here, it is once again a manner-based implicatum which the metalinguistic negation removes, that is, the expectation that the order in which the two conjoined elements were originally reported (as represented in the first conjunct) corresponds to their order of occurrence (in (24a)) or to their order of importance (in (24b)).[10]

But we have clearly come a long way from either the well-behaved ordinary internal negation operator or the semantic external negation operator of three-valued logics or the K & P analysis. What we are dealing with in the negative examples of (13)–(23) is an extended version of what Ducrot (1972) aptly terms metalinguistic negation—a formally negative utterance which is used to object to a previous utterance on any grounds whatever, including (as in (14)–(15)) the way it was pronounced.[11]

It remains to be shown that these examples all involve the same basic use of negation as that found in K & P's examples of canceled or rejected conventional implicata, as in (8). To this end, note first that, for the negative sentences of (13)–(23) as for (1′) and (8), a felicitous utterance involves contrastive intonation with a final rise within the negative clause (the 'contradiction contour' of Liberman and Sag [1974] or the fall-rise of Ladd [1980], as the case may be), followed by a continuation in which the offending item is replaced by the correct item in the appropriate lexical, morphological, and phonetic garb—a RECTIFICATION, to borrow the label of Anscombre and Ducrot (1977). These diagnostics fit not only the presuppositional cases of (1′) and (8) (as noted by Karttunen and Peters) and the scalar cases of (13) (as noted by Jespersen), but the remaining negative examples of (14)–(24), as diverse as they otherwise appear to be.

But it is not only the intonation and rectification which point to a kinship with external or contradiction negation; in the cases just discussed, like those of K & P, no negative polarity items are triggered by the marked use of negation. Thus, while (13b) is possible with metalinguistic negation, its NPI counterpart is not:

(25) *Chris didn't manage to solve <u>any</u> of the problems—he managed to solve <u>all</u> of them.

And parallel to (9c), in which *some / any* suppletion is ruled out, we find the contrast in (25):

(25′) I didn't [mí̄ʸənɨǰ] to solve {some / *any} of the problems—I [mǽnɨǰd] to solve some of the problems.

I shall return to the interaction of polarity and metalinguistic negation below.

Of course the principal resemblance between the instances of marked negation introduced here and the classical examples of presupposition-canceling negation discussed earlier is that both types occur naturally only as responses to utterances by other speakers earlier in the same discourse contexts, or as mid-course corrections after earlier utterances by the same speakers. It is for this reason that I seek to encompass all these examples under the general rubric of metalinguistic negation: they all involve the same extended use of negation as a way for speakers to announce their unwillingness to assert something in a given way, or to accept another's assertion of it in that way. Given the behavioral resemblances just cited (which I shall explore in more detail), as well as the prevailing Occamist considerations, there is no obvious reason not to collapse the presupposition-canceling negation of (1′) and (8) with the negation attaching to conversational implicature in (13), (21), (22′), (23), and (24), to pronunciation in (14a, b) and (15), to morphology or syntax in (14c) and (15), to register, speech level, or social attitudes in (16), (19), and (20), and to perspective or point of view in (17).

6.2.1 Pragmatic Ambiguity

The notion to which I am appealing here has a rich and controversial, if brief, history. Donnellan (1966) coined the term PRAGMATIC AMBIGUITY to describe the two understandings he associates with sentences like (26).

(26) Smith's murderer is insane.
  a. ATTRIBUTIVE: Whoever it may have been who murdered Smith is insane.
  b. REFERENTIAL: That individual [to whom I refer via the phrase *Smith's murderer*] is insane.

In (26a) the description *Smith's murderer* is used essentially, but in (26b) it is employed as a device for picking out a specific individual and predicating something of him or her, regardless of whether or not that individual did in fact murder Smith.

Similarly, Wertheimer (1972) argues persuasively that sentences containing modals, such as (27), are not semantically ambiguous, but have either of two uses, as paraphrased in (27a, b), depending on the system of rules which is implicitly invoked for the evaluation of the sentence.

(27) Lee {should/ought to} be in Chicago today.
  a. EPISTEMIC: According to my calculations, Lee is (probably) in Chicago today.
  b. ROOT OR DEONTIC: It would be {desirable/a good idea} for Lee to be in Chicago today.

Kratzer (1977) arrives independently at her own analysis, which also treats modals as pragmatically ambiguous, as does McCallum-Bayliss (1985). Palmer (1979) and Leech and Coates (1980) offer complex mixed treatments incorporating both polysemy and semantic indeterminacy in their analysis of modal 'ambiguities'.

Stalnaker (1972:389–95) discusses a number of cases under the heading 'Pragmatic Ambiguity', including not only reference and modality but conditionals and parentheticals, where the duality of use rests in each case on the distinction between contexts and possible worlds in the determination of propositional content.

Another instance of pragmatic ambiguity formed the centerpiece of chapter 4 above. As suggested by Mill and Grice and elaborated in Horn 1972, 1973 and Gazdar 1979a, 1979b, scalar predications like those in (28) and (29) can be considered to be pragmatically ambiguous as between the implicature-bearing version (the 'two-sided' understanding of Aristotle, with lower and upper bounds, conveying (28a) and (29a), respectively) and the implicature-free 'one-sided' version (with lower bound only, as in (28b) and (29b)).

(28) Some men are chauvinists.
  a. TWO-SIDED: Some but (for all I know) not all men are chauvinists.
  b. ONE-SIDED: {At least some / Some if not all} men are chauvinists.

(29) It is possible that the Yanks will win.
  a. It is possible but (for all I know) not {necessary / certain} that the Yanks will win.
  b. It is at least possible that the Yanks will win.

None of these proposed analyses is uncontroversial. Donnellan (1978) argues for a semantic treatment of the ambiguity in (26) (but cf. Kaplan 1978, Kripke 1977 for defenses of the pragmatic version in Donnellan 1966). The standard linguistic line on the modals treats them as semantically (and possibly syntactically) ambiguous between epistemic and root/deontic readings (cf. Hofmann 1966; Newmeyer 1969; Horn 1972; Jackendoff 1972). There are several recent accounts of weak scalar predications of the type illustrated in (28) and (29) which treat them as semantically (or at any rate truth-conditionally) ambiguous (cf. Cormack 1980; Lehrer and Lehrer 1982; Burton-Roberts 1984; Carston 1985a, 1985b; Kempson 1986), as we shall see in more detail below. And Searle (1979:146–50) not only rejects the claim that the referential/attributive distinction and indirect illocutionary force involve pragmatic ambiguity, he rejects this construct entirely. Other objections to the notion of pragmatic ambiguity, or to my application of it for an account of natural language negation, are leveled by Burton-Roberts (1987) and Moser (1987).[12]

Nevertheless, I believe that the pragmatic version of these ambiguities is largely correct, and that the line taken on such constructions is extendable to negation.[13] Apparent sentence negation represents either a descriptive truth-functional operator, taking a proposition **p** into a proposition **not-p** (or a predicate **P** into a predicate **not-P**), or a metalinguistic operator which can be glossed 'I object to **U**', where **U** is crucially a linguistic utterance or utterance type rather than an abstract proposition.[14]

In claiming that negation is pragmatically, rather than semantically, ambiguous, I am partly in accord with the classic monoguist position summarized by Gazdar (1979a:92): 'There are no grounds for thinking that natural language negation is semantically distinct from the bivalent operator found in the propositional calculus.' But the spirit, if not the letter, of this position is violated by the present approach, which takes a wide array of uses of natural language negation to be essentially non-truth-functional, and indeed nonsemantic.

6.2.2 Truth vs. Assertability

If we temporarily set aside the more extreme cases of metalinguistic negation (e.g., those affecting phonetic representation), the distinction drawn above recalls one made elsewhere whose import has been insufficiently appreciated: that between the truth of a proposition vs. the assertability of a statement or sentence. As Grice (1967) has pointed out, either truth or assertability can be affected by negation; it is up to the addressee to factor in the relevant contextual clues so as to determine just what the speaker intended to object to or deny in the use of a negative form at a given point in the conversation.

Grice defends the view that ordinary language *or* exhibits the truth-conditional semantics associated with the standard truth table for inclusive disjunction represented in the third column in the table in (30) (other columns are included for purposes of later reference),

(30)

p	**q**	**p ∨ q**	**~(p ∨ q)**	**p w q**	**~(p w q)**	**p → q**	**~(p → q)**
T	**T**	**T**	**F**	**F**	**T**	**T**	**F**
T	**F**	**T**	**F**	**T**	**F**	**F**	**T**
F	**T**	**T**	**F**	**T**	**F**	**T**	**F**
F	**F**	**F**	**T**	**F**	**T**	**T**	**F**

and responds to a potential objection to this claim as follows (Grice 1967: lecture 5, p. 9): 'If you say "X or Y will be elected", I may reply

"That's not so: X or Y or Z will be elected". Here . . . I am rejecting "X or Y will be elected" not as false but as unassertable'.

Grice puts this distinction to work in defense of his truth-functional line on conditionals. He begins by conceding that (31)

(31) It is not the case that if X is given penicillin, he will get better.

does not have the truth conditions which we should expect of a negated material conditional (as given in the last column of (30)). After all, this sentence does not normally commit the speaker to an assertion of the conjunction 'X will be given penicillin and X won't get better', as the truth table would lead us to expect. In the same way, I can deny Nietzsche's notorious conditional (32a) without committing myself to the conjunction in (32b):

(32) a. If God is dead, everything is permitted.
b. God is dead and something is forbidden.

Grice points out, however, that a speaker uttering (31)—or, even more clearly, (31′):

(31′) It is not the case that if X is given penicillin he will get better; it might very well have no effect on him at all.

is not so much negating the contained conditional proposition as asserting his unwillingness to assert that proposition (lecture 5, p. 5). In (31′), as elsewhere, a negation outside the scope of a conditional is to be interpreted as a refusal to assert ***if p then q*** rather than as a (descriptive) negation of a conditional whose truth value is determined in accordance with the material equivalence in (33):

(33) $\sim(\mathbf{p} \rightarrow \mathbf{q}) \equiv (\mathbf{p} \wedge \sim\mathbf{q})$

The same point is made by Dummett (1973:328–30), who distinguishes negation outside the scope of a Fregean assertion operator, '**not** (⊢**A**)', from the normal assertion of a negative proposition, '⊢(**not A**)'. The former interpretation (which violates the Fregean spirit; cf. §1.2) 'might be taken to be a means of expressing an unwillingness to assert "**A**"'. The clearest candidates for this species of negation, not surprisingly, are those where **A** is a conditional. Dummett cites exchanges like those in (34):

(34) X: If it rains, the match will be canceled.
Y: That's not so. (or, I don't think that's the case.)

Y's contribution here is not actually a negation of X's content (presumably a material conditional, although Dummett fails to make this explicit); rather, we can paraphrase Y as having conveyed (34′a) or (34′b):

(34′) a. If it rains, the match won't necessarily be canceled.
b. It may [epistemic] happen that it rains and yet the match is not canceled.

Dummett in fact goes beyond Grice, concluding (p. 330) that apparently 'we have no negation of the conditional of natural language, that is, no negation of its sense: we have only a form for expressing refusal to assent to its assertion'. (While Dummett offers no explanation for this curious state of affairs, Grice does present a pragmatic story for the failure of conditionals to undergo ordinary descriptive negation.)

It should be acknowledged that the notion ASSERTABLE, as employed by Grice, Dummett, and me, must be taken as elliptical for something like 'felicitously assertable' or 'appropriately assertable', where the adverbial hedge is broad enough to cover the wide range of examples under consideration here. I take 'assertability' to be on all fuzzy fours with other instances of linguistic shorthand, such as 'Can you say X?' or 'You can't say X' for judgments of syntactic (un)acceptability.

Another auto-cavil: I do not want to insist here on a wholesale defense of the Grice-Dummett line on conditionals. Grice's unabashed championing of the material conditional as an adequate representation for the semantics of natural language conditionals is particularly moot; indeed, the truth conditions for *if-then* statements have been passionately but inconclusively argued since at least the third century B.C., when Callimachus observed that 'even the crows on the rooftop are cawing about which conditionals are true' (Mates 1949:234). But the distinction drawn by Grice and Dummett between rejecting a claim as false and rejecting it as (perhaps true, but) unassertable suggests the proper approach for characterizing the two uses of negation.[15]

6.2.3 Other Metalinguistic Operators

If the approach suggested here is correct for negation, it is plausible that the natural language reflexes of other logical operators should come in similar pairs, exhibiting metalinguistic uses alongside descriptive ones. This is indeed what we find.

Consider, for example, the extension of logical inclusive disjunction to the examples in (35):

(35) a. Kim is bright, or {even/should I say} brilliant.
b. New Haven, or the Elm City, is the pearl of the Quinnipiac Valley.
c. Is the conductor Bernst[íʸ]n or Bernst[áʸ]n? (cf. The

formulator of relativity theory wasn't Einst[íʸ]n but Einst[áʸ]n.)

d. The current President has appointed more colored folks—or should I say blacks?—to prominent positions than any of his predecessors.

e. She deprived her students of a lecture—or (better) spared them a lecture—on the performative hypothesis. (after Wilson 1975: 149)

f. Did Elizabeth have a baby and get married, or did she get married and have a baby? (after Wilson 1975 and McCawley 1981; cf. (24))

As Du Bois (1974) notes, a principal source of nonlogical disjunction is the phenomenon of intentional midsentence correction, as in one reading of (35a, d, e), or in the examples in (36), from Du Bois (1974: 8), where the self-corrections have 'survived presumably careful editing':

(36) a. I can only very briefly set forth my own views, or rather my general atittudes. (from Sapir, *Language*)

b. Let us look at the racial, or rather racist, themes in the argument for population control. (from Pohlman, *Population: A Clash of Prophets*)

A more extensive analysis of metalinguistic disjunction is provided by Ball (1986).

Metalinguistic conditionals include those in (37):

(37) a. If you're thirsty, there's some beer in the fridge.

b. If you haven't already heard, Punxsutawny Phil saw his shadow this morning.

c. If I may say so, you're looking particularly lovely tonight.

In these AUSTIN CONDITIONALS (as they have been known since Austin 1956), each antecedent clause specifies a sufficient condition for the appropriateness or legitimacy of asserting the consequent, rather than for its truth. As with metalinguistic negation, we can find morphosyntactic diagnostics for these metalinguistic uses of disjunctions and conditionals: note, for example, that the disjunctions in (35) and (36) cannot be paraphrased by *either . . . or,* and that the consequent clauses in (37) exclude initial *then.*

To the Austin conditionals of (37), Ducrot (1972: 175–78) adds another metalinguistic conditional statement type, exemplified in sentences translating into (38):

(38) a. If the Cité is the heart of Paris, the Latin Quarter is its soul.

b. If the Bois de Boulogne is the lungs of Paris, the neighborhood square is its pores.

As Ducrot observes, the speaker in these cases is understood as proposing to justify the metaphor in the main clause by virtue of accepting the metaphor in the antecedent. The sense is 'If you're willing to grant **p**, you must equally grant **q**'.[16]

Perhaps the closest pragmatic doublet for negation, however, is offered by questions. What are generally labeled ECHO questions (or, following Perlmutter and Soames [1979:589–90], INCREDULITY questions) might, in the present context, be aptly renamed METALINGUISTIC questions. As with the most natural occurrences of metalinguistic negation, echo questions generally require a linguistic context in which the original utterance (be it a declarative, an imperative, or itself a question) has been previously uttered within the discourse. Consider the circumstances which might evoke the echo questions in (39):

(39) a. You did <u>what</u> with Sally and Bill?
b. Take out the <u>what</u>?
c. Do I <u>what</u> on the first date?

The distribution of echo questions is determined in accordance with the sentence type they are used to echo. Echoes of declaratives occur in declarative contexts, echoes of questions in interrogative environments, and so on:

(40) a. John thinks {Mary is dating {Fred / <u>who</u>?} / *who Mary is dating.}
b. John wonders {who Mary is dating. / *Mary is dating {Fred / <u>who</u>?} / where {Fred went / <u>who</u> went?}}

And just as metalinguistic negation is impotent to trigger negative polarity items or to incorporate prefixally as do descriptive negations (see below), echo questions, as is well known, fail to exhibit normal interrogative syntax; they neither exhibit *wh*-fronting nor trigger subject-auxiliary inversion. (Cf. Cooper 1983:148–50 for a complementary treatment of echo questions.)

Conjunction can be added to the brew as well. While *and* is standardly viewed as a concatenator for propositions (or subpropositional constituents), Dummett (1973:337) cites the double question *Has she gone away and has she stolen all the teaspoons?* as an instance in which speech acts, rather than propositions, are connected. Walker (1975:147) characterizes Dummett's conjunction in terms very much in keeping with my analysis of two-pronged negation: *and* here, Walker notes, has 'a special sense' which is however 'a natural extension of the truth-functional use', motivated by 'the striking analogy between the logical relations which can obtain between propositions and the relations which can hold between speech acts'.[17]

There is, then, reason to believe that the existence of parallel metalinguistic/descriptive splits for other logical operators, rather than supporting the strong monoguist line on negation (as Kempson [1975:184] suggests), in fact reinforces the line on negation urged here. If we are unwilling, in constructing the simplest syntactic and semantic theory for natural language, to collapse the *or* clauses of (35) and (36) with ordinary inclusive disjunctions, the *if* clauses of (37) and (38) with ordinary conditionals (whatever they are), and the echo questions of (39) and (40) with normal *wh*-questions, we must be equally unwilling to claim that all negations are one.

6.3 Metalinguistic Negation and Conversational Implicature

6.3.1 The Scalar Cases

I have noted that languages tend not to distinguish internal from external negation morphologically. It is thus especially significant that natural languages almost invariably allow a descriptive negation operator to double for metalinguistic use as a comment on the utterance, challenging what is, or potentially would be, its presupposition or implicature, as well as its conventional content. One frequent use of metalinguistic negation is as a means for disconnecting the implicated upper bound of relatively weak scalar predicates, as in (13) above. Parallel naturally occurring examples can easily be attested:

(41) a. Around here, we don't like coffee, we love it. (Lauren Bacall, in a TV commercial for High Point decaffeinated coffee)
 b. That wasn't a bad year, it was horrible. (Reggie Jackson, commenting on his 1983 season with the Angels)
 c. I'm not happy he's gone—I'm elated. Never has an assistant coach gotten so much credit. (Chicago Bears football coach Mike Ditka; reported reaction to the departure of former assistant Buddy Ryan to become head coach for Philadelphia)
 d. I have two homes and I don't dig my roots into one or the other. I dig them into both. (twelve-year-old Becky Margulies, of her joint custody, *New York Times*, 25 March 1984)

I have argued (cf. Horn 1972 and chapter 4 above) that the apparent ambiguity triggered with weak scalar predicates is pragmatic in character, traceable to whether or not a generalized **Q**-based upper-bounding implicature is associated with the assertion of the sentence containing that predicate within a given linguistic and extralinguistic context. If, for example, *possible* is semantically or lexically ambiguous between a one-sided (at least possible) reading and a two-sided (exactly possible) reading, a consistent account of scalar operators would find analogous ambiguities in such sentences as:

		ONE-SIDED READING	TWO-SIDED READING
(42)	a. Max has three children.	at least three.	exactly three.
	b. You ate some of the cookies.	some if not all.	some but not all.
	c. It's possible she'll win.	at least possible.	possible but not certain.
	d. John is patriotic or quixotic.	and perhaps both.	but not both.
	e. I'm happy.	if not ecstatic.	but not ecstatic.
	f. It's warm out.	at least warm.	but not hot.

But on this line, the putative ambiguity of *possible* extends to every scalar predicate, including each of the $\aleph_0$-many cardinal numbers substitutable into the frame of (42a). On the other hand, my pragmatically based account—utilizing an independently motivated principle, the Maxim of Quantity (whose domain, as Grice has observed, is not even purely linguistic, let alone restricted to the realm of the logical constants), or its generalization as the **Q** Principle—offers an intuitive treatment of the apparent discrepancies between the standard logical characterizations of the operators (quantifiers, binary connectives, modals, negation) and their natural language equivalents. This modular approach yields a simpler, more conservative account of the semantics of these operators than a holistic treatment which either intermingles the semantic and pragmatic facts, ignores the latter entirely, or embraces an infinititude of lexical ambiguities.

This is where negation enters the picture, in its usual befuddling way. While the negations of scalar predications like those in (42) are ordinarily understood as negating the one-sided values which are the putative logical forms of these sentences—cf. *He didn't eat three cookies* (= less than three), *It isn't possible she'll win* (= impossible), and so forth—the PARADOXICAL negations of (43) (following the terminology of Cormack

1980) must be understood as negating the corresponding two-sided understandings:

(43) a. He doesn't have three children, he has four.
b. You didn't eat some of the cookies, you ate all of them.
c. It isn't possible she'll win, it's downright certain she will.
d. John isn't patriotic or quixotic, he's both patriotic and quixotic.
e. I'm not happy—I'm ecstatic.
f. It's not warm out; it's downright hot.

This dual character of negation with scalar predications has been recognized at least since Jespersen (1917:81; 1924:325–26; cf. epigraph to chapter 4), but its resolution is crucial for a determination of the character of the division of labor between semantics and pragmatics. If there is only one negation in natural language, the truth-functional one-place propositional connective of Fregean semantics, the well-formedness of the examples in (43) apparently indicates that the upper bound must be built into the logical form, or at least the propositional content, associated with these scalar predications—contrary to the Gricean line I have argued for. This is precisely the conclusion reached by Kempson and her fellows within what I have elsewhere (Horn 1984a) called the London School of Parsimony; I shall return to their program in §6.5 below.

Under the alternative view I have suggested here, the negation appearing in (43) (and in (41)) represents one more instance of the metalinguistic negation operator which cannot be assimilated to ordinary negation. If this view proves correct, the orthodox Mill-Grice view on scalar predications will be vindicated.

Let us focus on the contrast between (44a, b), or more precisely, on their mutual consistency.

(44) a. Max has three children—indeed, he has four.
b. Max doesn't have three children—(*but) he has four.
c. Max doesn't have three children, (but) he has two.

How is it that the same state of affairs can be alternately described in terms of Max's having three children and of his not having three children? On the present account, the negation in (44b) does not negate the proposition that Max has three children; rather, it operates on a metalinguistic level to reject the implicatum that may be associated with the assertion of that proposition (viz., that he has only three children). By uttering (44b), the speaker conveys an unwillingness to assert a sentence that would induce a misleading implicatum, even though this sentence would be true under these circumstances (as (44a) makes clear). As for (44c), its negation is naturally taken descriptively as attaching to the proposition that Max has three children. (I return to the distribution of *but* in this and similar paradigms in §6.4.)

When negation is used metalinguistically to focus on a **Q**-based implicatum, it often appears to build in a covert *just* or *only* which can in fact be expressed directly without directly affecting what is communicated. Thus, compare the versions of (45) with and without the parenthesized adverb immediately following the negation:

(45) a. Max doesn't have (just) three children—he has four.
b. You didn't eat (only) some of the cookies—you ate all of them.
c. Around here, we don't (just) like coffee—we love it.
d. I don't (just) believe it—I know it.

But, as we shall see, the fact that *just* or *only* may be insertable into this frame without changing the conveyed meaning should not be taken as evidence that metalinguistic *not* can be analyzed as elliptical for *not just, not only* (pace Lehrer and Lehrer 1982; cf. §6.5 below), any more than the fact that *some* and *not all* can be substituted for each other in some contexts implies that they contribute the same meaning to the sentences in which they occur.

Examples analogous to those in (45) are adduced in Horn 1981a to support the claim that the "exhaustiveness premise" is not part of the meaning of cleft sentences, that is, that (46a) does not (pace Atlas and Levinson 1981) entail or (pace Halvorsen 1978) conventionally implicate (46b):

(46) a. It was a pizza that Mary ate.
b. Mary ate nothing other than a pizza.
c. It wasn't John that Mary kissed—it was John and Bill.
d. It wasn't a pizza that Mary ate—it was a pizza, a calzone, and a side of ziti.

It may seem that the felicity of (46c), from Atlas and Levinson, or of the analogous (46d), suggests that the failure of exhaustiveness does constitute sufficient grounds for denying the truth of a cleft. But in fact these sentences are acceptable only with metalinguistic negation, canceling the upper-bounding exhaustiveness implicatum. Note that *just* or *only* can be inserted into these examples, as in those of (45), without changing the conveyed message. Furthermore, the same interpretation is available in focus constructions without cleft syntax:

(46′) a. Mary didn't eat a pizza—she ate a pizza, a calzone, and a side of ziti.
b. I didn't spend the night with Mary—I spent the night with Mary and her husband.

c. 'You don't renounce a bigot's word, you renounce the bigot and his word'. (Sandra Goldstein, Gary Hart delegate from Stamford, Conn., reacting to Jesse Jackson's speech to the 1984 Democratic National Convention, as reported in the Stamford *Advocate*)

If Mary ate a pizza along with various other items, it is undeniably true that she ate a pizza; negation in (46′), as in (46c, d), must deny not truth but assertability.

One more instance of the interaction of metalinguistic negation with scalar implicature is worth citing here, given the attention this construction has received in the recent pragmatic literature. Klein (1980: 28ff.), in his discussion of **AP**s of the form *six feet tall* within his semantics for comparatives, cites the following two dialogue sequences as posing potential problems for a Gricean theory of measure phrases:

(47)	A: The minimum height for applicants is 6′.	A: How tall is Mona?
	B: Well, Mona is six foot tall; in fact she's six foot three.	B: Six foot tall.
		C: No, she's not, she's six foot three.

'Possibly', Klein concedes, 'measure phrases are ambiguous between the *at least* and the *exactly* reading'.

A bit further on, Klein assigns the equative construction in (48a) the same truth conditions as that in (48b),

(48) a. Jude is as tall as Mona.
b. Jude is at least as tall as Mona.

in effect (as in Horn 1972) treating equatives as lower-bounded (one-sided) by their literal meaning and potentially upper-bounded (two-sided) by Quantity implicature: 'The fact that an utterance of [(48a)] will often convey that Jude is exactly as tall as Mona is to be explained along standard Gricean lines: in the absence of conflicting factors, a speaker who utters the sentence will conversationally implicate that he is not in a position to make the stronger assertion that Jude is taller than Mona' (Klein 1980: 38). This plausible assumption is further supported by the fact that equatives and comparatives pattern in other respects like weak and strong scalar predicates, as observed in Horn 1972: §1.22:

(48′) a. Jude is as tall as, if not taller than, Mona.
b. Jude is not only as tall as Mona, he's taller than Mona.
c. Jude is not even as tall as Mona, let alone taller.

Yet the same evidence that led Klein to opt for an ambiguist line on absolute measure phrases could be adduced for the equatives:

(49) A: Jude is as tall as Mona.
B: Yes, in fact he's (even) taller.

(50) A: Jude is as tall as Mona.
B: No, he's not as tall as Mona, he's taller.

But it is now clear that both (47B) and (49B) correspond exactly to (44a) in affirming a weaker scalar judgment when a stronger judgment on the same scale could be truly asserted, while (47C) and (50B) correspond exactly to (44b) in metalinguistically rejecting a weaker assertion as unassertable when a stronger scalar statement is known to hold.

Of course the unmarked reading of negation associated with equatives will be the descriptive 'less than' understanding:

(47′) Mona isn't six feet tall. (she's under six feet)
(50′) Jude isn't as tall as Mona. (he's shorter)

And for both absolute measure phrases and equatives, the suspension of the upper-bounding implicature (e.g., by *at least;* cf. chapter 4) neutralizes the distinction between metalinguistic and descriptive negation, since there will now be only a lower bound to reject; the B responses in (47″) and (50″) can only be assigned a 'less than' reading:[18]

(47″) A: Mona is at least six feet tall.
B: No she's not, she's {5′10″/#6′2″}.

(50″) A: Jude is at least as tall as Mona.
B: No he's not, he's {shorter/#taller}.

While there has been a recent flurry of activity in the literature on equatives and comparatives, as I noted earlier (chapter 4, n. 27; cf. Anscombre and Ducrot 1976, 1978, 1983; Sadock 1981; Atlas 1984; and Cornulier 1984), I see no compelling argument from negation (or anywhere else) to abandon the position that measure phrases and equatives are pragmatically, rather than semantically, ambiguous, in exactly the manner of other scalar predications inducing **Q**-based implicatures.

6.3.2 **Q**-Based vs. **R**-Based Implicature

I have argued that metalinguistic negation applies to cancel the **Q**-based implicatum standardly associated with scalar predication. But what of **R**-based implicata? A number of instances of **R**-based implicature are linked to the negative strengthening rule that turns apparent contradictories into acting contraries, as we saw in detail in chapter 5; but testing these implicata against metalinguistic negation turns out to be somewhat impractical, given the accumulation of negatives that perforce results (and the mind-boggling effects when these are processed). Fortunately, there are more than a few cases of **R**-based implicata that do not directly involve negation.[19]

To say that someone was able to solve the problem may **R**-implicate that he in fact solved it (cf. Karttunen 1971). Similarly, to assert that someone was clever enough to do something will generally implicate that she did it. As I have also noted, my confiding that I broke a finger will (ceteris paribus) **R**-implicate that the unfortunate finger was one of mine (rather than **Q**-implicating that it was someone else's). And, as I also observed, Wittgenstein, Searle, and others take an assertion of the form *I believe that p* to count as an indirect assertion of **p**; this too can be subsumed under the general rubric of **R**-based inference, as I suggested in chapter 5.

Yet, as seen in (51), none of these implicata can be canceled by negation:

(51) a. He wasn't able to solve the problem. (≠ He was able to solve it, but he didn't)
b. She wasn't clever enough to figure out the solution. (≠ She was clever enough to do it, but she didn't do it)
c. I didn't break a finger yesterday. (≠ I broke a finger, but it wasn't one of mine)
d. I don't believe the Yanks will win the pennant. (≠ I believe [autobiographically] that they will, but I'm not indirectly asserting that they will)

Why should this difference in cancelability exist? The answer would seem to inhere in the logic of **Q**- and **R**-based inference. Let **S** represent a given scalar predication (based on the strong scalar value $\mathbf{P_S}$) and **W** the weaker proposition (based on the weak scalar value $\mathbf{P_W}$) which it unilaterally entails and from which the relevant implicata are to be drawn. In the case of **Q**-based implicata, the assertion of *W* will ordinarily **Q**-implicate ~**S**. As a scalar predicate, $\mathbf{P_W}$ is truth-conditionally defined by its lower bound and the ordinary negation of **W** (*not* P_W) negates that lower bound, denoting 'less than $\mathbf{P_W}$', a value incompatible with $\mathbf{P_S}$. Thus, the assertion that Kevin did not eat three cookies would be taken to amount to the assertion that he ate fewer than three (and hence not four, five, or more). But *not* P_W when uttered in a context in which $\mathbf{P_S}$ is affirmed, as in (43a–f), self-destructs on the unmarked, descriptive 'less than $\mathbf{P_W}$' understanding and must in effect be sent back through, whence the marked, metalinguistic quality of the negation in these examples.

In the case of **R**-based implicata, on the other hand, the assertion of ***W*** implicates, not the negation of **S**, but **S** itself: the proposition that he solved the problem unilaterally entails the proposition that he was able to solve it (**S** entails **W**, as before), but the assertion that he was able to solve it may implicate that he in fact solved it (***W*** **R**-implicates **S**). Once again, *not* $\boldsymbol{P_W}$ signifies 'less than $\mathbf{P_W}$' and hence licenses the inference of ~**S** (via modus tollens from the original **S** ⊩ **W** entailment). But crucially, there are no circumstances under which the implicatum **S** is canceled and *not* $\boldsymbol{P_W}$ cannot be interpreted consistently, as an ordinary descriptive negation. The negations in (51a–d) thus never spontaneously self-destruct, to get sent back through to be interpreted metalinguistically as cancelers of implicata. Schematically, we have the following situation:

(52)

	Q-based implicata:	**R**-based implicata:
1.	**S** entails **W**	**S** entails **W**
2.	~**W** entails ~**S**	~**W** entails ~**S**
3.	***W*** **Q**-implicates ~**S**	***W*** **R**-implicates **S**
4.	Normally, *not* $\boldsymbol{P_W}$ = 'less than $\mathbf{P_W}$', incompatible with $\mathbf{P_S}$	Normally, *not* $\boldsymbol{P_W}$ = 'less than $\mathbf{P_W}$', incompatible with $\mathbf{P_S}$
5.	*not* $\boldsymbol{P_W}$, asserted in context where **S** is given, reinterpreted as metalinguistic negation	*not* $\boldsymbol{P_W}$ never gets reinterpreted, since it's always compatible with ~**S** (the denial of **W**'s implicatum)

Now **R**-based implicata can be canceled without negation, simply by assigning the fall-rise contour (cf. Ladd 1980) and stressing the implicatum-inducing element:

(53) a. He was ˇable to solve the problem (but he didn't solve it).
b. She was ˇclever enough to figure out the solution (but she didn't do it).
c. I broke ˇa finger yesterday (but it wasn't one of mine).
d. I ˇbelieve the Mets will win the pennant (but I'm not saying they will).

Notice that we tend to get the opposite, **Q**-based implicatum in these contexts; the contour which cancels the **R**-based inference sets up a strong expectation for the kind of continuations exemplified in (53).

When we do appear to get the cancelation of an **R**-based implicatum by negation, the implicatum in question has, in fact, become conventionalized as part of the literal meaning of the expression (cf. Grice 1975:58; Morgan 1978; and the discussion in §5.3 above). Thus, for example, predicate expressions which denote various personal relationships may, as in (54a), take on a narrowed symmetric sense, by virtue of a partial conventionalization of an **R**-based inference; but, as in (54b), they do not always take on this sense.

(54) a. Pat and Leslie are {married/friends/lovers/in love}.
b. Pat and Leslie are spouses.

When the symmetric sense of these predicates is intended, negation may leave the more general sense unaffected:

(54′) They aren't {married/friends/lovers/in love}. (as in the country song title "When You're Married, but Not to Each Other")

To support the claim that only conventionalized **R**-based implicata can be canceled by negation, we can turn to a pair of examples from Atlas and Levinson (1981). As they note, there is a strong tendency for a predicate which is semantically nonspecific with respect to a given distinction to become pragmatically restricted so as to denote a certain proper subset of its semantic denotation. This 'inference to the best interpretation' is formulated as follows (Atlas and Levinson 1981:42):

> If a predicate **Q** is semantically nonspecific with respect to predicates $\mathbf{P}_i$, $\mathbf{l} \leq \mathbf{i} \leq \mathbf{n}$, but for some j, $\mathbf{l} \leq \mathbf{j} \leq \mathbf{n}$, $\mathbf{P}_j$ is stereotypical of **Q**s, then in saying **Qt** a speaker will convey $\mathbf{P}_j\mathbf{t}$.

The key notion here is the restriction of a more general predicate to a stereotypical instance, constituting an instantiation of Atlas and Levinson's 'Principle of Informativeness' (Read as much into an utterance as is consistent with what you know about the world—Levinson 1983:146–47) and of my **R**-based narrowing (cf. §5.3). Among Atlas and Levinson's examples of what they have in mind are those in (55):

(55) a. John had a drink --› John had an alcoholic drink.
b. The secretary smiled --› The female secretary smiled.

But, for our current purposes, it is crucial to recognize that (as both speakers' intuitions and lexicographers' practice suggest) the implicatum in the former case has become fossilized into conventional meaning, while that in the latter case has not. Thus, in the terms of Horn 1984a, *drink* has developed into an autohyponym (a word with two privatively related senses), while *secretary* continues to retain just one (relevant) meaning. In this light, negation behaves precisely as we would predict:

(55′) a. John didn't have a drink—that was a Shirley Temple.
b. #My secretary didn't smile—I have a male secretary.

A male secretary is still a secretary (although not a stereotypic one), but a nonalcoholic beverage may or may not count as a drink, just as two illicit married lovers (of each other) may truthfully admit that they are not married (to each other).

In arguing that nonconventionalized **R**-based implicata cannot be canceled by negation, I have implicitly built in the assumption that when negation can be read descriptively, it must be, or in other words, that metalinguistic negation is marked psycholinguistically, as well as structurally. Although I cannot support this claim by citing evidence that metalinguistic negation (as in (44b)) takes longer to process or verify than ordinary descriptive negation (as in (44c)), this difference does obtain intuitively, and not for the scalar cases alone.

In general, the effect of metalinguistic negation is either the rejection of a previous utterance (on any grounds whatever, as we have seen), or the rhetorical demolition of some feature of one's own utterance, often after considerable effort has been made precisely to set up that feature. Consider this opening passage from a newspaper op-ed piece subsequent to the 1984 elections: 'When Ronald Reagan carried 49 states and won 525 electoral votes, it was not an historic victory. Walter F. Mondale's poor showing wasn't an historic defeat. Mr. Mondale's choice of Geraldine A. Ferraro as his running mate wasn't an historic decision, either. None of these was an historic event. Each was a historic event' (John Chancellor, *New York Times* op-ed column). Mr. Chancellor is clearly expecting his reader to process the first four negations of his "Article about an Article" as descriptive operators, denying the historic significance of the events in question. It is only when the fifth sentence is encountered and that primary reading becomes untenable that these negations must be retroactively reprocessed as metalinguistic, functioning to reject an aspect of the allomorphy, not the content, of the sentences in which they occur.

The garden-path effect of this passage, and of the majority of metalinguistic negations whose discourse domain does not cross speakers (see the examples in (35)), is the rule rather than the exception. It might even be speculated (although I shall not try to support this speculation here) that there is an inherent ordering within the processing of metalinguistic negation: the addressee, on recognizing that a given negation cannot be coherently read descriptively (as denying a given predicate of a given subject or as an internal, predicate term negation), will try first to take it as a rejection of the conventional implicata (or presuppositions) associated with the negated utterance, then (if that fails) as a rejection of its potential conversational implicata, then (if that fails) as a rejection of the formal properties (grammar or phonology) of that utterance.

Whether or not this speculation can be supported by psycholinguistic or other evidence, the proposed asymmetry between descriptive and metalinguistic uses of negation seems plausible enough, especially since we often must go out of our way (intonationally and syntactically) to allow the latter understanding to emerge. Given the systematic nonexistence of contexts in which a negation both cannot be interpreted descriptively and can be read as a device for canceling a nonconventionalized **R**-based implicatum, we can see why the metalinguistic understanding is ruled out in the examples of (51).

6.4 Three Diagnostics for Metalinguistic Negation

6.4.1 Incorporated Negation vs. Metalinguistic Negation

One key correlate of the negative dichotomy resides in the inability of metalinguistic negation to incorporate prefixally. We saw in §2.3 that a negative particle interpreted as outside the scope of a presupposed element loses that potential interpretation when the negation is incorporated as a prefix.[20] Let us now consider the paradigms below:

(56) a. The king of France is {not happy / #unhappy}—there isn't any king of France.
b. The queen of England is {not happy / #unhappy}—she's ecstatic.

(57) a. {It isn't possible / It's not possible / #It's impossible} for you to leave now—it's necessary.
b. {not probable / #improbable} but certain
(vs. {not probable / improbable} but possible)
c. {not likely / #unlikely} but certain
(vs. {not likely / unlikely} if not impossible)
d. {not interesting / #uninteresting} but fascinating
(vs. {not interesting / uninteresting} but important)

(58) I {don't believe / #disbelieve} they'll win—I know they will.

Whenever negation is used metalinguistically to deny the appropriateness of using a predicate which would yield a true but misleading assertion (one which would induce false conventional or conversational implicata), it operates, in effect, on another level from that of the rest of the clause in which it is superficially situated, whence its impotence to trigger polarity items within that clause (as seen above and, in more detail, below). For the same reason, the metalinguistic operator cannot incorporate morphologically as the *un-* or *iN-* prefix; the acceptable incorporated negatives in (57) all involve ordinary, truth-functional uses of the negation.

We can thus correlate the disappearance of the metalinguistic understanding of incorporated negation with other lost readings of incorporated, narrow scope negation. Alongside the *not X but Y* construction illustrated in (57), we note that the ambiguity of the *X if not Y* construction, investigated in Horn 1972: §1.22 and Welte 1978:205, disappears when the negation is incorporated. While *The book is excellent if not perfect* can be read either as a concessive (cf. (59a)) or as an implicatum suspender (cf. (59b)),

(59) a. The book is excellent if not (exactly) ˇperfect. (cf. German *wenn (auch) nicht*)
b. The book is excellent if not (downright) `perfect. (cf. German *wenn nicht (sogar)*)

its incorporated counterpart *The book is excellent if imperfect* allows only the former interpretation. The difference in incorporability corresponds to a difference in bracketing: *X if [not Y]* for (59a) vs. *X [if not] Y* for (59b).

Similar examples (from the above sources) are given in (60):

(60) a.	Our victory is possible if not probable.	(concessive or suspensive)
	Our victory is possible if improbable.	(concessive only)
b.	All the men {were not/ weren't} happy.	(NEG-Q or NEG-V; cf. §4.3)
	All the men were unhappy.	(NEG-V only)
c.	She was not fortunate enough to lose her husband.	(unfortunately, he survived)
	She was unfortunate enough to lose her husband.	(unfortunately, he died)
d.	Sue didn't trust Bill or John.	(She trusted neither or not both)
	Sue distrusted Bill or John.	(≠ She trusted neither of them)

While these examples may differ in their details, they share with each other—and with the examples of (56)–(58)—one essential trait: the incorporation of a negative element as a prefix on the following element is only possible when those two elements are immediate constituents within a single level of analysis.[21]

The incorporability diagnostic for metalinguistic negation interacts in a significant way with an argument by Gazdar (1977, 1979a) on behalf of a monoguist analysis of natural language disjunction. On the view that *or* is semantically or lexically ambiguous (cf. the table in (30) above and the discussion in §4.2), a sentence like (61) will be assigned two distinct logical forms, corresponding to the inclusive (**p** ∨ **q**) and exclusive (**p** w **q**) interpretations of the connective and paraphrasable as in (61a, b), respectively.

(61) John is either patriotic or quixotic. [cf. (42d)]
a. INCLUSIVE: John is patriotic, quixotic, or both.
b. EXCLUSIVE: John is either patriotic or quixotic, but not both.

This ambiguist thesis is rejected by Gazdar (1979a: 81–82) on the grounds that it makes a 'bizarrely false prediction' when the disjunction is brought within the scope of negation. Thus the negation of (61), which Gazdar gives alternately as (61′a, b)

(61′) a. John isn't either patriotic or quixotic.
b. John is neither patriotic nor quixotic.

—which he takes to be mutual paraphrases—will be assigned two readings under the ambiguist analysis, with negation outside the scope of inclusive and exclusive disjunction, respectively. Given the values in the fourth column of (30), and the equivalence in (62),

(62) $\sim(\mathbf{p} \le \mathbf{q}) \leftrightarrow (\sim\mathbf{p} \wedge \sim\mathbf{q}) \vee (\mathbf{p} \wedge \mathbf{q})$

the exclusive reading of (61′a, b) will come out true if John is both patriotic and quixotic. Since (61′a, b) are 'patently false' in this state of affairs, Gazdar concludes that the ambiguist thesis is patently false as well.

It is true that when (61′a) is used descriptively as a negated disjunctive proposition, that proposition is indeed patently false if both disjuncts are true. But the same sentence can be used as a metalinguistic negation, to make an indirect assertion which may in fact be true in the same context. Consider (63a, b):

(63) a. Maggie isn't either patriotic or quixotic—she's both! [cf. (43d)]
b. #Maggie is neither patriotic nor quixotic—she's both!

Not only is (63a) a possible discourse utterance, it's also one with which most British subjects (at least those of the Tory persuasion) would cheerfully agree, for the most prominent referent of the name. However, this reading predictably disappears when the negation is incorporated as in (61′b) or (63b). Since such incorporated negation can only be descriptive, (61′b) is unambiguous and (63b) logically contradictory, as Gazdar predicts.

As further confirmation of the claim that the negation in (63a) is metalinguistic in the strong sense intended here (as a rejection of another real or possible utterance in the discourse, rather than as a propositional operator or predicate denial), consider the apparent graphemic contradiction in

(64a) (a 'paradoxical negation' if there ever was one), which can nevertheless be resolved in the appropriate phonetic context, as illustrated in (64b):

(64) a. Maggie isn't either patriotic or quixotic—she's either patriotic or quixotic.
  b. —Say, {Clive / Fiona}, you have to admit your Maggie is [íʸðr pèʸtriáDɪk ɔr kwɪksáDɪk]
  —No I haven't. Maggie isn't [íʸðr pèʸtriáDɪk ɔr kwɪksáDɪk]—she's [áʸðə pætrɔ́:tɪk ɔᵊ kwɪksɔ́:tɪk].

Curiously, Gazdar's argument was prefigured—flaw and all—in a parallel attack by Grice (1978) on a different ambiguist treatment of disjunction. In this case, the relevant strengthening property is not the truth-functional one of exclusivity, but a non-truth-functional, epistemically based condition on the appropriate use of disjunctive statements which Grice attributes to Strawson, but whose history goes back a bit further. On this view, a speaker who utters a statement of the form ***p or q*** licenses the inference (at least on the 'strong reading' of *or*) that he or she does not know for a fact that ***p*** is the case—or, of course, that ***q*** is. Thus I cannot felicitously utter (65),

(65) My wife is either in Oxford or in London.

on this strong reading, if I know for a fact that my wife is in Oxford (and if I presume that this fact is relevant to you when I utter (65)).

The argument against the position that (65) is (or has a reading which is) false or neither true nor false when my wife is, say, in Oxford, is straightforward and by now predictable. As in the scalar cases, the crucial point is that while a disjunction may be unassertable (without infelicity) when one of its disjuncts is known to be true, it is clearly true—that is, it expresses a true proposition. In the same Negation symposium I reviewed in §1.2, Mabbott in fact endorses the extreme position that it is always false that *This railway signal (round the next corner) is either red or green,* given that 'at this moment the signal . . . is not "either red or green"; it is [let us say] green' (Mabbott 1929:74).

To which Ryle retorts with his own railway example:

> I judge at Reading, say, 'That train is going either to Swindon or to Oxford'; and I do so without necessarily implying that the engine-driver, the passengers, or even I myself are in ignorance or doubt which its route actually is. Ordinarily, of course, I would not bother to make the statement if I was not in some doubt, since if I could identify its route it would be superfluous to mention such non-individuating facts about it. But facts do not cease to be

> facts or cease to be known when it becomes superfluous to mention them. (Ryle 1929:92–93)

While we may not be far from Oxford, it is clear that we are also back at the crossing of truth and assertability, as Ryle and Grice recognize, and as Mabbott and Strawson evidently do not.

But Grice (1978:116–18) goes on to observe that the 'strong' reading of disjunctions like (65) seems to disappear under negation, as in (66a). What he does not observe is that the relevant non-truth-functional aspect of the interpretation of such disjunctions may indeed be affected when the negation is used metalinguistically, as in (66b–d):

(66) a. Your wife isn't (either) in Oxford or in London.
b. Your wife isn't (either) in Oxford or in London, dammit, she's in London, as you bloody well know!
c. I didn't do it once or twice—I did it once and once only!
d. The signal is not 'either red or green', it's quite clearly green!

In these examples, a disjunction is disowned not because it is false, but (as in (63a)) because the utterance expressing it is too weak to be felicitously asserted. Once again, the metalinguistic reading disappears when the negative is incorporated into the disjunction:

(66′) a. Your wife is neither in Oxford nor in London (#, dammit, she's in London!)
b. The signal is neither red or green (#—it's quite clearly green!)

As we have observed, the incorporation of negation eliminates any possible metalinguistic understanding of that negative operator—provided that we are dealing with the sort of metalinguistic negation which takes scope over the entire utterance to focus on, and object to, a particular aspect of that utterance (e.g., a potential conversational implicatum). But constituent as well as sentence negation may be interpreted metalinguistically, as a rejection of a previous speaker's lexical choice. In this event, the metalinguistic operator may take the form of an affixal negation if the focused element admits a negative affix—and sometimes even if it doesn't.

Zimmer points out that while the (e-neg) adjective *sullen* does not normally occur with a negative prefix, his experimental subjects unanimously rejecting *unsullen*, 'It is not really very difficult to imagine a situation where *unsullen* might be used, e.g., one involving a kind of echo effect: "He's a sullen fellow, isn't he?"—"On the contrary, I think he's a remarkably unsullen fellow"' (Zimmer 1964:87; emphasis mine). Notice that the exigencies of English syntax leave this reply as the only possible form of objection; the unincorporated counterpart—**He's a remarkably not sullen fellow*—does not occur.

Thus it is not metalinguistic or echo negation per se which rules out incorporation, but the combination of metalinguistic negation with the wide scope readings associated with the earlier examples in this subsection. Crucially, while the respondent in Zimmer's dialogue is using negation metalinguistically, his utterance is nevertheless true as a descriptive negation (assuming the referent is indeed unsullen). In my earlier examples, on the other hand, ranging from (56)–(58) and (i)–(iv) of n. 20 to (63), (64), and (66), the statement used to reject a previous utterance is not true as a descriptive negation—it may be false, or in some cases nonbivalent or meaningless, depending on one's semantics of presupposition failure—since what is objected to here is not the content of the corresponding affirmative, but the implicata, presuppositions, or phonetic form associated with that content or its reification. In these circumstances, the negation may not incorporate.

6.4.2 Polarity as a Diagnostic

I have suggested that because metalinguistic negation does not operate on the same rhetorical or grammatical level as the clause in which it occurs—because it is in the clause but not of it—it will fail to display those traits which are characteristic of the more fully integrated object-level negator. One such trait, as we have just seen, is its failure to incorporate prefixally; a second is a two-part property relating to its interaction with polarity phenomena. First of all, as we have already observed, metalinguistic negation does not trigger negative polarity items.

But at the same time, affirmative or positive polarity items, expressions which normally do not occur felicitously inside the scope of an immediately commanding negation, can occur following an instance of metalinguistic negation. As Baker (1970:169) puts it, affirmative polarity items (*already, would rather, could just as well,* $_{\text{ADV}}$ *pretty, far ——er*) are acceptable 'when they represent, word by word, an emphatic denial of a preceding speaker's assertion'; he notes that while (67b) does not occur in isolation, it is felicitous as an emphatic denial of (67a).

(67) a. The Sox have already clinched the pennant.
b. The Sox haven't already clinched the pennant.

Thus metalinguistic negation is neither an active trigger of NPIs nor a passive countertrigger or inhibitor of PPIs.

The focus of the literature on this topic is the indefinite *some* (and its compounds), which is normally in suppletive variation with *any* (Jespersen 1917; Klima 1964; but see also R. Lakoff 1970; Bolinger 1977: §2; Sahlin 1979). While *some* is normally blocked after negation, there are exceptions. Thus Kruisinga (1931 : §1336): 'If the negation serves to contradict a statement made, or to contrast two things, *some* is used: *You cannot get something for nothing; A play isn't something you read, it's something actors do on a stage*'. One locus of contradiction and/or contrast which Kruisinga cites as especially favorable for the *not . . . some* pattern is the *not X but Y* construction ('*They produced not something new and distinctive but something closely resembling a ms*'. [ibid.]).

The intuitive sense of direct denial or contradiction that accompanies *not . . . some* is often accentuated by intonation or, in print, quotation marks, as in the following attested example (from a letter to the editor of *Ms.* magazine, May 1986, in response to an earlier article on yeast infections):

(68) Chlamydia is not "sometimes" misdiagnosed, it is frequently misdiagnosed.

The NPI version of this response is of course ruled out: *#Chlamydia is {not ever/never} misdiagnosed, it is frequently misdiagnosed.* Here, as in the parallel examples of §6.3, we are dealing with the use of metalinguistic negation to reject a weak scalar item for the upper-bounding implicatum it would induce and, as we have seen (cf. *Chris didn't manage to solve {some/*any} of the problems—he managed to solve all of them*), only PPIs are acceptable in this context.

A clear statement of the phenomenon in its French manifestation is offered by Tasmowski-De Ryck (1972: 199), who observes that in a 'pure et simple reprise', such as that in her example in (69),

(69)	—Si je t'avais caché quelque chose. . .	'If I had hidden something from you'
	—Tu ne peux pas me cacher quelque chose.	'You can't hide something from me'

we get the PPI *quelque chose* 'something', rather than its usual suppletive partner in negative and interrogative contexts, *rien* 'nothing, anything'. In such cases, 'Une proposition déjà énoncée est répétée telle quelle, et la négation s'y applique comme à un tout inanalysable, dans lequel elle ne s'intègre pas réellement'.

A similar characterization is offered by Bolinger (1977: 44), who cites 'the use of *some* rather than *any*' along with 'a special intonation, or punctuation, to show that the thing denied is quoted', as the two signs of what he calls EXTERNAL negation, in which 'the speaker denies something that has supposedly been affirmed'. His examples of this phenomenon, which would appear to bear greater kinship with metalinguistic negation than with any of the formal (two- or three-valued) accounts of semantic external negation, include the exchange in (70):

(70) You ate some mushrooms.—I did not 'eat some mushrooms'.

In his discussion of the failure of negation to trigger *some/any* suppletion in English sentences like (71),

(71) He doesn't have some assets hidden away.

Ladusaw (1980: 144) posits a semantically distinct operator $\textbf{not}_2$ which is 'used to deny that something is true, rather than to assert that something is false'. But while this may be a necessary condition for the appearance of marked ($\textbf{not}_2$) negation in (71) and other *not . . . some* examples, it is not a necessary condition in general for metalinguistic negation (compare the examples in (13)–(20), where the same operator, equally incapable of triggering NPIs or inhibiting PPIs, rejects a previous utterance on grounds irrelevant to truth or falsity), nor is it a sufficient condition. If you tell me he has some assets hidden away, I can respond either with (71) or with its NPI analogue (*No, you're wrong, he doesn't have any assets hidden away*) and in either case I would seem to be 'deny[ing] that something is true', namely, your statement. Perhaps we simply need better criteria for distinguishing denials of truth from assertions of falsity. (Leech 1981 invokes the same distinction, without equipping us any better for drawing it.)

Not all NPIs and PPIs occur in suppletive pairs like *some/any, sometimes/ever.* English, like other languages, employs a large variety of nominals to express weak scalar predications. Many of these items retain the standard 'less than' interpretation under descriptive negation, as in the pairs in (72) and (72′), discussed insightfully by Bolinger (1972).

(72) a. I'm a bit tired.	(72′) a. I'm not a bit tired.
b. I ate a bit.	b. I didn't eat a bit.
c. There was a trace of them.	c. There {wasn't a/was no} trace of them.

Quite often, items which originally participated in this pattern come to lose their positive use and become conventionalized (or, as Bolinger puts it, STEREOTYPED) as NPI equivalents of *any.* These strong negative scalar values, like *a bit* in (72′), seem to build in an implicit *even* (cf. Schmerling 1971; Fauconnier 1975a, 1975b; Heim 1984), but (unlike *a bit*) cannot occur with the relevant understanding—if they can occur at all—in the frame of (72):

(73) a. I didn't eat a thing.	(*??I ate a thing*)
b. I didn't {drink a drop/sleep a wink}.	(positive versions odd or jocular)
c. I don't {give a damn/give a hoot/care a fig}	(positive versions only jocular)

d. It's not worth a {nickel/thin dime/red cent}	(positive versions jocular or literal)
e. There isn't {the slightest chance/a ghost of a chance/the chance of a snowball in hell} that we'll succeed.	(positive versions only jocular)
f. There wasn't a sign of them.	(positive version only literal)

When these items occur in positive contexts (if they do), they denote a minimal quantity; when they occur in negative contexts, the negation denotes the absence of a minimal quantity, and hence the presence of no quantity at all. These are Bolinger's MINIMIZERS, which he contrasts (1972: 120ff.) with the DIMINISHERS, expressions denoting small quantities which either do not allow negation at all, or whose negation can only be interpreted via litotes (as 'more than' rather than 'less than'). Thus, while *a trace* can occur either with or without negation (cf. (72c), (72′c)), and *a sign* can occur (with the same meaning) only negatively (cf. (73f)), Bolinger finds that *an indication* occurs only positively (*There {was/*wasn't} an indication of them*). In my dialect, other positive polarity diminishers include *a tad* and, curiously, *a wee bit* (*He isn't a (#wee) bit tired*). Bolinger's most striking contrast between minimizers and diminishers is that between *a bit* and *a little:*

(74)	a. I ate a bit.	=	(74′)	a. I ate a little.
	I'm a bit tired	=		I'm a little tired.
	b. I didn't eat a bit.	≠		b. I didn't eat a little.
	I was not a bit tired.	≠		I was not a little tired.

Although Bolinger's labels may not be particularly mnemonic, his distinction is real enough: *neg* + minimizer = zero (as in (74)), while *neg* + diminisher, if it occurs at all, yields a higher quantity on the same scale (as in (74′b)).[22] But, as Bolinger observes, there is a difference between the two examples of (74′b). The uncontracted (*not a little*) form, as a stereotyped (short-circuited) instance of the rhetorical device of litotes within the formal register, need not be triggered by a previous utterance. But the contracted form (*I didn't eat a little*) normally requires a prior discourse context (and often a rectification); the sequence of auxiliary negation + *a little* is

> not normally found in utterances of first instance, but in response to real or imagined utterances already containing the item in question, which is then thrown into contrast:

'Were you a little worried?'
'I wasn't a little worried, my friend; I was worried sick.'
(Bolinger 1972: 122)

Thus, while *a little*—unlike *a bit*—is normally a PPI, it can be used as a metalinguistic negation, resulting in what Bolinger elsewhere (1952: 1123) characterizes as SECOND-INSTANCE SENTENCES, used for 'setting someone right' who had 'mistakenly asserted' something else, or (I might add) asserted the same thing in a mistaken way.

Notice, however, that this nonconventionalized, second-instance use of negated *a little* is not restricted to the 'more than' interpretation cited by Bolinger. If you tell me that I seem a little tired, I can reply '*I am not a little tired (I'm just thinking)*' as well as '*I'm not a little tired, {I'm exhausted / in fact I'm not at all tired}*. But this property is not restricted to *a little* in particular, or to the diminishers in general. Just as the minimizers, the minimum-quantity constructions in (73), represent the broadest and most productive class of NPIs in English and numerous other languages (cf. Horn 1978a: §2 and chapter 7 below for additional examples and references), the most productive class of positive polarity items consists of the qualifiers and mild intensifiers which Stoffel (1901) dubs DOWN-TONERS and Bolinger COMPROMISERS: *fairly, pretty, rather, somewhat, sort of, tolerably*, and their ilk. These are degree adverbs whose lower bound is above that of the minimizers but below that of the intensifiers I explored in §5.3 (*too, very, overly*). The members of this category, as exemplified in (75), do not ordinarily take well to negation:[23]

(75) a. He is {pretty / somewhat / rather / sort of / kind of} tired (ill, tall).
b. ??He isn't {pretty / somewhat / rather / sort of / kind of} tired (ill, tall).

The negations in (75b) become more plausible when the context permits a metalinguistic interpretation. One environment for such 'echo contradictories', as Bolinger (1972: 124) calls them, is of course in direct denials of the corresponding positive utterance, as in (75′):

(75′) a. He isn't {pretty / somewhat / rather} tall—he's humongous.
b. 'Still', Edwin concludes, 'I did rather like him, didn't you?'
'No,' Vinnie says . . . 'I didn't "rather like Chuck", if you want to know. I loved him'. (from Alison Lurie's novel, *Foreign Affairs*, p. 420)

But the same metalinguistic reading, for these and other examples of PPIs, can also be triggered by the kind of statement exemplified in (76), where the speaker uses the negative ironically to hint at the corresponding positive:

(76) a. You aren't {slightly/just the least bit} tipsy, are you? (= You are . . . , aren't you)
b. The salmon isn't just a wee bit off, is it? (= It is, isn't it)
c. You wouldn't be {sort of/ kind of} letting me down easy? (= You are, aren't you)

It is also worth noting that like descriptive negation, metalinguistic negation may be conventionally expressed by a morphologically affirmative expression.[24] And just as a superficially absent descriptive negation triggers NPIs—indeed, even strict NPIs (cf. the discussion of *I'll be damned if* and its Japanese analogues in §5.3)—metalinguistic negation, whether or not it is overtly expressed, is compatible only with PPIs:

(77) a. Like hell, I {still love you/*love you anymore}.
b. Like fudge, he's {already washed up/*washed up yet}.

6.4.3 Two *Buts* about Negation

One more correlate of the metalinguistic/descriptive split I am claiming for natural language negation can be found in the distribution of concessive and contrastive *but* conjunctions. We have seen that metalinguistic uses of negation tend to occur in contrastive environments, either across speakers in a given discourse context or within a single speaker's contribution, and the English representation par excellence of contrast is *but*. The archetypal frame for metalinguistic negation is the *not* ***X*** *but* ***Y*** construction, functioning as a single constituent within a sentence.[25] This construction provides a straightforward way to reject **X** (on any grounds) and to offer **Y** as its appropriate rectification. As with other forms of metalinguistic negation, it is irrelevant whether or not the rejected utterance [. . . **X** . . .] in fact expressed a true proposition.

As we have observed, this frame provides a favored type of opposition within the synoptic Gospels of the New Testament:

(78) Do not store up your riches on earth, where moths and rust destroy them . . . but store up your riches in heaven. . . .
I tell you not to resist injury, but if anyone strikes you on your right cheek, turn the other to him too.
Do not think that I have come to bring peace to the earth. I have not come to bring peace but a sword.

Bald (1971:10) cites several examples of this construction culled from Quirk's corpus, including those in (78′):

(78′) a. The apostles are not great men, but men filled with the Holy Spirit.
b. Their main business is not to survive, but to do well.
c. This is not mere convention, . . . but a consideration of a problem of reality.
d. He is not 'drab' but 'golden'.

As Bald observes, the rectification may not occur overtly, in which case the 'incompleteness' of what I am calling the metalinguistic understanding of *not* will still be 'made explicit through intonation', that is, the fall-rise (for which Bald employs the same notation as Ladd: *This is not mere con*ˇ*vention*). But while Bald is correct in pointing out that the construction in (78′) typically involves a contrast between two items from a given class which figure as alternatives within a given context, whence the oddity of (78″b) as against (78″a), he does not observe that semantic kinship is not a necessary criterion for the establishment of such contrast sets, as shown by misperception repairs like (78″c):

(78″) a. The plate is not red but green.
b. *The plate is not hot but green. (Bald 1971:26)
c. The plate is not 'hot' but 'hard'.

In some cases, the syntax requires that the rectification be overtly expressed. That is, *not X but Y* may occur in syntactic frames (e.g., nominal-internally and postverbally) where *not X* may not:[26]

(79) a. We have {?not three children/not three but four children/not three children but four}.
b. Negation is ambiguous {*not semantically/not semantically but pragmatically}.
c. {*Not John/Not John but Mary} supports the family. (from Klima 1964:302)
d. I saw {*not Chris/not Chris but Pat}.

A particularly striking example of a metalinguistic negation which could not occur without its *but* rectification appears in reviewer Elizabeth Stone's explanation of why it is that for Jessie—the protagonist of Marsha Norman's play *'night, Mother*—suicide counts as a positive act expressing not despair but autonomy:

(80) Not she chooses to die, but she chooses to die.

Like my earlier Anglo-American example (64b), (80) can only be a graphemic contradiction if the negation is taken truth-functionally.

It should be observed that, while the *not X but Y* construction appears to function as a single constituent—and, within it, *not X but* as a sub-constituent (cf. Klima 1964; Gates and Seright 1967)—it may often be paraphrased by a metalinguistic negation which is contracted onto the preceding copula or auxiliary element:

(79′) a. We don't have three children, but four. (cf. (79a))
b. Negation isn't ambiguous semantically, but pragmatically. (cf. (79b))

There are, in fact, two alternative canonical forms for rectifying a metalinguistic negation, and one canonical form for not doing so. Consider the following paradigm:

(81) a. It isn't hot, but scalding. (= It is not hot but scalding)
b. It isn't hot—it's scalding.
c. #It isn't hot, but it's scalding.

As in the related scalar examples in §6.3, *hot* is rejected here on the grounds that the predication it would yield (or yielded, earlier in the discourse frame), though true, is too weak. This interpretation is possible with *but* (as in (81a)), with a rectified full clause (as in (81b)), but not with both (as in (81c)). In effect, the relevant ('rather') reading of *but* renders gapping obligatory. The syntax of (81c) forces an interpretation on which *but* functions as a true sentential connective (rather than a rectifier, as in (78)–(81b)), and the negation as an ordinary descriptive operator.

The pragmatic deviance of (81c) on the descriptive reading of the negative stems, of course, from the fact that—given that anything scalding is also (at least) hot—it is inconsistent to assert of anything that it is scalding, yet not hot; crucially, the felicitous utterer of (81a, b) makes no such assertion. Similarly, the metalinguistic understanding of (82a, b) disappears with the unreduced syntax of the *but* clause in (82c) (recall the paradigm of (44) in the last section):

(82) a. We don't have three children { but four.
b. —we have four.
c. #but we (do) have four. }

When such sentential *but* conjunctions are acceptable as descriptive ('less than') negations, they tend to be assigned the intonation contour characteristic of concessions:[27]

(83) a. We don't have three children, but we do have two. (#but we do have four)
b. It isn't hot, but it is warm. (#but it is scalding)
c. Negation isn't ambiguous semantically, but it is pragmatically.

The acceptability contrast in (83a, b) hinges on what can count as a concession. The appearance of supportive (and apparently emphatic) *do* in (83a) and the stress on the auxiliary (as well as on the two underlined focal peaks) in (83a–c) are additional correlates of the concessive *but* clause.[28]

In the light of the differentiation of fall-rise (ˇ) from straight fall (ˋ) in Ladd 1980 and other work on intonation (cf. §4.3), in minimal pairs like that in (84B) vs. (84B′),

(84) A: Do you love me?
B: I ˇlike you.
B′: I aˋdore you.

the concessive structure in (83), exemplifying descriptive negation, is readily distinguished from the pattern in (81) and (82), exemplifying metalinguistic negation + rectification. The difference will involve not the first clause (which in both cases employs fall-rise) but the second:

(85) a. It isn't ˇhot, but it ˇis ˇwarm.
I don't (quite) ˇlove you, but I (ˇdo) ˇlike you.
I didn't ˇeat any apples, but I ˇdid ˇsniff one. [note NPI *any*]
b. It isn't ˇhot, (#but) it's ˋscalding.
I don't (just) ˇlove you, (#but) I aˋdore you.
I didn't ˇeat some apples—(#but) I deˋvoured them. [note PPI *some*]

The now-you-see-it-now-you-don't character of *but* in these paradigms is hardly a new discovery about English or, as we shall see, about other languages. Gates and Seright (1967) touch on the distribution of *but* in their study of what they call NEGATIVE-CONTRASTIVE constructions.[29] As we have seen, we can paraphrase the metalinguistic-rectification formula *not X but Y* as *Y (#but) not X*, while *Y but not X* can only be read as a descriptive negation.[30] This point is implicit for Gates and Seright, who provide the minimal pairs in (86):

(86) a. They had heard, but not seen, the intruder.
a′. They had run, (#but) not walked, to the station. [= not walked, but run]
b. He approached cautiously, but not furtively.
b′. He approached slowly, (#but) not quickly. [= not quickly, but slowly]

In each case, negation is used descriptively (in my terms) in the unprimed example and metalinguistically in its primed counterpart. I can add the adjectival pair below to fill out the paradigm; note in the light of my earlier discussion that the negation in the *not X but Y* version cannot be prefixally incorporated.

(87) a. He is resigned, but not happy.
b. He is sad, (#but) not happy. [= {not happy / #unhappy} but sad]

I have claimed that negation bears two distinct functions in natural language. The evidence I have been considering here suggests that there are two distinct functions for *but* as well. But here, the cross-linguistic evidence supports the hypothesis that there is a lexical rather than merely a pragmatic ambiguity involved. As discussed by Tobler (1896) and, in more detail, by Melander (1916), a language may either (as with English) contain one adversative particle with two functions or (as with German) display two particles differentiated for these two functions.[31]

The first function, to paraphrase Melander, is that of modifying or restricting the idea set forth in the preceding clause, which may or may not be negative; German *aber* and Swedish *men* illustrate this type of conjunction. The second function is that of excluding or suppressing the idea set forth in the preceding clause, which in this case must contain negation; German *sondern* and Swedish *utan* fill this role. Spanish *pero* and *sino* differ in essentially the same way, as observed by Anscombre and Ducrot (1977), as do Finnish *mutta* and *vaan* (cf. Whitney 1956).[32] But if German, Swedish, Spanish, and Finnish distinguish two *but*s, English does not—nor does French, the focus of attention on the part of Tobler and Melander (cf. also Anscombre and Ducrot 1977; Ducrot and Vogt 1979).

While Classical Latin employed *autem* for modification and *sed* for exclusion or modification, the comparative particle *magis* eventually subsumed both of these functions.[33] Its Old French heir was employed consistently to signal modification, and often for exclusion as well. In the former (= *aber*) use, as '*mais* modifiant' (Tobler's 'einschränkender *mais*'), it stood alone; in the latter (= *sondern*) use, as '*mais* excluant' (Tobler's 'ersetzender *mais*'), it competed with, and eventually prevailed over, the particle *ains* (*ainz, ançois*), whose rise and fall are chronicled by Melander, Sturel (1908), and Antoine (1952:1114–57).

Let us follow Anscombre and Ducrot (1977) in distinguishing these two functions of *mais* clauses as PA (for ***pero / aber***) and SN (for ***sondern / sino***), respectively. What we find in the historical records, as Tobler and Melander show, is that $mais_{PA}$ occurs in full clauses after positive or negative propositions, while $mais_{SN}$—from the earlier period on—occurs only in reduced (gapped) clauses, immediately after negation.[34]

While *ains* barely survived into the seventeenth century, when it was already being dismissed by prescriptivists as 'un vieil mot, qui ne vaut rien', the semantic differentiation between *mais* and *ains*, Sturel (1908: 385) points out, enabled the latter to function with 'un sens analogue à celui de l'allemand *sondern*', while retaining for *mais* the 'nuance étymologique' of *magis*. In the older language, the semantics of the adversatives leads to different collocation properties; we find *ains au contraire* and *ains seulement*, on the one hand, and *mais pourtant, mais cependant, mais néanmoins*, on the other. With the loss of the distinction, *mais* now co-occurs with both the 'on the contrary, rather' (SN) and the 'yet, however, nevertheless' (PA)-type adverbs (Antoine 1952: 1143).

But, as pointed out by Tobler and Melander and—apparently independently of them and of each other—by both Lang (1977) and Anscombre and Ducrot (1977), $mais_{PA}$ and $mais_{SN}$ remain distinct in their semantic and distributional properties, even if the morphological distinction has collapsed. The syntactic diagnostics are those I have already traced: $mais_{PA}$ occurs (like *aber* and *pero*) in full clauses after negative or positive propositions and collocates with *cependant, néanmoins, pourtant, en revanche*, or *par contre*, while $mais_{SN}$ appears (like *sondern* and *sino*) only in reduced clauses after unincorporated negation and collocates with *au contraire* or, in familiar style, *même que*.

The negation in the **neg-P** SN **Q** construction must be overt and the entire sequence must represent a single speech act (cf. Lang 1977: 237; Anscombre and Ducrot 1977: 25ff.). Anscombre and Ducrot provide the examples in (88), in which **Q** is presented as the motivation for rejecting **P**, and the negation is interpreted as 'polemic' (i.e., metalinguistic).[35]

(88) **S**: Eso {no es consciente / #es inconsciente}, sino totalmente automatico.
G: Das ist {nicht bewusst / #unbewusst}, sondern ganz automatisch.
F: {Ce n'est pas conscient / #c'est inconscient}, mais totalement automatique.
'It's {not conscious / #unconscious} but (rather) totally automatic'

But the **(neg-)P** PA **Q** construction necessarily involves the descriptive use of negation (when a negative is present, that is; unlike SN, PA is not restricted to follow negation). Here, following Anscombre and Ducrot's exposition (cf. also Ducrot and Vogt 1979), **P** and **Q** must have the same 'argumentative orientation' within a given scale, and **P** must be 'argumentatively superior' to **Q** (cf. §4.4 above on argumentative, quantitative, and pragmatic scales). Thus we get (89), but not (90):

(89) **S**: No es cierto, pero es probable.
G: Das ist nicht sicher, aber das ist wahrscheinlich.
F: Ce n'est pas certain, mais$_{PA}$ c'est (pourtant) probable.
'It's not certain, but it is probable' [cf. (83), (85a)]

(90) **S**: #No es probable, pero es cierto.
G: #Das ist nicht wahrscheinlich, aber das ist sicher.
F: #Ce n'est pas probable, mais$_{PA}$ c'est (pourtant) certain.
'#It's not probable, but it is certain' [cf. (83), (85b)]

Crucially, as Anscombre and Ducrot (1977) and Lang point out, *mais$_{SN}$* clauses block incorporated negation in the 'corrigendum' and trigger reduction in the 'corrigens', although paratactic SN clauses can (and indeed must) be unreduced.[36] Anscombre and Ducrot (1977:35–36) demonstrate this pattern with the paradigms in (91) and (92):

(91) Il n'est pas grand { mais$_{SN}$ très grand. / —il est très grand. / #mais il est très grand. }
'He's not tall, { but very tall / —he's very tall / #but he's very tall }

(92) Il {n'est pas intelligent/#est inintelligent}, mais seulement bûcheur.
'He's {not intelligent/#unintelligent} but just a grind'.

As I have already amply signaled, the distinction between the two types of adversative conjunctions manifested overtly in German, Swedish, and Spanish and covertly in French is attested in English as well.[37] My earlier examples which can be classed as instances of *but$_{PA}$* include (83), (85a), (86a, b), (87a), and the glosses to (89); instances of *but$_{SN}$* include those in (78)–(82), (85b), (86a′, b′), (87b), and the glosses to (88), (91), and (92). And Anscombre and Ducrot's conclusion (1977:40) that the PA connective represents a semantic coordinator and the SN connective (whether or not the two are lexically distinct) a semantic subordinator is reminiscent of Gates and Seright's (1967) similar characterization of the two uses of English *but*.[38]

Indeed, the hypothesis that English contains both *aber-* and *sondern*-type *buts*, made explicitly by Bald (1971:10) and Welte (1978:193), is implicit in the OED's entry under that conjunction, where we find (in abridged form) the following:[39]

(93) **but**[23]: 'on the contrary = Ger. *sondern*' [all examples follow negation] 'appending a statement contrary to, or incompatible

> with, one that is negatived. . . . In a compound sentence the second member is often greatly contracted—*Thou hast not lied unto men, but (thou hast lied) unto God*'
>
> **but**[24]: 'nevertheless, yet, however = Ger. *aber*'
>
> **but**[25]: 'however, on the other hand, moreover, yet'

It is clear that the OED's **but**[23] is but_{SN}, while both **but**[24] and **but**[25] (the distinction between which is not entirely clear to me) map onto but_{PA}.

My #'d examples from English (like the parallel examples from French) are ruled out because they are simultaneously disambiguated in both directions by their syntax and/or context of utterance and hence cannot be assigned either PA or SN reading felicitously. But the concessive PA examples are worth examining a bit more closely. Sentence (83) exemplifies the usual pattern: two scalar terms are juxtaposed in the construction (**neg**-)**P** PA **Q**, with **P** taken to be a stronger element than **Q** on a given scale. In the simple cases, such scales can be defined by unilateral entailment: *four* is stronger than *three* because any simple affirmation with the scalar element *four* entails the corresponding proposition with *three*, but not vice versa. The scale on which *scalding, hot,* and *warm* are situated and which is implicitly invoked in (81), (83b), and (85) can be similarly defined (cf. §4.4 for details).

But as we also saw in my earlier discussion of quantitative scales, the requisite notion of scale is far wider than logical or semantic entailment alone can accommodate. The entailment cases are special instances of what is more broadly a pragmatic relation defined as much by knowledge and beliefs about the world that are (assumed to be) shared by the speech participants as it is by the language itself (cf. Ducrot 1973; Fauconnier 1975a, 1975b, 1979a; and §4.4 above).

In this light, consider these additional examples of well-formed and ill-formed concessive but_{PA} conjunctions:

(94) a. I don't have my master's degree, but I do have my {bachelor's / #doctorate}.

b. I wasn't born in L.A., but { I did spend a few years there. / #I was born in New York. / (rather) in New York. [OK on SN reading] }

c. 'Of course it isn't cotton, but it is cottony soft'. (commercial for Cottonelle toilet paper)

(95) He isn't handsome, but he is {rich / presentable / a Catholic / a linguist}.
{#ugly / ?#mean}.

In the well-formed concessive examples of (94), it is still relatively straightforward to construct a scale on which the negated element outranks the item being affirmed. In (95), however, the concessive pattern expands to admit a case in which the two elements in opposition do not stand in an obvious scalar relation, but they do occur as fellow members of an implicitly evoked set of attributes. The examples in (95) might be paraphrased as, for example, *He isn't handsome <u>and</u> rich, but (at least) he is rich.*

Bearing in mind that concessive clauses are marked by a fall-rise contour (cf. (83) above), the examples in (94) and (95) fall into place if we recall Ladd's insightful characterization of the meaning associated with fall-rise (Ladd 1980: 153; cf. §4.3 above). Why is it, Ladd asks, that in the exchange in (96),

(96) A: Did you feed the animals?
B: I fed the ˇcat.

B 'clearly implies that he didn't feed the dog (or whatever)?' The key is the 'focus in a given set' meaning conventionally associated (as a conventional implicature?) with fall-rise. What (96B) contributes to the discourse context is thus 'I fed something [focus presupposition] from a set of things in the context [fall-rise nuance] and it was the cat [assertion]'. But, as Ladd also observes, the hierarchy evoked in (96), that is, (97),

(97) **animals**—from A's utterance
/ \
dog, etc. **cat**—from B's utterance

can also be invoked by the previous mention of a cohyponym:

(98) A: What would you think of getting a dog?
B: A ˇcat maybe.

Notice that Ladd's examples can be approximately translated into concessives, at the price of rendering explicit what had been implicit in the earlier exchanges:

(96′) I didn't feed the ˇanimals (ˇall the animals), but I ˇdid feed the cat.

(98′) I don't think {I/we} should get a ˇdog, but {I/we} ˇmight consider a cat.

By the same token, my earlier concessives can be elucidated within Ladd's framework. Given that cats are animals (but not vice versa), and that the speaker has, and is known to have, a cat, the fall-rise in (96) and (96′) works off the (unilateral) entailment in (96″); the concessive in (94c) depends on the parallel entailment in (99):

(96″) X fed the animals ⊩ X fed the cat

(99) Y is cotton ⊩ Y feels cottony soft

Similarly, being handsome and being rich (as in (95)) can be taken as subcases of desirable qualities, just as getting a dog and getting a cat (as in (98) and (98′)) are subcases of getting a pet.

Concessive structures of the form **neg-P** PA **Q** do seem to be unacceptable when the affirmation of **Q** is judged incompatible with the negation of **P**, whether this is because it constitutes a stronger item on the same scale (as in (81c), (82c), and (94a)), because **P** and **Q** are incompossible nonscalar expressions (e.g., being born in L.A. and being born in New York in (94b); but see discussion below), or because it is simply too mind-boggling to construct the superset of which the two items in question function as fellow members (e.g., the set of attributes containing *handsome* and *mean* for (95)). Even in the unlikely but attested example in (100),

(100) Tipping is not so common in Nepal. Tipping is not compulsory but it is obligatory. (*Nepal Travel Companion,* by S. D. Bista and Y. R. Satyal, cited in the *New Yorker,* 19 July 1982)

we infer that Messrs. Bista and Satyal are assuming a scale on which *compulsory* somehow outranks *obligatory*—that is, where anything compulsory is ipso facto obligatory, but not vice versa.

In fact, however, even the #-daubed examples can be rendered acceptable to the extent that ingenuity permits construction of the relevant pragmatic scale. Let us suppose that you have announced that you are looking for people with three children (to study the effects of large families, for example, or to offer them aid and solace). If I assume that my having four children qualifies me almost as well (or even better!) for your interests, I can nominate myself by uttering the now redeemed (82c) (*I don't have three children, but I do have four*).

Let us now turn to the infelicity of the latter version of (94b). Abbott (1972), citing some unpublished observations of Charles Fillmore, considers the related examples in (101):

(101) a. John was born, not in Boston, but in Philadelphia.
b. # John was born in Philadelphia, but not in Boston.
c. (#)John wasn't born in Boston, but he was born in Philadelphia.

While (101a) is good on the SN reading forced by the constituent negation (cf. (79)), the syntax of (101b, c) forces the PA interpretation, the former because its first clause lacks negation, and the latter because its second clause is unreduced and contains an overt *but*. As Fillmore and Abbott note, (101b) suggests the (unsatisfiable) expectation that John could have been born in both Philadelphia and Boston, while (101c) seems to have 'an associated assumption that there is a scale connected with places to be born in, and that Boston represents a more extreme point on that scale than Philadelphia' (Abbott 1972: 19). For me, one context which renders (101c) acceptable by commissioning the construction of just such a scale is the following: a casting director for a school play in a small town in Iowa or Mississippi, needing a fifth-grader to portray JFK, is being persuaded to settle for the only East Coaster in the class.

Of course if either the nonfocused material *he was born* or the conjunction *but* itself is deleted from (101c), we obtain the SN reading which, as in (101a), needs no special context:

(101c′) John wasn't born in Boston, {but/he was born} in Philadelphia.

Auxiliary negation allows both the PA reading (as in (101c)) and the SN reading (as in (101c′)), depending on the syntax of the second clause (and on the intonation contour), while postauxiliary constituent negation can only be taken metalinguistically and is thus incompatible with the concessive PA reading. (As we shall see in §6.6, the same pattern obtains in other languages, including French, Russian, and Hindi.) Thus, while (101c) is the PA version of (101c′), the latter's postauxiliary-negated paraphrase (101a) has no acceptable PA counterpart.

(101c″) #John was born not in Boston, but he was born in Philadelphia.

As seen in my reconsideration of (82c), the requisite pragmatic scale may force an inversion of the ordinary semantically based (entailment-generated) scale involving the same elements. Thus, too, the #-marked version of (94a) becomes acceptable if the speaker feels that the interlocutor is looking essentially for someone with a graduate degree, regardless of level, rather than specifically someone with a master's degree. Similarly, (102) is implausible in isolation, since it seems to allude to a rank ordering on which a private outranks a corporal.

(102) #He isn't a private, but he is a corporal.

Yet just such an ordering can be constructed if the context is fleshed out in the right way: the Colonel has ordered the Lieutenant to find a private to blame for a recent debacle. The Lieutenant reports back to the Colonel:

(102′) I've found a soldier we can volunteer for that last mission, sir.
He isn't a private, but he is a corporal. Will he do, sir?

Note that in this same context, the scalar terms *almost, barely, not even,* and so forth, reverse their normal distribution (*He's almost a private; He's a corporal, if not a PFC*), helping to confirm this ad hoc inversion of the standard rank ordering. Inspection seems to indicate that similar unusual (if not outlandish) contexts can be constructed to rescue the unacceptable PA/concessive examples from my earlier discussion, including those borrowed from Anscombre and Ducrot.

The English examples considered here are consistent with Anscombre and Ducrot's thesis that the negation which (optionally) figures in the concessive PA constructions is necessarily descriptive, while the negation required by the SN environments is typically understood as metalinguistic. Thus the contrast between the SN and PA types of *but* constructions in languages like English and French—as well as (if more subtly than) in languages like Spanish, German, Swedish, and Finnish—constitutes another diagnostic for metalinguistic vs. descriptive uses of negation.

6.5 Other Approaches to Metalinguistic Negation

6.5.1 Truth, TRUE, and Negation

As I have noted, the Kroch-Linebarger line on negative statements treats external negation as an ordinary truth-functional negative operator applied to a semantic TRUE operator predicated directly within its scope. Seeking to explain the unacceptability of negative polarity items in contexts like (103),

(103) a. *She did not lift a finger to help.
b. *We did not get up until 12:00.

read with rising 'denial' intonation, Linebarger (1981:35) cites Kroch's definition of 'external negation' as 'a "metalinguistic" usage in which the negative sentence NOT S does not directly comment on the state of affairs but instead denies the truth of the statement S previously uttered or implied. Sentence-external negation can be paraphrased as "The sentence S is not true"'. Linebarger proposes to formalize this account of metalinguistic external negation by representing the logical form of the 'denial' readings of (103) as in (103′):

(103′) a. NOT TRUE (she lifted a finger to help) . . .
b. NOT TRUE (we got up until 12:00) . . .

What rules these out as possible well-formed formulas is that the NPIs *lift a finger* and *until* are no longer within the immediate scope of negation, thus failing to meet what is for Linebarger a necessary (though not sufficient) condition for the acceptability of the relevant type of polarity trigger. In the same fashion, Linebarger (p. 36ff.) notes, the ill-formedness of

(104) is correctly predicted by assigning it the 'external' representation (104'):

(104) *The king of France didn't contribute one red cent, because there is no king of France.

(104') NOT TRUE (the king of France contributed one red cent). . .

One question raised by this characterization of marked negation is that, as we have seen at some length in earlier chapters, investigators as diverse as Leibniz, Hegel and the neo-Hegelians, Bergson, Wittgenstein, Wason, Ducrot, and Givón, have taken all instances of negation—including those of the NPI-triggering internal variety—as representing a way to deny, in Kroch's words, 'the truth of the statement S previously uttered or implied'.

Whether or not some such characterization of (descriptive) negation can be maintained, and if so which one (cf. §3.3), it is not clear how this line can serve as a way to differentiate Linebarger's versions of internal and external negation. In any event, a distinct problem remains for the Kroch-Linebarger position. I have highlighted many cases which pose insurmountable difficulties for any theory in which the special metalinguistic negation exemplified in (103) and (104) is directly associated with a denial of truth. It hardly seems plausible, for example, to analyze (13a) (*Some men aren't chauvinists—all men are chauvinists*) in terms of a Linebargerian representation like that in (105):

(105) NOT TRUE (some men are chauvinists). . .

Nor does this approach fare any better when applied to my examples of metalinguistic rejections of the register, grammar, phonetics, or Weltanschauung associated with an earlier utterance.

Even in the more clearly semantically based examples of conventional implicata considered by Karttunen and Peters (1979), such as (8) (*Chris didn't manage to solve the problem—it was quite easy for him*), we encounter a similar problem. An analysis of this external or contradiction negation along the lines of the Linebarger model yields (106):

(106) NOT TRUE (Chris managed to solve the problem). . .

Yet, as Karttunen and Peters observe, the simplest truth-conditional account of sentences like (8) is one in which the proposition corresponding to the parenthesized material in (106) is indeed true in any state of affairs in which Chris solved the problem.

Metalinguistic negation, as we have seen, is used to deny or object to any aspect of a previous utterance—from the conventional or conversational implicata that may be associated with it to its syntactic, morphological, or phonetic form. There can be no justification for inserting an operator TRUE into the logical form for a certain subclass of marked negative sentences, in order for negation to able to focus on it, if metalinguistic negation does not in principle have to do with truth conditions.

Perhaps in these cases of non-truth-functional negation, we could try placing the negative operator outside the scope of a semantic operator dubbed APPROPRIATE or CORRECT, rather than TRUE. But this solution merely shifts the problem back one step, given that metalinguistic negation—unlike ordinary descriptive negation, or the so-called external negation adopted in different guises by classical and multivalued logicians and by Kroch and Linebarger—is simply not an operator (truth-functional or otherwise) on propositions. Thus, representations like (107a) are essentially as inadequate as (107b):

(107) a. NOT {APPROPRIATE/CORRECT} (**p**)
b. NOT TRUE (**p**)

for the full range of cases under consideration here, given that those aspects of the utterance which metalinguistic negation is used to focus on may have nothing to do with the proposition expressed by that utterance. Conventional implicata or presuppositions may be analyzed as attributes (albeit non-truth-conditional attributes) of propositions; but conversational implicata, and—a fortiori—morphological and phonetic form, register, and so on, cannot be coherently treated in this way.

This key difference between descriptive and metalinguistic negation provides the most serious problem for the over-Occamistic claim of the strong monoguists that all uses of negation can be assimilated to one truth-functional analysis. It must not be overlooked that marked negation differs from descriptive negation not only phonologically, morphologically, and syntactically, but also in semantic function. In particular, metalinguistic negation, as an extralogical operator, plays no straightforward role with respect to such key laws of inference as double negation and modus tollendo ponens (MTP).

As a result, these laws would be unstatable if all uses of negation were to be treated identically. If we chose to tar descriptive negation with the same brush as metalinguistic negation, we could no longer draw such basic inferences as those in (108):

(108) a. I didn't manage to solve the problem.
∴ I didn't solve the problem. (cf. (8), (14))
b. Maggie isn't either patriotic or quixotic.
∴ Maggie isn't patriotic. (cf. (63a), (64a))

In the same vein, Wilson (1975: 149), citing disjunctive denials of the type earlier noted by Grice (cf. §6.2.2 above), observes that the two clauses of (109) seem to constitute premises in a disjunctive syllogism, that is, (109′).

(109) The next Prime Minister won't be Heath: it will be Heath or Wilson.

(109′) $\sim \mathbf{p}$
$\underline{\mathbf{p} \vee \mathbf{q}}$
$\therefore \mathbf{q}$ (via MTP)

Yet we don't in practice actually infer **q**—that is, *The next Prime Minister will be Wilson*—from an assertion of (109). Instances of descriptive negation, however, do license MTP: if I know that Heath or Wilson has been elected, and I hear Heath conceding on the BBC, I do conclude that Wilson (Harold, not Deirdre) was the winner. In short, forcing all instances of negation into a single Procrustean bed—however skillfully the bed may have been designed—accomplishes little beyond playing Pandar to some rather odd theoretical bedfellows.

But if metalinguistic uses of negation involve denial of assertability, rather than of truth, why is it that the syntax used to express this use of negation often seems to bring in some explicit reference to what is (and is not) true? Recall that in my discussion of (44b), *Max doesn't have three children—he has* <u>*four*</u>, I argued that negation attaches metalinguistically to the conversational implicatum associated with the utterance of *Max has three children,* rather than descriptively to the proposition expressed by that utterance. But some speakers can also get (110a), and sentences like (110b–d) are also heard and interpreted without undue difficulty.

(110) a. It isn't true that Max has <u>three</u> children—he has <u>four</u>.
b. It's not {true / the case} that <u>some</u> men are chauvinists—<u>all</u> men are chauvinists!
c. It's not so that the next Prime Minister will be Heath: it will be Heath or Wilson.
d. It's not the case that if X is given penicillin he will get better; it might very well have no effect on him at all. (= (31′) above, from Grice 1967: lecture 5, p. 5)

Does this mean we're on the wrong track? Do these examples involve a semantic external truth negation after all—so that, Occam to the contrary notwithstanding, we must acknowledge that conditionals, disjunctions, and weak scalar predications are all semantically ambiguous? No. Rather, what these sentences show is that the distribution of the English expressions *It is true that, It is the case that,* and *It is so that*—and their cross-linguistic counterparts—is a poor guide at best as to where the logical predicate TRUE is to be applied in the simplest, most elegant semantic/pragmatic theory of natural language meaning and communication. (Cf. Walker 1975: 138–40 for a similar point.)

We often say that something isn't true, meaning that it isn't felicitously assertable. This is not always possible: thus it strikes me as odd to insert *true* into those metalinguistic negations hinging on grammar, speech level, or phonetics:

(111) a. ?#It's not true that I [mɨ́ʸənɨǰd] to solve the problem—I [mǽnɨǰd] to solve the problem.
b. ?#It's not true that I managed to trap two mon<u>geese</u>—I managed to trap two mon<u>gooses</u>.
c. ?#It's not the case that the dog <u>shat</u> on the carpet—he <u>defecated</u> on it.
d. ?#*Ce n'est pas vrai que j'ai 'coo-pay luh vee-and'—(ce qui est vrai, c'est que) j'ai coupé la viande.*

It is true that the implicature-canceling examples of (110) remain problematic. But it is no less true that, in ordinary language, we often seem to be (literally) denying or ascribing truth to a given proposition when the simplest theory would represent us as doing something else entirely.

For one case in point, I can adapt the earlier Wilsonian example (24):

(112) It's not true that they had a baby and got married—they got married and had a baby.

Here the self-proclaimed truth negation focuses on an aspect of the use of conjunction, which, as we have seen (in n. 10), can be convincingly argued to be non-truth-conditional and in fact outside the domain of meaning proper: the interpretation of *and* in certain contexts as *and then.*

For an even more clear-cut (and more personal) example, I turn to some evidence involving the extended use of *true* in a nonnegative context. Several years ago, I was awakened for a pragmatics class I was teaching by the sound of my G.E. clock-radio cheerfully dispensing reveille:

(113)

After the familiar tune faded out, the announcer commented, 'Yes, it's true, it <u>is</u> time to wake up'. Now, what has been asserted to be true here?

The proposition (abbreviated) in (113)? Hardly: there is no proposition there, just a bunch of notes in search of a bugle.[40] Rather, the playing of reveille, given certain nonlinguistic conventions in our culture, can be performed (and is conventionally performed) with the intention of indirectly conveying the proposition that it is time for the reluctant hearer to awaken. It is this conveyed proposition which is being called true; the prior indirect assertion of this proposition is further illustrated by the anaphoric destressing in the radio announcer's utterance.[41]

Parallel to the conjunction and reveille examples just discussed is the phenomenon induced by rhetorical questions of the type which Sadock (1971) has labeled QUECLARATIVES, sentences of interrogative form but assertive force (cf. also §5.3). Notice that the *true* and *so* of B's alternate responses in (114) can only be directed at the proposition indirectly (but conventionally) conveyed by A:

(114) A: Who the hell buys that cockamamy line about pragmatic ambiguity?
B: (1) Yes, that's true. (= Nobody does)
(2) No, that's not necessarily so; there might be something to it.

Unlike the conjunction in (112), the queclarative case (like reveille) cannot survive embedding:

(115) a. I guess I'll have to settle for polyester, because where the hell can you find a 100 percent cotton jumpsuit anymore?
b. *It's not true that where the hell can you find a 100 percent cotton jumpsuit anymore.

But this is presumably caused by syntactic factors: neither bugle tunes nor *wh*-moved questions normally occur embedded.[42] In effect, the readings of the type described here constitute a root or main clause phenomenon in the sense of Emonds 1976 and Green 1976; hence the tension in (115) between *because* (whose status as a coordinator is borne out by interaction with other root phenomena) and the subordinating complementizer *that*. Crucially, however, what is being negated or affirmed, agreed with or disagreed with, in (114B) is not the question in (114A)—which, like the tune in (113), has no obvious truth value as such (but see Karttunen and Peters 1976)—but rather the proposition which A is taken to have pragmatically conveyed.

In its use as a validity assessor, the predicate *true* often picks out, not the entire proposition expressed by a previous utterance, but some subassertion within it. If you solemnly announce *I hereby state (declare, announce) that the performative hypothesis is dead,* I can reply *That's (not) true,* intending (unless I'm a neo-Jesuit—cf. Fauconnier 1979b) that my agreement or disagreement is to be applied to your embedded clause (--› *It's not all that dead*), rather than to your matrix, which is presumably automatically true and indeed self-verifying. Thus, recording a postgame interview with Miami Dolphin quarterback David Woodley, who had just helped lead his team to defeat in the 1983 Super Bowl game, journalist Malcolm Moran writes,

> It was suggested to Woodley that when many people remember Super Bowl XVII, they will say the Dolphins lost because David Woodley failed to complete his last nine passes.
>
> 'That's probably true', Woodley said.
>
> Woodley was not saying it was true that the critics will blame him. He was saying that the critics will be correct in saying the quarterback lost the game. (*New York Times,* 2 January 1983)

The cases we have been considering here suggest the line which I urge for (110) and related examples: what is denied is not the proposition actually asserted, if any, but the assertability of the proposition conveyed in the context of utterance. Adapting the important distinction drawn in Kripke (1977), we must acknowledge a divergence of speaker's meaning and sentence meaning.

We should note one additional way in which the use of the *It is true that* preface in ordinary discourse differs from the semantic value of truth predicates. Often the only felicitous discourse-initiating use of the affirmative formula *It is true that* is a concessive one. A sentence which begins by affirming *It is true that snow is white* tends to set up a later clause beginning with *but.* An instance of this usage can be found in the text above, immediately following (111d). I shall not dwell on this point here, except to suggest that it seems susceptible to a natural conversational account and to note that it gives us one more reason to dissociate the definition of the semantic truth predicate from the behavior of ordinary language *true.* (Cf. Strawson 1949; Grice 1978:125–27; and G. Lakoff 1975:259 for related discussion.)

6.5.2 Monoguists and Ambiguists Revisited

The analysis presented here, on which marked negation is taken to represent a metalinguistic use of the negative operator rather than (as with descriptive negation) a semantic operator within the logical form of a proposition, bears varying degrees of kinship to other accounts of negation which have been presented or defended over the last few years. I now turn to some of these accounts to investigate the relation between them and my own analysis.

We have seen in some detail that philosophers, linguists, and psychologists from Plato and Aristotle through Jespersen and Strawson to Greenberg, Clark, and Givón have argued for taking negative statements to be generally marked and/or complex relative to their corresponding affirmatives. We have also observed, in §1.2, the long-standing (but ultimately incoherent) philosophical practice, instantiated by Kant, Wittgenstein, Searle, Apostel, and Givón, to take negative statements as representing a speech act of denial, on a different level from their affirmative counterparts. On this view, it is maintained (in the words of Gale 1970:201) that 'negation signifies a person's mental act of denying, rejecting, or rebutting a statement that is actually made or envisioned as being made by someone', as expressed by the purported equivalence in (116).

(116) $\left\{\begin{array}{l}\text{It is not true that } \mathbf{p} \\ \text{It is not the case that } \mathbf{p} \\ \textbf{not-p}\end{array}\right\} \leftrightarrow$ I deny that **p**

I touched as well on the related tradition in which negation is identified with falsity, as represented in the work of Leibniz and the Idealists.

For Bergson, perhaps the most doctrinaire proponent of the position that negation is invariably a second-order operation, every negative statement is a subjective judgment about some actual or potential affirmative statement, not a description of reality per se. On this view (Bergson 1911, cited in §1.2.1), *not X* is inherently elliptical for *not X but Y;* in effect, all negation is metalinguistic.

But, as we also saw, Frege consistently treated negative sentences as simple assertions of negative propositions, explicitly warning against confusing a lexical form (negation) with a speech-act function (speaker denial), while Austin, Quine, Gale, and especially Geach level their own salvos against the asymmetricalist tenets, insisting on the logical parity of negation and affirmation. While such caveats are well taken, they seem to offer no explanation for why such luminaries as Kant, Bergson, and Russell might have been seduced into drawing their radical logical distinctions between negative and affirmative statements. Nor is any connection drawn between the treatment of negation as a propositional operator and the properties of morphosyntactic markedness characteristic of the form and function of negation in natural language.

One solution to this standoff, I suggest, is the recognition that while truth-conditional semantics does indeed (as argued by Gale (1970) and Geach [1972] 1980) contain a propositional- and/or term-level negative operator, corresponding to descriptive negation in the object language, not all occurrences of natural language negation can be represented in this way. As we have seen, a need clearly exists to accommodate the use of negation for (speaker) denial or rejection of an earlier statement, but once we have weaned ourselves from the strong monoguist thesis, there is no reason to expect the putative equivalence in (116) to hold.

In any case, *I deny that* **p**—as in (116)—is simply too restrictive a gloss for the metalanguage-level use of negation; we have observed a number of cases where a speaker uses metalinguistic negation not strictly to deny a proposition **p**, or to call **p** false, but rather more broadly to reject the utterance expressing a given proposition, or the implicata associated with that utterance, or the manner in which it was uttered. As remarked earlier, Dummett (1973:328–30) is on the right track in characterizing this use of negation as 'a means of expressing an unwillingness to assert "**A**"', without necessarily constituting a willingness to deny '**A**'. However, Dummett's neo-Fregean representations, utilizing scope distinctions to account for the difference between the two ways in which negation can be understood, may not be sufficiently general or generalizable. While '⊢ (**not A**)' may be unobjectionable for descriptive (propositional) negation, it is not clear that a representation like '**not** (⊢**A**)' can be interpreted coherently for all the cases cited in this chapter.

Some of the recent radically monoguist theories of negation suffer from the flaw noted by Austin, Gale, and Geach: the failure to distinguish negation from falsity and to recognize that to call a statement false is to say something (on a metalinguistic level) about that statement, but to apply (descriptive) negation to a proposition is simply to form another proposition which may itself be true or false. Here is Allwood (1972:43–45), offering a summary of his seminal univocal analysis of negation:

> We have in all cases taken negation to be the same basic semantic operation, indicating that a certain state of affairs is not a fact. We have taken negation to have exactly the properties of logical negation: always giving the predication it operates on an opposite truth value. . . . To negate a certain statement or to say of the same statement that it is false is logically to do the same thing, namely to claim that the state of affairs described in the statement does not obtain.

Allwood's identification of negation and falsity is precisely what Austin, Gale, and Geach warn against; his *prise de position* appears to mingle descriptive and metalinguistic uses of negation.

Kempson's more careful and thorough monoguist treatment of negation also identifies descriptive negation in natural language with 'the falsity operator of logic' (1975:95), but from the context of this identification, she may be referring elliptically to the propositional operator whose semantics corresponds to falsity, that is, a negative expression of the language rather than a negative comment about it. She goes on to summarize and challenge a variety of presuppositionalist views of ambiguous negation in which external or denial negation is taken as a semantic operator. I agree with Kempson that her 'denial negation' cannot be a semantic operator, and is instead—in her words (p. 99)—'one of the uses to which negative sentences could be put'. But she goes on to take this correct observation as a license to either ignore those cases of 'denial' negation whose behavior does not naturally fall within the proper bounds for logical negation or to subsume them within the general category of propositional negation, as Allwood does. Yet, as I have argued, no single logical notion of negation as a truth function can collect all natural language tokens of negation.

Kempson avers that 'marked (contrastive stress) interpretations of negative sentences' tend to function as denials, although she argues that 'this correspondence . . . does not carry over to compound sentences'. But, in general, Kempson's citations of marked, 'presupposition-cancelling' negation (1975:68, 78, 86–87) 'can only be conceived of as answers to a previous utterance', as Kiefer (1977:252–53) points out in his review. An example cited by Kiefer is (117):

(117) Edward didn't regret that Margaret had failed because he knew it wasn't true.

Kiefer's formulation is in keeping with the metalinguistic line on marked negation I have urged here.

The most sophisticated, as well as most radical, of the contemporary monoguists is Jay Atlas. His position has shifted perceptibly over the years (from Atlas 1974 through 1977, 1979 [cf. also Atlas and Levinson 1981] to 1980, 1981), as he has considered a progressively wider range of data. But he has consistently maintained that negation is ambiguous neither in scope nor in meaning, even when that position has pushed him into the sobering (or intoxicating?) conclusion that no set-theoretical semantic theory can do justice to negation—or, hence, to natural language in general. On the basis of a sampling of the kind of data I considered in §§6.2–6.4, Atlas summarizes his findings as follows: 'The range of interpretation includes statements that are internal negations, external negations, and metalinguistic predications. *Not*-sentences are semantically less specified, and theoretically more complex, than the tradition in logical theory has heretofore recognized' (Atlas 1981:127).

It should be clear that I share Atlas's misgivings about logical theories that either ignore metalinguistic uses of negation or take them as a subcase of a special semantic external negation operator; but I cannot agree that the appropriate solution lies in placing all our negative eggs into one 'radically underspecified' basket. To put it another way, the evidence I have cited here does not support the radical move of throwing out the model-theoretic baby with the ambiguist bathwater. The real bone of contention between Atlas's current view and my own concerns the proper treatment of descriptive negation, a question I shall consider in chapter 7.

Ambiguist treatments of negation have not entirely passed beyond the pale of modern logic. We have already encountered Bergmann's 'two-dimensional' theory (1977, 1981) of external negation, in which the truth/falsity axis intersects the anomaly/nonanomaly axis, producing four distinct assignments (cf. §2.4 above). But, as Atlas notes, Bergmann's system inherits empirical and theoretical problems from her ambiguist ancestors, in addition to some which are created by the innovations in her own account. Double negation no longer holds for Bergmann's internal negation; furthermore, given her projection rules, a conditional like *If there's a king of France, then he's bald* comes out true but anomalous. Yet 'intuitively there is no anomaly in this sentence at all' (Atlas 1981: 126–27).

But an equally fundamental flaw in Bergmann's account of negation is one not pointed out by Atlas: there is no obvious way to extend the formal 'anomaly' treatment from those negative statements which involve sortal incorrectness (Bergmann 1977) or referential vacuity (as in the *king of France* case) to those involving conversational implicata, grammar, style or register, phonetics, and so forth. It is these cases which most clearly demand a metalinguistic treatment outside the bounds of one- or two-dimensional logical semantics.

Similar problems arise in an account which is in some ways rather congenial to Bergmann's. Karttunen and Peters (1979:47) correctly describe their so-called contradiction negation as having 'a special function in discourse' of contradicting 'something that the addressee has just said, implied, or implicitly accepted'. While this is a necessary condition for a negative to be functioning metalinguistically, it is not sufficient: as we have seen, most ordinary (descriptive) negations can be characterized as serving the same 'special function' (cf. also Atlas 1980). And in stipulating that 'contradiction negation differs semantically from ordinary negation only by virtue of having a broader target', so that it 'pertains to the total meaning of its target sentence, ignoring the distinction between truth conditions and conventional implicatures', Karttunen and Peters fatally overlook just how broad a target marked, metalinguistic negation can have.

One additional contemporary account of negation, more neomonoguist than neoambiguist, is worth mentioning here. Lehrer and Lehrer (1982) distinguish two rival analyses of the relation between scalar operators like *good* and *excellent:* the HYPONYMY interpretation, on which *good* is a superordinate term for the category containing *excellent,* and the INCOMPATIBLE interpretation, on which the predicates *good* and *excellent* are mutually inconsistent. The Lehrers point out that (118a) seems to favor the former analysis and (118b) the latter. (The next five examples repeat Lehrer and Lehrer's (14)–(18).)

(118) a. This wine is good—it's even excellent.
  b. This wine is not good, it's excellent.

They opt for the hyponymy interpretation, based largely on the acceptability of (119) in a construction which excludes 'true incompatibles', as seen in (119′):

(119) That wine is not only good; it's excellent.

(119′) a. *That's not only a cat, it's a dog.
  b. That's not only a car, it's a Cadillac.

As ought to be clear from the discussion in chapter 4, I agree with the Lehrers' conclusion that *excellent* is a hyponym, rather than an incompatible, relation of *good.* But I cannot accept their implication that the negative predication *not good* in the first clause of (118b) is to be regarded as elliptical for *not only good* in (119). Given the scalar nature of the relation between *good* and *excellent*—that is, that ***a*** *is excellent* unilaterally entails ***a*** *is good*—(118b) and (119) will in fact convey the same information; the same point was made in connection with the examples of (45) above. But only those instances of metalinguistic negation which involve an upper-bounding **Q**-based implicatum will share this characteristic.

Thus, there is no way to extend the Lehrers' elliptical analysis of (118b) to conventional implicature cases like those in (1′) and (8), to phonetic cases like (14a, b), to morphological cases like (14c) and (15), to stylistic and connotative cases like those in (17)–(20), or even to those negations which focus on other varieties of conversational implicata; thus, alongside (119), compare:

(120) a. The king of France isn't (#just) bald—he's doesn't exist. [cf. (1′)]
  b. I didn't (#just) manage to trap two mongeese—I managed to trap two mongooses. [cf. (14c)]
  c. For a pessimist like him, the glass isn't (#only) half full—it's half empty. [cf. (17e)]
  d. He's not only meeting a woman this evening—he's meeting his wife. [OK, but ≠ (22′)]

Even among those cases which do involve the cancelation of a quantity-based scalar implicatum, the syntax may render a Lehrer and Lehrer–style paraphrase awkward or impossible:

(120′) ?#Maggie isn't just patriotic or quixotic—she's both! [cf. (63a)]

Lehrer and Lehrer correctly characterize the 'more than good' reading of the negation in (118b) as requiring that 'the intonation contour . . . remain high instead of dropping, signaling a clarification to follow'—but this same characterization applies across the board to all instances of metalinguistic negation, those which are paraphrasable in the manner of (119) and those which are not. In the final analysis, taking metalinguistic *not* to stand for *not only, not just* proves as inadequate as taking it to represent *not true*.

6.5.3 Metalinguistic Negation and 'négation métalinguistique'

As already acknowledged, I am indebted for both the (approximate) concept and the label of metalinguistic negation to Ducrot (1972, 1973). For Ducrot (1972: 37ff.), descriptive negation constitutes a comment on facts and preserves presuppositions.[43] Metalinguistic (aka. polemic) negation comments on utterances and challenges or rejects presuppositions.

Ducrot (1973: 240) defines metalinguistic negation as 'un rejet d'une affirmation préalable (implicite ou explicite)'; crucially, descriptive negation cannot then be characterized in the same way (as it often is, not least by Ducrot himself). While descriptive negation has the general property of reversing scales—cf. §4.4—metalinguistic negation may reverse or conserve scales, as determined by the rectification: *The ticket doesn't cost 10 francs, it costs* {*5*/*15*}.

In Ducrot's system, presuppositions (*présupposés*) are distinguished on the one hand from assertions (*posés*) and on the other hand from rhetorical implicata (sous-entendus). An intermediate formal language (which I shall dub $\mathbf{L_D}$) is introduced (Ducrot 1972: §5) for representing statements of ordinary language in such a way as to allow presuppositions and assertions to be distinguished in the predicate calculus translations of $\mathbf{L_D}$ formulas. The notation $\mathbf{X|Y}$ represents a 'predicative pair', where **X** and **Y** can be filled by atomic or complex predicates. Any $\mathbf{L_D}$ expression of the form $\mathbf{X|Y(a_1, \ldots, a_n)}$ will then correspond to two predicate calculus expressions: one, the translation of $\mathbf{X(a_1, \ldots, a_n)}$, for the presupposition, and the other, the translation of $\mathbf{Y(a_1, \ldots, a_n)}$, for the assertion.

Natural language operators (*only, some*) and negation are represented in $\mathbf{L_D}$ by boldface 'copulative operations' which convert one predicative pair into another (Ducrot 1972: 147). Two such copulative operations are **NEG** (presupposition-preserving descriptive negation) and **REF** (refutational, i.e., metalinguistic, negation). Their effect is indicated as follows (where paleface NEG eventually translates into predicate calculus '$\sim$' and ET into '$\wedge$'):

(121) a. **NEG**(**X**|**Y**) = **X** | NEG **Y**
b. **REF**(**X**|**Y**) = — | NEG (ET(**X**,**Y**))

It will be noticed that the distinction between (121a, b) directly (mutatis mutandis) prefigures that drawn between ordinary and 'contradiction' negation in Karttunen and Peters (1979), as represented in chapter 2 and in (11) above. More specifically, the marked negation of (121b), as in (11b), brings the presupposed (conventionally implicated) material within the logical scope of negation; but ordinary descriptive negation, in (121a) as in (11a), respects presupposed (conventionally implicated) material by according it a kind of logical transparency. (Note that the presuppositional component in the output of (121b) is empty, just as the conventional implicature component of (11b) is vacuous.)

But we have already seen that this scopal distinction does not generalize to the entire range of possible applications of metalinguistic negations discussed in this chapter; in particular, such foci of negation as phonetic form, allomorphy, syntax, conversational implicature, register, and connotative meaning are not part of logical form (in Ducrot's $\mathbf{L_D}$ any more than in K & P's version of Montague Grammar), and hence cannot be plugged into the format of (121b). Ducrot does acknowledge a 'rhetorical' function of marked negation, to deny the 'sous-entendus' associated with a given utterance, but his representations and account of 'la négation métalinguistique' do not do justice to the protean character of metalinguistic negation in French or English.

Nevertheless, the account of metalinguistic/polemic negation offered in various works by Ducrot and his colleagues (Ducrot 1972: 37ff.; 1973: 124–25; Anscombre and Ducrot 1977) is certainly helpful and suggestive for what a complete analysis must encompass. Thus, Ducrot correctly observes (echoing Grice and Dummett; see §6.2 above) that the negation associated with a conditional tends to be interpreted only as a metalinguistic device indicating the speaker's unwillingness to assert that conditional. Elsewhere, Ducrot points out that metalinguistic or polemic negation corresponds to a special negative speech act—a way of rebutting a previously uttered affirmation.

In her empirical study of the scope of negation in French, Heldner expands on the role of Ducrot's metalinguistic negation and its interaction with scalar predications (cf. Ducrot and Barbault's essay in Ducrot 1973). A sample citation involving metalinguistic negation is (122):

(122)	Jules ne chante pas bien, il chante comme un dieu.	'Jules doesn't sing well, he sings like a god'

where 'the speaker makes it clear that *bien* must be replaced by a more adequate term'—one not necessarily (as with descriptive negation) below *bien* on the relevant scale, but possibly higher or on another scale entirely (Heldner 1981 : 92).

As Heldner points out (p. 65), Ducrot and his colleagues originally took the descriptive/metalinguistic dichotomy to be morphologically neutralized in French, but more recent work has suggested a candidate for an unambiguous signal of the latter. For Gross 1977, the use of *non/non pas,* immediately preceding the negated item, can only be interpreted 'contrastively'—where Gross's CONTRASTIVE negation corresponds directly to the metalinguistic negation of Ducrot. (Anscombre and Ducrot [1977] independently cite *non* as an unambiguously polemic negation.) Thus the negation in (123a) may be interpreted 'contrastively', but that in (123b) must be:

(123) a. Max n'a pas abattu un if, mais (il a abattu) ce pin.
'Max didn't fell a yew, but (he felled) this pine'
b. Max a abattu non pas un if, mais (*il a abattu) ce pin.
'Max felled not a yew, but (*he felled) this pine'

Note that the reduction in the *mais* clause (as well as in its English gloss) is obligatory in (123b), while it is optional in (123a); as we saw in §6.4, this pattern is diagnostic for SN as opposed to PA *but.* Given that *non (pas)* can only be read as a metalinguistic negation, we predict correctly that it will occur only in environments which permit SN (rather than forcing PA) readings of *mais.* Anscombre and Ducrot (1977 : 37) provide the following minimal pairs:[44]

(124)	a.	Il n'est pas français mais il est belge.	'He isn't French, but$_{PA}$ he is Belgian'
	b.	Il est non pas français mais (*il est) belge.	'He is not French but$_{SN}$ Belgian'
(125)	a.	C'est non seulement vraisemblable, mais certain.	'It's not just likely but$_{SN}$ certain'
	b.	*C'est non pas certain, mais reste possible.	'It's not certain but$_{SN}$ remains possible'

Gross (1977:51) constructs another argument for distinguishing contrastive from ordinary negation, based on the distribution of partitive *de* + article vs. simple *de*. He takes (126a) to be necessarily contrastive, understood with a continuation (. . . *il boit autre chose*), while (126b) is understood noncontrastively.

(126)	a. Max ne boit pas du vin.	'Max doesn't drink wine, . . .'
	b. Max ne boit pas de vin.	'Max doesn't drink wine'

Gross finds that *non (pas)*, as expected, occurs only with *de* + article:

(127)	a. Max a bu du vin, non (pas) {de l'eau / *d'eau}.	'Max drank wine, not water'
	b. Max a bu non (pas) {du / *de} vin, mais de l'eau.	'Max drank not wine but water'

Clefts, too, force the contrastive reading on negation, and hence demand the article:

(128) Ce n'est pas {du / *de} vin qu'il boit, mais de l'eau.	'It isn't wine that he drinks, but water'

If, as is reasonable, we take the use of *de* without the article to constitute a negative polarity item in French (cf. Gaatone 1971; Horn 1978a, 1978b), then Gross's correlation of *de* + article with contrastive (i.e., metalinguistic) negation will define a diagnostic for French parallel to the observation for English (cf. Karttunen and Peters 1979; Linebarger 1981; and §6.4 above) that external or contradiction negation, and, by extension, the generalized metalinguistic operator, fail to trigger NPIs.

But in fact, the evidence is a bit murkier than Gross intimates. For Heldner (1981:77), both (129a) and (129b) are acceptable in isolation:

(129) a. Je ne bois pas du vin, (*mais) je bois de la grenadine.
b. Je ne bois pas de vin, mais je bois de la grenadine.

The former is interpreted as specific in time and space (= 'I am not drinking wine, I'm drinking grenadine'); the latter is taken as habitual (= 'I don't drink wine, but I drink grenadine'). In any case, however, Heldner does grant, with Gross, that negative polarity *de* is unacceptable in the unambiguously metalinguistic negation of (127b).[45]

6.5.4 Negation in London

The English-language account of negation bearing the greatest kinship to the approach taken here is probably that offered by Deidre Wilson. She includes in her primary data a wide variety of uses of negation, many derived from Grice, which are not reducible to garden-variety descriptive negation. Among these examples are the following (Wilson 1975: 149ff.):

(130) a. I'm not happy: I'm ecstatic.
b. The next Prime Minister won't be Heath: it will be Heath or Wilson. [= (109)]
c. I don't love Johnny: I love Johnny or Billy.

Although the passage that follows, inspired by such examples, does exhibit the error (decried by Quine, Austin, Geach, and others) of identifying ordinary descriptive negation with falsity, we are provided here with a clear description of why natural language negation cannot always be reduced to the familiar one-place logical connective:

> To assert that ***not-p*** (or to deny that ***p***) cannot be the same thing as to assert that ***p*** is false. It may also be to assert that ***p*** is inadequate to the facts without necessarily being false: it may be too weak, or too strong, or misleading. . . . Once negation and falsity are distinguished, semantic statements of entailment and contradiction could be made in terms of falsity, while the treatment of negation could include, but go beyond, relations of falsity alone.
> (Wilson 1975: 150)

Given the existence of cases like (130), there must be non-truth-functional aspects to the interpretation of (at least some uses of) negation—instances in which the value of ***not-p*** cannot be simply a function of the value of ***p***. We see from these examples that the falsity of ***p*** is a sufficient but not a necessary reason for asserting ***not-p***: given that uttering ***p*** might suggest ***q***, and that one does not wish to suggest ***q***, one might say '***not-p***' (Wilson 1975: 151).

What Wilson does not make clear is just how the fact that some instances of ***not-p*** count as refusals to assert ***p*** is to be related to the fact that other instances of ***not-p*** do contain negation as an object-language connective, translating into logical form as '∼'. What is lacking here is precisely a full characterization of the distinction between negation as a truth-functional connective (not equivalent to falsity) and negation as a metalinguistic objection to some aspect of a previous actual or implicit utterance.

In particular, just as not all uses of metalinguistic negation can be analyzed as semantic external negation—or as negation outside the scope of a semantic operator TRUE—it is also the case that not all the cases explored here can be taken as refusals to assert a given proposition (or sentence; Wilson is not entirely clear on just what sort of entity ***p*** is intended to stand for in the passages cited above). Her characterization collects those cases where negation attaches to conversational implicata, along with those arguably involving conventional implicata or presuppositions (notions whose utility Wilson challenges, but that's another story). But it does not directly generalize to examples like those in (14)–(15), where the objection is not to the assertion of a given proposition (much less to the truth of that proposition), but rather to the way that the proposition was reified into a sentence, or the way that sentence was uttered. The use of negation to signal that a speaker finds a given proposition unassertable (cf. Grice 1967; Dummett 1973; Ducrot 1973; and Grim 1981; along with Wilson) is more inclusive than the external negation operators of the ambiguists (the three-valued logicians, van Fraassen, Bergmann, Karttunen and Peters, Linebarger, and others), but is itself a proper subcase of the generalized use of negation as a metalinguistic operator.

Ironically, it is Wilson herself who cites and attacks two alternative views of marked negative statements, views which—while not fully fleshed out—more closely anticipate the notion of metalinguistic negation than anything in her own work or in that of other logicians, philosophers, and linguists. The relevant excerpts, from Fillmore 1969 and Kiparsky and Kiparsky 1971, emanate from that heady period immediately after the discovery by generative linguists of those great presuppositional vistas and swamps that philosophers had been uncertainly navigating for centuries. As is typical of the era, they combine keen insight with a certain lack of rigor and precision:

> Uses of the verb *chase* presuppose that the entity identified as the direct object is moving fast. Uses of the verb *escape* presuppose that the entity identified by the subject noun-phrase was contained somewhere by force previous to the time of focus. These presuppositions, as expected, are unaffected by sentence negation:
>
> (58) The dog {chased / didn't chase} the cat.
>
> (59) He {escaped / didn't escape} from the tower.
>
> It seems to me that sentences like (60) and (61) are partly comments on the appropriateness of the words *chase* and *escape* for the situations being described. These are sentences that would most naturally be used in contexts in which the word *chase* or *escape* had just been uttered:
>
> (60) I didn't 'chase' the thief; as it happened, he couldn't get his car started.
>
> (61) I didn't 'escape' from the prison; they released me.
>
> (Fillmore 1971:381–82)

> If you want to deny a presupposition, you must do it explicitly:
>
> Mary didn't clean the room; it wasn't dirty.
>
> Abe didn't regret that he had forgotten; he had remembered.

> The second clause casts the negative of the first into a different level; it's not the straightforward denial of an event or situation, but rather the denial of the appropriateness of the word in question [underlined above]. Such negations sound best with the inappropriate word stressed. (Kiparsky and Kiparsky 1971:351)

These passages are quoted by Wilson (1975:84) in the course of her blitz against all extant presuppositionalist theories, Fillmore's and the Kiparskys' included. Her objections to the views illustrated here have more to do, I think, with her skepticism about the viability of semantic (and pragmatic) notions of presupposition than with the metalinguistic line on so-called external negation; in attacking the proposed examples of 'lexical' presuppositions, Wilson also (quite properly) takes on some of the weakest candidates for presuppositional status ever proposed by philosophers or linguists.

In assuming that marked negation can only be used to deny presuppositions, Wilson may or may not be faithful to the intent of Fillmore and the Kiparskys. In any case, I have argued here for a different account of the metalinguistic use of negation—one which strikes me as entirely compatible with more recent theories of presuppositional phenomena, including the pragmatic presuppositions of Karttunen (1974) and Stalnaker (1974), the context-cancelable presuppositions of Gazdar (1979a, 1979b) and the ordered entailments of Wilson and Sperber (1979).

Note, however, that both the above excerpts specifically allude not only to the fact that metalinguistic negation is used to object to an earlier utterance as inappropriate—rather than to judge a proposition previously expressed as false—but also to the fact that it occurs (as does any metalinguistic operator, by definition) on a different level, that is, as a predication about the object language rather than a formal device within it. Moreover, while Wilson (1975:84–85) correctly recognizes that we cannot define all instances of external or presupposition-canceling negation as 'denials of appropriateness', as the Kiparskys seem to believe, their notion does provide a closer approximation to the general phenomenon of metalinguistic negation, in the complete range of exemplification I have attested, than does Wilson's own view of marked negation as a refusal to assert a given proposition.

More recently, Ruth Kempson and her fellow monoguists have drawn their own conclusions from the existence of Grice-Wilson negation (cf. (130) above). As Cormack (1980) points out, negations like (130a) (*I'm not happy: I'm ecstatic*) or (131):

(131) Justin didn't paint three squares, he painted four.

appear paradoxical: 'If Justin painted four squares, he certainly painted three; if someone is ecstatic, they are certainly happy, and so on'. Furthermore, as Burton-Roberts (1984:202–3) observes, (132)—while apparently paradoxical relative to standard modal systems—is nevertheless acceptable:

(132) It's not possible that mammals suckle their young, you ignoramus, it's downright necessary.[46]

I have already considered and rejected Lehrer and Lehrer's (1982) elliptical analysis of these 'paradoxical' negations. Burton-Roberts opts for a different approach, one in which the weak scalar element (*possible, three, happy*) is taken as lexically ambiguous between what I have called its one-sided (at least) and two-sided (exactly) understandings, given that 'as Cormack points out, the alternative to this is to invoke a special (denial, quotational) negation to handle the phenomenon . . . (an alternative that she rejects in favor of treating implicatures semantically)'. Having previously employed Occam's razor to shave off one of the forks of the purported ambiguity of negation, she does not consider this alternative as a live option.

The practitioners of the London School of Parsimony argue from the nonexistence of one ambiguity (that for negation) to the existence of infinitely many ambiguities, at either a semantic level (à la Cormack and Burton-Roberts) or a propositional level (à la Carston 1985a and Kempson 1986). If *Justin painted 3 squares* is ambiguous, so must be *Justin painted 4 squares, Justin painted 137 squares,* and so on. If these and other scalar predications are semantically or propositionally ambiguous in English, so are their translations into French, Basque, Swahili, and every other language in which a weak scalar predication may or may not (in my terms) induce an upper-bounding **Q**-based implicatum. Razor, where is thy sting? Parsimony, where is thy victory?

Having argued against the predecessors of the London School's ambiguist analyses (those of Aristotle on *possible,* Hamilton on *some,* and Smith on the cardinals), I admit a certain reluctance to abandon the Mill-Grice program for scalar predication (cf. Horn 1972, 1973 and chapter 4 above). But this reluctance is reinforced on the one hand by the argument (Horn 1984a, 1984c) that privative ambiguity cannot simply be argued away (à la Kempson 1980) and on the other hand by the arguments presented in this chapter. I have tried to show that a pragmatic ambiguity can be motivated for negation, not only in the scalar cases focused on by the London School but in a wide range of examples for which the considerations invoked by Cormack, Burton-Roberts, and Kempson are irrelevant. The alternative rejected by Cormack in the passage cited above remains the most general and elegant account of paradoxical uses of negation, while at the same time enabling us to preserve the simple Gricean line on scalar "ambiguities."

But is this line worth preserving? Kempson (1986) maintains that while scalar predications are not ambiguous either lexically or at the level of semantic representation, they are ambiguous propositionally, at the level of enriched logical form. For Kempson, utterance interpretation is radically underspecified by linguistic meaning; pragmatic principles—including the familiar Gricean implicata—may (contra Grice) influence propositional content and hence help determine truth conditions. If she is right (and see Atlas 1979, Carston 1985a, 1985b, and Sperber and Wilson 1986 for parallel arguments), no straightforward distinction between what is implicated and what is said (as defended or assumed by Grice, Gazdar, Karttunen and Peters, and of course Horn) will survive.

For the cases under consideration in this chapter, Kempson acknowledges the existence of metalinguistic negation but rejects its (necessary) application to the paradoxical scalar cases. More specifically, she questions the applicability of my polarity and incorporation diagnostics for metalinguistic vs. descriptive negation, although it strikes me as plausible that a sharpening of these tools (one which I shall not undertake here) can avoid the pitfalls she cites. But if sentences like (130)–(132) 'can and naturally are interpreted as straightforward cases of descriptive negation' (Kempson 1986:88), then the negated scalar predications themselves (based on *happy, or, three, possible*) are presumably available in a choice of truth conditions: 'We seem forced to the conclusion that there is much more ambiguity in natural languages than anyone has previously envisaged as part of the linguistic specification of natural languages, despite the fact that the variation in interpretation has a pragmatic basis' (Kempson 1986:82). While this conclusion may seem to necessitate biting the razor, we should note that it is not senses per se which undergo multiplication, but propositions. Note also that the same pragmatic principles I outlined in chapter 4 will be invoked for scalar predication on Kempson's account; they will just enter into the picture at an earlier stage of composition.

Carston (1985a, 1985b) offers an alternative view of paradoxical scalar negation. Her response to the positions of Horn 1985 (essentially reproduced in this chapter) and Kempson 1986 seems at first glance to represent a middle position between these two poles. While endorsing the Kempsonian propositional ambiguity for positive scalar predications, Carston (1985a: 14) supports my parallel between scalar negation and the use of a negative operator to object to phonetic, morphological, or stylistic aspects of a previous utterance. Thus, as against Kempson, Carston recognizes that 'we are not dealing with a straightforward descriptive use' in the cases of paradoxical negation: 'there is something metalinguistic here, whether it is the negation or some other aspect of the utterance'. Carston also observes correctly that the 'echoic use' of language involved in my examples is not limited to negative utterances alone (cf. §6.2.3 above).

So far so good. But Carston's landing site (1985a: 17) is the position that there is no metalinguistic use of negation per se: what our putative instances of metalinguistic negation illustrate is 'plain ordinary truth-functional negation operating over an echoic use of language'. We are now back to the ultimately incoherent view that negation is invariably a truth function—even when it takes as an argument the 'echoic use of language'. If there is no category mistake here, there is at the very least a good deal of explaining to do, since Carston is forced by her neomonoguism to propositionalize every target of metalinguistic negation, from grammatical usage to phonology, from register to musical technique. Occam's razor cuts more ways than one; when we bear in mind what a truth function must be a function of, we recognize the implausibility in the view that negation is invariably truth-functional.

6.6 Metalinguistic Negation and Surface Structure

While serving to reject an entire utterance, metalinguistic negation focuses on one particular aspect of that utterance, whence its superficial nature as a constituent or special (as opposed to sentential or nexal) negative. This is not to say that all constituent negation is metalinguistic; indeed, the vast majority of **VP**-scope negation is descriptive in function. Furthermore, as we have seen earlier in this chapter, certain instances of incorporated (and hence necessarily constituent or special) negation can only be descriptive, not metalinguistic. But other varieties of 'narrow-focus' negation (cf. Bolinger 1961; Jackendoff 1972; Ladd 1980) seem to involve a metalinguistic operator 'reaching down' into the hidden places of the sentence, penetrating even into direct quotes (cf. Bolinger 1961; L. Carlson 1983).

In these cases, it would appear that only the pair consisting of the element focused by negation and its rectification—even when this element is a single syllable corresponding to a bound morpheme or to no morpheme at all—represents the new information or rhematic material in the sentence, the remainder (corresponding to the earlier discourse token of the utterance under rejection) being old, given, or backgrounded. One example of this type appears (in both negative and disjunctive versions) in (35c); others are given in (133):

(133) a. This whiskey was not exported from Ireland, it was deported. (from Bolinger 1961: 83)

b. It's not your 'pronounciation' you need to work on—it's your pronunciation. (adapted from L. Carlson 1983: 196)

c. I'm advocating prosecution, not persecution.
d. I called for a policeman, not a policewoman.

As Bolinger (1961) points out, the focus of negation may undergo a shift to heavy, contrastive stress (and to the unreduced vowel entailed by this shift), reinforcing the sense that the rest of the sentence is treated as given information.

The cleft form in (133b) is another characteristic device for realizing a metalinguistic negation, an unsurprising fact given the focusing nature of both operations. This clearly applies to French, in which **AP**s can be clefted as well as **NP**s and **PP**s; Ducrot cites (134) as a standard form for the metalinguistic scalar negation (note also the SN *mais*):

(134) Ce n'est pas satisfait qu'il est, mais enthousiaste.
'He isn't satisfied, {but/he's} enthusiastic' [lit., 'It's not satisfied that he is . . .']

Cleft negation frequently takes propositional or sentential scope; the entire utterance being rejected is in the focus of the cleft, and the following rectification tells the addressee what aspect of the utterance is being rejected. One favorite form for this sentence type is *It's not that **p**, it's (just) that **q***. As in my earlier instances of metalinguistic negation, whether or not a given proposition is true is irrelevant to its rejection:

(135) a. It's not that she's rich and beautiful—{although, as heiresses go, she is quite lovely/in fact she's just a plain country girl}—it's that her heart is pure.
b. It's not that I don't want to go—{although I don't/in fact I'd love to/I hadn't really made up my mind one way or the other}—it's just that I've made plans to clean my bathtub.

The distribution of *although* and *in fact* in these examples suggests that descriptive and metalinguistic negation may be plotted on the same scale, with the former as the stronger item on that scale (*not only is it not that **p***, in fact *not-**p***).

Shakespeare seems to have been particularly fond of the multifarious potential of the related *not that . . . but that . . .* formula for rectification. Sentence (136a) is the more familiar citation, but (136b) the more revealing; Bianca here consciously exploits the metalinguistic function of the *not that* form to twist Cassio's assurance of love into a quibble over syntax:

(136) a. Not that I loved Caesar less but that I loved Rome more. (*Julius Caesar*, 3.2)
b. *Cassio:* Leave me for this time.
Bianca: Leave you? Wherefore?
Cassio: I do attend here on the general

And think it no addition, nor my wish
To have him see me womaned.
Bianca: Why, I pray you?
Cassio: Not that I love you not.
Bianca: But that you do not love me! (*Othello,* 3.4)

The words Bianca manages to put into Cassio's mouth can be expressed as: *'Tis not that I love you not, but that I do not love you.*

A more contemporary writer, David Mamet, calls upon a similar metalinguistic cleft negation to explain his recruitment of his wife Lindsay Crouse to star in his movie, *House of Games,* in the role of an expert on compulsive gambling: *It's not that she doesn't like poker—she hates it.*

A rather different surface form for metalinguistic negation which seems to be passing from the language is investigated by Arvid Smith. As Smith observes, a verb-focus contrastive context can consist of either *not* or *rather than* together with a finite verb; a nonfinite verb in the same environment is interpreted in a somewhat different way, as suggested by the following minimal pair (Smith 1933:79; glosses in original):

(137) a. He ran rather than walked. 'He progressed rather by running than by walking'
b. He ran rather than walk. 'He ran because he did not want to walk'

When the finite verb in the (137a)-type construction is in present tense, we get (or, into the early twentieth century, got) *not* as well as *rather than* in this frame. While preverbal nonauxiliary negation has for some centuries been unavailable in English, an exception seems to be made for this oppositional (metalinguistic) function of negation; we are reminded of the inability of modern metalinguistic negations to incorporate prefixally or to trigger NPIs. Smith attests the following examples:

(138) The wise mother suggests the duty, not commands it.
That is the fire that is lasting . . . that glows, not sparkles only; that comforts, not excites alone; that is certain, sure, and steady, . . . not thinly leaps and flutters and varies before a touch of gale.
Margery fills the world for you . . . —No, not fills it, said he.

If preverbal nonauxiliary negation can be used to express metalinguistic rejection in the context of (138), a far more common device cross-linguistically is the postverbal constituent negation we have already encountered in conjunction with but_{SN} in the English and French examples reproduced here:

(79) a. We have {?not three children / not three but four children / not three children but four}.
b. Negation is ambiguous {*not semantically / not semantically but pragmatically}.
d. I saw {*not Chris / not Chris but Pat}.

(101) a. John was born not in Boston, but in Philadelphia.

(123) b. Max a abattu non pas un if, mais (*il a abattu) ce pin. 'Max felled not a yew, but (*he felled) this pine'

(124) b. Il est non pas français mais (*il est) belge. 'He is not French but_{SN} Belgian'

Similarly, Bhatia (1977:25) observes that the basic negative markers in Hindi and Punjabi, *nahī̃* and *nai* respectively, take on a special contrastive function when they appear in postverbal position, as seen in (139):

(139) a.	**Hindi**:	Vo āyā nahī̃, āyegā.	'He did not come,
	Punjabi:	O āiā naī, āegā.	[but] will come'
		he came not will come	
b.	**Hindi**:	Usne patr nahī̃, kitāb parhī.	'He read not a
	Punjabi:	One xat naī, katāb parī.	letter, but a
		he letter not book read	book'

In Russian, metalinguistic negation typically involves the combination of a postverbal negative marker (the regular negation *ne*) combined with an obligatory rectification clause introduced by the SN connective *a*. Crockett (1977:241–42) provides the following minimal pairs; note especially her overtly metalinguistic gloss for the last example:

(140) a.	Boris umeet govorit' {*ne po-kitajski / ne po-kitajski a po-japonski}.	'Boris knows how to speak {*not Chinese / not Chinese but Japanese}'
b.	*Alik {byl ne doma / ne byl doma}.	'Alik was not home'
	Alik {byl ne doma / ne byl doma}, a budet.	'I'm not saying Alik was home; he will be'

In fact, as Crockett notes, each of the starred examples can surface without overt rectification, 'under the assumption that the addressee is able to complete it', as in *Alik byl ne doma* 'Where Alik was was not at home'.

But there is another feature associated with 'contrastive' constituent negation, as Babby (1980:106ff.) points out. Just as metalinguistic negation fails to trigger NPIs in English or the polarity *de* + **common noun** sequence in French, an **NP** in the scope of contrastive negation in Russian is never marked with the genitive of negation which is assigned to both objects and existential subjects in construction with descriptive negation. Babby provides the following instantiations of this pattern:

(141) a. U nego v rukax ne bylo slovarja.
his in hands not was dict.-GEN
'He didn't have a dictionary in his hands'
b. U nego v rukax byl
ne slovar' (a tom enciklopedii).
dict.-NOM *but volume encyc.*-GEN
'He didn't have in his hands a dictionary (but a volume of the encyclopedia)'

(141′) a. On ètogo ne delal. 'He didn't do that'
he-NOM *that*-GEN *not did*
b. Ne on èto delal. 'It wasn't he that did that'
that-ACC

(142) Ščuku nuzno ne uničtožat', a razvodit'
pike-ACC *nec. not destroy but breed*
'One must not destroy but breed pike' [D.O. 'pike' cannot be marked GEN]

Other factors are involved, however; while genitive marked objects occur only inside the scope of descriptive negation, accusative marking is triggered not only by metalinguistic negation with either verb or object focus—as in (143b, c) in this paradigm from Crockett 1977 (the bracketing is Babby's):

(143) a. Brat [ne est mjasa].
brother not eats meat-GEN
'My brother doesn't eat meat'
b. Brat est [ne mjaso].
meat-ACC
'It isn't meat my brother is eating'
c. Brat [ne est] mjaso (a zrët).
but gobbles
'My brother isn't eating the meat but gobbling it',
'My brother is not eating but gobbling the meat'

it may also signal, as Babby points out, that the object is specific and/or definite. If we have an actual meal in (143), involving 'a specific piece of meat on a specific occasion', we get an accusative-marked object within the scope of descriptive negation: *Brat [ne est mjaso]*. It has also frequently been observed that, ceteris paribus, the more formal genitive marking has been steadily losing ground to more informal accusative. (Cf. Magner

1955, Davidson 1967, Timberlake 1975, Crockett 1977, and Babby 1980 for additional commentary on case assignment and scope in Russian negative sentences.)

A predictably different word-order-based distinction between descriptive sentence negation and metalinguistic constituent (or focus) negation obtains in German (data from Payne 1985:232). Here, the normal position for matrix negation is sentence-final; as in Russian, a metalinguistic negation will immediately precede its target and will normally be rectified with the etymal SN connective *sondern:*

(144) a. Er besuchte uns gestern nicht. — 'He didn't seek us yesterday'
b. Er besuchte uns nicht gestern (sondern . . .) — 'He sought us not yesterday (but . . .)'

Payne also cites Persian (1985:232), in which the unmarked descriptive sentence negator, the prefix *na-/ne-*, gives way in 'contrastive' contexts to the construction *na . . . balke,* a freestanding negative particle plus a *sondern*-type rectification.

Metalinguistic negation in Hungarian may also be marked by word order, although the pattern again is different. As documented by Varga (1980:89–93), the focused element appears immediately to the right of the negation *nem* and is typically assigned the same fall-rise intonation contour we have already witnessed as a correlate of metalinguistic negation in English. Among Varga's examples are the following:

(145) a. \`Nem ˇPéter játszik hanem \`János.
not *plays* *but*
'Not Péter but János is playing'
b. \`Nem a ˇkonyvet olvassa, hanem az \`újságot.
the book-ACC *reads* *but* *the paper*-ACC
'He is reading not the book, but the paper'

With scalar values, as I predict, the descriptive reading yields only the 'less than' interpretation, while the metalinguistic use of negation, triggered by the fall-rise contour and appropriate rectification, is not restricted in this way.[47] Commenting on minimal pairs like that in (146),

(146) a. \`Nem \`olyan gyorsan gépel mint te.
as *fast* *types* *as* *you*
'He doesn't type as fast as you' (= he types more slowly)
b. \`Nem ˇolyan gyorsan gépel mint te, hanem \`gyorsabban.
'He doesn't type as fast as you, but faster'.

Varga (p. 90) comments that the latter sentence 'may have a surprising or humorous effect (because it contradicts our pragmatic expectations mobilized by the first part of the sentence), but is perfectly acceptable. . . . On the pragmatic level the lower-value ['less than'] interpretation prevails unless the higher-value interpretation is explicitly stated'. Similarly, in (146′) the result is 'a pragmatic contradiction', 'exploited to create a surprising and/or humorous effect' (Varga 1980:93).

(146′) \`Nem ˇolyan magas Ø, mint te, hanem \`sokkal \`magasabb Ø.
not as tall is as you but much taller is
'He isn't as tall as you, he's much taller'

This is of course precisely the effect I have copiously illustrated for metalinguistic negation in English.

I have argued that marked negation is a reflex of an extended metalinguistic use of the negative operator in English and other languages. Negative morphemes generally allow (in principle) both descriptive and metalinguistic functions; the syntactic environment—including such parameters as affixation, polarity, word order, case marking, and the presence of an SN or a PA connective—often helps select one of these uses as the more plausible or salient for a given negative token.

In some cases, however, a particular morphological realization of negation may in fact force or exclude a particular understanding. (Like the aforementioned contextual disambiguation, this too is a regular occurrence in the realm of pragmatic ambiguity; cf. Zwicky and Sadock 1975, Horn and Bayer 1984, and §5.3 above.) Thus, as we saw in the previous section, French *non (pas)*, positioned immediately before the item in the focus of negation, must be interpreted metalinguistically, while *ne . . . pas* allows both descriptive and metalinguistic functions.[48] Korean may offer an instance of the opposite state of affairs, in which one morphological variety of negation is unambiguously descriptive, while the other may be understood in either way.

The two constructions in question are, in the traditional descriptive terminology, the SHORT FORM negator *an(i)*, placed before the verb, and the LONG FORM *an(i) hada* (lit., 'not do'), placed after the verb stem suffixed by the nominalizer *ci-/ji-*. Thus, corresponding to a basic affirmative sentence like (147a), we have the short-form negative (147b) and the long-form negative (147c):

(147) a. Mica ka canta. 'Mica sleeps'
b. Mica ka an(i) canta. 'Mica does not sleep'
c. Mica ca-ci ani hanta. 'Mica does not sleep'

The issue is whether (147b, c), and members of similar pairs, differ in meaning or use, and if so how. Kuno (1980:162–63) maintains that the two constructions are either interchangeable or differ only in emphasis;

he notes that other commentators detect more of a real difference, for example, that the former is a 'verb negation' and the latter a 'sentence negation'. This distinction, as Kuno explicates it, is reminiscent of (but not identical to) the internal/external dichotomy discussed in chapter 2.

Other researchers have taken different and often conflicting (if not internally inconsistent) positions. Choi (1983) considers several possibilities raised in these studies and concludes that the closest match for the two Korean constructions within the Western literature on negation may be Aristotle's contrary vs. contradictory negations, corresponding to the short-form and long-form negative operators, respectively. In any case, Choi's data indicate that the preverbal short form is always used descriptively, while the long form is not restricted to metalinguistic uses—and indeed, often fills in suppletively for the distributionally defective short form when the syntax demands it. If the choice to use long-form negation is often interpreted metalinguistically in those contexts which would have permitted the short form, this interpretive tendency may well be grounded in what I have called the 'division of pragmatic labor' (Horn 1984b; cf. also McCawley 1978).

An additional factor relevant to the Korean case is the restricted scope often associated with the unmarked negative form in verb-final languages (cf. Davison 1978, Kuno 1980 for general discussion). Kuno notes that the scope of the Japanese negation *-na-i* is generally limited to the immediately preceding verb (although quantifiers can 'escape' this restriction and enter the scope of a noncontiguous negation). The normal Turkish negation *-mA-* is similarly restricted, with the suppletive periphrastic form *değil* surfacing in contrastive and other contexts.

Other considerations relevant to the interpretation of Japanese negative sentences, bearing particularly on the interaction of metalinguistic negation with the topic marker *-wa*, are brought out by McGloin (1982:57–58). Citing the following three-way distinction in English from Horn 1978a:137:

(148) a. She isn't pretty. (= less than pretty)
b. She isn't pretty, but she is intelligent. (= other than pretty)
c. She isn't (just) pretty, she is beautiful. (= more than pretty)

McGloin notes that the unmarked descriptive interpretation of (148a) is available in Japanese whether or not the scalar element is suffixed by *-wa*. Thus, both (149a, b) may be read as conveying that it is less than, that is, cooler than, hot:

(149) a. Atsuku na-i.
hot NEG-PRES
b. Atsuku wa na-i.
hot TOP NEG-PRES
} 'It isn't hot'

But only the former version may be assigned the nonscalar 'other than' interpretation of (148b) (e.g., 'It's not hot but it is dirty'). By contrast, McGloin reports that neither (149a) nor (149b) can be read as an English-style 'paradoxical' metalinguistic negation, as in (148c) above, or my earlier examples in this chapter ((13), (41), (43), and their ilk), where the negation focuses on the upper-bounding implicature associated with scalar predications. In order to get such a reading, a periphrastic form must be employed:

(150) a. Atsui dokoroka nietagit-te i-ru yo.
hot far from boiling be-PRES
'It's far from being hot: it's boiling'
b. Atsui nante yuu mon ja na-i. Nietagit-te i-ru yo.
say
'It's not something you can call hot. It's boiling'

But Japanese does contain a construction, *wake de wa nai,* which (as described by Kato 1985:180–84) seems to be specialized for expressing metalinguistic negation. In particular, as Kato shows, adverbs which cannot appear at all in negative sentences (either inside or outside the semantic scope of that negation) can occur inside the scope of *wake de wa nai:*

(151) a. *Kuruma ga totsuzen {tomat-ta/*tomar-anakat-ta}.
suddenly stop-PAST/*stop* -NEG-PAST
'Our car suddenly {stopped/*didn't stop}'
b. Kuruma ga totsuzen tomat-ta wake de wa na-i.
'It's not that our car stopped suddenly' [it stopped gently, etc.]

Whatever the details of the behavior of negation in specific verb-final languages (or recalcitrant languages of other typologies), the overall pattern seems confirmed: no language contains two negative operators corresponding exactly to descriptive and marked negation, whether the latter is to be characterized as an external semantic operator or (as urged here) a metalinguistic use of basic negation. At the same time, every language contains at least one negative morpheme which can be used either descriptively (to form a negative proposition) or metalinguistically (to reject a previous utterance), the choice between these two understandings often being made by the addressee in accordance with the grammatical properties of a particular negative token in a particular sentence type within a particular language.

One issue which remains is the directionality of the relationship between descriptive and metalinguistic negation: which use is primary and which derivative? Or do both uses branch off separately from some more basic, undifferentiated notion? I have little to contribute to this etiology, given that the connection is explicable in either direction and that a full answer would appear to be buried in the realms of speculation. My review of the literature of acquisition in §3.1 suggests that the metalinguistic use may be ontogenetically prior, in that the prohibitive or rejection/refusal negative of early child language predates and evolves into truth-functional negation (cf. Pea 1980b). On the other hand, for what it's worth, Rumbaugh and Gill (1977: 169–70) report that the chimpanzee Lana, having been taught the propositional, truth-functional, descriptive use of negation as part of her computer-based symbolic repertoire of 'Yerkish', spontaneously innovated what can only be viewed as metalinguistic uses of the same negative operator. Of course even if we conclude that the generalized metalinguistic use of negation as a sign of objection or refusal is learned earlier than its logical, truth-functional use, it does not follow that this order of development should be associated with any logical asymmetry in the account we give for negation (or, analogously, for the other operators) in an idealized model of the adult speaker's linguistic competence.

I have maintained in this chapter that conditions on truth must be kept distinct from conditions on assertability, and that more explanatory burden should be shifted from the former onto the latter. I have also argued that while there is indeed only one descriptive sentence-level negation operator in English and other languages, the ordinary truth-functional interpretation of this operator motivates it for an extended use as a general metalinguistic sign of rejection or objection, leveled against the choice of a particular object-language expression or the manner in which that expression was overtly realized.

In reply to the query posed in the title of Atlas 1981, 'Is *not* logical?', some have answered 'yes' and others (including Atlas himself) 'no'. I conclude that the only full and complete answer must be 'sometimes'—that is, when it functions descriptively rather than metalinguistically. Neither the strong monoguist nor the ambiguist approach to the data I have considered here can deal successfully with the unity and diversity of the phenomenon of metalinguistic negation.

While it is clear that natural language negation is not always logical, it should be recognized (contra Bergson and—in a different sense—contra Atlas) that it is not always nonlogical either. In particular, there is a procedural sense in which the descriptive use of negation is primary; the nonlogical metalinguistic understanding is typically available only on a 'second pass', when the descriptive reading self-destructs. While Reggie Jackson may have acknowledged that his 1983 season 'wasn't a bad year—it was horrible' (cf. (41b)), it is nevertheless undeniable that upon sending his cleats on to the Hall of Fame, Mr. October could hardly exclude 1983 from the ranks of his bad years on the grounds that it was (not bad but) horrible. And it is clearly harder to verify a metalinguistic negation (*The king of France is not bald*) than a descriptive one (*The queen of England is not bald*). Both semanticist and pragmatic/assimilationist theories of univocal negation (à la Kempson 1986 and Atlas 1981, respectively) must somehow come to terms with the fact that not all negative understandings are created equal.

One important question which I did not, and will not, directly address here is just how metalinguistic negation is to be represented within a formal theory of natural language discourse; this question, along with the larger issue of the relation between language and metalanguage in linguistic theory, would push us beyond the purview of this study. We must be content for now with the negative fact extracted from this chapter: some instances of negation in natural language are not formally representable in an interpreted propositional language.

But what conclusions do I finally draw from my analysis of descriptive and metalinguistic negation in this chapter for the question with which I began it, namely, the treatment of presuppositional phenonema and bivalency in a logic of natural language? If every negation is either descriptive or metalinguistic, how exclusive is this disjunction? Can some negations be both? And what of descriptive negation proper: what varieties of sentential- and constituent-scope negation must be recognized within the syntactic and semantic description of natural languages? These are the questions I shall address in the next (and final) chapter.

7 Negative Form and Negative Function

One issue I have stalked (and vice versa) in the six previous chapters of this study is the nature of the negative inventory. Just how many distinct species of negation must be admitted into our grammatical and semantic catalogues, and (if the answer exceeds one) on what basis are we to characterize the variants in our stock?

I began with an exploration of Aristotle's binary model, in which predicate denial—negation as a mode of predication—yields contradictory negation in singular statements, while term negation focuses on, and is often incorporated into, a particular subsentential constituent. This constituent is typically, but not invariably, the predicate term, whence predicate term negation with its contrary semantics.

In chapter 2, I moved on to consider several alternative ambiguist theories of negation. These theories were seen to differ with respect to whether the ambiguity was situated within the lexical semantics of negation or within the scope of the logical syntax of the negative expression, and with respect to the interaction of the posited negations with presuppositional phenomena, especially in contexts of reference failure or category mistakes.

In chapter 3, I determined that so-called sentential negation can indeed be regarded as the semantic counterpart of affirmation, with the objections of the asymmetricalists (see §1.2) satisfied within an independently motivated account of pragmatic inference. A functional model was proposed for the tendency of negation to appear to be (ceteris paribus) more presuppositional or context-bound than affirmation. Proceeding in chapters 4 and 5 to an examination of the semantic and pragmatic properties of negation in scalar predications, I argued that ordinary sentential negation yields formal contradictories which can nevertheless be filled in or pragmatically strengthened in certain specifiable scalar contexts to yield functional contraries.

I returned in chapter 6 to the inventory question proper and presented evidence for a dichotomy between descriptive sentential negation and the metalinguistic use of the negation operator as a sign of objection to some contextually present utterance. I suggested that the acceptance of a pragmatic ambiguity between descriptive and metalinguistic negation might vitiate the purported semantic ambiguity for negation and specifically the internal/external distinction approached from different directions within the models reviewed in chapter 2. This in turn led me to question the utility of the notion of semantic or logical presupposition, which is generally assumed to require the admission of just such a semantic ambiguity.

But I have not yet undertaken to investigate the logical nature of descriptive negation per se. In particular, assuming (with Aristotle, Jespersen, Klima, and Jackendoff, inter alia) that subsentential negation must be accepted—an assumption shared by most, though not all, linguists, and by some, though not many, philosophers—what becomes of sentential negation? Should this notion be identified with the one-place external proposition connective (the Stoics' *apophatikon*) or with the device for combining subjects with predicates, typically surfacing between the two expressions or within the latter one (Aristotle's predicate denial)? How does the cross-linguistic appearance and behavior of negation help to decide the question?

In this, my last chapter, I examine these issues, beginning with a survey of the forms of negation. In §7.1 I explore a number of parameters governing the surface expression of negation within a given language. I observe that the split between two negators may correlate with distinctions in mood (typically indicative vs. directive), with the tense and aspect of the verb, or with the category of the constituent in the focus of negation (typically verbal vs. nonverbal). In particular, I outline the tendency for a formally weak verb-based and specifically existential negation to be differentiated from a marked and often archaizing strong form characteristically used in nonverbal, identificational, and/or contrastive (metalinguistic) contexts. I also illustrate, discuss, and try to explain several typological tendencies affecting the form of negative sentences, including Jespersen's Cycle (the repeated pattern of successive weakening and restrengthening of the negative marker) and the Neg First principle (the preference for negation to precede its focus). I also summarize the results of Dahl 1979 demonstrating the extreme typological rarity of syntactic external negation.

In §7.2 I reinforce Dahl's findings with other arguments in support of my conclusion that Aristotle was right: wide-scope sentential negation is a mode of predication within a subject-predicate-based logical syntax, not an iterating unary connective within a propositional calculus. I develop an extended term logic emphasizing the connection between Aristotelian and Montagovian theories of negation (predicate denial), with the bridge formed by the theory of generalized quantifiers, spelling out the parallel assumed in term logic between singular and general predications. At the same time I depart from orthodox term logic in insisting on a distinction between one-place propositional connectives, which are excluded from my extension, and two-place truth-functional connectives (including conjunction and disjunction), which are endorsed within my model (as they are in natural language).

Finally, in §7.3, I attempt to draw together the remaining strands of my study by filling in some of the details in my picture of predicate denial: the effect of negation on presuppositional phenomena, the relation of predicate denial to constituent (predicate term) negation on the one hand and to metalinguistic negation on the other, and the proper description of the scope interaction between quantifiers and negation. We shall see, as we have seen in the preceding chapters, that the most natural and explanatory account of the complex properties of negative sentences requires the establishment of a division of labor between a formal theory of syntactic and semantic form and a pragmatic theory of language use and function.

7.1 Varieties of Negative Experience: A Typology of Descriptive Negation

Negative coexistence

In chapter 6, I touched on the typologically frequent distributional pattern in which two descriptive negators are differentiated, one occurring in indicative and/or main clauses and the other restricted to imperative contexts or to certain (typically subjunctive or nonfinite) embedded clauses. Among the languages investigated in their survey of syntactic realization types for speech act distinctions, Sadock and Zwicky (1985:175ff.) find that about half display a special negative imperative sentence type, the PROHIBITIVE, which differs significantly from other negative and/or other imperative types. In three-quarters of the languages surveyed there are no straightforward negative imperatives, the functional gap being filled by special negative markers, nonimperative verb forms, or both.

This pattern is also exemplified in the classical languages of Indo-European antiquity. In Vedic Sanskrit, for example (cf. Renou 1946), the sentential negation *ná,* appearing in clause-initial or preverbal position, does not co-occur with imperatives; instead, we get the prohibitive *mā* in such contexts. Negation is rare in embedded clauses of either subjunctive or indicative mood.

The two negators of Ancient Greek, *ou(k)* and *mē,* have been variously distinguished as objective and subjective, assertive and nonassertive, or independent and dependent, respectively (cf. Pott 1859; Mirambel 1946; Weinreich 1963). More specifically, as Mirambel (1946:58–60) shows, the former appears in main indicative clauses and in embedded infinitive clauses governed by verbs meaning 'say' or 'believe', the latter in imperative contexts, and one or the other elsewhere, depending on the interaction of various syntactic and semantic parameters. But the alternation between *ðen* (< *ouden*) and *mē(n)* in modern Greek represents—in the terminology of Dixon's (1979) description of case-marking systems—a SPLIT rather than FLUID pattern, determined by partially conventional or arbitrary grammatical considerations, rather than by the particular meaning or use of a given utterance token. Indicatives and conditionals always govern *ðen*, subjunctive and participle constructions govern *mē(n)*, and the imperative (the archgovernor of *mē* in Ancient Greek) cannot be negated at all (the subjunctive being used suppletively in negative environments, as in Swahili and many other languages). One result of this complementary distribution, as Mirambel points out, is that where the indicative and subjunctive have fallen together in positive contexts, the negative form preserves the distinction—a counterexample to the general tendency (cited by Givón, inter alia; cf. chapter 3) for syntactic distinctions to be neutralized under negation.

As Davies (1975) shows, the ancient Anatolian languages also differentiated two negative markers, along the same lines as Sanskrit, Greek, and Latin:

(1)

	Factual	Prohibitive
Hieroglyphic Luwian	na	ni
Cuneiform Luwian	nawa	nis
Lycian	ne, nepe	ni, nipe
Hittite	natta	lē

What is not clear is whether the two forms of negation in these and other ancient Indo-European languages descend from the same or different Proto-Indo-European sources, and—in the former event—when, how, and why the split might have occurred.

In addition to this frequently attested split along indicative vs. directive (or assertive vs. prohibitive) axes, the expression of sentential negation varies in many languages according to the tense or aspect of the sentence or the category membership of the predicator in negative focus. The negator used in nonexistence statements and other verbal environments is often formally distinct from the one used in negative identity statements and/or for constituent (especially nominal) negation. Classical Chinese, for example, featured sixteen or more negative markers, all glossed (at least sometimes) as 'no' or 'not'. These markers can be assigned to two basic categories, reconstructible with initial *p-* and *m-*, which have been seen as representing negations of identity and of existence, respectively (cf. Kennedy 1952; Graham 1959: 111). But this system proved to be unstable. In one daughter language, Cantonese, there are seven negative markers, all *m-* initial, one of which (*m hai*) seems to function exclusively as a metalinguistic operator (see chapter 6); it can negate any element of the sentence—which is then assigned contrastive stress—and is followed by a rectification (Yau 1980: 39–40).

Tense-based distinctions are also typically unstable. Classical Arabic contained two basic verbal negations, past *maa* and nonpast *laa*, but the reflexes of this formal dichotomy tend to be governed in the colloquial dialects of today by other syntactic and semantic parameters. In Gulf Arabic, *ma* negates verbs and *muu(b)* nouns, adjectives, and phrases; *la* co-occurs with the imperfect to convey a negative command (and nakedly for free-standing 'no!'). Iraqi Arabic distinguishes *la-* and *ma-*prefixed negations for imperative and nonimperative verbs, respectively, while *muu* is used elsewhere, to negate **NP**s, **AP**s, **PP**s, and so forth.[1]

In other modern vernaculars, category but not mood is relevant. Egyptian Colloquial Arabic differentiates a discontinuous verbal negative *ma . . . -š* from a constituent negator *muš/miš* used with nominal, adjectival, adverbial, and participial phrases and in copular (verbless) sentences. A similar dichotomy between a discontinuous verbal negation (*mɛ . . . -(ə)š*) and a nonverbal simple negation (*miš*) obtains in Tunisian Arabic: *mɛ ktibəš* 'he did not write' vs. *miš bɛ:hi* 'not good', *miš wildi* 'not my son', *miš inti* 'not you'. Similar patterns obtain in other North African Arabic vernaculars (cf. Comrie 1985).

Like some of the modern Arabic vernaculars, Temne (Nemer 1985) displays a three-way negative opposition. The standard indicative negation is *-hɛ*, attached to the auxiliary if there is one, and to the main verb if not. The negative particle *té* appears preverbally in imperatives, infinitives, and embedded finite clauses (including relatives). A third marker, *Tá*, occurs in nonverbal sentences and clefts; this form also co-occurs with other negative morphemes and may also signal metalinguistic negation in some contexts. Given the different syntactic properties of these negative morphemes, the negation will follow an indicative but precede an imperative:

(2) a. ɔ́bɔ́rkɔ̀ ɔ́ dí. 'The woman is eating'
ɔ́bɔ́rkɔ̀ ɔ́ díhɛ. 'The woman isn't eating'
DEF + *woman* 3P EAT-PRES±NEG
b. dīf kɔ́. 'Kill him!'
té dīf kɔ́. 'Don't kill him!'
±NEG *kill*-IMPER 3P

This ordering discrepancy is not surprising in the light of Jespersen's Neg First principle, 'the natural tendency, . . . for the sake of clearness, to place the negative first, or at any rate as soon as possible, very often immediately before the particular word to be negatived (generally the verb)' (Jespersen 1917:5). We invoked Neg First in chapter 5 to motivate both the tendency for affixal negation to manifest itself prefixally even when it does not function as a category-internal operation (cf. §5.1.2) and the tendency for semantically embedded negation to manifest itself in a higher clause in 'neg-raising' contexts, even when this introduces ambiguity (cf. §5.2).

The same principle can be held responsible for frequently attested restrictions on the distribution of sentential and constituent negation. Thus, for example, English pleonastic negative parentheticals may follow but not precede sentential negation (*She didn't, I don't believe, order pizza* vs. **She, I don't believe, didn't order pizza;* cf. Ross 1973a), while Italian requires sentential negation to be marked in the environment of postverbal but not preverbal negative **NP**s (*Non ha visto nessuno* 'He saw nobody' vs. *Nessuno (*non) l'ha visto* 'Nobody saw him'; thanks to Raffaella Zanuttini for the minimal pair). In each case, the sentence must be marked as negative before a secondary concordial negation can be marked on a nonverbal constituent.

But while Neg First is operative in both declarative and imperative contexts, there is a particularly strong motivation for avoiding postverbal negation in directive speech acts (imperatives and their functional equivalents). While a violation of Neg First by the postverbal negation in (2a) might result in temporary confusion, a similar transgression in the context of (2b) would literally constitute a matter of life and death (*Kill him—oops—not!*). Jespersen (1917:5–6) discusses parallel forms in Danish (*ikke spis det!*), German (*nicht hinauslehnen*), and Latin (*noli putare*), in which the negative marker appears preverbally in nonfinite directives 'where it is important to make the hearer realize as soon as possible that it is not a permission that is imparted'.

One particularly suggestive distributional pattern is found in Kannada, a Dravidian language of southern India (cf. Kittel 1903; Spencer 1914; Gowda 1970; Bhatia 1977). The two negators at issue here are *illa*, expressing nonexistence or straightforward predicate denial, and *alla*, variously analyzed as a negator of essence, quality, or identification. *Alla* often serves as a constituent negation (He who did the deed is not I) and characteristically appears in *sondern*-type environments, including the 'not **X** but **Y**' and 'not only . . . but also' constructions (cf. §6.4). We get contrastive pairs like that in (3a, b), from Spencer (1914:153):

(3) a. Ī grāmadalli maravu illa. 'In this village there is not a tree'

b. Idu maravu alla giḍavāgide. 'This is not a tree, it is a shrub'

The *alla* form is historically the third person singular of an obsolete verb *al-* 'to be fitting or proper', an etymology which seems particularly fitting and proper for its frequent use as a metalinguistic operator; indeed, as Kittel observes, *alla* is often glossed as 'not like that'. (In imperative sentences, a third negation is used: *bēda,* the historical negation of *bēku* 'want'.)

Similar patterns obtain in other Dravidian languages. Malayalam (Ravindran 1970) distinguishes the suffixes *-illa* and *-alla* along the same lines as the corresponding free forms in Kannada, while negative imperatives involve a periphrastic construction of special interest to us. In isolation, *vēṇṭa* is glossed as 'not needed', deriving from *vēṇ-* 'needed' + the negative remnant *-ta*. But when the same form occurs as a verbal suffix in negative imperatives, it can only be read as 'necessary not', as in *varavēṇṭa* 'do not come', *pōkavēṇṭa* 'do not go', *otavēṇṭa* 'do not run'. We have a clear instance here of the **O** --> **E** drift discussed in §4.5 above.

Telugu differentiates the *lē* negation of existentials, locatives, and other predications from the *kā* negation appearing in equational and other copular (verbless) sentences (Narasimharao 1970), while Tuḷu maps the *illa* and *alla* forms of Kannada and Malayalam onto *ijji* and *attɪ*, respectively (Madtha 1970). In keeping once again with the tendency to maximize the lexical expression of **E** negation and minimize that of **O** negation, we find in Tuḷu the portmanteau form *balli* which can be interpreted either as 'cannot, be unable to' (governing the genitive or instrumental) or as 'should not' (construed with the infinitive). The corresponding **O** values (= 'possible not', 'needn't') can only be expressed periphrastically.

In many languages, as touched on above, a special negative existential form can be isolated from both the general predicational negation and the special emphatic or constituent negator, if any. As noted by Schachter (1985:57–60), Hausa distinguishes the positive existential marker *akwai* from the negative existential *babu,* while ordinary predicate denial is expressed by *bà X bá,* where *X* 'very neatly' marks the scope of negation. Tagalog employs three unrelated morphemes, to mark negation in imperative and optative contexts (*huwag*), in existential contexts (*wala*), and elsewhere (*hindi*). But if there is a formal neutralization, the existential negation will typically fall together with the marker for negation in ordinary predicational frames; indeed, as in the Dravidian languages, we can often trace the standard negation back to an inflection of the existential verb.

This recurring morphosyntactic split between one negation employed for straightforward negative predications (predicate denials) and for nonexistence claims and another employed for negating identity statements or nonverbal constituents recalls the Hegelian dichotomy between significant and insignificant negation; cf. also Mabbott's view of negative identity statements as 'pseudo-judgments' (both in §1.2.2). We can also detect echoes of the dichotomy traced by Gebauer (1885) and other Indo-Europeanists between QUALITATIVE negation (realized as a negated finite verb, resulting in sentential scope for the negative operator) and QUANTITATIVE negation (realized as a negative focusing on or incorporated into some other sentence element, which may or may not equate to sentential negation). If we take the qualitative, predicational category to represent the prototype instance of ordinary descriptive sentence negation, we can see in quantitative and/or narrow-scope negation the contextually bound nature of descriptive constituent negation and of the characteristic metalinguistic/contrastive use of the negative operator.

Negation, word order, and Jespersen's Cycle

The Neg First principle expresses the strong tendency for negative markers to gravitate leftward so as to precede the finite verb or other possible foci of negation. As shown by the evidence collected in Dahl 1979 and Payne 1985, free particle negation is overwhelmingly likely to precede the verb in **SVO**, **VSO**, and **VOS** languages, and may do so in **SOV** languages as well (cf. Payne 1985:224). But this tendency is not absolute, and it interacts crucially with another basic principle, also associated with Jespersen: the cyclical pattern wherein the negative marker is gradually weakened into a verbal proclitic, then reinforced by the accretion of NPI minimizers or indefinites, and ultimately replaced by its reinforcement. This process has come to be known as JESPERSEN'S CYCLE: 'The history of negative expressions in various languages makes us witness the following curious fluctuation: the original negative adverb is first weakened, then found insufficient and therefore strengthened, generally through some additional word, and this in its turn may be felt as the negative proper and may then in course of time be subject to the same development as the original word' (Jespersen 1917:4).

Minimizers, those 'partially stereotyped equivalents of *any*' (Bolinger 1972:121; cf. §6.4 above), occur within the scope of a negation as a way of reinforcing that negation. As far back as Pott (1857:410), linguists have recognized this function of positive expressions denoting small or negligible quantities, often incorporating a sense of scorn or ridicule, which Pott sees as implicitly evoking the formula *nicht einmal das* 'not even . . .'; cf. also Schmerling 1971; Horn 1971; Fauconnier 1975a, 1975b; Heim 1984. Impressive, though hardly exhaustive, inventories of NPI minimizers specialized for this function are given by Pott (1857:410–11) and Wagenaar (1930:74–75). Their examples—from Sanskrit, Greek, Latin, French, Old Spanish, Italian, English, Dutch, German, and Slavic—include minimal quantities from the culinary domain (= 'not a cherrystone, a chestnut, a crumb, an egg, a fava, a fig, a garlic, a grain, a leek, an oyster, a parsnip, a pea'), coins of little value ('not a dinero, sou' [cf. not a red cent, plugged nickel, thin dime]), animals and body parts ('not a cat's tail, a hair, a mosquito, a lobster [*sic*], a sparrow'), and other objects of little value and/or salience ('not an accent, an atom, a nail, a pinecone, a point, a shred, a splinter, a straw').[2] Indeed, it would appear that any entity whose extension is small enough to be regarded as atomic in an accessible set of contexts can be used productively in this frame as a means of negative reinforcement.

Nor is this tendency by any means restricted to Indo-European. Negative-polarity minimizers occur as negation strengtheners in Basque (cf. Lafitte 1962), in Japanese (cf. McGloin 1976:397–419), and in many other languages.

But NPI minimizers are just one device, albeit the most colorful, for producing 'une négation energique' (Wagenaar 1930:75). The systematic use of indefinites of either positive or negative morphological character within the scope of negation also serves to reinforce that negation. We can trace the cyclical history of negative forms through the development of French negation. Preclassical Latin *ne dico* 'I do not say' is first strengthened by the addition of the indefinite *œnum* 'one (thing)', forming the preverbal particle *non (< noenum < ne-œnum),* hence Classical *non dico* (simple *ne* being retained in verbs with incorporated negation, e.g., *nescio* 'I do not know'). But *non* is then phonologically weakened through successive shifts back to the original Proto-Indo-European form *ne,* giving us Old French *jeo ne di.* (As we saw in chapter 6, another reflex of Latin *non* survives into modern French as the metalinguistic operator and free negation *non* [nõ].) Although there remain several isolated constructions with the simple proclitic negation *n(e)—je ne peux, je ne saurais dire, n'importe*—the standard expression of negation involves another round of strengthening, this time involving the accrual of such minimizers as *point* (from Latin *punctum,* 'a point'), *rien* (from *rem,* 'a thing'), *personne* (from *persona,* a person—cf. *not a soul*), and above all *pas* (from *passum,* 'a step', originally collocating with verbs of motion).

Bréal (1900:200–202) observes that while *pas, point, rien, personne, jamais, aucun,* and similar morphologically positive items originally 'served to reinforce the only genuine negative, to wit *ne*', they eventually 'by their association with the word *ne* became themselves negatives', to the point where they can now 'dispense with their companion' and signify negation in their own right. This holds particularly in verbless contexts: Bréal cites the exchange *Qui va là? —Personne* ('Who goes there? —Nobody') and the NP *pas d'argent* (no money), while Gaatone (1971:99) provides minimal pairs like *une chose jamais vue* (a thing never seen) vs. *Je n'ai jamais vue cette chose* (I've never seen that thing). When *ne* is retained in verbal environments, in the discontinuous EMBRACING negation of the standard dialect (*je ne dis pas*), it is no longer a negative marker per se, but rather 'une marque de redondance de la négation . . . conditionné par une autre term qui est le principal porteur de la valeur négative' (Gaatone, ibid.). This redundant *ne* is often absent in colloquial speech, where its evanescence is subject to a sociolinguistically conditioned variable rule; cf. Ashby 1981 for a valuable empirical study of the loss of the preverbal *ne* particle in contemporary French and the linguistic, stylistic, and social factors by which it is conditioned.

While Littré insisted on restricting *aucun* and *rien* to their etymological positive meanings *quelqu'un* and *quelque chose*, respectively, in the seventeenth century Racine was already using the formula *Je veux rien ou tout* to exclude his middles. Even the French Academy finally threw in the towel, conceding in the 1878 edition of its dictionary that *aucun* could be glossed as *nul, pas un* and *rien* as *néant, nulle chose*. Bréal (1900:201–2) attributes this semantic shift to CONTAGION, the new meaning of the indefinites arising from their 'long sojourn in negative phrases'.

Other Romance and Germanic languages have independently undergone analogous shifts in the representation of negatives (see Wagenaar 1930 for an especially thorough and insightful account of the history of negation in pre-fifteenth-century Spanish), but usually with one important difference from French. The standard pattern is exemplified by English *not,* Latin *nullus* (< *ne* + *ullus* 'something, anything'; cf. French *nul*) and *nemo* (< *ne* + *homo,* lit. 'not a man'), and Spanish *ningun* (cf. *algun* 'something'): the reinforcing postverbal indefinite incorporates a negative prefix through the process of negative concord or agreement (cf. Labov 1972), thereby facilitating its enthronement as the principal marker of negation in the sentence. While instances of positive reinforcers undergoing infection by exposure to negation exist outside of French—we have Spanish *nada* 'nothing' < Latin *(res) nata* 'insignificant [lit., born] thing' and *nadi(e)* 'nobody' < *(homo) natus*—it may be significant that their *n-* initial forms conveniently misrepresent their positive ancestry.

The evolution of English *not* starts out parallel to that of its Germanic cousins German *nicht* (< *ne* + *wicht*) and Dutch *niet:* an indefinite, used for postverbal reinforcement of a weak proclitic *ne,* incorporates prefixal negation and comes to supplant the preverbal element as the main conveyor of negative force. The English realization of the cycle is detailed by Jespersen (1917), Marchand (1938), and Joly (1972). As in Latin, the proclitic Indo-European negative *ne* proved too weak to survive unaccompanied in Old English, and the earlier *Ic ne secge* is strengthened into the Middle English *Ic ne seye not,* with embracing negation. Now the original intensifier or reinforcer is no longer perceived as emphatic, but reinterpreted as a simple marker of negation, and the vestigial proclitic is doomed to extinction (Stern 1937:263). By the Elizabethan period we find the simple *I say not.* This part of the cycle parallels that of French:

(4)

Old French:	Jeo ne dis	**Old English:**	Ic ne secge
Modern French (standard):	Je ne dis pas	**Middle English:**	Ic ne seye not
Modern French (colloq.):	Je dis pas	**Early Modern English:**	I say not

As Joly points out, we can see the beginning of this development in Old English itself, where (5b), with its embracing or reinforced negation, occurs alongside the simple (5a):

(5) a. ac hi ne cneowan hine. — 'but they did not know him'
b. & ne þenceaþ no hwæt hie don sceolde. — 'and they did not think of what they ought to do'

The 'négation composée' of the second example could take any of a number of forms.[3] Negation introduced by proclitic *ne* was consummated variously by *nō* or *nā* ('never' or 'not at all', from *ne* + *ō, ā* 'always, ever'; note that this combination predictably yields the stronger **E** meaning and not the **O** value 'not always'), by *ǣfre* (ever) or *nǣfre* (never), by *ǣnig* (any) or *nǣnig* (none), by *n(e)alles* ('not at all', from *ne* + *ealles* 'entirely'), and increasingly by *nowiht/nawiht* (nothing), from *nō/nā* (see above) + *wiht* 'person, creature, thing' (cf. *wight* and *not a whit*). This last form, representing a double incorporation, eventuated (through double contraction) as *noht/naht,* and finally *not.* Presumably facilitated by contexts allowing both interpretations, the negative existential pronoun is reanalyzed as a simple adverb (*I sowed nought* > *I sowed not*), spreading into forms containing no semantic indefinite (cf. Bossuyt 1983:311–12).

Some neg-incorporated forms from this period do survive as negative pronouns—*no, nought, none* (< OE *nān* 'not one', from *ne* + *ān* 'one')—while the neg-prefixed adverb *never* now patterns with other positive and negative adverbs in its preverbal position and its failure to trigger *do*-support. We have also encountered the neg-incorporated connectives *neither* and *nor.* But the neg-incorporated copulas and auxiliaries of Old English—*nis* 'not-is', *næs* 'not-was', *næbbe* 'not-have', *nolde* 'not-would', *nylle* 'not-want'—gradually disappeared (except in relics like *willy-nilly*), along with proclitic *ne,* as the reinforcing element began to take over the functional load of negation. A parallel history is displayed by the negative verbs of Latin—*neglego* 'disregard, not-heed', *nego* 'say no, deny', *nescio* 'not-know', *nolo* 'not-want'—which either passed out of Romance or lost their connection with overt negation. In both families, the nonverbal categories have retained their *n*-initial negative members, while the verbal categories have not.

Most of the Germanic dialects underwent the same shift as English, the preverbal particle (Gothic *ni*, OHG *ni*, Old Saxon *ni* / *ne*) undergoing reinforcement and gradually replacement by neg-incorporated indefinites (cf. Delbrück 1910; Coombs 1976).[4] But English—unlike German, Dutch, and colloquial French—did not stop there. The periphrastic *do*-supported negation of modern English was already extant in the fifteenth century and had become standard by the seventeenth: *I say not* → *I do not say.*[5] While postverbal negation (like solitary preverbal negation in French) survives in relic expressions (*I kid you not, She loves me not*) and is standard in modal and other auxiliary constructions, the result in ordinary main verb clauses is formally akin to the status quo ante: 'Henceforth the negative particle belongs to the verb again; it returns to the place which it had left' (Marchand 1938 : 198).

But why does negation need to be strengthened or reinforced in the first place? In particular, why must the preverbal Indo-European *ne* repeatedly sue for nonsupport? What triggers each round of postverbal reescalation? One recent suggestion is that the shift from preverbal to postverbal negation correlates with the more general shift from verb-final to verb-medial order. If negation is an adverb, and typological considerations dictate that adverbs tend to precede their verbs in **XV**-type languages but follow them in **VX**-type languages (cf. Greenberg 1963), the shift in negative positioning was to be expected as the **SOV** protolanguage and its verb-final classical daughters passed the torch to the verb-medial granddaughters alive today.[6] Embracing negation represents a transitional stage, a kind of halfway house on this relentless march. So argues Vennemann (1974).

But there are insurmountable difficulties for this approach. Ashby (1981) points out that Vennemann's approach, tailored to fit the facts of French, falters when confronted with other Romance languages no more verbfinal than French (e.g., Spanish and Italian) which have stubbornly retained their preverbal negations and show no signs of replacing them. Furthermore, as Bossuyt (1983 : 310) observes, Vennemann offers no explanation of why embracing negation should persist in standard French, a consistent **SVX**-type language, centuries after it has disappeared in the Germanic languages.

Lehmann (1974 : 11–17; 1978 : 181–83) casts a different typological net; for him, the negative element as a 'qualifier of sentential scope' is predicted to precede the verb in SVO languages. While this approach gets the predictions right for Spanish and Italian (and, if one is patient, for English), it runs into obvious difficulties when confronted with French, as Ashby notes, not to mention German, Dutch, and related languages. It is not clear why such languages should trouble to go against the grain of both their own typology and the Neg-First principle in order to innovate postverbal negation.

But if word order is not a crucial factor (or at least not the crucial factor) in motivating Jespersen's cycle, what is? Jespersen's answer is clear from his description of the cycle: an unstressed monosyllabic syllable consisting of nasal + neutral vowel is simply too weak to serve as the conduit for the vital function with which it has been entrusted, that of differentiating a positive statement from its contradictory:

> The negative notion, which is logically very important, is . . . made to be accentually subordinate to some other notion; and as this happens constantly, the negative gradually becomes a mere proclitic syllable (or even less than a syllable) prefixed to some other word. The incongruity between the notional importance and the formal insignificance of the negative may then cause the speaker to add something to make the sense perfectly clear to the hearer. (Jespersen 1917:5)

But once the phonetically weak negative particle has been reinforced by a postverbal indefinite (with or without incorporated concording negation) or other minimizer (*Let me make this perfectly clear: I ne knew it not!*), it is perceived as redundant. And so this barely audible and now nonfunctional linguistic form discreetly folds its tents and elides away.

We can see in this development an instance of the tension between a least-effort (**R**-based) tendency toward weakening and an information-preserving (**Q**-based) tendency toward strengthening (cf. Horn 1984b). As with all such dialectic processes, the new synthesis is never a resting place, but only the first step in a new cycle. The next stage of development is the movement of the new reinforced negation into a preverbal position (as with Latin *non* or the English *do* + *n't*), satisfying Neg First but leaving the door open to a new round of phonologically inspired weakening and semantically inspired restrengthening.

As we saw in my discussion of logical double negation (§5.1.3), the struggle between the two countervailing functional principles may leave in its wake a horde of enraged prescriptivists sputtering at the **Q**-motivated use of forms (*not uncommon, not inelegant*) which appear (from an **R**-based perspective) to be otiose or redundant. Similar attacks have been leveled against the **R**-violating but **Q**-affirming "redundant" reversatives (*unravel, uncork, unthaw, unskin, debone, depit*). The same vilification is incurred by embracing negation. Thus, as Bossuyt (1983) observes, forms with and without the preverbal negative particle coexisted in Early Middle Dutch, the no-longer-functional proclitic having become phonologically conditioned. But not everyone accepted this fait accompli: 'From a sociological point of view, particle-less negation was [from the thirteenth century] considered careless, and embracing negation [*ne* . . . *niet*] positively valued. In the seventeenth century, however, this evaluation of the negative particle had completely changed, and [the prescriptive grammarian] Leupenius . . . strongly reproves its use as illogical' (Bossuyt 1983:317). Such verdicts, not infrequently registered on the parallel shifts in French and English, are reminiscent in turn of the condemnation grammarians periodically direct to the analogous (**R**-violating but **Q**-motivated) appearance of 'pleonastic' or 'expletive' negation in comparative constructions and after verbs of forbidding, fearing, doubting, refusing, missing, and their ilk (cf. Horn 1978a: §3.2 for discussion and references).

While it may seem odd, notwithstanding the force of least effort, that an element as semantically crucial as negation could effectively disappear, the process is not as exotic as it may appear. As Marchand notes, the periphrastic preverbal negative in colloquial English is often realized as a non-nasal proclitic—*I də wanna go, I dunno*—while other instances of postauxiliary negation may be signaled more by vowel quality, stress, and rhythm than by the presence of a segmental element (Marchand 1938: 200–201; cf. Jespersen 1917:11). Thus, we distinguish *he can come* from *he can't come* largely by rhythmic structure: [hì kən kʌ́m] vs. [hì kǽnʔ kʌ́m]. (When the modal is contrastively stressed, the distinction tends to become neutralized, leading to some rather extreme repair sequences: *he can-yes or he can't?*).

In contexts not triggering *do*-support, English finite negation standardly appears—according to the standard analysis—as a weak contracted enclitic on a stressed auxiliary (*have, be, will, would*), its presence serving to block that auxiliary from itself contracting as it normally does in the corresponding affirmative examples. We thus get minimal pairs like *I've* vs. *I haven't, she's* vs. *she isn't, they'd* vs. *they wouldn't.*[7]

The result, as Marchand points out, is that the modern colloquial language contains two distinct morphemes for sentential negation: a weak contracted *n't* on tensed auxiliaries, often signaled indirectly by prosody, and a strong negative adverb *not* which occurs in nonfinite clauses, as a constituent or word negation, and in emphatic contexts (including, but not limited to, the *not X but Y* construction discussed in §6.4).[8] As we have seen, this state of affairs is by no means cross-linguistically unprecedented. Against the typological background I have sketched, we can view Jespersen's cycle as a turf war between a simple, descriptive negator which occurs predominantly and prototypically in noncontrastive indicative contexts and an emphatic negator which originates in nonfinite, nonverbal, and/or contrastive (metalinguistic) environments. Thus, the twin functions of negation in natural language, as outlined in chapter 6, are diachronically as well as synchronically interactive.

It will be recalled that the postverbal reinforcers of weak preverbal negation are first interpreted as intensifying the negative force of the sentence. Consider, for example, the distribution of such emphatic NPI adverbials in modern Indo-European languages, including English (*(not) . . . at all, in the world*), French (*(pas) . . . du tout, le moins du monde; absolument pas*), German (*durchaus (überhaupt) nicht*), and Spanish (*en mi vida, en absoluto*).[9] These are precisely the kind of intensifiers which develop into freestanding post-verbal negators. Indeed, the Spanish forms are used in that way, without an overt triggering negation, as Joly notes: *En mi {vida/dias} lo he visto* [lit., 'In my {life/days} I have seen that'] conveys that I have never seen that, and *en absoluto* can be used in a reply to indicate a categorical '*no*'.

Furthermore, the indefinite reinforcers—at least in the Germanic languages—have always permitted a contrastive interpretation. The *ne . . . na/no/naht* sequences of Old English, especially in their nontemporal uses, seem to have appeared most freely in contexts of contrast (*ne . . . no . . . ac . . .* 'not . . . but . . .', with the adversative *sondern*-type particle *ac*), of prohibition, or of affirmative/negative opposition. Joly provides the examples in (6); where I have signaled the embracing negatives:

(6)	a. Ne slapige no þin eage.	'Let not thine eyes sleep'
	b. Ne sohte ic na hine ac he sylf com to me.	'I sought him not but he himself came to me'
	c. He lange wiþstode ac hit ne forheol naht.	'He long resisted but it availed not'

The same is true for Dutch, as Bossuyt (1983:332) makes clear. Among the earliest citations of postverbal *niet/nit* as sole marker of negation (without the proclitic *ne* particle) is as a constituent negation in *mar* (*sondern*-type) adversative constructions:

(7) a. So ghinc hi na tire feeste, nit oppenbarlec mar
so went he to that party NEG *openly* *but*$_{SN}$
al heimelec.
secretly

b. Wart ic ben comen uan den hemele nit om minen wille
for I am come from the heavens NEG *to my will*
te doene mar den wille mijns uader.
to do *but*$_{SN}$ *the will my*-GEN *father*

These sentences closely resemble the modern English and Dutch forms of metalinguistic negation in the equivalent *not X but Y* environments, while the syntactic representation of descriptive "sentential" negation—especially in English—has undergone much more radical revision. In this sense, the metalinguistic use of the negative operator is formally more conservative than the descriptive.

But what do we mean, precisely, by descriptive "sentential" negation (scare quotes and all)? As we have seen in our whirlwind tour of the great negations of the world, finite noncontrastive, nonconstituent negation is typically associated with the main finite verb or **VP** of the sentence over which it has scope. Does ordinary language in fact possess a truly sentential or propositional contradictory negation operator, in the logical sense? Where does sentential negation show up in natural languages, and where doesn't it?

In a study of 240 languages from 40 separate families, Dahl (1979) finds that so-called sentence negation characteristically surfaces as an invariant adverb or adverbial particle in proximity with the finite verb, as a morphological category on the verb, or as an inflected auxiliary to the lexical verb. The first two of these alternatives are typologically widespread; the third represents the standard pattern in Uralic, although it is attested elsewhere (Payne 1985). Thus Finnish, for example, expresses negation by a negative verb inflected for person and number, followed by the notional main verb in a neutral form, yielding paradigms like that in (8) (from Bowerman 1973:234):

(8)	Minä en mene.	'I don't go'	[lit. 'I not go']
	Sinä et mene.	'you (sg.) don't go'	[lit. 'you not go']
	Hän ei mene.	'he / she / it doesn't go'	[lit. 'he, etc., nots go']

In colloquial speech, the Finnish negative verb moves farther leftward, yielding $\mathbf{V_{neg}}$**-S-X** order: *Et sinä mene* 'you don't go'. Negative imperatives involve a different negative verb which also appears before the main verb and its arguments: *Älä mene* 'don't go'.

Notice that the Neg First principle is respected in these forms as in the corresponding English examples with auxiliary negation; indeed, we might be tempted to analyze *do(es)n't* as a negative verb in English, were it not for the fact that *do* plays a similar supporting role in questions and emphatic contexts. If we take English contracted negatives to represent a set of lexically listed negative auxiliaries (*don't, can't, haven't*), as pointed out by Zwicky and Pullum (1983:510), then English will tend to resemble Japanese, Swahili, or Turkish more closely than it does its Indo-European cousins. But in fact Finnish might offer a better model for comparison.

Dahl's categories are of course not mutually disjoint. Negation in Japanese, for example, is marked by a bound verbal suffix *-nai* inflected for tense and aspect; cf. Ōta & Katō 1986:26. Often, as in Russian, negation immediately precedes the focus or new information, whatever its categorial status. The tendency for negation to precede its focus is of course Jespersen's Neg First principle, and Dahl's evidence strongly supports the application of Neg First to both **OV**- and **VO**-type languages. (As noted by Bossuyt [1983:310], this evidence constitutes another problem for Vennemann's typological explanation of Jespersen's Cycle.)

Where Dahl does not find negation is in the one place that the standard theories of transformational grammar—as in the model of Klima 1964—and of propositional logic would lead us to look for it: in sentence- or clause-peripheral position.[10] Dahl points out that even if we could motivate the negative-placement rule such approaches require, there would remain no explanation for why an underlyingly initial (or final) marker should be so consistently associated with the finite verb at surface structure.

Of course when a predicate happens to appear clause-initially (in VSO languages, pro-drop languages, or fronted constructions), a preverbal negative element will automatically accompany it to that position, but this is not true sentential external negation. A more plausible candidate for clause-peripheral negation can be found outside of Dahl's survey group, in the 'negative higher verbs' of Polynesian and Yuman languages (cf. Payne 1985). But even these entries in the propositional negation look-alike contest are subject to qualification.

Negation in Tongan (Churchward 1953:56, cited by Payne 1985:208) is marked by a verb which appears almost sentence-initially and takes a sentential complement:

(9) a. Na'e 'alu 'a Siale — 'Charlie went'
PAST *go* ABS *Charlie*
b. Na'e 'ikai [$_S$ ke 'alu 'a Siale] — 'Charlie didn't go'
PAST NEG ASP *go* ABS *Charlie*

But this state of affairs proves to be unstable; in Fijian, for example, the higher initial negation of the protolanguage has undergone auxiliarization and shows up within a verbal complex not unlike that cited in Finnish above (cf. Payne 1985:209–11). This reanalysis is represented by Payne as follows:

(10) [$_S$ **V$_{neg}$** [$_S$ **Comp V$_{lex}$ Obj Subj**]] (essentially as in Tongan) ⇒
[$_S$[$_{VG}$ **V$_{neg}$ Comp V$_{lex}$**] **Obj Subj**] (as in Fijian; VG = verbal group)

A similar development can be seen in the SOV Yuman family (Munro 1976:106–9), except that the sentence-peripheral higher negation reconstructible for Proto-Yuman is not strictly a one-place connective. Munro concedes that although—from the perspective of generative semantics—'a simple negative would seem to be a good example of a verb which should take just one sentential argument', there is in fact no direct evidence for the initial (pre-Proto-Yuman) stage she posits in the sequence in (11):

(11)

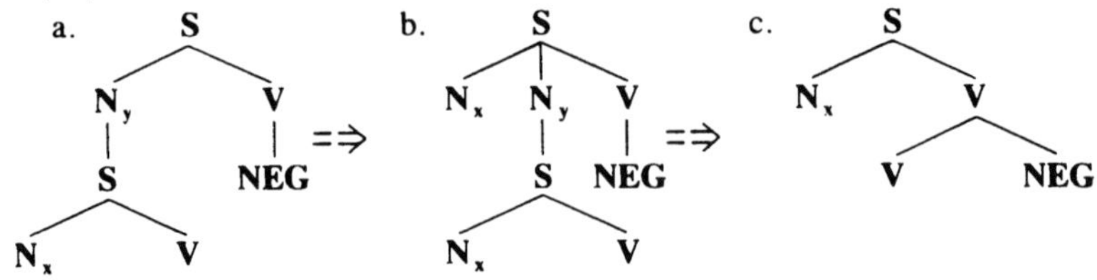

The Proto-Yuman (11b) stage, with an application of a subject-copying rule, is directly reflected in Diegueño, but this situation remains unstable. The Proto-Yuman negative verb is more typically reanalyzed as a suffix, through an application of Predicate Raising on Munro's analysis, resulting in the (11c) stage represented by Mojave:

(11′) **Diegueño**: ʔnʸa:-č ʔ-a:m- x ʔ- ma:w
/-SUBJ 1- *go*- IRR 1- NEG
Mojave: ʔinʸeč ʔ-iyem- mo- t- m
/ 1- *go*- NEG-EMPH-TNS
'I didn't go'

Patterns of this sort are significant, in that they tend to show that even in those few cases in which a descriptive negation connective seems to take a sentence as its surface argument, the motivation for such a conclusion is tenuous and the structures conforming to this analysis are relatively unstable.

In fact, the one natural locus of apparent sentence-external wide-scope negation appears to be child language. Bellugi (1967) and other developmental researchers attest a common but transient stage in the acquisition of English and other languages in which children express negation regularly or exclusively by a marker at sentence-initial position: *Not mommy go, No I want spinach.*[11] If my earlier speculations are correct, the negation at this stage is in fact still metalinguistic, so that the appearance and later disappearance of this paradigm reinforces the general observation, based on typologies like Dahl's and Payne's, that descriptive negation is not a sentential operator.

What is the significance of the (virtual) absence of sentence-peripheral descriptive negation for the standard theories of propositional logic and generative syntax which seem to assume the existence of such a chimera? How are we to represent the various forms descriptive negation takes within various languages? A program designed to address these questions is launched in §7.2.

7.2 Aristotle as a Montague Grammarian: Negation in Extended Term Logic

> All object language negation is internal. Either terms are negated (logical contrariety) or predicates are negated (predicate denial). Negation never applies to a sentence as a whole. Object language negation is never external. (Englebretsen 1981a:59)

The 2,300-year war between the term logic of Aristotle and the Peripetetics on the one hand and the propositional logic of the Stoics and post-Fregeans on the other seems to have ended in the complete rout of the former camp by the forces of the latter. Fifty years ago, Lukasiewicz (1934:79ff.) offered his postmortem to the conflict, which we already encountered in my earlier history of these events (§1.1): 'We know today that propositional logic . . . founded by the Stoics, carried on by the Scholastics, and axiomatized by Frege, . . . is logically prior to the logic of terms'. I shall maintain, in this necessarily programmatic introduction to what I shall dub Extended Term Logic, that the death of the Aristotelian system has been grossly exaggerated.[12]

Let us begin by reviewing some of the principal tenets of term logic, summarized in (12); cf. Aristotle and Englebretsen (1981a, 1981b) for details:

(12) a. Logic is the study of propositions. A proposition is a sentence which can be true or false; every proposition is either true or false (possibly excepting future contingents: see discussion in §2.1).

b. Every (simple) proposition is CATEGORICAL, containing something (the SUBJECT) about which something (the PREDICATE) is said.

c. There are no propositional operators as such.

d. Subject and predicate can be combined by two different MODES OF PREDICATION, (PREDICATE) AFFIRMATION (in which the predicate is AFFIRMED of its subject) and (PREDICATE) DENIAL (in which the predicate is denied of its subject). Affirmations and denials differ in QUALITY.

e. A predicate may be truly denied of its subject because the subject fails to exist, because the predicate fails to apply naturally to it, or because the predicate expresses a property which the subject accidentally fails to possess. Thus, *Socrates is not wise* (assuming a nonexistent Socrates), *The number 2 is not blue*, and *Reagan is not a Democrat* are all true instances of predicate denial.

f. Any term may be negated. *Not-man* (as in *Not-man is furry*) is an instance of subject term negation, *not-happy* (*Socrates is not-happy*) of predicate term negation. A predicate term negation affirms a negative term (*not-happy*) of the subject; a predicate denial (*Socrates is not happy*) denies a positive term (*happy*) of the subject. If Socrates does not exist, the former is automatically false and the latter true.
g. Predicate term negation yields CONTRARY oppositions: α *is* β and α *is not-*β cannot both be true but may both be false. Predicate denial yields CONTRADICTORY oppositions: γ *is* δ and γ *is not* δ necessarily differ in truth value in that one must be true and the other false.
h. Predications, whether affirmations or denials, may be UNIVERSAL (*All men are (not) happy, No men are (not) happy*), PARTICULAR (*Some men are (not) happy*), SINGULAR (*Socrates is (not) happy*), or INDEFINITE (*(A) man is (not) happy*). This is a distinction in QUANTITY.
i. Predications of whatever quality and quantity may be ASSERTORIC (*is/isn't*), APODEICTIC (*must be/needn't be*), or PROBLEMATIC (*may be/can't be*). This is a distinction in MODALITY.

Some of the essential differences between the term logic of Aristotle and the propositional and predicate calculus of the moderns are spelled out in (13):

(13)

Term Logic (Aristotle, Leibniz, Sommers, Englebretsen)	Propositional / Predicate Logic (Stoics, Cicero, Abelard, Frege, Russell, mathematical logic)
a. Every statement is categorical, of subject / predicate form.	No subject / predicate analysis; function / argument application is basic.
b. Any term may refer or characterize, depending on its position in the sentence.	Only logical functions characterize; only logical arguments refer.
c. Categorical syllogisms.[13]	Hypothetical syllogisms and laws of inference.
d. Parallel treatment of sin-	Differential treatment of

Term Logic (Aristotle, Leibniz, Sommers, Englebretsen)	Propositional / Predicate Logic (Stoics, Cicero, Abelard, Frege, Russell, mathematical logic)
gular and general statements: both of subject / predicate form.	singular and general statements (see below).
e. No sentential negations, hypotheticals, conjunctions, or disjunctions.	One- and two-place sentential connectives freely occur, as defined since Boole.
f. Neither of the two defined varieties of negation (cf. (12d–g)) reduces to an external sentential connective.	Negation is an external truth-functional propositional or sentential connective (*Not: it is day*).
g. No Law of Double Negation; predicate denial does not iterate, since it does not apply to a full proposition.	LDN accepted since Alexander of Aphrodisias (*Not: not: it is day* differs from *It is day* only in manner of speech). Negation freely iterates.

Of particular relevance here are (13a) the categorical (subject / predicate) nature of the term logic statement, rejected by Frege and his successors, (13d) the parallel treatment of singular and general statements (*Socrates is wise, Every man is wise*) in term logic and their differentiation within standard mathematical logic, (13e) the absence of propositional operators (negation, conjunction, disjunction, conditional) within term logic as against their fundamental role as one- and two-place connectives in propositional logic, and (13f, g) the differences between the two approaches with respect to negation. As spelled out in more detail in (12d–g), term logic accepts two varieties of negation, neither of which reduces to the external one-place connective of the Stoics and Fregeans. Subject and predicate can be combined by two distinct modes of predication, since a predicate (which may or may not itself contain a negative term) may be either affirmed or denied of its subject.

Since Aristotle and Leibniz, the agenda for a term logician has always crucially involved a demonstration of the claim that both singular and general statements are essentially of subject / predicate form. This claim is congenial to the enterprise of Montague Grammar, where (as in PTQ: Montague 1974: §8) both complex quantifier phrases and formally simple

NPs are GENERALIZED QUANTIFIERS denoting third-order entities: just as *every man* and *no man* denote the sets of properties which are properties of every man and of no man, respectively, so too *John* denotes the set of properties which are properties of John.

The theory of generalized quantifiers has been recently elaborated by Barwise and Cooper (1981), in the work I touched on in chapter 4. On this analysis every **NP**, singular or general—*John, a woman, the baby, every eel, most of the donkeys*—is taken to be a quantifier whose domain is a set of individuals. A quantifier of the form $_{\mathrm{Q}}[\mathbf{Det\ A}]$ denotes the set **B** of which (**Det A**) **B** holds. Quantifiers denote families of sets, that is, families (sets) of subsets of the domain of discourse; determiners denote functions from common noun denotations to quantifier denotations. Where **E** is the set of entities provided by the model, the extension of the determiners *all, most, some, and no,* for example, will be the functions assigning to each $\mathbf{A} \subseteq \mathbf{E}$ the families of sets picked out in (14a, b, c, d), respectively:

(14) a. *all A* denotes $\{\mathbf{X} \subseteq \mathbf{E}: [[\mathbf{A}]] \subseteq \mathbf{X}\}$
b. *most A* denotes $\{\mathbf{X} \subseteq \mathbf{E}: |[[\mathbf{A}]] \cap \mathbf{X}| > |[[\mathbf{A}]] - \mathbf{X}|\}$
c. *some A* denotes $\{\mathbf{X} \subseteq \mathbf{E}: [[\mathbf{A}]] \cap \mathbf{X} \neq \emptyset\}$
d. *no A* denotes $\{\mathbf{X} \subseteq \mathbf{E}: [[\mathbf{A}]] \cap \mathbf{X} = \emptyset\}$
(where $[[\alpha]]$ = denotation of α and $|\alpha|$ = cardinality of α)

While Barwise and Cooper's work is linked to an overall Montague Grammar program for the representation of the syntax and semantics of natural language, it is only a term-logic-based approach, with its insistence on the categorical nature of root sentences, which motivates and indeed forces this insightful analysis.

We have seen that standard versions of mathematical logic, by following Frege in his rejection of the two internal negations of Aristotelian logic (predicate denial and term negation) in favor of a single one-place propositional negation, cannot assign subject / predicate form to general statements. In the usual modern analysis, universal and particular statements are represented as universally quantified conditionals and existentially quantified conjunctions, respectively. In effect, then, a proposition like *All ravens are black,* as analyzed into $\forall \mathbf{x}\ (\mathbf{raven}\ (\mathbf{x}) \rightarrow \mathbf{black}\ (\mathbf{x}))$, is not about ravens at all. It is in fact about everything: it states of every individual **x** that if **x** is a raven, then **x** is black, that is, everything is either black or a nonraven. More generally, as Sommers (1970:38) observes, 'A statement of the form "All S is P" in quantificational transcription is not about all S or any S; it is about all things and affirms of them "is either un-S or P"'. A general statement is thus entirely distinct from a singular statement (e.g., *This raven is black*) in its logical form.

Barwise and Cooper, in their independent adumbration of the same point, point out that while the underlined portions of the English sentences in (15) clearly represent **NPs**—patterning here as subjects and elsewhere as direct and prepositional objects—these **NPs** are nowhere to be seen in the translations of these sentences into their predicate calculus representations in (15′):

(15)	a. Harry sneezes.	(15′)	a. **sneeze (Harry)**
	b. Some person sneezes.		b. $\exists$**x (person (x)** $\wedge$ **sneeze (x))**
	c. Every man sneezes.		c. $\forall$**x (man (x)** $\rightarrow$ **sneeze (x))**
	d. Most babies sneeze.		d. [no representation available in PC]

While the English **NPs** have mysteriously vanished into thin air, the conjunction and implication in (15′b, c) have mysteriously appeared out of it, the choice between the two connectives dictated by the choice of quantifier. An additional problem is posed by the nonstandard quantifiers (e.g., *most* α) which, as Barwise and Cooper and others have pointed out, cannot be adequately modeled by a quantified conditional or conjunction. Under the Montague–Barwise and Cooper analysis, as we have noted, each of the **NPs** in (15) is a generalized quantifier, allowing all four sentences to be assigned subject-predicate form, as in term logic. Each sentence in (15) is true iff the set of sneezers contains Harry, some person, every man, and most babies, respectively.[14]

The curious tradition of unpacking general predications into quantified conjunctions and conditionals is premised partly on the interaction of quantification with propositional negation. But the basis of this premise—the Stoic doctrine of *apophatikon*, the iterating external truth-functional negation connective—is a misrepresentation of natural language. In the last section, we observed that linguistic analogues of this logical connective are never (or hardly ever) found. The theoretical implications of this typological result must not be overlooked. We have already encountered Geach's verdict that the Stoic-Fregean propositional negation, however convenient it may prove for the purposes of formal logic, does violence to ordinary linguistic practice; I reprise his remarks after citing the parallel observations of Katz:

> In natural language, negation is not a mechanism for forming compound propositions. Logicians treat negation as a propositional connective even though it does not connect propositions, but in constructing artificial languages one is free to do what one wants, and furthermore, in the kind of artificial languages acceptable within the orthodox conception of logic there is no choice. . . .

> [In natural language] negative elements do not behave like the connectives 'and' and 'or' but like adverbs. (Katz 1977:238)

> Propositional negation was as foreign to ordinary Greek as to English, and [Aristotle] never attained to a distinct conception of it. The Stoics did reach such a conception, but in doing so they violated accepted Greek usage; their use of an initial *οὐχί* as the standard negation must have read just as oddly as sentences like 'Not: the sun is shining' do in English. . . . In ordinary language, it is rather rare to negate a statement by prefixing a sign of negation that governs the whole statement; negation is almost always applied primarily to some part of a statement, though this often has the effect of negating the statement as a whole.
> (Geach [1972] 1980:75)

Geach's last comment echoes Frege's observation that a constituent negation may have semantic scope over the entire sentence. The same point is reiterated by Cooper (1984:25): 'English uses the negation of constituents other than sentences to achieve the effect of what would be sentence negation in a predicate calculus–like representation'. What remains to be determined is precisely when the ordinary constituent-associated negation of natural language achieves this sentential effect. As Frege observes (1919:131), in (16c) as in (16b):

(16)	a. The man is celebrated.	(16′)	a. The man is happy.
	b. The man is not celebrated.		b. The man is not happy.
	c. The man is uncelebrated.		c. The man is unhappy.

we merely 'indicate the falsity of the thought' expressed in (16a). But, as we have seen, this result is far from general: in saying (16′c), we do not, as in (16′b), just indicate that the man is not happy (or that it is false that he is happy).[15] Assuming, contra Frege, that we do need to allow for a constituent negator, Aristotle's predicate term negation, for the strong contrariety operator of (16′c), one basic question remains: is the contradictory operator of (16b) and (16′b) the iterating propositional connective of the Stoics and Fregeans, or is it the noniterating predicate denial—negation as a mode of predication—of Aristotle and the Peripatetics?

Without acknowledging the Aristotelian connection, Montague quietly opts for the latter approach. In PTQ (Montague 1974:§8, 252–53), negation, like tense, is introduced syncategorematically. In addition to the basic positive subject-predicate rule **S4**, Montague offers an assortment of five distinct operations for forming negative sentences, constituting (in the words of Dowty, Wall, and Peters: 1981:244) 'alternative ways of putting subject with predicate to form a sentence'. The 'rules of tense-and-sign' in his **S17** produce analysis trees of the sort exemplified in (17):

(17)

```
a. the man seeks a unicorn, 4            b. the man doesn't seek a unicorn, 11
         /            \                              /            \
  the man, 2    seek a unicorn, 5            the man, 2       seek a unicorn, 5
      |           /        \                     |              /        \
     man        seek   a unicorn, 2             man           seek   a unicorn, 2
                            |                                             |
                         unicorn                                       unicorn

c. the man doesn't seek a unicorn, 10, 6
          /              \
  a unicorn, 2   the man doesn't seek him6, 11
      |                  /          \
   unicorn        the man, 2    seek him6, 5
                      |           /    \
                     man        seek   he6
```

There is no single operation by which auxiliary negation is introduced syncategorematically along with tense; rather, we are offered one subrule for the 'negative third person singular present', another for the 'negative third person singular present perfect', and so on. But this is an artifact of Montague's decision to intersperse agreement rules and other morphological processes with the bottom-up syntactic construction of sentences. What is crucial for our purposes is that ordinary (auxiliary) negation is treated as a mode of predication, a recipe for combining subject (a **T** or term phrase) and predicate (an **IV** phrase) to form a proposition or sentence (an expression of category **t**), rather than an operation on a fully formed proposition or sentence. But the **I**(ntensional) **L**(ogic) translation of syncategorematic negation is the external propositional connective, with the semantics of the contradictory operator (taking scope over the subject phrase, as in Aristotle). Thus, we get the semantic effect of Fregean negation without the syntactic commitment to an external presentential operator.

Notice that while it would have been technically straightforward for Montague to have treated contradictory negation in the manner of his de dicto modals, as a truth-functional connective of category **t/t**, he chose not to do so. In fact, in his earlier "English as a Formal Language," Montague (1974: §6, 190) does introduce *not* as a member of $\mathbf{B}_5$, the set of basic adformula phrases.[16] In "Universal Grammar" (Montague 1974: §7), on the other hand, *not* is introduced syncategorematically, as *n't* is in PTQ. It is my contention here that by the time of PTQ, Montague had come around to recognizing (at least implicitly) that—logical convenience notwithstanding—the appropriate treatment of English as a formal language requires negation to be analyzed syntactically as a mode of predication, not as an external propositional connective.

Since Montague's contradictory negation, like Aristotle's, forms a sentence out of a subject and a predicate, rather than converting one fully formed sentence into another, it will not iterate. Thus, we generate no sequences of the form of (18),

(18) not (not (not (not . . . (the cat is on the mat))))

which seems correct in the light of the behavior of natural languages. But the noniterating property of Montague negation does not follow systematically from his framework as it does from Extended Term Logic (ETL), which bars all one-place truth-functional propositional connectives and disallows any expression of category **t/t**.

Sentence negation as such is also difficult, although (as Cooper [1984] shows) not impossible, to accommodate within Situation Semantics, but this appears to be not so much a principled result of this framework as an epiphenomenon of it. Barwise and Perry (1983: 138), recognizing that 'the simplest form of negation in English is verb phrase negation, rather than sentence negation', point out that the former can be treated easily in Situation Semantics, while the latter proves (as perhaps it should) to be a complicated matter: 'If I say "A dog is not barking", this can describe any factual situation in which some dog is not barking at the location referred to. However, if I say "It's not true that a dog is barking", my new utterance doesn't mean that. Indeed, to ask what situation I am describing seems to miss the mark. Rather it seems that my utterance, if informative, serves primarily to preclude certain types of situations, namely, those with barking dogs'. Thus, sentence negation, for Barwise and Perry, operates on a different level from **VP**-negation. This characterization of **S**-negation as a metasituational operator is in some ways analogous to my metalinguistic negation; in both cases, syntactically external negation is identified as an unusual and linguistically marked device operating on a metalevel. But it is not clear from this passage whether Barwise and Perry's remarks are intended to carry against all instances of semantically wide-scope (contradictory) negation, or just (certain?) cases of quantifier-negation scope interaction. As we shall see, some instances of apparent sentence negation can in fact be reanalyzed as metalinguistic uses of the negative operator, but every wide scope negation cannot be so analyzed.

Aristotle's narrow-scope negator, predicate term negation, while not acknowledged by Montague, is partially recaptured in Bennett's and Dowty's extensions. Dowty (1979: 349) treats *un-* adjectives as instances of **IV** (predicate-internal) negation, but assigns them the semantics of contradictory negation, so that (16′b) and (16′c) will incorrectly come out truth-conditionally identical (as in the classical Fregean model). The 'internal' **IV** negation of Bennett 1976 (cf. his rules S29–32, T29–32) is also semantically contradictory. (See also Dowty, Wall, and Peters 1981: 94–96,

104–5, where the predicate negation **non** is formally defined within $\mathbf{L}_{type}$.) Clearly at least some instances of morphologically incorporated narrow-scope negation must be assigned the semantics of contrary opposition, so that (16′c) can come out false when (16′b) is true, given that a man may be neither happy nor unhappy.[17]

Within the generative tradition, **IV** or **VP** negation—while absent from Klima's grammar—is explicitly introduced by Jackendoff (1969), who distinguishes this operation from S-negation semantically in (more or less) the Aristotelian manner. The **S-neg** reading of *The arrow didn't hit the target* denies that the arrow hit the target; its **VP-neg** reading asserts of the arrow that it failed to hit the target. But Jackendoff, like Klima, introduces sentence negation in the initial position where it never surfaces; nor can I accept the **S-neg–VP-neg** boundary line determined by Jackendoff's semantic criterion, for reasons which emerge in §7.3.

In their post-Montagovian Boolean semantics for natural language, Keenan and Faltz (1978) allow negation to combine freely with expressions of any category, including proper names, thereby generating some 'admittedly unnatural combinations' (*Not* {*John*/*the man*} *left*); the semantics treats any expression of the form *not* α as denoting the set of properties which is the complement of that set of properties denoted by α. (Analogous cross-categorial semantics are provided for conjunction and disjunction, which are defined in terms of set intersection and union, respectively.)

A modified version of term logic, combining the dual-negation bivalent categorical logic of Aristotle with the formal rigor of the theory of generalized quantifiers due to Montague and his epigones, would seem to provide the best account of where descriptive negation surfaces in natural language, and of where it doesn't. As I have noted, citing Dahl 1979 and Payne 1985, syntactically external (clause-peripheral) negation, as an iterating one-place connective on propositions, never—or hardly ever—happens. Dahl's putative counterexamples to the nonoccurrence of syntactic external negation actually involve not clause-initial sentence negation at all but quantifier-internal term negation, as in *Not anybody can do it.* (I return to the nonsentential nature of quantifier negation in §7.3.)

Let us pause to take stock. My approach to negation in ETL is summarized in (19):

(19) a. The categorical (subject-predicate) nature of root sentences disallows all one-place, truth-functional (**t**/**t**) operators, including the external negation connective.

b. Ordinary (descriptive) negation is realized either as a narrow-scope term operator (internal to the predicate/**VP**, the quantifier/**NP**, etc.) or as a mode of predication by which the subject and the predicate combine to form a proposition. Only the latter corresponds semantically to wide-scope, contradictory negation.

c. Apparent residual instances of external negation are in fact manifestations of metalinguistic negation, a means of objecting to an utterance on any grounds whatever, including its grammatical or phonetic form; cf. chapter 6 for details.

My rejection of external negation as a sentential operator is motivated by many of the same factors that lead Bach (1980) and Enç (1981: chapter 3) to challenge traditional analyses of tense (including that within Montague Grammar) as a sentential operator whose scope varies with that of other operators. Enç points out that tense in natural languages tends to surface as a bound morpheme on the verb, a free morpheme immediately preceding the verb, an element morphophonemically incorporated into the verb (via ablaut, etc.), or as a marker on an auxiliary verb. On the basis of this distribution and of other considerations indicating that scopal analyses are simultaneously too strong and too weak, she concludes that any adequate treatment of natural language syntax and semantics must treat tense, not as a one-place sentential connective, but rather as an operator on **VPs** (as in Bach 1980) or on the predicate itself. As tense goes, so (I argue) goes sentential negation.

Another operator which bears some resemblance to, and is often treated like, negation is the question or *wh*-binding operator. Interrogative and negative sentences are generally classed together within accounts of sentence types in traditional grammars. In Chomsky 1957, negative and interrogative sentences are both analyzed as optional transforms of an unmarked positive declarative kernel, and by the time of Katz and Postal 1964 both negation and question formation are taken to be similarly triggered by base-generated markers, NEG and Q respectively. Other structural parallels can be cited, including the conditions on *do*- support in English or the relation between negative and interrogative parentheticals (discussed by J. R. Ross (1973a), inter alia). Kraak (1966: 90–91, 99) and Seuren (1967: 335) provide further evidence for what the latter calls a 'similarity between negative sentences and questions'. But just how well motivated is this analogy?

Notice first of all that while negations are not systematically assigned to sentence-initial position, question markers often are. Similarly, we find yes-no questions marked via global intonation contours, plausibly associated with an abstract sentence-level operator. In his typology of interrogative systems, Ultan (1978) points out that the two most widespread devices for marking yes-no questions, intonation and question particles, generally form a constituent with the entire clause. He also finds that interrogative particles (which appear in languages as diverse as Chinese, Cree, Finnish, Latin, Russian, Swahili, and Tagalog) are typically assigned to a fixed position within the sentence: sentence-initial, sentence- or clause-final, or as an enclitic to the clause-initial constituent. Contrarily, negation is assigned a fixed position with respect to the finite verb, as I have noted, and is never marked primarily by sentential intonation contours.[18]

This striking formal difference between interrogation and negation reflects a semantic asymmetry between the two categories. Crucially, the yes-no question operator applies to a fully formed proposition (and a constituent or *wh*-question to an open proposition), arguably in the manner of the illocutionary-force indicating devices of speech act theory (cf. Searle 1969). But descriptive negation does not constitute a speech act (pace Searle, Givón, and others; cf. §1.2), or indeed a propositional operator. While metalinguistic (echo) questions may parallel metalinguistic negations, as I have suggested (cf. §6.2.3), descriptive yes-no or constituent questions do not parallel descriptive negations.

It was Frege (1919) who first warned against taking ordinary negation to constitute a speech act on the order of asserting, questioning, or commanding. On the Fregean view, the latter operations apply to an integral proposition or thought which may or may not itself be negative, while negation is simply a one-place connective mapping one proposition into another. From my perspective, Frege was as right about what negation isn't as he was wrong about what it is.

7.2.1 Sommers and the Limits of Term Logic

In seeking to resuscitate term logic from oblivion, I am following the lead of Fred Sommers (1970, 1982) and George Englebretsen (1976, 1981a, 1981b). Sommers, in particular, has spent the last two decades in relentless pursuit of 'a term logic in which natural syntax and canonical [logical] form are in accord' (Sommers 1970:3). Sommers follows Aristotle—and, without recognizing it, Montague—in treating (20a, b) as essentially parallel to (20c) in terms of the logical form of their subjects (cf. Sommers 1982:339).

(20) a. Every man is wise.
b. {A man / Some man} is wise.
c. Socrates is wise.

Rather than sharing Frege's vision of natural language as incoherent in its disrespect for quantifier/variable notation, Sommers sees the neo-Stoic approach of mathematical logic as unnatural in its disrespect for the **NP-**

hood of the subjects of sentences like (20a, b). But Sommers's eloquent objections to the standard formal approach do not apply to Montague Grammar, where *every man, a man,* and *Socrates* are all taken to be term phrases.

Where I differ from Sommers, and from Leibniz before him, is in my toleration of deviations from the hard-line tenets of classical term logic. Sommers's absolutist stance leads him to define away not only propositional negation, where I follow him, but also the two-place connectives corresponding to *and, or,* and *if-then* (cf. (12b, c), (13e) above). For Sommers (1970:22), apparent noncategorical statements of the type in (21) are nominalized into the corresponding categorical, subject-predicate statements in (21′):

(21)	a. not-**p**	(21′)	a. The [**p**] does not obtain
	b. If **p** then **q**		b. All [**p**] is [**q**]
	c. **p** or **q**		c. All [non-**p**] isn't [non-**q**]
	d. **p** and **q**		d. Some [**p**] is [**q**] where [α] is read as 'case of α', 'state of affairs in which α'.

I would endorse the conversion of the sentential negation of (21a) into the second-order (i.e., metasituational or metalinguistic) statement of (21′a). But the translation proposed for *if-then* statements, defensible as it may appear to be for a nomic conditional like Sommers's example (22), where (22a) comes out as (22b) and hence as (22c), seems much less plausible for simple *if-then* statements like those in (23) and (24):

(22) a. If it rains, it pours.
b. All [it rains] is [it pours]
c. Every case of 'it is raining' is a case of 'it is pouring'.

(23) a. If Mary leaves before midnight, John will be sad.
b. All [Mary leaves before midnight] is [John will be sad]

(24) a. If the Yankees lose tonight, they fall into third place.
b. All [the Yankees lose tonight] is [the Yankees fall into third place]

The problem in Sommers's account (aside from some uncertainty over number agreement) is that while his critique of the standard predicate calculus translation of universal statements of the form *All F is G* into the logical form of a universally quantified hypothetical, $\forall \mathbf{x}(\mathbf{Fx} \rightarrow \mathbf{Gx})$, may be well taken, an approach which simply reverses the direction of translation by analyzing conditionals as disguised universal categoricals does not seem to represent much of an improvement.[19]

Nor do matters improve when we examine the Sommers line on disjunctions and conjunctions. The reduction of (21c) to (21′c) amounts to rendering an apparently simple sentential disjunction as a universal negated categorical whose terms are negated nominalizations; (25a) is spelled out as (25b).

(25) a. The Yankees will win or the Red Sox will win.
b. All [non-(the Yankees will win)] isn't [non-(the Red Sox will win)]

The appearance of a phantom universal determiner, a phantom subject/predicate construction, and (count 'em) three phantom negations makes those quaint deep structures of the generative semantics era (or, for that matter, of Russell's theory of descriptions) look trivial and surfacey by comparison.

There is no equivalent complication in the proposed categoricalization of apparent sentential conjunction, in the move from (21d) to (21′d), but there is nonetheless an oddity. If the conjunction in (26a) is unpacked into the particular categorical in (26b),

(26) a. The Yankees will win and the Mets will win.
b. Some [the Yankees will win] is [the Mets will win]

we not only predict a nonexistent parallel between *and* and *some*, we also fail to account for the well-attested syntactic, semantic, pragmatic, and intonational parallels between *or* and *some*, on the one hand, and between *and* and *all*, on the other; cf. Horn 1972 and chapter 4 above for some of these parallels.

In any case, even if it is possible to come up with a less problematic and more convincing method for translating apparent sentential hypotheticals (conditionals), disjunctions, and conjunctions into categorical subject/predicate sentences, it is neither necessary nor desirable to do so.[20] If sentential negation is distinguished by its absence from ordinary language, the same cannot be said for its binary cousins.

It is true that conjunction and disjunction occur most frequently as operations on subsentential constituents, typically on **NP**s or **VP**s. In this capacity they represent no difficulty for either Aristotle's or Sommers's version of term logic. But these operators undeniably connect sentences or propositions as well, typically (although not invariably) appearing between the two conjuncts or disjuncts. The realization of the if-then relation is more various, but it is not unusual to express conditionals by means of two connected clauses, of which one—the antecedent or protasis—may be (as in English) syntactically subordinate to the other.

The natural language reflexes of two-place truth-functional sentential connectives are in fact two (or **n** ≥ 2) –place connectives, while the one-place connective does not connect at all. Within Extended Term Logic, not all sentences, but all root sentences are categoricals.

If sentential conjunctions, disjunctions, and hypotheticals (conditionals) are countenanced, how do these sentence types interact with negation? The prediction would seem to be that they don't. Of course either conjunct or disjunct can itself be negative, as can the antecedent and/or consequent of a conditional. But if predicate denial is a mode of predication, a means for mapping a subject and predicate into a proposition, there is no way to apply it to a noncategorical sentence of the form ***p and q, p or q, if p then q***.

Thus, any negation which takes scope over a conjunction, disjunction, or conditional must be metalinguistic. This is a result I already argued for in the case of the negated (i.e., rejected) conditionals of chapter 6; cf. the analyses of Grice, Dummett, and Ducrot cited there. With the other connectives, there is nowhere for a wide-scope descriptive negation to surface, given that conjunctive and disjunctive sentence types have no main verb, **VP**, or auxiliary as such. The form of wide-scope negation we do find cross-linguistically (= *It's not ({true / the case}) that Chris won and Sandy lost*) are precisely what we would identify elsewhere in the same language as reflexes of the metalinguistic use of the negation operator. And it is striking that even these metalinguistic negations are expressed by gathering the rejected conjunction / disjunction into the predicate expression with a pseudo-logical predicate (*be true, be the case*) invoked (and denied) for the occasion. What we do not find even here is the straightforward *not: Chris won and Sandy lost,* expressed in the canonical sentence-negation form dear to the hearts of Stoics, Fregeans, and transformationalists alike.

If there is no site for wide-scope negation to roost within a conjoined sentence, negation is even more homeless in the case of paratactic conjunction, where the conjuncts in question are simply two consecutive independent sentences forming a single discourse frame:

(27) a. Chris won. Sandy lost.
b. They had a baby. They got married.

(As noted in chapter 6, paratactic and overt conjunctions induce similar temporal/causal implicata in similar contexts.) The only direct means for denying such paratactic conjunctions is a metalinguistic negation of the form *No, that's not true* (or *No, you're wrong,* or *That's not the way it happened*), in which the focus of negation is recovered from the context. Alternatively, the negative interlocutor can supply her own overt connective to draw the two sentences into one compound proposition; negation can then apply to this noncategorical proposition, but again only as a metalinguistic operator (*It's not the case that {Chris won and Sandy lost/they had a baby and got married*}). Either way, we have no ordinary negation, since conjoined sentences (whether connected or not) are not predications: where there is no single predicate to deny, there can be no predicate denial.

As pointed out by Avicenna, the negation of a categorical (subject-predicate) judgment, of either singular or general form, denies the association of the subject and predicate, the negation of a disjunctive judgment denies both terms, and the negation of a hypothetical judgment denies the logical connection between antecedent and consequent (cf. Madkour 1934: 168). But we can now see that only the first of these involves true predicate denial. In natural language, negated disjunctions virtually always take the form of a neg-incorporated subject or predicate term (i.e., a term or constituent negation, not a predicate denial), or of the denial of a predicate which itself incorporates a disjunction:

(28) a. *Neither* Aristotle *nor* Montague allowed iterating negation.
b. Aristotle {allowed *neither* sentential disjunctions *nor* propositional negation.
did*n't* allow *either* sentential disjunction *or* propositional negation.}
c. Aristotle {*neither* endorsed *nor* rejected Situation Semantics.
did*n't either* endorse *or* reject Situation Semantics.}

A true sentential disjunction can only be metalinguistically rejected (*It's not* {*true / the case*} *that Chris won or (that) Sandy lost*), not descriptively negated.

In endorsing binary sentential connectives, ETL differs crucially from orthodox term logic in which, as Englebretsen (1981a: 59) recognizes, all sentences are categorical and there can be no external operators: all operators (including modality and the binary connectives) must be internalized. But my position on negation is essentially Englebretsen's (1981a: 49ff.; cf. also epigraph to §7.2): '[So-called] sentential negation is either predicate denial or the metalinguistic predication of "untrue". . . . There is no genuine, object level, external, sentential negation'. What there is is predicate denial, term (constituent) negation, and metalinguistic negation.

Thus, alongside his two categories of 'internal' negation (predicate denial and logical contrariety), Englebretsen allows for the existence of metalinguistic negation. But we should note that his conception of this notion is more limited than mine. For Englebretsen, the metalinguistic operator (the 'predication of "untrue"') is used to assert the falsity or nontruth of a previous assertion. As we saw in chapter 6, this account is fundamentally unsatisfactory; potential targets of metalinguistic negation include conventional and conversational implicata, which by assumption have nothing to do with truth, as well as such clearly nonpropositional attributes of an utterance as register and morphological or phonetic form.

The view of negation presented here follows from a logic of natural language which—like Aristotle's or Montague's—respects the surface structure of natural languages. It should be stressed that the very criticisms so often leveled against Aristotle, addressing the Stagirite's purported confusion of formal logic and natural language, can be turned on their heads. Just as Montague has lent credibility to the analysis of English as a formal language, so too should Aristotle be seen as proposing to treat Ancient Greek as a formal language. To be sure, this practice may lead at times to excesses; consider Montague's notorious practice of bulldozing the actual relative clause structures of English in favor of the more tractable but rather less natural *such that* construction. Similarly, we recall (from chapter 1) Aristotle's decision to countenance the generation and interpretation of such unlikely sentences as (the Greek equivalent of) *Every not-recovers is not a not-man*. But in each instance a fundamental insight can be captured if the relevant generalization is allowed.

In the latter case, translating Aristotle into Montague English, the result is that both **IV** phrases and common nouns can be quantified over (i.e., turned into a **T** or term phrase—or a generalized quantifier, à la Barwise and Cooper—by combining with a determiner: *every, some*), both can be predicated of term phrases to form a sentence, and both can combine with negation to form a negative term, in the Aristotelian sense. Thus, the type-identical Montague categories **t / e** and **t // e**, both of which express properties and denote sets, as laid out in (29), tend to fall together.

(29)

Category Name	TG Category	Category Definition	Type	Denotation
IV: intransitive verb phrase	**VP** or **V**	**t / e**	⟨⟨**s**,**e**⟩,**t**⟩	set of individual concepts
CN: common noun	**NOM**, $\bar{\mathbf{N}}$	**t // e**		

This result would be congenial to some post-Aristotelian syntacticians. In the initial program for categorical syntax, Ajdukiewicz 1935, there was in fact no category for common nouns distinct from the S/N category of predicate phrases. Within generative semantics, G. Lakoff (1965) and Bach (1968) collapse set-denoting predicators of various surface categories within a single metacategory, Bach's CONTENTIVES. But, as Aristotle recognized, verbs differ formally from nouns in inflecting for tense, a difference which may be semantically grounded. Accounting for both the similarities and the differences between verbs and nouns as predicators remains a controversial question, one which I shall not attempt to address here. (Cf. G. Carlson 1977 and Parsons 1985 for two possible solutions utilizing a theory of stages and kinds and a Davidsonian theory of events, respectively.)

An Aristotelianized Montague Grammar also affords a more natural and coherent account of presuppositional and scope phenomena and their interaction with negation than any available within the Russellian theory of descriptions adopted in classical MG. I shall defer the discussion of this interaction to §7.3.2.

I have offered here a programmatic sketch of Extended Term Logic, my proposal for coming to terms with term logic. In this way, I have suggested, the treatment of singular and general expressions can be collapsed and a motivated distinction can be drawn between one-place truth-functional operators (which are excluded, ruling out any iterative syntactic external negation) and two-place truth-functional operators (which are permitted). I see this approach as a means for capturing the insights of Aristotle and Montague in taking surface form seriously as a mirror of, and guide to, logical form.

There remain some important open questions whose resolution will affect the development of a full-fledged ETL. Can instances of apparent de dicto modalities be assigned variously to a de re operator, on the one hand, and to a metalinguistic operator, on the other? Or should we follow Aristotle in taking de dicto modals, like negation, to constitute a mode of predication affecting the subject/predicate connection but taking scope over both? Secondly and more crucially, just what is the status of Aristotle's predicate denial? Is it possible to assimilate all instances of apparent predicate denial either to predicate term (constituent) negation or to metalinguistic negation, or must we retain predicate denial as syntactically and semantically distinct from both? The former question I cannot address here, but the latter I must.[21]

7.3 Scope, Presupposition, and the Grammar of Negation

—I have two coins which add up to fifty-five cents and one isn't
a nickel. How can that be?
—The other one is a nickel. (ancient childhood riddle)

Il faut savoir la grammaire lorsqu'on veut être roi de France.
(King Louis XVIII)

In this, the final (but by no means definitive) section of my study, I shall consider a proposal for representing the two descriptive negations of term logic—predicate denial and predicate term negation—within the grammar of English. I then turn to the unanswered question with which I ended §7.2:

given the need for accommodating both wide-scope metalinguistic negation and narrow-scope predicate term negation within natural language, what is the evidence that we need to allow for a representation of predicate denial? My pursuit of this evidence will lead me back to the reconsideration of the status of logical presupposition and to the investigation of the syntactic, semantic, and—as we shall see—pragmatic scope of negation in quantified sentences.

7.3.1 The Representation of ETL Negation: Gazdar, Pullum, Sag, and Aristotle

As I have noted, sentential or wide-scope descriptive negation in English can be realized either as the word *not* following the tensed auxiliary or, more colloquially, as a contraction on that auxiliary. In the latter case, as Lapointe (1980) and Zwicky and Pullum (1983) have convincingly argued, *-n't* is an inflectional suffix realizing a lexical feature on the auxiliary element, rather than (as usually assumed) a "mere" clitic freely attaching to a host. In its dual manifestations as free particle and inflection, the category of negation would thus parallel time reference (cf. the phrasal future *will kick* vs. the inflected past *kicked*) or, as Zwicky and Pullum observe, adjectival comparison (the phrasal *more sleepy* vs. the inflected *happier*).

This analysis is supported by distributional and morphological criteria, as well as by the semantics of contracted auxiliaries. As Zwicky and Pullum point out (1983:509), the treatment of the constraints on negative incorporation in Horn 1972:chapter 4 (and in §4.5 above) presupposes the characterization of negative auxiliaries as lexical items. Thus, for example, *can't* and *couldn't* can only be interpreted as involving wide-scope (modal) negation, while *mustn't* and (usually) *shouldn't* involve narrow-scope negation, affecting the following verb-phrasal constituent but not the preceding modal. If these and other expressions of the form ***Xn't*** are analyzed as lexical items, rather than simple host + clitic combinations formed by a postsyntactic attachment rule, such facts can be accounted for straightforwardly.

On my account, the asymmetry between *couldn't VP* (= **not′** (**could′** (**VP′**)) and *mustn't VP* (= **must′** (**not′** (**VP′**)) stems from the tendency for lexicalization rules to favor configurations which result in **E**-vertex rather than **O**-vertex logical forms. Other notorious irregularities in the distribution of contracted auxiliaries (cf. Boyd and Thorne 1969; Horn 1972; Givón 1978)—the marginality and/or nonoccurrence of *mayn't, mightn't,* and *oughtn't* within many dialects, the restriction of *mustn't* to root (deontic) rather than epistemic interpretations (*You mustn't work hard* vs. *You must not work hard*)—are also to be anticipated if these forms are all to be listed (or, as the case may be, unlisted) within inflectional paradigms.

Within ETL, it is natural to take *n't* forms as the canonical realization of predicate denial, a negative marker with scope over the entire predication (in the unmarked case) which surfaces within the predicate expression. Since the semantics associated with each of these forms is determined by the lexical item in question, nothing prevents the idiosyncratic association of a given ***Xn't*** auxiliary with the narrow scope *not-X* semantic interpretation.

Wide-scope predicate denial, of course, can also take the form of a *not* particle located between the auxiliary (Modal, *have, be,* or *do*) and the verbal element ($\bar{\mathbf{V}}$) for which it subcategorizes. But *not* also realizes nonfinite negation. In this capacity it can occur as the leftmost daughter of any nonfinite $\bar{\mathbf{V}}$. This results in narrow ($\bar{\mathbf{V}}$ or **IV**-phrase) scope negation, corresponding to Aristotle's predicate term negation (PTN) and to Jackendoff's **VP**-scope constituent negation. The semantics associated with this configuration may or may not be identical with that of the corresponding predicate denial (if any), depending on the semantic properties of the auxiliary verb involved. Gazdar, Pullum, and Sag (1982:604–5) offer the following characterization of the two rules for introducing *not* within a Generalized Phrase Structure grammar, vintage 1982. Sentences (30a, b) are the **PS** rule schemata which introduce *not* into tenseless and tensed $\bar{\mathbf{V}}$s respectively, (30a′, b′) instantiate two structures induced by these rules, and (30a″, b″) give the Montague-style semantic translations of these two structures.[22]

(30)

a. $\langle \alpha, [_{\bar{V}}\ \underset{\left[\begin{smallmatrix}+\text{AUX}\\ +\text{FIN}\end{smallmatrix}\right]}{V}\ \textit{not}\ \underset{[+\text{BSE}]}{\bar{V}}],\ \lambda\mathcal{P}[\sim V'({}^{\wedge}\bar{V}'(\mathcal{P}))]\rangle$

b. $\langle \beta, [_{\underset{[-\text{FIN}]}{\bar{V}}}\ \textit{not}\ \bar{V}],\ \lambda\mathcal{P}[\sim \bar{V}'(\mathcal{P})]\rangle$

a′.

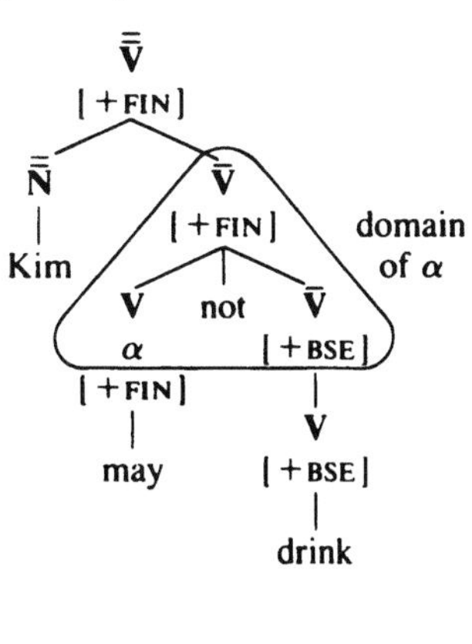

b′.

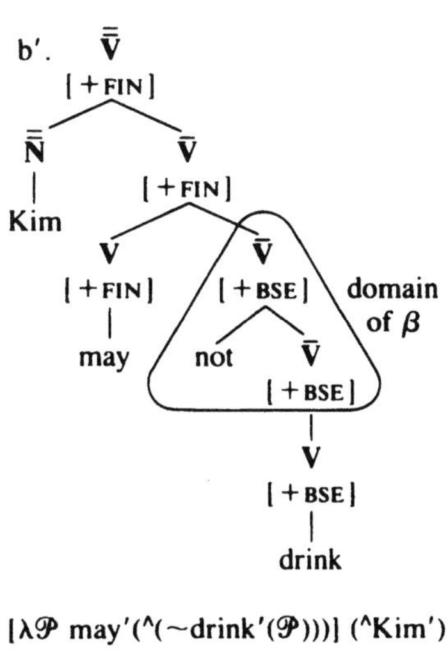

a″. $[\lambda\mathcal{P}[\sim\text{may}'({}^{\wedge}\text{drink}'(\mathcal{P}))]]\ ({}^{\wedge}\text{Kim}')$

b″. $[\lambda\mathcal{P}\ \text{may}'({}^{\wedge}(\sim\text{drink}'(\mathcal{P})))]\ ({}^{\wedge}\text{Kim}')$

The rule in (30a) is lexically restricted to those auxiliaries which, in effect, subcategorize for predicate denial. Modals like *can* and *could,* deontic *may,* and *need,* as well as the copula *be* and the perfect *have,* can be introduced by this rule; these auxiliaries will be assigned the appropriate subcategorization feature (the α in the tree in (30a′) above, keyed to the rule number introducing that item in the version of GPSG adopted by Gazdar, Pullum, and Sag 1982). Other modals, including *must* and epistemic *may* (cf. *It may not rain*) are not introduced by rule α, and hence will not co-occur with wide scope negation (predicate denial).

The right-hand rule and structure, on the other hand, are putatively available for all nonfinite $\bar{V}$s, although it is not entirely clear that the implicit claim here, that is, that all auxiliaries allow $\bar{V}$ negation within their scope, is tenable. In particular, it would appear that the unstressed *do* of *do*-support is never followed by a narrow-scope negation, at least in singular predications; note the following distribution of the Klima *neither/so*–tag diagnostics:

(31) He is not voting and neither is she.
He has not succeeded and neither has she.
He cannot attend and neither can she.
He did not come and neither did she.

(31′) He is (intentionally) not voting and so is she.
He has (often) not succeeded and so has she.
He can (always) not attend and so can she.
*He did (perhaps) not come and so did she.

The narrow ($\bar{V}$-) scope negations triggering *so-* rather than *neither*-clause reinforcement are in general possible after auxiliary verbs, provided of course that they are realized by a freestanding negative (à la rule (30b))—negative inflections (*isn't, hasn't, can't*) are compatible only with predicate denial (wide-scope) interpretations. But a negation following *do* is apparently restricted to the predicate denial operation of rule (30a). If, however, *do* is inverted (i.e., is generated with the [+INV] feature within the GPSG framework), narrow-scope ($\bar{V}$) negation is perfectly acceptable: *Did he ({possibly/ever}) not succeed?*

While a marriage of the GPSG syntax for English $\bar{V}$s with the bifurcated term logic analysis of negation (predicate denial as a mode of predication vs. term—narrow scope, $\bar{V}$—negation) strikes me as a promising match, I shall not pursue this coupling here. Nor shall I consider the less radical move of extending Montague's PTQ fragment to include particle, as well as contracted, negation and constituent, as well as wide-scope, negation (cf. Bennett 1976). In any case, the GPSG grammar of negation is very much in the Montagovian spirit. Indeed, with his syncategorematic insertion of of the negated auxiliary verb, Montague can be seen as indirectly anticipating the Lapointe–Zwicky and Pullum–GPSG inflection treatment of *n't*.

7.3.2 Scope and Presupposition for Russell, for Montague, and for Me

Whatever version of the grammar of negation is adopted, the analyst must eventually confront one of the most extensively studied and least-understood phenomena within the semantics of negation: the scope interaction of the negative operator with quantified subjects and with descriptions. In PTQ, definite descriptions and quantified subject phrases can be introduced directly or quantified in. In the latter case, the sentence resulting from an application of either the simple positive subject-predicate rule **S4** or one of the rules of negation-and-sign in **S17** contains an individual variable (*he*$_n$) in subject position. This variable is then bound by the application of the relevant quantification rule (Montague 1974:252ff.).

This procedure accounts for the scope interaction of quantified expressions with each other, with negation, and with other scope-defining operators. Thus, the two readings of the {*all*/*every*} . . . *not* constructions, as discussed in §4.3 above and exemplified in (32), result from two syntactically distinct analysis trees in the disambiguated language, as seen in (32a, b); the simplified **IL** translations that would be assigned are given in (32a′, b′) respectively:

(32)
Every fish doesn't sleep.

```
a.  every fish doesn't sleep, 10, 2              b.  every fish doesn't sleep, 11
           /            \                                  /          \
every fish, 2    he₂ doesn't sleep, 11            every fish, 2      sleep
      |               /      \                          |
    fish            he₂     sleep                      fish
```

a′. $\Lambda x(\text{fish}'(x) \rightarrow \neg\text{sleep}'(x))$ b′. $\neg\Lambda x(\text{fish}'(x) \rightarrow \text{sleep}'(x))$

In (32a), the subject phrase *every fish* is quantified into an open negative sentence which has been formed by the application of the negative–third person singular–present tense subject/predicate rule; it is thus assigned wide scope with respect to the negation, yielding the NEG-V reading (= 'Every fish is such that it doesn't sleep'). In (32b), the quantifier phrase *every fish* is built up first; it then combines with the **IV** phrase *sleep* by the application of the same tense-and-sign rule. Negation, inserted last, gets wide scope, whence the NEG-Q reading (= 'Not every fish is such that it sleeps').

This same technique generates an ambiguist account of definite descriptions in negative sentences à la Russell (1905). As we saw in §2.2, Russell's theory of descriptions assigns two alternative positions to the negative operator in the logical form of sentences like (33):

(33) The king of France is not bald.
  a. INTERNAL: $\exists x(Kx \wedge \forall y(Ky \rightarrow y = x) \wedge \sim Bx)$
    [= 'The king of France is not-bald']
  b. EXTERNAL: $\sim\exists x(Kx \wedge \forall y(Ky \rightarrow y = x) \wedge Bx)$
    [= 'not (The king of France is bald)']

with scope inside or outside that of the existential quantifier into which the description is unpacked. In the former case, the description is said to have a PRIMARY occurrence with respect to negation, and in the latter a SECONDARY occurrence.

The syntactic approach to scope ambiguities within MG automatically predicts the two readings detected by the ambiguists—Russellian (as Delacruz [1976] observes) or Aristotelian. Thus we have, inter alia, the two analysis trees in (34a, b), where I assume an extended version of PTQ in which predicate adjectives and adjective phrases are basic categories which combine with a subject to yield a sentence; the copula is introduced syncategorematically as a sign of affirmation or negation (as in traditional analyses; cf. Mill [1843] 1919):

(34)

The king of France isn't bald.

a. the king of France isn't bald, 10, 3
  the king of France — king of France
  he_3 isn't bald, 13 — he_3, bald

b. the king of France isn't bald, 13
  the king of France, 2 — king of France
  bald

a′. $\bigvee x(\text{king-of-France}'(x) \wedge \bigwedge y(\text{king-of-France}'(y) \rightarrow y = x) \wedge \neg \text{bald}'(x))$

b′. $\neg \bigvee x(\text{king-of-France}'(x) \wedge \bigwedge y(\text{king-of-France}(y) \rightarrow y = x) \wedge \text{bald}'(x))$

Sentences (34a, a′), corresponding to (33a) where the description has primary occurrence (wide scope) with respect to negation, asserts of the king of France that he isn't bald, while (34b, b′), corresponding to (33b) where the description has secondary occurrence (narrow scope), denies of the king of France that he is bald. For Montague, as for Russell (or Aristotle), the former is false and the latter true if there is no king of France.

Alternatively, one might superimpose a presuppositionalist analysis and treat the former case as neither true nor false, by assigning it a third value or no value at all if its existential presupposition fails (cf. Delacruz 1976).

Or again, one might assign the value **false** to this statement under such circumstances (following Aristotle and Russell), but at the same time associate it with a false conventional implicatum, whence the Strawsonian squeamishness upon evaluating it in kingless contexts (cf. Karttunen and Peters 1979). In any event, the contrast between (33a)/(34a), on the one hand, and (33b)/(34b), on the other, will be analyzed as a scope distinction in the syntax of logical form, directly analogous to the treatment of quantifier-negation ambiguities like (32).

But, as we saw in §2.2, Russell's analysis is far less convincing when it is generalized to names, which induce parallel presuppositional phenomena (cf. Frege 1892 on *Kepler did not die in misery*), yet which are not plausibly regarded as complex in the same sense as descriptions. Russell's and Quine's programs for unpacking names into descriptions permit a generalization of the appropriate sort, but at a heavy cost; as I noted in my earlier discussion, there are metaphysical and epistemological considerations militating against this move. We recall Henry's warning (1972:74) that when negative sentences are taken to be semantically ambiguous within a propositional rather than term-based logic, 'The distinction which has evidently been desirable all along is introduced in a tortuous and *ad hoc* fashion under the misleading guise of the "primary and secondary occurrence" of descriptions and all names have to be construed as disguised descriptions in order to be able to take advantage of this *ad hoc* distinction'.

The same dilemma arises within an MG-style scope analysis, although the deconstruction of definite descriptions in logical form is no longer necessary. But now proper names must be quantified in along with descriptions if the analysis tree in (34a) is to be adapted to Aristotle's *Socrates is not well* or Frege's *Kepler didn't die in misery,* which exhibit the same ambiguity as (33). While Montague does, in fact, allow Socrates, Kepler, and their ilk to be quantified in as variable-binding term phrases, along with *the king of France, every fish,* and *an eel,* there is no compelling evidence for—or benefit from—this move.

Whatever arguments one might muster for treating descriptions as variable-binding operators introducing scope ambiguities, no analogous arguments independent of the cases under discussion here are available for proper names. In any case, given Montague's endorsement of Kripke's (1972) analysis of names as rigid designators, the translations for the two analysis trees assigned by the PTQ grammar to sentences of the Aristotle-Frege variety come out identical, which doesn't help to provide the two different readings we apparently require.

If names do not enter into the quantificational ambiguities for which syntactic scope distinctions are designed to account, neither do all descriptions.[23] This leads us to the next problem for the MG approach to presuppositional phenomena. As we saw in chapter 2, there is a strong parallel, elucidated by Aristotle, between reference failure and category mistakes. Like (33), (35) allows two understandings.

(33) The king of France is not bald.

(35) The number 2 is not blue.

On its primary interpretation, (35) involves category-restriction-preserving negation and is a priori false (or, for those so inclined, neither true nor false). But negation may also be assigned wide scope with respect to the category or selection restriction; in this case, (35) is automatically true. For Russell, of course, category violations yield meaninglessness, and the parallel between (33) and (35) is implicitly rejected. But if we follow Aristotle, Quine, Lambert, and their fellow no-typers (cf. §2.3) in taking (35) to have a reading as the true contradictory of its a priori false counterpart *The number 2 is blue,* we will seek a parallel explanation of these parallel cases. But neither Russell's exponible analysis nor Montague's quantifying in extends from the flaccid king of France to the rigidly designating number 2.

There is one embarrassment for Montague's analysis which does not afflict Russell's. The principles which predict the ambiguity of a negative sentence like (33) automatically predict the same ambiguity for the corresponding positive sentence (*The king of France is bald*), yet there is no evidence for such an ambiguity. While the proliferation of multiple logically equivalent structures associated with an unambiguous surface sentence may not be unprecedented within grammatical theory, neither is it cause for pride. I return to this problem below.

Russell's dissection of the king of France is of course just one of many. Within the presuppositionalist analysis of Strawson (1950, 1952), only one, essentially internal, reading is assigned to (33), preserving the presuppositions associated with the corresponding affirmative singular predication (*The king of France is bald*). In multivalued presuppositional logics, Russell's (and Aristotle's) ambiguity resurfaces as a dichotomy between presupposition-preserving, nonbivalent choice (internal) negation and presupposition-free, bivalent exclusion (external) negation (cf. §2.4). In the monoguist theories of contemporary work in pragmatics, negative sentences are essentially unambiguous and general; the wide-scope reading of (33b) is assigned by semantic rules, and the narrow-scope force of (33a) is obtained through some variety of pragmatic strengthening (cf. Wilson 1975; Kempson 1975; Boër and Lycan 1976; Wilson and Sperber 1979; Gazdar 1979a; Grice 1981).

Cutting across the significant differences between these views, the current consensus is clearly that—pace Frege and Strawson—an external, nonentailing (and/or nonpresupposing) understanding is available as one of the versions, if not the unique version, of the logical form assigned to negative sentences like (33). But how compelling is this consensus? As we saw in chapter 6, the discourse oddness of those sentence tokens which effectively compel the external, presupposition-free reading of negation:

(33′) The king of France isn't bald—there is no king of France.

(35′) The number 2 isn't blue—integers aren't colored.

is essentially the same oddness that afflicts other sentence tokens where negation functions as a metalinguistic operator, associated with what are clearly (or arguably) extrasemantic properties of utterance meaning or use. As we further noted, the evidence from intonation, incorporation, and rectification underlines the kinship between these cases and supports a unified treatment in metalinguistic terms. The standard theories of semantic external negation, monoguist and ambiguist alike, fail to account for the double-take effect in the processing of these sentence tokens and for the linguistic correlates unifying the class of metalinguistic negations.

Should I not then conclude, with Burton-Roberts (1987), that the metalinguistic analysis of apparent wide-scope negation in fact requires me to adopt a Frege-Strawson version of presuppositional semantics, in which descriptive negation applies unambiguously so as to preserve truth-value gaps induced by reference failure (and category mistakes)?[24] If I do adopt this view, is there anything preventing me from subsuming all instances of wide-scope negation into the metalinguistic category? Having argued that so-called sentential negation, with scope over the subject, is to be reanalyzed as Aristotelian predicate denial rather than the Stoic-Fregean *apophatikon,* am I now to reanalyze it out of semantic existence?

To these rhetorical questions I must answer 'no', 'yes', and 'by no means', respectively.[25] The abandonment of semantic presupposition (cf. Karttunen 1974; Stalnaker 1974; Kempson 1975; Wilson 1975; Boër and Lycan 1976; Gazdar 1979a; Lycan 1984) was not occasioned by perversity or whimsy, or solely by the specter of the ambiguity of negation. The balance of the evidence leads me to agree with the prevailing view (expressed most forcefully in Lycan 1984: chapter 4) that the conceptual obscurities and implementational difficulties besetting the notion of logical presupposition render it at best otiose for the description of natural language semantics.

Among the problems to be faced by the would-be presupposition collector is the capricious nature of the prey. One nonlogical variable affecting the behavior of presuppositional phenomena is isolated by Strawson himself, who noted (1964:95) that while (36a), like (33), can be argued to presuppose the existence of the king of France, (36b) does not share this presupposition or the truth-value gap associated with reference failure.

(36) a. The king of France {visited/didn't visit} the exhibition.
b. The exhibition {was visited/wasn't visited} by the king of France.

On the revisionist account championed in Strawson 1964, it is not definite descriptions per se which induce existential presuppositions, but only those singular expressions which a sentence is understood as being about. But this is a pragmatic and not grammatical criterion, as we shall see in more detail later in this section.

One variable affecting presuppositionality not discussed by Strawson resides in the nature of the predicate. If what you announce to me is not (37a) but (37b),

(37) a. The king of France {is/isn't} bald.
b. The king of France {is/isn't} sitting in the chair to your right.

and if I note that the designated chair is clearly empty, I am far less inclined to grant that you have presupposed the existence of the king of France—and even less likely to grant that what you have said is neither true nor false if, in fact, France is a republic. What is at issue here is not the criterion of 'aboutness', but a distinction in verification procedures. In neither Strawson's case nor mine is the instability of the "presupposition" a matter of logical semantics.

Nor is it clear to me, as it seems to be to Burton-Roberts (1987), that the metalinguistic analysis of marked negation requires an endorsement of Strawson's monoguist presuppositional semantics. It is true that the wide-scope understanding of the negation in (33), corresponding to Russell's (33a) in failing to commit the speaker to the existence of a French monarch, does occur most naturally as a metalinguistic negation, in a context where the attribution of baldness to that monarch has just been entertained. But in precisely such contexts, the predicate denial—denying baldness of the king of France—is of course true: the very act of issuing the metalinguistic objection commits the speaker to the truth of the corresponding wide-scope descriptive negation, that is, to the predicate denial. This will occur whenever the focus of the metalinguistic negation—the existential or uniqueness presupposition in vacuous subject cases, the sortal presupposition in category mistakes—is a necessary condition for the truth of the positive corresponding affirmative. In contexts where the objection is not directed at a truth condition for the affirmative, the use of metalinguistic negation (*He didn't manage to pass the test—he was given the answers*) does not guarantee the truth of the predicate denial.[26]

But to say that a given predicate denial is true is not to guarantee that this predicate denial can be felicitously expressed as a descriptive negation. As we saw in chapter 6, truth is no guarantee of assertability. Each of the cases under discussion—*The king of France is not bald* (given that he doesn't exist), *Socrates is not well* (given that he's dead), *The number 2 is not blue*—is true (on the predicate denial reading), yet each is virtually unassertable, at least by someone who recognizes the presupposition failure. This follows from the criterion of plausibility of denial (cf. chapter 3): baldness arises only for those who exist, health for those who are alive, and blueness for objects which can be colored. If I assume you know that France is a republic, it is pointless for me to inform you of the truth of (33); if I assume you don't know it, (33) would be an inefficient way for me to inform you of the fact.[27] But if you have just claimed that the French king is bald, I can disabuse you of your assumptions by an apt metalinguistic application of (33)—or, more likely, of (33′). Of course if you are laboring under the misconception that Ted Kennedy reigns at the Elysée palace, you could perfectly well utter (33) as a predicate denial; your utterance would be plausible enough—for you—and indeed it would be, through no fault of your own, entirely true.

Thus a typical felicitous token of so-called external or presupposition-canceling negation, at least in vacuous singular expressions and category mistakes, is metalinguistic, whence the oddness and marked character of such examples, the behavior of such negations with respect to the diagnostics, and the failure of Frege and Strawson to take seriously these apparent counterexamples to their analyses. But these very tokens of metalinguistic negation necessitate the assignment of truth to the corresponding predicate denial, an assignment made by Aristotle in his analysis of the wide-scope reading of *Socrates is not wise*.

I follow Aristotle in taking predicate denials to be true in instances of reference failure or category mistakes, and predicate term negations to be false in the same contexts. In ETL, as in the Aristotelian, Russellian, and Montagovian systems, there is an existence-entailing narrow-scope (PTN) reading of negation, realized as nonfinite particle negation (cf. (30b)) or by a prefix on the predicate (*dislike, unwell, impossible*). But while predicate term negations (*The king of France dislikes pizza, Socrates is unwell, The number 2 is {not-blue / nonblue}*) necessarily share the existence and type commitments of the corresponding positive assertions, predicate denials do not necessarily lack these commitments, at least in practice. The normal use of (34), or of (33) as a predicate denial, does strongly suggest that

France has one and only one king, and the normal use of ***x** isn't blue* that **x** is colored. Indeed, the internal or presuppositional value of negation in singular predications is typically expressed with predicate denial, not with predicate term negation. Thus, the distinction between predicate denial and term (**IV**-phrase) negation does not provide us with the full scopal or presuppositional disambiguation that we might (based on the *Organon*) have desired and expected. How the presuppositional properties of predicate denial arise is a question I shall address below; we shall see that neither syntax nor semantics provides the appropriate language in which to frame its answer.

7.3.3 *Every* vs. *Some:* Different Scopes for Different Folks

Within ETL, we distinguish the predicate denials of (38a), in which the negation has scope over the subject and the NEG-Q reading is assigned, from the predicate term negations in (38b), where the nonfinite or prefixal negation is restricted to predicate-internal scope and the NEG-V reading emerges.[28]

(38) a. All the cookies {weren't / were not} eaten.
All things {aren't possible / are not possible}.
b. All the cookies were {not eaten / uneaten}.
All things are {not possible / impossible}.

The scope distinction in these examples correlates with Klima's diagnostics for sentence vs. constituent negation. Thus, the wide-scope (sentential) negation of the predicate denials determines positive tags, while the narrow-scope negation is associated with negative tags:[29]

(38′) a.	All the cookies weren't eaten, {were they / *weren't they}?	[only PD / NEG-Q reading]
	All the cookies [were not] eaten, were they?	[on PD / NEG-Q reading]
b.	All the cookies were [not eaten], weren't they?	[on PTN / NEG-V reading]
	All the cookies were uneaten, were they?	[only PTN / NEG-V reading]

So far so good. Indeed, the need for a wide-scope descriptive negation for characterizing the readings of (38a) supports the status of predicate denial as the default realization of semantically contradictory descriptive negation, despite the elusiveness of the predicted noncommittal readings in examples like (33) and (35). Unfortunately, however, many speakers obtain a narrow-scope, NEG-V reading for *all* + contracted (inflected) negation, either along with or instead of the wide-scope interpretation we expect. For some speakers, this NEG-V reading is possible only with a negative tag—*All the students didn't fail the test, didn't they*—in which context the otherwise salient NEG-Q reading of (38a)-type sentences is eliminated (Heringer 1970:293). (Other speakers reject both negative tags and NEG-V readings.) I must apparently give up my identification of auxiliary negation with predicate denial, weaken my claim that predicate denial invariably yields contradictory negation with *do* auxiliaries, and/or provide a mechanism for contradictory predicate denials, with certain quantified subjects, to be strengthened to contraries.

But an even more serious problem lurks on the paths of my approach to quantifier negation scope interaction, as well as on those of its competitors. In correctly predicting the scope ambiguity for universally quantified subjects in negative sentences, Montagovian and neo-Aristotelian frameworks predict a parallel ambiguity when the determiner in question is the existential or particular *some* or *a(n)*. Thus, merely by substituting *a* for every token of *every* in the trees of (32a, b), we obtain two readings for *A fish doesn't sleep,* the former with narrow-scope negation (= 'There is a fish which doesn't sleep'), and the latter an apparently unavailable NEG-Q interpretation with wide-scope negation (= 'Not a fish sleeps'). As is well known from the linguistic literature (cf., e.g., Givón 1978), the NEG-Q readings attested for the universal negations of (39a) tend to disappear in the corresponding particular negations of (39b).

(39)	a.	b.
	Every man didn't win.	{A/Some} man didn't win.
	Everybody isn't happy.	Somebody isn't happy.
	All the boys don't like you.	Some (of the) boys don't like you.
	All that glitters isn't gold.	Something that glitters isn't gold.
	(NEG-V[$\forall\sim$ . . .] or NEG-Q[$\sim\forall$ . . .])	(only NEG-V[$\exists\sim$ . . .], not NEG-Q[$\sim\exists$ . . .])

The analysis trees generated by Montague Grammar predict that in principle both universals and existentials in subject position can be either quantified into an open negative proposition (yielding the NEG-V reading with wide scope for the subject phrase) or combined with a CN to form a term expression which in turn combines negatively with a predicate to obtain a sentence (yielding the NEG-Q reading with wide scope for the negation). I have proposed jettisoning this line in favor of a distinction between finite inflected or particle negation (PD, taking scope over the entire predication) and nonfinite or prefixal negation (PTN, with scope confined to the predicate expression). But on either approach, the wide-scope reading for negation must apparently be filtered out when the subject is quantified by *some* or *a(n)*. How can this be accomplished—in MG, ETL, or any other framework?

My first step is to argue that both sets of examples in (39) do indeed involve predicate denial, despite the optional narrow-scope (contrary) NEG-V semantics of the sentences in (39a) and the apparently obligatory narrow-scope (subcontrary) NEG-V semantics of those in (39b). Besides the fact that the negations in these examples are realized as inflections, which I have taken to betoken predicate denial, the distribution of the Klima diagnostics is inconsistent with the treatment of the *some . . . not* constructions as instantiating constituent (predicate term) negation:

(40) a. Some of the arrows didn't hit the target and {?neither/*so} did some of the javelins. [adapted from Jackendoff 1972:363]
  b. Some of the arrows didn't hit the target and some of the javelins didn't {either/??too}.
  c. Some of the arrows didn't hit the target, {?did they/*didn't they}?

Crucially, this pattern is sharply distinguished from the one determined by clear cases of PTN, such as unlexicalized {*can/could*} . . . *not* on its **O**-vertex (M-NEG) reading:

(40′) a. She can [not attend], and {*so/neither} can he.
  b. She can [not attend], and he can [not attend] {*either/too}.
  c. She can [not attend], {*can she/can't she}?

(The S-neg diagnostics starred in (40′) are of course all grammatical when *can not* is read as wide-scope predicate denial, that is, as *cannot.*)

The fact that (for most speakers) neither set of diagnostics is impeccable in (40) is attributable to the clash between the syntax of these sentences, which I take to involve predicate denial, and their semantics, which is not that of the contradictory opposition normally associated with predicate denial (or Klima's sentential negation). For Jackendoff, who defines sentence negation as contradictory negation (via the *it is not so that S* paraphrase test), the NEG-V readings of (39a, b) must involve constituent (**VP**) negation, leaving us with no account of the difference between the patterns of (40) and (40′). Nor can I accept Jackendoff's claim (1972:332) that '**S** and **VP** negation differ in meaning exactly when there is a quantifier in the derived subject', given the clear syntactic and semantic evidence for narrow-scope **VP** negation in (40′) and other modal contexts (cf. Boyd and Thorne 1969, Horn 1972, and Palmer 1979 for additional examples and discussion).

But if both (39a, b) involve predicate denials, with narrow-scope NEG-V readings resulting from a delay in quantifying in *every* and *some*, as in (32a, a′), what prevents the alternative (32b, b′)-type analysis, with wide-scope negation, for the *some* . . . *not* examples of (39b)? Lee (1974) offers a creative but unsuccessful program for adapting Montague negation to the scope interaction facts obtaining in English. Foreshadowing Barwise and Cooper (1981), Lee takes 'prenominal negation'—both lexicalized (*nobody*) and free (*not everybody*)—to involve basic negative term phrases. For the range of sentences in (39b), where he rejects the MG-predicted ambiguity, Lee replaces the negation-as-a-mode-of-predication introduced by PTQ's **S17** with a strictly narrow-scope **IV/IV** operator. Thus, we are left with quantifier negation and verb phrase negation, but no remnant of wide-scope sentential negation or predicate denial. What then of the NEG-Q readings attested for universal subjects in negative sentences like those of (39a)? Lee acknowledges his failure to provide an account of such readings, which he excuses on the grounds that 'many speakers do not get these sentences at all', and that 'if they do, they only get the Neg-V reading' (Lee 1974:381). But in fact, the NEG-V-only dialect represents the smallest minority attested in Carden's (1970) study. Curiously, the same sentence cited as unambiguously NEG-V in Lee's earlier discussion, *Everyone didn't come*, turns out five pages later to be ambiguous after all; its NEG-Q reading is predicted by the good old tense-and-sign rule **S17** of PTQ, which Lee now sees as complementing, rather than replacing, his own rule (Lee 1974:386). But then the particular negations of (39b) must be ambiguous as well, and we are back where we began.

Cresswell (1973), recognizing the problematic asymmetry between universals and particulars in the context of (39), suggests an intriguing solution: since compositional semantic theories cannot predict the right scope assignments in these cases, all scope distinctions should be scrapped: logical form is essentially scope-free. This yields the full range of both attested and unattested readings as possible understandings of any sentence with two scope-wielding operators, whether one of these operators is negation (as in (32), (33), (35), (38), and (39)) or not (cf. the classic pair *Everyone loves someone, Someone is loved by everyone*). It will then be up to the pragmatics to filter out the nonoccurring readings. (Cf. Sadock 1976 for a similar proposal, independently arrived at.)

While this proposal may strike us as a counsel of despair, there is something to be said for overgeneration even in those theories which admit scope distinctions in logical form or semantic interpretation. In particular, it is worth reexamining the standard assumption that a particular or existentially quantified subject term never takes narrow scope with respect to a following negation. This assumption may appear justified when we come upon the examples of (39b) in isolation, but consider the following attested cases (emphasis added):

(41) a. A sociopath wouldn't get through the first ten minutes of my films. They are too slow. Someone isn't killed in the credits. (from a newspaper interview with Brian de Palma)
b. She swung round, she took two strides to him, waiting for someone to stop her, but someone didn't. (from John Le Carré's *The Little Drummer Girl*)
c. Neither Inspector Walker nor the book's readers can be entirely certain that an innocent man has not gone to the gallows. (from a book review in the *New York Times*)

Evidently, the appearance of the *some/a . . . not* construction within a context where the corresponding positive expectation has been explicitly established licenses a NEG-Q reading, neutralizing the asymmetry between particulars and universals.[30]

Thus, in (41a) Brian de Palma is implicitly contrasting his films with those of other directors (whose creations are in fact as sociopathic as his own are alleged to be); it is in those unspeakable celluloid horrors that someone invariably is killed while the opening credits roll. In (41b), the antiheroine Charlie expects someone to stop her. In (41c), the state-sanctioned hanging of an innocent is taken to be an unexpected, not to say disconcerting, possibility. Crucially, each case involves the disappointment of an expectation assumed to be shared by speaker and addressee (or by reader and writer).

These examples recall an observation by Baker (1970: 182ff.). Baker cites a number of (constructed) sentences in which *some . . . not* occurs embedded under one of a set of 'special predicates', inducing a reading in which the embedded negation seems to take wide scope with respect to its subject:

(42) a. I'm surprised that someone hasn't already said something to you.
b. John is relieved that someone didn't sign up ahead of him.

He attributes the acceptability of these complements to the nature of the governing predicate: 'Speaking intuitively, we can say that each of these predicates expresses a relation of contrariness between a certain fact and some mental or emotional state. For example, we say that we are *surprised* when a certain fact does not conform to our *expectations; relieved* when it does not conform to our *fears; disappointed* when it is not in line with our *hopes*'. Thus, (42a, b) are licensed by the propositions in (42′a, b), respectively:

(42′) a. I expected that someone would have already said something to you.
b. John was afraid that someone would sign up ahead of him.

Baker's account extends naturally to the reduced comparative in (43), warranted by (43′):

(43) It's {amazing/lucky} more people haven't already been killed at this intersection.

(43′) It was expected that more people would have already been killed at this intersection.

Note that the embedded clauses of (42a) and (43) cannot stand on their own with their multiple affirmative polarity items, nor can they occur in a non-"affective" environment:

(44) a. #I'm convinced that someone hasn't already said something to you.
b. Someone hasn't (#already) said {anything/#something} to you.
c. #I knew that more people haven't already been killed at this intersection.
d. ?More people haven't (#already) been killed at this intersection.

On Baker's transderivational theory of polarity, a given negative or positive polarity item **i** can be rendered acceptable in a sentence **S** if there is some other sentence **S′** such that **S** entails **S′** and **i** is well-formed within **S′**. Thus (42a) is acceptable because it entails (42′a), (43) because it entails (43′), and so on. But while some notion of allusion may well play a role in the description of polarity, Baker's theory is seriously flawed. Entailment relations prove to be neither a necessary nor a sufficient condition for the acceptability of polarity items, as recognized by Baker himself (1970: 182–84) and demonstrated in more detail by others. (Cf. Horn 1970; Fauconnier 1975a, 1975b, 1976; Ladusaw 1979; and Linebarger 1981, 1987 for criticisms and alternative proposals.)

For my present case, it suffices to note that Baker's framework fails to incorporate the correct generalizations about the set of entailments that license polarity shifts. What is crucial here is the dimension of emotivity Baker alludes to in the passage cited above; when this dimension is absent, the presence of an entailment does not in itself suffice. Thus (45a, b) are much less likely than (42a),

(45) a. #She denied that someone hasn't already said something to you.
b. #I doubt that someone hasn't already said something to you.

although each entails a positive sentence in which the positive polarity items are appropriate:

(45′) a. She said that someone has already said something to you.
b. I believe that someone has already said something to you.

(Of course simple negative polarity items are licensed by *deny* and *doubt,* as Baker and others have observed: *She {denied/doubts} that anyone has ever said anything about it.*)

Beyond the details of the constructions under consideration, three points need to be stressed. The first is that the negation in sentences like (42a, b) and (43) fails to interact with polarity items in the usual way; rather, the items in question are those which would have occurred in a corresponding proposition which is in some sense unmarked (expected) in the discourse context. We have encountered other instances of similar nonpolarizing negation in §6.4; as in those cases, we can take the present examples to represent METALINGUISTIC or second-instance negation.

A related point is that the fall-rise contours which tend to be associated with the NEG-Q readings of *{all/every}* . . . *not* (and of the related *both* . . . *not* and *and* . . . *not* constructions; cf. §4.3) is in fact a general characteristic of metalinguistic negation, as I noted in chapter 6. This supports the view that the wide-scope (NEG-Q) reading of negation in sentences with quantified subjects occurs most naturally in metalinguistic uses. On the other hand, the fact that no special intonation is required to bring out the wide-scope reading of negation in, for example, *All is not lost,* together with the typical absence of any rectification, suggests that this construction, with its $\sim\forall$ interpretation, must also be analyzable as realizing ordinary predicate denial.[31]

The third point is that, as we saw in the attested examples of (41), no 'special predicate' need be overtly present within the syntactic frame in order to trigger the polarity items in question (or the wide-scope reading of postexistential negation). Thus, alongside (42a, b) we can get the sequences in (46):

(46) a. This is incredible. You mean to tell me that someone hasn't already spoken to you about the party?!
b. What a relief—this must be my lucky day. Evidently someone hasn't signed up ahead of me after all!

Not surprisingly, metalinguistic *some . . . not* is also available in the context of the 'word by word, emphatic denial' (Baker 1970: 169) or the 'pure et simple reprise' (Tasmowski-De Ryck 1972: 199) of a previous assertion. We have already observed the appearance of postnegative *some* in such environments (see §6.4 for examples), and prenegative *some* is similarly attested in the exchange in (47a) from an episode of Sesame Street, while (47b) represents an exchange from the 1987 movie *Dirty Dancing* on the possibility of replacing a dancer who must miss the crucial mambo exhibition because she will be otherwise engaged (having an abortion) at the time:

(47) a. Forgetful Jones: 'Somebody broke my balloon, somebody broke my balloon.'
Maria: 'No, somebody didn't break your balloon.'
[It develops that Forgetful broke it himself.]
b. "Baby" Houseman: 'Can't someone else fill in?'
Johnny Castle: 'No, someone else can't fill in.'

But even when a wide-scope reading for the negation in a *some . . . not* construction is available in direct comebacks, it is often difficult to perceive, especially when this reading is not forced by the context or when the scornful flavor of the rebuttals of (47) is absent. Compare (47′):

(47′) A: Somebody spilled something.
B_1: {Nobody spilled anything/No, they [*sic*] didn't}—it's just the rain.
B_2: #Somebody didn't spill something, it's just the rain.

The same pattern holds for other weak or 'tolerant' determiners (cf. chapter 4). This point is illustrated nicely by an anecdote recalled by Katz (1972). When an outraged Parliament demanded that he withdraw the imputation in (47″a), Churchill responded by offering the "retraction" in (47″b):

(47″) a. Half of the ministers are asses.
b. Half of the ministers are not asses.

While an external metalinguistic reading is certainly available here (compare *ˇHalf of the ministers aren't asses, {but many/in fact all} of them are*), it is clearly not salient.

But what of the asymmetry between universals and particulars when no metalinguistic objection is involved? Why is it so much easier, both in English and cross-linguistically, to get the wide-scope reading for the negative in the context of (48a) than in the context of (48′a)?

(48) a. Everybody didn't come.
b. Not everybody came.

(48′) a. Somebody didn't come.
b. Nobody came.

For Jespersen (1924:327), the availability of NEG-Q readings for ordinary auxiliary (NEXAL) negation can be attributed to 'the result of two tendencies, to place the subject first, and to attract the negation to the verb', so that the negative which would 'logically' precede the universal is attracted instead to the unmarked nexal position on the finite auxiliary. But this latter tendency to 'use nexal negation whenever it is possible' (Jespersen 1917:44) is offset by a complementary tendency Jespersen invokes elsewhere. By this principle, our familiar Neg First, the preferred realization of wide-scope negation should be as a negative-incorporated quantified subject (*none, not every*), rather than as an auxiliary particle or inflection which must then "cross over" its subject term in the interpretation of the sentence. (The NEG-V reading, of course, involves no parallel interpretive crossover of scopes.) Furthermore, the prepositioning of the universal subject in (48a) 'for the sake of emphasis' (Jespersen 1917:87) would be equally satisfied by assigning the appropriate contour to the unambiguous *not*-initial version in (48b). And in any case, we must still distinguish the ambiguous nexal negation in (48a) from its (normally) unambiguous counterpart in (48′a).

The crucial difference between these two predicate denials is that a NEG-Q reading of the latter value could be alternately (and unambiguously) expressed by a fully lexicalized, inherently negative **E**-vertex quantifier or determiner, as in (48′b).[32] The only unambiguous alternative realization of the $\sim\forall$ value constituting the NEG-Q reading of examples like (48a), on the other hand, involves a relatively unlexicalized **O**-vertex negative quantifier or determiner, as in (48b): *not all (the) α, not everybody, not everything.* While the *some . . . not* configuration of (48′a) is of course logically equivalent to a negated universal, a particular negation too is less than fully lexicalized—as indeed any **O**-vertex quantifier must be, given my arguments in §4.5.

The relative availability of NEG-Q readings for predicate denials with quantified subjects is thus determined by the outcome of a rivalry among several functional tendencies. The missing link in Jespersen's account can be supplied by reformulating one of his principles: there is a preference for overt negation to surface in its unmarked (nexal, predicate denial) position, as a particle or inflection on the finite verb or auxiliary. This preference—call it NEXAL NOT—is satisfied in (48a) and (48′a) and is irrelevant to (48′b) with its inherently negative quantifier, but is contravened in (48b).

While Neg First predicts a general preference for the (b) over the (a) examples as representations of NEG-Q understandings, and the subject-first tendency predicts the opposite preference, it is Nexal Not which establishes the asymmetry between (48) and (48′). The marked character of negated quantifiers (*not everybody, not all the α*) as contrasted with inherently negative quantifiers (*nobody, none of the α, no α*) is underlined by the well-known distributional asymmetries between these two sets: the former expressions cannot occur freely as direct or prepositional objects, while the latter can.

But why should Nexal Not exist? What more general pattern of functional explanation can be invoked here? Notice that my Division of Pragmatic Labor (see §3.3) predicts a tendency for the (a) forms to become restricted to conveying their NEG-V meanings, given the existence of the alternative (b) expressions specialized for conveying their potential NEG-Q meanings, especially in the light of the preference for scope to correlate with surface order. As we see in appendix 2 (cf. also Horn 1984b), the strength of this blocking effect varies inversely with the markedness of the alternative expressions. Since *not everybody* is morphologically and syntactically more marked than *nobody,* it will have a relatively weak restrictive effect on the use of (48a) to convey its potential NEG-Q meaning.

If my proposed explanation for the asymmetry is correct, NEG-Q readings will be available for those predicate denials which do not have a lexicalized paraphrase. One indication that this is true is the pattern indicated in (49)-(49′), paralleling that in (48)-(48′).

(49) a. Kim and Lee didn't come.	(49′) a. Kim or Lee didn't come.
b. Not both Kim and Lee came.	b. Neither Kim nor Lee came.

As observed in Horn 1972, with acknowledgments to Barbara Partee, negation can be interpreted as outside the scope of the conjoined subject in (49a), but must be inside the disjoined subject in (49′a). But only the latter example has an unambiguous unmarked alternative connective (49′b) which expresses its potential (and hence nonexistent) NEG-CONJ value. (As with its quantificational analogue (48′a), (49′a) can be read as a wide-scope metalinguistic negation, given the right intonation contour, discourse context, and rectification.)

Another case in point is provided by data I borrow here from Ladusaw (1979:81), who notes (but does not explain) the contrast between the ambiguity of the sentences in (50) and the (virtual) univocality of those in (50′).

(50) a. Both seminar rooms aren't in use at the same time.
b. One TA can't always grade all of the homework assigned.
c. Three bricks won't be sufficient to prop up the bookcase.

(50′) a. Many students don't take classes after 4 P.M.
b. A lot of wine wasn't consumed.

The NEG-Q readings of (50′a, b) tend to be blocked by the existence of lexicalized negative quantifiers with the same meaning (*few students, little wine*).[33] But that in (50a) (as in the parallel (48a) and (49a)) can only be paraphrased by a complex negated quantifier, *not both seminar rooms.* And the wide-scope readings for the negations in (50b, c) do not allow even semilexicalized equivalents. Both *not n* and *less than n* are available only to express ordinary scalar negation, and in precisely these cases the NEG-Q readings associated with the grouplike subjects of (50b, c) instantly disappear:[34]

(51) a. {One / A} TA couldn't make it to the meeting. (≠ none / not one could make it)
b. Three bricks weren't the right color for the wall. (≠ fewer than three were)

Support for my functional approach to the *every / some* scope asymmetry is not limited to English. One source of evidence is provided by an **SOV** language, Turkish, where predicate denial is marked by a verbal suffix and there are fully lexicalized **E**-valued quantifiers (*hiç* 'none', *hiçbir* 'not one'), but no complex negated quantifiers corresponding to *not every, not all.* The example in (52) is cited by Payne (1985:234),

(52) a. Herkez cevab-ı bil-iyor.
everybody answer-OBJ *know*-PRES
'Everybody knows the answer'
b. Herkez cevab-ı bil-m-iyor.
everybody answer-OBJ *know*-NEG-PRES
preferred: 'Not everybody knows the answer'
dispreferred: 'Nobody knows the answer'

who points out that the salience of the NEG-Q interpretation here is 'undoubtedly' influenced by 'the fact that for the [NEG-V] *everybody not* reading, there is a universal tendency to prefer a quantifier of the *none or not any* type', for example, *hiç / hiçbir.*

It is significant in this connection that in a large number of verb-final languages where there is no incorporation of negation into quantifiers, it is more the rule than the exception for a negative to the right of an existential or particular quantifier to be assigned (at least optionally) wide scope with respect to that quantifier (cf. Davison 1978). The literal translations of the sentences of (39b) into Indo-Aryan, Dravidian, and Turkish can be, and generally are, interpreted as NEG-Q (= $\sim\exists$ . . .).

In accord with my **O** --> **E** semantic drift (cf. §4.5), then, a negative / quantifier configuration which can in principle be read either as a particular negative (**O**-vertex) or a negated particular (**E**-vertex) value will prefer the latter interpretation—when there is no lexicalized **E**-valued quantifier to block it. In at least one non-verb-final language, only the NEG-Q (**E**-vertex) interpretation is possible for particular negation. Foley (1975:145) observes that the ambiguity of (53) in English (in which, it will be recalled, the wide-scope reading of negation is anything but salient) disappears in Palauan, where the literal translation of such sentences can only be assigned the NEG-Q interpretation in (53b).

(53) One of his friends didn't sing.
  a. NEG-V: Of his friends, only [*sic*] one didn't sing.
  b. NEG-Q: None of his friends sang.

In Japanese, a somewhat more complex situation prevails. Both universals and existentials are normally assigned wide scope with respect to negation, unless the topic-marker *-wa* is suffixed to the quantifier, in which case the NEG-Q readings (= $\sim\forall$, $\sim\exists$) prevail. The contrasts in (54) and (55) are from Kato (1985:105ff.); similar pairs are provided by McGloin (1976, 1982) (but cf. Ōta and Katō 1986:34ff. for a different account, in which the behavior of topic *-wa* is distinguished from that of 'focus' or 'contrastive' *-wa*).

(54) a. Zen'in ga repooto o das -anakat- ta.
  all SM *report* OM *hand in* -NEG- PAST
  'None of them handed in the report'
  (NEG-V, $\forall\sim$ reading)
  b. Zen'in wa repooto o das -anakat- ta.
  TM
  'Not all of them handed in the report'
  (NEG-Q, $\sim\forall$ reading)

(55) a. Gakusei ga juu-nin ko -nakat- ta.
  student *ten* *come* -NEG- PAST
  'Ten students didn't come'
  (NEG-V, $\exists\sim$ reading)
  b. Gakusei ga juu-nin wa ko -nakat- ta.
  'Not more than ten students came'
  (NEG-Q, $\sim\exists$ reading)[35]

Other factors are involved as well in determining the scope possibilities of negation and quantifiers in Japanese, as discussed by McGloin, Kato, and Ōta and Katō. The external or wide-scope reading of negation is forced by the metalinguistic negator *wake de wa nai* (see §6.6), while lexical properties of certain existential and universal operators prevent them from entering the scope of negation even in the presence of *wa* (compare the behavior of *each* and *several* in English, which induce wide scope even with respect to a preceding negative). Furthermore, either the wide- or narrow-scope reading may be rendered more salient in a given example by properties of the discourse context.

In English, as we have seen, some explicitly quantified subjects may be understood as falling within the scope of a predicate denial marked on the finite verb, while others demand wide scope. The NEG-Q interpretation, yielding the semantics of contradictory negation, may or may not be filtered out by a functional principle (perhaps partly conventionalized) which blocks crossing scopes at surface structure. The availability of an unmarked alternative realization for the wide-scope reading of a given predicate denial tends to result in the restriction of that predicate denial to the NEG-V reading assigned to it—but assigned how? We have no difficulty associating NEG-V interpretations with predicate term negations, where the syntactic position of the negative element (incorporated or not) prevents it from taking scope over the subject. But in the case of predicate denials, where do the NEG-V readings come from?

I have explored and tentatively rejected Montague's approach to quantifier-scope ambiguities, partly on the basis of its tendency to overgenerate logical forms for ambiguous—and unambiguous—sentences. But this embarrassment is not restricted to MG; the same proliferation afflicts other theories, from Generative Semantics to the (Revised) Extended Standard Theory, which posit a disambiguated level of logical form as the site for the resolution of scope ambiguities. As Cooper (1975, 1983) and Ladusaw (1979) have argued, the ambiguities I am trying to predict are intrinsically semantic, not syntactic in nature. If we assume that syntactic form underdetermines the projection of meanings from the constituents of a sentence, we can sort out the different readings we require without assuming distinct derivations for distinct meanings in a hierarchically structured representation, as in the disambiguated language of Montague (1974), the predicate-calculus-like conceptual structure of G. Lakoff (1969) and McCawley (1972), or the LF of May (1977).

Let us follow Cooper and Ladusaw in taking scope ambiguities to result from the optional storage of **NP** meanings during the bottom-up interpretation of a sentence. As in MG, the point at which the **NP** meaning is retrieved and quantified in determines the reading for a given sentence, but as distinct from MG (and from other GEOMETRIC theories of scope, to borrow Ladusaw's term), there are no syntactic correlates of this distinction. Cooper storage involves a principled and limited relaxation of the compositionality requirement built into Montagovian (and other neo-Fregean) theories of formal semantics, but the payoff is clear. We no longer need to burden our logical syntax with an otherwise unmotivated rule of quantifier-lowering or Q-Magic (Carden 1970, 1973; G. Lakoff 1969), quantifier raising (May 1977), or **NP**-lowering (Barwise and Cooper 1981) to license the narrow-scope reading of auxiliary-based negation. Subject **NPs** get narrow-scope (NEG-Q) readings in predicate denials when their meanings are not stored; object **NPs** (*I didn't eat some of the cake*) get wide-scope readings (across negation) when their meanings are stored.

Now it may seem that this approach merely replaces unmotivated syntactic rules by the equally unmotivated semantic storage convention. But there is an important difference. Each **NP**, that is, each generalized quantifier, will indeed generate at least two distinct formal interpretations. But, as Ladusaw puts it, a distinct interpretation is not necessarily a distinct READING. If a reading is taken to be an equivalence class of interpretations, then two interpretations with the same truth conditions and the same conventional implicata will determine the same reading. In (56a, b), the two logically distinct equivalence classes of interpretations will correspond to two distinct readings for the single constituent structure (and logical syntax) of each sentence,

(56) a. Everybody ate something.
b. Somebody ate everything.
c. Somebody ate something.

but the various interpretations of (56c)—as determined by the various storage and retrieval possibilities for the **NP** meanings—all reduce to the same reading, with a unique set of truth conditions and conventional implicata (Ladusaw 1979:67). Thus (56c), unlike (56a, b), is predicted to be semantically univocal, which in fact it is.

Just as Lewis's account of logical double negation (Lewis 1970, discussed in §5.1.3 above) permits differences in meaning to be neutralized into a single intension, so too does the Cooper-Ladusaw line on scope ambiguities allow the multiple interpretations assigned to a single syntactic form to become neutralized into a single reading, in which case neither syntactic nor semantic ambiguity will be attested.

Without offering a detailed defense of the meaning-storage approach to scope ambiguities here, I shall assume—with Gazdar, Klein, Pullum, and Sag (1985:15)—that 'quantifier ambiguities should be handled by some variant of Cooper storage'.[36] On this assumption, the sentences of (39) are all taken to be syntactic predicate denials which receive multiple interpretations in the semantics. Whether two projected interpretations result in the assignment of two distinct and accessible readings for a given sentence will depend on the quantifier involved and on the utterance context of the sentence token.

7.3.4 Scope and Representation, Semantics and Pragmatics: A Final Conspectus

The complex interaction of the scope of negation with structural and extra-structural aspects of language (including focus, intonation, grammatical relations, new vs. old information, and the semantic and lexical properties of other scope-bearing operators, particularly quantifiers and modals) has been barely touched on in this study. For general or English-based treatments from various perspectives within the overall generative paradigm, see G. Lakoff 1969; Jackendoff 1969, 1971, 1972; Smith 1970; Partee 1970; Chomsky 1970, 1971; Bald 1971; Horn 1972, 1978a; Carden 1973; Stockwell, Schachter, and Partee 1973; Sgall, Hajičová, and Benešová 1973; Lasnik 1975; Welte 1978; C. Ross 1978; Linebarger 1981; Culicover 1981; and Payne 1985; cf. also Kraak 1966 and Seuren 1967 for Dutch (and English); Bhatia 1977 and Davison 1978 for South Asian languages; Crockett 1977 and Babby 1980 for Russian; Varga 1980 for Hungarian; Heldner 1981 for French; Kuno 1980, McGloin 1982, and Kato 1985 for Japanese, and so on.

But one of my promissory notes must now be redeemed. Following Aristotle, I have maintained that ordinary negation, that is, predicate denial, takes scope over the subject-predicate connection. Earlier in this chapter, I observed that the negation in *x isn't well* must have scope over the subject in order to come out false when *x* names something which does not exist (*Socrates isn't well*) or something which exists but is not the sort of thing of which *well* can be predicated (*The number 4 isn't well*). Indeed, such sentences cannot be disambiguated syntactically or semantically unless we are prepared to forego the treatment of names and numbers as rigid designators and, at the same time, to project the same ambiguity in the positive counterparts of these sentences ({*Socrates*/*The number 4*} *is well*). Why is it, then, that subject terms (and the presuppositional phenomena they induce) seem so often to take wide scope with respect to an unmarked, syntactically internal negation?

I have considered and rejected the position that negations whose apparent scope is within that of the subject are not sentential, that is, that they realize term (**IV**, **VP**) negation rather than predicate denial. In the classical languages, as I noted in chapter 1, word order distinguishes predicate denial (*Socrates well not is*) from predicate term (**IV**-phrase) negation (*Socrates not well is*) in copular sentences, even when the latter is not morphologically incorporated. But even in predicate denials, as the textual evidence makes clear, the normal interpretation of singular predicate expressions assumes the existence of the referent of the subject term. In other languages, especially those with verb-final word order, term or constituent negation may not exist; predicate denial is expressed by an Aux- (or **INFL**-) based negative morpheme which may receive various apparent scope assignments depending on the formal, semantic, and discourse context (see the Davison, Kuno, and Payne references above).

In English, as I observed in chapter 3, the class of sentence negations picked out by Klima's syntactic criteria (in particular, his four diagnostic tags based on *either, neither,* negative appositives, and simple positive question tags) overlaps with, but does not reduce to, the class of semantic sentence negations—that is, contradictories—identified by the *it is not so that S* paraphrase criterion favored by Jackendoff and others. Membership in the former class is neither a necessary nor a sufficient condition for membership in the latter. By the same token, most, but not all, predicate denials result in contradictory negations, while some term negations—those yielding **E**-vertex predicates like *impossible, unable*—may also result in contradictory negations.

Reviewing the evidence, we recall that the following sentence-types, all of which I take to involve predicate denial,

(57)	a.	I don't think the Yankees will win.	(=⇒ I think they won't win)
	b.	Chris isn't happy.	(=⇒ Chris is fairly unhappy)
	c.	You mustn't buy that car.	(You must [not buy that car])

tend to convey a stronger negative than the mere contradictory of the corresponding positive,

(57′) a. I think the Yankees will win.
b. Chris is happy.
c. You must buy that car.

That is, we seem to have—at least pragmatically—an excluded middle in each case:

(57″) a. I have no opinion one way or the other about whether the Yankees will win.
b. Chris is neither happy nor not happy (she's just feeling sort of blaah).
c. You can buy that car or not buy it, as you choose.

I maintained in chapter 5 that (57a) is the semantic contradictory of (57′a), with the stronger (NR) understanding (excluding (57″a)) filled in by a short-circuited conversational implicature; (57b) conveys a contrary negation of (57′b) by virtue of a related (but non-short-circuited) strengthening rule applying to the class of unmarked positive scalar predicates which includes *happy*. In the case of (57c), the strong internal reading is arguably a semantic, rather than pragmatic, fact; the same interpretation is assigned to the particle negation version, *You must [not buy that car],* where we are dealing with a **VP** negation. Crucially, however, (57c)—like (57a, b)—is an instance of Klima's sentence negation, as the diagnostics show:

(57‴) a. I don't think the Yankees will win, not even if the Sox take bribes.
b. Chris isn't happy, and Kim {isn't either/ *is too}.
c. You mustn't buy that car, and {neither/ *so} must your sister. [cf. (40′a)]

What we must conclude here, then, is that if (as I claim) a negatively inflected auxiliary always constitutes a predicate denial, then predicate denials—like sentence negations à la Klima (cf. §3.3)—do not invariably represent sentence negations à la Kraak, Seuren, and Jackendoff, that is, sentences which allow the *it is not so that* **S** paraphrase, where **S** represents the positive counterpart as in (57′). The contradictory reading will be assigned by the semantics, however, unless overridden by a lexical property of the predicate; in the case of (57c), the neg-inflected modal *mustn't* (unlike *needn't, can't, doesn't,* etc.) is lexically associated with the contrary, inner-neg reading.[37]

If contradictory opposition is not a necessary condition for a negation to be a predicate denial, neither is it sufficient. Each of the sentences in (58) constitutes a contradictory negation of the corresponding affirmative in (58′), allowing the *it is not so that* paraphrase, yet none of them represents a predicate denial.

(58) a. It's impossible for a bachelor to be married.
b. He ate nothing.
c. No bachelors are honest.
d. Not everyone reads Aristotle.
e. Not many children like war.

(58′) a. It's possible for a bachelor to be married.
b. He ate something.
c. Some bachelors are honest.
d. Everyone reads Aristotle.
e. Many children like war.

Sentence (58a) clearly involves Aristotle's term negation, Jespersen's special negation, or Klima's constituent negation, as indicated by its form and its behavior with respect to the diagnostics (cf. Klima 1964:291–92).

While (58b) displays a mixed pattern with respect to the same diagnostics, the internal position of the negative within the **VP** once again suggests a PTN analysis. Nevertheless, it is clear that a negative in this position can be interpreted with wide scope, although it need not be; cf. Klima (1964: 285) on the ambiguity of *I will force you to marry no one*. The availability of wide-scope, contradictory readings for **VP**-internal negations is further discussed by Bolinger (1977) and Jackendoff (1969).

Sentences (58c–e) generally conform to the syntactic diagnostics for S-negation, leading Klima (1964:271ff.) to assimilate such sentences to this category. These sentences also constitute logical contradictories of the corresponding positive general statements in (58′c–e). Yet they are not predicate denials. Sentence (58c) represents a prima facie instance of narrow-scope negation—not PTN, but quantifier (subject term) negation. The negative quantifiers of (58d, e) must also involve constituent negation if *not everyone, not many children* are themselves constituents in these sentences.

My line on these examples is prefigured by Attal (1971: 108), who argues that the negation in (58e) must be thought of as negating the quantifier rather than the sentence, especially given that *not many* can be paraphrased by an inherently negative quantifier: *Few children like war.* For Attal, as for me, Klima and Jackendoff must both yield here to Jespersen (1917:42), for whom (58e) exemplifies special negation, where 'the negative notion . . . belong[s] logically to one definite idea', rather than nexal negation, in which negation belongs 'to the combination of two ideas', typically the subject-predicate 'nexus'.[38]

The Jespersen–Attal–Barwise and Cooper–ETL conclusion that negative quantifiers do not express sentential negation is reached from different directions by Lee (1974) and Keenan and Faltz (1978: 132), as I noted earlier in this chapter. For Payne (1985:201–5), too, sentences with negated quantifiers (*Not everyone, Not many* α) or inherently negative quantifiers (*No* α, *Few* α) in subject position, while they may yield contradictory oppositions and obey Klima's diagnostics for sentential negation, fail to exemplify STANDARD negation. More recently, Hoeksema (1987) reconsiders Jacobs's (1982) analysis of German sentences like *Nicht jede Schwester bewundert einen Arzt* 'Not every nurse admires a doctor'. Where Jacobs takes the initial negation to constitute a sentential operator, Hoeksema cites evidence from Jacobs's own study for a Barwise and Cooper–type analysis in which *nicht jede Schwester* forms a constituent, that is, a **mon** $\downarrow$ generalized quantifier.

Within ETL, then, (48a) and (48′a), along with the sentences of (39a, b), express predicate denials, whatever the semantic scope of negation may be. The sentences of (48b) and (48′b), on the other hand, along with

(58c–e), realize narrow-scope quantifier negation. But where did the Stagirite stand? In fact, for Aristotle—hugging the surface of Ancient Greek syntax and semantics—no predicate denial reading is possible for negation in general (quantified) statements. This point is emphasized by Geach (1970), whose 'program for syntax' offers a neo-Ajdukiewiczian categorial syntax for natural language similar in spirit to Montague's (minus the IL translations) but quite different in its details.

Geach points out that in the *De Interpretatione,* Aristotle allows a predicate expression (*rhēma*) like *petetai* 'flies' to combine either with a simple name (*onoma*) or with a general expression like *pās anthrōpos,* yielding the sentences in (59a, b), respectively.

(59) a. Petetai Sōkratēs. 'Socrates flies'
b. Pās anthrōpos petetai. 'Every man flies'

But negation (by which Geach intends contradictory negation) affects these two formations differently, since the contradictory of (59a) is (59′a), while that of (59b) is (59′b):

(59′) a. Ou petetai Sōkratēs. 'Socrates does not fly'
b. Ou pās anthrōpos petetai. 'Not every man flies'

And far from constituting a mere fact about the logical syntax of Greek, 'this is a profound insight, ignored by those who would lump together proper names and phrases like "every man" as Noun Phrases; we have two different syntactical categories' (Geach 1970:484).

What Geach seeks is a way to assign *ou* 'not' to the category :**ss** (corresponding to the **s/s** category of Ajdukiewicz [1935] or to Montague's **t/t**), that is, to 'the category of a sentence-forming operator upon sentences', while allowing *ou petetai* to be a constituent of (59′a) and *ou pās anthrōpos* a constituent of (59′b). Geach satisfies his quest by developing a means of treating negation as a sentence-level operator semantically which is realized as a predicate negation in singular sentences but as a quantifier negation in general sentences. But the fundamental point is thereby obscured: syntactically speaking, proper names and quantified phrases <u>are</u> both **NP**s, precisely as Geach's ignorant lumpers maintain. Nor is it clear that Aristotle would endorse the view that a (semantically) contradictory negation invariably represents the canonical denial of a given general predication; whatever may be the correct approach to (59′b)—Ancient Greek allowed no NEG-Q reading for universal negations—Geach's line does not extend naturally to the corresponding particular statements. (Aristotle's views on how to deny general statements are the subject of interesting, if inconclusive speculation, in Sommers 1970 and Englebretsen 1981a, 1981b.)

In the alternative approach developed by Barwise and Cooper (1981), which I have adopted here, negative quantifiers are negative quantifiers, and indeed all subject terms—'singular' or 'general'—are set-denoting **NPs**, that is, generalized quantifiers. A given combination of a negative quantifier + predicate expression may be logically equivalent to a predicate denial, so that predicating something (e.g., *flying*) of not every man amounts to denying it of every man. In the same way, the proposition that no man flies is not identical with, but is logically equivalent to, the proposition that it is not the case that some man flies. This line yields more natural results for the analysis of negation in ordinary language—Ancient Greek or modern English.

But what of negative sentences with singular subjects? In the classical Aristotelian framework, and in my neoclassical extension of it, the scope of predicate denial logically includes that of its singular subject. But, as I suggested earlier in this section, this fact may be disguised when the subject term can be pragmatically presupposed to denote an existent. If *Kepler did not die in misery* and *The present king of France is not bald* seem to presuppose (imply, implicate, suggest) that the name *Kepler* and the description *the present king of France* denote objects whose existence is immune to the scope of the negation, this may be seen as an attribute of surface subjects in English (and evidently in Ancient Greek and German). The connection between apparent presuppositionality and subjecthood (or, more properly, topichood) has been stressed in work ranging from Strawson 1964 to Reinhart 1981, and I shall not dwell on this topic here (see Horn 1986 for a review of the evidence). But if this approach is correct, the apparent location of subjects outside the semantic scope of sentence negation (predicate denial) is a pragmatically induced mirage.

A similar functional approach would seem apt for the cases of category mistakes I explored in §2.3: the normal state of affairs in which I would feel called upon to deny an instance of (60a) is one in which I would intend the denial, (60b), to convey the corresponding PTN, (60c).

(60) a. **x** is red.
 b. **x** is not red.
 c. **x** is not-red. (i.e., **x** exists ∧ [(**x** is blue) ∨ (**x** is green) ∨ (**x** is yellow) ∨ . . .])

If a statement is taken to be about its subject (à la Strawson and Reinhart), that subject will normally be taken to exist and to be within the domain of the logical predicate, that is, of what is said about it. But if **x** happens not to exist, or to be outside the domain of the predicate, the statement made in uttering (60b) will remain meaningful and in fact come out, as Aristotle maintained in the *Categories*, automatically true (cf. §1.1.1). When such a statement is uttered—*The number 2 is not red, you dolt, numbers have no* {*color* / *extension*}—it will normally be as a metalinguistic negation, with the appropriate fall-rise contour and rectification identified in chapter 6.

What needs to be emphasized is the nonsemantic nature of the correlation I am drawing here. The relevant notion of topic, as defined by the aboutness criterion, often coincides with the grammatical subject, but not always. As Strawson himself recognizes, not all subjects are topics. Strikingly, some surface subjects are assigned contrastive stress and function as the sentence focus rather than topic. Under these conditions, when the sentence is not about its subject, the latter typically will not be associated with an existential presupposition and will be understood as falling within the scope of an aux-based negation. Examples of this phenomenon include (61a–d), borrowed from Strawson (1964:96), Grice (1975:122), Gabbay and Moravcsik (1978:255), and a Tom Wicker op-ed column in the *New York Times* (28 March 1987), respectively:

(61) a. —What bald notables are there?
—The king of France {is / isn't} bald.
b. Jónes didn't pay the bill; Smíth paid it.
c. The cat is not on the mat; the dog is.
d. God did not allow nor Satan force Jim Bakker to indulge in sex outside his marriage. Mr. Bakker decided that for himself.

However we explain (or explain away) presuppositional phenomena, a predicate denial, or a sentence negation in its unmarked position, sometimes appears to take scope over the subject and sometimes (in fact usually) does not, depending on what can be inferred in a given context. This distinction—reflecting, on my account, a difference, not in logical, but in pragmatic, scope—may be reinforced by syntactic and morphological correlates.

In a provocative and unfortunately overlooked paper, Kuroda (1972) seeks to assimilate a morphosyntactic differentiation in Japanese to the dichotomy drawn by Brentano and Marty almost a century earlier between CATEGORICAL and THETIC judgments. For Brentano and Marty, Aristotle's categorical sentences involve a DOUBLE judgment (*Doppelurteil*) involving two separate acts: the recognition of a subject and the affirmation or denial of the predicate with respect to that subject. A thetic judgment, on the other hand, is a SIMPLE judgment (*einfache Urteil*), representing 'simply the recognition or rejection of material of a judgment' (Kuroda 1972:154). Among those sentences realizing thetic judgments figure existentials and impersonals (*Gott ist, es gibt gelbe Blumen, es regnet*), where there is no logical subject, or at least no subject whose existence is typically taken for granted. But at least for Marty, the sentences realizing thetic judgments are still crucially assumed to be of subject-predicate form; the two judgment types differ in their 'inner speech form', but this difference is neutralized at the level of syntax ('ordinary speech form').

Kuroda sees in this distinction a means for explicating the distribution of *-wa* and *-ga* markers on subjects in Japanese: the former is taken as a signal of categorical (double) judgments, the latter of thetic (simple) judgments. Thus, the sentences in (62) contain a logical subject about which something is predicated, while their counterparts in (63) lack logical subjects, representing unpartitioned predications.

(62)	a.	Inu wa hasitte iru.	'The dog is running'
	b.	Inu wa neko o oikakete iru.	'The dog is chasing a cat'
(63)	a.	Inu ga hasitte iru.	'{A dog/The dog} is running'
	b.	Inu ga neko o oikakete iru.	'{A dog/The dog} is chasing a cat'

While these sentences permit both options, generic statements can only be categorical judgments, and specific indefinites only thetic.[39] The categorical judgments in (62) respond to the explicit or implicit question 'What is the dog doing?', while the thetic judgments in (63) answer the question 'What is happening?'

Notice that the notions of subject and of categorical judgment appealed to here are not Aristotle's. In term logic, the statements under discussion here are all categorical, of subject-predicate form semantically as well as syntactically. And indeed, the relevant notion of subject here (as a means for characterizing what is present in (62) and absent in (63)) is not that of LOGICAL subject, whether this notion is explicated in terms of Montagovian function-argument structure (subject = the **T** with which an **IV** phrase combines to form a sentence, as in Dowty 1982), semantic or thematic roles (agency, etc.), initial-stratum grammatical relations (cf. Perlmutter and Rosen 1984), or otherwise.

Rather, the Brentano-Marty characterization of subjects in terms of an entity's role within the structure of the judgment is PSYCHOLOGICAL, largely identifiable (as Kuroda recognizes) with the notion of THEME or (SENTENCE-)TOPIC developed within the Prague school and related functionalist paradigms. The aboutness criterion for themes or topics defined within Praguean work (cf. Firbas 1964, 1966) maps onto the Strawsonian notion cited above (Strawson 1964); both strands of thought are reviewed in recent work by Reinhart (1981) and van Oosten (1986). In fact the very distinction of *wa*- vs. *ga*-marked subjects in Japanese is explicated by Kuno (1972), in work contemporaneous with Kuroda's, via the Praguean notion of theme; the *-ga* sentences of (63) are NEUTRAL DESCRIPTIONS in which the subject (*inu*) is nonthematic (cf. Firbas 1966).

But then the categorical / thetic dichotomy must be represented neither in the syntactic analysis of the sentence nor in the combinatory semantics per se, but rather in the pragmatic relation of form and content to context. If a given term phrase is singled out by the speaker to represent not just the (logical) subject of the predication but the theme or psychological subject as well—referring to an entity whose existence (in reality and/or in the discourse frame) is pragmatically presupposed or inferrable from what is presupposed (cf. Prince 1981)—then that term phrase will effectively be outside the scope of assertion, and of course, of negation (predicate denial) as well.[40]

A consequence of my functional / pragmatic line on the apparent immunity of subjects to the scope of negation is that in languages like Russian, where the dichotomy I have outlined correlates with case marking in negative sentences, these morphosyntactic patterns must be seen as interacting crucially with extrasyntactic—and indeed extrasemantic—parameters.

As detailed by Babby (1980), the subjects of (certain) intransitives and the objects of (certain) transitives receive genitive case marking in Russian when they are in the scope of negation. In ordinary negative declarative sentences, an intransitive verb agrees with its nominative-marked subject, whose referent is assumed or presupposed to exist. But in negative existentials and other negative intransitive clauses with nonspecific indefinite subjects or impersonal syntax, no such presupposition holds, the subject (if any) is genitive-marked, and the verb takes neutral or unmarked (third person neuter singular) agreement. Contrasts include the following (Babby 1980:5ff.):

(64) a. Sobaka bol'še ne pokazalas'.
dog-NOM F.SG. *again* NEG *appeared*-F.SG.
'The dog did not reappear'

b. No ni odnoj sobaki ne pokazalos'.
but not single dog-GEN NEG *appeared*-N.SG.
'But not a single dog appeared'

(65) a. Dokumenty ne obnaružilis'.
documents-NOM.PL. NEG *were found*-PL.
'The documents were not found'

b. Dokumentov ne obnaružilos'.
documents-GEN.PL. NEG *was found*-N.SG.
'No documents were found'
[= 'There weren't documents found']

The same distinction is realized in transitive sentences via accusative vs. genitive case marking:

(66) a. My ne obnaružvili dokumenty.
We NEG *found* *documents*-ACC.PL.
'We didn't find the documents'

b. My ne obnaružvili documentov.
We NEG *found* *documents*-GEN.PL.
'We didn't find (any) documents'

But a similar contrast obtains in affirmative intransitive clauses, only reflected in word order rather than case marking: if the subject term refers to an entity whose existence is presupposed, **SV** order normally obtains, but in existential sentences in which existence is asserted rather than presupposed, we find **VS** order.[41] Like the impersonal (non-agreement-governing) character of negative existentials, obligatory **VS** order in affirmative existentials is taken to represent a syntactic 'adjustment' to signal the marked context; both these sentence types represent themeless 'neutral descriptions' or 'thetic judgments', as in Kuno 1972 and Kuroda 1972, respectively (Babby 1980:69ff.).

Thus, both the 'genitive of negation' and verb-subject order will occur in a past-tense sentence when that sentence is taken as an implicit continuation of *What happened (next) was. . . .* In such a discourse context, no **NP** is singled out as theme and placed outside the functional scope of negation or assertion. Rather, the entire predication counts as the rheme. The four possibilities are distinguished schematically by Babby (1980:72) as follows:[42]

(67)

	AFFIRMATIVE	NEGATIVE	
Existentials:	$[_{\text{S of A}}$ **VP NP**$]$	$[_{\text{ne}}$ **VP** $\mathbf{NP_{gen}}]$	(THETIC JUDGMENTS)
Declaratives:	**NP** $[_{\text{S of A}}$ **VP**$]$	$\mathbf{NP_{nom}}$ $[_{\text{ne}}$ **VP**$]$	(CATEGORICAL JUDGMENTS)

(where $[_{\text{S of A}}\ \alpha]$ and $[_{\text{ne}}\ \alpha]$ indicate scope of assertion and of negation, respectively)

While word order is a clear diagnostic for whether an **NP** is inside or outside the scope of assertion in a positive intransitive clause, according to Babby, it is a less faithful guide in negative sentences, where $\mathbf{NP_{gen}}$ *ne* **VP** order will occur when the subject (whose genitive marking places it within the scope of negation) denotes an entity already familiar within the discourse context, albeit one asserted not to exist with respect to the frame of predication, as in (65b). Sentences of this type demonstrate the necessity of separating theme, in the sense of aboutness, from old or given information (cf. Allerton 1978; Reinhart 1981; Prince 1981; Gundel 1985; van Oosten 1986; and Horn 1986 on this distinction).

In other languages, however, word order may be the only guide (in the absence of the genitive-of-negation phenomenon) to the effective scope of negation. In verb-initial and verb-medial languages, thematic material will typically occur to the left of negation; all material to the right of negation potentially occurs within its scope, although stress and intonation may serve to signal a particular item as the focus of negation (cf. Jackendoff 1972). In Hungarian, for instance, when the negative (*nem*) is proposition-initial and no material is anaphorically destressed as given, the result is a neutral (thetic) sentence (Varga 1980). But when some constituent precedes *nem*, that constituent is an 'uncontrasted topic' outside the scope of negation. (Contrastive stress brings in additional complications, including the possibility of metalinguistic readings in the presence of a fall-rise contour; cf. Varga 1980:89–93 and §6.6 above.)

Even in relatively fixed **SVO** languages like English, it is no accident that an ordinary sentential negation—serving to deny a given predicate of a given subject—tends to surface between subject and predicate, thereby placing the subject outside the pragmatic scope of negation, as a default, in terms of left-to-right processing.[43] I have maintained that any strengthening of this default guess to the status of an inference is dependent on the discourse properties of the particular negative token.

The same nonaccident is responsible for the fact, observed by Praguean and other scholars, that the grammatical subject—which in the unmarked case also represents the topic or theme (if any), the agent (if any), and the embodiment of old or predictable information (if any)—almost always precedes not only negation but the object **NP** and, in the vast majority of languages (i.e., those of **SVO** or **SOV** type), the other major constituent(s) as well. In verb-final languages the negative marker tends to be assigned a fixed position within the sentence, typically as a morphological affix attached to the verb stem. But there may be other devices available for distinguishing negative thetic judgments in which the subject, whether simple or quantified, is interpreted as within the scope of negation, from the default case in which the subject is interpreted as topic and hence automatically outside the scope of negation (cf. McGloin 1982, Kato 1985, and Ōta and Katō 1986 on the interaction of topichood and negation in Japanese).

I am arguing, then, for a distinction between the semantic scope of predicate denial and the pragmatic scope of speaker denial. Given the correlation between subjecthood and topichood and the dictates of information structuring (the old-to-new, theme-to-rheme principle which Firbas, Kuno et al. term FUNCTIONAL SENTENCE PERSPECTIVE), a subject potentially within the semantic scope of a sentential negation is typically interpreted as thematic and hence outside its pragmatic scope.

The result is that predicate denial—negation as a mode of predication, a rule for combining subject and predicate, usually (but not always) resulting in contradictory opposition—tends in practice to be functionally assimilated to **IV** (verb-phrase) negation, the (relatively) narrow scope operator allowed for by Aristotle, Jackendoff, and extenders of Montague. When an apparently sentential negation is not taken as a metalinguistic operator, an objection to an earlier utterance, typically intended as a rectification of one subpart of that utterance, it often mimics (without actually reducing to) constituent negation. The position in which predicate denial frequently finds itself, flanking subject and predicate or located within the predicate expression, facilitates this mimicry.[44]

The distinction between the grammatical scope of negation and the pragmatic scope of speaker denial also comes into play in connection with constituent negation. The linguistic literature is replete with negations which seem to be sentential (i.e., to instantiate predicate denial) according to their syntactic position and Klima effects, yet are constituentlike in their apparent restriction to a subpart of the sentence in which they occur.[45] Some writers have in fact sought to assimilate such cases to the general phenomenon of constituent negation. Jackendoff (1972:254), for example, treats (68) as 'multiply ambiguous' as among (inter alia) the readings brought out by the paraphrases in (68′),

(68) Maxwell didn't kill the judge with a silver hammer.

(68′) a. Maxwell didn't . . . (i.e., someone else did)
b. He didn't kill the judge . . . (i.e., he did something else to him)
c. He didn't kill the judge . . . (i.e., he killed someone else)
d. He didn't kill him with a silver hammer (but with something else)

while Gabbay and Moravcsik (1978:253) view (69a–d) on their 'natural readings' as exemplifying adjective negation, **PP** negation, adverb negation, and **VP** negation, respectively.

(69) a. This is not a large house.
b. He did not make this hole with a drill.
c. He did not run quickly.
d. He is not hunting lions.

But a simpler and more consistent approach would view these negative sentences as realizing wide-scope predicate denial; the negative element takes semantic scope over the entire predication, but will typically focus on a particular element as determined by the prior discourse frame and signaled by the stress pattern. The negation will be understood as associated with that rhematic constituent which receives the intonation peak (cf. Chomsky 1970). This analysis is consistent with the verdict of the Klima diagnostics, signaling in (68) and (69a–c) a primary or unique reading of sentential and not constituent negation.[46]

Furthermore, as indicated by the intonational contour and rectification triggered by felicitous utterances of the sentences in (68′), Jackendoff's example instantiates the same phenomenon we observed earlier with respect to predications denying baldness of nonexistent kings or blueness of numbers: a given utterance may constitute a metalinguistic objection which serves to guarantee the truth of the corresponding predicate denial.

It is clear that I cannot endorse the outright dismissal of the notion of constituent negation, as exemplified in a classic *prise de position* from the era of generative semantics—'The notion of negating a sentence element is suspicious. . . . It is hard to understand exactly what could be meant by negating a noun, an adverb, or the like' (Bach 1968: 97)—or in the opinion rashly offered elsewhere without argumentation that 'it is hard to see how this notion [i.e., constituent negation] can be made semantically coherent' (Horn 1978a: 135). In the case of **VP**-, **AP**-, or **QP**-internal negation, constituent negation is perfectly coherent, as semanticists from Aristotle to Barwise and Cooper and to Gazdar, Pullum and Sag have demonstrated. Further, the alternative approach—as advocated by monoguist theories of negation from Generative Semantics to MG—results in a theory of grammar which countenances an unrestricted and unmotivated proliferation of clauses simply in order to satisfy the one negation–one clause requirement.

At the same time, I do reject the characterization of (68) and (69a–c) as involving constituent negation, in the absence of any syntactic evidence for that position. Rather than assuming that sentences containing **n** constituents and descriptive auxiliary-based negation are **n** + 1 ways ambiguous, we can view them as general and unambiguous instantiations of predicate denial. The alternative specific understandings (as to just why the truth conditions for the corresponding affirmation fail to be satisfied) can be attributed to the different possible pragmatic presuppositions compatible with that negative sentence—the set of propositions and propositional functions which are taken as established, under discussion, or part of the common ground, and thus not available for direct denial.

I have argued here for a rather complex interrelation of form and function, on the grounds that no simpler account provides a fully adequate treatment of the syntactic, semantic, and pragmatic facts I have surveyed. For English copular sentences in particular, I have drawn a three-way distinction in the grammar of descriptive negation and a related but not isomorphic three-way distinction in the semantics. This conclusion is schematically represented in (70):

(70)

inflected negation (*Kim isn't happy*)
particle negation (*Kim is not happy, Kim's not happy*)
→ PREDICATE DENIAL (sentence negation)
semantics: CONTRADICTORY NEGATION[47]
(± existential commitment)

particle negation (*Kim is not happy, Kim's not happy*)
→ **VP** ($\bar{\mathbf{V}}$) constituent (TERM) NEGATION
semantics: IMMEDIATE CONTRARY
(+ existential commitment)

incorporated negation (*Kim is unhappy*)
→ **AP** constituent (TERM) NEGATION
semantics: MEDIATE CONTRARY[48]
(+ existential commitment)

The mediate contrary reading available for inflected and particle negation (with unmarked scalar predicate terms like *happy*) I have taken to be a fact about pragmatic inference, rather than semantic interpretation.

We saw, in the first chapters of this study, that there is ample cause for skepticism toward the skepticism of psychologists and functionally oriented linguists over the utility of the notion of logical negation. I have maintained that while there is indeed an asymmetry between affirmation and negation in natural language, this asymmetry is to be explicated in pragmatic rather than in semantic terms; I proposed an account of the negative markedness implicature in terms of my dualistic model for nonlogical inference. This same model was later invoked to account for the ordinary reading of scalar negation, for the interpretations associated with double negation, for the pragmatic strengthening of negative statements, and for the asymmetry in the availability of wide-scope readings for the negation in *every . . . not* vs. *some . . . not.*

I have further tried to show (in the line of argument begun in chapter 6 and reprised here) that by isolating the deviant behavior of marked, metalinguistic negation, we can prepare a place for a well-behaved wide-scope descriptive negation within a classically bivalent semantics for natural language. But, as I have argued in this chapter, the appropriate place of this operator at the logical table is not the one reserved for the standard unary propositional connective of the Stoics and Fregeans.

Cross-linguistic evidence on where negation does and does not surface lends support to the Aristotelian syntax for contradictory negation as a mode of predication within term logic, rather than as an external operator on fully formed propositions; one candidate for the representation of syntactically internal but semantically contradictory negation is offered by the GPSG analysis reviewed at the beginning of this section. As we have seen in chapter 5 and again in this last section, the contradictory semantics generally associated with predicate denial may itself be pragmatically overridden within the context of utterance, licensing the interlocutors to build in a stronger (presuppositional) understanding. The design of ordinary language once again confirms Bosanquet's dictum: the force of the contrary is invested in the form of the contradictory.

Appendix I: On LEM and the Purported Incoherence of Multivalued Logic

The Geach "proof" of the incoherence of MVL cited in §2.4 runs as follows:

(i)	A thing must not be both not-**F** and not not-**F**.	by LC
(ii)	A thing must not be both neither **F** nor not-**F**.	from (i) by De Morgan
(iii)	∴ A thing must be either **F** or not-**F**.	from (ii) by LEM.

From (i), a purported instance of the Law of Contradiction, we are claimed to be led (even the multivalued among us) inexorably to (iii), a statement of the Law of Excluded Middle, which MVL is usually characterized as necessarily repudiating. But, as Geach himself warns us elsewhere in the same section ([1972] 1980:78), *caveat lector:* 'The sleight of hand is performed where he [the lector] is not looking'.

Without a formalization it is difficult to determine what the premises in Geach's argument are meant to say. In particular, we cannot tell which negative operator (of the two permitted in every explicit variety of MVL) is involved in the steps along the way. But (i) only counts as an instantiation of LC if it has the form $\Box-(\mathbf{p}\wedge-\mathbf{p})$, and it is relevant to MVL only if the expression *not-F* employed by Geach in (i)–(iii) is read as internal negation. If the negations are read as external, the "proof" goes through but the conclusion, $\Box(\mathbf{F}\alpha \vee -\mathbf{F}\alpha)$, is as uncontroversial for a proponent of MVL as for any classical, two-valued logician. Thus, let us assume that (i) is to be read as

(i′) $\Box-(\neg\mathbf{F}\alpha \wedge -\neg\mathbf{F}\alpha)$

We must still decide how to take (ii), and the remainder of the proof hinges on that decision. There would seem to be two possible versions to choose between:

Version I	Version II
(i′) $\Box-(\neg\mathbf{F}\alpha \wedge -\neg\mathbf{F}\alpha)$	(i′) $\Box-(\neg\mathbf{F}\alpha \wedge -\neg\mathbf{F}\alpha)$
(ii′) $\Box--(\mathbf{F}\alpha \wedge \neg\mathbf{F}\alpha)$	(ii″) $\Box(--\neg\mathbf{F}\alpha \vee ---\neg\mathbf{F}\alpha)$
(iii′) $\Box(\mathbf{F}\alpha \vee \neg\mathbf{F}\alpha)$	(iii″) $\Box(--\neg\mathbf{F}\alpha \wedge \neg\mathbf{F}\alpha)$

Version I conveys us from a premise, (i′), which properly instantiates LC to a conclusion, (iii′), which properly instantiates LEM. But the application of De Morgan's Law to yield (ii′) from (i′) is illegitimate, since the parenthesized material in (i′) does not have the form of a conjoined negation, $(\mathbf{Np} \wedge \mathbf{Nq})$, which it must if it is to be converted into a negated disjunction of the form $\mathbf{N}(\mathbf{p} \vee \mathbf{q})$. On the other hand, while the proof in Version II is valid as far as it goes, the conclusion it yields is not the desired (iii′)—*α must be either **F** or not-**F***—but rather (iii″), *α must be either not not-**F** or not-**F***. Now it is notorious that within MVL *not not-**F*** does not reduce to ***F***. Even in Aristotle's and Russell's two-valued dual-negation models, as Geach is surely aware, we cannot infer from *Socrates is not not-wise, 2 is not not-red,* or *The king of France is not not-bald* that Socrates, 2, and the king of France *are* wise, red, and bald, respectively.

Finally, there is one last possible formulation of Geach's "proof" that might be entertained:

VERSION III

(i″) $\Box-(-\mathbf{F}\alpha \wedge -\neg\mathbf{F}\alpha)$

(ii′) $\underline{\Box--(\mathbf{F}\alpha \wedge \neg\mathbf{F}\alpha)}$

(iii′) $\Box\,(\mathbf{F}\alpha \vee \neg\mathbf{F}\alpha)$

As in Version I, we get to the desired conclusion; as in Version II, the reasoning is correct. But rather than conveying us to the best of all possible worlds, Version III transpires in an irrelevant world, since (i″) is not an instance of LC. What (i″) excludes is the possibility of α being neither **F** nor not-**F**. Yet, as just noted, in the bivalent logics of Aristotle and Russell, as in any system of MVL, Socrates can be neither wise nor not-wise, 2 neither red nor not-red, the king of France neither bald nor not-bald. (Note that Version III also requires us to treat Geach as unsportingly taking 'α is not **F**' to abbreviate $\neg\mathbf{F}\alpha$ in one place and $-\mathbf{F}\alpha$ in another.) Thus, however we read Geach's argument, we do not get a valid demonstration that LEM (for internal negation, as symbolized in (iii′)) logically follows from a proper instance of LC.

Appendix 2: Inherent negation revisited

Let us reconsider the psycholinguistics of inherent negation touched on in chapter 3. We shall focus on the question emerging from the work of Clark and his research group (cf. Clark 1971, 1974) and of Fodor, Fodor, and Garrett (1975): why is inherent (covert) negation less difficult to process and evaluate than overt negation? One possible avenue to an explanation passes through the markedness implicature discussed in §3.3 and its partial conventionalization.

Consider the case of *dissuade*. G. Lakoff (1969) attempts to support a coderivation of (i) and (ii),

(i) I persuaded Bill not to date many girls.
(ii) I dissuaded Bill from dating many girls.

based on the interaction of the scope of the negative element, explicit in the former and implicit (within the *dis-* prefix) in the latter, with quantifier scope. (Lakoff's original examples had Bill as the would-be dat<u>ee</u> rather than dat<u>er</u>, so that the quantified NP would occupy the matrix obj$\bar{\text{e}}$ct position, for reasons irrelevant to our discussion.)

The transformation substituting *dissuade* for the lexical complex otherwise resulting in *persuade . . . not* has been attacked on various empirical and theoretical grounds by Chomsky (1971), Schachter (1972), Hust (1975), and others. I will be concerned here only with Lakoff's argument (1969: 134) that (i) and (ii) are synonymous and with Chomsky's counterargument (1971: 143) that (ii), unlike (i), 'presupposes some sort of intention on the part of the person being dissuaded'. As Binnick spells it out,

> [(ii)] can be uttered only by a speaker who believes that Bill had the intention of dating many girls and that if Bill had not been talked out of it, he might have done so. It cannot be uttered by someone to whom Bill had come requesting advice on his future dating plans, with no intentions one way or the other. But [(i)] could certainly be used in either situation. (Binnick 1976: 221)

As Chomsky and Hust acknowledge, whether this apparent nonsynonymy is a problem for a decompositional analysis of *dissuade* depends on whether optional transformations can change meaning and/or whether they can be conditioned by presuppositions and other semantic and pragmatic information. But given the fact that (i) is usable felicitously in contexts where (ii) is inappropriate, is it in fact the case that the two sentences are nonsynonymous? We might take them to be truth-conditionally equivalent, differing in that (ii) conventionally implicates (cf. §2.5) or presupposes something not implicated or presupposed as strongly by (i). The implicature in question is evidently the markedness implicature for negation discussed in §3.3. What has happened here, then, is that the incorporation of the negative element of *persuade not* as the inherent negation of *dissuade* reinforces and conventionalizes the implicature associated with that negative element.

The *dissuade/persuade not* alternation recalls the cases triggering my division of pragmatic labor, where the speaker's avoidance of an **R**-specialized simple, unmarked, or more lexicalized item (*kill, pink, enjoyable, went to jail*) licenses the addressee to draw a **Q**-based inference that the speaker used the periphrastic alternative (*cause to die, pale red, capable of being enjoyed, went to the jail*), because the former term could not have been used appropriately. Here, however, the distribution of (i) is unlimited by the existence of the semantically more restricted (ii). It is suggested in Horn 1978c that (ii), while shorter than (i), is not a less marked alternative to its unincorporated counterpart (in the sense that *kill* is less marked, as well as shorter than, *cause to die*)—*dissuade* is historically younger, learned later, syntactically more specialized (in governing the marked *from . . . ing* complementizer rather than simple infinitivals and in barring finite complements), and less frequent in text tokens than *persuade not.*

Dowty (1979:291–92) traces the asymmetry in "presupposition" (scare quotes Dowty's) between (i) and (ii) to the fact that *intend,* the verb figuring in the most plausible decomposition of both *persuade* and *dissuade,* is a neg-raising trigger (cf. Horn 1978b; §5.2 above). Given that, as Dowty observes, any change-of-state verb implicates (conventionally or conversationally) that the opposite state obtained earlier (before the reference time of the predication), the implicatum associated with *dissuade* will then be not (iii) but (iv),

(iii) NOT (*intend*(NOT (**x**, **P**(**x**))))
(iv) NOT (NOT (*intend*(**x**, **P**(**x**))))

where the two contiguous negations will automatically cancel each other out, yielding the desired implicatum for *dissuade* sentences like (ii), namely, that **x** had intended to **P** before being dissuaded from **P**-ing.

But when we consider such overt/inherent dyads as those in (v)–(viii) (from Horn 1978c:204),

(v) **a** discouraged **b** from **X**ing.	(v′) **a** encouraged **b** not to **X**.
(vi) **a** {prevented/kept} **b** from **X**ing.	(vi′) **a** caused **b** not to **X**; **a** caused it not to be possible for **b** to **X**.
(vii) **a** denied that **p**.	(vii′) **a** asserted that **not-p**.
(viii) **a** doubts that **p**.	(viii′) **a** believes that **not-p**.

we see that the asymmetry between *dissuade* and *persuade not* is in fact quite general, extending freely to other pairs not involving change-of-state predications (cf. (vii), (viii)) or neg-raising predicates (cf. (vi), (vii)). In each instance, the use of the unprimed predication—in which the presence of an incorporated negation, morphological or inherent, is borne out by the ability of these predicates to trigger negative polarity items—is limited to contexts where the contained affirmative proposition is already understood. Thus, ***a** denied (doubts) that **p*** is appropriate only when ***p*** is a proposition evoked in (i.e., appearing in or directly inferable from) the earlier discourse. Similarly, you wouldn't speak of keeping someone from committing suicide unless you presupposed that he had some specific intention of so doing (away). The use of the corresponding primed form with unincorporated negation (in (v′)–(viii′)) does not signal the absence of this markedness implicature; it merely implicates it more weakly and less conventionally.

The notion that the incorporation of an implicature-bearing element tends to induce the partial or complete conventionalization of that implicature is supported by other lexical patterns. In Horn 1972: §1.2, it is argued that the incorporation of cardinal numbers regularly conventionalizes the upper-bounding scalar implicature associated with unincorporated numbers. A three-sided figure (or triangle) is semantically a figure with exactly (not at least) three sides; a square may count among those figures with (at least) three sides, but it does not thereby count as (at least) a triangle. Nor does a triple (a hit for exactly three bases) count among a player's total of two-base hits (or doubles), or a piece composed for eight instruments among a composer's quartets. Similarly, the incorporation of *able* into a verb stem yielding an adjective *V-able* reinforces the **R**-based strengthening inference associated with the free form *able* (Horn 1978c: 200–201): an *enjoyable* movie or *lovable* urchin is not just one which is *able to be (capable of being)* enjoyed or loved, but one which is or has in fact been (or would inevitably be) enjoyed or loved.

How does this asymmetry—which I have attributed to the degree of conventionalization of the markedness implicature—relate to the asymmetry in the processing of overt vs. inherent negation? If the inherent negatives of (ii) and of (v)–(viii) can be used appropriately only when their supposition—the action being constrained or the proposition being rejected—has been explicitly established in (and is thus recoverable from) the discourse context, the addressee need not take the (processing) time to reconstruct that supposition. In the more general, less constrained, cases of unincorporated negation (those of (i) and (v′)–(viii′) above), the supposition must be built into the context by the addressee. The time needed for this reconstruction would then constitute the difference in the latency for overt vs. inherent negation observed in the psycholinguistic literature. Thus, in effect, overt negation is psychologically harder or slower because it is semantically and/or pragmatically <u>less</u> complex than its more specialized inherent counterpart.

Notes

Chapter One

1. Privation as the absence of what would be expected by nature to be present is also discussed in the *Metaphysics* (1022b23–1023a8), where Aristotle—noting that privation can involve accidental removal or deliberate 'taking away of something by force'—extends the domain to include a 'sense' in which a mole can be said to be deprived of sight (privation 'with respect to its genus') as well as a blind man (privation 'with respect to himself'). In the end, Aristotle concedes that there are as many senses of 'privation' as there are *a-* prefixed terms in Greek (1022b33).

2. Note the subtle distinction between this case and that of privative/positive oppositions such as blindness and sight. Can sickness not equally well be taken as the privative of health by Aristotle's naturalness criterion? Is it not equally 'natural' for an old man to be ill as for him to be toothless?

3. In the more succinct language of the *Metaphysics*, 'In some cases there is something between (there are men who are neither good nor bad), but in other cases there is nothing between (a number must be odd or even)' (*Met.* 1055b23). I shall consider mediate and immediate contraries in greater detail in §1.1.5 below.

4. The negative nature of even nameable intermediaries between contraries can be illustrated in a number of ways. A journalist describing the plight of 'coloureds' within the convoluted system of apartheid notes that 'it is a race that is defined, in South African law, only by negatives: neither black nor white' (*New York Times*, 4 April 1985).

5. At *Metaphysics* 1023a7, Aristotle explicitly collapses privative and contrary opposition in his discussion of nonexcluded middles: 'It is not true that every man is either good or bad [contraries], either just or unjust [positive/privative], but there is also an intermediate state'. Indeed, the domain of privation in this section has so broadened that we are left with the impression that a privative opposition for Aristotle (in the *Metaphysics*, if not the *Categories*) is simply a contrary opposition in which the inherently marked member is marked with *a-*: 'the primary contrariety is that of possession and privation' (*Met.* 1055a34).

6. Ackrill's more literal rendering of this definition has Aristotle declaring, 'I speak of statements as opposite when they affirm and deny the same thing of the same thing'.

7. One remnant of Aristotle's syntactic approach to contradictory opposition emerges in his (rather confused) discussion of indefinites. He regards (i) and (ii) as contradictories:

(i) (A) man is white.
(ii) (A) man is not white.

despite the fact that, as he acknowledges, they can be true together (17b29–30). As Ackrill (1963: 129–30) points out, not only does this violate Aristotle's semantic criteria for contradiction—indeed, for any opposition—but it is also striking that (i) and (ii), unlike clear cases of contradictories, do not affirm and deny the same thing of the same thing, since their subject phrases (which in Greek consist of the bare noun with indefinite understanding) are referentially, although not formally, distinct. Indeed, (i) and (ii) appear to be best treated as neither contraries nor contradictories but rather as SUBCONTRARIES (to be discussed below). One can but sigh, with Ackrill, 'it is a pity that Aristotle introduces indefinite statements at all'.

8. Saint Thomas Aquinas, in his commentary on the Square, makes the same point on the particulars that Ackrill makes (cf. n. 7) on the indefinites: 'The particular affirmative and particular negative do not have opposition properly speaking, because opposition is concerned with the same subject' (lesson 11 in Oesterle 1962: 90). *Some men are bald* and *Some men are not bald* are about different subjects, Saint Thomas observes, and are thus not opposed, *proprie loquendo*.

9. One-sided possibility—$\Diamond_1$ in the scale in (7) below—is so-called because it is lower-bounded only (excluding the impossible); two-sided possibility—$\Diamond_2$—has both lower and upper sides, given its definition as 'neither necessary nor impossible'. Cf. chapter 4 for additional discussion.

10. The ambiguity mentioned here is that of *possible*, not of *necessary*.

11. 'The propositions "it may be" and "it may not be" appear each to imply the other' (De Int. 21b35); cf. also Pr. An. 32a29ff.

12. This word-order distinction is neutralized in noncopular sentences, but Aristotle argues that there are in fact readings of '*not-recovers*', '*not-sits*', and so forth as in(de)finite verbs (cf. Ackrill 1963: 120–21 for discussion).

13. For the Latin-based Scholastics, *S non est P* (as in (11′a) is *NEGATIO NEGANS*, and *S est non P* (as in (11′b)) *NEGATIO INFINITANS*, whence the label INFINITE NEGATION for the latter. Henry (1972: 37) offers the labels PROPOSITIONAL VS. NOMINAL negation, but these appear to evoke a different distinction.

14. At *De Int.* 19b24, Aristotle notes that the negative or infinite term *not-P* often functions analogously to the corresponding privative; thus *X is not-just* corresponds more closely to *X is unjust* than to the simple denial *X is not just*, in the sense that each of the first two propositions unilaterally entails the third. Stones are not *just*, but they are not *not-just* (or *unjust*) either. Aristotle may be assuming, however, that *Man is*

unjust also unilaterally entails *Man is not-just,* since the latter, but not the former, will be true in case man falls into the intermediate category, that which is neither just nor unjust (cf. Ackrill 1963: 143–44 for discussion).

15. While neither of these pairs represents a contradictory opposition, which is Aristotle's point here, the propositions in (13) are *contraries* (they can be simultaneously false but not simultaneously true), while those in (13′) are *subcontraries* (they can be simultaneously true but not simultaneously false).

16. Note that *equal/unequal* do not constitute mediate contraries in the sense of allowing a nonexcluded middle; nevertheless, as Aristotle maintains elsewhere (*Met.* 1055a10), 'Every object is equal or not equal, but not everything is equal or unequal, but only in the case of a thing which is receptive of the equal', namely, numbers or quantities, not arbitrary objects.

17. As I observed earlier, a similar inconsistency arises in the passage at *De Int.* 21a25–27, where the inference from (i) to (ii) is rejected.

    (i) Homer is a poet.
    (ii) Homer is.

    Yet this inference must be valid if we are to accept the standard line in the *Categories* and *De Interpretatione* that singular statements like (2a)—and, presumably, (i)—entail the existence of their subjects and are thus false when their subjects are vacuous. This discrepancy has been noted and discussed by, inter alia, Thompson (1953: 254–55), Ackrill (1963: 110–11), and Dancy (1975: 153–55), but never fully resolved. As Dancy observes, however, the weight of Aristotle's work lies on the side of the *Categories* treatment, in which singular terms do have existential import (cf. also §1.1.3 below).

18. The Law of Contradiction would more properly be labeled the Law of Noncontradiction, as in the practice of Dancy (1975) and Lear (1980). I shall follow the less logical but more widespread custom and retain the label LC.

19. Lukasiewicz ([1910] 1971) notes that in addition to this 'ontological formulation' of LC, Aristotle provides, elsewhere in the same Book Γ of the *Metaphysics,* both a 'logical formulation' ('The most certain of all basic principles is that contradictory propositions are not true simultaneously'—*Met.* 1011b13–14) and a 'psychological formulation' ('No one can believe that the same thing can [at the same time] be and not be'—*Met.* 1005b23–24) of the same law.

20. One application of this principle appears at *Categories* 14a10–13: 'Since the fact that Socrates is ill is the contrary of the fact that Socrates is well and two contrary conditions cannot both obtain in one and the same individual at the same time, both these contraries could not exist at once: for if that Socrates was well was a fact, than that Socrates was ill could not possibly be one.' (cf. also the argument at *De Int.* 24b8–10)

21. That LC is the basic, indemonstrable 'first principle' is affirmed by the greatest latter-day exponent of Aristotelian logic, Leibniz; notice that in

these passages LC is taken essentially as a relational variant of another of Aristotle's *axiomata*, the Law of Identity: 'The first of the truths of reason is the principle of contradiction, or, what comes to the same thing, that of identity' (Leibniz, cited by Russell [1900] 1937). 'Nothing should be taken as first principles but experiences and the axiom of identity or (what is the same thing) contradiction, which is primitive, since otherwise there would be no difference between truth and falsehood, and all investigation would cease at once, if to say yes or no were a matter of indifference' (Leibniz 1916:13–14). For Leibniz, everybody—even 'barbarians'—must tacitly assume LC as part of 'innate knowledge' which is implicitly called upon at every moment (p. 77).

22. Curiously, among those who have recently been cited as rejecting, or at least challenging, LC are 'some writers on Quantum Mechanics' (Dancy 1975:5, 9, 23).

23. Aristotle's impassioned defense of the indemonstrability of LC (and his attempt, notwithstanding this, to demonstrate it) establishes the standard position on this point, as exemplified in the passages from Leibniz cited above. But some of the details of the argumentation in *Metaphysics* Γ can be—and have been—questioned (cf. Lukasiewicz [1910] 1971, Barnes 1969, and Dancy 1975 for critical evaluations of the controversy). Lukasiewicz, for one, has sought to show that LC does not have the primacy argued for it in the *Metaphysics;* its value, he submits, is at best not 'logical' but 'ethical', serving as 'a weapon against error and falsehood', and in particular as an exceptionally useful tool for a defendant seeking to establish his innocence in a criminal proceeding (Lukasiewicz [1910] 1971:508). Even this lukewarm endorsement of LC would be rejected by skeptics, neo-Sophists, and others who have chosen (out of perversity or error, the Stagirite would no doubt insist) to follow the beat of a different drummer. In §1.3 I shall look more closely at the arguments of these conscientious objectors to the Law of Contradiction.

24. Compare Royce (1917:267):

> Nothing is both **X** and **Not X**
> Everything is either **X** or **Not X**

Rescher and Geach make their recommendations (on the superiority of (16) over (15) or (15′)) in the course of their respective evaluations of the status of LC and LEM within modern multivalued logics, on the need for which Rescher is neutral and Geach unsympathetic, as we shall see in chapter 2. I shall argue in later chapters, especially chapters 6 and 7, that propositional negation is not only less basic and more sophisticated than predicate denial but it may in fact not exist at all in an adequate logical representation for natural language.

25. And allowing as well its own group of dissenters, including Kierkegaard (cited in Dancy 1975), who defends the position that Christ both is a man and is not a man (at the same time, and presumably in the same respect). Yet this proposition can only constitute the serious paradox Kierkegaard intends if LC is otherwise (on the nontranscendental

plane) valid. In the same way, Tertullian's motto, endorsed by Kierkegaard, *Credo quia absurdum est,* can be interpreted only if there are laws whose violation creates the logical absurdity to be embraced.

26. In fact, the basic Stoic disjunction is exclusive, although some later members of the school allow an inclusive 'paradisjunction'; cf. Lukasiewicz 1934: 72ff.; Mates 1953: 33, 51–52; and the discussion in §4.3 below.

27. At least in its Western incarnation; early formulations of LDN appear in Indian and Chinese logic, as we shall see in §1.3.1. The connection between the propositional nature of the Stoic system and their discovery of LDN hinges on a point observed by Wittgenstein: 'That one can negate a negated proposition shows that what is negated is already a proposition' (Wittgenstein [1922] 1961: *Tractatus* 4.0641).

28. The Latin *omnis A non est B* could only get what Carden (1970) and others have called a NEG-V reading, with the scope of the negation inside that of the universal; compare the ambiguity of the corresponding sentences in other languages (*All pleasure is not good; Chaque plaisir n'est pas bon*). I return to this construction in §4.3 below.

29. By ~**A** ↔ **O**, I mean that the (contradictory) negation of an **A**-form predication is equivalent to the corresponding **O**-form predication. In the case of quantified (rather than modal) expressions, alternative notions have been employed for such equivalence statements, including −SaP = SoP (Moody 1953) and ~xAy = xOy (Strawson 1952), where *SaP* is read 'All S is/are P', and *xAy* as 'All x is/are y'. I shall employ the simpler and more general notation which is undifferentiated as between quantificational and modal statements—and indeed, as we shall see in chapter 4, extends readily to parallel oppositions among epistemic and deontic propositions as well.

30. An existence-free interpretation of the Aristotelian forms is also advocated by Vandamme (1972: 51–57).

31. Any decision on identifying Aristotle's class of negative propositions must also be tentative, because of what Ackrill (1963: 120) labels 'Aristotle's failure to distinguish between grammatical and logical analysis'. In chapter 7, I shall seek to explain, if not justify, this 'failure'.

32. As he must, given the differences between Greek and English syntax. We cannot take **Caesar not is dead* to be the canonical surface realization of predicate denial, as Aristotle was able to do for its Greek equivalent.

33. As Axinn (1964) points out, Ayer's notion of specificity 'always presupposes a definite, perhaps even finite or denumerable, universe of discourse', and even when this criterion is met Ayer's criteria may give the wrong predictions: compare (i), (ii) with (iii), (iv):

(i) Today's date is before the 5th of the month.
(ii) Today's date is not before the 5th of the month.
(iii) Today's date is before the 25th of the month.
(iv) Today's date is not before the 25th of the month.

Axinn (p. 75) also notes the role of 'pragmatic orientation' in determining whether a given statement is (psychologically) positive or negative: *That man is bareheaded* may count as positive or negative depending on whether we're interested in hairstyles or hat styles, respectively.

34. Kissin points out that Ayer's example was poorly chosen, since (29a) will be false and (29b) true if Everest is not a mountain in the world.

35. Sanford (1968:96), however, argues that we may need to allow contradictories to count as contraries (and as subcontraries) to avoid logical inconsistency when necessary truths and falsehoods are plotted on the Square.

36. More exactly, each term may have only one polar contrary with respect to a given scale on which it figures. As Aristotle observes (*Topica* 106a10ff.), *sharp* must be recognized as homonymous on the basis of the fact that each sense (in Greek as in English) determines a different polar contrary: *sharp* vs. *flat* for a musical note, *sharp* vs. *dull* for a solid edge. (It should be acknowledged that the criteria for determining ambiguity offered in the *Topica* (book I.15) tend to strike a modern reader as leading to the multiplication of senses beyond a clear necessity).

37. As I have defined contrariety or incompatibility, both immediate (C_2) and polar (C_3) contraries also satisfy the definition of the broader notion (C_1). That is,

$$\mathbf{C_2(F, G) \rightarrow C_1(F, G)}$$
$$\mathbf{C_3(F, G) \rightarrow C_1(F, G)}$$

For completeness, simple (i.e., Cajetan's reductive) contrariety could be defined negatively:

$$\mathbf{C_4(F, G) =_{df} C_1(F, G) \wedge \sim C_2(F, G) \wedge \sim C_3(F, G)}$$

but I shall in any case not be employing this notion directly in anything that follows.

38. As I noted above, the standard medieval labels for the two negations are respectively *negatio negans* and *negatio infinitans*. The great thirteenth-century logician William of Sherwood defines the former as negation outside a genus and the latter as negation within a genus, noting that the latter is best understood in accordance with the principle that 'speakers sometimes confine their discourse to determinate matter' (Kretzmann 1968:95)—an early version of the modern concept of a restricted universe of discourse.

39. The view that predicate term negation is an artifact also appears in Collinson (1937:89): 'We can take refuge in a form dear to logicians, such as "non-tall" or "non-fact"'.

40. Frege, unlike Russell, did not choose to side with Aristotle on this point: for him, no assertion can be made when the subject is empty (cf. Frege 1892, discussed in §2.2 below).

41. If *un-x* and *non-x (not-x)* predicates cannot be assimilated to contradictory negation, and if we share Frege's and Madkour's prejudice (now

the received view) against the existence of contrary negation, we are left with the result that such predicates must be taken as atomic. This approach, unlike Frege's, correctly predicts that (35b) patterns with affirmative sentences (e.g., *The man is obscure*), but at the cost of writing off the relationship between (35a) and (35b). A third approach to predicate term negation is that of Von Wright (1959), to which I return in §2.4.

42. The Fregean view is also reflected in the generative semantics tradition (represented by Bach [1968] and McCawley [1972], inter alia) of reducing all negation to a sentential operator with the semantics of falsity; cf. Englebretsen 1981a:24 for discussion. On the other hand, the separate treatment of constituent negation by interpretive semanticists (e.g., Jackendoff 1969, 1972) can be seen to reflect the Aristotelian approach. I return to these points in chapter 7.

43. Eventually, as indicated in the tree in (34) above, McCall introduces separate operators for weak and strong contrariety.

44. One possible rejoinder by a McCallian might exploit the latent ambiguity in (37), noted by Geach himself elsewhere in the same book ([1972] 1980:11–12). As recognized first by William of Sherwood (cf. Kretzmann 1968), (37) has two possible analyses differing according to which NP—*every cat* or *every dog*—'gets into the proposition first'. (This technique of disambiguation directly prefigures the distinction drawn within Montague Grammar according to the order in which different quantifier phrases are 'quantified in'; cf. Montague 1974.) Let me, following Geach, adopt a bracketing notation to distinguish these two analyses:

(i) Every cat (detests every dog)
 [asserts of every cat that it detests every dog]
(ii) (Every cat detests) every dog
 [asserts of every dog that every cat detests it]

In the case of (37), this makes for a distinction without a truth-conditional difference, but in closely related examples of the kind in (iii):

(iii) Every cat detests every dog except Lassie.

as William observes, the two analyses determine different truth conditions. If we take this approach, contrariety would be defined not on a proposition per se, but on a proposition-cum-analysis-tree; the two candidates for the contrary of (37), (37′a, b), might then each be the strong contrary of a different analyzed proposition, that is, (i) and (ii), respectively.

45. See the detailed and lucid exposition by Gale (1976:6ff.), from whom much of the subsequent discussion is adapted (cf. also Gale 1972: 469ff. on the circularity of Otherness and Incompatibility analyses of negation); other problems with Otherness are discussed by Wood (1933) and Toms (1972).

46. The view of negation as the rejection of an assumed affirmation will be discussed in §1.2.2 and in chapter 3.

47. W. D. Ross (1923:29) cites the 'attempt to reduce the negative to the affirmative by saying that "A is not B" really means "A is not-B"' as one of the 'two mistakes' of post-Aristotelian logicians never committed by the Stagirite himself. (The other is the admission, by Avicenna, Hegel, and others, of an "infinite judgment" alongside the affirmative and negative judgments; for Aristotle, as we have seen, *A is not-B* is simply an affirmation involving a negative or indefinite predicate term.)

48. It is perhaps not surprising that the confusion of language and metalanguage stressed by Austin and Quine was especially prevalent in the antiformalist works of the Idealists; note that the passages cited above reflect the very confusion of use and mention, especially in the random use of quotation, that is the object of Quine's calumny. But, as we shall see, negation and falsity are also treated as interchangeable by a number of contemporary linguists and philosophers of language who are otherwise sensitive to subtle formal distinctions.

49. If every determination is negation, then—as Royce (1917:266) points out—Goethe's description picks out not just Mephistopheles, but anyone who asserts anything.

50. This accomplishment may have inspired his friend's insightful observation (Russell 1948:520), cited above, that the universe can be described without 'not'. Note, however, that Professor Bumblowski's affirmative substitutions (Russell 1954:31) do not preserve full synonymy. It is unclear to what extent this point was recognized by his friend and confidant.

51. Bergson's contrast between the formal symmetry and the practical or "real" asymmetry of affirmation and negation is echoed in virtually the same words by García (1975) and Givón (1979:111–12); I shall return to their views later in this section.

52. In this passage, *subject* is either a mistake for *speaker* or is intended in some nonlinguistic sense; obviously *Snow is not black* implies no propositional attitude on the part of its grammatical subject.

53. Apostel's reductionist program also involves defining negation in terms of incompatibility, where two propositions are incompatible if it is impossible to assert them simultaneously. But *impossible* must then be taken as a basic modal operator, which is not a particularly desirable result.

54. This multiplicity is partially illustrated in chapter 2 below; we might say, given the role of Frege (1892) in initiating this confusion, that the term "presupposition" covers a multitude of *Sinns*.

55. In fact, Bosanquet's remarks would apply only to the passage at Sigwart 1895:122; the more measured Sigwart of three pages earlier would have had no trouble endorsing Bosanquet's conclusions.

56. Other problems arise for this analysis. It is questionable whether Demos would be happy with the analysis of (i) into (ii),

(i) God will not provide.
(ii) **not** (God will provide) = a contrary of (God will provide) is true.

Yet such an analysis follows from Demos's endorsement of the Aristotelian line on nonreferring singular terms, resulting in the ascription of truth to (i) in an atheist's universe of discourse.

57. Sigwart (1895: 123) had earlier pointed out that some true negative identities are bona fide negations serving to 'prohibit a threatened confusion'; his examples included *Apes are not men, Red is not blue,* and *Freedom is not license.* Mabbott's examples in (53), and his analysis of them, are closer to Hegel's on *The rose is not an elephant* and *Understanding is not a table,* but then there seems to be a significant difference between the Sigwart identities and those of Hegel and Mabbott: the latter class involves a cross-categorial error, while Sigwart's simply involve a wrong choice within the proper category.

58. The label is due to Gale (1976: 12–15), who provides a critique of this approach, which—as we shall see—retains its adherents even today.

59. Givón's analysis of negation is also exacerbated by an unidentified, idiosyncratic notion of 'internal negation' (and of assertion) on which negation 'applies only to the predicate phrase, i.e. the assertion itself' (1979: 115), whereby the negation of (i) is claimed to be (ii),

(i) Someone loves Mary.
(ii) Someone doesn't love Mary. (p. 113)

contrary to the well-established tradition dating back at least to Aristotle (cf. §1.1.1).

60. Along the same lines, Jespersen (1917: 4–5) writes of 'the chief use of a negative sentence being to contradict and to point a contrast'. Note, however, that in describing the 'use' or 'chief use' of a negative statement or sentence, Jespersen, Strawson, and Givón allow for the possibility (explicitly realized by the first of these scholars) that negation may have a meaning, as distinct from its (chief) use, which cannot be reduced to that of contradiction, correction, or contrast (however *that* is to be represented). I return to this question in chapter 6.

61. A subtler version of this equation appears in Sweet's *New English Grammar,* where the following three versions of a negative sentence are contrasted:

(i) He's not a fool.
(ii) He is not a fool.
(iii) He isn't a fool.

The negation in (i) is associated with the following noun, that in (ii) with either the noun or the copula. But the contracted negative in (iii) 'logically modifies the whole sentence', rendering it 'equivalent to "I deny that he is a fool"' (Sweet 1900: 126). Cf. my account of such contrasts in §7.3.

62. The same point is made by Quine (1952: 1): 'To deny a statement is to affirm another statement, known as the negation or contradictory of the first'.

63. Ayer could have added that some negative descriptions may well be <u>more</u> informative than the corresponding affirmatives—compare *No-*

body came to my party with *Somebody came to my party*, or *John is not currently breathing* with *John is currently breathing*. I return to this point in §3.3.

64. This suggestion is essentially identical to the line I have taken on Aristotle's future contingents (Horn 1981a), where what is a truth-value gap on some interpretations of *De Interpretatione*, chapter 9 reanalyzed as an 'assertability gap' (cf. the 'Farabian' reading summarized in §2.1 below, and the discussion of truth and assertability in chapter 6).

65. Cf. Korzybski (1933), Hayakawa (1949), or any issue of *Etc*.

66. So called, appropriately enough, after the Indian parable of the blind men and the elephant; of course we (all linguists) are the blind men, while the object of linguistic study, in all its ineffable splendor, is the variably perceived elephant (as presented in Ross's unpublished contribution to the 1978 Milwaukee Conference of Current Approaches to Syntax, aka the Syntax Bakeoff). Actually, the theories in question may best be characterized as endorsing the slogan "I'm OK, you're OK, he's not so hot", where *he* refers to Aristotle, Russell, Chomsky, or some other prototypical binary, dichotomous, formalist thinker.

67. The reluctance expressed by other Neo-Nyāya logicians with regard to endorsing LDN is discussed by Ingalls (1951:68–72) and Staal (1962: 65–66). The latter attributes this reluctance to the recognition that two negatives (in Sanskrit and elsewhere) may be used to reinforce rather than annul each other (see Delbrück's *Ergängungsnegation*, which has been discussed extensively by Jespersen (1917) and others; cf. Horn 1978a: §3 for a modern appraisal). This argument is unsupported, however, and I see no reason to credit it; Ancient Greek, Polish, and French logicians do not seem to have been particularly prone to view LDN with suspicion on the basis of the surface syntax of their respective native language.

68. In the West this restriction can be traced to the Aristotle of *De Interpretatione*, for whom only those sentences which are declarative and indicative can be propositions, and hence capable of being true or false; nondeclaratives are not the subject of logic, but of rhetoric.

69. This 'functor' must also be assumed to be noncompositional; otherwise it is hard to see how the expressions $(\sim\mathbf{N})[\mathbf{F}(\mathbf{x})]$ and $\sim(\mathbf{N}[\mathbf{F}(\mathbf{x})])$, which are crucially distinct for Staal (1962:62), can be kept from falling together. But if $\sim\mathbf{N}$ . . . is not the negation of $\mathbf{N}$ . . . , what is it?

70. Kaufmann (1965:192) also argues that the Hegelian dialectic 'is not meant to flout the law of contradiction'.

71. Elsewhere Hegel (1892:180 (§96)) notes 'the double meaning of *aufheben* . . . (1) to clear away or annul: Thus, we say, a law or a regulation is set aside; (2) to keep or preserve: in which sense we use it when we say: something is well put by'. This ambiguity represents 'the speculative spirit of our language rising above the mere "Either-or" of understanding'.

72. Negation remains a vital organizing principle in contemporary (neo-)Marxian thought. Marcuse (1968) titles a collection of his critical

essays *Negations*, assigning contradiction and negation center stage in the history of ideas. He argues that 'thought in contradiction' in today's post-Holocaust (and possibly pre-holocaust) world 'must become more negative and more utopian in opposition to the status quo'. (Cf. also Adorno's (1973) *Negative Dialectics*.)

73. In his essay on the origin of language, Abel (1882:225) detects in the coexistence of antonymic (quasi-)homophones in the 'highly cultivated tongues' of Europe the remnant of an earlier 'primeval dulness'. This stage of linguistic development, simultaneously more sensuous and less intelligible or abstract than that of familiar languages, is epitomized by the 'rank luxuriance of Old Egyptian synonymy', which 'takes us back into the childhood of mankind' (p. 237). Freud simply pushes one step beyond Abel, perceiving in this alleged psycholinguistic echo of the history of our human race a more distant echo of the predualistic unconscious mind in the history of the individual.

Chapter Two

1. The depth of Lukasiewicz's feelings on behalf of three-valued logic can be sensed in his 1918 plea for political and intellectual freedom in Poland and the world. On behalf of 'man's creative activity', Lukasiewicz 'declared a spiritual war' against the 'logical coercion' built into the classical two-valued system. The three-valued alternative, with an intermediate value characterized as representing 'objective possibility', lights the way for 'the ideal struggle for the liberation of the human spirit' (Lukasiewicz 1970:84).

2. As Rescher points out elsewhere (1963:55), besides being unpalatable to an al-Fārābi or an Aquinas ('How can there be divine foreknowledge if future-contingent statements are neither true nor false?'), the Boethian reading is philosophically unfortunate. It is not particularly compelling to assign propositions like (i) and (ii) different truth values,

   (i) It will rain tomorrow. (asserted on 12 April)
   (ii) It did rain yesterday. (asserted on 14 April)

   given that the two statements, uttered when indicated, make 'precisely the same claim about the facts, viz., rain on April 13'.

3. I shall adopt Rescher's updating of Aristotle's term-logic representations for ease of exposition; the distinction between a proposition **p** and a predication **F(a)** is not relevant to my discussion here.

4. In the light of the parallels I have already observed between Western and Indian logic (§1.3), it is not surprising that the issue of future contingents has its Eastern incarnation as well. Śaṇkara (eighth century) seems to recall the Sea Battle in his commentary on *vikalpa* (option) with respect to the mutual compatibility of the two Vedic injunctions (i) and (ii) when the outcome is not predetermined and both possibilities remain alive.

   (i) At the atirātra sacrifice he takes the sodaşin cup.
   (ii) At the atirātra sacrifice he does not take the sodaşin cup.

Śaṇkara's view, according to Staal (1962:61–62) is that 'the law of contradiction need not hold in the realm of what is to be established (*sādhya*) but holds in the realm of what is established (*siddha*)'. This corresponds to the Boethian interpretation of Aristotle on the Sea Battle.

5. Frege can be reckoned as the father of modern presupposition in a more negative sense as well. Careful study of Frege 1892 and later works reveals that there are at least three different relations collapsed under the rubric of *Voraussetzung:* sentences may have presuppositions, uses of sentences (i.e., assertions) may involve presuppositions, and speakers may make presuppositions (cf. Atlas 1975; Levinson 1983:170). This pernicious ambiguity was to fester and metastasize seventy-five years later in the treatment of presupposition within the generative semantics tradition, where (as in Kiparsky and Kiparsky 1971) the value of *x* in the formula *x presupposes y* may be filled in by a sentence, a set of sentences, a proposition, a speech act, a speaker, an utterance, or a verb, and the value of *y* by a sentence, a proposition, or a truth value.

6. The opposition between AMBIGUIST and MONOGUIST theories alluded to here and in later chapters is a terminological borrowing from Wertheimér (1972), who applies these labels to rival theories on the semantics of modals.

7. Strawson's remarks are in fact addressed not to (15) but to the analogous sentence *The king of France is wise*. The respective fixations of Messrs. Russell, Reichenbach, and Strawson on the hairlessness, age, and wisdom of the nonexistent monarch will not be dwelt upon here. At least we have mercifully put behind us the morbidity of the earlier examples with their allusions to illness, death, and destruction.

8. For Ewing, commands, wishes, and exclamations are also meaningless by definition, since the meanings of sentences are propositions, and nondeclaratives 'do not assert propositions'. This conclusion is only as compelling as its premises, which recent work on the semantics of nondeclaratives (e.g., Lewis 1979) has shown to be dubious at best.

9. It might be objected that Pap's gloss here is too confining: one can deny that x is kind without affirming that x is unkind, even if x is of the appropriate type to be evaluated for kindness. In my earlier terminology, Pap is confusing mediate with immediate contraries.

10. This may be partly a terminological dispute. Routley (1969:375–76) argues that it is the semantic version of LEM (what I have defined as LBV), that is, 'every proposition is either true or false', which must be violated by sentences with truth-value gaps; the syntactic version assumed by Lukasiewicz, van Fraassen, and Lambert (i.e., LEM proper, 'for every proposition **p**, **p** ∨ ~**p**') is compatible with sentences that are neither true nor false, assuming the appropriate definition of ~**p** as contradictory negation. I shall return in the next section to the differentiation of LEM and LBV within nonclassical logics.

11. I shall for the moment take trivalent logics like Lukasiewicz's, in which the classical values **T** and **F** do not exhaust the values assignable to a

given statement, as notational variants of the Fregean and Strawsonian theories, on which only the two classical values are available, but some meaningful sentences on some occasions of utterance are assigned neither of these values. (Within the tradition of mathematical logic represented by Kleene (1952: 344), a sentence is either true, false, or undefined; this view of partially defined truth functions can be regarded as a formal means for capturing the Frege-Strawson intuitions about Kepler and the king of France.) A logic allowing truth-value gaps is in fact conceptually distinct from, although related to, a genuine trivalent (or multivalued) logic. The distinction between truth-value gaps and nonclassical truth values is crucial for formal systems like van Fraassen's, which accept the former but reject the latter.

12. Internal and external negation seem to get relabeled more often than any Parisian boulevard. Among the hardier varieties from this appellation spring are (respectively) PRIMARY VS. SECONDARY negation (Smiley 1960), WEAK VS. STRONG negation (after Kleene 1938, 1952), STRONG VS. WEAK negation (Von Wright 1959; Zimmer 1964; Keenan 1969), CHOICE VS. EXCLUSION negation (van Fraassen 1969), and CHOICE NEGATION VS. COMPLEMENTATION (Herzberger 1970); cf. (70) below for further available options. In this section I shall retain Bochvar's terminology, which captures the link between the scopal ambiguity of Russell 1905 and that of MVL, by exploiting the definition in (51). When I am dealing with a logical system (with or without truth-value gaps and/or nonclassical values) that distinguishes two varieties of negation along either scopal or lexical lines, I shall use '$\neg$**p**' and '$-$**p**' to represent the internal and external varieties, respectively, since neither operator corresponds precisely to the sole negation operator of standard formal post-Fregean logic, that is, $\sim$**p**.

13. Carlson does not refer to Strawson, Pap, or Kuroda, and neither Kuroda nor Carlson refers to Von Wright, despite the strong affinities between the later analyses and their predecessors. This phenomenon is nonprejudicially alluded to by the dictum that great minds think alike.

14. In the earlier paper in which Grice introduced his two varieties of implicature, he explicitly dissociated the non-truth-conditional conventional implicatum induced by *but* from the (semantic) presupposition he took to be associated with *Smith has left off beating his wife* and similar classical examples (Grice 1961: §3). At least for the purposes of his argument in that paper, Grice accepted presupposition failure (but not implicatum failure) as a sufficient condition for truth-value gaps, à la Strawson. The distinction between presupposition and conventional implicature is reexamined in Grice 1981; cf. also Levinson 1983 for discussion.

15. Cf. Fraser 1971, Horn 1971 and Stalnaker 1974 for some earlier considerations, Horn 1979 and Langendoen 1981 for critiques of the K&P line on *even* and related particles, and Levinson 1983 for a general challenge to K&P's move of collapsing the conventional implicature-bearing particles like *even, too, only,* and *but* with the classic instances of presuppositional phenomena—definite descriptions, factives, category mistakes, et al. A promising recent treatment of *even,* synthesiz-

ing insights from earlier accounts and proposing a well-defined scalar model for the evaluation of *even* sentences and related constructions, is offered by Kay (1987).

16. At least one speaker, Andy Rogers has informed me, has little or no problem in accepting these sentences.

17. The peculiar semantics of (90c) vitiates one of Langendoen's misgivings (1981:217) about K&P's *even* analysis; Langendoen apparently assumes, without argument, that (90c) is the negation (or at least a negation) of (90a), which it simply cannot be.

18. Reasons for skepticism include, inter alia: (1) the fact that by any sane criteria, *even* is not a predicate; (2) the received modern verdict that neg-raising, aka negative transportation, is not a syntactic (or even semantic) rule—cf. Horn 1978b, Horn and Bayer 1984, and chapter 5; and (3) the calisthenics required on the NR analysis to deal with sentences like (i) and (ii):

(i) Not all the boys even kissed the maiden.
(ii) None of the boys even kissed the maiden.

Noting that *even* takes wide scope with respect to the negative quantifier in (i) and (ii), Horn (1971:130–31) advocates, apparently with a straight face, that these surface sentences are to be derived from sources directly underlying (i′) and (ii′) respectively:

(i′) Some of the boys didn't even (even didn't) kiss the maiden.
(ii′) All the boys didn't even (even didn't) kiss the maiden. (NEG-V reading on *all . . . not*)

This gambit requires the incorporation of the predicate calculus laws of Quantifier Negation into the grammar, but perhaps it is not surprising that such a suggestion should emanate from a scholar on record elsewhere (Horn 1972: §3.1) as advocating the importation of De Morgan's Laws into English syntax. Such proposals have received the acclaim they no doubt deserved.

Chapter Three

1. These logical contraries are *not-wise* and *not-happy*, rather than *unwise* and *unhappy*, given that one can be neither wise nor unwise, and neither happy nor unhappy; cf. the discussion below and in §5.1.

2. The first allusion to the function of *signe zero* may well have been Vātsysāyana's sūtra (A.D. 300) on the marked and unmarked cloths (cf. discussion in §1.3.1 above).

3. As an affixal parallel to Greenberg's Vietnamese example, we might note that languages often mark stems for both positive and negative categories (*useful* vs. *useless*) or—more frequently—for the negative only (*please/displease*, *fair/unfair*, *possible/impossible*, *terminal/nonterminal*), but never just for the positive (cf. Zimmer 1964 and §5.1 below).

4. Indeed, in some instances negativity is marked only semantically, as with the 'affectives' or 'inherent negatives' of English, for example, *doubt* (vs. *believe*), *deny* (vs. *assert*), *lack* (vs. *have*), *reluctant* (vs. *eager*), *useless* (vs. *useful*), and the antonymic adjective pairs (e.g., *short/tall, bad/good*). Cf. Klima (1964) and Osgood and Richards (1973:386) for additional examples and discussion.

5. Polarity, especially negative polarity, has been the focus of many studies in recent years, ranging from Buyssens (1959) and Klima (1964) to Baker (1970), Borkin (1971), Fauconnier (1975a, 1975b), Horn (1978a: §2; 1978b), Ladusaw (1979, 1980), Linebarger (1981, 1987) and Horn and Bayer (1984; recapitulated in §5.3 below).

6. I shall argue in chapter 6 that a 'nonlogical', utterance-level use of negation does in fact need to be recognized, but the metalinguistic operator posited there is found in neither all nor only nondeclarative environments.

7. The sequence of negative categories would then parallel that in the acquisition of modals, where the more subjective root and deontic readings are acquired before the more objective epistemic values.

8. Fraiberg (1959:62–66) writes eloquently of the power and autonomy associated with the first use of the magic *no*, often uttered even as the child performs the very prohibited act his verbal or gestural negation rejects. For the toddler, this negation counts as 'a political gesture', indeed a 'declaration of independence'.

9. This is not the null hypothesis it seems to be; what is being claimed here is that the crucial distinction between negatives and positives is not a formal asymmetry of derivational structure or length, but an asymmetry of meaning.

10. Trabasso and his colleagues (e.g., in Trabasso, Rollins, and Shaughnessy 1971) independently developed analogous models for the processing of negation, RESPONSE CHANGE (Clark's true model) and (especially when the predicates are contradictories or immediate contraries) OPTIONAL RECODING (Clark's conversion model). Cf. Wason 1972:25 for discussion.

11. The emotive component of the positive/negative dichotomy has, of course, a much longer history than this; as I have noted, it can be taken back—and is, by Osgood and Richards (1973)—four thousand years to the codification of the *I Ching*.

12. Actually, while Wason determines that (17a) requires less reaction time, Herb Clark points out that this result is a property of a particular experimental design; in other studies, it is the less plausible but more obviously true (17b) which is easier to verify.

13. Wason's notion of plausible denial must be distinguished from the homonymous principle recently popularized in the testimony of the President's men on their commitment to the protection, and redefinition, of national security.

14. In one more instance of great minds thinking alike, Wason's circle study was set in print just as a short reply by Axinn (1964) to Ayer (1952) containing almost the identical example (albeit as a thought experiment) appeared in a philosophy journal. Axinn points out that Ayer's identification of the negative member of an opposed pair of statements with the one that is 'more specific' (cf. §1.1.4) must be qualified to allow for 'pragmatic choice': 'If the universe of discourse is ten red objects and a thousand others, "red" is more specific than "not red". But if the universe of discourse is ten red objects and one blue, "red" is less specific than "not red"' (Axinn 1964:75). In these terms, what Wason and his colleagues have shown is that as the context shifts to render a negative predicate more specific relative to its positive counterpart, it automatically renders a negation based on that predicate more natural and easier to process. The asymmetry of affirmation and negation remains, however, since a relatively nonspecific affirmative (*Circle 4 is red* in Wason's example) retains its naturalness and psycholinguistic simplicity.

15. It may be worth noting that at least some young children exhibit, in practice, a blissful ignorance of others' familiarity with these conditions. My family lore includes an incident in which my then three-year-old brother announced—out of a clear blue sky—'There are no marbles in my milk'. He appeared genuinely surprised when his assertion was not taken at face value but was further investigated (and disconfirmed) by an adult reasoning from the apparent violation of the plausibility requirement.

16. The context in which a sentence of the form *X is not all y* would be maximally implausible, while nevertheless true, would presumably be that in which x is entirely non-y, as in the evaluation of (20) against a circle which is entirely blue. This fact, and the general tendency to read *not all* here as 'almost but not quite all' (rather than as, say, 'almost not at all'), correlate with the scalar properties of quantifiers discussed in chapter 4.

17. Wason might have gone back further to Kant's verdict that 'the task peculiar to negative judgments is that of rejecting error' and to similar observations by other pre-Strawsonian asymmetricalists cited in §1.2.2.

18. Furthermore, as Eifermann (1961:266) points out, the two terms of this archetype of nongradable contradictory opposition may differ with respect to negatability. In Hebrew, *lo-zuġi*, literally 'not even', is a common expression for 'odd' in describing numbers, while *lo-pirdi* ('not odd') is an extremely unlikely periphrasis for 'even'. It would appear that *zuġi* 'even' is somehow unmarked with respect to *pirdi* 'odd', possibly (although this is not Eifermann's suggestion) because oddness for integers is naturally thought of as parasitic on evenness. That is, an even number is divisible by 2, and hence a member of the set {x: x = 2n for some integer n}, while an odd number is one not divisible by 2, a member of the set {x: x = 2n ± 1}. Anticlimactically, Eifermann finds no difference in processing time or error rates for the two expressions *lo-zuġi* and *lo-pirdi* in her study.

19. For an Aristotelian or a Russellian (cf. §2.2), (22) could also be true if there is no drink at all.

20. More accurately, this is what I implicate in a neutral context, one in which a stronger proposition, were I in a position to assert it, would have been relevant to the interests of the addressee; the operation of the maxim of quantity is constrained by the operation of the maxims of quality and relation, along with other nonconversational principles. (Cf. Grice 1975, Harnish 1976, Levinson 1983, Horn 1984b, and other recent works on pragmatics for discussion of maxim clash, which I shall touch on in §3.3 below.)

21. Both Wason and Clark at times recognize the need for this looser formulation of the conditions on how the positive counterpart of negation gets into the discourse model; cf. for example, Cornish and Wason 1970:113.

22. Compare these sequences:

 (i) A: If I were you, I wouldn't go there tonight. But then, I'm not you; you'll have to make up your own mind.
 (ii) A: If I were you, I wouldn't marry him.
 B: But you're not me.

 Neither A's continuation in (i) nor B's retort in (ii) contains a globally plausible denial, yet each negation is perfectly appropriate in the given context.

23. It should be clear that Clark's explicit denial is a linguistic category, one probably better labeled 'explicit negation'. Tottie's explicit denial, on the other hand, is a pragmatically defined category picking out those negatives with overt positive counterparts in the discourse frame.

24. Delbrück (1910:6) offers (i) as a Modern German instance of qualitative negation, and (ii) and (iii) as instances of quantitative negation.

(i) Meine Gäste sind nicht gekommen.	'My guests have not come'.
(ii) Keiner meine Gäste ist gekommen.	'None of my guests has come'.
(iii) Ich sehe nirgend einen meiner Gäste.	'Nowhere do I see one of my guests'.

25. The role of the diagnostics in these and other contexts will be reconsidered in §7.3, when I present my own version of sentential negation, corresponding more closely to Jespersen's nexal negation (and Aristotle's predicate denial) than to the Klima and Jackendoff models.

26. The phonological and phonetic reflexes of the two Zipfian forces are under investigation in an ongoing research project by Michael Studdert-Kennedy, Björn Lindblom, and Peter McNeilage. The characterization in this paragraph is adapted from an oral presentation by Studdert-Kennedy at Yale in the spring of 1987.

27. An earlier version of the Quantity maxim is implicitly bounded by **R** and Quality: 'One should not make a weaker statement than a stronger

one unless there is a good reason for so doing' (Grice 1961:132; emphasis mine; I return to the context of this formulation in chapter 4).

28. The key property involved here is informational strength as defined by unilateral entailment (in a given context); P_j may be stronger than P_i without there being a simple quantitative or pragmatic scale definable over the two values (cf. Hirschberg 1985 for discussion). Thus, we can think of *thumb* as informationally stronger than *finger,* in the sense that all thumbs are fingers but not vice versa. We will then correctly predict that a sentence like *I broke my finger* will (normally) **Q**-implicate that the finger I broke was not a thumb. (Thumbs, however, are fingers, since humans are reckoned as having ten fingers and not eight.)

29. In Leech 1983, this principle of population imbalance is dubbed the SUBMAXIM of negative informativeness, but it is never given in the directive form of a (sub)maxim, and it is unclear just why it should be elevated to this status. I should note in passing that Leech seems to endorse the proliferation of maxims to the point where any principle of language use seems to qualify as a (sub)maxim. Besides the Occamistic considerations militating against this approach, the development of an algorithm for the resolution of maxim clash under these conditions becomes formidable indeed, as pointed out in unpublished work by Steve Levinson.

30. I follow here the currently standard practice exemplified by Barnes (1969) and H. Clark (1974), in which *odd/even, healthy/sick, possible/impossible,* and so forth, are defined as contradictory pairs, rather than (following Aristotle) as immediate contraries. Here and in later chapters (especially chapter 5) I shall be more directly concerned with whether a given opposition between two terms allows an unexcluded middle than with those cases in which the terms in question simultaneously fail to hold of the subject, through a category mistake or the subject's nonexistence (see chapters 1 and 2).

31. As Aristotle has taught us, there is one sense in which (46a) is informationally weaker than (46b): it can be logically concluded from the latter, but not from the former, that we have a cat. It is not clear whether this asymmetry is relevant to the case at hand.

32. Or, alternatively, of trying to get the functional tail to wag the logical dog.

Chapter Four

1. Actually, this may not be quite so clear as is asserted here. As we shall see in chapter 5, there is a strong pragmatic tendency in natural language to take (6b) to convey that C_4, rather than C_3, obtains. I shall ignore this factor within the present discussion.

2. Significantly, for Apuleius, the first geometer of the Square, the **I** and **O** terms are not *subcontraria* but *subpares,* 'nearly [but not quite] equals'. Cf. Sullivan 1967:64–67 for discussion.

3. Hamilton evidently overlooked the parallel between his two *somes* and Aristotle's two *possibles*; in fact, he has little to say about the modalities.

4. Coincidentally, Grice's conversational implicata map directly onto the sous-entendus of Ducrot's parallel system (1972, 1973). Ducrot's analogue of MQ is the *Loi d'exhaustivité,* which 'exige que le locuteur donne, sur le thème dont il parle, les renseignments les plus forts qu'il possède, et qui sont susceptible d'intéresser le destinataire' (1972: 134).

5. That is, what I am objecting to is the institutionalization of the move from ~**K**(**p**) (speaker does not know for a fact that **p**, e.g., (15c), is true) to **K**~(**p**) (speaker knows that **p**, e.g., (15c), is not true). This practice, initiated by Gazdar (1979a) and adopted by Levinson (1983: 135–36), leads to unfortunate complications and loss of generality, as Levinson concedes, and seems in any case empirically unmotivated. (Cf. also Soames 1982 and Hirschberg 1985 for related discussion.)

6. I shall return to the Law of Double Negation and associated problems in §5.1.3. The morpho-syntactic consequences of linguistic double negation, particularly its effect on polarity phenomena and mood, are discussed in Baker 1970; Stockwell, Schachter, and Partee: 1973: 257ff.; and Horn 1978a: §3.1.

7. Curiously, Jespersen omits any statement of an equivalence between **B** and **B**~, although such an equivalence is clearly derivable within his system, since bilateral *some (possible)* is identical to its own inner negation, bilateral *some (possible) . . . not.*

8. Pott also develops a notion of Gradation or scale which prefigures both Wittgenstein's use of Gradation (1922: §4.464) to describe the scalar structure of the epistemic modalities and my own advocacy of the quantity scale to account for the logical and pragmatic principles relating the quantifiers and analogous notions (Horn 1972; cf. Ducrot 1973; Fauconnier 1975a, 1975b, 1976; Gazdar 1979a; Levinson 1983; Hirschberg 1985).

9. Likewise, in Cantonese (Yau 1980: 26–27), the flanking of a **B**-category modal (the epistemic possibility operator *wui* or the root ability operator *hɔji*) by two negations amounts to an assertion of the corresponding **A**-category modal:

(i) Kœy wui hœy. *he poss. go*	'It is possible that he will go'
(ii) Kœy m wui m hœy. NEG NEG	'He will definitely go' (lit. 'It is not possible that he won't go')
(iii) Kœy hɔji hœy. *he able go*	'He can go'
(iv) Kœy m hɔji m hœy.	'He must go' (lit. 'He can't not go')

Yau lists several other **B**-category predicates which can be flanked by negation to yield a strong modal (must, have to), while there seem to be no instances of the reverse process, instantiating the second equivalence in (26). This asymmetry—which is by no means restricted to Cantonese—is functionally motivated, as we shall see in chapter 5.

10. The complex interactions of modality, negation, and scope are explored in greater detail by Leech (1969); Seuren (1969); and Boyd and Thorne (1969), inter alia. Cf. also Leech 1969, 1974 for an informal account of the properties of duals (which he terms INVERSE OPPOSITES) in ordinary language.

11. Sapir's student and collaborator Collinson was later to offer his own treatment of positive and negative 'indication'. His point of departure (1937: 90–92) is similar to Sapir's: 'We say "*some (certain)* people would not like that" usually implying that there are others who would. . . . "*Not-every*" might, in theory, be expected to include total negation and mean "none or some though less than all". In practice, however, *not-every* means one or some but less than all'. Despite this appropriately pragmatic characterization of the relation between the subcontraries, Collinson follows Jespersen in taking the usual conveyed meaning of these operators to constitute their literal meaning. He thus (p. 92–93) offers us such (23i)-type equivalences as those in (i),

(i)		
	not everywhere	'in some places only'
	not always	'sometimes'
	not everyone	'someone', 'some people'
	not everything	'some thing(s)'
	not all	'some', 'at least one'

and a neo-Jespersenian chart (pp. 108–9) in which the tripartition of Totalizers (**A**), Indefinites (**B**), and Negatives (**C**) is extended to German, Latin, Romance, and even a number of International Auxiliary Languages; this last set of data is reproduced in (ii) for curiosity value:

(ii)	**Latino sino flexione**	**Occidental**	**Novial**	**Ido**	**Esperanto**
A	omni	omni	omni	omna	ĉiu
B	aliquo	alcun	kelki	ula	iu, kelkiu
C	nullo	nul	nuli	nula	neniu

12. By the same token, as Cicero notes (*Topica* 56–57), the Stoics tended to treat the inference from *not both* to *at least one* as logical rather than pragmatic; this led them to promote their questionable 'seventh mode' of inference, that is, (i), alongside their unarguably valid 'sixth mode', given in (ii).

(i)		(ii)	
	Not both this and that		Not both this and that
	Not this		This
	Therefore, that		Therefore, not that

See Mates (1953: 125) and Lukasiewicz (1934) for discussion.

13. Elsewhere, Quine (1951: 12) cites 'the prevalent use of the expressions "or both" [as in *p or q or both*] and "and/or"' as 'a presumption in favor of the exclusive interpretation, since otherwise these expressions would always be superfluous'. But the use of such forms is precisely what we should expect if the exclusive understanding is filled out from

the inclusive logical form via **Q**-based implicature. Furthermore, as Quine observes, the inclusive sense for natural language is itself supported by the nonsuperfluity of the addendum in *p or q but not both*. If the former (*or both*) rider is more 'prevalent' than the latter (*but not both*), this is due to the fact that the upper-bounding (exclusivity) of *or* results from a generalized conversational implicature, one which by definition (cf. Grice 1961, 1975) applies in a default or neutral context, where it is not explicitly canceled. (It may be worth noting that the 'prevalent use' of expressions like *some if not all* and *some or all* does not lead Quine to conclude (with Hamilton) that the standard representation for ordinary language *some* must be upper-bounded so as to exclude *all.*)

14. The pragmatic account of the two *or*s offered here (and in Horn 1972:97–99) is also prefigured in Joseph's observation (1916:187ff.) that the context ('the nature of the case') determines whether the two alternatives are mutually exclusive; when it does not, it is 'perhaps safer to assume exclusivity, unless the contrary is stated', as in the typical **Q**-implicature cancelation frames 'A or B or both', 'A and/or B' (see note 13).

15. Cf. Kraak (1966:177) and Seuren (1967:358) on the nonambiguity of Dutch *Alle jongens lopen niet* 'All the boys were [not walking]'.

16. In French, where the NEG-Q reading is likewise salient in the absence of a forcing context and where Sara Lee products are unknown, the NEG-V reading tends to be forced in a different example (as noted by François Latraverse): *Tout ce que tu fais ne vaut pas de la merde.*

17. The logically analogous **A**-category modals do not take a following negative interpretable as outside their scope. Sentences (i) and (ii) are unambiguously M-NEG (**A**~) rather than NEG-M (~**A**).

 (i) {Necessarily/Certainly} John isn't a bachelor.
 (ii) 2 + 2 {necessarily/certainly} are not 5.

 In these examples, unlike those involving quantifiers and binary connectives, the **A** operator is not part of the subject phrase; for Tobler—and for the analysis I defend in chapter 7—this makes all the difference.

18. Cf. Jespersen 1917:47 on the ambiguity of *I didn't go because I was afraid*. This two-faced construction has been the focus of much (relatively) recent attention; cf. inter alia G. Lakoff 1965; Kraak 1966; Seuren 1967; Stockwell, Schachter, and Partee 1973; Sgall, Hajičova, and Benešová 1973; Linebarger 1987.

19. This intuition is supported by the fact that fall-rise, though a sufficient condition for triggering the NEG-Q reading, is not a necessary condition, as a thought experiment on the sentences in (36′) bears out. Note further that it is harder to avoid the fall-rise contour when selecting the NEG-CONJ reading on *and* in examples like (39c), although even here it may be obviated by contrastive stress: *Leslie and Kim didn't come, just Leslie did.*

20. Hintikka's actual notation for epistemic possibility is **P**. I substitute POSS here, having preempted **P** for use as a predicate variable.

21. The difference between these two types of frames emerges especially clearly when we look, not at scales, but at a related class of partially ordered sets, those which (following Lehrer 1974:29 and Lyons 1977:289) I shall term RANKS. Unlike elements within a scale, ranked values are mutually incompatible. Thus, in such canonical ranks as those of (i)–(iii), discussed in Horn 1972 and Hirschberg 1985,

(i) «general, major, . . . , lieutenant, sergeant, corporal, private»
(ii) «felony, misdemeanor, tort»
(iii) «win, place, show»

it is not the case that a sergeant is (a fortiori) a corporal, nor is a felony a misdemeanor. The fact that the members of a rank are not related by entailment is reflected by the failure of the (50b)-type cancelation frames to apply:

(i′) He's not (#only) a corporal, he's a sergeant.
(ii′) #Smoking marijuana is a misdemeanor, in fact it's a felony.

But the suspenders of (50a) are possible here: I can state that all of my sons are at least corporals even if two are sergeants and one a lieutenant, and I can assert (with partial command of the facts) that smoking marijuana is at least a misdemeanor. Similarly, consider the rank of poker hands in (iv), on which we can superimpose a number of overlapping scales (inter alia those in (v)):

(iv) «Five of a kind [in wild-card games], royal flush, straight flush, four of a kind, full house, flush, straight, three of a kind, two pairs, one pair (two of a kind)»
(v) ⟨Five of a kind, four of a kind, three of a kind, two of a kind⟩
⟨royal flush, straight flush, flush⟩
⟨royal flush, straight flush, straight⟩

While a full house is not ipso facto a flush (#*Not only is it a flush, it's a full house*), it outranks a flush (cf. (iv)). If I deduce from your betting pattern that you have at least a flush, my subsequent discovery that you in fact have a full house does not invalidate this deduction.

22. A similar conception of an 'extended square of opposition' is utilized by McCall (1967a:123ff.), where (sub)contraries are similarly related by horizontals, and contradictories by diagonals. McCall uses his extended squares for plotting the relations of English sentences like those in (i):

(i) John adores Mary
John loves Mary
John does not hate Mary

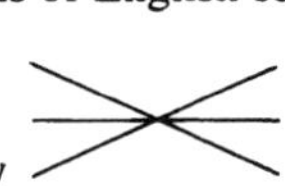

John hates Mary
John does not love Mary
John does not adore Mary

23. When we enter the subjective domain, the waters become murkier. A country/pop song of a few years back ('You Really Got a Hold on Me') begins with the line in (i).

(i) I don't like you, but I love you

In this context *love* (eros) and *like* (philia) cannot be situated on a scale and all bets are off; indeed, the singer—after a change of heart—might now (extrametrically) declare *I not only love you—I like you!* But it is less plausible that someone could love, say, a brand of pasta without (a fortiori) liking it. Similarly, one may be judged beautiful and yet (to a given perceiver) not attractive, but it is harder to imagine a day which is hot without being (at least) warm. (It should be pointed out that I am continuing to defer the examination of sentences like those in (ii) and (iii):

(ii) I don't (just) like you, I love you.
(iii) It's not warm out—it's (downright) hot!

to chapter 6, where I shall argue that they do not in fact counter-exemplify the scalar analysis.)

24. Hirschberg (1985) generalizes the set of scale- (and scalar-implicature-) inducing contexts further still, bringing set/subset, set/member, entity/attribute, and other types within the purview of her definition. Essentially, for Hirschberg, any partially ordered set (POSET) will induce a scale and support scalar implicature.

25. Notice that *lukewarm* (like its synonym *tepid*) appears to figure on the ⟨cold, cool⟩ rather than the ⟨hot, warm⟩ scale, at least for the figurative sense of the terms; cf. Horn 1972:48 for related discussion and Ducrot 1972:284–85 for an analogous treatment of French *tiède*. But when actual temperature distinctions are literally involved, the context (including the goals or desiderata of the interlocutors) plays a crucial role:

(i) My beer is { lukewarm, if not downright {warm/#cool}. / more lukewarm [warmer] than yours. }
(ii) My coffee is { lukewarm, if not downright {cold/#hot}. / more lukewarm [cooler] than yours. }

Notice also that these terms when used literally can refer only to the temperature of substances, not to the weather; it cannot be lukewarm or tepid out.

26. Ducrot (1972:274–75) sets out with a pragmatic treatment of the weak scalar (subcontrary) "ambiguities", for example, the two understandings of *quelques* 'some' (the unilateral *au moins quelques* and the bilateral *quelques seulement*). On this treatment, essentially identical to that traced above, *quelques* says *quelques au moins* (at least some), which is its representation in the linguistic component, but a simple assertion with *quelques*, given the *Loi d'exhaustivité*, licenses the inference that the speaker couldn't have used a stronger statement (e.g., one with *tous* 'all'). Thus the conveyed meaning of . . . *quelques* . . . provided by the rhetorical component (*le composant rhétorique*) is bilateral—'only some', 'some but not all'. But, Ducrot observes, a sentence like (i):

(i) Some students have read some of Chomsky's books; others have read all of them.

is interpreted as if *some of Chomsky's books* says or means 'not all'. Similarly,

(ii) Pierre a lu quelques livres de Chomsky. 'Pierre has read some of Chomsky's books'

may be taken to be false if Pierre has in fact read all of Chomsky's books. In the light of such data, Ducrot concludes that the output of the rhetorical component must somehow get back into the linguistic component in order to affect truth conditions. But since (ii) may also be judged true if Pierre has read all of Chomsky's books, Ducrot ends up endorsing the (Hamiltonian) position that *quelques*, and presumably all weak and intermediate scalar operators, are linguistically ambiguous, precisely the unparsimonious position the pragmatic analysis (equipped with the Maxim of Quantity or the *Loi d'exhaustivité*) was set up to avoid. I return to this issue in chapter 6.

27. Fauconnier's point has since been reinforced by Cornulier (1984) in his critique of the argumentation-based account of comparatives and equatives offered by Anscombre and Ducrot (1976, 1978, 1983). Curiously enough, the debate over the proper treatment of equatives like *John is as tall as Brian*—with Anscombre and Ducrot on one side and Fauconnier and Cornulier on the other—has been simultaneously raging in a parallel, trans-Atlantic universe as well: cf. Klein 1980; Sadock 1981; and especially Atlas 1984. Even the same examples (e.g., *John is as tall as Brian but Brian is not as tall as John*), with the same judgment differences, figure in the two debates, albeit in different languages. While essentially supporting the (Mill-Grice-Horn) minimalist or nonrestrictive thesis, Cornulier, like Fauconnier before him and Kempson (1986) after him, shows that whatever the ultimate account of scalar predication may be, the interaction of pragmatics and logical form is more complex (and more interesting) than envisaged in earlier neo-Gricean treatments (e.g., Horn 1972).

28. These contrasts reflect a general tendency to assign a special, often affective, ingredient to constituent negations (*no* . . .) as against the unmarked (and hence scalar) readings associated with ordinary *not* negation. Thus, Kruisinga (1931 : §1248) distinguishes the 'word modifier' *no Xer* from the 'sentence modifier' *not Xer* in frames like *The patient is {no/not} better today.* Minimal pairs demonstrating a similar asymmetry in nominal constructions can be found in the descriptive literature; compare *I'm {no/not a} spring chicken, She's {no/not an} angel.* (Cf. Bolinger (1977) and Welte (1978 : 165–85) for further illustration and insightful discussion of the notional contrast between *not (a)* and *no.*)

29. As Löbner (1985) points out, any two incompatible first-order predicates, like *hot/cold* or even nonscalars like *cat/dog,* will generate a version of the Square of Opposition, but it will always be a version which lacks duality and inner negation. His examples include (i):

(i)

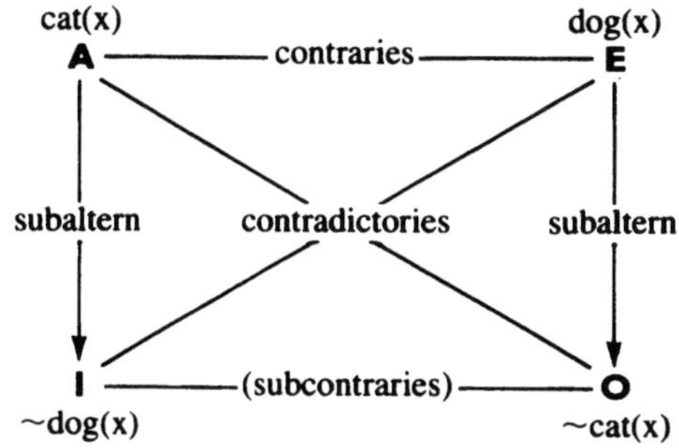

The **I** and **O** values—~*dog* and ~*cat*—are subcontraries, since something can be both a nondog and a noncat (e.g., a chipmunk), but nothing can be neither. Notice that the two contraries (**A** and **E** values) in this defective square do not differ in quality (positive-negative polarity) as the contraries in the standard (second-order operator) squares do; in fact the relevant placement of the **A** and **E** values here (the arranging of *cats* and *dogs*) is entirely arbitrary.

30. *But* conjunctions tend to be possible when—and only when—the two quantifiers or determiners are heterogeneous in monotonicity type, as substitution of *but* for *and* in the frames of (73)–(75) demonstrates. B & C note, as others have, that *but* is in various respects not a true conjunction of the class represented by *and* and *or;* among other differences, it does not iterate.

31. Actually, the **M**-class quantifiers of Horn 1969 do not map onto the entire **mon** ↑ set, but rather onto its 'persistent' subset; see Barwise and Cooper 1981 : 193 for an explication of persistence. (**M**- and **L**-class quantifiers are discussed further in Horn 1972 : §2.1.)

32. Compare also the nonmonotone quantifier *a single CN,* in which the upper-bounding implicature is not only strengthened but conventionalized; as Hirschberg (1985 : 35) observes, (i) can be truly (if misleadingly) uttered by a far-wider class of individuals than (ii) can:

(i) I have one leg.
(ii) I have a single leg.

33. If a speaker specifies that Kim earns $33,726.97 a year, she will be taken as implicating (if not asserting) the exact value; why would it be relevant to specify the value so exactly unless all those figures are literally significant? On the other hand, if the salary is given as $34,000, where there are only two significant figures, the upper-bounding implicature will be far weaker and more readily cancelable. The semantics and pragmatics of approximation are discussed further in Sadock 1977, Wachtel 1980, and Channell 1980.

34. As noted in Horn 1972 : §1.21, the context may reverse the direction of a scale even when no negation is overtly present; in the example cited there, a bowler can shoot *200 if not 210 (# if not 190),* while a golfer

can card a *70 if not 68 (#if not 72)*. Context plays an equally crucial role in the case of ordinals: a doctor may diagnose her patient as afflicted with *second- if not third- (#first-) degree* burns, whereas a prosecutor may portray the defendant as culpable of *second- if not first- (#third-) degree* murder. Cf. Hirschberg 1985: §5.1.4 for related discussion.

35. As argued in Horn 1972: §1.21 (cf. also Horn 1978c; Hirschberg 1985: 93–94), and noted in appendix 2, the lexical incorporation of cardinals tends to strengthen their upper-bounding implicatum into an aspect of conventional meaning. While squares have (at least) three (and in fact four) sides, they are not three-sided figures. Nor is a triple (i.e., a hit for exactly three bases) a two-base hit or double, although it may be listed among a batter's hits for (at least) two bases. But Wade Boggs might report the intuition that he's about to hit at least a double, and possibly a triple or home run: as with the military, legal, and poker nomenclature cited above, terms with lexically incorporated cardinals may in general be ranked, but not scaled.

36. It is clearly an oversimplification to suppose that a language gets only the lexical items it truly needs. Yet there is equally clearly a functional tendency in that direction, reflected not only in the story of **O** being related in this section, but also in the 'blocking' effect considered by Aronoff (1976), Kiparsky (1983), and others, wherein the meaning, use, or very existence of a given (potential) word is affected by the existence and range of a related (and more basic or specific) entry in the lexicon. Some ramifications of blocking are discussed in chapter 5.

37. In fact, while we do find examples like this in the older Germanic dialects—Coombs (1976: 176, 209) cites (i), also from Beowulf, and parallel examples, including (ii) and (iii), from the Old Icelandic Eddas:

(i) þæt ðǣr ǣnig mon / wordum nē worcum wǣre ne brǣce
that there any man words nor deeds were NEG *break*
'that no one there broke the treaty with words nor with deeds'

(ii) . . . mat þú villat né mannzcis gamar
food you want-NEG *nor of-anyone joys*
'you don't want food nor anyone's joys'

(iii) hlyra henni borcr né barr
shield-NEG *it bark nor foliage*
'bark nor foliage (do not) shield it'

it will be noticed that in each of these citations, the main verb itself is negated (with preverbal *ne* in Old English or the *-a(t)* enclitic in Old Icelandic; cf. §7.1 below). Similarly, the full context of Jespersen's (*sūð nē nōrð*) example reveals the main verb *nǣre* 'were not', which incorporates its own negation. Thus the entire conjoined phrase is commanded by, or is within the scope of, a sentence negation, rendering this construction less exotic than it may at first appear.

38. On the theories of negation investigated in chapter 3, of course, all instances of negation introduce such an implicature or suggestion, so that the difference between ***p** and not **q*** and ***p** rather than **q*** would seem

to collapse. But, if my account of negation in §3.3.1 is combined with Dieterich and Napoli's treatment of *rather than*, these expressions will still differ in the degree to which the 'markedness implicature' is conventionalized in the two cases.

39. A revealing case in point is Latin, which contained two forms combining a negative in the first syllable with a connective enclitic in the second. Crucially, both such forms, *neque* (lit., 'not-and') and *neve* (lit., 'not-or'), allowed only the 'and not' meaning. Forms equivalent to, and indeed cognate with, *neque* are also to be found in the early Germanic dialects: see the discussion of Gothic *nih* (< *ni* 'not' + *-uh* 'and') and OHG *noh* in Delbrück 1910 and Coombs 1976.

40. Greenbaum (1974) observes that Americans tend to prefer 'lexical' negation with *need* ({*doesn't*/*don't*} *need to*) over 'modal' negation (*need not, needn't*). Greenbaum further remarks on what he sees as an 'antipathy in American English toward the contracted form of the modal', excluding *mayn't, daren't,* and presumably *mightn't*. But this antipathy clearly does not extend to *can't, couldn't, shouldn't,* and (deontic) *mustn't,* a fact which Greenbaum's approach, ignoring the semantics and pragmatics of the modal notions, cannot accommodate. (I do not mean to suggest that an approach which <u>is</u> equipped with the proper semantics and pragmatics can sail home scot-free; the domain of modal contraction, like the wider domain of the modal auxiliaries, is notoriously rife with synchronic irregularity. It is this very irregularity which leads Zwicky and Pullum (1983) to argue that *-n't* represents an inflection rather than a negative enclitic.)

41. The existence of languages like Korean (Na 1981), in which the words for 'unnecessary' (*philyo-əp*) and 'uncertain' (*pul-hwaksilha*) are as lexicalized and as free as their **E** counterparts (*halsu-əp* 'unable', *pul-kaniŋha* 'impossible') does not counterexemplify this implicational universal.

42. Within Barwise and Cooper's logic for generalized quantifiers, the positive (**mon** ↑) values are in fact simpler on one level than their negative (**mon** ↓) counterparts, although this is merely stipulated. B&C predict that the response latencies for the quantifiers would conform to a schema like that in (i):

(i) **mon** ↑ < **mon** ↓ < nonmonotone

Verification tests involving the nonmonotone, nonscalar values (e.g., *exactly* **n**) involve the application of two different subroutines (one for confirming *at least **n*** and one for *at most **n***) and so should prove the most complex of all.

Chapter Five

1. While simple contraries (Cajetan's reductive contraries) will continue to count as contraries, this category will not figure crucially in the following discussion.

2. The prevalence of this metaphor, playing off what is, of course, Aristotle's favorite example of polar contraries, tells us much about our awareness of the nature of polarization, just as our tacit understanding of just which side (in a moral or ethical opposition) counts as white and which as black reveals much about our cultural heritage.

3. In the Hungary of the late 1960s, the liberalizer Janos Kadar transformed the party credo *Those who are not with us are against us* (attributed, however, to Lenin rather than Jesus, Donka Farkas informs me) into its inclusive counterpart *Those who are not against us are with us*. This still-productive pattern is invoked in Bishop Desmond Tutu's disjunction from a January 1986 speech at Stanford University, 'You are either for us or against us'.

4. The school of General Semantics (Korzybski 1933; Hayakawa 1949; cf. also Leech 1974:39–40) has been especially critical of the polarizing tendency, repeatedly excoriating the mindless endorsement of a two-valued orientation, which they trace back to Aristotle. Korzybski, Hayakawa, et al. argue passionately for a new 'non-Aristotelian' mode of thinking, a multivalued approach with its place reserved for the neither-nor—Sapir's zone of indifference. Yet Aristotle, as we have had ample occasion to observe, was himself a non-Aristotelian in the General Semanticists' sense, since he consistently defined predicate contrariety (the relation of *good* to *bad, black* to *white, tall* to *short*) in such a way as to explicitly allow for an unexcluded middle. The polarizing enemy may be real, but his identity has been mistaken; Jesus and Hitler, Brecht and Cleaver, are more appropriate villains than the Stagirite.

5. The more lexicalized and less productive *iN-* forms always have the potential to 'restrict the domain of *un-*': hence, we have no **unpossible* or **unactive* (Zimmer 1964:30); cf. OED (**un-**[1], 7), Jespersen (1942: §26.1), Marchand (1969), and—on the general phenomenon of BLOCKING OF PREEMPTION BY SYNONYMY—Bréal (1900:27), Aronoff (1976:43ff.), Clark and Clark (1979:§4.4) and Horn (1984b:25ff.). Aronoff's formulation of the blocking principle, as Broselaw (1977:60) observes, stipulates that 'no more than one item may be listed in a meaning slot of any particular stem'. Thus, *iN-* and *un-* doublets will be tolerated if they maintain or develop different meanings; we shall see examples of such doublets below.

6. Cf. also Marchand (1960:150–53) on *undisquieted, undishonored,* and so forth. As Zimmer recognizes, there remain some problematic cases involving e-pos derivatives of e-neg stems which involve no productive pattern. He attests a number of *-ous* derivatives (*unmalicious, unobnoxious, unvicious*) and assorted other examples (*uncorrupt, undegenerate, unguilty, unselfish, unsordid, unvulgar*). Zimmer points out, however, that an attested form is not necessarily a natural one. Standard dictionaries, after all, include such apparent simplex counterexamples to (7) as *uncruel, unignorant, unsick,* and *unstupid,* virtual nonce forms which must be taken with an untiny grain of salt. This is not to dismiss the existence of real counterexamples, that is, contradictory, e-pos or neu-

tral formations with *iN*- and unproductive *un*-, relatively rare though they may be: *unworthy* (from Jespersen), *unselfish* (from Zimmer), *impartial, inanimate, incredible, inexpensive* (less e-neg than *cheap*), *infrequent, unafraid, unapologetic,* and *unsusceptible* (from Funk 1971).

7. The contradictory reading assigned to *in-x-able, un-x-able* derivatives must have another source as well, however; note that fully lexicalized entries with this form (*impossible, incapable, unfeasible*) take contradictory prefixal negation, as do other adjectives of the same (roughly defined) semantic class (*impractical, unable, unapt*). I shall return to the proper characterization of this class in §5.2.

8. One attested violator of this constraint, involving the attachment of *un*- to a base containing not *dis*- but *in*-, is *uninfallible;* this is largely a term of art within Catholic theology.

9. Of course it does not follow that words containing unproductive affixes necessarily are unanalyzable by native speakers; the reinterpretation of *inflammable* as 'not flammable', with negative *in*-, instead of (or alongside) the etymologically "correct" 'capable of becoming inflamed', reveals (as does the occasional creative use) that speakers are fully aware of the negative character of the prefix. In the same way, a jocular innovation like *coolth* suggests that the nominalizer *-th* is analyzed (or analyzable) as a derivational formative, albeit a dead one. As Zimmer (1964:86) notes for the lexicalized prefixal negations of *untrue, unhappy, unkind,* 'We need a grammar which will analyze such morpheme combinations but will not generate them'.

10. This is not to deny the existence of cases, especially in the technical domain, where *un-X* and *non-X* forms do seem to constitute simple free variants; cf. *unprefixable/nonprefixable, unproductive/nonproductive, unvoiced/nonvoiced.*

11. In fact, *uneven* is attested for numbers, as is its former synonym *unequal.* The increasing oddity of *uneven* seems in fact to be correlated with the frequency of *odd* in the relevant sense, exactly as predicted by the theory of preemption by synonymy I am assuming here. (Nor is it odd that **unodd,* the parallel term in the paradigm, is entirely unattested; recall the discussion of *lo-zuǧi* vs. *lo-pirdi* in §3.2.) While *unequal* no longer bears this sense, its continued existence is an affront to the overly strong claim (by, e.g., Selkirk 1982:108) that *un*- never attaches to ungradables.

12. Of course phonologically transparent *iN*- adjectives may also drift into noncompositionality; cf. *indifferent, invaluable* (Jespersen 1917:145).

13. Allen (1978) actually situates *non*- within the grammar of compounds, but her position is supported primarily by the observation that *non*- attaches outside compounds, a property which (as Selkirk (1982) notes) is shared by *un*-. Marchand (1960) cites such forms as *un-cross-examined* and *un-self-conscious* (cf. (10a) above), and Selkirk adds *un-self-sufficient, un-germ-resistant,* and *un-laid-back* to the inventory. Al-

len's treatment of these forms as ungrammatical but acceptable is as unconvincing as such claims generally are (cf. §5.1.3 below).

14. These forms also result in what we might term unrecoverable haplology; for many speakers of American English, *unpacked* may be read as 'not unpacked', especially in the frame of (i),

    (i) I'm still unpacked.

    which is nevertheless not quite equivalent to the simple positive ((i) ≠ *I'm still packed*).

15. In one of his "On Language" columns (16 March 1986) for the *New York Times Magazine,* William Safire lambastes the 'linguistic evasion' and 'sheer laziness on the part of the categorizer' which leads to the (over-) use of exclusionary *non-*. His dire warning to mend our ways 'lest *yin* and *yang* become *yin* and *nonyin*' is misplaced, however: it is *yang*/*nonyang* that would ensue (see chapter 3).

16. It should also be noted that even prefixes which don't affect the category of their stem do affect subcategorial frames. Prefixes may allow or impose transitivity on an intransitive stem: one can *outthink* an adversary or *rethink* a problem, but one cannot *think* anything (except perhaps a thought). At the same time, it is generally true that 'addition of prefixes to verbs rules out non-nominal complements' (Carlson and Roeper 1980: 123; cf. Ross 1974): one can *think,* but not *outthink* or *rethink,* that it's going to rain. Under certain conditions, the prefixation of category-respecting negative *un-* also affects subcategorization. For many speakers, *It is uncertain that it will rain* and/or *It is uncertain to rain* are at best marginal, where the unprefixed equivalent *not certain* is impeccable. As Welte (1978: §4.2.1.5) shows in considerable detail, similar subcategorization differences obtain between verbal pairs like *disagree* and *don't agree, disbelieve* and *don't believe, discontinue* and *don't continue,* and so on: in each case, the negatively prefixed verb differs from its unprefixed counterpart (with or without particle negation) in its unwillingness to take sentential complements.

17. Zimmer also remarks on the failure of Russian *ne-* to combine with members of multiple-opposition sets; we have no adjectives like **ne-krasnyj* 'nonred'. In this respect, *ne-* is closer to *un-* than to *non-*.

18. Notice that these cases need not conform to Jespersen's rule for the weakening effect of double negation; if *X* is a weak scalar operator, *not . . . X . . . not* will convey a stronger affirmative than *X*. Typical examples of this pattern, as observed in §4.2, involve sequences of the form *not . . . can . . . not,* as one (or the only) way of expressing necessity (= *must*) in a given language.

19. Whereas two affirmatives never reduce to a single negative—to which one is tempted to respond, in the immortal riposte of Sidney Morgenbesser, 'Yeah, yeah'. Seriously, though, this difference in the cancelation properties of negation and affirmation has often been taken as a diagnostic for notional negativity in predicates and propositions: one who is bad at being bad ends up being good, but one who is good at being

good is not bad. Cf. Givón (1970) and Cruse (1980, 1986) for additional examples and discussion.

20. Aside from the question with which I am concerned here, it should be noted that *bad, evil, silly, dangerous,* and other e-neg adjectives which bar negative prefixes are not necessarily analyzable in terms of denoting 'the absence of something'; cf. Zimmer 1964 for elaboration.

21. Langendoen retains the ungrammatical-but-acceptable line for those L&B examples involving, not DNAAs, but negated unprefixed adjectives inside NPs. Bolinger (1980) noted that such examples improve dramatically with the intercalation of material between the *not* and the adjective: cf. **a not sad person* vs. *a not, shall we say, sad turn of events.* The efficacy of 'prosodic schmaltz' of this sort suggests to Langendoen that we are dealing here (unlike in the case of DNAAs under his reanalysis) with true ungrammaticality.

22. See Bolinger 1972: 115–25 ("Litotes and negation") for an insightful treatment of logical double negation in terms of litotes. Bolinger points out (p. 116) that logically speaking, 'the denial of the negative leaves the entire positive range open to whatever degree is appropriate', the context narrowing down the intended range actually intended by the speaker. Thus, to cite two of his examples, *I was not unaware of the problem* may convey that 'I was damn well aware of it', and *It was a not unkindly meant remark* [note the adverbial equivalent of a DNAA] suggests that the remark in question was not intended as particularly kind or unkind.

23. While I have been concentrating here on those double negations which focus on a single constituent or, in Jespersen's terms, 'refer to the same idea or word', some of the same remarks carry over to the use of flanking double negations ($\sim\Diamond\sim$ for $\Box$) cited in §4.2. Thus Yau (1980: 83), discussing the affective value of Cantonese sentences with two nonconsecutive negations, finds that this construction 'permet d'exprimer l'affirmative d'une manière tactique et prudente'. Indeed, when double negation flanks a **B**-category (**I**-vertex) verb or modal (like Cant. *hɔji, hɔ nang* 'be able to') so as to yield an **A**-category affirmative (*m hɔji m V* 'can't not, ought to, must'), the effect does seem to be that of cushioning the iron fist of necessity or obligation within the velvet glove of double negation. It is striking that the opposite practice of enclosing a strong, **A**-category modal by double negation to produce a **B**-category value (e.g., 'not obligatory not', 'needn't not' = 'permitted') is a far rarer and altogether less conventional practice. While the logical equivalences are parallel in the two cases (cf. (26) in chapter 4), the periphrasis is functionally motivated only in the former case; there is no reason to avoid the direct expression of permission or ability. (Cf. §4.2 above for more on nonconsecutive, or flanking, double negation, and §5.3 for further discussion of the attenuating effect of negation.)

24. Hintikka ignores those uses of the double negative prompted by a previous use (by the same or another speaker in or relevant to the discourse context) of a single negative:

That looks impossible.—No, it's not impossible.
Contrary to popular opinion (previous claims, etc.), this task is not impossible.

These examples represent one more instance in which a double negation is more appropriate within a specific, nonstereotypic context than the simple affirmative logically equivalent to it. As we saw in chapter 3, even such apparently pointless negative statements as *5 is not even* may occur felicitiously in place of the briefer and logically equivalent *5 is odd* when the context contains either the false proposition that 5 is even, or the open proposition that x is even.

25. The claim that two logically equivalent expressions which differ in logical form may count as nonsynonymous is also exploited in Atlas and Levinson's (1981) analysis of the cleft construction.

26. For extensive discussion and references, see Horn 1978b and Horn and Bayer 1984, both of which I shall draw freely on for the present account. As in the latter of these papers, the phrase NEG-RAISING PHENOMENON (NRP) will be employed expositorily and nonprejudicially, to designate the correlation in question, with no assumptions made about its ultimate treatment within linguistic theory.

27. Traditional linguists, from Tobler on (see below), have also tended to assume, without argument, that the NRP is 'illogical', while at the same time striving to find a general explanation for its existence. Here, for example, is Grevisse (1969:884; emphasis mine): 'Il arrive que par un déplacement curieux, des verbes tels que falloir, vouloir, devoir, aller, etc. prennent la négation qui logiquement porte sur la proposition ou l'infinitif qui les accompagne'.

28. Spitzer (1927:69) notes that the same *Zusammenfall* of the reduced forms Kalepky cites for *hoffen*—but not *espérer*—does obtain for a true neg-raiser in French: *je ne pense pas* and *je pense que non* can both bear the stronger, contrary interpretation (as, I might add, can their English equivalents, *I don't think so/I think not*).

29. Spitzer's *Gefühlslogik* is independently invoked by Le Bidois and Le Bidois ([1935] 1968:§985) in their defense of the *déplacement de la négation* affecting *vouloir* and *falloir:* 'A la défaut . . . de la logique rationelle et abstraite, c'est la logique autrement intime, autrement puissante du sentiment et de la vie qui est ici en jeu'. But while Le Bidois and Le Bidois do not share Spitzer's loony views on negative predispositions, they incorrectly take the NRP to be restricted to prohibitive contexts, that is, verbs which denote will and necessity and which govern the subjunctive.

30. Bolinger invokes the same metaphor in his early treatment of the substitution of *I don't think I want it* for *I think I don't want it,* in which 'a negative logically belonging to the subordinate verb passes to the preceding governing verb' (1957:94). But in labeling this process 'ABSORBED negation', Bolinger anticipates both Klima's (1964) account of 'negative absorption' and his own later view (Bolinger 1968:23–24)

that the negative in question belongs semantically to both subordinate and governing verbs, a claim to which I return in §5.3.

31. As seen in the contrast between the third and fourth versions, and in that between the last two, the first of these principles takes precedence over the second. Thus, a negatively inflected auxiliary *X-n't* will be weaker than the corresponding postauxiliary negation *X not*. This is borne out by the standard scalar diagnostics:

 (i) She isn't happy; in fact (I'd go so far as to say) she's not happy.
 (ii) #She's not happy; in fact (I'd go so far as to say) she isn't happy.

32. As we observed in earlier chapters, and shall see in more detail below, a similar observation could be made about the relation between the stronger (predicate-term or internal) reading vs. the weaker (predicate denial or external) reading of the negation in *The book is not white* or *The king of France is not bald*. Whether these examples involve true privative ambiguity is an open question, one to which I shall return in chapters 6 and 7.

33. Note also the distributional correlates of this fact (cf. (42) above and the detailed discussion in Horn 1978b): *Kim {doesn't believe / #didn't claim / #isn't certain} that Pat will arrive until midnight.*

34. The assertion of Lehmann (1980) alleging the nonexistence of the NRP in verb-final languages cannot be substantiated, given the data on Hindi (and related Indo-Aryan languages), Korean, Japanese, and Turkish cited in Horn 1978b and papers referred to therein; cf. also Wali 1972 and Bhatia 1977 on South Asian languages and McGloin 1976, 1982 on Japanese.

35. It is argued in Horn 1978b: 205–7 that certain apparent verbs of communication, including Greek *phēmi* (cited by Bosanquet ([1888] 1911: 319) to illustrate the NRP, in the passage reproduced above) and English *would say*, along with related predicates in Basque and Hebrew, actually constitute opinion predicates of Category (**a**), thereby qualifying as NR triggers.

36. This class may be more open than is generally recognized. When the Philadelphia 76ers' team doctor Michael Clancy was quoted as assuring the team and its fans that an injured player, Caldwell Jones, would be OK because 'He did not examine as though it was a fracture', he clearly suggests that Jones examined as though it wasn't a fracture. Apparently any verb **X**, including patient-subject predicates like *examine* or *read* (*This book doesn't read as though the author knew anything about formal semantics*), can occur in the frame *not **X** {as though / like} **Y*** and convey the lower-neg understanding ***X** {as though / like} not **Y***.

37. Escure (1974) argues unconvincingly that French allows NR over factives; the problem is that her examples of "factives" (*désirer, souhaiter*) aren't. Shnukal (1980) asserts that there can be no semantic criteria for neg-raisability, contra Horn (1975, 1978b), since 'even some factives, traditionally excluded from the list of Neg-Raising predicates (Horn

1975:287) may, on occasion, allow NR'. But her evidence for this claim is illustrated by examples like (i):

(i) I don't {see/know} that there would be any problem at all, would there?

But whatever else is going on here, the NRP is not directly involved, since—as noted by Cattell (1973:623–24) and Horn (1978b:154–55)—(i) is paraphrased not by (ii) but by something like (iii):

(ii) I {see/know} that there wouldn't any problem at all.
(iii) There won't be any problem at all {that I can see/as far as I know}.

And in any case, the verbs in question here, as in Escure's example, are not factive when they occur in the frame of (i), which clearly does not presuppose that there would (or will) be a problem.

38. Actually, all we are licensed to infer is that the negations are not sub-**VP**-level. We shall see in chapter 7 that there is reason to doubt that sentential negation exists as such within the syntax of natural language. In Aristotelian terms, what I am claiming here is that the negations in the sentences under consideration are predicate denials, not predicate term negations.

39. Modulo the blocking phenomenon cited in my discussion in §5.1.1 which precludes the formation of **ungood*, **untall*, **unrich*, and so forth. The failure of *sad* to block the formation of *unhappy* may be attributed to the fact, noticed by Lehrer and Lehrer (1982:20), that the former is not simply a stronger version of the latter: while one who is very unhappy could be described as sad, 'one could also be unhappy because one is frustrated, unsatisfied, disappointed, or angry'.

40. SCIs, as conventionalized conversational implicatures, are not to be confused with Grice's (or Karttunen and Peters's) conventional implicatures (cf. §2.5). Unlike conversational implicata (short-circuited or not), conventional implicata are—as their name betokens—part of an expression's conventional meaning, although they do not affect the truth conditions of the sentence in which that expression occurs. Like SCIs, conventional implicata are detachable; unlike SCIs—or indeed any conversational implicata—they are noncalculable and noncancelable. This does not mean that it is always obvious when a given implicatum ceases being an SCI and acquires the status of a conventional implicatum; these are precisely the issues which a pragmatic theory must help decide. (See Cole 1975 and Sadock 1978 for related discussion.)

41. The treatment of the NRP as an instance of short-circuited conversational implicature was suggested in 1980 in a paper by Samuel Bayer for a course of mine at Yale; the present version of this analysis is essentially a revision of Horn and Bayer 1984. The notion of short-circuited implicature is also invoked by Caplan (1978) to account for some of the material dealt with in this section involving partly conventionalized litotes.

42. Cf. Brown and Levinson 1978: §5 for an especially thorough and insightful cross-linguistic and theoretical study of the interaction of implicature, indirectness, and politeness, with a focus (in their §5.4) on negative politeness in particular.

43. The two effects are not only related but interactive. Sentence (99′b) in fact represents a doubly hedged version of the proposition being (indirectly) asserted here, that is, (i)

    (i) Your jumpsuit is {not entirely appropriate/rather inappropriate}.

    Thus the addressee is doubly blessed, at least conventionally, since the disagreeable proposition is phrased so as to incorporate both forms of indirection. The actual content of the assertion in such cases is signaled by tag questions, as R. Lakoff (1969) points out in connection with her notorious example (ii), where the tagged proposition must be (iii):

    (ii) I don't suppose the Yankees will win, will they?
    (iii) The Yankees will not win.

    (Cf. also Cattell 1973 on tag questions and the NRP.)

44. Unless it is forced by the presence of the relevant linguistic diagnostics. I have already touched on the role of strict NPIs as forcing the NR (lower-clause) understanding for higher-clause negation (cf. Horn 1978b for more details), the role of preverbal *please* as a signal of indirect requests, and of *but* (= 'but not') for universally quantified assertions. In the same way, the indirect assertion or parenthetical function of *I believe, I suppose,* and *I guess,* etc. (insightfully discussed in the analytic philosophy literature by Urmson [1952] and Wittgenstein [1953: 190]) can be signaled or forced by postponing this adverbial-like qualifier as in (i), or 'niching' it (cf. Ross 1973a) as in (ii):

    (i) Your answer is not wholly satisfactory, I believe.
    (ii) Your jumpsuit, I think, is not entirely appropriate.

45. Compare this (anti-LEM) verdict delivered by a *New Yorker* cartoon forewoman of the jury:

    'We find the defendant not guilty but not all that innocent, either'.

46. A clear minimal pair is provided by *anxious to,* which is e-pos and supports the process in (103), and *anxious about,* which is e-neg and blocks it:

(i) I'm not that anxious to see the movie.	[--› contrary (I'd rather <u>not</u> see it)]
(ii) I'm not that anxious about the exam.	[simple contradictory]

    While *anxious to* is not an NR trigger (à la *likely to*), its negation is typically strengthened to a contrary, whether it is intensified, as in (i), or not, as in (iii), courtesy of Edgar Rice Burroughs (*Tarzan's First Love*):

(iii) Tarzan knew that once the great bulls were aroused none of the jungle, not even Numa, the lion, was anxious to match fangs with them.

47. Even when neither negated value is unmarked, the irony seems to be somewhat deader when it is an e-pos evaluation that is negated. Both responses in (i) are possible, but more calculation is required to extract the ironic (litotic) interpretation in the latter case.

(i) A: How are you feeling?
B_1: Not great. [--› pretty bad]
B_2: Not lousy. [? --› pretty good]

48. Bolinger (1972:121ff.) sees this 'complimentary use' of *not half bad* (= 'more than half good') as nonquantitative: he observes that the *even* tacitly built into a standard example of *not half* (with its 'less than' interpretation characteristic of scalar negation; cf. chapter 4 above) is obligatorily absent here:

(i) The show isn't (even) half over yet.
(ii) As a pianist, he's not (*even) half bad.

Similarly, we might note that describing the wine as not half bad is quite different from saying that the bottle is not [= less than] half full (or not half empty).

49. As Ladd notes, his theory of intonational meaning correctly rules out a fall-rise contour with what I have termed extreme scalar values (i.e., those compatible with *absolutely*):

(i) A: How do you like my new color scheme?
B_1: #{ˇHorrible/Fanˇtastic} B_2: √{`Horrible/Fan`tastic}

The interaction of intonation contours with implicature will be considered again in chapter 6.

50. It should also be pointed out that the availability of the litotic reading of evaluative *not bad* depends on the choice of intensifer inserted after the negation, while no such variability obtains for its counterpart, *not good:*

(i) A: How are you feeling?
B: Not too good. [--› fairly bad]
Not too bad. [--› fairly good]
(ii) A: How are you feeling?
B: Not very good. [--› fairly bad]
(?) Not very bad. [-/-› fairly good]

Conventionalization of usage strikes again.

51. Other attenuations of the positive occurring as possible responses to a general *How's it goin'* question are *Can't complain* (slightly more upbeat than *Not so bad*) and *It could be worse* (cited as the appropriate response of a farmer interviewed after a tornado touchdown has leveled all his buildings and sent the boards of his barn flying into the next county) (Mohr 1987:7–8).

Chapter Six

1. Thus the purported ambiguity of (3)—and that of (1)—cannot be demonstrated by an application of the ellipsis (identity-of-sense) transformations which block crossed readings in the case of non-privative ambiguities:

(i) Tracy left a deposit at the bank, and so did Lee.
(ii) I saw her duck, {and you did too/but you didn't}.

As Zwicky and Sadock (1975) note, the identification of crossed readings in cases of privative opposition is problematic, since the more inclusive understanding will always be available, for example, in (iii) and (iv):

(iii) Fido is a dog, and so is Queenie.
(iv) The king of France is not bald, and neither is the queen of England.

Thus, pace Kempson (1975:99–100), the acceptability of (iv)—where both conjuncts permit the more inclusive external understanding—has no bearing either way on the purported ambiguity of negation. By the ambiguity criterion classical employed by philosophers, according to which a statement is ambiguous if it can be simultaneously true and false relative to the same possible world, context, or state of affairs (cf. Quine 1960:27 on the ambiguity of *light;* see also Martin 1982 and Kempson 1986), (3) comes out ambiguous, since it can be either true or false if I bought a bitch. Sentence (1) seems to be ambiguous by the same test: in the world of 1905 or today, Russell's classic example may well be simultaneously true and—depending on one's semantic persuasion—either false or neither true nor false (i.e., in any case, not true). The a priori objection to an ambiguous negation operator thus comes to rest largely on the metatheoretical desideratum of parsimony, rather than on the nature of ambiguity in natural language. (Cf. Horn 1984a, c for a general defense of privative ambiguity, and Atlas 1974, 1977, 1979, 1980, 1981 for a wholesale rejection of the ambiguity of natural language negation.)

2. Sources for data include Andrew 1940 for Old English, Banjọ 1974 for Yoruba, Marilyn Vihman (pers. com.) for Estonian, and Yau 1980 for Cantonese.

3. Linebarger (1981) also explores the failure of external negation to trigger NPIs. Following Kroch (1974), she attributes this failure to the intervention of an abstract TRUE predicate immediately within the scope of negation, blocking the relation between the NPI and its potential trigger. This approach to external negation will be evaluated in §6.5 below.

4. This comment applies primarily to the seminal work on presupposition in Strawson 1950, 1952. As Dahl (1981:197) points out, Strawson later (1964:95) acknowledges that the negation of a sentence guilty of reference failure may well be true (contra the misrepresentation of his

position in Wilson 1975:19). Dahl's neo-Strawsonian account of negation in PPCDs (purported presupposition-canceling discourses) is in fact consistent with the treatment of metalinguistic negation I shall undertake here. Cf. also Burton-Roberts (1987) for a neo-Strawsonian theory of presupposition which explicitly invokes my notion of metalinguistic negation to account for such PPCDs.

5. Atlas (1977, 1978, 1979, 1980, 1981) differs from his fellows in the monoguist cohort in rejecting the view that negation is invariably an external, truth-functional operator, a view which he himself endorsed in Atlas 1974. I shall return to his current position in §6.5 below.

6. The former example may be hard to process with wide scope on the negation, as desired. There is a strong tendency to take auxiliary negation as inside the scope of a preceding indefinite (**B**-category quantifier), a tendency I shall investigate further in chapter 7. But (13a) is read with wide-scope negation if it immediately follows the assertion that some men are chauvinists.

7. While the response in (15) is possible, a more natural correction would be that in (i), as suggested to me by Benoît de Cornulier (pers. com.).

(i) Ce n'est pas que j'ai 'coo-pay luh vee-and', c'est que j'ai <u>coupé la viande</u>.

I shall return below to the correlation between cleft syntax and metalinguistic negation.

8. For the view from the other side, compare this plaintive comment from Jim Garrett, during his brief tenure as football coach at Columbia University, on his then 0–8 team:

(i) I really believe we don't lose. We just don't win.

Notice that Coach Garrett was not alluding to ties.

9. Quantity-based implicata are also generated by tautologies (e.g., *war is war, boys will be boys*), which can be interpreted as informative only on the assumption that the speaker is obeying the Cooperative Principle and in particular the maxim of Quantity (cf. Grice 1975; Levinson 1983:110–11). Of course, just which informative proposition will be taken as having been conveyed by a given tautology will depend on contextual factors, some of which are clearly conventionalized: compare the respective implicata of *When it's over, it's over* and *It ain't over till it's over.* These implicata too are susceptible to metalinguistic rejection, as is clear from the following scenario, due to Quaker State Motor Oil:

(i) A: What brand of motor oil do you use?
B[starting car engine]: Motor oil is motor oil.
[Smoke belches out of B's exhaust.]
Voice-over: Motor oil is definitely <u>not</u> motor oil.

10. The presumed trigger of (24a), *They had a baby and got married,* might be taken to be semantically ambiguous between atemporal (symmetric or logical, *and also*) and temporal (*and then*) readings. Evidence against this position includes the following:

1. On the two- *and* theory, conjunction in virtually every language would be described as ambiguous in just the same way as in English.
2. No natural language contains a single conjunction ambiguous between 'and also' and 'and earlier' readings: no language could be just like English except for containing a conjunction *shmand* such that *They had a baby shmand got married* would be interpreted either atemporally or as 'They had a baby and, before that, they had gotten married'.
3. The same ambiguity exhibited by *and* arises in paratactic constructions in which two clauses describing related events are juxtaposed without any overt connective (*They had a baby. They got married.*).

Grice's alternative position, which I take to be correct, is that conjunctions are semantically univocal, but may conversationally implicate (through the exploitation of the Manner submaxim 'Be orderly') that the events occurred in the order in which they were described. The impossibility of *shmand* on this account is pegged to the absence of any maxim enjoining the speaker to 'Be disorderly'. Notice that, as with scalar implicature, the asymmetric implicatum may be canceled or suspended: *They had a baby and got married, but not necessarily in that order.* (Grice's line on asymmetric conjunction is supported by Wilson (1975), Schmerling (1975), Gazdar (1979a), and Levinson (1983), while it is rejected in favor of a semantic account by Bar-Lev and Palacas (1980), McCawley (1981), Carston (1985b), and Kempson (1986).)

11. In later work by Ducrot and his colleagues, beginning with Ducrot 1973, 'négation métalinguistique' is rechristened 'négation polémique', prompted by an increased emphasis on the structure of argumentation. I consider the earlier term more felicitous, expecially in the light of examples not considered by Ducrot, for example, (14)–(16). In these cases, negation constitutes a means for rejecting the language used by an earlier speaker, and is therefore indeed metalinguistic; but the notion of polemics or argumentation must be stretched to the snapping point to treat the negation here as polemic.

12. Moser points out that a more exact model of pragmatic ambiguity for our purposes is the dual-function analysis of left dislocation offered by Prince (1983), in which a single construction or lexical form is seen as serving two distinct (but systematically related) discourse functions.

13. Note, however, that the dual functions of negation are not entirely on a par with the instances of pragmatic ambiguity just cited, where the distinction between the two understandings in each case is neutralized at the level of logical form.

14. As Barbara Abbott has pointed out to me, U need not even involve a specifically linguistic utterance, as seen by the function of metalinguistic negation in the following musical scenario:

(i) Piano student plays passage in manner μ.
Teacher: It's not [plays passage in manner μ]—it's [plays same passage in manner μ'].

The teacher's use of *not* is clearly not assimilable to anything remotely resembling truth-functional propositional negation. (Cf. also Partee 1973 for an account of belief contexts in which utterances as well as propositions can be the object of belief, and in which ***a** believes that U* $\neq$ ***a** believes that **p***).

15. This same distinction can be applied successfully to the problem of future contingents I touched on in §2.1. Aristotle's dilemma—how do we assign truth values to (i) and (ii) today without embracing determinism?

    (i) There will be a sea battle tomorrow.
    (ii) There will not be a sea battle tomorrow.

    —can be resolved without adopting the Boethian position championed by Lukasiewicz, in which a third, indeterminate value is assigned to both of these statements. What we need to recognize is that (i) and (ii) are simply true or false, according to what the future holds, although neither (i) nor (ii) is assertable in the present, in the absence of foreknowledge or clairvoyance. The same approach generalizes naturally to the other varieties of present unknowables cited in §2.1, such as *Aristotle ate no breakfast on the day he died,* or *The number of stars is even.*

16. For those who lack the appropriate Parisian frame of reference to evaluate the propositions in (38), (i) is a rough equivalent based on my own hometown:

    (i) If the docks are the burly forearms of New York, the subways are the pits.

    Nongeographic instantiations of this construction, especially in the heart-and-soul frame of (38a), are frequently attested in journalistic prose, whence cometh (ii):

    (ii) If Patrick Ewing was the defensive heart of Georgetown's 1984 championship team, Gene Smith was unquestionably its soul.

17. While it begins to seem as though all logical operators may well have an extended metalinguistic function in addition to (and motivated by) their conventional logical function, I do not mean to suggest that it is only the connectives and variable binders which have this dual character. It might be maintained, for example, that definite pronouns are used metalinguistically in contexts like (i) and (ii):

    (i) What do you mean the king of France is bald—he doesn't exist.

    (ii) She told me a story about Santa Claus, but I know he doesn't really exist.

18. The same judgments will apply to intrasentential negations:

    (i) Mona is not at least six feet tall, she's {5′10″/#6′2″}.
    (ii) Jude is not at least as tall as Mona, he's {shorter/#taller}.

    The same pattern, of course, applies to other scalar negations:

    (iii) Max doesn't have at least three children, he has {two/#four}.

(iv) It isn't at least possible she'll win (#—it's downright certain she will).
(v) It's not at least warm out, it's downright {chilly / #hot}.

19. The subsequent discussion is lifted essentially verbatim from Horn 1984b: 20–22; the reader is referred to that paper for additional details.

20. My examples of this phenomenon, discussed in chapter 2, included the sortal incorrectness pairs in (i) (from Zimmer 1964), (ii) (from Drange 1966), and (iii) (from Bergmann 1977).

(i) Triangles are {not intelligent / #unintelligent}.
(ii) The number 4 is {not tolerant / #intolerant} of carelessness.
(iii) The theory of relativity is {not interested / #uninterested} in classical music.

Sentence (56a) in the text echoes (iv) (from Zimmer 1964: 23),

(iv) The king of France is {not intelligent / #unintelligent}—there isn't any.

in which the negation outside the scope of the existential presupposition is similarly immune to prefixal co-optation. My point in this subsection is simply that this immunity is not limited to the negation affecting presuppositions or conventional implicata, providing one more argument for the unity—and the nonsemantic nature—of marked negation.

21. This account of the incorporation diagnostic must draw a sharp distinction between the lexicalized prefixal negations of (56)–(58) and the "contracted" *-n't* in examples like (57a) and numerous other sentences scattered throughout this chapter. Evidently nothing constrains metalinguistic negation from contracting as an enclitic onto a preceding copula. If (as has been traditionally assumed) the *-n't* forms are produced by postlexical syntactic and/or phonological rules of encliticization, rather than in the lexicon, the distinction is made for us automatically. However, as Jerry Sadock has pointed out to me, the adoption of Zwicky and Pullum's (1983) analysis of ***Xn't*** as an inflection of the auxiliary element **X**, generated by the morphology, would require a different account here, perhaps one exploiting the distinction between the derivational and inflectional morphology. I shall simply assume that the grammar has some way of distinguishing the lexical prefixes *un-*, *iN-*, and so forth, which are incompatible with metalinguistic negation, from the *-n't* forms, which are not.

22. As Bolinger also recognizes, however, a diminisher may be turned into a minimizer by the insertion of *even;* compare (74′b) to (i) and (ii):

(i) I didn't eat even a little. (= nothing at all)
(ii) I was not even a little tired. (= not a bit tired)

23. That is, the descriptive negations corresponding to the qualified affirmations in (75a) are not those in (75b), but must take the form of (i), or—even more simply—in (ii):

(i) He isn't at all tired (ill, tall).
(ii) He isn't tired (ill, tall).

The positive polarity status of the examples in (75a) is obviously not a coincidence, given the high degree of nondetachability of whatever pragmatic process is responsible for this restriction. But just why is it that while negating a minimum quantity results—not surprisingly—in a strong negative, negating a moderate quantity results in anomaly (as in (75b))? To what factors do we owe the PPI-hood of this class of adverbs? While I cannot pursue an adequate answer here, I would suggest that the lack of straightforward negation in these cases is functionally motivated. There may well be sufficient reasons for qualifying or toning down a strong affirmative (especially when the modified adjective is e-neg, as in (75a); cf. Stoffel 1901 on 'down-toners'), and there are—as we saw in chapter 5—sufficient reasons for toning down a strong negative by expressing it indirectly (as in (72′) or (73), or by the *not* {*too*/*very*/*overly*} construction discussed in §5.3), even when the indirectness is only pro forma. But there is no similar motivation for expressing a negated qualified statement; if the true negations of (75a) are really those in (i) and (ii), the economy (**R**) Principle blocks the use of the more cumbersome forms in (75b) to convey that same meaning, and (disregarding the metalinguistic use, which has its own motivation) there is no other proposition these negations would usefully serve to express.

24. Bolinger (1977:45) points out that at least for some speakers, these 'formulaic external negations'—he includes *poppycock, fiddlesticks, like hell, like fun,* and the functionally related overt negation *nothing*—can 'reach down' across clause boundaries, or even island boundaries, to zero in on the one item they are used (metalinguistically) to reject:

(i) —He found proofs that clinched the argument.
—He found proofs that clinched {nothing/fiddlesticks}.
(ii) —You'll tell me when you get the reply.
—I'll tell you when I get nothing.
(iii) —He has a dog that bit a policeman.
—He has a dog that bit your old man.

Judgments may differ on these, but perhaps the reader will share my own intuition that Bolinger's examples become more plausible if *nothing* gives way to *nothing of the sort,* or if we employ in each case the all-purpose 'formulaic external negation' *par excellence,* namely, *yo' mama.*

25. Sgall, Hajičová, and Benešová (1973:21ff.) treat instances of the *not X but Y* construction as 'second-instance sentences', that is, as metalinguistic rejections of an earlier explicit or implicit assertion.

26. The difference between these examples and Bald's is that a negation immediately following *be* (as in (78′), (78″a), and for that matter (78)) may be a descriptive or a metalinguistic operator, since the syntax of negative placement obscures the difference. But in modern English, a negation placed after a main verb must be understood metalinguistically, with narrow focus, and hence must normally be completed by a rectification. (I return to postverbal negation below.)

27. Concessive *but* clauses are also discussed by Bolinger (1972:118), who differentiates the 'adversative *but*' of the concessive clause in (i) and (ii) from the 'verifying *in fact*' of the implicatum-suspending clause in (iii).

(i) I'm not very happy, but I am happy.
(ii) He doesn't have many friends, but he does have a few.
(iii) He doesn't have many friends, in fact he hardly has any at all.

(Cf. also the discussion of concessive vs. suspensive *if not* clauses in Horn 1972 and Welte 1978, cited earlier in this section.)

28. The same contours are determined by concessive structures which do not involve negation, as in the attested example in (i):

(i) 'For a cat he's a pretty good guy, but he is a cat'. (one mouse to another on TV cartoon show)

29. Klima (1964:302–3) and (apparently independently) Gates and Seright (1967) cite additional linguistic correlates of the negative-contrastive construction, both of which reinforce its status as a noncoordinate structure. First, the negative element is nonrestrictive and hence optional:

(i) a. (Not pleasure but) business is the purpose of my visit.
b. Business (not pleasure) is the purpose of my visit.

Secondly, it is always the positive element which controls agreement:

(ii) a. Not this book but those {are / *is} acceptable.
b. This book but not those {is / *are} acceptable.

Thirdly, it is only the positive element on which a tag can be based:

(iii) a. Not the father but the mother supports the family, doesn't {she / *he}?
b. The father {and / but} not the mother supports the family, doesn't {he / *she}?

Thus, as Klima observes, the subject of the *not **X** but **Y*** construction is ***Y***.

30. That this restriction did not always exist is clear from a sixteenth-century example of finite-verb-focus metalinguistic negation which Curme (1931) cites from Nashe:

(i) They deafe men's eares, but not edifie.

While Curme finds 'something unnatural' about (i) vis-à-vis the modern language, Nashe's sentiment could be equivalently expressed without the offending particle, as in (ii),

(ii) They deafen, not edify, men's ears.

Cf. Smith (1933:80) for related discussion.

31. Cf. König (forthcoming) for a comprehensive cross-linguistic study of adversative constructions in a framework partly based on that of Horn 1984b, 1985.

32. Having offered two equivalents for 'but' in Finnish, Whitney (1956: 187) goes on to observe the following:

> After a negative statement *mutta* introduces a mild difference or a concessive statement, while *vaan* begins a stronger or complete contradiction:
>
> *Hän ei ole sairas, mutta heikko hän on.* 'He is not ill, but he is weak'
>
> *Hän ei ole sairas, vaan aivan terre.* 'He is not ill, but quite well'

(The gapping in the latter construction is characteristic of the metalinguistic / contrastive *but* pattern in Finnish as well as English—and French, as we are about to see.)

33. As pointed out by Ducrot and Vogt (1979: 318–19), there is clear evidence that within the Classical period *magis* could already serve as a rectifying, *sondern*-type adversative. They cite the following examples from Virgil and Catullus, respectively, in which the negation must be understood metalinguistically:

(i) Non equidem invideo, magis miror. 'I am not envious, but (rather) astonished'
(ii) Id, Manli, non est turpe, magis miserum. 'It's not shameful, Manlius, but unhappy'

The metalinguistic nature of the latter example is brought out by Ducrot and Vogt by their free rendering, 'Il ne faut pas dire que c'est honteux, il faut dire plutôt que c'est malheureux'. The compound particle *nisi* (the reverse cognate of Spanish *sino*) was also employed as an adversative; cf. Tobler [1896] 1908: 70–71, Melander 1916: 122–23, and Wagenaar 1930: 133–34 for examples and discussion.

34. Actually, as Melander (1916: 90–94) points out, both *mais* and *ains* do occur without an overt negation immediately preceding, provided that the understood negative proposition is recoverable from the context of rectification:

(i) —Tu menz. 'You lie'
—Mes tu, dex te confond. 'You [do], God confound you'
(ii) —Vous avés tort. 'You're wrong'
—Ains ai bien droit. '[On the contrary] I'm quite right'

Notice that unlike *mais*$_{SN}$, the SN particle *ains*—whose closest modern equivalent may be *au contraire* 'on the contrary'—is attested in unreduced clauses. It is also worth noting that both *mais* and *ains* were systematically used in scalar contexts to detach the upper-bounding implicatum, in the fashion of *indeed* (cf. §4.4). Melander provides the following citations, inter alia, from Ronsard and his contemporaries:

(iii) "Estes vous . . . blechies?"—"Blechies? Ains sui tues"
(iv) Quelle plaisir est-ce, ains quelle merveille.

(v) Lustre de ta patrie, ains de tout l'univers.
(vi) Jusqu'aux Rois (ô ma Muse), ains jusqu'aux Dieux tu pousses.
(vi) Que tout ton sens envers le mien / Vault moult petit, mais ne vault rien.
(vii) Promis? Mes doné quitement.

In each case, a scalar value (wounded, pleasure, the fatherland, the Kings, little, promised) is rejected in favor of a stronger item on the same scale (killed, wonder, the entire universe, the Gods, nothing, given). This 'augmentative' use of *ains* and *mais*$_{SN}$, clearly akin to metalinguistic negation, is now expressed by the specialized adverb *voire*, which—like its English counterpart *nay* (*Wounded? Nay, killed*)—retains a palpable *soupçon* of the literary.

35. While the typical use of **neg-P** SN **Q** directly follows a previous assertion of P, this is not a necessary condition on the felicity of SN-*but*. The sequence in (i) is well formed in both Spanish and German versions,

(i) A: Pierre is nice.
B: He's not just nice, but$_{SN}$ quite generous.

Here, the object of B's denial (*he's just nice*) has not been actually asserted, but is inferrable via the 'loi d'exhaustivité' (or the Maxim of Quantity; cf. §4.2) from A's assertion.

36. As Tobler ([1896] 1908:94) recognizes, however, the distinction between the *aber*-type *mais* of (i) and the *sondern*-type *mais* of (ii) is neutralized in (iii), where the negation focuses on the verb (more specifically, on its tense).

(i) Il n'est pas riche, mais il est sain.	'He isn't rich, but$_{PA}$ he is healthy'
(ii) Il n'est pas riche mais pauvre.	'He is not rich but$_{SN}$ poor'
(iii) Il n'est pas riche, mais il l'a été.	'He is not rich, but$_{PA/SN}$ he was'

37. Extensive discussion of the German and Swedish facts is given in Lang 1977 (cf. also Welte 1978:193ff. on German and English), while Anscombre and Ducrot (1977) and Ducrot and Vogt (1979) provide detailed analyses of data from French and other Romance languages from the perspective of a theory of argumentation.

38. The proposal that English possesses both *aber*-type and *sondern*-type *buts* would appear to be inconsistent with Klima's analysis (1964: 302–3) on which (i) is derived from the source directly underlying (ii);

(i) Not John but Mary supports the family. [= (79c)]
(ii) Mary, {and / but} not John, supports the family.

note that the former contains an instance of *but*$_{SN}$ and the latter an instance of *but*$_{PA}$.

39. Welte shows that even when the syntax of PA clauses is compatible with gapping, the incorporation criterion still permits us to distinguish the two *buts* of English as well as their German counterparts. Consider

the following paradigms, from Welte (1978:193–94) (cf. Lang 1977, 1984):

(iG) Bill machte Sue nicht glücklich, {aber/sondern} reich.
(iE) Bill made Sue not happy, but (he made her) rich.
(iiG) Bill machte Sue unglücklich, {aber/*sondern} reich.
(iiE) Bill made Sue unhappy but rich.

Welte observes that the *aber*-type constructions in all four sentences involve two facts, triggering in each case the entailment that Bill made Sue both rich and not (or un-) happy, while a syntactically negated *sondern*-type 'Korrektursatz' like (i) involves just one fact; the only inference warranted here being that Bill made Sue rich. If we replace *r(e)ich* with its antonym, the PA/*aber* connective becomes semantically implausible, while its SN/*sondern* counterpart—if available—is fine:

(iiiG) Bill machte Sue nicht glücklich, {#aber/sondern} arm.
(iiiE) Bill made Sue not happy, but (#he made her) poor.
(ivG) Bill machte Sue unglücklich, {#aber/*sondern} arm.
(ivE) #Bill made Sue unhappy but poor.

The implausibility of the necessary presuppositions renders (ivG/E) anomalous on the PA version (cf. discussion in text below), while the lexical negation rules it out as a possible Korrektursatz.

40. As Georgia Green has reminded me, reveille does have words—indeed, several alternate sets of words; but we cannot infer from the announcer's comment that he is transderivationally alluding to any particular set of words, or in fact that he even knows any set of words.

41. This phenomenon, in which a proposition that is pragmatically presupposed as part of the common ground can serve as the basis for anaphoric-type destressing of material actually uttered for the first time, is amply instantiated in the literature. The contrasts in (i), (ii), (iii), and (iv) are discussed in Morgan 1969, Akmajian and Jackendoff 1970, Horn 1981b, and Ladd 1980, respectively.

(i) a. How does it feel to be a beautiful GIRL? (no pragmatic presupposition necessary)
b. How does it FEEL to be a beautiful girl? (speaker presupposes addressee is a beautiful girl)

(ii) John washed the car—
a. I was afraid someone ELSE would do it. (hope fulfilled: I wanted John to wash it)
b. I was AFRAID someone else would do it. (disappointment: I wanted to do it myself)

(iii) a. I thought you were COMING. (but you didn't)
b. I THOUGHT you were coming. (and sure enough, you did)

(iv) —How did your operation go?
—Don't talk to me about it.
a. ?#I'd like to strangle the BUTCHER. (only literal reference to the butcher; OK after *How's the filet mignon?*)
b. I'd like to STRANGLE the butcher. (epithet reading, with pragmatic presupposition).

As in my musical example (113), the material following the stressed element in the (b) examples here is treated as though it had been asserted earlier in the discourse context, and hence is destressed. A frequently encountered example of pragmatically triggered destressing is found in sportscasters's updating of scores; which numeral receives the major stress in (v) depends on which team has just scored one or more runs.

(v) {And that makes it / And the score is now} Red Sox 6, Yankees 3.

Ladd (1980) offers a persuasive account of anaphoric destressing as default accent, although other proposals abound in the literature on intonation.

42. Compare, however, the following remark by a gloomy drive-in movie owner in suburban Atlanta, quoted in Newsweek (9 August 1982):

(i) Things are so much more open sexually today that who needs a drive-in?

43. It is often difficult to determine just how a given expression, for example, those in (i),

(i) John, too, is coming to the party.
Even John passed the exam. [cf. my discussion of *not . . . even* in §2.5 above]

may be (descriptively) negated. Ducrot suggests we can determine intuitively what an expression presupposes, and thereby discover its descriptive, presupposition-preserving negation. Thus the descriptive negations of (ii)–(iii) can be given as (ii′)–(iii′), respectively (Ducrot 1972:105):

(ii) We have finally arrived.
(ii′) We haven't arrived yet.
(iii) For a Frenchman, he knows a lot of logic.
(iii′) Even for a Frenchman, he doesn't know much logic.

44. Anscombre and Ducrot (1977:40) point out that *non (pas)* requires an overt rectification:

(i) #Pierre viendra non pas demain.	'Pierre will come not tomorrow'
(ii) Pierre viendra non pas demain mais la semaine prochaine.	'Pierre will come not tomorrow but next week'

As seen in the glosses, and earlier in (79), the same property is shared by postverbal metalinguistic negation in English.

45. As carefully documented by Danell (1974), much more is going on here than meets the eye. The interplay of factors determining the distribution of *pas du X* (or *pas un X*) vs. *pas de X* is extremely complex, hinging on such variables as the scope of negation, the modality of the sentence, the nature of the complement(s), and the 'degree of existence' of the object focused by negation.

46. Note that incorporation is impossible here, as predicted (cf. my minimal pairs in §6.4):

(i) #It's impossible that mammals suckle their young, you ignoramus, it's downright necessary.

47. With cardinal values, the metalinguistic reading is available without fall-rise intonation or rectification, if the focused (objected-to) value immediately follows the negative: *Nem tizet akar* (literally, 'Not ten [he] wants') can convey that he wants less or more than ten, while *Nem akar tizet* can only be interpreted descriptively as signaling that he wants less than ten.

48. Cf. also Stickel (1972) on the functions of *nein* (freestanding *no*) as a '*Kontroll- und Korrektursignale*' (i.e., metalinguistic negation) in German.

Chapter Seven

1. The same pattern is attested in the unrelated language Thadou (a Kuki-Chin tongue of the Tibeto-Burmese group, spoken in the Manipur Hills of India): *hih* occurs in imperative contexts, *po* with nonimperative verbs, and *lou* to mark constituent (phrasal, word, or affixal) negation, as in *apha-lou* 'not good' (cf. Thirumalai 1970 for details). Similarly, in the Bantu language Kinyarwanda, sentential or verbal negation is expressed by *nti-* / *-ta-*, nominal negation by *nta,* and constituent or focus negation (often functioning metalinguistically, as in the *not X but Y* construction) by *si* (Kimenyi 1973).

2. 'Not an accent' meaning not even a diacritic, presumably the linguistic mark of least value. Similar minimizers from the linguistic domain are *tittle* and *jot* (often appearing in tandem: *not depart one tittle or jot from*), Latin *titulus,* Greek *stigmē,* German *(kein) Iota,* and so forth. It is hard to find anything much smaller or less significant than the dot over an *i.*

3. The stage represented by (5b), in which finite-verb negation co-occurs with one or more neg-incorporated indefinites to express a negative proposition, is of course still extant in nonstandard English (cf. Labov 1972), and is realized as well in a wide variety of the world's languages. Jespersen (1917:64–80) illustrates 'cumulative', 'resumptive', and 'paratactic' varieties of multiple negation with data from Latin, Old and Modern English, Middle High German, Danish, Spanish, Serbian, Greek, Hungarian, and Bantu. (Cf. also Coombs 1976 for extensive examples from the older Germanic dialects and Wagenaar 1930 for Romance.)

4. The history of negation in Old Icelandic is somewhat different. The Old Icelandic preverbal particle *né* seems to have dropped out fairly early on in clause-initial position; finite negation is marked instead by the historically positive reinforcer *eigi* 'not at all' in prose and by the suffixes *-gi* and *-a(t)* (cognate with OE *ā* and Gothic *aiw* 'ever') in poetry. Cf. Delbrück 1910:22–24 and Coombs 1976:§7 for elaboration.

5. By the turn of the sixteenth century, Shakespeare was able to exploit the interchangeability of the two alternants, as we saw in §6.6.

6. Or more accurately, in the structures of the respective types within a single language; like other adverbs, *nicht* appears postverbally in verb-second main clauses (or root sentences) in German, but preverbally in verb-final embedded clauses.

7. As I shall discuss in more detail later in this chapter, *n't* is in fact not an enclitic but an inflectional suffix on tensed auxiliaries: cf. Lapointe 1980; Gazdar, Pullum, and Sag 1982; and especially Zwicky and Pullum 1983. As Zwicky and Pullum (1983:507) point out, if clitics (including the auxiliary forms *'ve, 's,* and *'d*) are attached postsyntactically, and inflectional affixes are associated with stems in the lexicon, there is no way to get an inflectional suffix (e.g., *-n't*) outside an enclitic. Thus, while no general constraint blocks the accumulation of clitics per se (cf. *I'd've, she'd'a*), we automatically rule out **I'ven't*, **she'sn't* as variants of *I haven't/I've not, she isn't/she's not.*

8. Fortunately, Marchand's insightful analysis of the forms and functions of modern English negation does not depend crucially on his *Sprachgefühl*-ish account (1938:204) of the predominance of contracted negation: 'As a rule, English does not like to emphasize the idea of negativity, being a language which naturally avoids emotionality.' Given the notorious description of English as a 'masculine' tongue in Jespersen 1922, we can only conclude that non-native grammarians are lamentably susceptive to the stiff upper lips of their informants.

9. The foregoing inventory is from Joly 1972:39; other forms could easily be added to the list. And once again, this type of negative reinforcement occurs widely among non-Indo-European languages as well. Examples include Japanese *tittomo* '(not) at all', Basque *(ez) deus* '(not) at all', and Swahili *(si) kamwe* '(not) at all, (not) ever'.

10. The generative semantics version of negative sentence structure (cf., e.g., G. Lakoff 1970; McCawley 1972) differs from the standard theory by assigning the negative marker of status of an aunt, rather than sister, of the clause it eventually negates. That is, in place of the sister adjunction of the Klima-Jackendoff model illustrated in (i), we get the Chomsky-adjoined structures of (ii) or (iii). A close relative is the Kraak (1966) model shown in (iv).

Whatever the virtues of these analyses for moving the deep structure of English closer to the logical forms of the propositional calculus, and notwithstanding the much-debated syntactic evidence for these structures, (ii)–(iv) are no better than (i) at approximating the surface appearance of negation in English or, as is clear from Dahl, in natural languages in general.

11. At least one 2-year-old (my niece) is on record as subjecting this initial

negation to category-neutral procliticization in vowel-initial contexts like that in (i):

(i) Nabby wan' dat one. 'Abby [the speaker] does not want that one'

One presumes that such speakers have not yet fully accepted the **Q**-based clarity constraint.

12. This section is an elaboration of a paper (Horn and Farkas 1985) co-authored by my esteemed colleague Donka Farkas, delivered orally at the LSA but not heretofore committed to print. The initial premise and well as the more egregious errors and overstatements are my own.

13. The categorical syllogism has the canonical form of (i), the hypothetical syllogism that of (ii):

(i) All **A** is **B**	(ii) If ϕ then ψ
All **B** is **C**	If ψ then χ
∴ All **A** is **C**	∴ If ϕ then χ

14. While it may seem counterintuitive to treat the subject of (15a) as a third-order entity, Barwise and Cooper distinguish the generalized quantifier consisting of a name, for example, $[_{NP}$ **Harry**], which does indeed denote the family of sets of which Harry is a member, from the name *Harry* itself, which simply denotes an individual.

15. Of course as we saw in §5.3, the simple unincorporated version may itself be pragmatically strengthened to a contrary negation, so that both (16′b) and (16′c) result in communicating something more than the statement that a certain proposition (that the man is happy) is false. This does not affect the current point, which is that the latter example cannot be assimilated (as (16c) may be) to propositional negation.

16. Dowty, Wall, and Peters (1981:87–88) find it convenient to introduce an iterating one-place sentential negation of category ⟨**t**/**t**⟩ into their proto-Montagovian type-theoretic language $\mathbf{L}_{\mathbf{type}}$.

17. It will be recalled that in the classical framework, as summarized in chapter 1, (16′a) and (16′c) are MEDIATE contraries, since they allow an unexcluded middle. But (16a) and (16c) are IMMEDIATE (or LOGICAL) contraries: immediate because no man can be neither celebrated nor uncelebrated and contraries because neither (16a) nor (16c) will be true if no is man referred to.

18. On the other hand, negative force may be carried by lexical tone, as Payne (1985:229) indicates.

19. Conditionals are also claimed to incorporate implicit universal quantifiers within the analysis proposed by Lewis (1975) and adopted into discourse representation theory by Kamp (1981) and Heim (1982). But Lewis's approach, though independently conceived and motivated, is beset by some of the same difficulties as Sommers's. Besides the intuitive implausibility of taking the conditionals in (23a) and (24a) as universally quantifying over cases, neither account of conditionals as universals predicts the salient (one-time-only) readings of (i) and (ii):

(i) If you find a dime, give it to me. [thanks to Cleo Condoravdi for this example]
(ii) If my daughter has a baby I will be pleased.

20. In fact, Sommers eventually retreats from his proposal for the conversion of disjunctions. He later (1982: 153) rejects the formula in (21′c), suggesting that '**p** or **q**' might be represented as 'every [−**p**] is a [**q**]', effectively translating the disjunction into the corresponding conditional, which is in turn converted as in (21′b). On the other hand, *every* is now (1982: 170) implausibly translated into a double negation (actually the contradictory of a contrary), so that *every S is P* (e.g., *Every child of mine was fed*) comes out as *no S is not-P* (*No child of mine was unfed*).

21. In any case, de dicto modality, unlike negation, is never simply a truth function, since (as with all modalities) it is in principle intensional. Thus, by allowing for the representation of de dicto epistemic modals (*It is {possible/likely/necessary} that*) as external operators on a proposition, we would still not be ipso facto endorsing the need for any truth function of the category **t/t**.

22. $\mathcal{P}$ here is a variable over nonfinite $\bar{\mathbf{V}}$-type meanings. The feature [±FIN] determines tensed vs. tenseless inflection on the head of a $\bar{\mathbf{V}}$, while [+BSE] determines a bare stem on the head. [+AUX] is the feature assigned to just those verbs in the class of auxiliaries. These are all head features, shared between mothers and heads in accordance with the Head Feature Convention of GPSG. Rule β, providing for negation as an optional left sister of nonfinite (tenseless) verbal expressions, is not restricted to [+BSE] complements of modals, extending as well to [+PSP] (i.e., past participial) complements of perfect *have,* [+PRP] (i.e., present participial) complements of progressive *be,* and so on. (Cf. Gazdar, Pullum, and Sag 1982 for details.)

23. Wide-scope negation has also been traditionally employed to unravel de dicto–de re ambiguities in intensional (opaque) contexts, along the lines of Quine 1960. And it is true that names do participate in such ambiguities, so that, for example, (i) entails (ii) on the wide-scope (de re, transparent) reading but not on the narrow-scope (de dicto, opaque) reading of the names, given that Superman = Clark Kent.

(i) Lois believes she loves Superman.
(ii) Lois believes she loves Clark Kent.

But there is reason to suspect that such ambiguities are conceptually and analytically distinct from the quantificational ambiguities for which wide-scope readings are appropriately invoked; cf. Barwise and Perry 1983 for discussion.

24. For Burton-Roberts, Strawson (presumably along with Frege) was correct to ignore the cancelation effects associated with external negation, since no such cancelation is possible with descriptive negation. Presumably those external, presupposition-canceling negations occasionally countenanced by Strawson (see the discussion of *He neither cares nor doesn't care; he's dead* in §2.2 above) can be reanalyzed as involving the metalinguistic use of the negative operator.

25. Attal (1971:107–8) entertains the possibility that all sentence negation might reduce to (Ducrot's) refutational or polemic negation; he ultimately (and in my view, correctly) rejects this hypothesis.

26. Notice that the metalinguistic use of negation in (33) guarantees the truth of the predicate denial only when it does focus on a truth condition of the corresponding affirmation. In contexts like (i) and (ii), the existence assumption goes through in the normal way.

(i) The king of France isn't [bæld]—the king of France is [bɔ:ld].
(ii) The king of France is not bald but galled.

27. Cf. Lycan (1984:103) for a similar attempt to derive presuppositional effects—at least in the existential and sortal cases—from the application of Grice's Strength (= Quantity) maxim, that is, my **Q** Principle. Although Lycan's framework, like mine, is bivalent and nonpresuppositional, it assumes a standard external propositional connective and a Russellian approach to scope disambiguation for (33); nondenoting subjects and category mistakes must thus be treated as either decomposable or nonsensical, an approach I have rejected here.

28. The ambiguity predicted for unincorporated particle negation in these examples is also associated with distributive quantification and with mass expressions displaying similar semantics:

(i) Every cookie was not eaten.
(ii) All the cake was not eaten.
(iii) The whole cake was not eaten.

29. Actually, the facts are somewhat more complex, especially with respect to the distribution of tag questions; cf. Carden (1970), Heringer (1970), and Stokes (1974) for data and discussion.

30. The simple indefinite article can be interpreted as a generic determiner, rather than a true indefinite or particular, in which case (as with other generics and universals) a wide-scope reading for Aux-based negation is readily available:

(i) A tiger doesn't eat zucchini. (cf. *The tiger isn't a vegetarian, Tigers don't eat apples*)

Other indefinite subjects may be taken as nonspecifics within the scope of negation when the NEG-V (specific) reading is pragmatically deviant:

(ii) A .44 calibre pistol wasn't available in the shop. (from Ōta and Katō 1986:33)
(iii) A child hasn't been born in this town for two years. (from Cooper 1984:25)

In none of these examples is *some* substitutable for *a* (salva sensu).

31. It is worth noting tha there remains a metalinguistic tinge to the NEG-Q *All . . . not* construction. *All that glitters is not gold* is a natural candidate for the wide-scope interpretation of the negative precisely because of the natural expectation that is thereby disappointed (viz., that all that glitters should be gold). If no such expectation is likely to be entertained, as in (i)–(iii),

(i) All philosophers don't ignore the complications of natural language.
(ii) All my friends aren't cokeheads.
(iii) The party was a disaster. First, all the people who said they were coming didn't come. Then, all the soufflés didn't rise. Then, everybody didn't remember the words to 'Happy Birthday'.

the NEG-Q readings may still be (at least marginally) available, but only when the specific expectation to the contrary can be built into the context; it is as if each *all* . . . *not* sequence were prefaced by 'Contrary to what you {said/assumed}'. No such expectation is evoked in the negated universal (*Not all* . . .) counterparts of these sentences.

32. This account of the *some not/every not* asymmetry is an elaboration of the one suggested in Horn 1978a: 142.

33. NEG-Q interpretations are possible here with metalinguistic negation, but in this case the unmarked scalar (less than) reading is not forced:

(i) ˇMany students don't take classes after 4:00 P.M.—{all/just some} do.

Indeed, metalinguistic negation can override even the scalar value of the negative quantifier:

(ii) Not ˇmany but most students don't take classes after 4:00 P.M.

The normally unavailable wide-scope reading for (50′a, b) also reemerges in Baker's 'double negative contexts'. Ladusaw cites the NEG-Q reading of the embedded clause in (iii)

(iii) I'm surprised that a lot of wine wasn't consumed.

as constituting 'one phenomenon which points to the necessity of distinguishing assertion of a negation from negation of an assertion'—or, in my terms, descriptive from metalinguistic negation.

34. Additional evidence supporting my transderivational/functionalist line on NEG-Q readings is provided by an attested example (from the *New York Times*, 25 December 1986), where *not one* or *none* cannot felicitously appear and the negation receives the wide-scope interpretation unavailable in (51a).

(i) The Giants . . . have signed a handful of free agents for next year. One of them is not Steve Ciskowski. [and, as the context makes clear, neither are any of the others]

35. I am assuming, with Kato, that cardinal determiners are to be considered as special cases of the existential; cf. also Barwise and Cooper 1981. Notice the 'less than' interpretation associated with the denial of the scalar operator *ten*, as predicted by my account in chapter 4.

36. Cf. Ladusaw (1979) for a formal definition of the scope of an expression with respect to an interpretation and for additional evidence for this framework (and against its geometric rivals) based on polarity phenomena in English.

37. Note that unlike the cases of the positive midscalars *shouldn't* and *oughtn't, mustn't* cannot be analyzed as an instance of my pragmatic rule of NR strengthening. (i) shares a reading with (i′) and (i″), but (ii), if it is acceptable at all, cannot be read as conveying (ii′) or (ii″):

(i) I don't think you two {should/ought to} split up.
(i′) I think you two {shouldn't/oughtn't to} split up.
(i″) I think you two {should/ought to} [not split up].
(ii) ?I don't think you two must split up.
(ii′) I think you two mustn't split up.
(ii″) I think you two must [not split up].

38. In this respect, Jespersen provides a more appropriate icon for ancestor worship at the ETL shrine than Aristotle himself, since the latter did not clearly recognize the Greek counterparts of (58c–e) as predicate affirmations with negative subject terms (see Geach 1970 and discussion below), while the former did regard such sentences as instances of special negation, despite their association with the semantics of contradictory opposition.

39. The facts are slightly more complicated than suggested here. Kuno (1972:271) observes that while the 'theme marker' *-wa* cannot occur in nongeneric sentences with indefinite subjects, as in (i), the same strings are acceptable in contexts like that in (ii), where *-wa* marks contrast rather than ordinary topic.

(i) Ame {ga/*wa} hutte imasu.
rain falling is
'Rain is falling'.
(ii) Ame wa hutte imasu, yuki wa hutte imasen.
rain falling is snow falling is-not
'Rain is falling, but snow is not falling'.

It seems plausible to take the latter example as involving metalinguistic rather than descriptive negation. I shall not pursue the relation between thematic and contrastive *-wa* here.

40. The recasting of the Brentano-Marty-Kuroda distinction in pragmatic terms is advocated on independent grounds by Lambrecht (1987).

41. Similar facts obtain in other languages conforming to the pro-drop parameter, including Romance languages like Italian and Rumanian; cf. also Firbas (1966) on the effect of functional sentence perspective on word order in Czech.

42. I am skipping over a number of details which are not directly relevant to our present concerns. As I noted earlier, following Crockett (1977) and Babby, a certain species of constituent and/or contrastive negation—which I took in §6.6 to represent my metalinguistic operator—never permits genitive marking on an **NP** within the apparent scope of negation. (Cf. Magner 1955, R. M. Davidson 1967, and Timberlake 1975 for other factors involved in determination of GEN/NOM and GEN/ACC case marking in Russian negative sentences.)

43. When this left-to-right principle is to be violated, the violation is typically signaled by marked intonation. As shown by G. Lakoff (1969) and Jackendoff (1972), inter alia, the correlation between scope and surface word order is not a simple fact about negation but a general tendency affecting the preferred interpretation of logical operators in natural language.

44. This conclusion, according to which semantically distinct linguistic expressions often merge syntactically and/or pragmatically within a given language, may seem peculiar, attributing to the linguistic system an apparently wasteful duplication of effort. Yet this is precisely what we expect to find (and do find, elsewhere) if we view language as defined by the interaction of parallel but autonomous components. This view has been persuasively expounded in recent work by Sadock (1983, 1985).

45. I am intentionally leaving aside here the question of embedded finite negation (*She said that he wasn't leaving*), where the negation technically has narrow scope with respect to the matrix clause, but is otherwise an instance of ordinary propositional negation or predicate denial (but not, as I noted in earlier chapters, of speaker's denial). Embedded nonfinite negation, on the other hand, does represent true constituent negation (PTN).

46. Note, for example, the contrast in Klima effects between (69a–c) and an unambiguous PTN, for example,

(i) This is not a large house, and {neither/??so} is that one.
(ii) This is a not large house, and {so/*neither} is that one.

Sentence (64d) is more easily interpretable as an instance of PTN (*He is busy not hunting lions*), whence the split diagnostics of (iii) and (iv):

(iii) He [is not] hunting lions, and neither is she. [on PD reading]
(iv) He is [not hunting lions], and so is she. [on PTN reading]

The purported semantic ambiguity of sentences like (68) and (69a–c) is presupposed or defended by Kraak (1966), as well as Jackendoff and Gabbay and Moravcsik; the alternative position that such sentences are semantically general or unspecified is argued for in Seuren 1967, Atlas 1974, Cornulier 1974, Kempson 1975, and Horn 1978a.

47. As I argued above, following the Zwicky and Pullum and GPSG inflectional analyses of contracted negation, the semantics associated with a given *Xn't* form (especially when *X* is a modal) may not be contradictory.

48. Whether prefixal negation is interpreted as a mediate or immediate contrary depends on the lexical item in question; cf. §5.1 for discussion.

Bibliography

Abbott, B. 1972. "The Conjunction *but*." Unpublished. University of California, Berkeley.

Abel, C. 1882. *Linguistic Essays*. London: Trübner.

Abelard, Peter. 1956. *Dialectica*. L. M. DeRijk, ed. Assen, Netherlands: Van Gorcum.

Ackrill, J., ed. and trans. 1963. *Aristotle's Categories and De Interpretatione*. Oxford: Clarendon.

Adorno, T. 1973. *Negative Dialectics*. E. B. Ashton, trans. New York: Seabury.

Aitchison, J., and G. Bailey. 1979. "Unhappiness about Not Unhappy People." *Journal of Linguistics* 15:245–66.

Ajdukiewicz, K. 1935. "Syntactic Connexion." Translated in McCall, ed. (1967b), 1–27.

Akmajian, A., and R. Jackendoff. 1970. "Coreferentiality and Stress." *Linguistic Inquiry* 1:124–26.

Algeo, T. 1971. "The Voguish Uses of *non*." *American Speech* 46:87–105.

Allen, M. R. 1978. Morphological Investigations. Ph.D. diss., University of Connecticut.

Allerton, D. 1978. "The Notion of 'Givenness' and Its Relations to Presupposition and to Theme." *Lingua* 44:133–68.

Allwood, J. 1972. "Negation and the Strength of Presuppositions." Reprinted in *Logic, Pragmatics, and Grammar*, Ö. Dahl, ed., 11–57. University of Götebörg Department of Linguistics.

Allwood, J., L.G. Andersson, and Ö. Dahl. 1977. *Logic in Linguistics*. Cambridge: Cambridge University Press.

Altmann, S. A. 1967. "Structure of Social Communication." In *Social Communication among Primates*, S. A. Altmann, ed., 325–62. Chicago: University of Chicago Press.

Anderson, S., and P. Kiparsky, eds. 1973. *A Festschrift for Morris Halle*. New York: Holt.

Andrew, S. O. 1940. *Syntax and Style in Old English*. Cambridge: Cambridge University Press.

Andrews, E. 1986. "A Synchronic Semantic Analysis of *de-* and *un-* in American English." *American Speech* 61:221–32.

Anscombe, G. E. M. 1956. "Aristotle and the Sea Battle." *Mind* 65:1–15.

Anscombre, J.-C., and O. Ducrot. 1976. "L'argumentation dans la langue." *Langages* 42:5–27.

———. 1977. "Deux *mais* en français?" *Lingua* 43:23–40.

———. 1978. "Echelles argumentatives, échelles implicatives, et lois de discours." *Semantikos* 2:43–66.

———. 1983. *L'argumentation dans la langue*. Brussels: Pierre Mardaga.

Antoine, G. 1952. *La coordination en français, Tome 2*. Paris: Editions d'Artrey.

Apostel, L. 1972a. "Negation: The Tension between Ontological Positivity and Anthropological Negativity." *Logique et Analyse* 15:209–317.

——. 1972b. "The Relation between Negation in Linguistics, Logic, and Psychology." *Logique et Analyse* 15:333–401.

Apostle, H., ed. and trans. 1966. *Aristotle's Metaphysics*. Bloomington: Indiana University Press.

Saint Thomas Aquinas. 1496. *Expositio super libros Posteriorum et De Interpretatione Aristotelis*. Completed by Thomas Caietanus. Venice: Otinus de Cuna, for Alexander Calcedonius.

———. 1945. *Basic Writings of Saint Thomas Aquinas*. A. C. Pegis, ed. and trans. New York: Random House.

Aristotle. *Works*. 1961–66. Including *Categories, De Interpretatione, Prior and Posterior Analytics*. Translated under the editorship of W. D. Ross. London: Oxford University Press.

Arlen, H., and J. Mercer. 1944. "Ac-Cent-Tchu-Ate the Positive." Song from *Here Come the Waves*, film musical of 1944.

Aronoff, M. 1976. *Word Formation in Generative Grammar*. Cambridge: MIT Press.

Ashby, W. 1981. "The Loss of the Negative Particle *ne* in French." *Language* 57:674–87.

Atlas, J. 1974. "Presupposition, Ambiguity, and Generality: A Coda to the Russell-Strawson Debate on Referring." Unpublished. Pomona College.

———. 1975. "Frege's Polymorphous Concept of Presupposition and its Role in a Theory of Meaning." *Semantikos* 1:29–44.

———. 1977. "Negation, Ambiguity, and Presupposition." *Linguistics and Philosophy* 1:321–36.

———. 1978. "On Presupposing." *Mind* 87:396–411.

———. 1979. "How Linguistics Matters to Philosophy: Presupposition, Truth, and Meaning." In Oh and Dinneen, eds., 265–81.

———. 1980. "A Note on a Confusion of Pragmatic and Semantic Aspects of Negation. *Linguistics and Philosophy* 3:411–14.

———. 1981. "Is *not* Logical?" *Proceedings of the Eleventh International Symposium on Multiple-Valued Logic*, 124–28. New York: IEEE.

———. 1984. "Comparative Adjectives and Adverbials of Degree." *Linguistics and Philosophy* 7:347–77.

Atlas, J., and S. Levinson. 1981. "*It*-clefts, Informativeness, and Logical Form." In Cole, ed. (1981), 1–61.

Attal, P. 1971. "Négation de phrase et négation de constituent." *Langue française* 12:98–111.

Austin, J. L. [1950] 1970. "Truth." In Austin (1970), 117–33.

———. 1956. "Ifs and Cans." In Austin (1970), 205–32.

———. 1970. *Philosophical Papers*, J. O. Urmson and G. J. Warnock, eds. 2d ed. London: Oxford University Press.

Avicenna. 1971. *Treatise on Logic*, F. Zabeeh, ed. and trans. The Hague: Martinus Nijhoff.

Axinn, S. 1964. "Ayer on Negation." *Journal of Philosophy* 61:74–75.

Ayer, A. J. 1936. *Language, Truth, and Logic*. London: Victor Gallancz.

———. [1952] 1963. "Negation." *Journal of Philosophy* 49:797–815. Reprinted in A. J. Ayer, *Philosophical Essays*, 36–65. London: Macmillan.

Babby, L. 1980. *Existential Sentences and Negation in Russian*. Ann Arbor: Karoma.

Bach, E. 1968. "Nouns and Noun Phrases." in Bach and Harms, eds., 90–122.

———. 1980. "Tenses and Aspects as Functions on Verb-Phrases." In *Time, Tense, and Quantifiers*, C. Rohrer, ed., 19–37. Tübingen: Niemeyer.

Bach, E., and R. Harms, eds. 1968. *Universals in Linguistic Theory*. New York: Holt.

Bach, K., and R. Harnish. 1979. *Linguistic Communication and Speech Acts*. Cambridge: MIT Press.

Bacon, F. [1620] 1853. *Novum Organum*. J. Devey, trans. London: Henry G. Bohn.

Bacri, N. 1976. *Fonctionnement de la négation*. Paris: Mouton.

Baker, C. L. 1970. "Double Negatives." *Linguistic Inquiry* 1:169–186.

Bald, W.-F. 1971. "The Scope of Negation and Copula Sentences in English." *Journal of English Linguistics* 5:1–28.

Baldwin, J. M., ed. 1928. Entries for "Negation" and "Negative," *Dictionary of Philosophy and Psychology*, vol. 2, 146–49. New York: Macmillan. (Entries written by C. S. Peirce et al.)

Ball, C. 1986. "Metalinguistic Disjunction." *Penn Review of Linguistics*. Tenth Penn Linguistics Colloquium.

Baltin, M. 1977. "Quantifier Negative Interaction." In *Studies in Linguistic Variation*, R. Shuy and R. Fasold, eds., 30–36. Washington, Georgetown University Press.

Banjọ, A. 1974. "Sentence Negation in Yoruba." *Studies in African Linguistics* 5:35–47.

Bar-Lev, Z., and A. Palacas. 1980. "Semantic Command over Pragmatic Priority." *Lingua* 51:137–46.

Barnes, J. 1969. "The Law of Contradiction." *Philosophical Quarterly* 19:302–9.

Barrett, R., and A. J. Stenner. 1971. "On the Myth of Exclusive 'Or'." *Mind* 79:116–21.

Bartsch, R. 1973. "'Negative Transportation' Gibt es Nicht." *Linguistische Berichte* 27:1–7.

Barwise, J., and R. Cooper. "Generalized Quantifiers and Natural Language." *Linguistics and Philosophy* 4:159–219.

Barwise, J., and J. Perry. 1983. *Situations and Attitudes*. Cambridge: MIT Press.

Baudry, L. 1950. *La querelle des futurs contingents*. Paris: J. Vrin.

Bauer, L. 1983. *English Word-formation*. Cambridge: Cambridge University Press.

Baugh, A. 1935. *A History of the English Language*. London: Routledge & Kegan Paul.

Bellugi, U. 1967. The Acquisition of the System of Negation in Children's Speech. Ph.D. diss., Harvard University.

Bennett, M. 1976. "A Variation and Extension of a Montague Fragment of English." In Partee, ed., 119–63.

Bergmann, M. 1977. "Logic and Sortal Incorrectness." *Review of Metaphysics* 31:61–79.

———. 1981. "Presupposition and Two-Dimensional Logic." *Journal of Philosophical Logic* 10:27–53.

Bergson, H. 1911. *Creative Evolution*. A. Mitchell, trans. New York: Modern Library.

Bhatia, T. 1977. *A Syntactic and Semantic Description of Negation in South Asian Languages*. Ph.D. diss., University of Illinois.

Bierwisch, M. 1971. "On Classifying Semantic Features." In Steinberg and Jakobovitz, eds., 410–35.

Biligiri, H. S., ed. 1970. *Papers and Talks*. Presented and delivered at the Summer School of Linguistics. Mysore: Wesley Press.

Binnick, R. 1976. "The Iffyness of Transitive Verbs." In *Syntax and Semantics 6: The Grammar of Causative Constructions*, M. Shibatani, ed., 217–27. New York: Academic Press.

Black, M., ed. 1965. *Philosophy in America*. Ithaca: Cornell University Press.

Blanché, R. 1969. *Structures intellectuelles*. 2d ed. Paris: Librairie philosophique J. Vrin.

Bloom, L. 1970. *Language Development: Form and Function in Emerging Grammars*. Cambridge: MIT Press.

Bloomfield, L. 1933. *Language*. New York: Henry Holt.

Bochvar, D. A. 1938. [On a three-valued logical calculus and its application to the analysis of contradictions.] *Matematičeskij Sbornik* 4:287–308.

Bodine, A. 1975. "Androcentrism in Prescriptive Grammar." *Language in Society* 4:129–46.

Boër, S., and W. Lycan. 1976. "The Myth of Semantic Presupposition." Distributed by IULC.

Bolinger, D. 1952. "Linear Modification." *PMLA* 67:1117–44.

———. 1957. *Interrogative Structures of American English.* Publication of the American Dialect Society, no. 28. University: University of Alabama Press.

———. 1961. "Contrastive Accent and Contrastive Stress." *Language* 37:83–96.

———. 1968. "Postposed Main Phrases: An English Rule for the Romance Subjunctive." *Canadian Journal of Linguistics* 14:3–30.

———. 1972. *Degree Words.* The Hague: Mouton.

———. 1977. *Meaning and Form.* London: Longman.

———. 1980. "A Not Impartial Review of a Not Unimpeachable Theory." In *Language Use and the Uses of Language,* R. W. Shuy and A. Shnukal, eds. Washington: Georgetown University Press.

Borkin, A. 1971. "Polarity Items in Questions." *CLS 7,* 53–62.

Bosanquet, B. [1888] 1911. *Logic, vol. 1.* 2d edition. Oxford: Clarendon.

———. 1895. *The Essentials of Logic.* London: Macmillan.

Bosley, R. 1975. *Aspects of Aristotle's Logic.* Assen, Netherlands: Van Gorcum.

Bossuyt, A. 1983. "Historical Functional Grammar: An Outline of an Integrated Theory of Language Change." In *Advances in Functional Grammar,* S. Dik, ed., 301–25. Dordrecht: Foris.

Bottomore, T., ed. 1983. *A Dictionary of Marxist Thought.* Oxford: Basil Blackwell.

Boucher, J., and C. Osgood. 1969. "The Pollyanna Hypothesis." *Journal of Verbal Learning and Verbal Behavior* 8:1–8.

Bowerman, M. 1973. *Early Syntactic Development.* Cambridge: Cambridge University Press.

Boyd, J., and J. Thorne. 1969. "The Semantics of Modal Verbs." *Journal of Linguistics* 5:57–74.

Bradley, F. H. 1883. *Principles of Logic.* London: K. Paul, Trench.

Bréal, M. 1900. *Semantics.* Trans. Mrs. H. Cust. New York: Henry Holt.

Broselaw, E. 1977. "Language Change and Theories of the Lexicon." *CLS 13,* 58–63.

Brouwer, L. E. J. 1908. "The Unreliability of the Logical Principles." Translated in *Collected Works, I,* A. Heyting, ed., 107–11. Amsterdam: North-Holland, 1975.

———. 1923. "On the Significance of the Principle of the Excluded Middle in Mathematics." Translated in van Heijenoort, ed., 334–45.

Brown, M. 1984. "Generalized Quantifiers and the Square of Opposition." *Notre Dame Journal of Formal Logic* 25:303–22.

Brown, P., and S. Levinson. 1978. "Universals in Language Usage: Politeness Phenomena." In *Questions and Politeness,* E. Goody, ed., 56–311. Cambridge: Cambridge University Press.

Buelens, J. 1972. "Negation in Freud." *Logique et Analyse* 15:319–31.

Buridan, John. 1966. *John Buridan: Sophisms on Meaning and Truth*. T. R. Scott, trans. New York: Appleton-Century-Crofts.

Burton-Roberts, N. 1984. "Modality and Implicature." *Linguistics and Philosophy* 7:181–206.

———. 1987. "On Horn's Dilemma: Presupposition and Negation." University of Newcastle upon Tyne.

Burtt, E. A. 1955. "What Can Western Philosophy Learn from India?" *Philosophy East and West* 5:195–210.

Buyssens, E. 1959. "Negative Contexts." *English Studies* 40:163–69.

Campbell, R. N. 1981. "Language Acquisition, Psychological Dualism, and the Definition of Pragmatics." In Parret et al., eds., 93–103.

Caplan, I. 1978. "Understatement." Unpublished. LSA Linguistic Institute, University of Illinois.

Carden, G. 1970. "A Note on Conflicting Idiolects." *Linguistic Inquiry* 1:281–90.

———. 1973. *English Quantifiers*. Tokyo: Taishukan.

Carlson, G. 1977. *Reference to Kinds in English*. Ph.D. diss., University of Massachusetts. Distributed by IULC.

Carlson, G., and T. Roeper. 1980. "Morphology and Subcategorization." In Hoekstra et al., eds., 123–64.

Carlson, L. 1983. *Dialogue Games*. Dordrecht: D. Reidel.

Carroll, J., and M. Tanenhaus. 1975. "Prolegomena to a Theory of Word Formation." In Grossman, San, and Vance, eds., 47–62.

Carston, R. 1985a. "A Reanalysis of Some 'Quantity Implicatures.'" University College, London.

———. 1985b. "Saying and Implicating." University College, London.

Castañeda, H.-N. 1976. "Leibniz' Syllogistico-Propositional Calculus." *Notre Dame Journal of Formal Logic* 17:481–500.

Cattell, R. 1973. "Negative Transportation and Tag Questions." *Language* 49: 612–39.

Channell, J. 1980. "More on Approximations: A Reply to Wachtel." *Journal of Pragmatics* 4:461–76.

Chao, Y. R. 1955. "Notes on Chinese Grammar and Logic." *Philosophy East and West* 5:31–41.

Charniak, E. 1972. *Towards a Model of Children's Story Comprehension*. MIT Artificial Intelligence Laboratory Monographs, No. 226.

Chattopadhyaya, D., and M. Gangopadhyaya. 1968. *Nyāya Philosophy: Literal Translation of Gautama's Nyāya-Sūtra and Vātsyāyana's Bhāṣya*. Part 2: Second Adhyāya. Calcutta: Indian Studies: Past and Present.

Choi, S. 1983. "Some Aspects of Negation in Korean." Unpublished. Yale University.

Chomsky, N. 1957. *Syntactic Structures*. The Hague: Mouton.

———. 1965. *Aspects of the Theory of Syntax*. Cambridge: MIT Press.

———. 1970. "Deep Structure, Surface Structure, and Semantic Interpretation." Reprinted in Steinberg and Jakobovitz, eds., 183–216. Also in Chomsky (1972), 62–119.

———. 1971. "Some Empirical Issues in the Theory of Transformational Grammar." Reprinted in Chomsky (1972), 120–202.

———. 1972. *Studies on Semantics in Generative Grammar.* The Hague: Mouton.

———. 1981. *Lectures on Government and Binding.* Dordrecht: Foris.

Churchward, C. M. 1953. *Tongan Grammar.* Oxford University Press.

Clark, E. V. Forthcoming. "The Principle of Contrast." In *Proceedings of the Twentieth Annual Carnegie Symposium on Cognition: Mechanisms of Language Acquisition,* B. MacWhinney, ed.

Clark, E. V., and H. H. Clark. 1979. "When Nouns Surface as Verbs." *Language* 55:767–811.

Clark, H. H. 1971. "The Chronometric Study of Meaning Components." In *Colloques Internationaux du C.N.R.S.*, no. 206, Problèmes actuels en Psycholinguistique, 489–505.

———. 1974. "Semantics and Comprehension." In *Current Trends in Linguistics,* vol. 12, T. Sebeok, ed., 1291–1428. The Hague: Mouton. Also published separately; The Hague: Mouton, 1976.

Clark, H. H., P. A. Carpenter, and M. A. Just. 1973. "On the Meeting of Semantics and Perception." In *Visual Information Processing,* W. G. Chase, ed., 311–81. New York: Academic Press.

Clark, H. H., and E. V. Clark. 1977. *Psychology of Language: An Introduction to Psycholinguistics.* New York: Harcourt, Brace.

Clark, H., and S. Haviland. 1977. "Comprehension and the Given-New Contract." In *Discourse Production and Comprehension,* R. Freedle, ed., 1–40. Hillside, N.J.: Lawrence Erlbaum.

Clark, R. 1974. "Ontology and the Philosophy of Mind in Sellars' Critique of Russell." In *Bertrand Russell's Philosophy,* G. Nakhnikian, ed., 101–16. London: Duckworth.

Cohen, L. J. 1971. "The Logical Particles of Natural Language." In *Pragmatics of Natural Language,* Y. Bar-Hillel, ed., 50–68. Dordrecht: D. Reidel.

Cole, P. 1975. "The Synchronic and Diachronic Status of Conversational Implicature." In Cole and Morgan, eds., 257–88.

Cole, P. ed. 1978. *Syntax and Semantics 9: Pragmatics.* New York: Academic Press.

———. 1981. *Radical Pragmatics.* New York: Academic Press.

Cole, P., and J. Morgan, eds. 1975. *Syntax and Semantics 3: Speech Acts.* New York: Academic Press.

Collinson, W. 1937. *Indication.* Language Monograph Number 17.

Comrie, B. 1985. "The Possessive Predicate in North African Vernacular Arabic." Paper presented at the Sixteenth Annual Conference on African Linguistics, New Haven.

Coombs, V. 1976. *A Semantic Syntax of Grammatical Negation in the Older Germanic Dialects.* Göppingen Arbeiten zur Germanistik, no. 177. Göppingen: Verlag Alfred Kümmerle.

Cooper, R. 1983. *Quantification and Syntactic Theory.* Dordrecht: D. Reidel.

———. 1984. "Sentence Negation in Situation Semantics." *CLS 20,* 25–33.

———. 1975. *Montague's Semantic Theory and Transformational Grammar.* Ph.D. diss., University of Massachusetts.

Cooper, W., and J. R. Ross. 1975. "World Order." In Grossman, Vance, and San, eds., 63–111.

Cormack, A. 1980. "Negation, Ambiguity, and Logical Form." Unpublished. University of London.

Cornish, E. R. 1971. "Pragmatic Aspects of Negation in Sentence Evaluation and Completion Tasks." *British Journal of Psychology* 62:505–11.

Cornish, E. R., and P. C. Wason. 1970. "The Recall of Affirmative and Negative Sentences in an Incidental Learning Task." *Quarterly Journal of Experimental Psychology* 22:109–14.

Cornulier, B. de. 1973. "Sur une règle de déplacement de négation." *Le français moderne* 41:43–57.

———. 1974. "Remarques à propos de la negation anticipée." *Le français moderne* 42:206–16.

———. 1984. "Pour l'analyse minimaliste de certaines expressions de quantité; Réponse à des objections d'Anscombre et Ducrot." *Journal of Pragmatics* 8:661–91.

Covington, M. 1981. "Evidence for Lexicalism: A Critical Review." Distributed by IULC.

Cresswell, M. 1973. *Logics and Languages.* London: Methuen.

Crockett, D. 1977. "The Scope of Denial in Russian Negative Sentences." *Lingua* 43:229–45.

Cruse, D. 1980. "Antonyms and Gradable Complementaries." In *Perspektiven der lexikalischen Semantik,* D. Kastovsky, ed., 14–25. Bonn: Bouvier Verlag Herbert Grundmann.

———. 1986. *Lexical Semantics.* Cambridge: Cambridge University Press.

Culicover, P. 1981. "Negative Curiosities." Distributed by IULC.

Curme, G. 1931. *A Grammar of the English Language.* New York: D. C. Heath.

Cutler, A. 1977. "The Context-Dependence of 'Intonational Meanings'" *CLS 13,* 104–15.

Dahl, Ö. 1979. "Typology of Sentence Negation." *Linguistics* 17:79–106.

———. 1981. "In Defense of a Strawsonian Approach to Presupposition." In *Crossing the Boundaries in Linguistics,* W. Klein and W. Levelt, eds., 191–200. Dordrecht: D. Reidel.

Dancy, R. M. 1975. *Sense and Contradiction: A Study in Aristotle.* Dordrecht: D. Reidel.

Danell, K. J. 1974. "La concurrence *pas de vin—pas du vin.*" Studia Neophilologica 46:409–25.

Davidson, D., and G. Harman, eds. 1972. *Semantics of Natural Language.* Cambridge: Cambridge University Press.

Davidson, R. M. 1967. "The Use of the Genitive in Negative Constructions." *Studies in the Modern Russian Language,* 2. Cambridge: Cambridge University Press.

Davies, A. M. 1975. "Negation and Disjunction in Anatolian—and Elsewhere." *Anatolian Studies* 25:157–68.

Davison, A. 1978. "Negative Scope and Rules of Conversation: Evidence from an OV Language." In Cole, ed. (1978), 23–45.

Delacruz, E. 1976. "Factives and Proposition Level Constructions in Montague Grammar." In Partee, ed., 177–99.

Delbrück, B. 1910. *Germanische Syntax I. Zu den negativen Sätzen.* Leipzig: B. G. Teubner.

De Mey, M. 1972. "The Psychology of Negation and Attention." *Logique et Analyse* 15:137–53.

De Morgan, A. 1847. *Formal Logic.* London: Taylor & Walton.

———. 1858. "On the Syllogism: III, and on Logic in General." Reprinted in *On the Syllogism and Other Logical Writings,* P. Heath, ed., 74–146. London: Routledge & Kegan Paul.

Demos, R. 1917. "A Discussion of a Certain Type of Negative Proposition." *Mind* 26:188–96.

Deutscher, M. 1965. "A Note on Saying and Disbelieving." *Analysis* 105:53–57.

de Villiers, J. G., and H. B. T. Flusberg. 1975. "Some Facts One Simply Cannot Deny." *Journal of Child Language* 2:279–86.

Dieterich, T. G., and D. J. Napoli. 1982. "Comparative *rather.*" *Journal of Linguistics* 18:137–65.

Dixon, R. M. W. 1979. "Ergativity." *Language* 55:59–138.

Donnellan, K. 1966. "Reference and Definite Descriptions." *Philosophical Review* 75:281–304.

———. 1970. "Categories, Negation, and the Liar Paradox." In Martin, ed., 113–20.

———. 1978. "Speaker Reference, Descriptions and Anaphora." In Cole, ed. (1978), 47–68.

Dowty, D. 1979. *Word Meaning and Montague Grammar.* Dordrecht: D. Reidel.

———. 1982. "Grammatical Relations and Montague Grammar." In *The Nature of Syntactic Representation,* P. Jacobson and G. Pullum, eds., 79–130. Dordrecht: D. Reidel.

Dowty, D., R. Wall, and S. Peters. 1981. *Introduction to Montague Semantics.* Dordrecht: D. Reidel.

Drange, T. 1966. *Type Crossings.* The Hague: Mouton.

DuBois, J. 1974. "Syntax in Mid-Sentence." *Berkeley Studies in Syntax and Semantics,* vol. 1, 3:1–25. Berkeley: Department of Linguistics and Institute of Human Learning, University of California.

Ducrot, O. 1972. *Dire et ne pas dire.* Paris: Hermann.

———. 1973. *La preuve et le dire.* Paris: Maison Mame.

Ducrot, O., and C. Vogt. 1979. "De *magis* à *mais:* Une hypothèse sémantique," *Revue de linguistique romane* 43:317–41.

Dummett, M. 1973. *Frege: Philosophy of Language.* London: Duckworth.

Edgerton, F., trans. 1929. *The Mīmāṅsā Nyāya Prakāsa.* New Haven: Yale University Press.

Eifermann, R. 1961. "Negation: A Linguistic Variable." *Acta Psych.* 18:258–73.

Emonds, J. 1976. *A Transformational Approach to Syntax.* New York: Academic Press.

Enç, M. 1981. Tense without Scope: An Analysis of Nouns as Indexicals. Ph.D. diss., University of Wisconsin-Madison.

Englebretsen, G. 1974. "A Note on Contrariety." *Notre Dame Journal of Formal Logic* 15:613–14.

———. 1976. "The Square of Opposition." *Notre Dame Journal of Formal Logic* 17:531–41.

———. 1981a. *Logical Negation.* Assen, Netherlands: Van Gorcum.

———. 1981b. *Three Logicians: Aristotle, Leibniz, and Sommers and the Syllogistic.* Assen, Netherlands: Van Gorcum.

———. 1984. "Opposition." *Notre Dame Journal of Formal Logic* 25:79–85.

Epstein, S. 1976. Investigations in Pragmatic Theory. Ph.D. diss., University of California, San Diego.

Erasmus, D. [1519] 1650. *Colloquia.* Amsterdam: L. Elzeverius.

Escure, G. 1974. "Negation and Dialect Variation in French." *Papers in Linguistics* 7:403–34.

Ewing, A. C. 1937. "Meaningless." *Mind* 46:347–64.

Fauconnier, G. 1975a. "Polarity and the Scale Principle." *CLS 11,* 188–99.

———. 1975b. "Pragmatic Scales and Logical Structures." *Linguistic Inquiry* 6:353–75.

———. 1976. "Remarque sur la théorie des phénomèmes scalaires." *Semantikos* 1:13–36.

———. 1979a. "Implication Reversal in a Natural Language." In *Formal Semantics and Pragmatics for Natural Languages,* F. Guenther and S. J. Schmidt, eds., 289–301. Dordrecht: D. Reidel.

———. 1979b. "Comment contrôler la vérité; Remarques illustrés par des assertions dangereuses et pernicieuses en tout genre." *Actes de la recherche en sciences sociales,* 25.

Fillmore, C. 1963. "The Position of Embedding Transformations in a Grammar." *Word* 19:208–31.

———. 1968. "The Case for Case." In Bach and Harms, eds., 1–88.

———. 1971. "Types of Lexical Information." In Steinberg and Jakobovitz, eds., 370–92.

Fillmore, C., and D. T. Langendoen, eds. 1971. *Studies in Linguistic Semantics*. New York: Holt.

Firbas, J. 1964. "On Defining the Theme in Functional Sentence Analysis." *Travaux Linguistiques de Prague 1*, 267–80.

———. 1966. "Non-Thematic Subjects in Contemporary English." *Travaux Linguistiques de Prague 2*, 239–56.

Fodor, J. A., and M. Garrett. 1966. "Competence and Performance." In Lyons and Wales, eds., 135–73.

Fodor, J. A., and J. Katz, eds. 1964. *The Structure of Language*. Englewood Cliffs, N.J.: Prentice-Hall.

Fodor, J. D. 1979. "In Defense of the Truth-Value Gap." In Oh and Dinneen, eds., 199–244.

Fodor, J. D., J. A. Fodor, and M. F. Garrett. 1975. "The Psychological Unreality of Semantic Representations." *Linguistic Inquiry* 6:515–31.

Foley, W. 1975. "How to Say 'No' in Palauan." *Proceedings of the Berkeley Linguistic Society 1*, 132–46.

Fraiberg, S. 1959. *The Magic Years*. New York: Scribner.

Fraser, B. 1971. "An Analysis of 'Even' in English." In Fillmore and Langendoen, eds., 150–78.

———. 1975. "Hedged Performatives." In Cole & Morgan, eds., 187–210.

Frege, G. 1892. "Über Sinn und Bedeutung." *Zeitschrift für Philosophie und philosophische Kritik*, NF 100:25–50. Reprinted as "On Sense and Reference" in Geach and Black, eds., 56–78.

———. 1919. "Negation." *Beiträge zur Philosophie des Deutschen Idealismus* 1:143–57. Reprinted in Geach and Black, eds., 117–35.

Freud, S. 1910. "The Antithetical Meaning of Primal Words." In Strachey, ed., vol. 11, 155–61.

———. [1924] 1960. *A General Introduction to Psychoanalysis*. J. Riviere, trans. New York: Washington Square Press.

———. 1925. "Negation." In Strachey, ed., vol. 19, 235–9.

Friedrich, C., ed. 1953. *The Philosophy of Hegel*. New York: Modern Library.

Funk, W.-P. 1971. "Adjectives with Negative Prefixes in Modern English and the Problem of Synonymy." *Zeitschrift für Anglistik und Amerikanistik* 19:364–86.

Gaatone, E. 1971. *Etude descriptive du système de la négation en français contemporain*. Geneva: Droz.

Gabbay, D., and J. Moravcsik. 1978. "Negation and Denial." In *Studies in Formal Semantics*, F. Guenther and C. Rohrer, eds., 251–65. Amsterdam: North-Holland.

Gale, R. M. 1970. "Negative Statements." *American Philosophical Quarterly* 7:206–17.

———. 1972. "On What There Isn't." *Review of Metaphysics* 25:459–88.

———. 1976. *Negation and Non-Being*. American Philosophical Quarterly Monograph no. 10. Oxford: Basil Blackwell.

García, E. 1975. "*Other than* Ambiguity." Unpublished. Lehmann College, CUNY.

Garner, R. T. 1970. "Some Doubts about Illocutionary Negation." *Analysis* 31:106–12.

Gates, D. L., Jr., and O. D. Seright. 1967. "Negative-Contrastive Constructions in Standard Modern English." *American Speech* 42:136–42.

Gazdar, G. 1977. "Univocal *or*." In *CLS Book of Squibs*, S. Fox et al., eds., 44–45. Chicago: Chicago Linguistic Society.

———. 1979a. *Pragmatics: Implicature, Presupposition, and Logical Form*. New York: Academic Press.

———. 1979b. "A Solution to the Projection Problem." In Oh and Dinneen, eds., 57–89.

Gazdar, G., E. Klein, G. Pullum, and I. Sag. 1985. *Generalized Phrase Structure Grammar*. Cambridge, MA: Cambridge University Press.

Gazdar, G., and G. Pullum. 1976. "Truth-Functional Connectives in Natural Language." *CLS 12*, 220–34.

Gazdar, G., G. Pullum, and I. Sag. 1982. "Auxiliaries and Related Phenomena in a Restrictive Theory of Grammar." *Language* 58:591–638.

Geach, P. T. 1962. *Reference and Generality*. Ithaca: Cornell University Press.

———. 1970. "A Program for Syntax." *Synthèse* 22:3–17. Also in Davidson and Harman, eds., 483–97.

———. [1972] 1980. *Logic Matters*. Reprint. Berkeley and Los Angeles: University of California Press.

Geach, P. T., and M. Black, eds. 1952. *Translations from the Philosophical Writings of Gottlob Frege*. Oxford: Blackwell.

Gebauer, J. 1885. "Über die Negation insbesondere im Altböhmischen." *Archiv für slavische Philologie* 8:177–93.

Geis, M., and A. M. Zwicky. 1971. "On Invited Inferences." *Linguistic Inquiry* 2:561–65.

Givón, T. 1970. "Notes on the Semantic Structure of English Adjectives." *Language* 46:816–37.

———. 1978. "Negation in Language: Pragmatics, Function, Ontology." In Cole, ed. (1978), 69–112.

———. 1979. *On Understanding Grammar*. New York: Academic Press.

Goddard, L. 1960. "The Exclusive 'or'." *Analysis* 20:97–105.

Gordon, D., and G. Lakoff. 1971. "Conversational Postulates." In *CLS 7*, 63–84.

Gowda, K. S. G. 1970. "Negation in Kannada." In Biligiri, ed., 1–4.

Graham, A. C. 1959. "'Being' in Western Philosophy compared with *shih/fei* and *yu/wu*." *Asia Major* 7:79–112.

Green, G. 1974. *Semantics and Syntactic Regularity*. Bloomington: Indiana University Press.

———. 1975. "How to Get People to Do Things with Words: The Whimperative Question." In Cole and Morgan, eds., 107–42.

———. 1976. "Main Clause Phenomena in Subordinate Clauses." *Language* 52:382–97.

Greenbaum, S. 1974. "Problems in the Negation of Modals." *Moderna Språk* 68:244–55.

Greenbaum, S., G. Leech, and J. Svartvik, eds. 1980. *Studies in English Linguistics*. London: Longman.

Greenberg, J. 1963. "Some Universals of Grammar with Particular Reference to the Order of Meaningful Elements." In Greenberg, ed., 73–113.

———. 1966. *Language Universals, with Special Reference to Feature Hierarchies*. The Hague: Mouton.

Greenberg, J., ed. 1963. *Universals of Language*. Cambridge: MIT Press.

Greenberg, J., C. Ferguson, and E. Moravcsik, eds. 1978. *Universals of Human Language*. vol. 4, *Syntax*. Stanford: Stanford University Press.

Greene, J. 1970a. "The Semantic Function of Negatives and Passives." *British Journal of Psychology* 61:17–22.

———. 1970b. "Syntactic Form and Semantic Function." *Quarterly Journal of Experimental Psychology* 22:14–27.

Greene, J., and P. C. Wason. 1970. "Negation: A Rejoinder to Wales and Grieve." *Perception and Psychophysics* 8:238–9.

Grevisse, M. 1969. *Le bon usage*. 9th ed. Gembloux, Belgium: J. Duculot.

Grice, H. P. 1961. "The Causal Theory of Perception." *Proceedings of the Aristotelian Society* Supplementary volume 35:121–52.

———. 1967. Logic and Conversation. Unpublished. William James lectures at Harvard University.

———. 1975. "Logic and Conversation." In Cole and Morgan, eds., 41–58. Lecture 2 of William James lectures.

———. 1978. "Further Notes on Logic and Conversation." In Cole, ed. (1978), 113–28. Lecture 3 of Grice (1967).

———. 1981. "Presupposition and Conversational Implicature." In Cole, ed. (1981), 183–98.

Grim, P. 1981. "A Note on the Ethics of Theories of Truth." In Vetterling-Braggin, ed., 290–98.

Gross, M. 1977. "Une analyse non présuppositionelle de l'effet contrastif." *Lingvisticae Investigationes* 1:39–62.

Grossman, R., L. San, and T. Vance, eds. 1975. *Papers from the Parasession on Functionalism*. Chicago: Chicago Linguistic Society.

Guérard, A. L. 1922. *A Short History of the International Language Movement*. London: T. Fisher Unwin.

Gundel, J. 1985. "'Shared Knowledge' and Topicality." *Journal of Pragmatics* 9:83–107.

Halpern, R. 1976. "The Bivalence of Neg Raising Predicates." *Studies in the Linguistic Sciences* 6:69–81.

Halvorsen, P.-K. 1978. *The Syntax and Semantics of Cleft Constructions*. Ph.D. diss., University of Texas. *Texas Linguistics Forum* 11.

Hamilton, Sir William, of Edinburgh. 1860. *Lectures on Logic*. Edinburgh: Blackwood and Sons.

Hare, R. M. 1970. "Meaning and Speech Acts." *Philosophical Review* 79:3–24.

Harnish, R. 1976. "Logical Form and Implicature." In *An Integrated Theory of Linguistic Ability*, T. G. Bever, J. J. Katz, and D. T. Langendoen, eds., 464–79. New York: Crowell.

Harries, H. 1973. "A Note on Double Negation and Modals as Main Verbs." *Working Papers on Language Universals* 11:211–17.

Hart, H. L. A. 1951. "A Logician's Fairy Tale." *Philosophical Review* 60:98–212.

Hawkins, J. 1978. *Definiteness and Indefiniteness*. London: Croom Helm.

Hayakawa, S. I. 1949. *Language in Thought and Action*. New York: Harcourt, Brace.

Hegel, G. W. F. [1812–16] 1929. *Science of Logic*, vol. 1. W. H. Johnston and L. G. Struthers, ed. and trans. London: George Allen & Unwin.

———. 1892. *The Logic of Hegel: Translated from the Encyclopædia of the Philosophical Sciences*, W. Wallace, trans. 2d ed. London: Oxford University Press.

Heim, I. 1982. The Semantics of Definite and Indefinite Noun Phrases. Ph.D. diss., University of Massachusetts.

———. 1984. "A Note on Negative Polarity and Downward Entailingness." In *Proceedings of NELS 14*, 98–107.

Heinemann, F. H. 1944. "The Meaning of Negation." *Proceedings of the Aristotelian Society* 44:127–52.

Heldner, C. 1981. *La portée de la négation*. Stockholm: Norstedts Tryckeri.

Henry, D. 1967. *The Logic of Saint Anselm*. Oxford: Oxford University Press.

———. 1972. *Medieval Logic and Metaphysics*. London: Hutchinson University Library.

Heringer, J. 1970. "Research on Quantifier-Negative Idiolects." *CLS 6*, 287–96.

Herzberger, H. 1970. "Truth and Modality in Semantically Closed Languages." In Martin, ed., 25–46.

———. 1971. "Setting Russell Free." Unpublished. University of Toronto.

Heyting, A. 1971. *Intuitionism, an Introduction*. 3d ed. Amsterdam: North-Holland.

Hilbert, D. 1927. "The Foundations of Mathematics." Translated in van Heijenoort, ed., 464–74.

Hintikka, J. 1960. "Aristotle's Different Possibilities." *Inquiry* 3:18–28. Reprinted as chapter 2 of Hintikka (1973).

———. 1962. *Knowledge and Belief*. Ithaca: Cornell University Press.

———. 1968. "Epistemic Logic and the Methods of Philosophical Analysis." *Australasian Journal of Philosophy* 46:37–51.

———. 1973. *Time and Necessity: Studies in Aristotle's Theory of Modality.* Dordrecht: D. Reidel.

Hirschberg, J. 1985. A Theory of Scalar Implicature. Ph.D. diss., University of Pennsylvania. Distributed by Department of Computer and Information Science, Moore School of Electrical Engineering, University of Pennsylvania.

Hoeksema, J. 1987. Review of Jacobs (1982). *Language* 63:432.

Hoekstra, T., H. van der Hulst, and M. Moortgat, eds. 1980. *Lexical Grammar.* Dordrecht: Foris.

Hofmann, T. R. 1966. "Past Tense Replacement and the Modal System." *Harvard University Computation Lab Report 17.* Cambridge, MA.

Hogben, L. 1943. *Interglossa.* New York: Penguin.

Hoosain, R., and C. Osgood. 1975. "Response Time for Yang (Positive) and Yin (Negative) Words." Unpublished. University of Illinois-Urbana.

Horn, L. 1969. "A Presuppositional Analysis of *only* and *even.*" *CLS 5,* 97–108.

———. 1970. "Ain't it hard (anymore)." *CLS 6,* 318–327.

———. 1971. "Negative Transportation: Unsafe at Any Speed?" *CLS 7,* 120–133.

———. 1972. *On the Semantic Properties of Logical Operators in English.* Distributed by IULC, 1976.

———. 1973. "Greek Grice: A Brief Survey of Proto-Conversational Rules in the History of Logic." *CLS 9,* 205–214.

———. 1975. "Neg-Raising Predicates: Towards an Explanation." *CLS 11,* 279–94.

———. 1978a. "Some Aspects of Negation." In Greenberg et al., eds., 127–210.

———. 1978b. "Remarks on Neg-Raising." *Syntax and Semantics 9: Pragmatics,* P. Cole, ed., 129–220. New York: Academic Press.

———. 1978c. "Lexical Incorporation, Implicature, and the Least Effort Hypothesis." *Papers from the Parasession on the Lexicon,* D. Farkas et al., eds., 196–209. Chicago: Chicago Linguistic Society.

———. 1979. "*Only, even,* and Conventional Implicature." Paper presented at the LSA annual meeting, Los Angeles.

———. 1980. "Affixation and the Unaccusative Hypothesis." *CLS 16,* 134–146.

———. 1981a. "Exhaustiveness and the Semantics of Clefts." *Proceedings of NELS 11,* 125–142.

———. 1981b. "A Pragmatic Approach to Certain Ambiguities." *Linguistics and Philosophy* 4:321–58.

———. 1984a. "Ambiguity, Negation, and the London School of Parsimony." *Proceedings of NELS 14,* 108–131.

———. 1984b. "Toward a New Taxonomy for Pragmatic Inference: Q-Based and R-Based Implicature." In Schiffrin, ed., 11–42.

———. 1984c. "In Defense of Privative Ambiguity." *Proceedings of the Berkeley Linguistics Society 10,* 141–56.

———. 1985. "Metalinguistic Negation and Pragmatic Ambiguity." *Language* 61:121–74.

———. 1986. "Presupposition, Theme and Variations." *Papers from the Parasession on Pragmatics and Grammatical Theory,* A. M. Farley et al., eds., 168–192. Chicago: Chicago Linguistic Society.

Horn, L., and S. Bayer. 1984. "Short-Circuited Implicature: A Negative Contribution." *Linguistics and Philosophy* 7:397–414.

Horn, L., and M. Covington. 1987. "Informativeness, Entropy, and the Redundant *un*-verb." Paper presented at the LSA annual meeting, San Francisco.

Horn, L., and D. Farkas. 1985. "Aristotle as a Montague Grammarian." Paper presented at the LSA annual meeting, Seattle.

Hust, J. 1975. "Dissuaded." *Linguistic Analysis* 1:173–89.

———. 1977. "The Syntax of the Unpassive Construction in English." *Linguistic Analysis* 3:31–63.

Ingalls, D. H. H. 1951. *Materials for the Study of Navya-Nyāya Logic.* Cambridge: Harvard University Press.

Jackendoff, R. 1969. "An Interpretive Theory of Negation." *Foundations of Language* 5:218–41.

———. 1971. "On Some Questionable Arguments about Quantifiers and Negation." *Language* 47:282–97.

———. 1972. *Semantic Interpretation in Generative Grammar.* Cambridge: MIT Press.

Jacobs, J. 1982. *Syntax und Semantik der Deutschen.* Munich: Fink.

Jacobsson, B. 1970. "Notes on Negation in English." *Moderna Språk* 64:18–29.

Jakobson, R. 1939. "Signe zéro." In *Mélanges de linguistique, offerts à Charles Bally,* 143–52. Geneva: Georg.

Jespersen, O. 1917. *Negation in English and Other Languages.* Copenhagen: A. F. Høst.

———. 1922. *Language: Its Nature, Development and Origin.* New York: Henry Holt.

———. 1924. *The Philosophy of Grammar.* London: Allen & Unwin.

———. 1928. *An International Auxiliary Language.* London: Allen & Unwin.

———. [1933] 1964. *Essentials of English Grammar.* Reprint. University: University of Alabama Press.

———. 1942. *A Modern English Grammar on Historical Principles. Part 6: Morphology.* Copenhagen: Einar Munksgaard.

———. 1949. *A Modern English Grammar on Historical Principles. Part 2: Syntax, First Volume.* Copenhagen: Einar Munksgaard. Written in 1918.

Jhering, R. von. [1883] 1923. *Der Zweck im Recht,* vol. 2, 8th ed. Leipzig: Breitkopf und Härtel.

Johnson-Laird, P. N., and J. M. Tridgell. 1972. "When Negation is Easier than Affirmation." *Quarterly Journal of Experimental Psychology* 24:87–91.

Joly, A. 1972. "La négation dite 'explétive' en vieil anglais et dans d'autres langues indo-européennes.' *Études anglaises* 25:30–44.

Joseph, H. W. B. 1916. *An Introduction to Logic*. 2d ed. Oxford: Clarendon Press.

Just, M. A., and P. A. Carpenter. 1971. "Comprehension of Negation with Quantification." *Journal of Verbal Learning and Verbal Behavior* 12:21–31.

Kadmon, N. 1984. "Indefinite Noun Phrases with Cardinality Indication." Unpublished. University of Massachusetts.

Kalepky, Th. 1891. *Von der Negation im Provenzalischen*. Berlin: R. Gaertners Verlagbuchhandlung.

Kamp, H. 1981. "A Theory of Truth and Semantic Representation." In *Formal Methods in the Study of Language*, J. Groenendijk, T. Janssen, and M. Stokhof, eds., 277–322. Amsterdam: Mathematical Centre Tracts.

Kant, I. [1787] 1964. *Critique of Pure Reason*. 2d ed. N. K. Smith, trans. London: Macmillan.

Kaplan, D. 1978. "DTHAT." In Cole, ed. (1978), 221–43.

Karttunen, L. 1971. "Implicative Verbs." *Language* 47:340–58.

———. 1974. "Presupposition and Linguistic Context." *Theoretical Linguistics* 1:181–93.

Karttunen, L., and S. Peters. 1976. "What Indirect Questions Conventionally Implicate." *CLS 12*, 351–68.

———. 1979. "Conventional Implicature." In Oh and Dinneen, eds., 1–56.

Kato, Y. 1985. *Negative Sentences in Japanese*. Sophia Linguistica number 19. Tokyo: Linguistic Institute for International Communication, Sophia University.

Katz, J. 1964. "Antonymy and Contradiction in Natural Language." In Fodor and Katz, eds., 519–43.

———. 1972. *Semantic Theory*. New York: Harper & Row.

———. 1977. *Propositional Structure and Illocutionary Force*. Cambridge: MIT Press.

Katz, J., and P. Postal. 1964. *An Integrated Theory of Linguistic Descriptions*. Cambridge: MIT Press.

Kaufmann, W. 1965. *Hegel*. Garden City, NY: Doubleday.

Kay, P. 1987. "Scalar Semantics and the Word EVEN." Paper presented at the Symposium on Lexical Semantics, Linguistic Institute, Stanford, CA.

Keenan, E. L. 1969. A Logical Base for English. Ph.D. diss., University of Pennsylvania.

———. 1971. "Two Kinds of Presupposition in Natural Language." In Fillmore and Langendoen, eds., 45–54.

Keenan, E. L., and B. Comrie, 1977. "Noun Phrase Accessibility and Universal Grammar." *Linguistic Inquiry* 8:63–98.

Keenan, E. L., and L. M. Faltz. 1978. *Logical Types for Natural Language*. UCLA Occasional Papers in Linguistics number 3. Los Angeles: UCLA Department of Linguistics.

Keenan, E. O. 1976. "The Universality of Conversational Postulates." *Language in Society* 5:67–80.

Keller-Cohen, D., K. C. Chalmer, and J. Remler. 1979. "The Development of Discourse Negation in the Nonnative Child." In Ochs and Schieffelin, eds., 305–22.

Kempson, R. 1975. *Presupposition and the Delimitation of Semantics*. Cambridge: Cambridge University Press.

———. 1977. *Semantic Theory*. Cambridge: Cambridge University Press.

———. 1979. "Presupposition, Opacity, and Ambiguity." In Oh and Dinneen, eds., 283–97.

———. 1980. "Ambiguity and Word Meaning." In Greenbaum, et al., eds., 7–16.

———. 1986. "Ambiguity, and the Semantics-Pragmatics Distinction." In *Meaning and Interpretation*, C. Travis, ed., 77–103. Oxford: Blackwell.

Kempson, R., and A. Cormack, 1979. "Ambiguity and Quantification." *Linguistics and Philosophy* 4:259–309.

Kennedy, G. A. 1952. "Negatives in Classical Chinese." *Wennti* 1:1–46.

Keynes, J. N. 1906. *Studies and Exercises in Formal Logic*. London: Macmillan.

Kiefer, F. 1977. "Two Studies on Presupposition." *Lingua* 43:247–71. Review article on Wilson (1975) and Kempson (1975).

Kimball, J., ed. 1975. *Syntax and Semantics 4*. New York: Academic Press.

Kimenyi, A. 1973. "Ambiguity in Neg-Constructions." Unpublished. UCLA.

Kiparsky, C., and P. Kiparsky. 1971. "Fact." In Steinberg and Jakobovitz, 345–69.

Kiparsky, P. 1982. "Lexical Morphology and Phonology." In *Linguistics in the Morning Calm*, P. Kiparsky, ed., 3–91. Seoul: Hanshin.

———. 1983. "Word Formation and the Lexicon." In *Proceedings of the 1982 Mid-America Linguistics Conference*. Lawrence, KS: University of Kansas Department of Linguistics, 47–78.

Kissin, P. 1969. *Negation in Natural Language*. Ph.D. diss., University of Wisconsin-Madison. Ann Arbor: University Microfilms, 1970.

Kittel, F. 1903. *A Grammar of the Kannaḍa Language*. Mangalore: Basel Mission Book and Tract Depository.

Kleene, S. C. 1938. "On a Notation for Ordinal Numbers." *Journal of Symbolic Logic* 3:150–55.

———. 1952. *Introduction to Metamathematics*. New York: Van Nostrand.

Klein, E. 1980. "A Semantics for Positive and Comparative Adjectives." *Linguistics and Philosophy* 4:1–45.

Klein, S. 1985. "On Ergativity and the Maturing Grammar." Paper presented at the annual LSA meeting, Seattle.

Klima, E. 1964. "Negation in English." In Fodor and Katz, eds., 246–323.

Klima, E., and U. Bellugi. 1966. "Syntactic Regularities in the Speech of Children." In Lyons and Wales, eds., 183–208.

Kneale, W., and M. Kneale. 1962. *The Development of Logic*. Oxford: Clarendon.

König, E. (Forthcoming). "Concessive Relations and Concessive Connectives: Some Cross-Linguistic Observations." In *Explanation for Language Universals*, J. Hawkins, ed.

Korzybski, A. 1933. *Science and Sanity*. Lancaster, PA: The International Non-Aristotelian Library Publishing Co.

Kraak, A. 1966. *Negatieve Zinnen*. Hilversum: W. de Haan.

Kratzer, A. 1977. "What 'must' and 'can' Must and Can Mean." *Linguistics and Philosophy* 1:337–55.

Kretzmann, N. 1968. *William of Sherwood's Treatise on Syncategorematic Words*. Minneapolis: University of Minnesota Press.

Kripke, S. 1972. "Naming and Necessity." In Davidson and Harman, eds., 253–355.

———. 1977. "Speaker's Reference and Semantic Reference." In *Midwest Studies in Philosophy 2*, 255–76.

Kroch, A. 1974. *The Semantics of Scope in English*. New York: Garland.

Kruisinga, E. 1931. *A Handbook of Present-Day English*. 5th ed. *English Accidence and Syntax I*. Groningen: P. Noordhoff.

Kuno, S. 1972. "Functional Sentence Perspective." *Linguistic Inquiry* 3:269–320.

———. 1980. "The Scope of the Question and Negation in Some Verb-Final Languages." *CLS 16*, 155–69.

Kuroda, S.-Y. 1972. "The Categorical and the Thetic Judgment (Evidence from Japanese)." *Foundations of Language* 2:153–85.

———. 1977. "Description of Presuppositional Phenomena from a Non-Presuppositionist Point of View." *Lingvisticae Investigationes* 1:63–162.

Kurrik, M. J. 1979. *Literature and Negation*. New York: Columbia University Press.

Labov, W. 1972. "Negative Attraction and Negative Concord in English Grammar." *Language* 48:773–818.

Ladd, D. 1980. *The Structure of Intonational Meaning*. Bloomington: Indiana University Press.

Ladusaw, W. 1979. *Polarity Sensitivity as Inherent Scope Relations*. Ph.D. diss., University of Texas. Distributed by IULC. Also published, New York: Garland Press, 1980.

———. 1980. "Affective *or*, Factive Verbs, and Negative Polarity Items." *CLS 20*, 170–84.

Lafitte, P. 1962. *Grammaire basque*. Rev. and corr. ed. Bayonne: Editions des Amis du Musée basque.

Lakoff, G. 1965. On the Nature of Syntactic Irregularity. Ph.D. diss., Indiana University. Published as *Irregularity in Syntax*, New York: Holt, 1970.

———. 1969. "On Derivational Constraints." *CLS 5*, 117–39.

———. 1970. "Pronominalization, Negation, and the Analysis of Adverbs." In *Readings in English Transformational Grammar,* R. Jacobs and P. Rosenbaum, eds., 145–65. Waltham, MA: Ginn.

———. 1971. "Presuppositions and Relative Grammaticality." In Steinberg and Jakobovitz, eds., 329–40.

———. 1975. "Pragmatics in Natural Logic." In *Formal Semantics of Natural Language,* E. L. Keenan, ed., 253–86. Cambridge: Cambridge University Press.

Lakoff, R. 1969. "A Syntactic Argument for Negative Transportation." *CLS 5,* 140–47.

———. 1970. "Some Reasons Why There Can't Be Any *some-any* Rule." *Language* 45:608–15.

———. 1975. *Language and Woman's Place.* New York: Harper & Row.

Lambert, K. 1968. "On the No Type Theory of Significance." *Australasian Journal of Philosophy* 46:79–86.

Lambrecht, K. 1987. "Sentence Focus, Information Structure, and the Thetic-Categorical Distinction." *Papers from the Parasession on Grammar and Cognition.* Berkeley: BLS.

Lang, E. 1977. *Semantik der koordinativen Verknüpfung.* Berlin: Akademie-Verlag.

———. 1984. *The Semantics of Coordination.* Trans. by J. Phelby of Lang (1977). Amsterdam: John Benjamins.

Langendoen, D. T. 1981. Review of Oh and Dinneen, eds. (1979). *Language* 57:214–20.

———. 1982. "On a Class of Not Ungrammatical Constructions." *Journal of Linguistics* 18:107–12.

Langendoen, D. T., and T. G. Bever. 1973. "Can a Not Unhappy Person Be Called a Not Sad One?" In Anderson and Kiparsky, eds., 392–409.

Lapointe, S. 1980. "A Lexical Analysis of the English Auxiliary Verb System." In Hoekstra et al., eds., 215–54.

Lasnik, H. 1975. "On the Semantics of Negation." In *Contemporary Research in Philosophical Logic and Linguistic Semantics,* D. Hockney, W. Harper, and B. Freed, eds., 279–311. Dordrecht: D. Reidel.

Lear, J. 1980. *Aristotle and Logical Theory.* Cambridge: Cambridge University Press.

Le Bidois, G., and R. Le Bidois. [1935] 1968. *Syntaxe du français moderne.* 2d ed. Paris: Auguste Picard.

Lee, K. 1974. "Negation in Montague Grammar." *CLS 10,* 378–89.

Leech, G. 1969. *Towards a Semantic Description of English.* London: Longman.

———. 1974. *Semantics.* Harmondsworth: Penguin.

———. 1981. "Pragmatics and Conversational Rhetoric." In Parret et al., eds., 413–42.

———. 1983. *Principles of Pragmatics*. London: Longman.

Leech, G., and J. Coates. 1980. "Semantic Indeterminacy and the Modals." In Greenbaum, et al., eds., 79–90.

Lehmann, W. P. 1974. *Proto-Indo-European Syntax*. Austin: University of Texas Press.

———. ed. 1978. *Syntactic Typology*. Austin: University of Texas Press.

———. 1980. Review of Greenberg, Ferguson, and Moravcsik, eds. (1978), *Language* 56:838–47.

Lehrer, A. 1974. *Semantic Fields and Lexical Structure*. Amsterdam: North-Holland.

———. 1985. "Markedness and Antonymy." *Journal of Linguistics* 21:397–429.

Lehrer, A., and K. Lehrer. 1982. "Antonymy." *Linguistics and Philosophy* 5:483–501.

Leibniz, G. W. 1916. *New Essays Concerning Human Understanding by Gottfried Wilhelm Leibnitz*. A. G. Langley, ed. and trans. 2d ed. Chicago: Open Court.

———. 1966. *Leibniz: Logical Papers*. G. H. R. Parkinson, ed. Oxford: Clarendon.

Lerner, J.-Y., and W. Sternefeld. 1984. "Zum Skopus der Negation im komplexen Satz des Deutschen." *Zeitschrift für Sprachwissenschaft* 3:159–202.

Levergood, B. 1984. "'Only' as an Operator over Conversational Implicatures." Unpublished. University of Texas.

Levi, J. 1978. *The Syntax and Semantics of English Complex Nominals*. New York: Academic Press.

Levinson, S. 1983. *Pragmatics*. Cambridge: Cambridge University Press.

———. 1987a. "Minimization and Conversational Inference." In *The Pragmatic Perspective*, M. Papi and J. Verschueren, eds., 61–129. Amsterdam: Benjamins.

———. 1987b. "Pragmatics and the Grammar of Anaphora." *Journal of Linguistics* 23:379–434.

Lewis, C. I. 1946. *An Analysis of Knowledge and Valuation*. La Salle: Open Court.

Lewis, D. 1969. *Convention*. Cambridge: Harvard University Press.

———. 1970. "General Semantics." *Synthèse* 22:18–67. Also appears in Davidson and Harman, eds., (1972), 169–218.

———. 1975. "Adverbs of Quantification." In *Formal Semantics of Natural Language*, E. Keenan, ed., 3–15. Cambridge: Cambridge University Press.

———. 1979. "Scorekeeping in a Language Game." *Journal of Philosophical Logic* 8:339–59.

Li, C., ed. 1976. *Subject and Topic*. New York: Academic Press.

Liberman, M., and I. Sag. 1974. "Prosodic Form and Discourse Function." *CLS 10*, 402–15.

Lieber, R. 1981. *On the Organization of the Lexicon*. Ph.D. diss., MIT. Distributed by IULC.

Linebarger, M. 1981. *The Grammar of Negative Polarity*. Ph.D. diss., MIT. Distributed by IULC.

———. 1987. "Negative Polarity and Grammatical Representation." *Linguistics and Philosophy* 10:325–87.

Ljung, M. 1974. "Some Remarks on Antonymy." *Language* 50:74–88.

Löbner, S. 1985. "Quantification as a Major Module of Natural Language Semantics." Unpublished. University of Düsseldorf. To appear in *Information, Interpretation, and Inference*, J. Groenendijk, D. de Jong, and M. Stokhof, eds. Dordrecht: D. Reidel.

Lörcher, E. 1900. "Die unechte Negation bei Otfrid und im *Heliand*." *Beiträge zur Geschichte der deutschen Sprache und Literature* 25:543–66.

Lowth, Bishop R. 1762. *A Short Introduction to English Grammar*. London: J. Hughs.

Luborsky, P. 1972. Neg-Hopping in English and Russian: A Comparative Description. Honors thesis, Harvard University.

Lukasiewicz, J. [1910]. 1971. "On the Principle of Contradiction in Aristotle." Trans. by V. Wedin in *Review of Metaphysics* 24:485–509.

———. 1922. "On Determinism." Trans. in McCall, ed., 19–39.

———. 1930. "Philosophical Remarks on Many-Valued systems of Propositional Logic." Trans. in McCall, ed., 40–65.

———. 1934. "On the History of the Logic of Propositions." Trans. in McCall, ed., 66–87.

———. 1970. *Jan Lukasiewicz: Selected Works*. L. Borkowski, ed. Amsterdam: North-Holland.

Lycan, W. 1984. *Logical Form in Natural Language*. Cambridge: MIT Press.

Lyons, J. 1977. *Semantics*. Cambridge: Cambridge University Press.

Lyons, J., and R. J. Wales, eds. 1966. *Psycholinguistics Papers*. Edinburgh: Edinburgh University Press.

Mabbott, J. D. 1929. "Negation." *Aristotelian Society* Supplementary vol. 9, 67–79.

McCall, S. 1967a. "Contrariety." *Notre Dame Journal of Formal Logic* 8:121–32.

McCall, S., ed. 1967b. *Polish Logic 1920–1939*. Oxford: Clarendon.

McCallum-Bayliss, H. 1985. The Modal Verbs: Univocal Lexical Items. Ph.D. diss., Georgetown University.

McCawley, J. D. 1968. "The Role of Semantics in a Grammar." In Bach and Harms, eds., 125–69.

———. 1972. "A Program for Logic." In Davidson and Harman, 496–544.

———. 1978. "Conversational Implicature and the Lexicon." In Cole, ed. (1978), 245–59.

———. 1979. "Presupposition and Discourse Structure." In Oh and Dinneen, 225–34.

———. 1980. "An Un-Syntax." In *Syntax and Semantics 13: Current Approaches to Syntax,* E. Moravcsik and J. Wirth, eds. New York: Academic Press.

———. 1981. *Everything That Linguists Always Wanted to Know about Logic (but Were Ashamed to Ask).* Chicago: University of Chicago Press.

McGloin, N. H. 1976. "Negation." In *Syntax and Semantics 5: Japanese Generative Grammar,* M. Shibatani, ed., 371–419. New York: Academic Press.

———. 1982. Negation in a Verb-Final Language. Unpublished. University of Wisconsin-Madison.

McNeill, D., and N. B. McNeill. 1968. "What Does a Child Mean When He Says No?" In *Proceedings of the Conference on Language and Language Behavior,* E. M. Zale, ed., 51–62. New York: Appleton-Century-Crofts.

McTaggart, J. McT. E. 1922. *Studies in the Hegelian Dialectic.* 2d ed. Cambridge: Cambridge University Press.

Madkour, I. 1934. *L'Organon d'Aristote dans le monde arabe.* Paris: Les Presses modernes.

Madtha, W. 1970. "Negation in Tuḷu." In Biligiri, ed., 25–28.

Magner, T. 1955. "Negation and Case Selection in Russian." *Slavic Word* 11: 531–541.

Malkiel, Y. 1959. "Studies in Irreversible Binomials." *Lingua* 8: 113–60.

Marchand, H. 1938. "Remarks about English Negative Sentences." *American Studies* 20: 198–204.

———. 1960. *The Categories and Types of Present-Day English Word Formation.* Wiesbaden: O. Harrassowitz.

———. 1969. Marchand (1960), 2d ed. Munich: Beck.

———. 1972. "Reversative, Ablative, and Privative Verbs in English, French, and German." In *Issues in Linguistics: Papers in Honor of Henry and Renee Kahane,* B. Kachru, R. Lees, Y. Malkiel, A. Pietrangeli, and S. Saporta, eds., 636–43. Urbana: University of Illinois Press.

Marcuse, H. 1954. *Reason and Revolution: Hegel and the Rise of Social Theory.* New York: Humanities Press.

———. 1968. *Negations: Essays in Critical Theory.* Boston: Beacon.

Martin, J. 1979. Some Misconceptions in the Critique of Semantic Presupposition. Distributed by IULC.

———. 1982. "Negation, Ambiguity, and the Identity Test." *Journal of Semantics* 1: 251–74.

Martin, R. M., ed. 1970. *The Paradox of the Liar.* Reseda, CA: Ridgeview.

Martinet, A. 1962. *A Functional View of Language.* Oxford: Clarendon.

Martinon, P. 1927. *Comment on parle en français.* Paris: Larousse.

Martyna, W. 1983. "Beyond the He/Man Approach: The Case for Non-Sexist Language." In *Language, Gender, and Society,* B. Thorne, C. Kramarae, and N. Henley, eds., 25–37. Rowley, MA: Newbury House.

Mates, B. 1949. "Diodorean Implication." *Philosophical Review* 58:234–52.

———. 1953. *Stoic Logic*. Berkeley and Los Angeles: University of California Press.

Matilal, B. K. 1968. *The Navya-Nyāya Doctrine of Negation*. Cambridge: Harvard University Press.

May, R. 1977. The Grammar of Quantification. Ph.D. diss., MIT.

Maynor, N. 1979. "The Morpheme *un*." *American Speech* 54:310–11.

Melander, J. 1916. *Etudes sur magis et les expressions adversatives dans les langues romanes*. Uppsala: Almqvist & Wiksell.

Menner, R. 1936. "The Conflict of Homonyms in English." *Language* 12: 229–44.

Michalski, K. 1937. *Le problème de la volonté à Oxford et à Paris au XIVe siècle*. Leopoli: Studia Philosophica.

Mill, J. S. [1843] 1919. *A System of Logic*. 8th ed. London: Longman.

———. 1867. *An Examination of Sir William Hamilton's Philosophy*. 3d ed. London: Longman.

Miller, C., and K. Swift. 1976. *Words and Women*. New York: Anchor.

Mirambel, A. 1946. "Négation et mode en grec moderne." *Bulletin de la société linguistique de Paris* 43:57–66.

Mohr, H. 1987. *How to Talk Minnesotan*. New York: Penguin.

Montague, R. 1974. *Formal Philosophy: Selected Papers*. R. Thomason, ed. New Haven: Yale University Press.

Moody, E. 1953. *Truth and Consequence in Mediæval Logic*. Amsterdam: North-Holland.

Morgan, J. 1969. "On Arguing about Semantics." *Papers in Linguistics* 1:49–70.

———. 1978. "Two Types of Convention in Indirect Speech Acts." In Cole, ed. (1978), 261–80.

Morris, C. W. 1938. *Foundations of the Theory of Signs*. Chicago: University of Chicago Press.

Morpurgo-Tagliabue, G. 1981. "Grammar, Logic, and Rhetoric in a Pragmatic Perspective." In Parret et al., eds., 493–508.

Moser, M. 1987. "Pragmatic Ambiguity of Negation." Linguistic Institute, Stanford, CA.

Mullally, J. P. 1945. *The Summulae Logicales of Peter of Spain*. Notre Dame, IN: University of Notre Dame Press.

Munro, P. 1976. "Subject Copying, Auxiliarization and Predicate Raising: The Mojave Evidence," *IJAL* 42:99–112.

Murray, L. 1803. *An English Grammar*. Philadelphia: B. Johnson.

———. 1814. *An English Grammar, 2d edition, improved*. New York: Collins.

Na, Y.-H. 1981. "A Constraint on the Lexicalization of Negation." *CLS 17*, 239–48.

Narasimharao, K. V. V. L. 1970. "Negation in Telugu." In Biligiri, ed., 21–24.

Nelson, J. O. 1954. "In Defense of the Traditional Interpretation of the Square." *Philosophical Review* 63:401–13.

Nemer, J. 1985. "Negation and Clefting in Temne." Paper presented at the Sixteenth Annual Conference on African Linguistics, New Haven.

Newmeyer, F. 1969. *English Aspectual Verbs.* Studies in Linguistics and Language Learning, 6. Seattle: Department of Linguistics, University of Washington.

———. 1980. *Linguistic Theory in America.* New York: Academic Press.

Noreen, A. 1904. *Vårt Språk: Nysvensk Grammatik,* vol. 5. Lund: C. W. K. Gleerup.

Nyrop, K. 1930. *Grammaire historique de la langue française, Tome sixième.* Copenhagen: Gyldendalske Boghandel.

Ochs, E., and B. Schieffelin, eds. 1979. *Developmental Pragmatics.* New York: Academic Press.

O'Donnell, J. R. 1941. "The Syncategoremata of William of Sherwood." *Mediæval Studies* 3:46–93.

Oesterle, J., ed. and trans. 1962. *Aristotle: On Interpretation. Commentary by Saint Thomas and Cajetan.* Milwaukee: Marquette University Press.

Oh, C.-K., and D. Dinneen, eds. 1979. *Syntax and Semantics 11: Presupposition.* New York: Academic Press.

Olshewsky, T., ed. 1969. *Problems in the Philosophy of Language.* New York: Holt.

Orwell, G. 1946. "Politics and the English Language." Reprinted in *Collected Essays,* 353–67. London: Secker & Warburg, 1961.

———. 1949. *Nineteen Eighty-Four.* New York: Harcourt, Brace.

Osgood, C., and M. M. Richards. 1973. "From Yang and Yin to *and* or *but.*" *Language* 49:380–412.

Ōta, A., and Y. Katō. 1986. "The Relative Scope of NEG and Quantifiers in English and Japanese." *Sophia Linguistica* 20/21:25–40.

Otero, C. 1972. "Acceptable Ungrammatical Sentences in Spanish." *Linguistic Inquiry* 2:233–42.

Palmer, F. R. 1979. *Modality and the English Modals.* London: Longman.

Pap, A. 1960. "Types and Meaninglessness." *Mind* 69:41–54.

Parmenides. 1965. *Parmenides. A Text with Translation, Commentary, and Critical Essays.* L. Taran, ed. Princeton: Princeton University Press.

Parret, H., M. Sbisà, and J. Verschueren, eds. 1981. *Possibilities and Limitations of Pragmatics.* Amsterdam: Benjamins.

Parsons, T. 1985. "Underlying Events in English." In *Actions and Events,* E. LePore and B. McLaughlin, eds., 235–67. Oxford: Blackwell.

Partee, B. H. 1970. "Negation, Conjunction, and Quantifiers: Syntax vs. Semantics." *Foundations of Language* 6:153–65.

———. 1973. "The Semantics of Belief Sentences." In *Approaches to Natural Languages,* J. Hintikka, J. Moravcsik, and P. Suppes, eds., 309–36. Dordrecht: D. Reidel.

———. 1979. "Montague Grammar and the Well-Formedness Constraint." In *Syntax and Semantics 10: Selections from the Third Groningen Round Table,* F. Heny and H. Schnelle, eds., 275–313. New York: Academic Press.

Partee, B. H., ed. 1976. *Montague Grammar.* New York: Academic Press.

Patton, T. E. 1968. "Katz on the Semantics of Negation." *Journal of Philosophy* 65:213–31.

Paul, H. 1898. *Principien der Sprachgeschichte.* 3d ed. Halle: M. Niemeyer.

Payne, J. 1985. "Negation." In Shopen, ed., 197–242.

Pea, R. 1980a. "Logic in Early Child Language." *Annals of the New York Academy of Sciences* 27–43.

———. 1980b. "The Development of Negation in Early Child Language." In *The Social Foundations of Language and Thought: Essays in Honor of Jerome S. Bruner,* D. R. Olson, ed., 156–86. New York: W. W. Norton.

Peirce, C. S. 1933. *The Simplest Mathematics.* Reprinted as vol. 4, *Collected Papers,* C. Hartshorne and P. Weiss, eds., Cambridge: Harvard University Press, 1960.

Pelletier, J. 1977. "Or." *Theoretical Linguistics* 4:61–74.

Perlmutter, D., and C. Rosen, eds. 1984. *Studies in Relational Grammar 2.* Chicago: University of Chicago Press.

Perlmutter, D., and S. Soames. 1979. *Syntactic Argumentation and the Structure of English.* Berkeley and Los Angeles: University of California Press.

Pesetsky, D. 1985. "Morphology and Logical Form." *Linguistic Inquiry* 16: 193–246.

Plato. 1937. *The Sophist.* B. Jowett, trans. New York: Random House.

Poldauf, J. 1964. "Some Points on Negation in Colloquial English." In *A Prague School Reader in Linguistics,* J. Vachek, ed., 366–74. Bloomington: Indiana University Press.

Pollack, J. 1974. A Re-Analysis of Neg-Raising in English. Master's thesis, Ohio State University. Appears in *OSU Working Papers in Linguistics 21,* 1976.

Pope, E. 1972. Questions and Answers in English. Ph.D. diss., MIT.

Postal, P. 1971. *Cross-over Phenomena.* New York: Holt.

Pott, A. F. 1859. *Etymologische Forschungen auf dem Gebiete der Indo-Germanischen Sprachen.* vol. 1. Lemgo and Detmold: Meyer.

Poutsma, H. 1928. *A Grammar of Late Modern English.* Groningen: P. Noordhoff.

Price, H. H. 1929. "Negation." *Aristotelian Society* Supplementary vol. 9, 97–111.

Prince, E. 1976. "The Syntax and Semantics of NEG-Raising, with Evidence from French." *Language* 52:404–26.

———. 1981. "Toward a Taxonomy of Given-New Information." In Cole, ed. (1981), 223–56.

———. 1983. "Topicalization and Left Dislocation: A Functional Analysis." In *Discourses in Reading and Linguistics,* S. White and V. Teller, eds. Annals of the New York Academy of Sciences, vol. 433.

Prior, A. N. 1954. "Entities." *Australasian Journal of Philosophy* 32:159–60.
———. 1967. "Negation." Entry in *Encyclopedia of Philosophy,* vol. 5, 458–63. New York: Macmillan.
Putnam, H. 1973. "Meaning and Reference." *Journal of Philosophy* 70:699–711.
Quine, W. V. O. 1948. "On What There Is." Reprint in *From a Logical Point of View,* 2d ed. Cambridge: Harvard University Press.
———. 1951. *Mathematical Logic.* New York: Harper & Row.
———. 1952. *Methods of Logic.* London: Routledge & Kegan Paul.
———. 1953a. "On a So-Called Paradox." *Mind* 62:65–67.
———. 1953b. "Mr. Strawson on Logical Theory." *Mind* 62:433–51.
———. 1960. *Word and Object.* Cambridge: MIT Press.
———. 1981. "What Price Bivalence?" *Journal of Philosophy* 78:90–95.
Raju, R. T. 1941. "The Reality of Negation," *Philosophical Review* 50:585–601.
———. 1954. "The Principle of Four-Cornered Negation in Indian Philosophy." *Review of Metaphysics* 7:694–713.
Randle, H. N. 1930. *Indian Logic in the Early Schools.* London: Oxford University Press.
Ravindran, N. 1970. "Negation in Malayalam." In Biligiri, ed., 5–8.
Reichenbach, H. 1947. *Elements of Symbolic Logic.* New York: Macmillan.
Reinhart, T. 1981. "Pragmatics and Linguistics: An Analysis of Sentence Topics." *Philosophica* 27:53–94.
Renou, L. 1946. "Sur la phrase négative dans le Rgveda." *Bulletin de la société linguistique de Paris* 43:43–49.
———. 1957. *Terminologie grammaticale du Sanskrit.* Paris: Librairie Ancienne.
Rescher, N. 1963. *Studies in the History of Arabic Logic.* Pittsburgh: University of Pittsburgh Press.
———. 1969. *Many-Valued Logic.* New York: McGraw-Hill.
Richter, F. K. 1944. "Adolf Hitler and Words." *Etc.* 1:256–58.
Robinson, R. H. 1956. "Some Logical Aspects of Nāgārjuna's System." *Philosophy East and West* 6:291–308.
Rodman, R. 1976. "Scope Phenomena, 'Movement Transformations,' and Relative Clauses." In Partee, ed., 165–76.
Rosch, E. 1977. "Human Categorization." In *Advances in Cross-Cultural Psychology,* N. Warren, ed., 1–72. New York: Academic Press.
Rosenbaum, P. 1967. *The Grammar of English Predicate Complement Constructions.* Cambridge: MIT Press.
Ross, A. C. 1976. "Meaningless '*De-*' in English." *Notes and Queries* 23:310–11.
Ross, C. 1978. "The Rightmost Principle of Sentence Negation." *CLS 14,* 416–29.
Ross, J. R. 1967. Constraints on Variables in Syntax. Ph.D. diss., MIT.
———. 1973a. "Slifting." In *The Formal Analysis of Natural Languages,* M.

Gross, M. Halle, and M.-P. Schützenberger, eds., 133–69. The Hague: Mouton.

———. 1973b. "Negginess." Paper presented at the annual LSA meeting, San Diego.

———. 1974. "Wording Up." Handout for unpublished talk, MIT.

Ross, W. D. 1923. *Aristotle*. London: Methuen.

Routley, R. 1966. "On a Significance Theory." *Australasian Journal of Philosophy* 44:172–209.

———. 1969. "The Need for Nonsense." *Australasian Journal of Philosophy* 47:367–84.

Royce, J. 1917. "Negation." Entry in *Encyclopedia of Religion and Ethics*, J. Hastings, ed., vol. 9, 264–71. New York: Charles Scribner's Sons.

Rumbaugh, D., and T. Gill. 1977. "Lana's Acquisition of Language Skills." In *Language Learning by a Chimpanzee: The Lana Project*, D. Rumbaugh, ed., 165–92. New York: Academic Press.

Russell, B. [1900] 1937. *A Critical Exposition of the Philosophy of Leibniz*. 2d ed. London: George Allen & Unwin.

———. 1905. "On Denoting." *Mind* 14:479–93.

———. [1908] 1967. "Mathematical Logic as Based on the Theory of Types." Reprinted in van Heijenoort, ed., 150–82.

———. 1918. *The Philosophy of Logical Atomism*, excerpted in Russell (1964), 177–281.

———. 1940. *An Inquiry into Meaning and Truth*. London: George Allen & Unwin.

———. 1948. *Human Knowledge, its Scope and Limits*. New York: Simon and Schuster.

———. 1954. "The Metaphysician's Nightmare." In *Nightmares of Eminent Persons*, 31–35. London: Bodley Head.

———. 1957. "Mr. Strawson on Referring." *Mind* 66:385–89.

———. 1964. *Logic and Knowledge: Essays 1901–1950*. R. C. Marsh, ed. London: Macmillan.

Ryle, G. 1929. "Negation." *Aristotelian Society*, Supplementary vol. 9, 80–86.

Sadock, J. 1971. "Queclaratives." *CLS 7*, 223–31.

———. 1972. "Speech Act Idioms." *CLS 8*, 329–39.

———. 1974. *Toward a Linguistic Theory of Speech Acts*. New York: Academic Press.

———. 1976. "Larry Scores a Point." *Pragmatics Microfiche*, Fiche 1.4.

———. 1977. "Truth and Approximations." *BLS 3*, 430–39.

———. 1978. "On Testing for Conversational Implicature." In Cole, ed. (1978), 281–97.

———. 1981. "Almost." In Cole, ed. (1981), 257–71.

———. 1983. "The Necessary Overlapping of Grammatical Components." In

Papers from the Parasession on the Interplay of Phonology, Morphology, and Syntax, 198–221. Chicago: CLS.

———. 1984. "Whither Radical Pragmatics?" In Schiffrin, ed., 139–49.

———. 1985. "Autolexical Syntax." *Natural Language and Linguistic Theory* 3:379–439.

Sadock, J., and A. Zwicky. 1985. "Speech Act Distinctions in Syntax." In Shopen, ed., 155–96.

Sag, I., and M. Liberman. 1973. "The Intonational Disambiguation of Indirect Speech Acts." *CLS 11,* 487–97.

Sahlin, E. 1979. Some *and* any *in Spoken and Written English.* Uppsala: Almqvist & Wiksell.

Sanford, D. H. 1968. "Contraries and Subcontraries." *Noûs* 2:95–96.

Sapir, E. 1930. *Totality.* Language Monograph Number 6.

———. 1944. "Grading: A Study in Semantics." Reprinted in *Selected Writings,* D. Mandelbaum, ed., 122–49. Berkeley: University of California Press, 1951.

Schachter, P. 1972. "On Post-Transformational Lexical Insertion." In *UCLA Papers in Syntax I,* P. Schachter and G. Bedell, eds. Los Angeles: UCLA Dept. of Linguistics.

———. 1985. "Parts-of-Speech Systems." In Shopen, ed., 3–61.

Schiffrin, D., ed. 1984. *Meaning, Form, and Use in Context: Linguistic Applications (GURT '84).* Washington: Georgetown University Press.

Schmerling, S. 1971. "A Note on Negative Polarity." *Papers in Linguistics* 4:200–206.

———. 1975. "Asymmetric Conjunction and Rules of Conversation." In Cole and Morgan, eds., 211–31.

Scriblerus, M. [1727] 1952. *The Art of Sinking.* Pseudonymous publication attributed to A. Pope, J. Swift, and J. Arbuthnot, appearing in first edition of the Swift-Pope Miscellany. Reproduced under editorship of E. L. Steeves. New York: King's Crown Press.

Searle, J. 1965. "What is a Speech Act?" In Black, ed., 221–39.

———. 1969. *Speech Acts.* Cambridge: Cambridge University Press.

———. 1975. "Indirect Speech Acts." In Cole and Morgan, eds., 59–82.

———. 1979. "Referential and Attributive." In *Expression and Meaning: Studies in the Theory of Speech Acts,* 137–61. Cambridge: Cambridge University Press.

Sebeok, T. A. 1962. "Evolution of Signalling Behavior." *Behavioral Science* 7:430–42.

Selkirk, E. O. 1982. *The Syntax of Words.* Cambridge: MIT Press.

Seright, O. D. 1966. "Double Negatives in Standard Modern English." *American Speech* 41:123–26.

Seuren, P. 1967. "Negation in Dutch" (revised article on Kraak [1966]). *Neophilologus* 51:327–63.

———. 1969. *Operators and Nucleus*. Cambridge: Cambridge University Press.

Sgall, P., E. Hajičová, and E. Benešová. 1973. *Topic, Focus, and Generative Semantics*. Kronberg: Scripton Verlag GmbH.

Shanon, B. 1981. "What's in the Frame? Linguistic Indicators." *Journal of Pragmatics* 5:35–44.

Sharma, Dh. 1970. *The Negative Dialectics of India*. (no publication data avaliable.)

Sheffer, H. M. 1913. "A Set of Five Independent Postulates for Boolean Algebra." *Transactions of the American Mathematical Society* 14:481–88.

Sheintuch, G., and K. Wise 1976. "On the Pragmatic Unity of the Rules of Neg-Raising and Neg-Attraction." *CLS 12*, 548–57.

Shnukal, A. 1980. "Syntactic Indications of Tentativeness in Discourse." In *Contemporary Studies in Romance Languages*, F. Nuessel, Jr., ed., 280–99. Distributed by IULC.

Shopen, T., ed. 1985. *Language Typology and Syntactic Description I: Clause Structure*. Cambridge: Cambridge University Press.

Siegel, D. 1971. "Some Lexical Transderivational Constraints in English." Unpublished. MIT.

———. 1973. "Some Non-Sources for Unpassives." In *Syntax and Semantics 2*, J. Kimball, ed., 301–17.

———. [1974] 1979. *Topics in English Morphology*. New York: Garland.

———. 1977. "The Adjacency Condition and the Theory of Morphology." *NELS 7*, 189–97.

Sigwart, C. 1895. *Logic*, vol. 1, 2d ed., H. Dendy, trans. New York: Macmillan. German edition published 1889.

Sloman, A. 1969. "Transformations of Illocutionary Acts." *Analysis* 30:56–59.

Smiley, T. 1960. "Sense without Denotation." *Analysis* 20:125–35.

Smith, A. 1933. "Två Bidrg Till Engelsk Ordföljd." *Moderna Språk* 27:79–82.

Smith, S. 1970. *Meaning and Negation*. Ph.D. diss., UCLA. Revised and published, The Hague: Mouton, 1975.

Soames, S. 1979. "A Projection Problem for Speaker Presuppositions." *Linguistic Inquiry* 10:623–66.

———. 1982. "How Presuppositions are Inherited: A Solution to the Projection Problem." *Linguistic Inquiry* 13:483–545.

Sommers, F. 1963. "Types and Ontology." *Philosophical Review* 72:327–63.

———. 1965. "Predicability." In Black, ed., 262–81.

———. 1970. "The Calculus of Terms." *Mind* 79:1–39.

———. 1982. *The Logic of Natural Language*. Oxford: Clarendon.

Spencer, H. 1914. *Kanarese Grammar*. Mysore: Wesleyan Mission Press.

Sperber, D., and D. Wilson. 1986. *Relevance*. Cambridge: Harvard University Press.

Spinoza, B. [1934] 1965. *The Philosophy of Spinoza*, vol. 1. H. Wolfson, ed. and trans. Reprint. New York: Meridian Books.

Spitzer, L. 1927. "Frz. *il ne faut pas que tu meures* 'du darfst nicht sterben'." *Germanisch-romanische Monatsschrift* 15:69–73.

Staal, F. 1962. "Negation and the Law of Contradiction in Indian Thought: A Comparative Study." *Bulletin of the School of Oriental and African Studies* 25:52–71.

Stalnaker, R. 1972. "Pragmatics." In Davidson and Harman, eds., 380–97.

———. 1974. "Pragmatic Presuppositions." In *Semantics and Philosophy*, M. Munitz and P. Unger, eds., 197–214. New York: New York University Press.

Steinberg, D. 1970. "Negation, Analyticity, Amphigory, and the Semantic Interpretation of Sentences." *Journal of Experimental Psychology* 84:417–23.

Steinberg, D., and L. Jakobovitz, eds. 1971. *Semantics: An Interdisciplinary Reader.* Cambridge: Cambridge University Press.

Stenner, A. J. 1981. "A Note on Logical Truth and Non-Sexist Semantics." In Vetterling-Braggin, ed., 299–306.

Stern, G. 1937. *Meaning and Change of Meaning*. Bloomington: Indiana University Press.

Stickel, G. 1972. "'Ja' und 'Nein' als Kontoll- und Korrektursignale." *Linguistische Berichte* 17:12–17.

Stockwell, R., P. Schachter, and B. H. Partee, 1973. *The Major Syntactic Structures of English.* Chapter 5, "Negation." New York: Holt.

Stoffel, C. 1901. *Intensives and Down-toners*. Anglistische Forschungen, part 1. Heidelberg: Carl Winter's Universitätsbuchhandlung.

Stokes, W. 1974. "All of the Work on Quantifier-Negation isn't Convincing." *CLS 10*, 692–700.

Strachey, J., ed. 1957. *The Standard Edition of the Complete Psychological Works of Sigmund Freud.* London: Hogarth Press.

Strang, C. 1960. "Aristotle and the Sea-Battle." *Mind* 69:447–65.

Strawson, P. F. 1949. "Truth." *Analysis* 9:83–97.

———. 1950. "On Referring." *Mind* 59:320–44.

———. 1952. *Introduction to Logical Theory.* London: Methuen.

———. [1964] 1971. "Identifying Reference and Truth-Values." Reprinted in Steinberg and Jakobovitz, eds., 86–99.

Sturel, R. 1908. *Jacques Amyot, traducteur des vies parallèles de Plutarque*. Paris: Honoré Champion.

Sullivan, M. W. 1967. *Apuleian Logic*. Amsterdam: North-Holland.

Sweet, H. 1900. *A New English Grammar.* Oxford: Clarendon.

Tannen, D. 1975. "Communication Mix and Mixup." *San Jose State Occasional Papers in Linguistics.*

———. 1983. *Communicative Style*. Norwood, NJ: Ablex.

Tasmowski-De Ryck, L. 1972. "La négation en français et la formalisation de la grammaire." *Logique et Analyse* 15:171–207.

Taylor, K. 1981. "Reference and Truth: The Case of Sexist and Racist Utterances." In Vetterling-Braggin, eds., 307–18.

Tesnière, L. 1959. *Eléments de syntaxe structurale*. Paris: C. Klincksieck.

Thirumalai, M. S. 1970. "Some Aspects of Negation in Thadou." In Biligiri, ed., 55–60.

Thomason, R. 1973. "Semantics, Pragmatics, Conversation, Presupposition." Unpublished. University of Pittsburgh.

Thomason, R., and R. Stalnaker. 1973. "A Semantic Theory of Adverbs." *Linguistic Inquiry* 4:195–220.

Thompson, M. 1953. "On Aristotle's Square of Opposition." *Philosophical Review* 62:251–65.

———. 1954. "A Reply to Mr. Nelson." *Philosophical Review* 63:414–19.

Timberlake, A. 1975. "Hierarchies in the Genitive of Negation." *Slavic and East European Journal* 19:123–38.

Tobler, A. [1882a] 1921. "Tout ce qui reluit n'est pas or." Reprinted in *Vermischte Beiträge zur französischen Grammatik 1*, 3d ed., 195–201. Leipzig: S. Hirzel.

———. [1882b] 1921. "Il ne faut pas que tu meures 'du darfst nicht sterben'." Reprinted in *Vermischte Beiträge zur französischen Grammatik 1*, 3d ed., 201–5. Leipzig: S. Hirzel.

———. [1896] 1908. "*ne . . . se . . . non, mais, fors, que*." Reprinted in *Vermischte Beiträge zur französischen Grammatik 3*, 2d ed., 78–111. Leipzig: S. Hirzel.

Toms, E. 1972. "The Problem of Negation." *Logique et Analyse* 15:1–38.

Tottie, G. 1980. "Affixal and Non-Affixal Negation in English—Two Systems in (Almost) Complementary Distribution." *Studia Linguistica* 34:101–23.

———. 1982. "Where Do Negative Sentences Come From?" *Studia Linguistica*. 36:88–105.

Trabasso, T., H. Rollins & E. Shaughnessy. 1971. "Storage and Verification Stages in Processing Concepts." *Cognitive Psychology* 2:239–89.

Ultan, R. 1978. "Some General Characteristics of Interrogative Systems." In Greenberg et al., eds., 211–48.

Urmson, J. O. 1952. "Parenthetical Verbs." *Mind* 61:480–96.

Vandamme, F. 1972. "On Negation, An Interdisciplinary Study." *Logique et Analyse* 15:39–101.

Van Dongen, W. A. 1918. "At All." *Neophilologus* 3:62–64.

van Fraassen, B. 1966. "Singular Terms, Truth-Value Gaps, and Free Logic." *Journal of Philosophy* 63:481–95.

———. 1968. "Presupposition, Implication, and Self-Reference." *Journal of Philosophy* 65:136–57.

———. 1969. "Presuppositions, Super-Valuations, and Free Logic." In *The Logical Way of Doing Things*, K. Lambert, ed., 67–91. New Haven: Yale University Press.

van Ginneken, J. 1907. *Principes de linguistique psychologique*. Paris: M. Rivière.

van Heijenoort, J., ed. 1967. *From Frege to Gödel*. Cambridge: Harvard University Press.

Van Oosten, J. 1986. *The Nature of Subjects, Topics, and Agents*. Bloomington: IULC.

Varga, L. 1980. "Observations on Negation in Hungarian and English." *Acta Linguistica* 30:67–96.

Vendler, Z. 1967. *Philosophy in Linguistics*. Ithaca: Cornell University Press.

Vennemann, T. 1974. "Topics, Subjects, and Word Order: From SXV to SVX via TVX." In *Historical Linguistics*, J. M. Anderson and C. Jones, eds., 1:339–76. Amsterdam: North-Holland.

Vetterling-Braggin, M., ed. 1981. *Sexist Language: A Modern Philosophical Analysis*. Totowa, NJ: Littlefield, Adams.

Volterra, V., and F. Antinucci. 1979. "Negation in Child Language: A Pragmatic Study." In Ochs and Schieffelin, eds., 281–303.

Von Wright, G. H. 1951. *An Essay in Modal Logic*. Amsterdam: North-Holland.

———. 1959. "On the Logic of Negation." *Societas Scientiarum Fennica, Commentationes Physico-Mathematicae* 22:1–30.

Wachtel, T. 1980. "Pragmatic Approximations." *Journal of Pragmatics* 4:201–12.

Wagenaar, K. 1930. *Etude sur la négation en ancien espagnol jusqu'au XVe siècle*. Groningen: J. B. Wolters.

Wales, R. J., and R. Grieve. 1969. "What Is So Difficult about Negation?" *Perception and Psychophysics* 6:327–32.

Wali, K. 1972. "Negation Dependent Idioms, and the Structure of Marathi Modals." *Papers in Linguistics* 5:382–401.

Walker, R. 1975. "Conversational Implicatures." In *Meaning, Reference, and Necessity*, S. Blackburn, ed., 133–81. Cambridge: Cambridge University Press.

Ward, G., and J. Hirschberg. 1985. "Implicating Uncertainty." *Language* 61: 747–76.

Warner, W. L. 1946. "The Absolute Negative Comparative." *American Speech* 21:302–3.

Warren, H. C. 1896. *Buddhism in Translations*. Cambridge: Harvard University Press.

Wason, P. C. 1959. "The Processing of Positive and Negative Information." *Quarterly Journal of Experimental Psychology* 11:92–107.

———. 1961. "Response to Affirmative and Negative Binary Statements." *British Journal of Psychology* 52:133–42.

———. 1965. "The Contexts of Plausible Denial." *Journal of Verbal Learning and Verbal Behavior* 4:7–11.

———. 1972. "In Real Life Negatives Are False." *Logique et Analyse* 15: 17–38.

Wason, P. C., and S. Jones. 1963. "Negation: Denotation and Connotation." *British Journal of Psychology* 54:299–307.

Wasow, T. 1972. "A Note on Not-Hopping." *Foundations of Language* 8:274–75.

———. 1977. "Transformations and the Lexicon." In *Formal Syntax*, P. Culicover, T. Wasow, and A. Akmajian, eds., 327–60. New York: Academic Press.

Waterlow, S. 1982. *Passage and Possibility: A Study of Aristotle's Modal Concepts*. Oxford: Clarendon.

Weinreich, U. 1963. "On the Semantic Structure of Language." In Greenberg, ed., 114–71.

Welte, W. 1978. *Negationslinguistik*. Munich: Wilhelm Fink.

Wertheimer, R. 1972. *The Significance of Sense: Meaning, Modality, and Morality*. Ithaca: Cornell University Press.

Wheeler, S. 1983. "Megarian Paradoxes as Eleatic Arguments." *American Philosophical Quarterly* 20:287–95.

Whitehead, A. N., and B. Russell. 1910. *Principia Mathematica*. Cambridge: Cambridge University Press.

Whitney, A. H. 1956. *Teach Yourself Finnish*. London: Teach Yourself Books, Hodden and Stoughton.

Williams, C. J. F. 1964. "Saint Anselm and His Biographers." *Downside Review* 82:124–40.

———. 1968. Review of Henry, *Logic of Saint Anselm*. *Mind* 77:609–11.

Williams, E. 1981. "On the Notions 'Lexically Related' and 'Head of a Word'." *Linguistic Inquiry* 12:245–74.

Williams, E. R. 1944. *The Conflict of Homonyms in English*. New Haven: Yale University Press.

Wilson, D. 1975. *Presupposition and Non-Truth-Conditional Semantics*. New York: Academic Press.

Wilson, D., and D. Sperber. 1979. "Ordered Entailments: An Alternative to Presuppositional Theories." In Oh and Dinneen, eds., 299–323.

———. 1981. "Irony and the Use-Mention Distinction." In Cole, ed. (1981), 295–318.

Wittgenstein, L. [1922] 1961. *Tractatus Logico-Philosophicus*. D. F. Pears and B. F. McGuinness, trans. London: Routledge & Kegan Paul.

———. 1953. *Philosophical Investigations*. Anscombe, trans. New York: Macmillan.

Wood, L. 1933. "The Paradox of Negative Judgment." *Philosophical Review* 42:412–23.

Wundt, W. 1886. "Das Sittliche in der Sprache." *Deutsche Rundschau* 47:70–92.

Yau, S. C. 1980. *Le système de la négation en Cantonais*. Kowloon, Hong Kong: Éditions Langages Croisés.

Ziff, P. 1960. *Semantic Analysis*. Ithaca: Cornell University Press.

Zimmer, K. 1964. *Affixal Negation in English and Other Languages.* Supplement to *Word,* Monograph No. 5.

———. 1971. "Some General Observations about Nominal Compounds." *Stanford University Working Papers on Language Universals* 5:C1–C21.

Zipf, G. K. 1935. *The Psycho-Biology of Language.* Cambridge: MIT Press.

———. 1949. *Human Behavior and the Principle of Least Effort.* Cambridge, MA: Addison-Wesley.

Zwicky, A. 1973. "Linguistics as Chemistry: The Substance Theory of Semantic Primes." In Anderson and Kiparsky, eds., 467–85.

Zwicky, A., and G. Pullum. 1983. "Cliticization vs. Inflection: English *n't.*" *Language* 59:502–13.

Zwicky, A., and J. Sadock. 1975. "Ambiguity Tests and How to Fail Them." In Kimball, ed., 1–36.

Index of Names

Subject Index